Algorithms on text (strings) have long been studied in computer science, and computation on molecular sequence data (strings) is at the heart of computational molecular biology. Existing and emerging algorithms for string computation provide a significant intersection between computer science and molecular biology. This book is a general, rigorous treatment of algorithms that operate on character strings and sequences. It covers a wide spectrum of string algorithms from classical computer science to modern molecular biology and, when possible, integrates those two fields.

In addition to explaining current algorithms, the book emphasizes fundamental ideas and techniques that are central in today's applications and should lead to new techniques, in the future. The book contains new approaches developed for complex material, simplifying methods that have been previously for the specialist alone. Biological problems are discussed in detail to explain the reasons that many biological questions have been productively cast as string problems.

The book is written for graduate or advanced undergraduate students in computer science, or computational biology, or bio-informatics. It can be used as a main text for courses on string algorithms, or for computer science oriented courses on computational biology and is also a reference for professionals. The book contains over 400 exercises to reinforce presented material, and to develop additional topics.

Algorithms on Strings, Trees, and Sequences

Algorithms on Strings, Trees, and Sequences

COMPUTER SCIENCE AND COMPUTATIONAL BIOLOGY

Dan Gusfield

University of California, Davis

CAMBRIDGE
UNIVERSITY PRESS

PUBLISHED BY THE PRESS SYNDICATE OF THE UNIVERSITY OF CAMBRIDGE
The Pitt Building, Trumpington Street, Cambridge CB2 1RP
40 West 20th Street, New York, NY 10011-4211, USA
10 Stamford Road, Oakleigh, Melbourne 3166, Australia

First published 1997
Reprinted 1999 (with corrections)

Printed in the United States of America

Typeset in Times Roman

Library of Congress Cataloging-in-Publication Data
Gusfield, Dan.
Algorithms on strings, trees, and sequences : computer science and
computational biology / Dan Gusfield.l
p. cm.
ISBN 0-521-58519-8 (hc)
1. Computer algorithms. 2. Molecular biology – Data processing.
I. Title.
QA76.9.A43387 1997
005.7'3 – dc21   86-46612

A catalog record for this book is available from the British Library.

ISBN 0 521 58519 8 hardback

Dedicated to the Memory of Gene Lawler:
Teacher, Colleague, and Friend.

Contents

Preface

History and motivation

Although I didn't know it at the time, I began writing this book in the summer of 1988 when I was part of a computer science (early bioinformatics) research group at the Human Genome Center of Lawrence Berkeley Laboratory.[1] Our group followed the standard assumption that biologically meaningful results could come from considering DNA as a one-dimensional character string, abstracting away the reality of DNA as a flexible three-dimensional molecule, interacting in a dynamic environment with protein and RNA, and repeating a life-cycle in which even the classic linear chromosome exists for only a fraction of the time. A similar, but stronger, assumption existed for protein, holding, for example, that all the information needed for correct three-dimensional folding is contained in the protein sequence itself, essentially independent of the biological environment the protein lives in. This assumption has recently been modified, but remains largely intact [297].

For nonbiologists, these two assumptions were (and remain) a god send, allowing rapid entry into an exciting and important field. Reinforcing the importance of sequence-level investigation were statements such as:

> The digital information that underlies biochemistry, cell biology, and development can be represented by a simple string of G's, A's, T's and C's. This string is the root data structure of an organism's biology. [352]

and

> In a very real sense, molecular biology is all about sequences. First, it tries to reduce complex biochemical phenomena to interactions between defined sequences . . . .[449]

and

> The ultimate rationale behind all purposeful structures and behavior of living things is embodied in the sequence of residues of nascent polypeptide chains . . . In a real sense it is at this level of organization that the secret of life (if there is one) is to be found. [330]

So without worrying much about the more difficult chemical and biological aspects of DNA and protein, our computer science group was empowered to consider a variety of biologically important problems defined primarily on *sequences*, or (more in the computer science vernacular) on *strings*: reconstructing long strings of DNA from overlapping string fragments; determining physical and genetic maps from probe data under various experimental protocols; storing, retrieving, and comparing DNA strings; comparing two or more strings for similarities; searching databases for related strings and substrings; defining and exploring different notions of string relationships; looking for new or ill-defined patterns occurring frequently in DNA; looking for structural patterns in DNA and

---

[1] The other long-term members were William Chang, Gene Lawler, Dalit Naor, and Frank Olken.

protein; determining secondary (two-dimensional) structure of RNA; finding conserved, but faint, patterns in many DNA and protein sequences; and more.

We organized our efforts into two high-level tasks. First, we needed to learn the relevant biology, laboratory protocols, and existing algorithmic methods used by biologists. Second we sought to canvass the computer science literature for ideas and algorithms that weren't already used by biologists, but which *might plausibly* be of use either in current problems or in problems that we could anticipate arising when vast quantities of sequenced DNA or protein become available.

Our problem

None of us was an expert on string algorithms. At that point I had a textbook knowledge of Knuth-Morris-Pratt and a deep confusion about Boyer–Moore (under what circumstances it was a linear time algorithm and how to do *strong* preprocessing in linear time). I understood the use of dynamic programming to compute edit distance, but otherwise had little exposure to specific string algorithms in biology. My general background was in combinatorial optimization, although I had a prior interest in algorithms for building evolutionary trees and had studied some genetics and molecular biology in order to pursue that interest.

What we needed then, but didn't have, was a comprehensive cohesive text on string algorithms to guide our education. There were at that time several computer science texts containing a chapter or two on strings, usually devoted to a rigorous treatment of Knuth-Morris-Pratt and a cursory treatment of Boyer–Moore, and possibly an elementary discussion of matching with errors. There were also some good survey papers that had a somewhat wider scope but didn't treat their topics in much depth. There were several texts and edited volumes from the biological side on uses of computers and algorithms for sequence analysis. Some of these were wonderful in exposing the potential benefits and the pitfalls of using computers in biology, but they generally lacked algorithmic rigor and covered a narrow range of techniques. Finally, there was the seminal text *Time Warps, String Edits, and Macromolecules: The Theory and Practice of Sequence Comparison* edited by D. Sankoff and J. Kruskal, which served as a bridge between algorithms and biology and contained many applications of dynamic programming. However, it too was much narrower than our focus and was a bit dated.

Moreover, most of the available sources from either community focused on string *matching*, the problem of searching for an exact or "nearly exact" copy of a pattern in a given text. Matching problems are central, but as detailed in this book, they constitute only a part of the many important computational problems defined on strings. Thus, we recognized that summer a need for a rigorous and fundamental treatment of the *general* topic of algorithms that operate on strings, along with a rigorous treatment of *specific* string algorithms of greatest current and potential import in computational biology. This book is an attempt to provide such a dual, and integrated, treatment.

Why mix computer science and computational biology in one book?

My interest in computational biology began in 1980, when I started reading papers on building evolutionary trees. That side interest allowed me an occasional escape from the hectic, hyper competitive "hot" topics that theoretical computer science focuses on. At that point, computational molecular biology was a largely undiscovered area for computer science, although it was an active area for statisticians and mathematicians (notably Michael Waterman and David Sankoff who have largely framed the field). Early on, seminal papers on computational issues in biology (such as the one by Buneman [83]) did not appear in mainstream computer science venues but in obscure places such as conferences on *computational archeology* [226]. But seventeen years later, computational biology *is* hot, and many computer scientists are now entering the (now more hectic, more competitive) field [280]. What should they learn?

The problem is that the emerging field of computational molecular biology is not well defined and its definition is made more difficult by rapid changes in molecular biology itself. Still, algorithms that operate on molecular sequence data (strings) are at the heart of computational molecular biology. The big-picture question in computational molecular biology is how to "do" as much "real biology" as possible by exploiting molecular sequence data (DNA, RNA, and protein). Getting sequence data is relatively cheap and fast (and getting more so) compared to more traditional laboratory investigations. The use of sequence data is already central in several subareas of molecular biology and the full impact of having extensive sequence data is yet to be seen. Hence, algorithms that operate on strings will continue to be the area of closest intersection and interaction between computer science and molecular biology. Certainly then, computer scientists need to learn the string techniques that have been most successfully applied. But that is not enough.

Computer scientists need to learn *fundamental* ideas and techniques that will endure long after today's central motivating applications are forgotten. They need to study methods that prepare them to frame and tackle future problems and applications. Significant contributions to computational biology might be made by extending or adapting algorithms from computer science, even when the original algorithm has no clear utility in biology. This is illustrated by several recent sublinear-time approximate matching methods for database searching that rely on an interplay between exact matching methods from computer science and dynamic programming methods already utilized in molecular biology.

Therefore, the computer scientist who wants to enter the general field of computational molecular biology, and who learns string algorithms with that end in mind, should receive a training in string algorithms that is much broader than a tour through techniques of known present application. Molecular biology and computer science are changing much too rapidly for that kind of narrow approach. Moreover, theoretical computer scientists try to develop effective algorithms somewhat differently than other algorithmists. We rely more heavily on correctness proofs, worst-case analysis, lower bound arguments, randomized algorithm analysis, and bounded approximation results (among other techniques) to *guide* the development of practical, effective algorithms. Our "relative advantage" partly lies in the mastery and use of those skills. So even if I were to write a book for computer scientists who only want to do computational biology, I would still choose to include a broad range of algorithmic techniques from pure computer science.

In this book, I cover a wide spectrum of string techniques – well beyond those of established utility; however, I have selected from the many possible illustrations, those techniques that seem to have the greatest *potential application* in future molecular biology. Potential application, particularly of ideas rather than of concrete methods, and to *anticipated* rather than to existing problems is a matter of judgment and speculation. No doubt, some of the material contained in this book will never find direct application in biology, while other material will find uses in surprising ways. Certain string algorithms that were generally deemed to be irrelevant to biology just a few years ago have become adopted by practicing biologists in both large-scale projects and in narrower technical problems. Techniques previously dismissed because they originally addressed (exact) string problems where *perfect data* were assumed have been incorporated as *components* of more robust techniques that handle imperfect data.

What the book is

Following the above discussion, this book is a general-purpose rigorous treatment of the entire field of deterministic algorithms that operate on strings and sequences. Many of those algorithms utilize trees as data-structures or arise in biological problems related to evolutionary trees, hence the inclusion of "trees" in the title.

The model reader is a research-level professional in computer science or a graduate or advanced undergraduate student in computer science, although there are many biologists (and of course mathematicians) with sufficient algorithmic background to read the book. The book is intended to serve as both a reference and a main text for courses in pure computer science and for computer science–oriented courses on computational biology.

Explicit discussions of biological applications appear throughout the book, but are more concentrated in the last sections of Part II and in most of Parts III and IV. I discuss a number of biological issues in detail in order to give the reader a deeper appreciation for the reasons that many biological problems have been cast as problems on strings and for the variety of (often very imaginative) technical ways that string algorithms have been employed in molecular biology.

This book covers all the classic topics and most of the important advanced techniques in the field of string algorithms, with three exceptions. It only lightly touches on probabilistic analysis and does not discuss parallel algorithms or the elegant, but very theoretical, results on algorithms for infinite alphabets and on algorithms using only constant auxiliary space.[2] The book also does not cover stochastic-oriented methods that have come out of the machine learning community, although some of the algorithms in this book are extensively used as subtools in those methods. With these exceptions, the book covers all the major styles of thinking about string algorithms. The reader who absorbs the material in this book will gain a deep and broad understanding of the field and sufficient sophistication to undertake original research.

Reflecting my background, the book rigorously discusses each of its topics, usually providing complete proofs of behavior (correctness, worst-case time, and space). More important, it emphasizes the *ideas* and *derivations* of the methods it presents, rather than simply providing an inventory of available algorithms. To better expose ideas and encourage discovery, I often present a complex algorithm by introducing a naive, inefficient version and then successively apply additional insight and implementation detail to obtain the desired result.

The book contains some new approaches I developed to explain certain classic and complex material. In particular, the preprocessing methods I present for Knuth-Morris-Pratt, Boyer–Moore and several other linear-time pattern matching algorithms differ from the classical methods, both unifying and simplifying the preprocessing tasks needed for those algorithms. I also expect that my (hopefully simpler and clearer) expositions on linear-time suffix tree constructions and on the constant-time least common ancestor algo-

---

[2] Space is a very important practical concern, and we will discuss it frequently, but constant space seems too severe a requirement in most applications of interest.

rithm will make those important methods more available and widely understood. I connect theoretical results from computer science on sublinear-time algorithms with widely used methods for biological database search. In the discussion of multiple sequence alignment I bring together the three major objective functions that have been proposed for multiple alignment and show a continuity between approximation algorithms for those three multiple alignment problems. Similarly, the chapter on evolutionary tree construction exposes the commonality of several distinct problems and solutions in a way that is not well known. Throughout the book, I discuss many computational problems concerning repeated substrings (a very widespread phenomenon in DNA). I consider several different ways to define repeated substrings and use each specific definition to explore computational problems and algorithms on repeated substrings.

In the book I try to explain in complete detail, and at a reasonable pace, many complex methods that have previously been written exclusively for the specialist in string algorithms. I avoid detailed code, as I find it rarely serves to explain interesting ideas,[3] and I provide over 400 exercises to both reinforce the material of the book and to develop additional topics.

What the book is not

Let me state clearly what the book is not. It is not a *complete* text on computational molecular biology, since I believe that field concerns computations on objects other than strings, trees, and sequences. Still, computations on strings and sequences form the heart of computational molecular biology, and the book provides a deep and wide treatment of sequence-oriented computational biology. The book is also not a "how to" book on string and sequence analysis. There are several books available that survey specific computer packages, databases, and services, while also giving a general idea of how they work. This book, with its emphasis on ideas and algorithms, does not compete with those. Finally, at the other extreme, the book does not attempt a definitive history of the field of string algorithms and its contributors. The literature is vast, with many repeated, independent discoveries, controversies, and conflicts. I have made some historical comments and have pointed the reader to what I hope are helpful references, but I am much too new an arrival and not nearly brave enough to attempt a complete taxonomy of the field. I apologize in advance, therefore, to the many people whose work may not be properly recognized.

In summary

This book is a general, rigorous text on deterministic algorithms that operate on strings, trees, and sequences. It covers the full spectrum of string algorithms from classical computer science to modern molecular biology and, when appropriate, connects those two fields. It is the book I wished I had available when I began learning about string algorithms.

Acknowledgments

I would like to thank The Department of Energy Human Genome Program, The Lawrence Berkeley Laboratory, The National Science Foundation, The Program in Math and Molec-

---

[3] However, many of the algorithms in the book have been coded in C and are available at
http://wwwcsif.cs.ucdavis.edu/~gusfield/strpgms.html.

ular Biology, and The DIMACS Center for Discrete Mathematics and Computer Science special year on computational biology, for support of my work and the work of my students and postdoctoral researchers.

Individually, I owe a great debt of appreciation to William Chang, John Kececioglu, Jim Knight, Gene Lawler, Dalit Naor, Frank Olken, R. Ravi, Paul Stelling, and Lusheng Wang.

I would also like to thank the following people for the help they have given me along the way: Stephen Altschul, David Axelrod, Doug Brutlag, Archie Cobbs, Richard Cole, Russ Doolittle, Martin Farach, Jane Gitschier, George Hartzell, Paul Horton, Robert Irving, Sorin Istrail, Tao Jiang, Dick Karp, Dina Kravets, Gad Landau, Udi Manber, Marci McClure, Kevin Murphy, Gene Myers, John Nguyen, Mike Paterson, William Pearson, Pavel Pevzner, Fred Roberts, Hershel Safer, Baruch Schieber, Ron Shamir, Jay Snoddy, Elizabeth Sweedyk, Sylvia Spengler, Martin Tompa, Esko Ukkonen, Martin Vingron, Tandy Warnow, and Mike Waterman.

PART I

Exact String Matching: The Fundamental String Problem

Exact matching: what's the problem?

Given a string P called the *pattern* and a longer string T called the *text*, the **exact matching** problem is to find all occurrences, if any, of pattern P in text T.

For example, if $P = aba$ and $T = bbabaxababay$ then P occurs in T starting at locations 3, 7, and 9. Note that two occurrences of P may overlap, as illustrated by the occurrences of P at locations 7 and 9.

Importance of the exact matching problem

The practical importance of the exact matching problem should be obvious to anyone who uses a computer. The problem arises in widely varying applications, too numerous to even list completely. Some of the more common applications are in word processors; in utilities such as *grep* on Unix; in textual information retrieval programs such as Medline, Lexis, or Nexis; in library catalog searching programs that have replaced physical card catalogs in most large libraries; in internet browsers and crawlers, which sift through massive amounts of text available on the internet for material containing specific keywords;[1] in internet news readers that can search the articles for topics of interest; in the giant digital libraries that are being planned for the near future; in electronic journals that are already being "published" on-line; in telephone directory assistance; in on-line encyclopedias and other educational CD-ROM applications; in on-line dictionaries and thesauri, especially those with cross-referencing features (the *Oxford English Dictionary* project has created an electronic on-line version of the *OED* containing 50 million words); and in numerous specialized databases. In molecular biology there are several hundred specialized databases holding raw DNA, RNA, and amino acid strings, or processed patterns (called motifs) derived from the raw string data. Some of these databases will be discussed in Chapter 15.

Although the practical importance of the exact matching problem is not in doubt, one might ask whether the problem is still of any research or educational interest. Hasn't exact matching been so well solved that it can be put in a black box and taken for granted? Right now, for example, I am editing a ninety-page file using an "ancient" shareware word processor and a PC clone (486), and every exact match command that I've issued executes faster than I can blink. That's rather depressing for someone writing a book containing a large section on exact matching algorithms. So is there anything left to do on this problem?

The answer is that for typical word-processing applications there probably is little left to do. The exact matching problem is solved for those applications (although other more so-phisticated string tools might be useful in word processors). But the story changes radically

---

[1] I just visited the Alta Vista web page maintained by the Digital Equipment Corporation. The Alta Vista database contains over 21 billion words collected from over 10 million web sites. A search for all web sites that mention "Mark Twain" took a couple of seconds and reported that twenty thousand sites satisfy the query. For another example see [392].

for other applications. Users of Melvyl, the on-line catalog of the University of California library system, often experience long, frustrating delays even for fairly simple matching requests. Even *grepping* through a large directory can demonstrate that exact matching is not yet trivial. Recently we used GCG (a very popular interface to search DNA and protein databanks) to search Genbank (the major U.S. DNA database) for a thirty-character string, which is a small string in typical uses of Genbank. The search took over four hours (on a local machine using a local copy of the database) to find that the string was not there.[2] And Genbank today is only a fraction of the size it will be when the various genome programs go into full production mode, cranking out massive quantities of sequenced DNA. Certainly there are faster, common database searching programs (for example, BLAST), and there are faster machines one can use (for example, an e-mail server is available for exact and inexact database matching running on a 4,000 processor MasPar computer). But the point is that the exact matching problem is not so effectively and universally solved that it needs no further attention. It will remain a problem of interest as the size of the databases grow and also because exact matching will continue to be a *subtask* needed for more complex searches that will be devised. Many of these will be illustrated in this book.

But perhaps the most important reason to study *exact* matching in detail is to understand the various ideas developed for it. Even assuming that the exact matching problem itself is sufficiently solved, the entire field of string algorithms remains vital and open, and the education one gets from studying exact matching may be crucial for solving less understood problems. That education takes three forms: specific algorithms, general algorithmic styles, and analysis and proof techniques. All three are covered in this book, but style and proof technique get the major emphasis.

Overview of Part I

In Chapter 1 we present naive solutions to the exact matching problem and develop the fundamental tools needed to obtain more efficient methods. Although the classical solutions to the problem will not be presented until Chapter 2, we will show at the end of Chapter 1 that the use of fundamental tools alone gives a simple linear-time algorithm for exact matching. Chapter 2 develops several classical methods for exact matching, using the fundamental tools developed in Chapter 1. Chapter 3 looks more deeply at those methods and extensions of them. Chapter 4 moves in a very different direction, exploring methods for exact matching based on arithmetic-like operations rather than character comparisons.

Although exact matching is the focus of Part I, some aspects of inexact matching and the use of wild cards are also discussed. The exact matching problem will be discussed again in Part II, where it (and extensions) will be solved using suffix trees.

Basic string definitions

We will introduce most definitions at the point where they are first used, but several definitions are so fundamental that we introduce them now.

Definition   A *string* S is an ordered list of characters written contiguously from left to right. For any string S, $S[i..j]$ is the (contiguous) *substring* of S that starts at position

---

[2]   We later repeated the test using the Boyer–Moore algorithm on our own raw copy of Genbank. The search took less than ten minutes, most of which was devoted to movement of text between the disk and the computer, with less than one minute used by the actual text search.

i and ends at position j of S. In particular, $S[1..i]$ is the *prefix* of string S that ends at position i, and $S[i..|S|]$ is the *suffix* of string S that begins at position i, where $|S|$ denotes the number of characters in string S.

Definition $S[i..j]$ is the empty string if $i > j$.

For example, *california* is a string, *lifo* is a substring, *cal* is a prefix, and *ornia* is a suffix.

Definition   A *proper* prefix, suffix, or substring of S is, respectively, a prefix, suffix, or substring that is not the entire string S, nor the empty string.

Definition   For any string S, $S(i)$ denotes the ith character of S.

We will usually use the symbol S to refer to an arbitrary fixed string that has no additional assumed features or roles. However, when a string is known to play the role of a pattern or the role of a text, we will refer to the string as P or T respectively. We will use lower case Greek characters (α, β, γ, δ) to refer to variable strings and use lower case roman characters to refer to single variable characters.

Definition   When comparing two characters, we say that the characters *match* if they are equal; otherwise we say they *mismatch*.

Terminology confusion

The words "string" and "word" are often used synonymously in the computer science literature, but for clarity in this book we will never use "word" when "string" is meant. (However, we do use "word" when its colloquial English meaning is intended.)

More confusing, the words "string" and "sequence" are often used synonymously, particularly in the biological literature. This can be the source of much confusion because "*sub*strings" and "*sub*sequences" are very different objects and because algorithms for substring problems are usually very different than algorithms for the analogous subsequence problems. The characters in a substring of S must occur *contiguously* in S, whereas characters in a subsequence might be interspersed with characters not in the subsequence. Worse, in the biological literature one often sees the word "sequence" used in place of "subsequence". Therefore, for clarity, in this book we will always maintain a distinction between "subsequence" and "substring" and never use "sequence" for "subsequence". We will generally use "string" when pure computer science issues are discussed and use "sequence" or "string" interchangeably in the context of biological applications. Of course, we will also use "sequence" when its standard mathematical meaning is intended.

The first two parts of this book primarily concern problems on strings and substrings. Problems on subsequences are considered in Parts III and IV.

1

Exact Matching: Fundamental Preprocessing and First Algorithms

1.1. The naive method

Almost all discussions of exact matching begin with the *naive method*, and we follow this tradition. The naive method aligns the left end of P with the left end of T and then compares the characters of P and T left to right until either two unequal characters are found or until P is exhausted, in which case an occurrence of P is reported. In either case, P is then shifted one place to the right, and the comparisons are restarted from the left end of P. This process repeats until the right end of P shifts past the right end of T.

Using n to denote the length of P and m to denote the length of T, the worst-case number of comparisons made by this method is $\Theta(nm)$. In particular, if both P and T consist of the same repeated character, then there is an occurrence of P at each of the first $m - n + 1$ positions of T and the method performs exactly $n(m - n + 1)$ comparisons. For example, if $P = aaa$ and $T = aaaaaaaaaa$ then $n = 3, m = 10$, and 24 comparisons are made.

The naive method is certainly simple to understand and program, but its worst-case running time of $\Theta(nm)$ may be unsatisfactory and can be improved. Even the practical running time of the naive method may be too slow for larger texts and patterns. Early on, there were several related ideas to improve the naive method, both in practice and in worst case. The result is that the $\Theta(n \times m)$ worst-case bound can be reduced to $O(n + m)$. Changing "$\times$" to "$+$" in the bound is extremely significant (try $n = 1000$ and $m = 10,000,000$, which are realistic numbers in some applications).

1.1.1. Early ideas for speeding up the naive method

The first ideas for speeding up the naive method all try to shift P by more than one character when a mismatch occurs, but never shift it so far as to miss an occurrence of P in T. Shifting by more than one position saves comparisons since it moves P through T more rapidly. In addition to shifting by larger amounts, some methods try to reduce comparisons by skipping over parts of the pattern after the shift. We will examine many of these ideas in detail.

Figure 1.1 gives a flavor of these ideas, using $P = abxyabxz$ and $T = xabxyabxyabxz$. Note that an occurrence of P begins at location 6 of T. The naive algorithm first aligns P at the left end of T, immediately finds a mismatch, and shifts P by one position. It then finds that the next seven comparisons are matches and that the succeeding comparison (the ninth overall) is a mismatch. It then shifts P by one place, finds a mismatch, and repeats this cycle two additional times, until the left end of P is aligned with character 6 of T. At that point it finds eight matches and concludes that P occurs in T starting at position 6. In this example, a total of twenty comparisons are made by the naive algorithm.

A smarter algorithm might realize, after the ninth comparison, that the next three

5

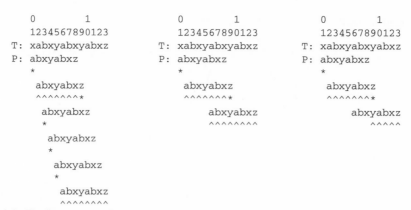

Figure 1.1: The first scenario illustrates pure naive matching, and the next two illustrate smarter shifts. A caret beneath a character indicates a match and a star indicates a mismatch made by the algorithm.

comparisons of the naive algorithm will be mismatches. This smarter algorithm skips over the next three shift/compares, immediately moving the left end of P to align with position 6 of T, thus saving three comparisons. How can a smarter algorithm do this? After the ninth comparison, the algorithm knows that the first seven characters of P match characters 2 through 8 of T. If it also knows that the first character of P (namely a) does not occur again in P until position 5 of P, it has enough information to conclude that character a does not occur again in T until position 6 of T. Hence it has enough information to conclude that there can be no matches between P and T until the left end of P is aligned with position 6 of T. Reasoning of this sort is the key to shifting by more than one character. In addition to shifting by larger amounts, we will see that certain aligned characters do not need to be compared.

An even smarter algorithm knows the next occurrence in P of the first three characters of P (namely abx) begin at position 5. Then since the first seven characters of P were found to match characters 2 through 8 of T, this smarter algorithm has enough information to conclude that when the left end of P is aligned with position 6 of T, the next three comparisons must be matches. This smarter algorithm avoids making those three comparisons. Instead, after the left end of P is moved to align with position 6 of T, the algorithm compares character 4 of P against character 9 of T. This smarter algorithm therefore saves a total of six comparisons over the naive algorithm.

The above example illustrates the kinds of ideas that allow some comparisons to be skipped, although it should still be unclear how an algorithm can efficiently implement these ideas. Efficient implementations have been devised for a number of algorithms such as the Knuth-Morris-Pratt algorithm, a real-time extension of it, the Boyer–Moore algorithm, and the Apostolico–Giancarlo version of it. All of these algorithms have been implemented to run in linear time ($O(n + m)$ time). The details will be discussed in the next two chapters.

1.2. The preprocessing approach

Many string matching and analysis algorithms are able to efficiently skip comparisons by first spending "modest" time learning about the internal structure of either the pattern P or the text T. During that time, the other string may not even be known to the algorithm. This part of the overall algorithm is called the *preprocessing* stage. Preprocessing is followed by a *search* stage, where the information found during the preprocessing stage is used to reduce the work done while searching for occurrences of P in T. In the above example, the smarter method was assumed to know that character a did not occur again until position 5, and the even smarter method was assumed to know that the pattern abx was repeated again starting at position 5. This assumed knowledge is obtained in the preprocessing stage.

For the exact matching problem, all of the algorithms mentioned in the previous section preprocess pattern P. (The opposite approach of preprocessing text T is used in other algorithms, such as those based on suffix trees. Those methods will be explained later in the book.) These preprocessing methods, as originally developed, are similar in spirit but often quite different in detail and conceptual difficulty. In this book we take a different approach and do not initially explain the originally developed preprocessing methods. Rather, we highlight the similarity of the preprocessing *tasks* needed for several different matching algorithms, by first defining a *fundamental preprocessing* of P that is independent of any particular matching algorithm. Then we show how each specific matching algorithm uses the information computed by the fundamental preprocessing of P. The result is a simpler more uniform exposition of the preprocessing needed by several classical matching methods and a simple linear time algorithm for exact matching based only on this preprocessing (discussed in Section 1.5). This approach to linear-time pattern matching was developed in [202].

1.3. Fundamental preprocessing of the pattern

Fundamental preprocessing will be described for a general string denoted by S. In specific applications of fundamental preprocessing, S will often be the pattern P, but here we use S instead of P because fundamental preprocessing will also be applied to strings other than P.

The following definition gives the key values computed during the fundamental preprocessing of a string.

Definition   Given a string S and a position $i > 1$, let $Z_i(S)$ be the *length* of the longest substring of S that *starts* at i and matches a prefix of S.

In other words, $Z_i(S)$ is the length of the longest *prefix* of $S[i..|S|]$ that matches a prefix of S. For example, when $S = aabcaabxaaz$ then

$$Z_5(S) = 3 \ (aabc...aabx...),$$

$$Z_6(S) = 1 \ (aa...ab...),$$

$$Z_7(S) = Z_8(S) = 0,$$

$$Z_9(S) = 2 \ (aab...aaz).$$

When S is clear by context, we will use Z_i in place of $Z_i(S)$.

To introduce the next concept, consider the boxes drawn in Figure 1.2. Each box starts at some position $j > 1$ such that Z_j is greater than zero. The length of the box starting at j is meant to represent Z_j. Therefore, each box in the figure represents a maximal-length

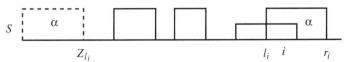

Figure 1.2: Each solid box represents a substring of S that matches a prefix of S and that starts between positions 2 and i. Each box is called a *Z-box*. We use r_i to denote the *right-most* end of any Z-box that begins at or to the left of position i and α to denote the substring in the Z-box ending at r_i. Then l_i denotes the left end of α. The copy of α that occurs as a prefix of S is also shown in the figure.

substring of S that matches a prefix of S and that does not start at position one. Each such box is called a *Z-box*. More formally, we have:

> **Definition** For any position $i > 1$ where Z_i is greater than zero, the *Z-box* at i is defined as the interval starting at i and ending at position $i + Z_i - 1$.

> **Definition** For every $i > 1$, r_i is the right-most endpoint of the Z-boxes that begin at or before position i. Another way to state this is: r_i is the largest value of $j + Z_j - 1$ over all $1 < j \leq i$ such that $Z_j > 0$. (See Figure 1.2.)

We use the term l_i for the value of j specified in the above definition. That is, l_i is the position of the *left end* of the Z-box that ends at r_i. In case there is more than one Z-box ending at r_i, then l_i can be chosen to be the left end of any of those Z-boxes. As an example, suppose $S = aabaabcaxaabaabcy$; then $Z_{10} = 7$, $r_{15} = 16$, and $l_{15} = 10$.

The linear time computation of Z values from S is the *fundamental* preprocessing task that we will use in all the classical linear-time matching algorithms that preprocess P. But before detailing those uses, we show how to do the fundamental preprocessing in linear time.

1.4. Fundamental preprocessing in linear time

The task of this section is to show how to compute all the Z_i values for S in linear time (i.e., in $O(|S|)$ time). A direct approach based on the definition would take $\Theta(|S|^2)$ time. The method we will present was developed in [307] for a different purpose.

The preprocessing algorithm computes Z_i, r_i, and l_i for each successive position i, starting from $i = 2$. All the Z values computed will be kept by the algorithm, but in any iteration i, the algorithm only needs the r_j and l_j values for $j = i - 1$. No earlier r or l values are needed. Hence the algorithm only uses a single variable, r, to refer to the most recently computed r_j value; similarly, it only uses a single variable l. Therefore, in each iteration i, if the algorithm discovers a new Z-box (starting at i), variable r will be incremented to the end of that Z-box, which is the right-most position of any Z-box discovered so far.

To begin, the algorithm finds Z_2 by explicitly comparing, left to right, the characters of $S[2..|S|]$ and $S[1..|S|]$ until a mismatch is found. Z_2 is the length of the matching string. If $Z_2 > 0$, then $r = r_2$ is set to $Z_2 + 1$ and $l = l_2$ is set to 2. Otherwise r and l are set to zero. Now assume inductively that the algorithm has correctly computed Z_i for i up to $k - 1 > 1$, and assume that the algorithm knows the current $r = r_{k-1}$ and $l = l_{k-1}$. The algorithm next computes Z_k, $r = r_k$, and $l = l_k$.

The main idea is to use the already computed Z values to accelerate the computation of Z_k. In fact, in some cases, Z_k can be deduced from the previous Z values without doing any additional character comparisons. As a concrete example, suppose $k = 121$, all the values Z_2 through Z_{120} have already been computed, and $r_{120} = 130$ and $l_{120} = 100$. That means that there is a substring of length 31 starting at position 100 and matching a prefix of S (of length 31). It follows that the substring of length 10 starting at position 121 must match the substring of length 10 starting at position 22 of S, and so Z_{22} may be very helpful in computing Z_{121}. As one case, if Z_{22} is three, say, then a little reasoning shows that Z_{121} must also be three. Thus in this illustration, Z_{121} can be deduced without any additional character comparisons. This case, along with the others, will be formalized and proven correct below.

Figure 1.3: String $S[k..r]$ is labeled β and also occurs starting at position k' of S.

Figure 1.4: Case 2a. The longest string starting at k' that matches a prefix of S is shorter than $|\beta|$. In this case, $Z_k = Z_{k'}$.

Figure 1.5: Case 2b. The longest string starting at k' that matches a prefix of S is at least $|\beta|$.

The Z algorithm

Given Z_i for all $1 < i \le k - 1$ and the current values of r and l, Z_k and the updated r and l are computed as follows:

Begin

1. If $k > r$, then find Z_k by explicitly comparing the characters starting at position k to the characters starting at position 1 of S, until a mismatch is found. The length of the match is Z_k. If $Z_k > 0$, then set r to $k + Z_k - 1$ and set l to k.

2. If $k \le r$, then position k is contained in a Z-box, and hence $S(k)$ is contained in substring $S[l..r]$ (call it α) such that $l > 1$ and α matches a prefix of S. Therefore, character $S(k)$ also appears in position $k' = k - l + 1$ of S. By the same reasoning, substring $S[k..r]$ (call it β) must match substring $S[k'..Z_l]$. It follows that the substring beginning at position k must match a prefix of S of length at least the *minimum* of $Z_{k'}$ and $|\beta|$ (which is $r - k + 1$). See Figure 1.3.

   We consider two subcases based on the value of that minimum.

   2a. If $Z_{k'} < |\beta|$ then $Z_k = Z_{k'}$ and r, l remain unchanged (see Figure 1.4).

   2b. If $Z_{k'} \ge |\beta|$ then the entire substring $S[k..r]$ must be a prefix of S and $Z_k \ge |\beta| = r - k + 1$. However, Z_k might be strictly larger than $|\beta|$, so compare the characters starting at position $r + 1$ of S to the characters starting a position $|\beta| + 1$ of S until a mismatch occurs. Say the mismatch occurs at character $q \ge r + 1$. Then Z_k is set to $q - k$, r is set to $q - 1$, and l is set to k (see Figure 1.5).

End

Theorem 1.4.1. *Using Algorithm Z, value Z_k is correctly computed and variables r and l are correctly updated.*

PROOF In Case 1, Z_k is set correctly since it is computed by explicit comparisons. Also (since $k > r$ in Case 1), before Z_k is computed, no Z-box has been found that starts between positions 2 and $k - 1$ and that ends at or after position k. Therefore, when $Z_k > 0$ in Case 1, the algorithm does find a new Z-box ending at or after k, and it is correct to change r to $k + Z_k - 1$. Hence the algorithm works correctly in Case 1.

In Case 2a, the substring beginning at position k can match a prefix of S only for length $Z_{k'} < |\beta|$. If not, then the next character to the right, character $k + Z_{k'}$, must match character $1 + Z_{k'}$. But character $k + Z_{k'}$ matches character $k' + Z_{k'}$ (since $Z_{k'} < |\beta|$), so character $k' + Z_{k'}$ must match character $1 + Z_{k'}$. However, that would be a contradiction to the definition of $Z_{k'}$, for it would establish a substring longer than $Z_{k'}$ that starts at k' and matches a prefix of S. Hence $Z_k = Z_{k'}$ in this case. Further, $k + Z_k - 1 < r$, so r and l remain correctly unchanged.

In Case 2b, β must be a prefix of S (as argued in the body of the algorithm) and since any extension of this match is explicitly verified by comparing characters beyond r to characters beyond the prefix β, the full extent of the match is correctly computed. Hence Z_k is correctly obtained in this case. Furthermore, since $k + Z_k - 1 \geq r$, the algorithm correctly changes r and l. $\square$

Corollary 1.4.1. Repeating Algorithm Z for each position $i > 2$ correctly yields all the Z_i values.

Theorem 1.4.2. *All the $Z_i(S)$ values are computed by the algorithm in $O(|S|)$ time.*

PROOF   The time is proportional to the number of iterations, $|S|$, plus the number of character comparisons. Each comparison results in either a match or a mismatch, so we next bound the number of matches and mismatches that can occur.

Each iteration that performs any character comparisons at all ends the first time it finds a mismatch; hence there are at most $|S|$ mismatches during the entire algorithm. To bound the number of matches, note first that $r_k \geq r_{k-1}$ for every iteration k. Now, let k be an iteration where $q > 0$ matches occur. Then r_k is set to $r_{k-1} + q$ at least. Finally, $r_k \leq |S|$, so the total number of matches that occur during any execution of the algorithm is at most $|S|$. $\square$

1.5. The simplest linear-time exact matching algorithm

Before discussing the more complex (classical) exact matching methods, we show that fundamental preprocessing alone provides a simple linear-time exact matching algorithm. This is the simplest linear-time matching algorithm we know of.

Let $S = P\$T$ be the string consisting of P followed by the symbol "$\$$" followed by T, where "$\$$" is a character appearing in neither P nor T. Recall that P has length n and T has length m, and $n \leq m$. So, $S = P\$T$ has length $n + m + 1 = O(m)$. Compute $Z_i(S)$ for i from 2 to $n + m + 1$. Because "$\$$" does not appear in P or T, $Z_i \leq n$ for every $i > 1$. Any value of $i > n + 1$ such that $Z_i(S) = n$ identifies an occurrence of P in T starting at position $i - (n + 1)$ of T. Conversely, if P occurs in T starting at position j of T, then $Z_{(n+1)+j}$ must be equal to n. Since all the $Z_i(S)$ values can be computed in $O(n + m) = O(m)$ time, this approach identifies all the occurrences of P in T in $O(m)$ time.

The method can be implemented to use only $O(n)$ space (in addition to the space needed for pattern and text) independent of the size of the alphabet. Since $Z_i \leq n$ for all i, position k' (determined in step 2) will always fall inside P. Therefore, there is no need to record the Z values for characters in T. Instead, we only need to record the Z values for the n characters in P and also maintain the current l and r. Those values are sufficient to compute (but not store) the Z value of each character in T and hence to identify and output any position i where $Z_i = n$.

There is another characteristic of this method worth introducing here: The method is considered an *alphabet-independent* linear-time method. That is, we never had to assume that the alphabet size was finite or that we knew the alphabet ahead of time – a character comparison only determines whether the two characters match or mismatch; it needs no further information about the alphabet. We will see that this characteristic is also true of the Knuth-Morris-Pratt and Boyer–Moore algorithms, but not of the Aho–Corasick algorithm or methods based on suffix trees.

1.5.1. Why continue?

Since function Z_i can be computed for the pattern in linear time and can be used directly to solve the exact matching problem in $O(m)$ time (with only $O(n)$ additional space), why continue? In what way are more complex methods (Knuth-Morris-Pratt, Boyer–Moore, real-time matching, Apostolico–Giancarlo, Aho–Corasick, suffix tree methods, etc.) deserving of attention?

For the exact matching problem, the Knuth-Morris-Pratt algorithm has only a marginal advantage over the direct use of Z_i. However, it has historical importance and has been generalized, in the Aho–Corasick algorithm, to solve the problem of searching for a *set* of patterns in a text in time linear in the size of the text. That problem is not nicely solved using Z_i values alone. The real-time extension of Knuth-Morris-Pratt has an advantage in situations when text is input on-line and one has to be sure that the algorithm will be ready for each character as it arrives. The Boyer–Moore method is valuable because (with the proper implementation) it also runs in linear worst-case time but typically runs in *sublinear* time, examining only a fraction of the characters of T. Hence it is the preferred method in most cases. The Apostolico–Giancarlo method is valuable because it has all the advantages of the Boyer–Moore method and yet allows a relatively simple proof of linear worst-case running time. Methods based on suffix trees typically preprocess the text rather than the pattern and then lead to algorithms in which the search time is proportional to the size of the pattern rather than the size of the text. This is an extremely desirable feature. Moreover, suffix trees can be used to solve much more complex problems than exact matching, including problems that are not easily solved by direct application of the fundamental preprocessing.

1.6. Exercises

The first four exercises use the fact that fundamental processing can be done in linear time and that all occurrences of *P* in *T* can be found in linear time.

1. Use the existence of a linear-time exact matching algorithm to solve the following problem in linear time. Given two strings α and β, determine if α is a circular (or cyclic) rotation of β, that is, if α and β have the same length and α consists of a suffix of β followed by a prefix of β. For example, *defabc* is a circular rotation of *abcdef*. This is a classic problem with a very elegant solution.

2. Similar to Exercise 1, give a linear-time algorithm to determine whether a linear string α is a substring of a *circular string* β. A circular string of length n is a string in which character n is considered to precede character 1 (see Figure 1.6). Another way to think about this

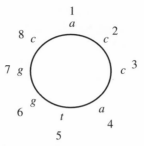

Figure 1.6: A circular string β. The linear string $\bar{\beta}$ derived from it is *accatggc*.

problem is the following. Let $\bar{\beta}$ be the *linear* string obtained from β starting at character 1 and ending at character n. Then α is a substring of circular string β if and only if α is a substring of some circular rotation of $\bar{\beta}$.

A digression on circular strings in DNA

The above two problems are mostly exercises in using the existence of a linear-time exact matching algorithm, and we don't know any critical biological problems that they address. However, we want to point out that *circular DNA* is common and important. Bacterial and mitochondrial DNA is typically circular, both in its genomic DNA and in additional small double-stranded circular DNA molecules called *plasmids*, and even some true eukaryotes (higher organisms whose cells contain a nucleus) such as yeast contain plasmid DNA in addition to their nuclear DNA. Consequently, tools for handling circular strings may someday be of use in those organisms. Viral DNA is not always circular, but even when it is linear some virus genomes exhibit circular properties. For example, in some viral populations the linear order of the DNA in one individual will be a circular rotation of the order in another individual [450]. Nucleotide mutations, in addition to rotations, occur rapidly in viruses, and a plausible problem is to determine if the DNA of two individual viruses have mutated away from each other only by a circular rotation, rather than additional mutations.

It is very interesting to note that the problems addressed in the exercises are actually "solved" in nature. Consider the special case of Exercise 2 when string α has length n. Then the problem becomes: Is α a circular rotation of $\bar{\beta}$? This problem is solved in linear time as in Exercise 1. Precisely this matching problem arises and is "solved" in *E. coli* replication under the certain experimental conditions described in [475]. In that experiment, an enzyme (RecA) and ATP molecules (for energy) are added to *E. coli* containing a single strand of one of its plasmids, called string β, and a double-stranded linear DNA molecule, one strand of which is called string α. If α is a circular rotation of $\bar{\beta}$ then the strand opposite to α (which has the DNA sequence complementary to α) hybridizes with β creating a proper double-stranded plasmid, leaving α as a single strand. This transfer of DNA may be a step in the replication of the plasmid. Thus the problem of determining whether α is a circular rotation of $\bar{\beta}$ is solved by this natural system.

Other experiments in [475] can be described as *substring* matching problems relating to circular and linear DNA in *E. coli*. Interestingly, these natural systems solve their matching problems faster than can be explained by kinetic analysis, and the molecular mechanisms used for such rapid matching remain undetermined. These experiments demonstrate the role of enzyme RecA in *E. coli* replication, but do not suggest immediate important computational problems. They do, however, provide indirect motivation for developing computational tools for handling circular strings as well as linear strings. Several other uses of circular strings will be discussed in Sections 7.13 and 16.17 of the book.

3. **Suffix-prefix matching.** Give an algorithm that takes in two strings α and β, of lengths n

and m, and finds the longest suffix of α that exactly matches a prefix of β. The algorithm should run in $O(n + m)$ time.

4. **Tandem arrays.** A substring α contained in string S is called a *tandem array* of β (called the base) if α consists of more than one consecutive copy of β. For example, if $S = xyzabcabcabcabcpq$, then $\alpha = abcabcabcabc$ is a tandem array of $\beta = abc$. Note that S also contains a tandem array of $abcabc$ (i.e., a tandem array with a longer base). A *maximal* tandem array is a tandem array that cannot be extended either left or right. Given the base β, a tandem array of β in S can be described by two numbers (s, k), giving its starting location in S and the number of times β is repeated. A tandem array is an example of a repeated substring (see Section 7.11.1).

Suppose S has length n. Give an example to show that two maximal tandem arrays of a given base β can overlap.

Now give an $O(n)$-time algorithm that takes S and β as input, finds every maximal tandem array of β, and outputs the pair (s, k) for each occurrence. Since maximal tandem arrays of a given base can overlap, a naive algorithm would establish only an $O(n^2)$-time bound.

5. If the Z algorithm finds that $Z_2 = q > 0$, all the values $Z_3, \ldots, Z_{q+1}, Z_{q+2}$ can then be obtained immediately without additional character comparisons and without executing the main body of Algorithm Z. Flesh out and justify the details of this claim.

6. In Case 2b of the Z algorithm, when $Z_{k'} \geq |\beta|$, the algorithm does explicit comparisons until it finds a mismatch. This is a reasonable way to organize the algorithm, but in fact Case 2b can be refined so as to eliminate an unneeded character comparison. Argue that when $Z_{k'} > |\beta|$ then $Z_k = |\beta|$ and hence no character comparisons are needed. Therefore, explicit character comparisons are needed only in the case that $Z_{k'} = |\beta|$.

7. If Case 2b of the Z algorithm is split into two cases, one for $Z_{k'} > |\beta|$ and one for $Z_{k'} = |\beta|$, would this result in an overall speedup of the algorithm? You must consider all operations, not just character comparisons.

8. Baker [43] introduced the following matching problem and applied it to a problem of software maintenance: "The application is to track down duplication in a large software system. We want to find not only exact matches between sections of code, but parameterized matches, where a parameterized match between two sections of code means that one section can be transformed into the other by replacing the parameter names (e.g., identifiers and constants) of one section by the parameter names of the other via a one-to-one function".

Now we present the formal definition. Let Σ and Π be two alphabets containing no symbols in common. Each symbol in Σ is called a *token* and each symbol in Π is called a *parameter*. A string can consist of any combinations of tokens and parameters from Σ and Π. For example, if Σ is the upper case English alphabet and Π is the lower case alphabet then $XYabCaCXZddW$ is a legal string over Σ and Π. Two strings S_1 and S_2 are said to *p-match* if and only if

a. Each token in S_1 (or S_2) is opposite a matching token in S_2 (or S_1).

b. Each parameter in S_1 (or S_2) is opposite a parameter in S_2 (or S_1).

c. For any parameter x, if one occurrence of x in S_1 (S_2) is opposite a parameter y in S_2 (S_1), then every occurrence of x in S_1 (S_2) must be opposite an occurrence of y in S_2 (S_1). In other words, the alignment of parameters in S_1 and S_2 defines a one-one correspondence between parameter names in S_1 and parameter names in S_2.

For example, $S_1 = XYabCaCXZddbW$ p-matches $S_2 = XYdxCdCXZccxW$. Notice that parameter a in S_1 maps to parameter d in S_2, while parameter d in S_1 maps to c in S_2. This does not violate the definition of *p-matching*.

In Baker's application, a token represents a part of the program that cannot be changed, whereas a parameter represents a program's variable, which can be renamed as long as all occurrences of the variable are renamed consistently. Thus if S_1 and S_2 p-match, then the variable names in S_1 could be changed to the corresponding variable names in S_2, making the two programs identical. If these two programs were part of a larger program, then they could both be replaced by a call to a single subroutine.

The most basic p-match problem is: Given a text T and a pattern P, each a string over Σ and Π, find all substrings of T that p-match P. Of course, one would like to find all those occurrences in $O(|P|+|T|)$ time. Let function Z_i^p for a string S be the length of the longest string starting at position i in S that p-matches a prefix of $S[1..i]$. Show how to modify algorithm Z to compute all the Z_i^p values in $O(|S|)$ time (the implementation details are slightly more involved than for function Z_i, but not too difficult). Then show how to use the modified algorithm Z to find all substrings of T that p-match P, in $O(|P|+|T|)$ time.

In [43] and [239], more involved versions of the p-match problem are solved by more complex methods.

The following three problems can be solved without the Z algorithm or other fancy tools. They only require thought.

9. You are given two strings of n characters each and an additional parameter k. In each string there are $n-k+1$ substrings of length k, and so there are $\Theta(n^2)$ pairs of substrings, where one substring is from one string and one is from the other. For a pair of substrings, we define the *match-count* as the number of opposing characters that match when the two substrings of length k are aligned. The problem is to compute the match-count for each of the $\Theta(n^2)$ pairs of substrings from the two strings. Clearly, the problem can be solved with $O(kn^2)$ operations (character comparisons plus arithmetic operations). But by better organizing the computations, the time can be reduced to $O(n^2)$ operations. (From Paul Horton.)

10. A DNA molecule can be thought of as a string over an alphabet of four characters $\{a, t, c, g\}$ (nucleotides), while a protein can be thought of as a string over an alphabet of twenty characters (amino acids). A gene, which is physically embedded in a DNA molecule, typically encodes the amino acid sequence for a particular protein. This is done as follows. Starting at a particular point in the DNA string, every three consecutive DNA characters encode a single amino acid character in the protein string. That is, three DNA nucleotides specify one amino acid. Such a coding triple is called a *codon*, and the full association of codons to amino acids is called the *genetic code*. For example, the codon *ttt* codes for the amino acid Phenylalanine (abbreviated in the single character amino acid alphabet as *F*), and the codon *gtt* codes for the amino acid Valine (abbreviated as *V*). Since there are $4^3 = 64$ possible triples but only twenty amino acids, there is a possibility that two or more triples form codons for the same amino acid and that some triples do not form codons. In fact, this is the case. For example, the amino acid Leucine is coded for by six different codons.

Problem: Suppose one is given a DNA string of n nucleotides, but you don't know the correct "reading frame". That is, you don't know if the correct decomposition of the string into codons begins with the first, second, or third nucleotide of the string. Each such "frameshift" potentially translates into a different amino acid string. (There are actually known genes where each of the three reading frames not only specifies a string in the amino acid alphabet, but each specifies a functional, yet different, protein.) The task is to produce, for each of the three reading frames, the associated amino acid string. For example, consider the string *atggacgga*. The first reading frame has three complete codons, *atg*, *gac*, and *gga*, which in the genetic code specify the amino acids *Met*, *Asp*, and *Gly*. The second reading frame has two complete codons, *tgg* and *acg*, coding for amino acids *Trp* and *Thr*. The third reading frame has two complete codons, *gga* and *cgg*, coding for amino acids *Gly* and *Arg*.

The goal is to produce the three translations, using the fewest number of character examinations of the DNA string and the fewest number of indexing steps (when using the codons to look up amino acids in a table holding the genetic code). Clearly, the three translations can be done with $3n$ examinations of characters in the DNA and $3n$ indexing steps in the genetic code table. Find a method that does the three translations in at most n character examinations and n indexing steps.

Hint: If you are acquainted with this terminology, the notion of a finite-state transducer may be helpful, although it is not necessary.

11. Let T be a text string of length m and let S be a *multiset* of n characters. The problem is to find all substrings in T of length n that are formed by the characters of S. For example, let $S = \{a, a, b, c\}$ and $T = abahgcabah$. Then *caba* is a substring of T formed from the characters of S.

Give a solution to this problem that runs in $O(m)$ time. The method should also be able to state, for each position i, the length of the longest substring in T starting at i that can be formed from S.

Fantasy protein sequencing. The above problem may become useful in sequencing protein from a particular organism after a large amount of the genome of that organism has been sequenced. This is most easily explained in prokaryotes, where the DNA is not interrupted by introns. In prokaryotes, the amino acid sequence for a given protein is encoded in a contiguous segment of DNA – one DNA codon for each amino acid in the protein. So assume we have the protein molecule but do not know its sequence or the location of the gene that codes for the protein. Presently, chemically determining the amino acid sequence of a protein is very slow, expensive, and somewhat unreliable. However, finding the multiset of amino acids that make up the protein is relatively easy. Now suppose that the whole DNA sequence for the genome of the organism is known. One can use that long DNA sequence to determine the amino acid sequence of a protein of interest. First, translate each codon in the DNA sequence into the amino acid alphabet (this may have to be done three times to get the proper frame) to form the string T; then chemically determine the multiset S of amino acids in the protein; then find all substrings in T of length $|S|$ that are formed from the amino acids in S. Any such substrings are candidates for the amino acid sequence of the protein, although it is unlikely that there will be more than one candidate. The match also locates the gene for the protein in the long DNA string.

12. Consider the two-dimensional variant of the preceding problem. The input consists of two-dimensional text (say a filled-in crossword puzzle) and a multiset of characters. The problem is to find a *connected* two-dimensional substructure in the text that matches all the characters in the multiset. How can this be done? A simpler problem is to restrict the structure to be rectangular.

13. As mentioned in Exercise 10, there are organisms (some viruses for example) containing intervals of DNA encoding not just a single protein, but three viable proteins, each read in a different reading frame. So, if each protein contains n amino acids, then the DNA string encoding those three proteins is only $n + 2$ nucleotides (characters) long. That is a very compact encoding.

(**Challenging problem?**) Give an algorithm for the following problem: The input is a protein string S_1 (over the amino acid alphabet) of length n and another protein string of length $m > n$. Determine if there is a string specifying a DNA encoding for S_2 that contains a substring specifying a DNA encoding of S_1. Allow the encoding of S_1 to begin at any point in the DNA string for S_2 (i.e., in any reading frame of that string). The problem is difficult because of the degeneracy of the genetic code and the ability to use any reading frame.

2

Exact Matching:
Classical Comparison-Based Methods

2.1. Introduction

This chapter develops a number of classical comparison-based matching algorithms for the exact matching problem. With suitable extensions, all of these algorithms can be implemented to run in linear worst-case time, and all achieve this performance by preprocessing pattern P. (Methods that preprocess T will be considered in Part II of the book.) The original preprocessing methods for these various algorithms are related in spirit but are quite different in conceptual difficulty. Some of the original preprocessing methods are quite difficult.[1] This chapter does not follow the original preprocessing methods but instead exploits fundamental preprocessing, developed in the previous chapter, to implement the needed preprocessing for each specific matching algorithm.

Also, in contrast to previous expositions, we emphasize the Boyer–Moore method over the Knuth-Morris-Pratt method, since Boyer–Moore is the practical method of choice for exact matching. Knuth-Morris-Pratt is nonetheless completely developed, partly for historical reasons, but mostly because it generalizes to problems such as real-time string matching and matching against a set of patterns more easily than Boyer–Moore does. These two topics will be described in this chapter and the next.

2.2. The Boyer–Moore Algorithm

As in the naive algorithm, the Boyer–Moore algorithm successively aligns P with T and then checks whether P matches the opposing characters of T. Further, after the check is complete, P is shifted right relative to T just as in the naive algorithm. However, the Boyer–Moore algorithm contains three clever ideas not contained in the naive algorithm: the right-to-left scan, the bad character shift rule, and the good suffix shift rule. Together, these ideas lead to a method that typically examines fewer than $m + n$ characters (an expected sublinear-time method) and that (with a certain extension) runs in linear worst-case time. Our discussion of the Boyer–Moore algorithm, and extensions of it, concentrates on *provable* aspects of its behavior. Extensive experimental and practical studies of Boyer–Moore and variants have been reported in [229], [237], [409], [410], and [425].

2.2.1. Right-to-left scan

For any alignment of P with T the Boyer–Moore algorithm checks for an occurrence of P by scanning characters from *right to left* rather than from left to right as in the naive

---

[1] Sedgewick [401] writes "Both the Knuth-Morris-Pratt and the Boyer–Moore algorithms require some complicated preprocessing on the pattern that is difficult to understand and has limited the extent to which they are used". In agreement with Sedgewick, I still do not understand the original Boyer–Moore preprocessing method for the *strong* good suffix rule.

algorithm. For example, consider the alignment of P against T shown below:

```
              1         2
              12345678901234567
    T:        xpbctbxabpqxctbpq
    P:          tpabxab
```

To check whether P occurs in T at this position, the Boyer–Moore algorithm starts at the *right* end of P, first comparing $T(9)$ with $P(7)$. Finding a match, it then compares $T(8)$ with $P(6)$, etc., moving right to left until it finds a mismatch when comparing $T(5)$ with $P(3)$. At that point P is shifted *right* relative to T (the amount for the shift will be discussed below) and the comparisons begin again at the right end of P.

Clearly, if P is shifted right by one place after each mismatch, or after an occurrence of P is found, then the worst-case running time of this approach is $O(nm)$ just as in the naive algorithm. So at this point it isn't clear why comparing characters from right to left is any better than checking from left to right. However, with two additional ideas (the *bad character* and the *good suffix* rules), shifts of more than one position often occur, and in typical situations large shifts are common. We next examine these two ideas.

2.2.2. Bad character rule

To get the idea of the bad character rule, suppose that the last (right-most) character of P is y and the character in T it aligns with is $x \neq y$. When this initial mismatch occurs, if we know the right-most position in P of character x, we can safely shift P to the right so that the right-most x in P is below the mismatched x in T. Any shorter shift would only result in an immediate mismatch. Thus, the longer shift is correct (i.e., it will not shift past any occurrence of P in T). Further, if x never occurs in P, then we can shift P completely past the point of mismatch in T. In these cases, some characters of T will never be examined and the method will actually run in "sublinear" time. This observation is formalized below.

Definition For each character x in the alphabet, let $R(x)$ be the position of right-most occurrence of character x in P. $R(x)$ is defined to be zero if x does not occur in P.

It is easy to preprocess P in $O(n)$ time to collect the $R(x)$ values, and we leave that as an exercise. Note that this preprocessing does not require the fundamental preprocessing discussed in Chapter 1 (that will be needed for the more complex shift rule, the good suffix rule).

We use the R values in the following way, called the *bad character shift rule*:

Suppose for a particular alignment of P against T, the right-most $n - i$ characters of P match their counterparts in T, but the next character to the left, $P(i)$, mismatches with its counterpart, say in position k of T. The *bad character rule* says that P should be shifted right by $\max[1, i - R(T(k))]$ places. That is, if the right-most occurrence in P of character $T(k)$ is in position $j < i$ (including the possibility that $j = 0$), then shift P so that character j of P is below character k of T. Otherwise, shift P by one position.

The point of this shift rule is to shift P by more than one character when possible. In the above example, $T(5) = t$ mismatches with $P(3)$ and $R(t) = 1$, so P can be shifted right by two positions. After the shift, the comparison of P and T begins again at the right end of P.

Extended bad character rule

The bad character rule is a useful heuristic for mismatches near the right end of P, but it has no effect if the mismatching character from T occurs in P to the right of the mismatch point. This may be common when the alphabet is small and the text contains many similar, but not exact, substrings. That situation is typical of DNA, which has an alphabet of size four, and even protein, which has an alphabet of size twenty, often contains different regions of high similarity. In such cases, the following *extended bad character rule* is more robust:

> When a mismatch occurs at position i of P and the mismatched character in T is x, then shift P to the right so that the closest x to the left of position i in P is below the mismatched x in T.

Because the extended rule gives larger shifts, the only reason to prefer the simpler rule is to avoid the added implementation expense of the extended rule. The simpler rule uses only $O(|\Sigma|)$ space (Σ is the alphabet) for array R, and one table lookup for each mismatch. As we will see, the extended rule can be implemented to take only $O(n)$ space and at most one extra step per character comparison. That amount of added space is not often a critical issue, but it is an empirical question whether the longer shifts make up for the added time used by the extended rule. The original Boyer–Moore algorithm only uses the simpler bad character rule.

Implementing the extended bad character rule

We preprocess P so that the extended bad character rule can be implemented efficiently in both time and space. The preprocessing should discover, for each position i in P and for each character x in the alphabet, the position of the closest occurrence of x in P to the left of i. The obvious approach is to use a two-dimensional array of size n by $|\Sigma|$ to store this information. Then, when a mismatch occurs at position i of P and the mismatching character in T is x, we look up the (i, x) entry in the array. The lookup is fast, but the size of the array, and the time to build it, may be excessive. A better compromise, below, is possible.

During preprocessing, scan P from right to left collecting, for each character x in the alphabet, a list of the positions where x occurs in P. Since the scan is right to left, each list will be in decreasing order. For example, if $P = abacbabc$ then the list for character a is 6, 3, 1. These lists are accumulated in $O(n)$ time and of course take only $O(n)$ space. During the search stage of the Boyer–Moore algorithm if there is a mismatch at position i of P and the mismatching character in T is x, scan x's list from the top until we reach the first number less than i or discover there is none. If there is none then there is no occurrence of x before i, and all of P is shifted past the x in T. Otherwise, the found entry gives the desired position of x.

After a mismatch at position i of P the time to scan the list is at most $n - i$, which is roughly the number of characters that matched. So in worst case, this approach at most doubles the running time of the Boyer–Moore algorithm. However, in most problem settings the added time will be vastly less than double. One could also do binary search on the list in circumstances that warrant it.

2.2.3. The (strong) good suffix rule

The bad character rule by itself is reputed to be highly effective in practice, particularly for English text [229], but proves less effective for small alphabets and it does not lead to a linear worst-case running time. For that, we introduce another rule called the *strong*

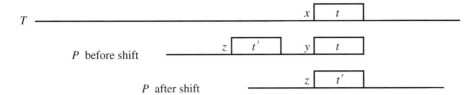

Figure 2.1: Good suffix shift rule, where character x of T mismatches with character y of P. Characters y and z of P are guaranteed to be distinct by the good suffix rule, so z has a chance of matching x.

good suffix rule. The original preprocessing method [278] for the strong good suffix rule is generally considered quite difficult and somewhat mysterious (although a weaker version of it is easy to understand). In fact, the preprocessing for the strong rule was given incorrectly in [278] and corrected, without much explanation, in [384]. Code based on [384] is given without real explanation in the text by Baase [32], but there are no published sources that try to fully explain the method.[2] Pascal code for strong preprocessing, based on an outline by Richard Cole [107], is shown in Exercise 24 at the end of this chapter.

In contrast, the fundamental preprocessing of P discussed in Chapter 1 makes the needed preprocessing very simple. That is the approach we take here. The *strong good suffix rule* is:

Suppose for a given alignment of P and T, a substring t of T matches a suffix of P, but a mismatch occurs at the next comparison to the left. Then find, if it exists, the right-most copy t' of t in P such that t' is not a suffix of P and *the character to the left of t' in P differs from the character to the left of t in P*. Shift P to the right so that substring t' in P is below substring t in T (see Figure 2.1). If t' does not exist, then shift the left end of P past the left end of t in T by the least amount so that a prefix of the shifted pattern matches a suffix of t in T. If no such shift is possible, then shift P by n places to the right. If an occurrence of P is found, then shift P by the least amount so that a *proper* prefix of the shifted P matches a suffix of the occurrence of P in T. If no such shift is possible, then shift P by n places, that is, shift P past t in T.

For a specific example consider the alignment of P and T given below:

```
         0         1
         123456789012345678
      T: prstabstubabvqxrst
                 *
      P:    qcabdabdab
            1234567890
```

When the mismatch occurs at position 8 of P and position 10 of T, $t = ab$ and t' occurs in P starting at position 3. Hence P is shifted right by *six* places, resulting in the following alignment:

---

[2] A recent plea appeared on the internet newsgroup comp. theory:

> I am looking for an elegant (easily understandable) proof of correctness for a part of the Boyer–Moore string matching algorithm. The difficult-to-prove part here is the algorithm that computes the dd_2 (good-suffix) table. I didn't find much of an understandable proof yet, so I'd much appreciate any help!

```
             0            1
             123456789012345678
      T:  prstabstubabvqxrst
      P:             qcabdabdab
```

Note that the extended bad character rule would have shifted P by only one place in this example.

Theorem 2.2.1. *The use of the good suffix rule never shifts P past an occurrence in T.*

PROOF Suppose the right end of P is aligned with character k of T before the shift, and suppose that the good suffix rule shifts P so its right end aligns with character $k' > k$. Any occurrence of P ending at a position l strictly between k and k' would immediately violate the selection rule for k', since it would imply either that a closer copy of t occurs in P or that a longer prefix of P matches a suffix of t. □

The original published Boyer–Moore algorithm [75] uses a simpler, weaker, version of the good suffix rule. That version just requires that the shifted P agree with the t and does not specify that the next characters to the left of those occurrences of t be different. An explicit statement of the weaker rule can be obtained by deleting the italics phrase in the first paragraph of the statement of the strong good suffix rule. In the previous example, the weaker shift rule shifts P by three places rather than six. When we need to distinguish the two rules, we will call the simpler rule the *weak* good suffix rule and the rule stated above the *strong* good suffix rule. For the purpose of proving that the search part of Boyer–Moore runs in linear worst-case time, the weak rule is not sufficient, and in this book the strong version is assumed unless stated otherwise.

2.2.4. Preprocessing for the good suffix rule

We now formalize the preprocessing needed for the Boyer–Moore algorithm.

Definition For each i, $L(i)$ is the largest position less than n such that string $P[i..n]$ matches a suffix of $P[1..L(i)]$. $L(i)$ is defined to be zero if there is no position satisfying the conditions. For each i, $L'(i)$ is the largest position less than n such that string $P[i..n]$ matches a suffix of $P[1..L'(i)]$ and such that the character preceding that suffix is not equal to $P(i-1)$. $L'(i)$ is defined to be zero if there is no position satisfying the conditions.

For example, if $P = cabdabdab$, then $L(8) = 6$ and $L'(8) = 3$.

$L(i)$ gives the right end-position of the right-most copy of $P[i..n]$ that is not a suffix of P, whereas $L'(i)$ gives the right end-position of the right-most copy of $P[i..n]$ that is not a suffix of P, with the stronger, added condition that its preceding character is unequal to $P(i-1)$. So, in the strong-shift version of the Boyer–Moore algorithm, if character $i-1$ of P is involved in a mismatch and $L'(i) > 0$, then P is shifted right by $n - L'(i)$ positions. The result is that if the right end of P was aligned with position k of T before the shift, then position $L'(i)$ is now aligned with position k.

During the preprocessing stage of the Boyer–Moore algorithm $L'(i)$ (and $L(i)$, if desired) will be computed for each position i in P. This is done in $O(n)$ time via the following definition and theorem.

Definition For string P, $N_j(P)$ is the length of the longest *suffix* of the substring $P[1..j]$ that is also a *suffix* of the full string P.

For example, if $P = cabdabdab$, then $N_3(P) = 2$ and $N_6(P) = 5$.

Recall that $Z_i(S)$ is the length of the longest substring of S that starts at i and matches a prefix of S. Clearly, N is the reverse of Z, that is, if P^r denotes the string obtained by reversing P, then $N_j(P) = Z_{n-j+1}(P^r)$. Hence the $N_j(P)$ values can be obtained in $O(n)$ time by using *Algorithm Z* on P^r. The following theorem is then immediate.

Theorem 2.2.2. *$L(i)$ is the largest index j less than n such that $N_j(P) \geq |P[i..n]|$ (which is $n-i+1$). $L'(i)$ is the largest index j less than n such that $N_j(P) = |P[i..n]| = (n-i+1)$.*

Given Theorem 2.2.2, it follows immediately that all the $L'(i)$ values can be accumulated in linear time from the N values using the following algorithm:

Z-based Boyer–Moore

for $i := 1$ to n do $L'(i) := 0$;
for $j := 1$ to $n - 1$ do
    begin
    $i := n - N_j(P) + 1$;
    $L'(i) := j$;
    end;

The $L(i)$ values (if desired) can be obtained by adding the following lines to the above pseudocode:

$L(2) := L'(2)$;
for $i := 3$ to n do $L(i) := \max[L(i - 1), L'(i)]$;

Theorem 2.2.3. *The above method correctly computes the L values.*

PROOF $L(i)$ marks the right end-position of the right-most substring of P that matches $P[i..n]$ and is not a suffix of $P[1..n]$. Therefore, that substring begins at position $L(i)-n+i$, which we will denote by j. We will prove that $L(i) = \max[L(i - 1), L'(i)]$ by considering what character $j - 1$ is. First, if $j = 1$ then character $j - 1$ doesn't exist, so $L(i - 1) = 0$ and $L'(i) = 1$. So suppose that $j > 1$. If character $j - 1$ equals character $i - 1$ then $L(i) = L(i - 1)$. If character $j - 1$ does not equal character $i - 1$ then $L(i) = L'(i)$. Thus, in all cases, $L(i)$ must either be $L'(i)$ or $L(i - 1)$.

However, $L(i)$ must certainly be greater than or equal to both $L'(i)$ and $L(i - 1)$. In summary, $L(i)$ must either be $L'(i)$ or $L(i - 1)$, and yet it must be greater or equal to both of them; hence $L(i)$ must be the maximum of $L'(i)$ and $L(i - 1)$.   □

Final preprocessing detail

The preprocessing stage must also prepare for the case when $L'(i) = 0$ or when an occurrence of P is found. The following definition and theorem accomplish that.

Definition   Let $l'(i)$ denote the length of the largest suffix of $P[i..n]$ that is also a prefix of P, if one exists. If none exists, then let $l'(i)$ be zero.

Theorem 2.2.4. *$l'(i)$ equals the largest $j \leq |P[i..n]|$, which is $n - i + 1$, such that $N_j(P) = j$.*

We leave the proof, as well as the problem of how to accumulate the $l'(i)$ values in linear time, as a simple exercise. (Exercise 9 of this chapter)

2.2.5. The good suffix rule in the search stage of Boyer–Moore

Having computed $L'(i)$ and $l'(i)$ for each position i in P, these preprocessed values are used during the search stage of the algorithm to achieve larger shifts. If, during the search stage, a mismatch occurs at position $i - 1$ of P and $L'(i) > 0$, then the good suffix rule shifts P by $n - L'(i)$ places to the right, so that the $L'(i)$-length prefix of the shifted P aligns with the $L'(i)$-length suffix of the unshifted P. In the case that $L'(i) = 0$, the good suffix rule shifts P by $n - l'(i)$ places. When an occurrence of P is found, then the rule shifts P by $n - l'(2)$ places. Note that the rules work correctly even when $l'(i) = 0$.

One special case remains. When the first comparison is a mismatch (i.e., $P(n)$ mismatches) then P should be shifted one place to the right.

2.2.6. The complete Boyer–Moore algorithm

We have argued that neither the good suffix rule nor the bad character rule shift P so far as to miss any occurrence of P. So the Boyer–Moore algorithm shifts by the largest amount given by either of the rules. We can now present the complete algorithm.

The Boyer–Moore algorithm

{Preprocessing stage}
    Given the pattern P,
    Compute $L'(i)$ and $l'(i)$ for each position i of P,
    and compute $R(x)$ for each character $x \in \Sigma$.
{Search stage}
    $k := n$;
    while $k \le m$ do
        begin
        $i := n$;
        $h := k$;
        while $i > 0$ and $P(i) = T(h)$ do
            begin
            $i := i - 1$;
            $h := h - 1$;
            end;
        if $i = 0$ then
            begin
            report an occurrence of P in T ending at position k.
            $k := k + n - l'(2)$;
            end
        else
            shift P (increase k) by the maximum amount determined by the
            (extended) bad character rule and the good suffix rule.
        end;

Note that although we have always talked about "shifting P", and given rules to determine by how much P should be "shifted", there is no shifting in the actual implementation. Rather, the index k is increased to the point where the right end of P would be "shifted". Hence, each act of shifting P takes constant time.

We will later show, in Section 3.2, that by using the strong good suffix rule alone, the

Boyer–Moore method has a worst-case running time of $O(m)$ provided that the pattern does not appear in the text. This was first proved by Knuth, Morris, and Pratt [278], and an alternate proof was given by Guibas and Odlyzko [196]. Both of these proofs were quite difficult and established worst-case time bounds no better than $5m$ comparisons. Later, Richard Cole gave a much simpler proof [108] establishing a bound of $4m$ comparisons and also gave a difficult proof establishing a tight bound of $3m$ comparisons. We will present Cole's proof of $4m$ comparisons in Section 3.2.

When the pattern does appear in the text then the original Boyer–Moore method runs in $\Theta(nm)$ worst-case time. However, several simple modifications to the method correct this problem, yielding an $O(m)$ time bound in all cases. The first of these modifications was due to Galil [168]. After discussing Cole's proof, in Section 3.2, for the case that P doesn't occur in T, we use a variant of Galil's idea to achieve the linear time bound in all cases.

At the other extreme, if we only use the bad character shift rule, then the worst-case running time is $O(nm)$, but assuming randomly generated strings, the expected running time is sublinear. Moreover, in typical string matching applications involving natural language text, a sublinear running time is almost always observed in practice. We won't discuss random string analysis in this book but refer the reader to [184].

Although Cole's proof for the linear worst case is vastly simpler than earlier proofs, and is important in order to complete the full story of Boyer–Moore, it is not trivial. However, a fairly simple extension of the Boyer–Moore algorithm, due to Apostolico and Giancarlo [26], gives a "Boyer–Moore–like" algorithm that allows a fairly direct proof of a $2m$ worst-case bound on the number of comparisons. The Apostolico–Giancarlo variant of Boyer–Moore is discussed in Section 3.1.

2.3. The Knuth-Morris-Pratt algorithm

The best known linear-time algorithm for the exact matching problem is due to Knuth, Morris, and Pratt [278]. Although it is rarely the method of choice, and is often much inferior in practice to the Boyer–Moore method (and others), it can be simply explained, and its linear time bound is (fairly) easily proved. The algorithm also forms the basis of the well-known Aho–Corasick algorithm, which efficiently finds all occurrences in a text of any pattern from a *set* of patterns.[3]

2.3.1. The Knuth-Morris-Pratt shift idea

For a given alignment of P with T, suppose the naive algorithm matches the first i characters of P against their counterparts in T and then mismatches on the next comparison. The naive algorithm would shift P by just *one* place and begin comparing again from the left end of P. But a larger shift may often be possible. For example, if $P = abcxabcde$ and, in the present alignment of P with T, the mismatch occurs in position 8 of P, then it is easily deduced (and we will prove below) that P can be shifted by four places without passing over any occurrences of P in T. Notice that this can be deduced without even knowing what string T is or exactly how P is aligned with T. Only the location of the mismatch in P must be known. The Knuth-Morris-Pratt algorithm is based on this kind of reasoning to make larger shifts than the naive algorithm makes. We now formalize this idea.

---

[3] We will present several solutions to that set problem including the Aho–Corasick method in Section 3.4. For those reasons, and for its historical role in the field, we fully develop the Knuth-Morris-Pratt method here.

Definition   For each position i in pattern P, define $sp_i(P)$ to be the *length* of the longest proper *suffix* of $P[1..i]$ that matches a prefix of P.

Stated differently, $sp_i(P)$ is the length of the longest proper substring of $P[1..i]$ that ends at i and that matches a prefix of P. When the string is clear by context we will use sp_i in place of the full notation.

For example, if $P = abcaeabcabd$, then $sp_2 = sp_3 = 0, sp_4 = 1, sp_8 = 3$, and $sp_{10} = 2$. Note that by definition, $sp_1 = 0$ for any string.

An optimized version of the Knuth-Morris-Pratt algorithm uses the following values.

Definition   For each position i in pattern P, define $sp_i'(P)$ to be the length of the longest proper suffix of $P[1..i]$ that matches a prefix of P, *with the added condition that characters $P(i + 1)$ and $P(sp_i' + 1)$ are unequal.*

Clearly, $sp_i'(P) \leq sp_i(P)$ for all positions i and any string P. As an example, if $P = bbccaebbcabd$, then $sp_8 = 2$ because string bb occurs both as a proper prefix of $P[1..8]$ and as a suffix of $P[1..8]$. However, both copies of the string are followed by the same character c, and so $sp_8' < 2$. In fact, $sp_8' = 1$ since the single character b occurs as both the first and last character of $P[1..8]$ and is followed by character b in position 2 and by character c in position 9.

The Knuth-Morris-Pratt shift rule

We will describe the algorithm in terms of the sp' values, and leave it to the reader to modify the algorithm if only the weaker sp values are used.[4] The Knuth-Morris-Pratt algorithm aligns P with T and then compares the aligned characters from *left to right*, as the naive algorithm does.

For any alignment of P and T, if the first mismatch (comparing from left to right) occurs in position $i + 1$ of P and position k of T, then shift P to the right (relative to T) so that $P[1..sp_i']$ aligns with $T[k - sp_i'..k - 1]$. In other words, shift P exactly $i + 1 - (sp_i' + 1) = i - sp_i'$ places to the right, so that character $sp_i' + 1$ of P will align with character k of T. In the case that an occurrence of P has been found (no mismatch), shift P by $n - sp_n'$ places.

The shift rule guarantees that the prefix $P[1..sp_i']$ of the shifted P matches its opposing substring in T. The next comparison is then made between characters $T(k)$ and $P[sp_i' + 1]$. The use of the stronger shift rule based on sp_i' guarantees that the same mismatch will not occur again in the new alignment, but it does not guarantee that $T(k) = P[sp_i' + 1]$.

In the above example, where $P = abcxabcde$ and $sp_7' = 3$, if character 8 of P mismatches then P will be shifted by $7 - 3 = 4$ places. This is true even without knowing T or how P is positioned with T.

The advantage of the shift rule is twofold. First, it often shifts P by more than just a single character. Second, after a shift, the left-most sp_i' characters of P are guaranteed to match their counterparts in T. Thus, to determine whether the newly shifted P matches its counterpart in T, the algorithm can start comparing P and T at position $sp_i' + 1$ of P (and position k of T). For example, suppose $P = abcxabcde$ as above, $T = xyabcxabcxadcdqfeg$, and the left end of P is aligned with character 3 of T. Then P and T will match for 7 characters but mismatch on character 8 of P, and P will be shifted

---

[4] The reader should be alerted that traditionally the Knuth-Morris-Pratt algorithm has been described in terms of *failure functions*, which are related to the sp_i values. Failure functions will be explicitly defined in Section 2.3.3.

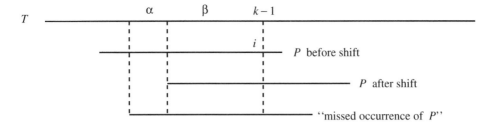

Figure 2.2: Assumed missed occurrence used in correctness proof for Knuth-Morris-Pratt.

by 4 places as shown below:

```
1         2
123456789012345678
xyabcxabcxadcdqfeg
   abcxabcde
     abcxabcde
```

As guaranteed, the first 3 characters of the shifted P match their counterparts in T (and their counterparts in the unshifted P).

Summarizing, we have

Theorem 2.3.1. *After a mismatch at position $i + 1$ of P and a shift of $i - sp_i'$ places to the right, the left-most sp_i' characters of P are guaranteed to match their counterparts in T.*

Theorem 2.3.1 partially establishes the correctness of the Knuth-Morris-Pratt algorithm, but to fully prove correctness we have to show that the shift rule never shifts too far. That is, using the shift rule no occurrence of P will ever be overlooked.

Theorem 2.3.2. *For any alignment of P with T, if characters 1 through i of P match the opposing characters of T but character $i + 1$ mismatches $T(k)$, then P can be shifted by $i - sp_i'$ places to the right without passing any occurrence of P in T.*

PROOF   Suppose not, so that there is an occurrence of P starting strictly to the left of the shifted P (see Figure 2.2), and let α and β be the substrings shown in the figure. In particular, β is the prefix of P of length sp_i', shown relative to the shifted position of P. The unshifted P matches T up through position i of P and position $k - 1$ of T, and all characters in the (assumed) missed occurrence of P match their counterparts in T. Both of these matched regions contain the substrings α and β, so the unshifted P and the assumed occurrence of P match on the entire substring $\alpha\beta$. Hence $\alpha\beta$ is a suffix of $P[1..i]$ that matches a proper prefix of P. Now let $l = |\alpha\beta| + 1$ so that position l in the "missed occurrence" of P is opposite position k in T. Character $P(l)$ cannot be equal to $P(i + 1)$ since $P(l)$ is assumed to match $T(k)$ and $P(i + 1)$ does not match $T(k)$. Thus $\alpha\beta$ is a proper suffix of $P[1..i]$ that matches a prefix of P, and the next character is unequal to $P(i + 1)$. But $|\alpha| > 0$ due to the assumption that an occurrence of P starts strictly before the shifted P, so $|\alpha\beta| > |\beta| = sp_i'$, contradicting the definition of sp_i'. Hence the theorem is proved.   □

Theorem 2.3.2 says that the Knuth-Morris-Pratt shift rule does not miss any occurrence of P in T, and so the Knuth-Morris-Pratt algorithm will correctly find all occurrences of P in T. The time analysis is equally simple.

Theorem 2.3.3. *In the Knuth-Morris-Pratt method, the number of character comparisons is at most 2m.*

PROOF   Divide the algorithm into compare/shift phases, where a single phase consists of the comparisons done between successive shifts. After any shift, the comparisons in the phase go left to right and start either with the last character of T compared in the previous phase or with the character to its right. Since P is never shifted left, in any phase at most one comparison involves a character of T that was previously compared. Thus, the total number of character comparisons is bounded by $m + s$, where s is the number of shifts done in the algorithm. But $s < m$ since after m shifts the right end of P is certainly to the right of the right end of T, so the number of comparisons done is bounded by $2m$.   □

2.3.2. Preprocessing for Knuth-Morris-Pratt

The key to the speed up of the Knuth-Morris-Pratt algorithm over the naive algorithm is the use of sp' (or sp) values. It is easy to see how to compute all the sp' and sp values from the Z values obtained during the fundamental preprocessing of P. We verify this below.

> **Definition**   Position $j > 1$ *maps to* i if $i = j + Z_j(P) - 1$. That is, j maps to i if i is the right end of a Z-box starting at j.

Theorem 2.3.4. *For any* $i > 1$, $sp'_i(P) = Z_j = i - j + 1$, *where* $j > 1$ *is the smallest position that maps to* i. *If there is no such* j *then* $sp'_i(P) = 0$. *For any* $i > 1$, $sp_i(P) = i - j + 1$, *where* j *is the smallest position in the range* $1 < j \leq i$ *that maps to* i *or beyond. If there is no such* j, *then* $sp_i(P) = 0$.

PROOF   If $sp'_i(P)$ is greater than zero, then there is a proper suffix α of $P[1..i]$ that matches a prefix of P, such that $P[i + 1]$ does not match $P[|\alpha| + 1]$. Therefore, letting j denote the start of α, $Z_j = |\alpha| = sp'_i(P)$ and j maps to i. Hence, if there is no j in the range $1 < j \leq i$ that maps to i, then $sp'_i(P)$ must be zero.

Now suppose $sp'_i(P) > 0$ and let j be as defined above. We claim that j is the smallest position in the range 2 to i that maps to i. Suppose not, and let j^* be a position in the range $1 < j^* < j$ that maps to i. Then $P[j^*..i]$ would be a proper suffix of $P[1..i]$ that matches a prefix (call it β) of P. Moreover, by the definition of mapping, $P(i + 1) \neq P(|\beta|)$, so $sp'_i(P) \geq |\beta| > |\alpha|$, contradicting the assumption that $sp'_i = \alpha$.

The proofs of the claims for $sp_i(P)$ are similar and are left as exercises.   □

Given Theorem 2.3.4, all the sp' and sp values can be computed in linear time using the Z_i values as follows:

Z-based Knuth-Morris-Pratt

```
for i := 1 to n do
     sp'_i := 0;
for j := n downto 2 do
begin
     i := j + Z_j(P) - 1;
     sp'_i := Z_j;
end;
```

The sp values are obtained by adding the following:

$sp_n(P) := sp'_n(P);$
for $i := n - 1$ downto 2 do
    $sp_i(P) := \max[sp_{i+1}(P) - 1, sp'_i(P)]$

2.3.3. A full implementation of Knuth-Morris-Pratt

We have described the Knuth-Morris-Pratt algorithm in terms of shifting P, but we never accounted for time needed to implement shifts. The reason is that shifting is only conceptual and P is never explicitly shifted. Rather, as in the case of Boyer–Moore, pointers to P and T are incremented. We use pointer p to point into P and one pointer c (for "current" character) to point into T.

> **Definition** For each position i from 1 to $n + 1$, define the failure function $F'(i)$ to be $sp'_{i-1} + 1$ (and define $F(i) = sp_{i-1} + 1$), where sp'_0 and sp_0 are defined to be zero.

We will only use the (stronger) failure function $F'(i)$ in this discussion but will refer to $F(i)$ later.

After a mismatch in position $i + 1 > 1$ of P, the Knuth-Morris-Pratt algorithm "shifts" P so that the next comparison is between the character in position c of T and the character in position $sp'_i + 1$ of P. But $sp'_i + 1 = F'(i + 1)$, so a general "shift" can be implemented in constant time by just setting p to $F'(i + 1)$. Two special cases remain. When the mismatch occurs in position 1 of P, then p is set to $F'(1) = 1$ and c is incremented by one. When an occurrence of P is found, then P is shifted right by $n - sp'_n$ places. This is implemented by setting $F'(n + 1)$ to $sp'_n + 1$.

Putting all the pieces together gives the full Knuth-Morris-Pratt algorithm.

Knuth-Morris-Pratt algorithm

begin
Preprocess P to find $F'(k) = sp'_{k-1} + 1$ for k from 1 to $n + 1$.
    $c := 1;$
    $p := 1;$
    While $c + (n - p) \le m$
    do begin
        While $P(p) = T(c)$ and $p \le n$
        do begin
            $p := p + 1;$
            $c := c + 1;$
        end;
    if $p = n + 1$ then
        report an occurrence of P starting at position $c - n$ of T.
    if $p := 1$ then $c := c + 1$
    $p := F'(p);$
    end;
end.

2.4. Real-time string matching

In the search stage of the Knuth-Morris-Pratt algorithm, P is aligned against a substring of T and the two strings are compared left to right until either all of P is exhausted (in which case an occurrence of P in T has been found) or until a mismatch occurs at some positions $i+1$ of P and k of T. In the latter case, if $sp'_i > 0$, then P is shifted right by $i - sp'_i$ positions, guaranteeing that the prefix $P[1..sp'_i]$ of the shifted pattern matches its opposing substring in T. No explicit comparison of those substrings is needed, and the next comparison is between characters $T(k)$ and $P(sp'_i + 1)$. Although the shift based on sp'_i guarantees that $P(i + 1)$ differs from $P(sp'_i + 1)$, it does not guarantee that $T(k) = P(sp'_i + 1)$. Hence $T(k)$ might be compared several times (perhaps $\Omega(|P|)$ times) with differing characters in P. For that reason, the Knuth-Morris-Pratt method is not a *real-time* method.

To be real time, a method must do at most a *constant* amount of work between the time it first examines any position in T and the time it last examines that position. In the Knuth-Morris-Pratt method, if a position of T is involved in a match, it is never examined again (this is easy to verify) but, as indicated above, this is not true when the position is involved in a mismatch. Note that the definition of real time only concerns the search stage of the algorithm. Preprocessing of P need not be real time. Note also that if the search stage is real time it certainly is also linear time.

The utility of a real-time matcher is two fold. First, in certain applications, such as when the characters of the text are being sent to a small memory machine, one might need to guarantee that each character can be fully processed before the next one is due to arrive. If the processing time for each character is constant, independent of the length of the string, then such a guarantee may be possible. Second, in this particular real-time matcher, the shifts of P may be longer but never shorter than in the original Knuth-Morris-Pratt algorithm. Hence, the real-time matcher may run faster in certain problem instances.

Admittedly, arguments in favor of real-time matching algorithms over linear-time methods are somewhat tortured, and the real-time matching is more a theoretical issue than a practical one. Still, it seems worthwhile to spend a little time discussing real-time matching.

2.4.1. Converting Knuth-Morris-Pratt to a real-time method

We will use the Z values obtained during fundamental preprocessing of P to convert the Knuth-Morris-Pratt method into a real-time method. The required preprocessing of P is quite similar to the preprocessing done in Section 2.3.2 for the Knuth-Morris-Pratt algorithm. For historical reasons, the resulting real-time method is generally referred to as a *deterministic finite-state string matcher* and is often represented with a finite state machine diagram. We will not use this terminology here and instead represent the method in pseudo code.

> **Definition** Let x denote a character of the alphabet. For each position i in pattern P, define $sp'_{(i,x)}(P)$ to be the length of the longest proper suffix of $P[1..i]$ that matches a prefix of P, *with the added condition that character $P(sp'_i + 1)$ is x.*

Knowing the $sp'_{(i,x)}$ values for each character x in the alphabet allows a shift rule that converts the Knuth-Morris-Pratt method into a real-time algorithm. Suppose P is compared against a substring of T and a mismatch occurs at characters $T(k) = x$ and $P(i+1)$. Then P should be shifted right by $i - sp'_{(i,x)}$ places. This shift guarantees that the prefix $P[1..sp'_{(i,x)}]$ matches the opposing substring in T and that $T(k)$ matches the next character in P. Hence, the comparison between $T(k)$ and $P(sp'_{(i,x)} + 1)$ can be skipped. The next needed comparison is between characters $P(sp'_{(i,x)} + 2)$ and $T(k + 1)$. With this shift rule, the method becomes real time because it still never reexamines a position in T involved in a match (a feature inherited from the Knuth-Morris-Pratt algorithm), and it now also never reexamines a position involved in a mismatch. So, the search stage of this algorithm never examines a character in T more than once. It follows that the search is done in real time. Below we show how to find all the $sp'_{(i,x)}$ values in linear time. Together, this gives an algorithm that does linear preprocessing of P and real-time search of T.

It is easy to establish that the algorithm finds all occurrences of P in T, and we leave that as an exercise.

2.4.2. Preprocessing for real-time string matching

Theorem 2.4.1. *For* $P[i+1] \neq x$, $sp'_{(i,x)}(P) = i - j + 1$, *where* j *is the smallest position such that* j *maps to* i *and* $P(Z_j + 1) = x$. *If there is no such* j *then* $sp'_{(i,x)}(P) = 0$.

The proof of this theorem is almost identical to the proof of Theorem 2.3.4 (page 26) and is left to the reader. Assuming (as usual) that the alphabet is finite, the following minor modification of the preprocessing given earlier for Knuth-Morris-Pratt (Section 2.3.2) yields the needed $sp'_{(i,x)}$ values in linear time:

Z-based real-time matching

for $i := 1$ to n do
    $sp'_{(i,x)} := 0$ for every character x;
for $j := n$ downto 2 do
    begin
        $i := j + Z_j(P) - 1$;
        $x := P(Z_j + 1)$;
        $sp'_{(i,x)} := Z_j$;
    end;

Note that the linear time (and space) bound for this method require that the alphabet Σ be finite. This allows us to do $|\Sigma|$ comparisons in constant time. If the size of the alphabet is explicitly included in the time and space bounds, then the preprocessing time and space needed for the algorithm is $O(|\Sigma|n)$.

2.5. Exercises

1. In "typical" applications of exact matching, such as when searching for an English word in a book, the simple bad character rule seems to be as effective as the extended bad character rule. Give a "hand-waving" explanation for this.

2. When searching for a single word or a small phrase in a large English text, brute force (the naive algorithm) is reported [184] to run faster than most other methods. Give a hand-waving explanation for this. In general terms, how would you expect this observation to hold up with smaller alphabets (say in DNA with an alphabet size of four), as the size of the pattern grows, and when the text has many long sections of similar but not exact substrings?

3. "Common sense" and the $\Theta(nm)$ worst-case time bound of the Boyer–Moore algorithm (using only the bad character rule) both would suggest that empirical running times increase with increasing pattern length (assuming a fixed text). But when searching in actual English

texts, the Boyer–Moore algorithm runs faster in practice when given longer patterns. Thus, on an English text of about 300,000 characters, it took about five times as long to search for the word "Inter" as it did to search for "Interactively".

Give a hand-waving explanation for this. Consider now the case that the pattern length increases without bound. At what point would you expect the search times to stop decreasing? Would you expect search times to start increasing at some point?

4. Evaluate empirically the utility of the extended bad character rule compared to the original bad character rule. Perform the evaluation in combination with different choices for the two good-suffix rules. How much more is the average shift using the extended rule? Does the extra shift pay for the extra computation needed to implement it?

5. Evaluate empirically, using different assumptions about the sizes of P and T, the number of occurrences of P in T, and the size of the alphabet, the following idea for speeding up the Boyer–Moore method. Suppose that a phase ends with a mismatch and that the good suffix rule shifts P farther than the extended bad character rule. Let x and y denote the mismatching characters in T and P respectively, and let z denote the character in the shifted P below x. By the suffix rule, z will not be y, but there is no guarantee that it will be x. So rather than starting comparisons from the right of the shifted P, as the Boyer–Moore method would do, why not first compare x and z? If they are equal then a right-to-left comparison is begun from the right end of P, but if they are unequal then we apply the extended bad character rule from z in P. This will shift P again. At that point we must begin a right-to-left comparison of P against T.

6. The idea of the bad character rule in the Boyer–Moore algorithm can be generalized so that instead of examining characters in P from right to left, the algorithm compares characters in P in the order of how unlikely they are to be in T (most *unlikely* first). That is, it looks first at those characters in P that are least likely to be in T. Upon mismatching, the bad character rule or extended bad character rule is used as before. Evaluate the utility of this approach, either empirically on real data or by analysis assuming random strings.

7. Construct an example where fewer comparisons are made when the bad character rule is used alone, instead of combining it with the good suffix rule.

8. Evaluate empirically the effectiveness of the strong good suffix shift for Boyer–Moore versus the weak shift rule.

9. Give a proof of Theorem 2.2.4. Then show how to accumulate all the $l'(i)$ values in linear time.

10. If we use the weak good suffix rule in Boyer–Moore that shifts the closest copy of t under the matched suffix t, but doesn't require the next character to be different, then the pre-processing for Boyer–Moore can be based directly on sp_i values rather than on Z values. Explain this.

11. Prove that the Knuth-Morris-Pratt shift rules (either based on sp or sp') do not miss any occurrences of P in T.

12. It is possible to incorporate the bad character shift rule from the Boyer–Moore method to the Knuth-Morris-Pratt method or to the naive matching method itself. Show how to do that. Then evaluate how effective that rule is and explain why it is more effective when used in the Boyer–Moore algorithm.

13. Recall the definition of l_i on page 8. It is natural to conjecture that $sp_i = i - l_i$ for any index i, where $i \geq l_i$. Show by example that this conjecture is incorrect.

14. Prove the claims in Theorem 2.3.4 concerning $sp_i'(P)$.

15. Is it true that given only the sp values for a given string P, the sp' values are completely determined? Are the sp values determined from the sp' values alone?

Using *sp* values to compute *Z* values

In Section 2.3.2, we showed that one can compute all the *sp* values knowing only the *Z* values for string *S* (i.e., not knowing *S* itself). In the next five exercises we establish the converse, creating a linear-time algorithm to compute all the *Z* values from *sp* values alone. The first exercise suggests a natural method to accomplish this, and the following exercise exposes a hole in that method. The final three exercises develop a correct linear-time algorithm, detailed in [202]. We say that *sp_i maps to k* if $k = i - sp_i + 1$.

16. Suppose there is a position i such that sp_i maps to k, and let i be the largest such position. Prove that $Z_k = i - k + 1 = sp_i$ and that $r_k = i$.

17. Given the answer to the previous exercise, it is natural to conjecture that Z_k always equals sp_i, where i is the largest position such that sp_i maps to k. Show that this is not true. Given an example using at least three distinct characters.

    Stated another way, give an example to show that Z_k can be greater than zero even when there is *no* position i such that sp_i maps to k.

18. Recall that r_{k-1} is known at the start of iteration k of the Z algorithm (when Z_k is computed), but r_k is known only at the end of iteration k. Suppose, however, that r_k is known (somehow) at the start of iteration k. Show how the Z algorithm can then be modified to compute Z_k using *no* character comparisons. Hence this modified algorithm need not even know the string S.

19. Prove that if Z_k is greater than zero, then r_k equals the largest position i such that $k \geq i - sp_i$. Conclude that r_k can be deduced from the sp values for every position k where Z_k is not zero.

20. Combine the answers to the previous two exercises to create a linear-time algorithm that computes all the Z values for a string S given only the sp values for S and not the string S itself.

    Explain in what way the method is a "simulation" of the Z algorithm.

21. It may seem that $l'(i)$ (needed for Boyer–Moore) should be sp_n for any i. Show why this is not true.

22. In Section 1.5 we showed that all the occurrences of P in T could be found in linear time by computing the Z values on the string $S = P\$T$. Explain how the method would change if we use $S = PT$, that is, we do not use a separator symbol between P and T. Now show how to find all occurrences of P in T in linear time using $S = PT$, but with sp values in place of Z values. (This is not as simple as it might at first appear.)

23. In Boyer–Moore and Boyer–Moore–like algorithms, the search moves right to left in the pattern, although the pattern moves left to right relative to the text. That makes it more difficult to explain the methods and to combine the preprocessing for Boyer–Moore with the preprocessing for Knuth-Morris-Pratt. However, a small change to Boyer–Moore would allow an easier exposition and more uniform preprocessing. First, place the pattern at the *right* end of the text, and conduct each search *left to right* in the pattern, shifting the pattern *left* after a mismatch. Work out the details of this approach, and show how it allows a more uniform exposition of the preprocessing needed for it and for Knuth-Morris-Pratt. Argue that on average this approach has the same behavior as the original Boyer–Moore method.

24. Below is working Pascal code (in Turbo Pascal) implementing Richard Cole's preprocessing, for the strong good suffix rule. It is different than the approach based on fundamental preprocessing and is closer to the original method in [278]. Examine the code to extract the algorithm behind the program. Then explain the idea of the algorithm, prove correctness of the algorithm, and analyze its running time. The point of the exercise is that it is difficult to convey an algorithmic idea using a program.

```pascal
program gsmatch(input,output);
{This is an implementation of Richard Cole's
preprocessing for the strong good suffix rule}
type
tstring = string[200];
indexarray = array[1..100] of integer;

const
zero = 0;

var
p:tstring;
bmshift,matchshift:indexarray;
m,i:integer;

procedure readstring(var p:tstring; var m:integer);

begin
read(p);

m:=Length(p);
writeln('the length of the string is ', m);

end;

procedure gsshift(p:tstring; var
gs_shift:indexarray;m:integer);

var
i,j,j_old,k:integer;
kmp_shift:indexarray;
go_on:boolean;

begin {1}
  for j:= 1 to m do
    gs_shift[j] := m;
  kmp_shift[m]:=1;

{stage 1}
  j:=m;

  for k:=m-1 downto 1 do
    begin {2}
        go_on:=true;
        while (p[j] <> p[k]) and go_on do
           begin {3}
           if (gs_shift[j] > j-k) then gs_shift[j] := j-k;
           if (j < m) then j:= j+kmp_shift[j+1]
           else go_on:=false;
           end; {3}
```

```
        if (p[k] = p[j]) then
            begin {3}
            kmp_shift[k]:=j-k;
          j:=j-1;
            end {3}
        else
        kmp_shift[k]:=j-k+1;
      end; {2}

  {stage 2}
   j:=j+1;
   j_old:=1;

   while (j <= m) do
     begin {2}
       for i:=j_old to j-1 do
       if (gs_shift[i] > j-1) then gs_shift[i]:=j-1;

       j_old:=j;
       j:=j+kmp_shift[j];
     end; {2}
  end; {1}

begin {main}

writeln('input a string on a single line');

readstring(p,m);
gsshift(p,matchshift,m);
writeln('the value in cell i is the number of positions to shift');
writeln('after a mismatch occurring in position i of the pattern');

for i:= 1 to m do
write(matchshift[i]:3);
writeln;
end. {main}
```

25. Prove that the shift rule used by the real-time string matcher does not miss any occurrences of P in T.

26. Prove Theorem 2.4.1.

27. In this chapter, we showed how to use Z values to compute both the sp_i' and sp_i values used in Knuth-Morris-Pratt and the $sp_{i,x}'$ values needed for its real-time extension. Instead of using Z values for the $sp_{i,x}'$ values, show how to obtain these values from the sp_i and/or sp_i' values in linear $[O(n|\Sigma|)]$ time, where n is the length of P and $|\Sigma|$ is the length of the alphabet.

28. Although we don't know how to simply convert the Boyer–Moore algorithm to be a real-time method the way Knuth-Morris-Pratt was converted, we can make similar changes to the strong shift rule to make the Boyer–Moore shift more effective. That is, when a mismatch occurs between $P(i)$ and $T(h)$ we can look for the right-most copy in P of $P[i+1..n]$ (other than $P[i + 1..n]$ itself) such that the preceding character is $T(h)$. Show how to modify

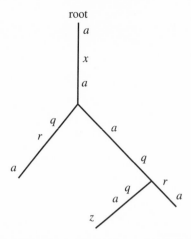

Figure 2.3: The pattern $P = aqra$ labels two subpaths of paths starting at the root. Those paths start at the root, but the subpaths containing $aqra$ do not. There is also another subpath in the tree labeled $aqra$ (it starts above the character z), but it violates the requirement that it be a subpath of a path starting at the root. Note that an edge label is displayed from the top of the edge down towards the bottom of the edge. Thus in the figure, there is an edge labeled "qra", not "arq".

the Boyer–Moore preprocessing so that the needed information is collected in linear time, assuming a fixed size alphabet.

29. Suppose we are given a tree where each edge is labeled with one or more characters, and we are given a pattern P. The label of a subpath in the tree is the concatenation of the labels on the edges in the subpath. The problem is to find all subpaths of paths starting at the root that are labeled with pattern P. Note that although the subpath must be part of a path directed from the root, the subpath itself need not start at the root (see Figure 2.3). Give an algorithm for this problem that runs in time proportional to the total number of characters on the edges of the tree plus the length of P.

3

Exact Matching: A Deeper Look at Classical Methods

3.1. A Boyer–Moore variant with a "simple" linear time bound

Apostolico and Giancarlo [26] suggested a variant of the Boyer–Moore algorithm that allows a fairly simple proof of linear worst-case running time. With this variant, no character of T will ever be compared after it is first matched with any character of P. It is then immediate that the number of comparisons is at most $2m$: Every comparison is either a match or a mismatch; there can only be m mismatches since each one results in a nonzero shift of P; and there can only be m matches since no character of T is compared again after it matches a character of P. We will also show that (in addition to the time for comparisons) the time taken for all the other work in this method is linear in m.

Given the history of very difficult and partial analyses of the Boyer–Moore algorithm, it is quite amazing that a close variant of the algorithm allows a simple linear time bound. We present here a further improvement of the Apostolico–Giancarlo idea, resulting in an algorithm that simulates *exactly* the shifts of the Boyer–Moore algorithm. The method therefore has all the rapid shifting advantages of the Boyer–Moore method as well as a simple linear worst-case time analysis.

3.1.1. Key ideas

Our version of the Apostolico–Giancarlo algorithm simulates the Boyer–Moore algorithm, finding exactly the same mismatches that Boyer–Moore would find and making exactly the same shifts. However, it infers and avoids many of the explicit matches that Boyer–Moore makes.

We take the following high-level view of the Boyer–Moore algorithm. We divide the algorithm into *compare/shift phases* numbered 1 through $q \leq m$. In a compare/shift phase, the right end of P is aligned with a character of T, and P is compared right to left with selected characters of T until either all of P is matched or until a mismatch occurs. Then, P is shifted right by some amount as given in the Boyer–Moore shift rules.

Recall from Section 2.2.4, where preprocessing for Boyer–Moore was discussed, that $N_i(P)$ is the length of the longest suffix of $P[1..i]$ that matches a suffix of P. In Section 2.2.4 we showed how to compute N_i for every i in $O(n)$ time, where n is the length of P. We assume here that vector N has been obtained during the preprocessing of P.

Two modifications of the Boyer–Moore algorithm are required. First, during the search for P in T (after the preprocessing), we maintain an m length vector M in which at most one entry is updated in every phase. Consider a phase where the right end of P is aligned with position j of T and suppose that P and T match for l places (from right to left) but no farther. Then, set $M(j)$ to a value $k \leq l$ (the rules for selecting k are detailed below). $M(j)$ records the fact that a suffix of P of length k (at least) occurs in T and ends exactly

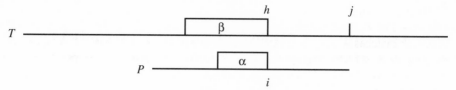

Figure 3.1: Substring α has length N_i and substring β has length $M(h) > N_i$. The two strings must match from their right ends for N_i characters, but mismatch at the next character.

at position j. As the algorithm proceeds, a value for $M(j)$ is set for every position j in T that is aligned with the right end of P; $M(j)$ is undefined for all other positions in T.

The second modification exploits the vectors N and M to speed up the Boyer–Moore algorithm by inferring certain matches and mismatches. To get the idea, suppose the Boyer–Moore algorithm is about to compare characters $P(i)$ and $T(h)$, and suppose it knows that $M(h) > N_i$ (see Figure 3.1). That means that an N_i-length substring of P ends at position i and matches a suffix of P, while an $M(h)$-length substring of T ends at position h and matches a suffix of P. So the N_i-length suffixes of those two substrings must match, and we can conclude that the next N_i comparisons (from $P(i)$ and $T(h)$ moving leftward) in the Boyer–Moore algorithm would be matches. Further, if $N_i = i$, then an occurrence of P in T has been found, and if $N_i < i$, then we can be sure that the next comparison (after the N_i matches) would be a mismatch. Hence in simulating Boyer–Moore, if $M(h) > N_i$ we can avoid at least N_i explicit comparisons. Of course, it is not always the case that $M(h) > N_i$, but all the cases are similar and are detailed below.

3.1.2. One phase in detail

As in the original Boyer–Moore algorithm, when the right end of P is aligned with a position j in T, P is compared with T from right to left. When a mismatch is found or inferred, or when an occurrence of P is found, P is shifted according to the original Boyer–Moore shift rules (either the strong or weak version) and the compare/shift phase ends. Here we will only give the details for a single phase. The phase begins with h set to j and i set to n.

Phase algorithm

1. If $M(h)$ is undefined or $M(h) = N_i = 0$, then compare $T(h)$ and $P(i)$ as follows:

    If $T(h) = P(i)$ and $i = 1$, then report an occurrence of P ending at position j of T, set $M(j) = n$, and shift as in the Boyer–Moore algorithm (ending this phase).

    If $T(h) = P(i)$ and $i > 1$, then set h to $h - 1$ and i to $i - 1$ and repeat the phase algorithm.

    If $T(h) \neq P(i)$, then set $M(j) = j - h$ and shift P according to the Boyer–Moore rules based on a mismatch occurring in position i of P (this ends the phase).

2. If $M(h) < N_i$, then P matches its counterparts in T from position n down to position $i - M(h) + 1$ of P. By the definition of $M(h)$, P might match more of T to the left, so set i to $i - M(h)$, set h to $h - M(h)$, and repeat the phase algorithm.

3. If $M(h) \geq N_i$ and $N_i = i > 0$, then declare that an occurrence of P has been found in T ending at position j. $M(j)$ must be set to a value less than or equal to n. Set $M(j)$ to $j - h$, and shift according to the Boyer–Moore rules based on finding an occurrence of P ending at j (this ends the phase).

4. If $M(h) > N_i$ and $N_i < i$, then P matches T from the right end of P down to character $i - N_i + 1$ of P, but the next pair of characters mismatch [i.e., $P(i - N_i) \neq T(h - N_i)$]. Hence P matches T for $j - h + N_i$ characters and mismatches at position $i - N_i$ of P. $M(j)$ must be set to a value less than or equal to $j - h + N_i$. Set $M(j)$ to $j - h$. Shift P by the Boyer–Moore rules based on a mismatch at position $i - N_i$ of P (this ends the phase).

5. If $M(h) = N_i$ and $0 < N_i < i$, then P and T must match for at least $M(h)$ characters to the left, but the left end of P has not yet been reached, so set i to $i - M(h)$ and set h to $h - M(h)$ and repeat the phase algorithm.

3.1.3. Correctness and linear-time analysis

Theorem 3.1.1. *Using M and N as above, the Apostolico–Giancarlo variant of the Boyer–Moore algorithm correctly finds all occurrences of P in T.*

PROOF   We prove correctness by showing that the algorithm simulates the original Boyer–Moore algorithm. That is, for any given alignment of P with T, the algorithm is correct when it declares a match down to a given position and a mismatch at the next position. The rest of the simulation is correct since the shift rules are the same as in the Boyer–Moore algorithm.

Assume inductively that $M(h)$ values are valid up to some position in T. That is, wherever $M(h)$ is defined, there is an $M(h)$-length substring in T ending at position h in T that matches a suffix of P. The first such value, $M(n)$, is valid because it is found by aligning P at the left of T and making explicit comparisons, repeating rule 1 of the phase algorithm until a mismatch occurs or an occurrence of P is found. Now consider a phase where the right end of P is aligned with position j of T. The phase simulates the workings of Boyer–Moore except that in cases 2, 3, 4, and 5 certain explicit comparisons are skipped and in case 4 a mismatch is inferred, rather than observed. But whenever comparisons are skipped, they are certain to be matches by the definition of N and M and the assumption that the M values are valid thus far. Thus it is correct to skip these comparisons. In case 4, a mismatch at position $i - N_i$ of P is correctly inferred because N_i is the maximum length of any substring ending at i that matches a suffix of P, whereas $M(h)$ is less than or equal to the maximum length of any substring ending at h that matches a suffix of P. Hence this phase correctly simulates Boyer–Moore and finds exactly the same mismatch (or an occurrence of P in T) that Boyer–Moore would find. The value given to $M(j)$ is valid since in all cases it is less than or equal to the length of the suffix of P shown to match its counterpart in the substring $T[1..i]$.   □

The following definitions and lemma will be helpful in bounding the work done by the algorithm.

Definition   If j is a position where $M(j)$ is greater than zero then the interval $[j - M(j) + 1..j]$ is called a *covered interval* defined by j.

Definition   Let $j' < j$ and suppose covered intervals are defined for both j and j'. We say that the covered intervals for j and j' *cross* if $j - M(j) + 1 \leq j'$ and $j' - M(j') + 1 < j - M(j) + 1$ (see Figure 3.2).

Lemma 3.1.1. *No covered intervals computed by the algorithm ever cross each other. Moreover, if the algorithm examines a position h of T in a covered interval, then h is at the right end of that interval.*

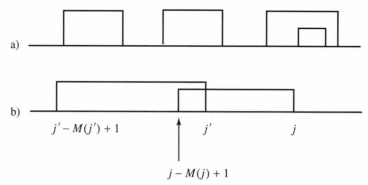

Figure 3.2: a. Diagram showing covered intervals that do not cross, although one interval can contain another. b. Two covered intervals that do cross.

PROOF   The proof is by induction on the number of intervals created. Certainly the claim is true until the first interval is created, and that interval does not cross itself. Now assume that no intervals cross and consider the phase where the right end of P is aligned with position j of T.

Since $h = j$ at the start of the phase, and j is to the right of any interval, h begins outside any interval. We consider how h could first be set to a position inside an interval, other than the right end of the interval. Rule 1 is never executed when h is at the right end of an interval (since then $M(h)$ is defined and greater than zero), and after any execution of Rule 1, either the phase ends or h is decremented by one place. So an execution of Case 1 cannot cause h to move beyond the right-most character of a covered interval. This is also true for Cases 3 and 4 since the phase ends after either of those cases. So if h is ever moved into an interval in a position other than its right end, that move must follow an execution of Case 2 or 5. An execution of Case 2 or 5 moves h from the right end of some interval $I = [k..h]$ to position $k - 1$, one place to the left of I. Now suppose that $k - 1$ is in some interval I' but is not at its right end, and that this is the first time in the phase that h (presently $k - 1$) is in an interval in a position other than its right end. That means that the right end of I cannot be to the left of the right end of I' (for then position $k - 1$ would have been strictly inside I'), and the right ends of I and I' cannot be equal (since $M(h)$ has at most one value for any h). But these conditions imply that I and I' cross, which is assumed to be untrue. Hence, if no intervals cross at the start of the phase, then in that phase only the right end of any covered interval is examined.

A new covered interval gets created in the phase only after the execution of Case 1, 3, or 4. In any of these cases, the interval $[h + 1..j]$ is created after the algorithm examines position h. In Case 1, h is not in any interval, and in Cases 3 and 4, h is the right end of an interval, so in all cases $h + 1$ is either not in a covered interval or is at the left end of an interval. Since j is to the right of any interval, and $h + 1$ is either not in an interval or is the left end of one, the new interval $[h + 1..j]$ does not cross any existing interval. The previously existing intervals have not changed, so there are no crossing intervals at the end of the phase, and the induction is complete.   □

Theorem 3.1.2. *The modified Apostolico–Giancarlo algorithm does at most $2m$ character comparisons and at most $O(m)$ additional work.*

PROOF   Every phase ends if a comparison finds a mismatch and every phase, except the last, is followed by a nonzero shift of P. Thus the algorithm can find at most m mismatches. To bound the matches, observe that characters are explicitly compared only in Case 1, and if the comparison involving $T(h)$ is a match then, at the end of the phase, $M(j)$ is set at least as large as $j - h + 1$. That means that all characters in T that matched a character of P during that phase are contained in the covered interval $[j - M(j) + 1..j]$. Now the algorithm only examines the right end of an interval, and if h is the right end of an interval then $M(h)$ is defined and greater than 0, so the algorithm never compares a character of T in a covered interval. Consequently, no character of T will ever be compared again after it is first in a match. Hence the algorithm finds at most m matches, and the total number of character comparisons is bounded by $2m$.

To bound the amount of additional work, we focus on the number of accesses of M during execution of the five cases since the amount of additional work is proportional to the number of such accesses. A character comparison is done whenever Case 1 applies. Whenever Case 3 or 4 applies, P is immediately shifted. Hence Cases 1, 3, and 4 can apply at most $O(m)$ times since there are at most $O(m)$ shifts and compares. However, it is possible that Case 2 or Case 5 can apply without an immediate shift or immediate character comparison. That is, Case 2 or 5 could apply repeatedly before a comparison or shift is done. For example, Case 5 would apply twice in a row (without a shift or character comparison) if $N_i = M(h) > 0$ and $N_{i-N_i} = M(h - M(h))$. But whenever Case 2 or 5 applies, then $j > h$ and $M(j)$ will certainly get set to $j - h + 1$ or more at the end of that phase. So position h will be in the strict interior of the covered interval defined by j. Therefore, h will never be examined again, and $M(h)$ will never be accessed again. The effect is that Cases 2 and 5 can apply at most once for any position in T, so the number of accesses made when these cases apply is also $O(m)$.   □

3.2. Cole's linear worst-case bound for Boyer–Moore

Here we finally present a linear worst-case time analysis of the *original* Boyer–Moore algorithm. We consider first the use of the (strong) good suffix rule by itself. Later we will show how the analysis is affected by the addition of the bad character rule. Recall that the good suffix rule is the following:

> Suppose for a given alignment of P and T, a substring t of T matches a suffix of P, but a mismatch occurs at the next comparison to the left. Then find, if it exists, the right-most copy t' of t in P such that t' is not a suffix of P and *such that the character to the left of t' differs from the mismatched character in P*. Shift P to the right so that substring t' in P is below substring t in T (recall Figure 2.1). If t' does not exist, then shift the left end of P past the left end of t in T by the least amount so that a prefix of the shifted pattern matches a suffix of t in T. If no such shift is possible, then shift P by n places to the right.
>
> If an occurrence of P is found, then shift P by the least amount so that a *proper* prefix of the shifted pattern matches a suffix of the occurrence of P in T. If no such shift is possible, then shift P by n places.

We will show that by using the good suffix rule alone, the Boyer–Moore method has a worst-case running time of $O(m)$, provided that the pattern does not appear in the text. Later we will extend the Boyer–Moore method to take care of the case that P does occur in T.

As in our analysis of the Apostolico–Giancarlo algorithm, we divide the Boyer–Moore algorithm into *compare/shift* phases numbered 1 through $q \leq m$. In compare/shift phase i, a suffix of P is matched right to left with characters of T until either all of P is matched or until a mismatch occurs. In the latter case, the substring of T consisting of the matched characters is denoted t_i, and the mismatch occurs just to the left of t_i. The pattern is then shifted right by an amount determined by the good suffix rule.

3.2.1. Cole's proof when the pattern does not occur in the text

Definition   Let s_i denote the amount by which P is shifted right at the end of phase i.

Assume that P does not occur in T, so the compare part of every phase ends with a mismatch. In each compare/shift phase, we divide the comparisons into those that compare a character of T that has previously been compared (in a previous phase) and those comparisons that compare a character of T for the first time in the execution of the algorithm. Let g_i be the number of comparisons in phase i of the first type (comparisons involving a previously examined character of T), and let g_i' be the number of comparisons in phase i of the second type. Then, over the entire algorithm the number of comparisons is $\sum_{i=1}^{q}(g_i + g_i')$, and our goal is to show that this sum is $O(m)$.

Certainly, $\sum_{i=1}^{q} g_i' \leq m$ since a character can be compared for the first time only once. We will show that for any phase i, $s_i \geq g_i/3$. Then since $\sum_{i=1}^{q} s_i \leq m$ (because the total length of all the shifts is at most m) it will follow that $\sum_{i=1}^{q} g_i \leq 3m$. Hence the total number of comparisons done by the algorithm is $\sum_{i=1}^{q}(g_i + g_i') \leq 4m$.

An initial lemma

We start with the following definition and a lemma that is valuable in its own right.

Definition   For any string β, β^i denotes the string obtained by concatenating together i copies of β.

Lemma 3.2.1. *Let γ and δ be two nonempty strings such that $\gamma\delta = \delta\gamma$. Then $\delta = \rho^i$ and $\gamma = \rho^j$ for some string ρ and positive integers i and j.*

This lemma says that if a string is the same before and after a circular shift (so that it can be written both as $\gamma\delta$ and $\delta\gamma$, for some strings γ and δ) then γ and δ can both be written as concatenations of some single string ρ.

For example, let $\delta = abab$ and $\gamma = ababab$, so $\delta\gamma = ababababab = \gamma\delta$. Then $\rho = ab$, $\delta = \rho^2$, and $\gamma = \rho^3$.

PROOF   The proof is by induction on $|\delta| + |\gamma|$. For the basis, if $|\delta| + |\gamma| = 2$, it must be that $\delta = \gamma = \rho$ and $i = j = 1$. Now consider larger lengths. If $|\delta| = |\gamma|$, then again $\delta = \gamma = \rho$ and $i = j = 1$. So suppose $|\delta| < |\gamma|$. Since $\delta\gamma = \gamma\delta$ and $|\delta| < |\gamma|$, δ must be a prefix of γ, so $\gamma = \delta\delta'$ for some string δ'. Substituting this into $\delta\gamma = \gamma\delta$ gives $\delta\delta\delta' = \delta\delta'\delta$. Deleting the left copy of δ from both sides gives $\delta\delta' = \delta'\delta$. However, $|\delta| + |\delta'| = |\gamma| < |\delta| + |\gamma|$, and so by induction, $\delta = \rho^i$ and $\delta' = \rho^j$. Thus, $\gamma = \delta\delta' = \rho^k$, where $k = i + j$.   □

Definition   A string α is *semiperiodic* with *period* β if α consists of a nonempty suffix of a string β (possibly the entire β) followed by one or more copies of β. String α is called *periodic* with *period* β if α consists of two or more *complete* copies of β. We say that string α is *periodic* if it is periodic with some period β.

For example, $bcabcabc$ is semiperiodic with period abc, but it is not periodic. String $abcabc$ is periodic with period abc. Note that a periodic string is by definition also semiperiodic. Note also that a string cannot have itself as a period although a period may itself be periodic. For example, *abababab* is periodic with period *abab* and also with shorter period *ab*. An alternate definition of a semiperiodic string is sometimes useful.

Definition   A string α is *prefix semiperiodic* with period γ if α consists of one or more copies of string γ followed by a nonempty prefix (possibly the entire γ) of string γ.

We use the term "prefix semiperiodic" to distinguish this definition from the definition given for "semiperiodic", but the following lemma (whose proof is simple and is left as an exercise) shows that these two definitions are really alternate reflections of the same structure.

Lemma 3.2.2. *A string α is semiperiodic with period β if and only if it is prefix semiperiodic with the same length period as β.*

For example, the string *abaabaabaabaabaab* is semiperiodic with period *aab* and is prefix semiperiodic with period *aba*.

The following useful lemma is easy to verify, and its proof is typical of the style of thinking used in dealing with overlapping matches.

Lemma 3.2.3. *Suppose pattern P occurs in text T starting at positions p and $p' > p$, where $p' - p \leq \lfloor n/2 \rfloor$. Then P is semiperiodic with period $p' - p$.*

The following lemma, called the *GCD Lemma*, is a very powerful statement about periods of strings. We won't need the lemma in our discussion of Cole's proof, but it is natural to state it here. We will prove it and use it in Section 16.17.5.

Lemma 3.2.4. *Suppose string α is semiperiodic with both a period of length p and a period of length q, and $|\alpha| \geq p + q$. Then α is semiperiodic with a period whose length is the greatest common divisor of p and q.*

Return to Cole's proof

Recall that the key thing to prove is that $s_i \geq g_i/3$ in every phase i. As noted earlier, it then follows easily that the total number of comparisons is bounded by $4m$.

Consider the ith compare/shift phase, where substring t_i of T matches a suffix of P and then P is shifted s_i places to the right. If $s_i \geq (|t_i| + 1)/3$, then $s_i \geq g_i/3$ even if all characters of T that were compared in phase i had been previously compared. Therefore, it is easy to handle phases where the shift is "relatively" large compared to the total number of characters examined during the phase. Accordingly, for the next several lemmas we consider the case when the shift is relatively small (i.e., $s_i < (|t_i| + 1)/3$ or, equivalently, $|t_i| + 1 > 3s_i$).

We need some notation at this point. Let α be the suffix of P of length s_i, and let β be the smallest substring such that $\alpha = \beta^l$ for some integer l (it may be that $\beta = \alpha$ and $l = 1$). Let $\bar{P} = P[n - |t_i|..n]$ be the suffix of P of length $|t_i| + 1$, that is, that portion of P (including the mismatch) that was examined in phase i. See Figure 3.3.

Lemma 3.2.5. *If $|t_i| + 1 > 3s_i$, then both t_i and $\bar{P}$ are semiperiodic with period α and hence with period β.*

PROOF   Starting from the right end of $\bar{P}$, mark off substrings of length s_i until less than s_i characters remain on the left (see Figure 3.4). There will be at least three full substrings since $|\bar{P}| = |t_i| + 1 > 3s_i$. Phase i ends by shifting P right by s_i positions. Consider how $\bar{P}$ aligns with T before and after that shift (see Figure 3.5). By definition of s_i and α, α is the part of the shifted $\bar{P}$ to the right of the original $\bar{P}$. By the good suffix rule, the portion

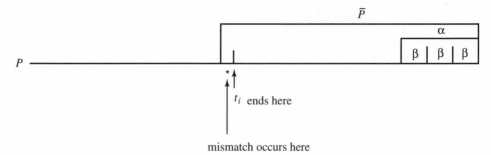

Figure 3.3: String α has length s_i; string $\bar{P}$ has length $|t_i| + 1$.

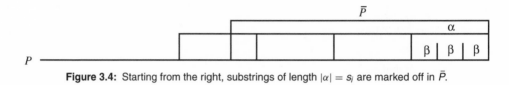

Figure 3.4: Starting from the right, substrings of length $|\alpha| = s_i$ are marked off in $\bar{P}$.

Figure 3.5: The arrows show the string equalities described in the proof.

of the shifted $\bar{P}$ below t_i must match the portion of the unshifted $\bar{P}$ below t_i, so the second marked-off substring from the right end of the shifted $\bar{P}$ must be the same as the first substring of the unshifted $\bar{P}$. Hence they must both be copies of string α. But the second substring is the same in both copies of $\bar{P}$, so continuing this reasoning we see that all the s_i-length marked substrings are copies of α and the left-most substring is a suffix of α (if it is not a complete copy of α). Hence $\bar{P}$ is semiperiodic with period α. The right-most $|t_i|$ characters of $\bar{P}$ match t_i, and so t_i is also semiperiodic with period α. Then since $\alpha = \beta^l$, $\bar{P}$ and t_i must also be semiperiodic with period β. □

Recall that we want to bound g_i, the number of characters compared in the ith phase that have been previously compared in earlier phases. All but one of the characters compared in phase i are contained in t_i, and a character in t_i could have previously been examined only during a phase where P overlaps t_i. So to bound g_i, we closely examine in what ways P could have overlapped t_i during earlier phases.

Lemma 3.2.6. *If $|t_i| + 1 > 3s_i$, then in any phase $h < i$, the right end of P could not have been aligned opposite the right end of any full copy of β in substring t_i of T.*

PROOF By Lemma 3.2.5, t_i is semiperiodic with period β. Figure 3.6 shows string t_i as a concatenation of copies of string β. In phase h, the right end of P cannot be aligned with the right end of t_i since that is the alignment of P and T in phase $i > h$, and P must have moved right between phases h and i. So, suppose, for contradiction, that in phase h the right end of P is aligned with the right end of some other full copy of β in t_i. For

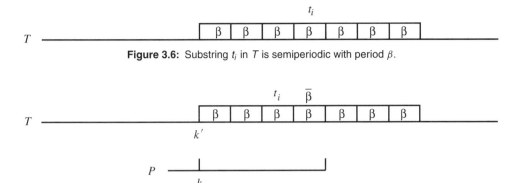

Figure 3.6: Substring t_i in T is semiperiodic with period β.

Figure 3.7: The case when the right end of P is aligned with a right end of $\bar{\beta}$ in phase h. Here $q = 3$. A mismatch must occur between $T(k')$ and $P(k)$.

concreteness, call that copy $\bar{\beta}$ and say that its right end is $q|\beta|$ places to the left of the right of t_i, where $q \geq 1$ (see Figure 3.7). We will first deduce how phase h must have ended, and then we'll use that to prove the lemma.

Let k' be the position in T just to the left of t_i (so $T(k')$ is involved in the mismatch ending phase i), and let k be the position in P opposite $T(k')$ in phase h. We claim that, in phase h, the comparison of P and T will find matches until the left end of t_i but then mismatch when comparing $T(k')$ and $P(k)$. The reason is the following: Strings $\bar{P}$ and t_i are semiperiodic with period β, and in phase h the right end of P is aligned with the right end of some β. So in phase h, P and T will certainly match until the left end of string t_i. Now $\bar{P}$ is semiperiodic with β, and in phase h, the right end of P is exactly $q|\beta|$ places to the left of the right end of t_i. Therefore, $\bar{P}(1) = \bar{P}(1 + |\beta|) = \cdots = \bar{P}(1 + q|\beta|) = P(k)$. But in phase i the mismatch occurs when comparing $T(k')$ with $\bar{P}(1)$, so $P(k) = \bar{P}(1) \neq T(k')$. Hence, if in phase h the right end of P is aligned with the right end of a β, then phase h must have ended with a mismatch between $T(k')$ and $P(k)$. This fact will be used below to prove the lemma.[1]

Now we consider the possible shifts of P done in phase h. We will show that every possible shift leads to a contradiction, so no shifts are possible and the assumed alignment of P and T in phase h is not possible, proving the lemma.

Since $h < i$, the right end of P will not be shifted in phase h past the right end of t_i; consequently, after the phase h shift a character of $\bar{P}$ is opposite character $T(k')$ (the character of T that will mismatch in phase i). Consider where the right end of P is after the phase h shift. There are two cases to consider: 1. Either the right end of P is opposite the right end of another full copy of β (in t_i) or 2. The right end of P is in the interior of a full copy of β.

Case 1 If the phase h shift aligns the right end of P with the right end of a full copy of β, then the character opposite $T(k')$ would be $P(k - r|\beta|)$ for some r. But since $\bar{P}$ is

---

[1] Later we will analyze the Boyer–Moore algorithm when P is in T. For that purpose we note here that when phase h is assumed to end by finding an occurrence of P, then the proof of Lemma 3.2.6 is complete at this point, having established a contradiction. That is, on the assumption that the right end of P is aligned with the right end of a β in phase h, we proved that phase h ends with a mismatch, which would contradict the assumption that h ends by finding an occurrence of P in T. So even if phase h ends by finding an occurrence of P, the right end of P could not be aligned with the right end of a β block in phase h.

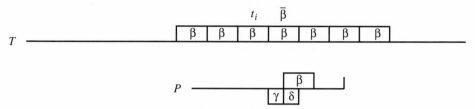

Figure 3.8: Case when the right end of P is aligned with a character in the interior of a β. Then t_i would have a smaller period than β, contradicting the definition of β.

semiperiodic with period β, $P(k)$ must be equal to $P(k - r|\beta|)$, contradicting the good suffix rule.

Case 2   Suppose the phase h shift aligns P so that its right end aligns with some character in the interior of a full copy of β. That means that, in this alignment, the right end of some β string in P is opposite a character in the interior of $\bar{\beta}$. Moreover, by the good suffix rule, the characters in the shifted P below $\bar{\beta}$ agree with $\bar{\beta}$ (see Figure 3.8). Let $\gamma\delta$ be the string in the shifted P positioned opposite $\bar{\beta}$ in t_i, where γ is the string through the end of β and δ is the remainder. Since $\bar{\beta} = \beta$, γ is a suffix of β, δ is a prefix of β, and $|\gamma| + |\delta| = |\bar{\beta}| = |\beta|$; thus $\gamma\delta = \delta\gamma$. By Lemma 3.2.1, however, $\beta = \rho^t$ for $t > 1$, which contradicts the assumption that β is the smallest string such that $\alpha = \beta^l$ for some l.

Starting with the assumption that in phase h the right end of P is aligned with the right end of a full copy of β, we reached the conclusion that no shift in phase h is possible. Hence the assumption is wrong and the lemma is proved. □

Lemma 3.2.7. *If $|t_i| + 1 > 3s_i$, then in phase $h < i$, P can match t_i in T for at most $|\beta| - 1$ characters.*

PROOF   Since P is not aligned with the end of any β in phase h, if P matches t_i in T for β or more characters then the right-most β characters of P would match a string consisting of a suffix (γ) of β followed by a prefix (δ) of β. So we would again have $\beta = \gamma\delta = \delta\gamma$, and by Lemma 3.2.1, this again would lead to a contradiction to the selection of β. □

Note again that this lemma holds even if phase h is assumed to find an occurrence of P. That is, nowhere in the proof is it assumed that phase h ends with a mismatch, only that phase i does. This observation will be used later.

Lemma 3.2.8. *If $|t_i| + 1 > 3s_i$, then in phase $h < i$ if the right end of P is aligned with a character in t_i, it can only be aligned with one of the left-most $|\beta| - 1$ characters of t_i or one of the right-most $|\beta|$ characters of t_i.*

PROOF   Suppose in phase h that the right end of P is aligned with a character of t_i other than one of the left-most $|\beta| - 1$ characters or the right-most $|\beta|$ characters. For concreteness, say that the right end of P is aligned with a character in copy β' of string β. Since β' is not the left-most copy of β, the right end of P is at least $|\beta|$ characters to the right of the left end of t_i, and so by Lemma 3.2.7 a mismatch would occur in phase h before the left end of t_i is reached. Say that mismatch occurs at position k'' of T. After that mismatch, P is shifted right by some amount determined by the good suffix rule. By Lemma 3.2.6, the phase-h shift cannot move the right end of P to the right end of β', and we will show that the shift will also not move the end of P past the right end of β'.

Recall that the good suffix rule shifts P (when possible) by the smallest amount so that all the characters of T that matched in phase h again match with the shifted P and so that the two characters of P aligned with $T(k'')$ before and after the shift are unequal. We claim these conditions hold when the right end of P is aligned with the right end of β'. Consider that alignment. Since $\bar{P}$ is semiperiodic with period β, that alignment of P and T would match at least until the left end of t_i and so would match at position k'' of T. Therefore, the two characters of P aligned with $T(k'')$ before and after the shift cannot be equal. Thus if the end of P were aligned with the end of β' then all the characters of T that matched in phase h would again match, and the characters of P aligned with $T(k'')$ before and after the shift would be different. Hence the good suffix rule would not shift the right end of P past the right of the end of β'.

Therefore, if the right end of P is aligned in the interior of β' in phase h, it must also be in the interior of β' in phase $h + 1$. But h was arbitrary, so the phase-$h + 1$ shift would also not move the right end of P past β'. So if the right end of P is in the interior of β' in phase h, it remains there forever. This is impossible since in phase $i > h$ the right end of P is aligned with the right end of t_i, which is to the right of β'. Hence the right end of P is not in the interior of β', and the Lemma is proved. $\square$

Note again that Lemma 3.2.8 holds even if phase h is assumed to end by finding an occurrence of P in T. That is, the proof only needs the assumption that phase i ends with a mismatch, not that phase h does. In fact, when phase h finds an occurrence of P in T, then the proof of the lemma only needs the reasoning contained in the first two paragraphs of the above proof.

Theorem 3.2.1. *Assuming P does not occur in T, $s_i \geq g_i/3$ in every phase i.*

PROOF This is trivially true if $s_i \geq (|t_i| + 1)/3$, so assume $|t_i| + 1 > 3s_i$. By Lemma 3.2.8, in any phase $h < i$, the right end of P is opposite either one of the left-most $|\beta| - 1$ characters of t_i or one of the right-most $|\beta|$ characters of t_i (excluding the extreme right character). By Lemma 3.2.7, at most $|\beta|$ comparisons are made in phase $h < i$. Hence the only characters compared in phase i that could possibly have been compared before phase i are the left-most $|\beta| - 1$ characters of t_i, the right-most $2|\beta|$ characters of t_i, or the character just to the left of t_i. So $g_i \leq 3|\beta| = 3s_i$ when $|t_i| + 1 > 3s_i$. In both cases then, $s_i \geq g_i/3$. $\square$

Theorem 3.2.2. [108] *Assuming that P does not occur in T, the worst-case number of comparisons made by the Boyer–Moore algorithm is at most $4m$.*

PROOF As noted before, $\sum_{i=1}^{q} g_i' \leq m$ and $\sum_{i=1}^{q} s_i \leq m$, so the total number of comparisons done by the algorithm is $\sum_{i=1}^{q}(g_i + g_i') \leq (\sum_i 3s_i) + m \leq 4m$. $\square$

3.2.2. The case when the pattern does occur in the text

Consider P consisting of n copies of a single character and T consisting of m copies of the same character. Then P occurs in T starting at every position in T except the last $n - 1$ positions, and the number of comparisons done by the Boyer–Moore algorithm is $\Theta(mn)$. The $O(m)$ time bound proved in the previous section breaks down because it was derived by showing that $g_i \leq 3s_i$, and that required the assumption that phase i ends with a mismatch. So when P does occur in T (and phases do not necessarily end with mismatches), we must modify the Boyer–Moore algorithm in order to recover the linear running time. Galil [168] gave the first such modification. Below we present a version of his idea.

The approach comes from the following observation: Suppose in phase i that the right end of P is positioned with character k of T, and that P is compared with T down to character s of T. (We don't specify whether the phase ends by finding a mismatch or by finding an occurrence of P in T.) If the phase-i shift moves P so that its left end is to the right of character s of T, then in phase $i + 1$ a *prefix* of P definitely matches the characters of T up to $T(k)$. Thus, in phase $i + 1$, if the right-to-left comparisons get down to position k of T, the algorithm can conclude that an occurrence of P has been found even without explicitly comparing characters to the left of $T(k + 1)$. It is easy to implement this modification to the algorithm, and we assume in the rest of this section that the Boyer–Moore algorithm includes this rule, which we call the *Galil rule*.

Theorem 3.2.3. *Using the Galil rule, the Boyer–Moore algorithm never does more than $O(m)$ comparisons, no matter how many occurrences or P there are in T.*

PROOF   Partition the phases into those that do find an occurrence of P and those that do not. Let Q be the set of phases of the first type and let d_i be the number of comparisons done in phase i if $i \in Q$. Then $\sum_{i \in Q} d_i + \sum_{i \notin Q}(|t_i| + 1)$ is a bound on the total number of comparisons done in the algorithm.

The quantity $\sum_{i \notin Q}(|t_i| + 1)$ is again $O(m)$. To see this, recall that the lemmas of the previous section, which proved that $g_i \leq 3s_i$, only needed the assumption that phase i ends with a mismatch and that $h < i$. In particular, the analysis of how P of phase $h < i$ is aligned with P of phase i did not need the assumption that phase h ends with a mismatch. Those proofs cover both the case that h ends with a mismatch and that h ends by finding an occurrence of P. Hence it again holds that $g_i \leq 3s_i$ if phase i ends with a mismatch, even though earlier phases might end with a match.

For phases in Q, we again ignore the case that $s_i \geq (n + 1)/3 \geq (d_i + 1)/3$, since the total number of comparisons done in such phases must be bounded by $\sum 3s_i \leq 3m$. So suppose phase i ends by finding an occurrence of P in T and then shifts by less than $n/3$. By a proof essentially the same as for Lemma 3.2.5 it follows that P is semi-periodic; let β denote the shortest period of P. Hence the shift in phase i moves P right by exactly $|\beta|$ positions, and using the Galil rule in the Boyer–Moore algorithm, no character of T compared in phase $i + 1$ will have ever been compared previously. Repeating this reasoning, if phase $i + 1$ ends by finding an occurrence of P then P will again shift by exactly $|\beta|$ places and no comparisons in phase $i + 2$ will examine a character of T compared in any earlier phase. This cycle of shifting P by exactly $|\beta|$ positions and then identifying another occurrence of P by examining only $|\beta|$ new characters of T may be repeated many times. Such a succession of overlapping occurrences of P then consists of a concatenation of copies of β (each copy of P starts exactly $|\beta|$ places to the right of the previous occurrence) and is called a *run*. Using the Galil rule, it follows immediately that in any single run the number of comparisons used to identify the occurrences of P contained in that run is exactly the length of the run. Therefore, over the entire algorithm the number of comparisons used to find those occurrences is $O(m)$. If no additional comparisons were possible with characters in a run, then the analysis would be complete. However, additional examinations are possible and we have to account for them.

A run ends in some phase $k > i$ when a mismatch is found (or when the algorithm terminates). It is possible that characters of T in the run could be examined again in phases after k. A phase that reexamines characters of the run either ends with a mismatch or ends by finding an occurrence of P that overlaps the earlier run but is not part of it. However, all comparisons in phases that end with a mismatch have already been accounted for (in the accounting for phases not in Q) and are ignored here.

Let $k' > k > i$ be a phase in which an occurrence of P is found overlapping the earlier run but is not part of that run. As an example of such an overlap, suppose $P = axaaxa$ and T contains the substring *axaaxaaxaaxaxaaxaaxa*. Then a run begins at the start of the substring and ends with its twelfth character, and an overlapping occurrence of P (not part of the run) begins with that character. Even with the Galil rule, characters in the run will be examined again in phase k', and since phase k' does not end with a mismatch those comparisons must still be counted.

In phase k', if the left end of the new occurrence of P in T starts at a left end of a copy of β in the run, then contiguous copies of β continue past the right end of the run. But then no mismatch would have been possible in phase k since the pattern in phase k is aligned exactly $|\beta|$ places to the right of its position in phase $k - 1$ (where an occurrence of P was found). So in phase k', the left end of the new P in T must start with an interior character of some copy of β. But then if P overlaps with the run by more than $|\beta|$ characters, Lemma 3.2.1 implies that β is periodic, contradicting the selection of β. So P can overlap the run only by part of the run's left-most copy of β. Further, since phase k' ends by finding an occurrence of P, the pattern is shifted right by $s_{k'} = |\beta|$ positions. Thus any phase that finds an occurrence of P overlapping an earlier run next shifts P by a number of positions larger than the length of the overlap (and hence the number of comparisons). It follows then that over the entire algorithm the total number of such additional comparisons in overlapping regions is $O(m)$.

All comparisons are accounted for and hence $\sum_{i \in Q} d_i = O(m)$, finishing the proof of the lemma.   □

3.2.3. Adding in the bad character rule

Recall that in computing a shift after a mismatch, the Boyer–Moore algorithm uses the largest shift given by either the (extended) bad character rule or the (strong) good suffix rule. It seems intuitive that if the time bound is $O(m)$ when only the good suffix rule is used, it should still be $O(m)$ when both rules are used. However, certain "interference" is plausible, and so the intuition requires a proof.

Theorem 3.2.4. *When both shift rules are used together, the worst-case running time of the modified Boyer–Moore algorithm remains $O(m)$.*

PROOF   In the analysis using only the suffix rule we focused on the comparisons done in an arbitrary phase i. In phase i the right end of P was aligned with some character of T. However, we never made any assumptions about how P came to be positioned there. Rather, given an arbitrary placement of P in a phase ending with a mismatch, we deduced bounds on how many characters compared in that phase could have been compared in earlier phases. Hence all of the lemmas and analyses remain correct if P is arbitrarily picked up and moved some distance to the right at any time during the algorithm. The (extended) bad character rule only moves P to the right, so all lemmas and analyses showing the $O(m)$ bound remain correct even with its use.   □

3.3. The original preprocessing for Knuth-Morris-Pratt

3.3.1. The method does not use fundamental preprocessing

In Section 1.3 we showed how to compute all the sp_i values from Z_i values obtained during fundamental preprocessing of P. The use of Z_i values was conceptually simple and allowed a uniform treatment of various preprocessing problems. However, the classical preprocessing method given in Knuth-Morris-Pratt [278] is not based on fundamental preprocessing. The approach taken there is very well known and is used or extended in several additional methods (such as the Aho–Corasick method that is discussed next). For those reasons, a serious student of string algorithms should also understand the classical algorithm for Knuth-Morris-Pratt preprocessing.

The preprocessing algorithm computes $sp_i(P)$ for each position i from $i = 2$ to $i = n$ (sp_1 is zero). To explain the method, we focus on how to compute sp_{k+1} assuming that sp_i is known for each $i \le k$. The situation is shown in Figure 3.9, where string α is the prefix of P of length sp_k. That is, α is the longest string that occurs both as a proper prefix of P and as a substring of P ending at position k. For clarity, let α' refer to the copy of α that ends at position k.

Let x denote character $k+1$ of P, and let $\beta = \bar{\beta}x$ denote the prefix of P of length sp_{k+1} (i.e., the prefix that the algorithm will next try to compute). Finding sp_{k+1} is equivalent to finding string $\bar{\beta}$. And clearly,

*) $\bar{\beta}$ is the longest proper prefix of $P[1..k]$ that matches a suffix of $P[1..k]$ and that is followed by character x in position $|\bar{\beta}| + 1$ of P. See Figure 3.10.

Our goal is to find sp_{k+1}, or equivalently, to find $\bar{\beta}$.

3.3.2. The easy case

Suppose the character just after α is x (i.e., $P(sp_k + 1) = x$). Then, string αx is a prefix of P and also a proper suffix of $P[1..k + 1]$, and thus $sp_{k+1} \ge |\alpha x| = sp_k + 1$. Can we then end our search for sp_{k+1} concluding that sp_{k+1} equals $sp_k + 1$, or is it possible for sp_{k+1} to be strictly greater than $sp_k + 1$? The next lemma settles this.

Lemma 3.3.1. *For any k, $sp_{k+1} \le sp_k + 1$. Further, $sp_{k+1} = sp_k + 1$ if and only if the character after α is x. That is, $sp_{k+1} = sp_k + 1$ if and only if $P(sp_k + 1) = P(k + 1)$.*

PROOF Let $\beta = \bar{\beta}x$ denote the prefix of P of length sp_{k+1}. That is, $\beta = \bar{\beta}x$ is the longest proper suffix of $P[1..k + 1]$ that is a prefix of P. If sp_{k+1} is strictly greater than

sp_k k $k+1$

Figure 3.9: The situation after finding sp_k.

k $k+1$

Figure 3.10: sp_{k+1} is found by finding $\bar{\beta}$.

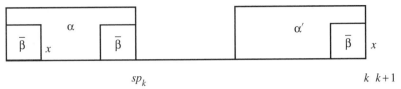

$$sp_k \qquad\qquad\qquad\qquad\qquad\qquad\qquad\qquad k \;\; k+1$$

Figure 3.11: $\bar{\beta}$ must be a suffix of α.

$sp_k + 1 = |\alpha| + 1$, then $\bar{\beta}$ would be a prefix of P that is longer than α. But $\bar{\beta}$ is also a proper suffix of $P[1..k]$ (because βx is a proper suffix of $P[1..k + 1]$). Those two facts would contradict the definition of sp_k (and the selection of α). Hence $sp_{k+1} \leq sp_k + 1$.

Now clearly, $sp_{k+1} = sp_k + 1$ if the character to the right of α is x, since αx would then be a prefix of P that also occurs as a proper suffix of $P[1..k + 1]$. Conversely, if $sp_{k+1} = sp_k + 1$ then the character after α must be x. $\quad\square$

Lemma 3.3.1 identifies the largest "candidate" value for sp_{k+1} and suggests how to initially look for that value (and for string β). We should first check the character $P(sp_k+1)$, just to the right of α. If it equals $P(sp_k + 1)$ then we conclude that $\bar{\beta}$ equals α, β is αx, and sp_{k+1} equals $sp_k + 1$. But what do we do if the two characters are not equal?

3.3.3. The general case

When character $P(k + 1) \neq P(sp_k + 1)$, then $sp_{k+1} < sp_k + 1$ (by Lemma 3.3.1), so $sp_{k+1} \leq sp_k$. It follows that β must be a prefix of α, and $\bar{\beta}$ must be a *proper* prefix of α. Now substring $\beta = \bar{\beta} x$ ends at position $k + 1$ and is of length at most sp_k, whereas α' is a substring ending at position k and is of length sp_k. So $\bar{\beta}$ is a suffix of α', as shown in Figure 3.11. But since α' is a copy of α, $\bar{\beta}$ is also a suffix of α.

In summary, when $P(k + 1) \neq P(sp_k + 1)$, $\bar{\beta}$ occurs as a suffix of α and also as a proper prefix of α followed by character x. So when $P(k + 1) \neq P(sp_k + 1)$, $\bar{\beta}$ is the longest proper prefix of α that matches a suffix of α and that is followed by character x in position $|\bar{\beta}| + 1$ of P. See Figure 3.11.

However, since $\alpha = P[1..sp_k]$, we can state this as

**) $\bar{\beta}$ is the longest proper prefix of $P[1..sp_k]$ that matches a suffix of $P[1..k]$ and that is followed by character x in position $|\bar{\beta}| + 1$ of P.

The general reduction

Statements $*$ and $**$ differ only by the substitution of $P[1..sp_k]$ for $P[1..k]$ and are otherwise exactly the same. Thus, when $P(sp_k + 1) \neq P(k + 1)$, the problem of finding $\bar{\beta}$ reduces to another instance of the original problem but on a smaller string ($P[1..sp_k]$ in place of $P[1..k]$). We should therefore proceed as before. That is, to search for $\bar{\beta}$ the algorithm should find the longest proper prefix of $P[1..sp_k]$ that matches a suffix of $P[1..sp_k]$ and then check whether the character to the right of that prefix is x. By the definition of sp_k, the required prefix ends at character sp_{sp_k}. So if character $P(sp_{sp_k}+1) = x$ then we have found $\bar{\beta}$, or else we recurse again, restricting our search to ever smaller prefixes of P. Eventually, either a valid prefix is found, or the beginning of P is reached. In the latter case, $sp_{k+1} = 1$ if $P(1) = P(k + 1)$; otherwise $sp_{k+1} = 0$.

The complete preprocessing algorithm

Putting all the pieces together gives the following algorithm for finding $\bar{\beta}$ and sp_{k+1}:

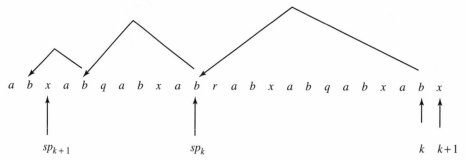

Figure 3.12: "Bouncing ball" cartoon of original Knuth-Morris-Pratt preprocessing. The arrows show the successive assignments to the variable v.

How to find sp_{k+1}

> $x := P(k + 1)$;
> $v := sp_k$;
> While $P(v + 1) \neq x$ and $v \neq 0$ do
> > $v := sp_v$;
>
> end;
> If $P(v + 1) = x$ then
> > $sp_{k+1} := v + 1$
>
> else
> > $sp_{k+1} := 0$;

See the example in Figure 3.12.

The entire set of sp values are found as follows:

Algorithm SP(P)

$sp_1 = 0$
For $k := 1$ to $n - 1$ do
begin
> $x := P(k + 1)$;
> $v := sp_k$;
> While $P(v + 1) \neq x$ and $v \neq 0$ do
> > $v := sp_v$;
>
> end;
> If $P(v + 1) = x$ then
> > $sp_{k+1} := v + 1$
>
> else
> > $sp_{k+1} := 0$;

end;

Theorem 3.3.1. *Algorithm SP finds all the $sp_i(P)$ values in $O(n)$ time, where n is the length of P.*

PROOF Note first that the algorithm consists of two nested loops, a *for* loop and a *while* loop. The *for* loop executes exactly $n - 1$ times, incrementing the value of k each time. The *while* loop executes a variable number of times each time it is entered.

The work of the algorithm is proportional to the number of times the value of v is assigned. We consider the places where the value of v is assigned and focus on how the value of v changes over the execution of the algorithm. The value of v is assigned once each time the *for* statement is reached; it is assigned a variable number of times inside the *while* loop, each time this loop is reached. Hence the number of times v is assigned is $n - 1$ plus the number of times it is assigned inside the *while* loop. How many times that can be is the key question.

Each assignment of v inside the *while* loop must decrease the value of v, and each of the $n - 1$ times v is assigned at the *for* statement, its value either increases by one or it remains unchanged (at zero). The value of v is initially zero, so the total amount that the value of v can increase (at the *for* statement) over the entire algorithm is at most $n - 1$. But since the value of v starts at zero and is never negative, the total amount that the value of v can *decrease* over the entire algorithm must also be bounded by $n - 1$, the total amount it can increase. Hence v can be assigned in the *while* loop at most $n - 1$ times, and hence the total number of times that the value of v can be assigned is at most $2(n - 1) = O(n)$, and the theorem is proved. □

3.3.4. How to compute the optimized shift values

The (stronger) sp'_i values can be easily computed from the sp_i values in $O(n)$ time using the algorithm below. For the purposes of the algorithm, character $P(n + 1)$, which does not exist, is defined to be different from any character in P.

Algorithm SP'(P)

$sp'_1 = 0$;
For $i := 2$ to n do
begin
    $v := sp_i$;
    If $P(v + 1) \neq P(i + 1)$ then
        $sp'_i := v$
    else
        $sp'_i := sp'_v$;
end;

Theorem 3.3.2. *Algorithm $SP'(P)$ correctly computes all the sp'_i values in $O(n)$ time.*

PROOF The proof is by induction on the value of i. Since $sp_1 = 0$ and $sp'_i \leq sp_i$ for all i, then $sp'_1 = 0$, and the algorithm is correct for $i = 1$. Now suppose that the value of sp'_i set by the algorithm is correct for all $i < k$ and consider $i = k$. If $P[sp_k + 1] \neq P[k + 1]$ then clearly sp'_k is equal to sp_k, since the sp_k length prefix of $P[1..k]$ satisfies all the needed requirements. Hence in this case, the algorithm correctly sets sp'_k.

If $P(sp_k + 1) = P(k + 1)$, then $sp'_k < sp_k$ and, since $P[1..sp_k]$ is a suffix $P[1..k]$, sp'_k can be expressed as the length of the longest proper prefix of $P[1..sp_k]$ that also occurs as a suffix of $P[1..sp_k]$ with the condition that $P(k + 1) \neq P(sp'_k + 1)$. But since $P(k + 1) = P(sp_k + 1)$, that condition can be rewritten as $P(sp_k + 1) \neq P(sp'_k + 1)$. By the induction hypothesis, that value has already been correctly computed as sp'_{sp_k}. So when $P(sp_k + 1) = P(k + 1)$ the algorithm correctly sets sp'_k to sp'_{sp_k}.

Because the algorithm only does constant work per position, the total time for the algorithm is $O(n)$. □

It is interesting to compare the classical method for computing sp and sp' and the method based on fundamental preprocessing (i.e., on Z values). In the classical method the (weaker) sp values are computed first and then the more desirable sp' values are derived

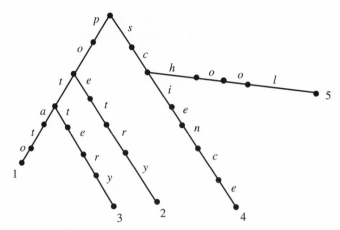

Figure 3.13: Keyword tree $\mathcal{K}$ with five patterns.

from them, whereas the order is just the opposite in the method based on fundamental preprocessing.

3.4. Exact matching with a set of patterns

An immediate and important generalization of the exact matching problem is to find all occurrences in text T of any pattern in a *set* of patterns $\mathcal{P} = \{P_1, P_2, \ldots, P_z\}$. This generalization is called the *exact set matching* problem. Let n now denote the total length of all the patterns in $\mathcal{P}$ and m be, as before, the length of T. Then, the exact set matching problem can be solved in time $O(n + zm)$ by separately using any linear-time method for each of the z patterns.

Perhaps surprisingly, the exact set matching problem can be solved faster than, $O(n + zm)$. It can be solved in $O(n + m + k)$ time, where k is the number of occurrences in T of the patterns from $\mathcal{P}$. The first method to achieve this bound is due to Aho and Corasick [9].[2] In this section, we develop the Aho–Corasick method; some of the proofs are left to the reader. An equally efficient, but more robust, method for the exact set matching problem is based on suffix trees and is discussed in Section 7.2.

Definition The *keyword tree* for set $\mathcal{P}$ is a rooted directed tree $\mathcal{K}$ satisfying three conditions: 1. each edge is labeled with exactly one character; 2. any two edges out of the same node have distinct labels; and 3. every pattern P_i in $\mathcal{P}$ maps to some node v of $\mathcal{K}$ such that the characters on the path from the root of $\mathcal{K}$ to v exactly spell out P_i, and every leaf of $\mathcal{K}$ is mapped to by some pattern in $\mathcal{P}$.

For example, Figure 3.13 shows the keyword tree for the set of patterns {*potato, poetry, pottery, science, school*}.

Clearly, every node in the keyword tree corresponds to a prefix of one of the patterns in $\mathcal{P}$, and every prefix of a pattern maps to a distinct node in the tree.

Assuming a fixed-size alphabet, it is easy to construct the keyword tree for $\mathcal{P}$ in $O(n)$ time. Define $\mathcal{K}_i$ to be the (partial) keyword tree that encodes patterns $P_1, \ldots, P_i$ of $\mathcal{P}$.

---

[2] There is a more recent exposition of the Aho–Corasick method in [8], where the algorithm is used just as an "acceptor", deciding whether or not there is an occurrence in T of at least one pattern from $\mathcal{P}$. Because we will want to explicitly find all occurrences, that version of the algorithm is too limited to use here.

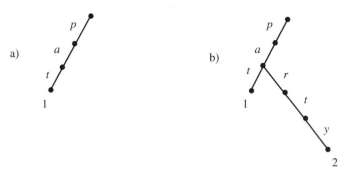

Figure 3.14: Pattern P_1 is the string *pat*. a. The insertion of pattern P_2 when P_2 is *pa*. b. The insertion when P_2 is *party*.

Tree $\mathcal{K}_1$ just consists of a single path of $|P_1|$ edges out of root r. Each edge on this path is labeled with a character of P_1 and when read from the root, these characters spell out P_1. The number 1 is written at the node at the end of this path. To create $\mathcal{K}_2$ from $\mathcal{K}_1$, first find the longest path from root r that matches the characters of P_2 in order. That is, find the longest prefix of P_2 that matches the characters on some path from r. That path either ends by exhausting P_2 or it ends at some node v in the tree where no further match is possible. In the first case, P_2 already occurs in the tree, and so we write the number 2 at the node where the path ends. In the second case, we create a new path out of v, labeled by the remaining (unmatched) characters of P_2, and write number 2 at the end of that path. An example of these two possibilities is shown in Figure 3.14.

In either of the above two cases, $\mathcal{K}_2$ will have at most one branching node (a node with more than one child), and the characters on the two edges out of the branching node will be distinct. We will see that the latter property holds inductively for any tree $\mathcal{K}_i$. That is, at any branching node v in $\mathcal{K}_i$, all edges out of v have distinct labels.

In general, to create $\mathcal{K}_{i+1}$ from $\mathcal{K}_i$, start at the root of $\mathcal{K}_i$ and follow, as far as possible, the (unique) path in $\mathcal{K}_i$ that matches the characters in P_{i+1} in order. This path is unique because, at any branching node v of $\mathcal{K}_i$, the characters on the edges out of v are distinct. If pattern P_{i+1} is exhausted (fully matched), then number the node where the match ends with the number $i+1$. If a node v is reached where no further match is possible but P_{i+1} is not fully matched, then create a new path out of v labeled with the remaining unmatched part of P_{i+1} and number the endpoint of that path with the number $i+1$.

During the insertion of P_{i+1}, the work done at any node is bounded by a constant, since the alphabet is finite and no two edges out of a node are labeled with the same character. Hence for any i, it takes $O(|P_{i+1}|)$ time to insert pattern P_{i+1} into $\mathcal{K}_i$, and so the time to construct the entire keyword tree is $O(n)$.

3.4.1. Naive use of keyword trees for set matching

Because no two edges out of any node are labeled with the same character, we can use the keyword tree to search for all occurrences in T of patterns from $\mathcal{P}$. To begin, consider how to search for occurrences of patterns in $\mathcal{P}$ that begin at character 1 of T: Follow the unique path in $\mathcal{K}$ that matches a prefix of T as far as possible. If a node is encountered on this path that is numbered by i, then P_i occurs in T starting from position 1. More than one such numbered node can be encountered if some patterns in $\mathcal{P}$ are prefixes of other patterns in $\mathcal{P}$.

In general, to find all patterns that occur in T, start from each position l in T and follow the unique path from r in $\mathcal{K}$ that matches a substring of T starting at character l.

Numbered nodes along that path indicate patterns in $\mathcal{P}$ that start at position l. For a fixed l, the traversal of a path of $\mathcal{K}$ takes time proportional to the minimum of m and n, so by successively incrementing l from 1 to m and traversing $\mathcal{K}$ for each l, the exact set matching problem can be solved in $O(nm)$ time. We will reduce this to $O(n + m + k)$ time below, where k is the number of occurrences.

The dictionary problem

Without any further embellishments, this simple keyword tree algorithm efficiently solves a special case of set matching, called the *dictionary problem*. In the dictionary problem, a set of strings (forming a dictionary) is initially known and preprocessed. Then a sequence of individual strings will be presented; for each one, the task is to find if the presented string is contained in the dictionary. The utility of a keyword tree is clear in this context. The strings in the dictionary are encoded into a keyword tree $\mathcal{K}$, and when an individual string is presented, a walk from the root of $\mathcal{K}$ determines if the string is in the dictionary. In this special case of exact set matching, the problem is to determine if the text T (an individual presented string) completely matches some string in $\mathcal{P}$.

We now return to the general set matching problem of determining which strings in $\mathcal{P}$ are *contained* in text T.

3.4.2. The speedup: generalizing Knuth-Morris-Pratt

The above naive approach to the exact set matching problem is analogous to the naive search we discussed before introducing the Knuth-Morris-Pratt method. Successively incrementing l by one and starting each search from root r is analogous to the naive exact match method for a single pattern, where after every mismatch the pattern is shifted by only one position, and the comparisons are always begun at the left end of the pattern. The Knuth-Morris-Pratt algorithm improves on that naive algorithm by shifting the pattern by more than one position when possible and by never comparing characters to the left of the current character in T. The Aho–Corasick algorithm makes the same kind of improvements, incrementing l by more than one and skipping over initial parts of paths in $\mathcal{K}$, when possible. The key is to generalize the function sp_i (defined on page 27 for a single pattern) to operate on a set of patterns. This generalization is fairly direct, with only one subtlety that occurs if a pattern in $\mathcal{P}$ is a proper substring of another pattern in $\mathcal{P}$. So, it is very helpful to (temporarily) make the following assumption:

Assumption No pattern in $\mathcal{P}$ is a proper substring of any other pattern in $\mathcal{P}$.

3.4.3. Failure functions for the keyword tree

Definition Each node v in $\mathcal{K}$ is *labeled* with the string obtained by concatenating in order the characters on the path from the root of $\mathcal{K}$ to node v. $\mathcal{L}(v)$ is used to denote the label on v. That is, the concatenation of characters on the path from the root to v spells out the string $\mathcal{L}(v)$.

For example, in Figure 3.15 the node pointed to by the arrow is labeled with the string *pott*.

Definition For any node v of $\mathcal{K}$, define $lp(v)$ to be the length of the longest proper suffix of string $\mathcal{L}(v)$ that is a prefix of some pattern in $\mathcal{P}$.

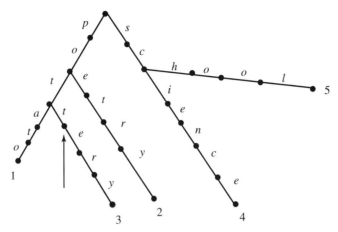

Figure 3.15: Keyword tree to illustrate the label of a node.

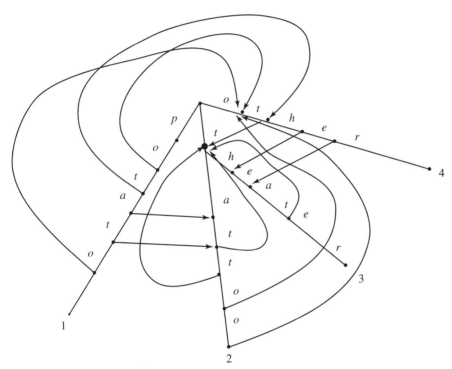

Figure 3.16: Keyword tree showing the failure links.

For example, consider the set of patterns $\mathcal{P} = \{$*potato, tattoo, theater, other*$\}$ and its keyword tree shown in Figure 3.16. Let v be the node labeled with the string *potat*. Since *tat* is prefix of *tattoo*, and it is the longest proper suffix of *potat* that is a prefix of any pattern in $\mathcal{P}$, $lp(v) = 3$.

Lemma 3.4.1. *Let α be the $lp(v)$-length suffix of string $\mathcal{L}(v)$. Then there is a unique node in the keyword tree that is labeled by string α.*

PROOF $\mathcal{K}$ encodes all the patterns in $\mathcal{P}$ and, by definition, the $lp(v)$-length suffix of $\mathcal{L}(v)$ is a prefix of some pattern in $\mathcal{P}$. So there must be a path from the root in $\mathcal{K}$ that spells out string α. By the construction of $\mathcal{T}$ no two paths spell out the same string, so this path is unique and the lemma is proved. □

Definition For a node v of $\mathcal{K}$ let n_v be the unique node in $\mathcal{K}$ labeled with the suffix of $\mathcal{L}(v)$ of length $lp(v)$. When $lp(v) = 0$ then n_v is the root of $\mathcal{K}$.

Definition We call the ordered pair (v, n_v) a *failure link*.

Figure 3.16 shows the keyword tree for $\mathcal{P} = \{potato, tatoo, theater, other\}$. Failure links are shown as pointers from every node v to node n_v where $lp(v) > 0$. The other failure links point to the root and are not shown.

3.4.4. The failure links speed up the search

Suppose that we know the failure link $v \mapsto n_v$ for each node v in $\mathcal{K}$. (Later we will show how to efficiently find those links.) How do the failure links help speed up the search? The Aho–Corasick algorithm uses the function $v \mapsto n_v$ in a way that directly generalizes the use of the function $i \mapsto sp_i$ in the Knuth-Morris-Pratt algorithm. As before, we use l to indicate the starting position in T of the patterns being searched for. We also use pointer c into T to indicate the "current character" of T to be compared with a character on $\mathcal{K}$. The following algorithm uses the failure links to search for occurrences in T of patterns from $\mathcal{P}$:

Algorithm AC search

    $l := 1$;
    $c := 1$;
    $w :=$ root of $\mathcal{K}$;
    repeat
        While there is an edge (w, w') labeled character $T(c)$
            begin
            if w' is numbered by pattern i then
                report that P_i occurs in T starting at position l;
            $w := w'$ and $c := c + 1$;
            end;
        $w := n_w$ and $l := c - lp(w)$;
    until $c > m$;

To understand the use of the function $v \mapsto n_v$, suppose we have traversed the tree to node v but cannot continue (i.e., character $T(c)$ does not occur on any edge out of v). We know that string $\mathcal{L}(v)$ occurs in T starting at position l and ending at position $c - 1$. By the definition of the function $v \mapsto n_v$, it is guaranteed that string $\mathcal{L}(n_v)$ matches string $T[c - lp(v)..c - 1]$. That is, the algorithm could traverse $\mathcal{K}$ from the root to node n_v and be sure to match all the characters on this path with the characters in T starting from position $c - lp(v)$. So when $lp(v) \geq 0$, l can be increased to $c - lp(v)$, c can be left unchanged, and there is no need to actually make the comparisons on the path from the root to node n_v. Instead, the comparisons should begin at node n_v, comparing character c of T against the characters on the edges out of n_v.

For example, consider the text $T = xxpotattooxx$ and the keyword tree shown in Figure 3.16. When $l = 3$, the text matches the string *potat* but mismatches at the next character. At this point $c = 8$, and the failure link from the node v labeled *potat* points to the node n_v labeled *tat*, and $lp(v) = 3$. So l is incremented to $5 = 8 - 3$, and the next comparison is between character $T(8)$ and character t on the edge below *tat*.

With this algorithm, when no further matches are possible, l may increase by more than one, avoiding the reexamination of characters of T to the left of c, and yet we may be sure that every occurrence of a pattern in $\mathcal{P}$ that *begins* at character $c - lp(v)$ of T will be correctly detected. Of course (just as in Knuth-Morris-Pratt), we have to argue that there are no occurrences of patterns of $\mathcal{P}$ starting strictly between the old l and $c - lp(v)$ in T, and thus l can be incremented to $c - lp(v)$ without missing any occurrences. With the given assumption that no pattern in $\mathcal{P}$ is a proper substring of another one, that argument is almost identical to the proof of Theorem 2.3.2 in the analysis of Knuth-Morris-Pratt, and it is left as an exercise.

When $lp(v) = 0$, then l is increased to c and the comparisons begin at the root of $\mathcal{K}$. The only case remaining is when the mismatch occurs at the root. In this case, c must be incremented by 1 and comparisons again begin at the root.

Therefore, the use of function $v \mapsto n_v$ certainly accelerates the naive search for patterns of $\mathcal{P}$. But does it improve the worst-case running time? By the same sort of argument used to analyze the search time (not the preprocessing time) of Knuth-Morris-Pratt (Theorem 2.3.3), it is easily established that the search time for Aho–Corasick is $O(m)$. We leave this as an exercise. However, we have yet to show how to precompute the function $v \mapsto n_v$ in linear time.

3.4.5. Linear preprocessing for the failure function

Recall that for any node v of $\mathcal{K}$, n_v is the unique node in $\mathcal{K}$ labeled with the suffix of $\mathcal{L}(v)$ of length $lp(v)$. The following algorithm finds node n_v for each node v in $\mathcal{K}$, using $O(n)$ total time. Clearly, if v is the root r or v is one character away from r, then $n_v = r$. Suppose, for some k, n_v has been computed for every node that is exactly k or fewer characters (edges) from r. The task now is to compute n_v for a node v that is $k + 1$ characters from r. Let v' be the parent of v in $\mathcal{K}$ and let x be the character on the v' to v edge, as shown in Figure 3.17.

We are looking for the node n_v and the (unknown) string $\mathcal{L}(n_v)$ labeling the path to it from the root; we know node $n_{v'}$ because v' is k characters from r. Just as in the explanation

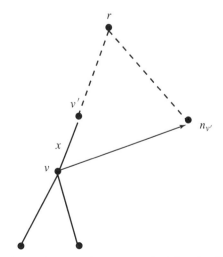

Figure 3.17: Keyword tree used to compute the failure function for node v.

of the classic preprocessing for Knuth-Morris-Pratt, $\mathcal{L}(n_v)$ must be a suffix of $\mathcal{L}(n_{v'})$ (not necessarily proper) followed by character x. So the first thing to check is whether there is an edge $(n_{v'}, w')$ out of node $n_{v'}$ labeled with character x. If that edge does exist, then n_v is node w' and we are done. If there is no such edge out of $n_{v'}$ labeled with character x, then $\mathcal{L}(n_v)$ is a *proper* suffix of $\mathcal{L}(n_{v'})$ followed by x. So we examine $n_{n_{v'}}$ next to see if there is an edge out of it labeled with character x. (Node $n_{n_{v'}}$ is known because $n_{v'}$ is k or fewer edges from the root.) Continuing in this way, with exactly the same justification as in the classic preprocessing for Knuth-Morris-Pratt, we arrive at the following algorithm for computing n_v for a node v:

Algorithm n_v

> v' is the parent of v in $\mathcal{K}$;
> x is the character on the edge (v', v);
> $w := n_{v'}$;
> While there is no edge out of w labeled x and $w \neq r$
>> do $w := n_w$;
> end (while);
> If there is an edge (w, w') out of w labeled x then
>> $n_v := w'$;
> else
>> $n_v := r$;

Note the importance of the assumption that n_u is already known for every node u that is k or fewer characters from r.

To find n_v for every node v, repeatedly apply the above algorithm to the nodes in $\mathcal{K}$ in a breadth-first manner starting at the root.

Theorem 3.4.1. *Let n be the total length of all the patterns in $\mathcal{P}$. The total time used by Algorithm n_v when applied to all nodes in $\mathcal{K}$ is $O(n)$.*

PROOF   The argument is a direct generalization of the argument used to analyze time in the classic preprocessing for Knuth-Morris-Pratt. Consider a single pattern P in $\mathcal{P}$ of length t and its path in $\mathcal{K}$ for pattern P. We will analyze the time used in the algorithm to find the failure links for the nodes on this path, as if the path shares no nodes with paths for any other pattern in $\mathcal{P}$. That analysis will overcount the actual amount of work done by the algorithm, but it will still establish a linear time bound.

The key is to see how $lp(v)$ varies as the algorithm is executed on each successive node v down the path for P. When v is one edge from the root, then $lp(v)$ is zero. Now let v be an arbitrary node on the path for P and let v' be the parent of v. Clearly, $lp(v) \leq lp(v')+1$, so over all executions of Algorithm n_v for nodes on the path for P, $lp()$ is increased by a total of at most t. Now consider how $lp()$ can decrease. During the computation of n_v for any node v, w starts at $n_{v'}$ and so has initial node depth equal to $lp(v')$. However, during the computation of n_v, the node depth of w decreases every time an assignment to w is made (inside the *while* loop). When n_v is finally set, $lp(v)$ equals the current depth of w, so if w is assigned k times, then $lp(v) \leq lp(v') - k$ and $lp()$ decreases by at least k. Now $lp()$ is never negative, and during all the computations along path P, $lp()$ can be increased by a total of at most t. It follows that over all the computations done for nodes on the path for P, the number of assignments made inside the *while* loop is at most t. The total time used is proportional to the number of assignments inside the loop, and hence all failure links on the path for P are set in $O(t)$ time.

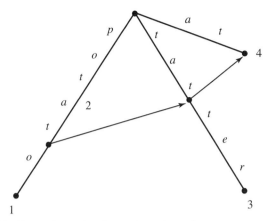

Figure 3.18: Keyword tree showing a directed path from *potat* to *at* through *tat*.

Repeating this analysis for every pattern in $\mathcal{P}$ yields the result that all the failure links are established in time proportional to the sum of the pattern lengths in $\mathcal{P}$ (i.e., in $O(n)$ total time). $\square$

3.4.6. The full Aho–Corasick algorithm: relaxing the substring assumption

Until now we have assumed that no pattern in $\mathcal{P}$ is a substring of another pattern in $\mathcal{P}$. We now relax that assumption. If one pattern is a substring of another, and yet *Algorithm AC search* (page 56) uses the same keyword tree as before, then the algorithm may make l too large. Consider the case when $\mathcal{P} = \{acatt, ca\}$ and $T = acatg$. As given, the algorithm matches T along a path in $\mathcal{K}$ until character g is the current character. That path ends at the node v with $\mathcal{L}(v) = acat$. Now no edges out of v are labeled g, and since no proper suffix of *acat* is a prefix of *acatt* or *ac*, n_v is the root of $\mathcal{K}$. So when the algorithm gets stuck at node v it returns to the root with g as the current character, and it sets l to 5. Then after one additional comparison the current character pointer will be set to $m + 1$ and the algorithm will terminate without finding the occurrence of *ca* in T. This happens because the algorithm shifts (increases l) so as to match the longest *suffix* of $\mathcal{L}(v)$ with a prefix of some pattern in $\mathcal{P}$. Embedded occurrences of patterns in $\mathcal{L}(v)$ that are not suffixes of $\mathcal{L}(v)$ have no influence on how much l increases.

It is easy to repair this problem with the following observations whose proofs we leave to the reader.

Lemma 3.4.2. *Suppose in a keyword tree $\mathcal{K}$ there is a directed path of failure links (possibly empty) from a node v to a node that is numbered with pattern i. Then pattern P_i must occur in T ending at position c (the current character) whenever node v is reached during the search phase of the Aho–Corasick algorithm.*

For example, Figure 3.18 shows the keyword tree for $\mathcal{P} = \{potato, pot, tatter, at\}$ along with some of the failure links. Those links form a directed path from the node v labeled *potat* to the numbered node labeled *at*. If the traversal of $\mathcal{K}$ reaches v then T certainly contains the patterns *tat* and *at* end at the current c.

Conversely,

Lemma 3.4.3. *Suppose a node v has been reached during the algorithm. Then pattern*

P_i occurs in T ending at position c only if v is numbered i or there is a directed path of failure links from v to the node numbered i.

So the full search algorithm is

Algorithm full AC search

> $l := 1$;
> $c := 1$;
> $w := $ root;
> repeat
> > While there is an edge (w, w') labeled $T(c)$
> > > begin
> > > if w' is numbered by pattern i or there is
> > > a directed path of failure links from w' to a node numbered with i
> > > > then report that P_i occurs in T ending at position c;
> > > $w := w'$ and $c := c + 1$;
> > > end;
> > $w := n_w$ and $l := c - lp(w)$;
> until $c > n$;

Implementation

Lemmas 3.4.2 and 3.4.3 specify at a high level how to find all occurrences of the patterns in the text, but specific implementation details are still needed. The goal is to be able to build the keyword tree, determine function $v \mapsto n_v$, and be able to execute the full AC search algorithm all in $O(m + k)$ time. To do this we add an additional pointer, called the *output link*, to each node of $\mathcal{K}$.

The output link (if there is one) at a node v points to that numbered node (a node associated with the end of a pattern in $\mathcal{P}$) other than v that is reachable from v by the fewest failure links. The output links can be determined in $O(n)$ time during the running of the preprocessing algorithm n_v. When the n_v value is determined, the possible output link from node v is determined as follows: If n_v is a numbered node then the output link from v points to n_v; if n_v is not numbered but has an output link to a node w, then the output link from v points to w; otherwise v has no output link. In this way, an output link points only to a numbered node, and the path of output links from any node v passes through all the numbered nodes reachable from v via a path of failure links. For example, in Figure 3.18 the nodes for *tat* and *potat* will have their output links set to the node for *at*. The work of adding output links adds only constant time per node, so the overall time for algorithm n_v remains $O(n)$.

With the output links, all occurrences in T of patterns of $\mathcal{P}$ can be detected in $O(m + k)$ time. As before, whenever a numbered node is encountered during the full AC search, an occurrence is detected and reported. But additionally, whenever a node v is encountered that has an output link from it, the algorithm must traverse the path of output links from v, reporting an occurrence ending at position c of T for each link in the path. When that path traversal reaches a node with no output link, it returns along the path to node v and continues executing the full AC search algorithm. Since no character comparisons are done during any output link traversal, over both the construction and search phases of the algorithm the number of character comparisons is still bounded by $O(n+m)$. Further, even though the number of traversals of output links can exceed that linear bound, each traversal of an output link leads to the discovery of a pattern occurrence, so the total time for the algorithm is $O(n+m+k)$, where k is the total number of occurrences. In summary we have,

Theorem 3.4.2. *If $\mathcal{P}$ is a set of patterns with total length n and T is a text of total length m, then one can find all occurrences in T of patterns from $\mathcal{P}$ in $O(n)$ preprocessing time plus $O(m+k)$ search time, where k is the number of occurrences. This is true even without assuming that the patterns in $\mathcal{P}$ are substring free.*

In a later chapter (Section 6.5) we will discuss further implementation issues that affect the practical performance of both the Aho–Corasick method, and suffix tree methods.

3.5. Three applications of exact set matching

3.5.1. Matching against a DNA or protein library of known patterns

There are a number of applications in molecular biology where a relatively stable library of interesting or distinguishing DNA or protein substrings have been constructed. The *Sequence-tagged sites* (*STSs*) and *Expressed sequence tags* (*ESTs*) provide our first important illustration.

Sequence-tagged-sites

The concept of a Sequence-tagged-site (STS) is one of the most useful by-products that has come out of the Human Genome Project [111, 234, 399]. Without going into full biological detail, an STS is intuitively a DNA string of length 200–300 nucleotides whose right and left ends, of length 20–30 nucleotides each, occur only once in the entire genome [111, 317]. Thus each STS occurs uniquely in the DNA of interest. Although this definition is not quite correct, it is adequate for our purposes. An early goal of the Human Genome Project was to select and map (locate on the genome) a set of STSs such that any substring in the genome of length 100,000 or more contains at least one of those STSs. A more refined goal is to make a map containing ESTs (expressed sequence tags), which are STSs that come from genes rather than parts of intergene DNA. ESTs are obtained from mRNA and cDNA (see Section 11.8.3 for more detail on cDNA) and typically reflect the protein coding parts of a gene sequence.

With an STS map, one can locate on the map any sufficiently long string of anonymous but sequenced DNA – the problem is just one of finding which STSs are contained in the anonymous DNA. Thus with STSs, map location of anonymous sequenced DNA becomes a string problem, an exact set matching problem. The STSs or the ESTs provide a computer-based set of indices to which new DNA sequences can be referenced. Presently, hundreds of thousands of STSs and tens of thousands of ESTs have been found and placed in computer databases [234]. Note that the total length of all the STSs and ESTs is very large compared to the typical size of an anonymous piece of DNA. Consequently, the keyword tree and the Aho–Corasick method (with a search time proportional to the length of the anonymous DNA) are of direct use in this problem for they allow very rapid identification of STSs or ESTs that occur in newly sequenced DNA.

Of course, there may be some errors in either the STS map or in the newly sequenced DNA causing trouble for this approach (see Section 16.5 for a discussion of STS maps). But in this application, the number of errors should be a small percentage of the length of the STS, and that will allow more sophisticated exact (and inexact) matching methods to succeed. We will describe some of these in Sections 7.8.3, 9.4, and 12.2 of the book.

A related application comes from the "BAC-PAC" proposal [442] for sequencing the human genome (see page 418). In that method, 600,000 strings (patterns) of length 500 would first be obtained and entered into the computer. Thousands of times thereafter, one would look for occurrences of any of these 600,000 patterns in text strings of length 150,000. Note that the total length of the patterns is 300 million characters, which is two-thousand times as large as the typical text to be searched.

3.5.2. Exact matching with wild cards

As an application of exact set matching, we return to the problem of exact matching with a single pattern, but complicate the problem a bit. We modify the exact matching problem by introducing a character ϕ, called a *wild card*, that matches any single character. Given a pattern P containing wild cards, we want to find all occurrences of P in a text T. For example, the pattern $ab\phi\phi c\phi$ occurs twice in the text $xabvccbababcax$. Note that in this version of the problem no wild cards appear in T and that each wild card matches only a single character rather than a substring of unspecified length.

The problem of matching with wild cards should need little motivating, as it is not difficult to think up realistic cases where the pattern contains wild cards. One very important case where simple wild cards occur is in DNA *transcription factors*. A transcription factor is a protein that binds to specific locations in DNA and regulates, either enhancing or suppressing, the transcription of the DNA into RNA. In this way, production of the protein that the DNA codes for is regulated. The study of transcription factors has exploded in the past decade; many transcription factors are now known and can be separated into families characterized by specific substrings containing wild cards. For example, the *Zinc Finger* is a common transcription factor that has the following signature:

$$CYS\phi\phi CYS\phi\phi\phi\phi\phi\phi\phi\phi\phi\phi\phi\phi HIS\phi\phi HIS,$$

where CYS is the amino acid cysteine and HIS is the amino acid histidine. Another important transcription factor is the *Leucine Zipper*, which consists of four to seven leucines, each separated by six wild card amino acids.

If the number of permitted wild cards is unbounded, it is not known if the problem can be solved in linear time. However, if the number of wild cards is bounded by a fixed constant (independent of the size of P) then the following method, based on exact set pattern matching, runs in linear time:

Exact matching with wild cards

0. Let C be a vector of length $|T|$ initialized to all zeros.

1. Let $\mathcal{P} = \{P_1, P_2, \ldots, P_k\}$ be the (multi-)set of maximal substrings of P that do not contain any wild cards. Let $l_1, l_2, \ldots, l_k$ be the starting positions in P of each of these substrings.

{For example, if $P = ab\phi\phi c\phi ab\phi\phi$ then $\mathcal{P} = \{ab, c, ab\}$ and $l_1 = 1, l_2 = 5, l_3 = 7.$}

2. Using the Aho–Corasick algorithm (or the suffix tree approach to be discussed later), find for each string P_i in $\mathcal{P}$, all starting positions of P_i in

text T. For each starting location j of P_i in T,
increment the count in cell $j - l_i + 1$ of C by one.

{For example, if the second copy of string ab is found in T
starting at position 18, then cell 12 of C is incremented by one.}

3. Scan vector C for any cell with value k. There is an occurrence of P
in T starting at position p if and only if $C(p) = k$.

Correctness and complexity of the method

Correctness   Clearly, there is an occurrence of P in T starting at position p if and only if, for each i, subpattern $P_i \in \mathcal{P}$ occurs at position $j = p + l_i - 1$ of T. The above method uses this idea in reverse. If pattern $P_i \in \mathcal{P}$ is found to occur starting at position j of T, and pattern P_i starts at position l_i in P, then this provides one "witness" that P occurs at T starting at position $p = j - l_i + 1$. Hence P occurs in T starting at p if and only if similar witnesses for position p are found for each of the k strings in $\mathcal{P}$. The algorithm counts, at position p, the number of witnesses that observe an occurrence of P beginning at p. This correctly determines whether P occurs starting at p because each string in $\mathcal{P}$ can cause at most one increment to cell p of C.

Complexity   The time used by the Aho–Corasick algorithm to build the keyword tree for $\mathcal{P}$ is $O(n)$. The time to search for occurrences in T of patterns from $\mathcal{P}$ is $O(m + z)$, where $|T| = m$ and z is the number of occurrences. We treat each pattern in $\mathcal{P}$ as being distinct even if there are multiple copies of it in $\mathcal{P}$. Then whenever an occurrence of a pattern from $\mathcal{P}$ is found in T, exactly one cell in C is incremented; furthermore, a cell can be incremented to at most k. Hence z must be bounded by km, and the algorithm runs in $O(km)$ time . Although the number of character comparisons used is just $O(m)$, km need not be $O(m)$ and hence the number of times C is incremented may grow faster than $O(m)$, leading to a nonlinear $O(km)$ time bound. But if k is assumed to be bounded (independent of $|P|$), then the method does run in linear time. In summary,

Theorem 3.5.1. *If the number of wild cards in pattern P is bounded by a constant, then the exact matching problem with wild cards in the Pattern can be solved in $O(n + m)$ time.*

Later, in Sections 9.3, we will return to the problem of wild cards when they occur in either the pattern, text, or both.

3.5.3. Two-dimensional exact matching

A second classic application of exact set matching occurs in a generalization of string matching to two-dimensional exact matching. Suppose we have a *rectangular* digitized picture T, where each point is given a number indicating its color and brightness. We are also given a smaller rectangular picture P, which also is digitized, and we want to find all occurrences (possibly overlapping) of the smaller picture in the larger one. We assume that the bottom edges of the two rectangles are parallel to each other. This is a two-dimensional generalization of the exact string matching problem.

Admittedly, this problem is somewhat contrived. Unlike the one-dimensional exact matching problem, which truly arises in numerous practical applications, compelling applications of two-dimensional exact matching are hard to find. Two-dimensional matching that is inexact, allowing some errors, is a more realistic problem, but its solution requires more complex techniques of the type we will examine in Part III of the book. So for now, we view two-dimensional exact matching as an illustration of how exact set matching can be used in more complex settings and as an introduction to more realistic two-dimensional problems. The method presented follows the basic approach given in [44] and [66]. Since then, many additional methods have been presented since that improve on those papers in various ways. However, because the problem as stated is somewhat unrealistic, we will not discuss the newer, more complex, methods. For a sophisticated treatment of two-dimensional matching see [22] and [169].

Let m be the total number of points in T, let n be the number of points in P, and let n' be the number of rows in P. Just as in exact string matching, we want to find the smaller picture in the larger one in $O(n + m)$ time, where $O(nm)$ is the time for the obvious approach. Assume for now that each of the rows of P are distinct; later we will relax this assumption.

The method is divided into two phases. In the first phase, search for all occurrences of each of the rows of P among the rows of T. To do this, add an end of row marker (some character not in the alphabet) to each row of T and concatenate these rows together to form a single text string T' of length $O(m)$. Then, treating each row of P as a separate pattern, use the Aho–Corasick algorithm to search for all occurrences in T' of any row of P. Since P is rectangular, all rows have the same width, and so no row is a proper substring of another and we can use the simpler version of Aho–Corasick discussed in Section 3.4.2. Hence the first phase identifies all occurrences of complete rows of P in complete rows of T and takes $O(n + m)$ time.

Whenever an occurrence of row i of P is found starting at position (p, q) of T, write the number i in position (p, q) of another array M with the same dimensions as T. Because each row of P is assumed to be distinct and because P is rectangular, at most one number will be written in any cell of M.

In the second phase, scan each *column* of M, looking for an occurrence of the string $1, 2, \ldots, n'$ in consecutive cells in a single column. For example, if this string is found in column 6, starting at row 12 and ending at row $n' + 12$, then P occurs in T when its upper left corner is at position (6,12). Phase two can be implemented in $O(n' + m) = O(n + m)$ time by applying any linear-time exact matching algorithm to each column of M.

This gives an $O(n + m)$ time solution to the two-dimensional exact set matching problem. Note the similarity between this solution and the solution to the exact matching problem with wild cards discussed in the previous section. A distinction will be discussed in the exercises.

Now suppose that the rows of P are not all distinct. Then, first find all identical rows and give them a common label (this is easily done during the construction of the keyword tree for the row patterns). For example, if rows 3, 6, and 10 are the same then we might give them all the label of 3. We do a similar thing for any other rows that are identical. Then, in phase one, only look for occurrences of row 3, and not rows 6 and 10. This ensures that a cell of M will have at most one number written in it during phase 1. In phase 2, don't look for the string $1, 2, 3, \ldots, n'$ in the columns of M, but rather for a string where 3 replaces 6 and 10, etc. It is easy to verify that this approach is correct and that it takes just $O(n + m)$ time. In summary,

Theorem 3.5.2. *If T and P are rectangular pictures with m and n cells, respectively, then all exact occurrences of P in T can be found in $O(n + m)$ time, improving upon the naive method, which takes $O(nm)$ time.*

3.6. Regular expression pattern matching

A *regular expression* is a way to specify a set of related strings, sometimes referred to as a *pattern*.[3] Many important sets of substrings (patterns) found in biosequences, particularly in proteins, can be specified as regular expressions, and several databases have been constructed to hold such patterns. The PROSITE database, developed by Amos Bairoch [41, 42], is the major regular expression database for significant patterns in proteins (see Section 15.8 for more on PROSITE).

In this section, we examine the problem of finding substrings of a text string that match one of the strings specified by a given regular expression. These matches are computed in the Unix utility *grep*, and several special programs have been developed to find matches to regular expressions in biological sequences [279, 416, 422].

It is helpful to start first with an example of a simple regular expression. A formal definition of a regular expression is given later. The following PROSITE expression specifies a set of substrings, some of which appear in a particular family of granin proteins:

$$[ED]\text{-}[EN]\text{-}L\text{-}[SAN]\text{-}x\text{-}x\text{-}[DE]\text{-}x\text{-}E\text{-}L.$$

Every string specified by this regular expression has ten positions, which are separated by a dash. Each capital letter specifies a single amino acid and a group of amino acids enclosed by brackets indicates that exactly one of those amino acids must be chosen. A small x indicates that any one of the twenty amino acids from the protein alphabet can be chosen for that position. This regular expression describes 192,000 amino acid strings, but only a few of these actually appear in any known proteins. For example, ENLSSEDEEL is specified by the regular expression and is found in human granin proteins.

3.6.1. Formal definitions

We now give a formal, recursive definition for a regular expression formed from an alphabet Σ. For simplicity, and contrary to the PROSITE example, assume that alphabet Σ does not contain any symbol from the following list: $*, +, (,), \epsilon$.

> **Definition**   A single character from Σ is a regular expression. The symbol ϵ is a regular expression. A regular expression followed by another regular expression is a regular expression. Two regular expressions separated by the symbol "+" form a regular expression. A regular expression enclosed in parentheses is a regular expression. A regular expression enclosed in parentheses and followed by the symbol "*" is a regular expression. The symbol * is called the Kleene closure.

These recursive rules are simple to follow, but may need some explanation. The symbol ϵ represents the empty string (i.e., the string of length zero). If R is a parenthesized regular expression, then R^* means that the expression R can be repeated any number of times (including zero times). The inclusion of parentheses as part of a regular expression (outside of Σ) is not standard, but is closer to the way that regular expressions are actually specified in many applications. Note that the example given above in PROSITE format does not conform to the present definition but can easily be converted to do so.

As an example, let Σ be the alphabet of lower case English characters. Then $R = (a + c + t)ykk(p + q)^* \, vdt(l + z + \epsilon)(pq)$ is a regular expression over Σ, and $S =$

---

[3] Note that in the context of regular expressions, the meaning of the word "pattern" is different from its previous and general meaning in this book.

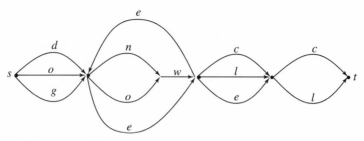

Figure 3.19: Directed graph for the regular expression $(d + o + g)((n + o)w)^*(c + l + \epsilon)(c + l)$.

aykkpqppvdtpq is a string specified by R. To specify S, the subexpression $(p+q)$ of R was repeated four times, and the empty string ϵ was the choice specified by the subexpression $(l + z + \epsilon)$.

It is very useful to represent a regular expression R by a directed graph $G(R)$ (usually called a nondeterministic, finite state automaton). An example is shown in Figure 3.19. The graph has a start node s and a termination node t, and each edge is labeled with a single symbol from $\Sigma \cup \epsilon$. Each s to t path in $G(R)$ specifies a string by concatenating the characters of Σ that label the edges of the path. The set of strings specified by all such paths is exactly the set of strings specified by the regular expression R. The rules for constructing $G(R)$ from R are simple and are left as an exercise. It is easy to show that if a regular expression R has n symbols, then $G(R)$ can be constructed using at most $2n$ edges. The details are left as an exercise and can be found in [10] and [8].

Definition   A substring T' of string T *matches* the regular expression R if there is an s to t path in $G(R)$ that specifies T'.

Searching for matches

To search for a *substring* in T that matches the regular expression R, we first consider the simpler problem of determining whether some (unspecified) *prefix* of T matches R. Let $N(0)$ be the set of nodes consisting of node s plus all nodes of $G(R)$ that are reachable from node s by traversing edges labeled ϵ. In general, a node v is in set $N(i)$, for $i > 0$, if v can be reached from some node in $N(i - 1)$ by traversing an edge labeled $T(i)$ followed by zero or more edges labeled ϵ. This gives a constructive rule for finding set $N(i)$ from set $N(i - 1)$ and character $T(i)$. It easily follows by induction on i that a node v is in $N(i)$ if and only if there is path in $G(R)$ from s that ends at v and generates the string $T[1..i]$. Therefore, prefix $T[1..i]$ matches R if and only if $N(i)$ contains node t.

Given the above discussion, to find all prefixes of T that match R, compute the sets $N(i)$ for i from 0 to m, the length of T. If $G(R)$ contains e edges, then the time for this algorithm is $O(me)$, where m is the length of the text string T. The reason is that each iteration i [finding $N(i)$ from $N(i - 1)$ and character $T(i)$] can be implemented to run in $O(e)$ time (see Exercise 29).

To search for a *nonprefix* substring of T that matches R, simply search for a prefix of T that matches the regular expression $\Sigma^* R$. Σ^* represents any number of repetitions (including zero) of any character in Σ. With this detail, we now have the following:

Theorem 3.6.1. *If T is of length m, and the regular expression R contains n symbols, then it is possible to determine whether T contains a substring matching R in $O(nm)$ time.*

3.7. Exercises

1. Evaluate empirically the speed of the Boyer–Moore method against the Apostolico–Giancarlo method under different assumptions about the text and the pattern. These assumptions should include the size of the alphabet, the "randomness" of the text or pattern, the level of periodicity of the text or pattern, etc.

2. In the Apostolico–Giancarlo method, array M is of size m, which may be large. Show how to modify the method so that it runs in the same time, but in place of M uses an array of size n.

3. In the Apostolico–Giancarlo method, it may be better to compare the characters first and then examine M and N if the two characters match. Evaluate this idea both theoretically and empirically.

4. In the Apostolico–Giancarlo method, $M(j)$ is set to be a number less than or equal to the length of the (right-to-left) match of P and T starting at position j of T. Find examples where the algorithm sets the value to be strictly less than the length of the match. Now, since the algorithm learns the exact location of the mismatch in all cases, $M(j)$ could always be set to the full length of the match, and this would seem to be a good thing to do. Argue that this change would result in a correct simulation of Boyer–Moore. Then explain why this was not done in the algorithm.

   Hint: It's the time bound.

5. Prove Lemma 3.2.2 showing the equivalence of the two definitions of semiperiodic strings.

6. For each of the n prefixes of P, we want to know whether the prefix $P[1..i]$ is a periodic string. That is, for each i we want to know the largest $k > 1$ (if there is one) such that $P[1..i]$ can be written as α^k for some string α. Of course, we also want to know the period. Show how to determine this for all n prefixes in time linear in the length of P.

   Hint: Z-algorithm.

7. Solve the same problem as above but modified to determine whether each prefix is semiperiodic and with what period. Again, the time should be linear.

8. By being more careful in the bookkeeping, establish the constant in the $O(m)$ bound from Cole's linear-time analysis of the Boyer–Moore algorithm.

9. Show where Cole's worst-case bound breaks down if only the *weak* Boyer–Moore shift rule is used. Can the argument be fixed, or is the linear time bound simply untrue when only the weak rule is used? Consider the example of $T = abababababababababab$ and $P = xaaaaaaaaa$ without also using the bad character rule.

10. Similar to what was done in Section 1.5, show that applying the classical Knuth-Morris-Pratt preprocessing method to the string $P\$T$ gives a linear-time method to find all occurrence of P in T. In fact, the search part of the Knuth-Morris-Pratt algorithm (after the preprocessing of P is finished) can be viewed as a slightly optimized version of the Knuth-Morris-Pratt preprocessing algorithm applied to the T part of $P\$T$. Make this precise, and quantify the utility of the optimization.

11. Using the assumption that $\mathcal{P}$ is *substring free* (i.e., that no pattern $P_i \in \mathcal{P}$ is a substring of another pattern $P_j \in \mathcal{P}$), complete the correctness proof of the Aho–Corasick algorithm. That is, prove that if no further matches are possible at a node v, then l can be set to $c - lp(v)$ and the comparisons resumed at node n_v without missing any occurrences in T of patterns from $\mathcal{P}$.

12. Prove that the search phase of the Aho–Corasick algorithm runs in $O(m)$ time if no pattern in $\mathcal{P}$ is a proper substring of another, and otherwise in $O(m + k)$ time, where k is the total number of occurrences.

13. The Aho–Corasick algorithm can have the same problem that the Knuth-Morris-Pratt algorithm

has when it only uses sp values rather than sp' values. This is shown, for example, in Figure 3.16 where the edge below the character a in *potato* is directed to the character a in *tattoo*. A better failure function would avoid this situation. Give the details for computing such an improved failure function.

14. Give an example showing that k, the number of occurrences in T of patterns in set $\mathcal{P}$, can grow faster than $O(n+m)$. Be sure you account for the input size n. Try to make the growth as large as possible.

15. Prove Lemmas 3.4.2 and 3.4.3 that relate to the case of patterns that are not substring free.

16. The time analysis in the proof of Theorem 3.4.1 separately considers the path in $\mathcal{K}$ for each pattern P in $\mathcal{P}$. This results in an overcount of the time actually used by the algorithm. Perform the analysis more carefully to relate the running time of the algorithm to the number of nodes in $\mathcal{K}$.

17. Discuss the problem (and solution if you see one) of using the Aho–Corasick algorithm when **a.** wild cards are permitted in the text but not in the pattern and **b.** when wild cards are permitted in both the text and pattern.

18. Since the nonlinear time behavior of the wild card algorithm is due to duplicate copies of strings in $\mathcal{P}$, and such duplicates can be found and removed in linear time, it is tempting to "fix up" the method by first removing duplicates from $\mathcal{P}$. That approach is similar to what is done in the two-dimensional string matching problem when identical rows were first found and given a single label. Consider this approach and try to use it to obtain a linear-time method for the wild card problem. Does it work, and if not what are the problems?

19. Show how to modify the wild card method by replacing array C (which is of length $m > n$) by a list of length n, while keeping the same running time.

20. In the wild card problem we first assumed that no pattern in $\mathcal{P}$ is a substring of another one, and then we extended the algorithm to the case when that assumption does not hold. Could we instead simply reduce the case when substrings of patterns are allowed to the case when they are not? For example, perhaps we just add a new symbol to the end of each string in $\mathcal{P}$ that appears nowhere else in the patterns. Does it work? Consider both correctness and complexity issues.

21. Suppose that the wild card can match any length substring, rather than just a single character. What can you say about exact matching with these kinds of wild cards in the pattern, in the text, or in both?

22. Another approach to handling wild cards in the pattern is to modify the Knuth-Morris-Pratt or Boyer–Moore algorithms, that is, to develop shift rules and preprocessing methods that can handle wild cards in the pattern. Does this approach seem promising? Try it, and discuss the problems (and solutions if you see them).

23. Give a complete proof of the correctness and $O(n+m)$ time bound for the two-dimensional matching method described in the text (Section 3.5.3).

24. Suppose in the two-dimensional matching problem that Knuth-Morris-Pratt is used once for each pattern in $\mathcal{P}$, rather than Aho–Corasick being used. What time bound would result?

25. Show how to extend the two-dimensional matching method to the case when the bottom of the rectangular pattern is not parallel to the bottom of the large picture, but the orientation of the two bottoms is known. What happens if the pattern is not rectangular?

26. Perhaps we can omit phase two of the two-dimensional matching method as follows: Keep a counter at each cell of the large picture. When we find that row i of the small picture occurs in row j of the large picture starting at position (i', j), increment the counter for cell

$(i', j - i + 1)$. Then declare that P occurs in T with upper left corner in any cell whose counter becomes n' (the number of rows of P). Does this work?

Hint: No.

Why not? Can you fix it and make it run in $O(n + m)$ time?

27. Suppose we have $q > 1$ small (distinct) rectangular pictures and we want to find all occurrences of any of the q small pictures in a larger rectangular picture. Let n be the total number of points in all the small pictures and m be the number of points in the large picture. Discuss how to solve this problem efficiently. As a simplification, suppose all the small pictures have the same width. Then show that $O(n + m)$ time suffices.

28. Show how to construct the required directed graph $G(R)$ from a regular expression R. The construction should have the property that if R contains n symbols, then $G(R)$ contains at most $O(n)$ edges.

29. Since the directed graph $G(R)$ contains $O(n)$ edges when R contains n symbols, $|N(i)| = O(n)$ for any i. This suggests that the set $N(i)$ can be naively found from $N(i - 1)$ and $T(i)$ in $O(ne)$ time. However, the time stated in the text for this task is $O(e)$. Explain how this reduction of time is achieved. Explain that the improvement is trivial if $G(R)$ contains no ϵ edges.

30. Explain the importance, or the utility, of ϵ edges in the graph $G(R)$. If R does not contain the closure symbol "*", can ϵ edges always be avoided? Biological strings are always finite, hence "*" can always be avoided. Explain how this simplifies the searching algorithm.

31. Wild cards can clearly be encoded into a regular expression, as defined in the text. However, it may be more efficient to modify the definition of a regular expression to explicitly include the wild card symbol. Develop that idea and explain how wild cards can be efficiently handled by an extension of the regular expression pattern matching algorithm.

32. PROSITE patterns often specify the number of times that a substring can repeat as a finite range of numbers. For example, CD(2–4) indicates that CD can repeat either two, three, or four times. The formal definition of a regular expression does not include such concise range specifications, but finite range specifications can be expressed in a regular expression. Explain how. How much do those specifications increase the length of the expression over the length of the more concise PROSITE expression? Show how such range specifications are reflected in the directed graph for the regular expression (ϵ edges are permitted). Show that one can still search for a substring of T that matches the regular expression in $O(me)$ time, where m is the length of T and e is the number of edges in the graph.

33. Theorem 3.6.1 states the time bound for determining if T contains a substring that matches a regular expression R. Extend the discussion and the theorem to cover the task of explicitly finding and outputting all such matches. State the time bound as the sum of a term that is independent of the number of matches plus a term that depends on that number.

4

Seminumerical String Matching

4.1. Arithmetic versus comparison-based methods

All of the exact matching methods in the first three chapters, as well as most of the methods that have yet to be discussed in this book, are examples of *comparison-based* methods. The main primitive operation in each of those methods is the comparison of two characters. There are, however, string matching methods based on *bit operations* or on *arithmetic*, rather than character comparisons. These methods therefore have a very different flavor than the comparison-based approaches, even though one can sometimes see character comparisons hidden at the inner level of these "seminumerical" methods. We will discuss three examples of this approach: the *Shift-And* method and its extension to a program called *agrep* to handle inexact matching; the use of the Fast Fourier Transform in string matching; and the random fingerprint method of Karp and Rabin.

4.2. The *Shift-And* method

R. Baeza-Yates and G. Gonnet [35] devised a simple, bit-oriented method that solves the exact matching problem very efficiently for relatively small patterns (the length of a typical English word for example). They call this method the *Shift-Or* method, but it seems more natural to call it *Shift-And*. Recall that pattern P is of size n and the text T is of size m.

Definition Let M be an n by $m + 1$ *binary* valued array, with index i running from 1 to n and index j running from 1 to m. Entry $M(i, j)$ is 1 if and only if the first i characters of P exactly match the i characters of T ending at character j. Otherwise the entry is zero.

In other words, $M(i, j)$ is 1 if and only if $P[1..i]$ exactly matches $T[j - i + 1..j]$. For example, if $T = california$ and $P = for$, then $M(1, 5) = M(2, 6) = M(3, 7) = 1$, whereas $M(i, j) = 0$ for all other combinations of i, j. Essentially, the entries with value 1 in row i of M show all the places in T where a copy of $P[1..i]$ ends, and column j of M shows all the prefixes of P that end at position j of T.

Clearly, $M(n, j) = 1$ if and only if an occurrence of P ends at position j of T; hence computing the last row of M solves the exact matching problem. For the algorithm to compute M it first constructs an n-length binary vector $U(x)$ for each character x of the alphabet. $U(x)$ is set to 1 for the positions in P where character x appears. For example, if $P = abacdeab$ then $U(a) = 10100010$.

Definition Define *Bit-Shift*$(j-1)$ as the vector derived by shifting the vector for *column* $j - 1$ down by one position and setting that first to 1. The previous bit in position n disappears. In other words, *Bit-Shift*$(j - 1)$ consists of 1 followed by the first $n - 1$ bits of column $j - 1$.

For example, Figure 4.1 shows a column $j - 1$ before and after the bit-shift.

```
        0                   1
        0                   0
        1                   0
        0                   1
        1                   0
        1                   1
        0                   1
        1                   0
```

Figure 4.1: Column $j-1$ before and after operation *Bit-Shift*($j-1$).

4.2.1. How to construct array M

Array M is constructed column by column as follows: Column one of M is initialized to all zero entries if $T(1) \neq P(1)$. Otherwise, when $T(1) = P(1)$ its first entry is 1 and the remaining entries are 0. After that, the entries for column $j > 1$ are obtained from column $j-1$ and the U vector for character $T(j)$. In particular, the vector for column j is obtained by the bitwise **AND** of vector *Bit-Shift*($j-1$) with the U vector for character $T(j)$. More formally, if we let $M(j)$ denote the jth column of M, then $M(j) = $ *Bit-Shift*($j-1$) **AND** $U(T(j))$. For example, if $P = abaac$ and $T = xabxabaaxa$ then the eighth column of M is

```
        1
        0
        1
        0
        0
```

because prefixes of P of lengths one and three end at position seven of T. The eighth character of T is character a, which has a U vector of

```
        1
        0
        1
        1
        0
```

When the eighth column of M is shifted down and an **AND** is performed with $U(a)$, the result is

```
        1
        0
        0
        1
        0
```

which is the correct ninth column of M.

To see in general why the *Shift-And* method produces the correct array entries, observe that for any $i > 1$ the array entry for cell (i, j) should be 1 if and only if the first $i-1$ characters of P match the $i-1$ characters of T ending at character $j-1$ and character $P(i)$ matches character $T(j)$. The first condition is true when the array entry for cell $(i-1, j-1)$ is 1, and the second condition is true when the ith bit of the U vector for character $T(j)$ is 1. By first shifting column $j-1$, the algorithm **AND**s together entry $(i-1, j-1)$ of column $j-1$ with entry i of the vector $U(T(j))$. Hence the algorithm computes the correct entries for array M.

4.2.2. *Shift-And* is effective for small patterns

Although the *Shift-And* method is very simple, and in worst case the number of bit operations is clearly $\Theta(mn)$, the method is very efficient if n is less than the size of a single computer word. In that case, every column of M and every U vector can be encoded into a single computer word, and both the *Bit-Shift* and the **AND** operations can be done as single-word operations. These are very fast operations in most computers and can be specified in languages such as C. Even if n is several times the size of a single computer word, only a few word operations are needed. Furthermore, only two columns of M are needed at any given time. Column j only depends on column $j-1$, so all previous columns can be forgotten. Hence, for reasonable sized patterns, such as single English words, the *Shift-And* method is very efficient in both time and space regardless of the size of the text. From a purely theoretical standpoint it is not a linear time method, but it certainly is practical and would be the method of choice in many circumstances.

4.2.3. *agrep*: The *Shift-And* method with errors

S. Wu and U. Manber [482] devised a method, packaged into a program called *agrep*, that amplifies the *Shift-And* method by finding *inexact* occurrences of a pattern in a text. By inexact we mean that the pattern either occurs exactly in the text or occurs with a "small" number of *mismatches* or *inserted* or *deleted* characters. For example, the pattern *atcgaa* occurs in the text *aatatccacaa* with two mismatches starting at position four; it also occurs with four mismatches starting at position two. In this section we will explain *agrep* and how it handles mismatches. The case of permitted insertions and deletions will be left as an exercise. For a small number of errors and for small patterns, *agrep* is very efficient and can be used in the core of more elaborate text searching methods. Inexact matching is the focus of Part III, but the ideas behind *agrep* are so closely related to the *Shift-And* method that it is appropriate to examine *agrep* at this point.

> **Definition** For two strings P and T of lengths n and m, let M^k be a binary-valued array, where $M^k(i, j)$ is 1 if and only if at least $i - k$ of the first i characters of P match the i characters up through character j of T.

That is, $M^k(i, j)$ is the natural extension of the definition of $M(i, j)$ to allow up to k mismatches. Therefore, M^0 is the array M used in the *Shift-And* method. If $M^k(n, j) = 1$ then there is an occurrence of P in T ending at position j that contains at most k mismatches. We let $M^k(j)$ denote the jth column of M^k.

In *agrep*, the user chooses a value of k and then the arrays $M, M^1, M^2, \ldots, M^k$ are computed. The efficiency of the method depends on the size of k – the larger k is, the slower the method. For many applications, a value of k as small as 3 or 4 is sufficient, and the method is extremely fast.

4.2.4. How to compute M^k

Let k be the fixed maximum permitted number of mismatches specified by the user. The method will compute M^l for all values of l between 0 and k. There are several ways to organize the computation and its description, but for simplicity we will compute column j of each array M^l before any columns past j will be computed in any array. Further, for every j we will compute column j in arrays M^l in increasing order of l. In particular, the zero column of each array is again initialized to all zeros. Then the jth column of M^l is computed by:

$$M^l(j) = M^{l-1}(j) \text{ OR } [Bit\text{-}Shift(M^l(j-1)) \text{ AND } U(T(j))] \text{ OR } M^{l-1}(j-1).$$

Intuitively, this just says that the first i characters of P will match a substring of T ending at position j, with at most l mismatches, if and only if one of the following three conditions hold:

- The first i characters of P match a substring of T ending at j, with at most $l-1$ mismatches.
- The first $i-1$ characters of P match a substring of T ending at $j-1$, with at most l mismatches, and the next pair of characters in P and T are equal.
- The first $i-1$ characters of P match a substring of T ending at $j-1$, with at most $l-1$ mismatches.

It is simple to establish that these recurrences are correct, and over the entire algorithm the number of bit operations is $O(knm)$. As in the *Shift-And* method, the practical efficiency comes from the fact that the vectors are bit vectors (again of length n) and the operations are very simple – shifting by one position and **AND**ing bit vectors. Thus when the pattern is relatively small, so that a column of any M^l fits into a few words, and k is also small, *agrep* is extremely fast.

4.3. The match-count problem and Fast Fourier Transform

If we relax the requirement that only bit operations are permitted and allow each entry of array M to hold an integer between 0 and n, then we can easily adapt the *Shift-And* method to compute for each pair i, j the number of characters of $P[1..i]$ that match $T[j-i+1..j]$. This computation is again a form of inexact matching, which is the focus of Part III. However, as was true of *agrep*, the solution is so connected to the *Shift-And* method that we consider it here. In addition, it is a natural introduction to the next topic, match-counts. For clarity, let us define a new matrix MC.

> **Definition** The matrix MC is an n by $m+1$ integer-valued matrix, where entry $MC(i, j)$ is the number of characters of $P[1..i]$ that match $T[j-i+1..j]$.

A simple algorithm to compute matrix MC generalizes the *Shift-And* method, replacing the **AND** operation with the *increment by one* operation. The zero column of MC starts with all zeros, but each $MC(i, j)$ entry now is set to $MC(i-1, j-1)$ if $P(i) \neq T(j)$, and otherwise it is set to $MC(i-1, j-1) + 1$. Any entry with value n in the last row again indicates an occurrence of P in T, but values less than n count the *exact number* of characters that match for each of different alignments of P with T. This extension uses $\Theta(nm)$ additions and comparisons, although each addition operation is particularly simple, just incrementing by one.

If we want to compute the entire MC array then $\Theta(nm)$ time is necessary, but the most important information is contained in the last row of MC. For each position $j \geq n$ in T, the last row indicates the number of characters that match when the right end of P is aligned with character j of T. The problem of finding the last row of MC is called the *match-count* problem. Match-counts are useful in several problems to be discussed later.

4.3.1. A fast worst-case method for the match-count problem?

Can any of the linear-time exact matching methods discussed in earlier chapters be adapted to solve the match-count problem in linear time? That is an open question. The extension of the *Shift-And* method discussed above solves the match-count problem but uses $\Theta(nm)$ arithmetic operations in all cases.

Surprisingly, the match-count problem can be solved with only $O(m \log m)$ arithmetic operations if we allow multiplication and division of complex numbers. The numbers remain small enough to permit the unit-time model of computation (that is, no number requires more than $O(\log m)$ bits), but the operations are still more complex than just incrementing by one. The $O(m \log m)$ method is based on the *Fast Fourier Transform* (*FFT*). This approach was developed by Fischer and Paterson [157] and independently in the biological literature by Felsenstein, Sawyer, and Kochin [152]. Other work that builds on this approach is found in [3], [58], [59], and [99]. We will reduce the match-count problem to a problem that can be efficiently solved by the FFT, but we treat the FFT itself as a black box and leave it to any interested reader to learn the details of the FFT.

4.3.2. Using Fast Fourier Transform for match-counts

The match-count problem is essentially that of finding the last row of the matrix MC. However, we will not work on MC directly, but rather we will solve a more general problem whose solution contains the desired information. For this more general problem the two strings involved will be denoted α and β rather than P and T, since the roles of the two strings will be completely symmetric. We still assume, however, that $|\alpha| = n \leq m = |\beta|$.

Definition Define $V(\alpha, \beta, i)$ to be the number of characters of α and β that match when the *left end* of string α is opposite position i of string β. Define $V(\alpha, \beta)$ to be the vector whose ith entry is $V(\alpha, \beta, i)$.

Clearly, when $\alpha = P$ and $\beta = T$ the vector $V(\alpha, \beta)$ contains the information needed for the last row of MC. But it contains more information because we allow the left end of α to be to the left of the left end of β, and we also allow the right end of α to be to the right of the right end of β. Negative numbers specify positions to the left of the left end of β, and positive numbers specify the other positions. For example, when α is aligned with β as follows,

```
             21123456789
B:              accctgtcc
A:              aactgccg
```

then the left end of α is aligned with position -2 of β.

Index i ranges from $-n + 1$ to m. Notice that when $i > m - n$, the right end of α is right of the right end of β. For any fixed i, $V(\alpha, \beta, i)$ can be directly computed in $O(n)$ time (for any i, just directly count the number of resulting matches and mismatches), so $V(\alpha, \beta)$ can be computed in $O(nm)$ total time.

We now show how to compute $V(\alpha, \beta)$ in $O(m \log m)$ total time by using the Fast Fourier Transform. For most problems of interest $\log m \ll n$, so this technique yields a large speedup. Further, there is specialized hardware for FFT that is very fast, suggesting a way to solve these problems quickly with hardware. The solution will work for any alphabet, but it is easiest to explain it on a small alphabet. For concreteness we use the four-letter alphabet a, t, c, g of DNA.

The high-level approach

We break up the match-count problem into four problems, one for each character in the alphabet.

Definition Define $V_a(\alpha, \beta, i)$ to be the number of matches of character a that occur when the start of string α is positioned opposite position i of string β. $V_a(\alpha, \beta)$ is the $(n + m)$-length vector holding these values.

Similar definitions apply for the other three characters. With these definitions,

$$V(\alpha, \beta, i) = V_a(\alpha, \beta, i) + V_t(\alpha, \beta, i) + V_c(\alpha, \beta, i) + V_g(\alpha, \beta, i)$$

and

$$V(\alpha, \beta) = V_a(\alpha, \beta) + V_t(\alpha, \beta) + V_c(\alpha, \beta) + V_g(\alpha, \beta).$$

The problem then becomes how to compute $V_a(\alpha, \beta, i)$ for each i. Convert the two strings into binary strings $\bar\alpha_a$ and $\bar\beta_a$, respectively, where every occurrence of character a becomes a 1, and all other characters become 0s. For example, let α be *acaacggaggtat* and β be *accacgaag*. Then the binary strings $\bar\alpha_a$ and $\bar\beta_a$ are 1011000100010 and 100100110. To compute $V_a(\alpha, \beta, i)$, position $\bar\beta_a$ to start at position i of $\bar\alpha_a$ and count the number of columns where both bits are equal to 1. For example, if $i = 3$ then we get

```
1011000100010
   10010110
```

and the $V_a(\alpha, \beta, 3) = 2$. If $i = 9$ then we have

```
1011000100010
         10010110
```

and $V_a(\alpha, \beta, 9) = 1$.

Another way to view this is to consider each space opposite a bit to be a 0 (so both binary strings are the same length), do a bitwise **AND** operation with the strings, and then add the resulting bits.

To formalize this idea, pad the right end of $\bar\beta$ (the larger string) with n additional zeros and pad the right end of $\bar\alpha$ with m additional zeros. The two resulting strings then each have length $n + m$. Also, for convenience, renumber the indices of both strings to run from 0 to $n + m - 1$. Then

$$V_a(\alpha, \beta, i) = \sum_{j=0}^{j=n+m-1} \bar\alpha_a(j) \times \bar\beta_a(i + j),$$

where the indices in the expression are taken modulo $n + m$. The extra zeros are there to handle the cases when the left end of α is to the left end of β and, conversely, when the right end of α is to the right end of β. Enough zeros were padded so that when the right end of α is right of the right end of β, the corresponding bits in the padded $\bar\alpha_a$ are all opposite zeros. Hence no "illegitimate wraparound" of α and β is possible, and $V_a(\alpha, \beta, i)$ is correctly computed.

So far, all we have done is to recode the match-count problem, and this recoding doesn't suggest a way to compute $V_a(\alpha, \beta)$ more efficiently than before the binary coding and padding. This is where correlation and the FFT come in.

Cyclic correlation

Definition Let X and Y be two z-length vectors with real number components indexed from 0 to $z - 1$. The *cyclic correlation* of X and Y is an z-length real vector $W(i) = \sum_{j=0}^{j=z-1} X(j) \times Y(i + j)$, where the indices in the expression are taken modulo z.

Clearly, the problem of computing vector $V_a(\alpha, \beta)$ is exactly the problem of computing the cyclic correlation of padded strings $\bar{\alpha}_a$ and $\bar{\beta}_a$. In detail, $X = \bar{\alpha}_a$, $Y = \bar{\beta}_a$, $z = n + m$, and $W = V_a(\alpha, \beta)$.

Now an algorithm based only on the definition of cyclic correlation would require $\Theta(z^2)$ operations, so again no progress is apparent. But cyclic correlation is a classic problem known to be solvable in $O(z \log z)$ time using the Fast Fourier Transform. (The FFT is more often associated with the *convolution* problem for two vectors, but cyclic correlation and convolution are very similar. In fact, cyclic correlation is solved by reversing one of the input vectors and then computing the convolution of the two resulting vectors.)

The FFT method, and its use in the solution of the cyclic correlation problem, is beyond the scope of this book, but the key is that it solves the cyclic correlation problem in $O(z \log z)$ arithmetic operations, for two vectors each of length z. Hence it solves the match-count problem using only $O(m \log m)$ arithmetic operations. This is surprisingly efficient and a definite improvement over the $\Theta(nm)$ bound given by the generalized *Shift-And* approach. However, the FFT requires operations over complex numbers and so each arithmetic step is more involved (and perhaps more costly) than in the more direct *Shift-And* method.[1]

Handling wild cards in match-counts

Recall the wild card discussion begun in Section 3.5.2, where the wild card symbol ϕ matches any other *single* character. For example, if $\alpha = ag\phi\phi ct\phi a$ and $\beta = agctctgt$, then $V(\alpha, \beta, 1) = 7$ (i.e., all positions are counted as a match except the last). How do we incorporate these wild card symbols into the FFT approach computing match-counts? If the wild cards only occur in one of the two strings, say β, then the solution is very direct. When computing $V_x(\alpha, \beta, i)$ for every position i and character x, simply replace each wild card symbol in β with the character x. This works because for any fixed starting point i and any position j in β, the jth position will contribute a 1 to $V_x(\alpha, \beta, i)$ for at most one character x, depending on what character is in position $i + j - 1$ of α (i.e., what character in α is opposite the j position in β).

But if wild cards occur in both α and β, then this direct approach will not work. If two wild cards are opposite each other when α starts at position i, then $V_x(\alpha, \beta, i)$ would be too large, since those two symbols will be counted as a match when computing $V_x(\alpha, \beta, i)$ for each $x = a, t, c$, and g. So if for a fixed i there are k places where two wild cards line up, then the computed $\sum_x V_x(\alpha, \beta, i)$ will be $3k$ larger than the correct $V(\alpha, \beta, i)$ value. How can we avoid this overcount?

The answer is to find what k is and then correct the overcount. The idea is to treat ϕ as a real character, and compute $V_\phi(\alpha, \beta, i)$ for each i. Then

$$V(\alpha, \beta, i) = \sum_{x \neq \phi} V_x(\alpha, \beta, i) - 3V_\phi(\alpha, \beta, i),$$

---

[1] A related approach [58] attempts to solve the match-count problem in $O(m \log m)$ integer (noncomplex) operations by implementing the FFT over a finite field. In practice, this approach is probably superior to the approach based on complex numbers, although in terms of pure complexity theory the claimed $O(m \log m)$ bound is not completely kosher because it uses a precomputed table of numbers that is only adequate for values of m up to a certain size.

where for each character x, $V_x(\alpha, \beta, i)$ is computed by replacing each wild card with character x. In summary,

Theorem 4.3.1. *The match-count problem can be solved in $O(m \log m)$ time even if an unbounded number of wild cards are allowed in either P or T.*

Later, after discussing suffix trees and common ancestors, we will present in Section 9.3 a different, more comparison-based approach to handling wild cards that appear in both strings.

4.4. Karp–Rabin fingerprint methods for exact match

The *Shift-And* method assumes that we can efficiently shift a vector of bits, and the generalized *Shift-And* method assumes that we can efficiently increment an integer by one. If we treat a (row) bit vector as an integer number then a left shift by one bit results in the doubling of the number (assuming no bits fall off the left end). So it is not much of an extension to assume, in addition to being able to increment an integer, that we can also efficiently multiply an integer by two. With that added primitive operation we can turn the exact match problem (again without mismatches) into an arithmetic problem. The first result will be a simple linear-time method that has a very small probability of making an error. That method will then be transformed into one that never makes an error, but whose running time is only expected to be linear. We will explain these results using a binary string P and a binary text T. That is, the alphabet is first assumed to be just $\{0, 1\}$. The extension to larger alphabets is immediate and will be left to the reader.

4.4.1. Arithmetic replaces comparisons

Definition   For a text string T, let T_r^n denote the n-length substring of T starting at character r. Usually, n is known by context, and T_r^n will be replaced by T_r.

Definition   For the binary pattern P, let

$$H(P) = \sum_{i=1}^{i=n} 2^{n-i} P(i).$$

Similarly, let

$$H(T_r) = \sum_{i=1}^{i=n} 2^{n-i} T(r + i - 1).$$

That is, consider P to be an n-bit binary number. Similarly, consider T_r^n to be an n-bit binary number. For example, if $P = 0101$ then $n = 4$ and $H(P) = 2^3 \times 0 + 2^2 \times 1 + 2^1 \times 0 + 2^0 \times 1 = 5$; if $T = 101101010$, $n = 4$, and $r = 2$, then $H(T_r) = 6$.

Clearly, if there is an occurrence of P starting at position r of T then $H(P) = H(T_r)$. However, the converse is also true, so

Theorem 4.4.1. *There is an occurrence of P starting at position r of T if and only if $H(P) = H(T_r)$.*

The proof, which we leave to the reader, is an immediate consequence of the fact that every integer can be written in a unique way as the sum of positive powers of two.

Theorem 4.4.1 converts the exact match problem into a numerical problem, comparing the two numbers $H(P)$ and $H(T_r)$ rather than directly comparing characters. But unless the pattern is fairly small, the computation of $H(P)$ and $H(T_r)$ will not be efficient.[2] The problem is that the required powers of two used in the definition of $H(P)$ and $H(T_r)$ grow large too rapidly. (From the standpoint of complexity theory, the use of such large numbers violates the unit-time *random access machine (RAM)* model. In that model, the largest allowed numbers must be represented in $O[\log(n + m)]$ bits, but the number 2^n requires n bits. Thus the required numbers are exponentially too large.) Even worse, when the alphabet is not binary but say has t characters, then numbers as large as t^n are needed.

In 1987 R. Karp and M. Rabin [266] published a method (devised almost ten years earlier), called the *randomized fingerprint* method, that preserves the spirit of the above numerical approach, but that is extremely efficient as well, using numbers that satisfy the *RAM* model. It is a *randomized* method where the *only if* part of Theorem 4.4.1 continues to hold, but the *if* part does not. Instead, the *if* part will hold *with high probability*. This is explained in detail in the next section.

4.4.2. Fingerprints of P and T

The general idea is that, instead of working with numbers as large as $H(P)$ and $H(T_r)$, we will work with those numbers *reduced modulo* a relatively small integer p. The arithmetic will then be done on numbers requiring only a small number of bits, and so will be efficient. But the really attractive feature of this method is a proof that the probability of error can be made small if p is chosen randomly in a certain range. The following definitions and lemmas make this precise.

> **Definition**   For a positive integer p, $H_p(P)$ is defined as $H(P)$ mod p. That is $H_p(P)$ is the remainder of $H(P)$ after division by p. Similarly, $H_p(T_r)$ is defined as $H(T_r)$ mod p. The numbers $H_p(P)$ and $H_p(T_r)$ are called *fingerprints* of P and T_r.

Already, the utility of using fingerprints should be apparent. By reducing $H(P)$ and $H(T_r)$ modulo a number p, every fingerprint remains in the range 0 to $p - 1$, so the size of a fingerprint does not violate the *RAM* model. But if $H(P)$ and $H(T_r)$ must be computed before they can be reduced modulo p, then we have the same problem of intermediate numbers that are too large. Fortunately, modular arithmetic allows one to reduce at any time (i.e., one can never reduce too much), so that the following generalization of Horner's rule holds:

Lemma 4.4.1. $H_p(P) = \{[\ldots(\{[P(1) \times 2 \bmod p + P(2)] \times 2 \bmod p + P(3)\} \times 2 \bmod p + P(4))\ldots] \bmod p + P(n)\} \bmod p$, *and no number ever exceeds* $2p$ *during the computation of* $H_p(P)$.

---

[2] One can more efficiently compute $H(T_{r+1})$ from $H(T_r)$ than by following the definition directly (and we will need that later on), but the time to do the updates is not the issue here.

For example, if $P = 101111$ and $p = 7$, then $H(P) = 47$ and $H_p(P) = 47 \bmod 7 = 5$. Moreover, this can be computed as follows:

$$1 \times 2 \bmod 7 + 0 = 2$$
$$2 \times 2 \bmod 7 + 1 = 5$$
$$5 \times 2 \bmod 7 + 1 = 4$$
$$4 \times 2 \bmod 7 + 1 = 2$$
$$2 \times 2 \bmod 7 + 1 = 5$$
$$5 \bmod 7 = 5.$$

The point of Horner's rule is not only that the number of multiplications and additions required is linear, but that the intermediate numbers are always kept small.

Intermediate numbers are also kept small when computing $H_p(T_r)$ for any r, since that computation can be organized the way that $H_p(P)$ was. However, even greater efficiency is possible: For $r > 1$, $H_p(T_r)$ can be computed from $H_p(T_{r-1})$ with only a small *constant* number of operations. Since

$$H_p(T_r) = H(T_r) \bmod p$$

and

$$H(T_r) = 2 \times H(T_{r-1}) - 2^n T(r-1) + T(r+n-1),$$

it follows that

$$H_p(T_r) = [(2 \times H(T_{r-1}) \bmod p) - (2^n \bmod p) \times T(r-1) + T(r+n-1)] \bmod p.$$

Further,

$$2^n \bmod p = 2 \times (2^{n-1} \bmod p) \bmod p.$$

Therefore, each successive power of two taken mod p and each successive value $H_p(T_r)$ can be computed in constant time.

Prime moduli limit false matches

Clearly, if P occurs in T starting at position r then $H_p(P) = H_p(T_r)$, but now the converse does not hold for every p. That is, we cannot necessarily conclude that P occurs in T starting at r just because $H_p(P) = H_p(T_r)$.

Definition   If $H_p(P) = H_p(T_r)$ but P does not occur in T starting at position r, then we say there is a *false match* between P and T at position r. If there is *some* position r such that there is a false match between P and T at r, then we say there is a false match between P and T.

The goal will be to choose a modulus p small enough that the arithmetic is kept efficient, yet large enough that the probability of a false match between P and T is kept small. The key comes from choosing p to be a *prime* number in the proper range and exploiting properties of prime numbers. We will state the needed properties of prime numbers without proof.

Definition   For a positive integer u, $\pi(u)$ is the *number* of primes that are less than or equal to u.

The following theorem is a variant of the famous *prime number theorem*.

Theorem 4.4.2. $\frac{u}{\ln(u)} \leq \pi(u) \leq 1.26 \frac{u}{\ln(u)}$, *where* $\ln(u)$ *is the base e logarithm of u* [383].

Lemma 4.4.2. *If $u \geq 29$, then the product of all the primes that are less than or equal to u is greater than 2^u* [383].

For example, for $u = 29$ the prime numbers less than or equal to 29 are 2, 5, 7, 11, 13, 17, 19, 23, and 29. Their product is 2,156,564,410 whereas 2^{29} is 536,870,912.

Corollary 4.4.1. *If $u \geq 29$ and x is any number less than or equal to 2^u, then x has fewer than $\pi(u)$ (distinct) prime divisors.*

PROOF   Suppose x does have $k > \pi(u)$ distinct prime divisors $q_1, q_2, \ldots, q_k$. Then $2^u \geq x \geq q_1 q_2 \ldots q_k$ (the first inequality is from the statement of the corollary, and the second from the fact that some primes in the factorization of x may be repeated). But $q_1 q_2 \ldots q_k$ is at least as large as the product of the smallest k primes, which is greater than the product of the first $\pi(u)$ primes (by assumption that $k > \pi(u)$). However, the product of the primes less than or equal to u is greater than 2^u (by Lemma 4.4.2). So the assumption that $k > \pi(u)$ leads to the contradiction that $2^u > 2^u$, and the lemma is proved.   □

The central theorem

Now we are ready for the central theorem of the Karp–Rabin approach.

Theorem 4.4.3. *Let P and T be any strings such that $nm \geq 29$, where n and m are the lengths of P and T, respectively. Let I be any positive integer. If p is a randomly chosen prime number less than or equal to I, then the probability of a false match between P and T is less than or equal to $\frac{\pi(nm)}{\pi(I)}$.*

PROOF   Let R be the set of positions in T where P does *not* begin. That is, $s \in R$ if and only if P does not occur in T beginning at s. For each $s \in R$, $H(P) \neq H(T_s)$. Now consider the product $\Pi_{s \in R}(|H(P) - H(T_s)|)$. That product must be at most 2^{nm} since for any s, $H(P) - H(T_s) \leq 2^n$ (recall that we have assumed a binary alphabet). Applying Corollary 4.4.1, $\Pi_{s \in R}(|H(P) - H(T_s)|)$ has at most $\pi(nm)$ distinct prime divisors.

Now suppose a false match between P and T occurs at some position r of T. That means that $H(P) \bmod p = H(T_r) \bmod p$ and that p evenly divides $H(P) - H(T_r)$. Trivially then, p evenly divides $\Pi_{s \in R}(|H(P) - H(T_s)|)$, and so p is one of the prime divisors of that product. If p allows a false match to occur between P and T, then p must be one of a set of at most $\pi(nm)$ numbers. But p was chosen randomly from a set of $\pi(I)$ numbers, so the probability that p is a prime that allows a false match between P and T is at most $\frac{\pi(nm)}{\pi(I)}$.   □

Notice that Theorem 4.4.3 holds for any choice of pattern P and text T such that $nm \geq 29$. The probability in the theorem is not taken over choices of P and T but rather over choices of prime p. Thus, this theorem does not make any (questionable) assumptions about P or T being random or generated by a Markov process, etc. It works for any P and T! Moreover, the theorem doesn't just bound the probability that a false match occurs at a fixed position r, it bounds the probability that there is even a single such position r in T. It is also notable that the analysis in the proof of the theorem feels "weak". That is, it only develops a very weak property of a prime p that allows a false match, namely being one of at most $\pi(nm)$ numbers that divide $\Pi_{s \in R}(|H(P) - H(T_s)|)$. This suggests that the true probability of a false match occurring between P and T is much less than the bound established in the theorem.

Theorem 4.4.3 leads to the following random fingerprint algorithm for finding all occurrences of P in T.

Random fingerprint algorithm

1. Choose a positive integer I (to be discussed in more detail below).
2. Randomly pick a prime number less than or equal to I, and compute $H_p(P)$. (Efficient randomized algorithms exist for finding random primes [331].)
3. For each position r in T, compute $H_p(T_r)$ and test to see if it equals $H_p(P)$. If the numbers are equal, then either declare a probable match or check explicitly that P occurs in T starting at that position r.

Given the fact that each $H_p(T_r)$ can be computed in constant time from $H_p(T_{r-1})$, the fingerprint algorithm runs in $O(m)$ time, excluding any time used to explicitly check a declared match. It may, however, be reasonable not to bother explicitly checking declared matches, depending on the probability of an error. We will return to the issue of checking later. For now, to fully analyze the probability of error, we have to answer the question of what I should be.

How to choose I

The utility of the fingerprint method depends on finding a good value for I. As I increases, the probability of a false match between P and T decreases, but the allowed size of p increases, increasing the effort needed to compute $H_p(P)$ and $H_p(T_r)$. Is there a good balance? There are several good ways to choose I depending on n and m. One choice is to take $I = nm^2$. With that choice the largest number used in the algorithm requires at most $4(\log n + \log m)$ bits, satisfying the *RAM* model requirement that the numbers be kept small as a function of the size of the input. But, what of the probability of a false match?

Corollary 4.4.2. When $I = nm^2$, the probability of a false match is at most $\frac{2.53}{m}$.

PROOF   By Theorem 4.4.3 and the prime number theorem (Theorem 4.4.2), the probability of a false match is bounded by

$$\frac{\pi(nm)}{\pi(nm^2)} \leq 1.26 \frac{nm}{nm^2} \frac{\ln(nm^2)}{\ln(nm)} = 1.26 \frac{1}{m} \left[\frac{\ln(n) + 2\ln(m)}{\ln(n) + \ln(m)} \right] \leq \frac{2.53}{m}. \qquad \square$$

A small example from [266] illustrates this bound. Take $n = 250$, $m = 4000$, and hence $I = 4 \times 10^9 < 2^{32}$. Then the probability of a false match is at most $\frac{2.53}{4000} < 10^{-3}$. Thus, with just a 32-bit fingerprint, for any P and T the probability that even a single one of the algorithm's declarations is wrong is bounded by 0.001.

Alternately, if $I = n^2m$ then the probability of a false match is $O(1/n)$, and since it takes $O(n)$ time to determine whether a match is false or real, the *expected* verification time would be constant. The result would be an $O(m)$ *expected* time method that never has a false match.

Extensions

If one prime is good, why not use several? Why not pick k primes $p_1, p_2, \ldots, p_k$ randomly and compute k fingerprints? For any position r, there can be an occurrence of P starting at r only if $H_{p_i}(P) = H_{p_i}(T_r)$ for *every* one of the k selected primes. We now define a false match between P and T to mean that there is an r such that P does not occur in T starting at r, but $H_{p_i}(P) = H_{p_i}(T_r)$ for *each* of the k primes. What now is the probability of a false match between P and T? One bound is fairly immediate and intuitive.

Theorem 4.4.4. *When k primes are chosen randomly between 1 and I and k fingerprints are used, the probability of a false match between P and T is at most* $[\frac{\pi(nm)}{\pi(I)}]^k$.

PROOF We saw in the proof of Theorem 4.4.3 that if p is a prime that allows $H_p(P) = H_p(T_r)$ at some position r where P does not occur, then p is in a set of at most $\pi(nm)$ integers. When k fingerprints are used, a false match can occur only if each of the k primes is in that set, and since the primes are chosen randomly (independently), the bound from Theorem 4.4.3 holds for each of the primes. So the probability that all the primes are in the set is bounded by $[\frac{\pi(nm)}{\pi(I)}]^k$, and the theorem is proved.   □

As an example, if $k = 4$ and n, m, and I are as in the previous example, then the probability of a false match between P and T is at most by 10^{-12}. Thus, the probability of a false match is reduced dramatically, from 10^{-3} to 10^{-12}, while the computational effort of using four primes only increases by four times. For typical values of n and m, a small choice of k will assure that the probability of an error due to a false match is less than the probability of error due to a hardware malfunction.

Even lower limits on error

The analysis in the proof of Theorem 4.4.4 is again very weak, because it just multiplies the probability that each of the k primes allows a false match *somewhere* in T. However, for the algorithm to actually make an error at some specific position r, each of the primes must *simultaneously* allow a false match at the same r. This is an even less likely event. With this observation we can reduce the probability of a false match as follows:

Theorem 4.4.5. *When k primes are chosen randomly between 1 and I and k fingerprints are used, the probability of a false match between P and T is at most* $m[\frac{\pi(n)}{\pi(I)}]^k$.

PROOF Suppose that a false match occurs at some fixed position r. That means that each prime p_i must evenly divide $|H(P) - H(T_r)|$. Since $|H(P) - H(T_r)| \leq 2^n$, there are at most $\pi(n)$ primes that divide it. So each p_i was chosen randomly from a set of $\pi(I)$ primes and by chance is part of a subset of $\pi(n)$ primes. The probability of this happening at that fixed r is therefore $[\frac{\pi(n)}{\pi(I)}]^k$. Since there are m possible choices for r, the probability of a false match between P and T (i.e., the probability that there is such an r) is at most $m[\frac{\pi(n)}{\pi(I)}]^k$, and the theorem is proved.   □

Assuming, as before, that $I = nm^2$, a little arithmetic (which we leave to the reader) shows

Corollary 4.4.3. When k primes are chosen randomly and used in the fingerprint algorithm, the probability of a false match between P and T is at most $(1.26)^k m^{-(2k-1)}(1 + 0.6 \ln m)^k$.

Applying this to the running example of $n = 250$, $m = 4000$, and $k = 4$ reduces the probability of a false match to at most 2×10^{-22}.

We mention one further refinement discussed in [266]. Returning to the case where only a single prime is used, suppose the algorithm explicitly checks that P occurs in T when $H_p(P) = H_p(T_r)$, and it finds that P does not occur there. Then one may be better off by picking a new prime to use for the continuation of the computation. This makes intuitive sense. Theorem 4.4.3 randomizes over the choice of primes and bounds the probability that a randomly picked prime will allow a false match anywhere in T. But once the prime has been shown to allow a false match, it is no longer random. It may well be a prime that allows numerous false matches (a *demon seed*). Theorem 4.4.3 says nothing about how bad a particular prime can be. But by picking a new prime after each error is detected, we can apply Corollary 4.4.2 to each prime, establishing

Theorem 4.4.6. *If a new prime is randomly chosen after the detection of an error, then for any pattern and text the probability of t errors is at most $(\frac{2.53}{m})^t$.*

This probability falls so rapidly that one is effectively protected against a long series of errors on any particular problem instance. For additional probabilistic analysis of the Karp–Rabin method, see [182].

Checking for error in linear time

All the variants of the Karp–Rabin method presented above have the property that they find *all true* occurrences of P in T, but they may also find *false matches* – locations where P is declared to be in T, even though it is not there. If one checks for P at each declared location, this checking would seem to require $\Theta(nm)$ worst-case time, although the expected time can be made smaller. We present here an $O(m)$-time method, noted first by S. Muthukrishnan [336], that determines if any of the declared locations are false matches. That is, the method either verifies that the Karp–Rabin algorithm has found no false matches or it declares that there is at least one false match (but it may not be able to find all the false matches) in $O(m)$ time.

The method is related to Galil's extension of the Boyer–Moore algorithm (Section 3.2.2), but the reader need not have read that section. Consider a list $\mathcal{L}$ of (starting) locations in T where the Karp–Rabin algorithm declares P to be found. A *run* is a maximal interval of consecutive starting locations $l_1, l_2, \ldots, l_r$ in $\mathcal{L}$ such that every two successive numbers in the interval differ by at most $n/2$ (i.e., $l_{i+1} - l_i \leq n/2$). The method works on each run separately, so we first discuss how to check for false matches in a single run.

In a single run, the method explicitly checks for the occurrence of P at the first two positions in the run, l_1 and l_2. If P does not occur in both of those locations then the method has found a false match and stops. Otherwise, when P does occur at both l_1 and l_2, the method learns that P is semiperiodic with period $l_2 - l_1$ (see Lemma 3.2.3). We use d to refer to $l_2 - l_1$, and we show that d is the smallest period of P. If d is not the smallest period, then d must be a multiple of the smallest period, say d'. (This follows easily from the GCD Theorem, which is stated in Section 16.17.1.) (page 431). But that implies that there is an occurrence of P starting at position $l_1 + d' < d_2$, and since the Karp–Rabin method *never* misses any occurrence of P, that contradicts the choice of l_2 as the second occurrence of P in the interval between l_1 and l_r. So d must be the smallest period of P, and it follows that if there are no false matches in the run, then $l_{i+1} - l_i = d$ for each i in the run. Hence, as a first check, the method verifies that $l_{i+1} - l_i = d$ for each i; it declares a false match and stops if this check fails for some i. Otherwise, as in the Galil method, to check each location in $\mathcal{L}$, it suffices to *successively* check the last d characters in each declared occurrence of P against the last d characters of P. That is, for position l_i, the method checks the d characters of T starting at position $l_i + n - d$. If any of these successive checks finds a mismatch, then the method has found a false match in the run and stops. Otherwise, P does in fact occur starting at each declared location in the run.

For the time analysis, note first that no character of T is examined more than twice during a check of a single run. Moreover, since two runs are separated by at least $n/2$ positions and each run is at least n positions long, no character of T can be examined in more than two consecutive runs. It follows that the total time for the method, over all runs, is $O(m)$.

With the ability to check for false matches in $O(m)$ time, the Karp–Rabin algorithm can be converted from a method with a small probability of error that runs in $O(m)$ worst-case time, to one that makes no error, but runs in $O(m)$ *expected* time (a conversion from a Monte Carlo algorithm to a Las Vegas algorithm). To achieve this, simply (re)run and (re)check the Karp–Rabin algorithm until no false matches are detected. We leave the details as an exercise.

4.4.3. Why fingerprints?

The Karp–Rabin fingerprint method runs in linear worst-case time, but with a nonzero (though extremely small) chance of error. Alternatively, it can be thought of as a method that never makes an error and whose expected running time is linear. In contrast, we have seen several methods that run in linear worst-case time and never make errors. So what is the point of studying the Karp–Rabin method?

There are three responses to this question. First, from a practical standpoint, the method is simple and can be extended to other problems, such as two-dimensional pattern matching with odd pattern shapes – a problem that is more difficult for other methods. Second, the method is accompanied by concrete proofs, establishing significant properties of the method's performance. Methods similar in spirit to fingerprints (or filters) predate the Karp–Rabin method, but, unlike the Karp–Rabin method, they generally lack any theoretical analysis. Little has been proven about their performance. But the main attraction is that the method is based on very different *ideas* than the linear-time methods that guarantee no error. Thus the method is included because a central goal of this book is to present a diverse collection of ideas used in a range of techniques, algorithms, and proofs.

4.5. Exercises

1. Evaluate empirically the *Shift-And* method against methods discussed earlier. Vary the sizes of P and T.

2. Extend the *agrep* method to solve the problem of finding an "occurrence" of a pattern P inside a text T, when a small number of insertions and deletions of characters, as well as mismatches, are allowed. That is, characters can be inserted into P and characters can be deleted from P.

3. Adapt *Shift-And* and *agrep* to handle a set of patterns. Can you do better than just handling each pattern in the set independently?

4. Prove the correctness of the *agrep* method.

5. Show how to efficiently handle wild cards (both in the pattern and the text) in the *Shift-And* approach. Do the same for *agrep*. Show that the efficiency of neither method is affected by the number of wild cards in the strings.

6. Extend the *Shift-And* method to efficiently handle regular expressions that do not use the Kleene closure. Do the same for *agrep*. Explain the utility of these extensions to collections of biosequence patterns such as those in PROSITE.

7. We mentioned in Exercise 32 of Chapter 3 that PROSITE patterns often specify a range for the number of times that a subpattern repeats. Ranges of this type can be easily handled by the $O(nm)$ regular expression pattern matching method of Section 3.6. Can such range

specifications be efficiently handled with the *Shift-And* method or *agrep*? The answer partly depends on the number of such specifications that appear in the expression.

8. (Open problem) Devise a purely comparison-based method to compute match-counts in $O(m \log m)$ time. Perhaps one can examine the FFT method in detail to see if complex arithmetic can be replaced with character comparisons in the case of computing match-counts.

9. Complete the proof of Corollary 4.4.3.

10. The random fingerprint approach can be extended to the two-dimensional pattern matching problem discussed in Section 3.5.3. Do it.

11. Complete the details and analysis to convert the Karp–Rabin method from a Monte-Carlo–style randomized algorithm to a Las Vegas–style randomized algorithm.

12. There are improvements possible in the method to check for false matches in the Karp–Rabin method. For example, the method can find in $O(m)$ time all those runs containing no false matches. Explain how. Also, at some point, the method needs to explicitly check for P at only l_1 and not l_2. Explain when and why.

PART II

Suffix Trees and Their Uses

5

Introduction to Suffix Trees

A suffix tree is a data structure that exposes the internal structure of a string in a deeper way than does the fundamental preprocessing discussed in Section 1.3. Suffix trees can be used to solve the exact matching problem in linear time (achieving the same worst-case bound that the Knuth-Morris-Pratt and the Boyer–Moore algorithms achieve), but their real virtue comes from their use in linear-time solutions to many string problems more complex than exact matching. Moreover (as we will detail in Chapter 9), suffix trees provide a bridge between *exact* matching problems, the focus of Part I, and *inexact* matching problems that are the focus of Part III.

The classic application for suffix trees is the *substring problem*. One is first given a text T of length m. After $O(m)$, or linear, preprocessing time, one must be prepared to take in any unknown string S of length n and in $O(n)$ time either find an occurrence of S in T or determine that S is not contained in T. That is, the allowed preprocessing takes time proportional to the length of the text, but thereafter, the search for S must be done in time proportional to the length of S, *independent* of the length of T. These bounds are achieved with the use of a suffix tree. The suffix tree for the text is built in $O(m)$ time during a preprocessing stage; thereafter, whenever a string of length $O(n)$ is input, the algorithm searches for it in $O(n)$ time using that suffix tree.

The $O(m)$ preprocessing and $O(n)$ search result for the substring problem is very surprising and extremely useful. In typical applications, a long sequence of requested strings will be input after the suffix tree is built, so the linear time bound for each search is important. That bound is *not* achievable by the Knuth-Morris-Pratt or Boyer–Moore methods – those methods would preprocess each requested string on input, and then take $\Theta(m)$ (worst-case) time to search for the string in the text. Because m may be huge compared to n, those algorithms would be impractical on any but trivial-sized texts.

Often the text is a fixed *set* of strings, for example, a collection of STSs or ESTs (see Sections 3.5.1 and 7.10), so that the substring problem is to determine whether the input string is a substring of any of the fixed strings. Suffix trees work nicely to efficiently solve this problem as well. Superficially, this case of multiple text strings resembles the *dictionary* problem discussed in the context of the Aho–Corasick algorithm. Thus it is natural to expect that the Aho–Corasick algorithm could be applied. However, the Aho–Corasick method does not solve the substring problem in the desired time bounds, because it will only determine if the new string is a *full* string in the dictionary, not whether it is a substring of a string in the dictionary.

After presenting the algorithms, several applications and extensions will be discussed in Chapter 7. Then a remarkable result, *the constant-time least common ancestor method*, will be presented in Chapter 8. That method greatly amplifies the utility of suffix trees, as will be illustrated by additional applications in Chapter 9. Some of those applications provide a bridge to inexact matching; more applications of suffix trees will be discussed in Part III, where the focus is on inexact matching.

5.1. A short history

The first linear-time algorithm for constructing suffix trees was given by Weiner [473] in 1973, although he called his tree a position tree. A different, more space efficient algorithm to build suffix trees in linear time was given by McCreight [318] a few years later. More recently, Ukkonen [438] developed a conceptually different linear-time algorithm for building suffix trees that has all the advantages of McCreight's algorithm (and when properly viewed can be seen as a variant of McCreight's algorithm) but allows a much simpler explanation.

Although more than twenty years have passed since Weiner's original result (which Knuth is claimed to have called "the algorithm of 1973" [24]), suffix trees have not made it into the mainstream of computer science education, and they have generally received less attention and use than might have been expected. This is probably because the two original papers of the 1970s have a reputation for being extremely difficult to understand. That reputation is well deserved but unfortunate, because the algorithms, although nontrivial, are no more complicated than are many widely taught methods. And, when implemented well, the algorithms are practical and allow efficient solutions to many complex string problems. We know of no other single data structure (other than those essentially equivalent to suffix trees) that allows efficient solutions to such a wide range of complex string problems.

Chapter 6 fully develops the linear-time algorithms of Ukkonen and Weiner and then briefly mentions the high-level organization of McCreight's algorithm and its relationship to Ukkonen's algorithm. Our approach is to introduce each algorithm at a high level, giving simple, *inefficient* implementations. Those implementations are then incrementally improved to achieve linear running times. We believe that the expositions and analyses given here, particularly for Weiner's algorithm, are much simpler and clearer than in the original papers, and we hope that these expositions result in a wider use of suffix trees in practice.

5.2. Basic definitions

When describing how to build a suffix tree for an arbitrary string, we will refer to the generic string S of length m. We do not use P or T (denoting pattern and text) because suffix trees are used in a wide range of applications where the input string sometimes plays the role of a pattern, sometimes a text, sometimes both, and sometimes neither. As usual the alphabet is assumed finite and known. After discussing suffix tree algorithms for a single string S, we will generalize the suffix tree to handle sets of strings.

> **Definition** A suffix tree $\mathcal{T}$ for an m-character string S is a rooted directed tree with exactly m leaves numbered 1 to m. Each internal node, other than the root, has at least two children and each edge is labeled with a nonempty substring of S. No two edges out of a node can have edge-labels beginning with the same character. The key feature of the suffix tree is that for any leaf i, the concatenation of the edge-labels on the path from the root to leaf i exactly spells out the suffix of S that starts at position i. That is, it spells out $S[i..m]$.

For example, the suffix tree for the string *xabxac* is shown in Figure 5.1. The path from the root to the leaf numbered 1 spells out the full string $S = xabxac$, while the path to the leaf numbered 5 spells out the suffix *ac*, which starts in position 5 of S.

As stated above, the definition of a suffix tree for S does not guarantee that a suffix tree for any string S actually exists. The problem is that if one *suffix* of S matches a *prefix*

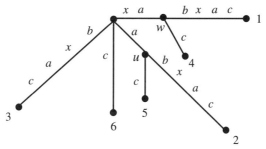

Figure 5.1: Suffix tree for string *xabxac*. The node labels *u* and *w* on the two interior nodes will be used later.

of another suffix of S then no suffix tree obeying the above definition is possible, since the path for the first suffix would not end at a leaf. For example, if the last character of *xabxac* is removed, creating string *xabxa*, then suffix *xa* is a prefix of suffix *xabxa*, so the path spelling out *xa* would not end at a leaf.

To avoid this problem, we assume (as was true in Figure 5.1) that the last character of S appears nowhere else in S. Then, no suffix of the resulting string can be a prefix of any other suffix. To achieve this in practice, we can add a character to the end of S that is not in the alphabet that string S is taken from. In this book we use $ for the "termination" character. When it is important to emphasize the fact that this termination character has been added, we will write it explicitly as in S. Much of the time, however, this reminder will not be necessary and, unless explicitly stated otherwise, every string S is assumed to be extended with the termination symbol $, even if the symbol is not explicitly shown.

A suffix tree is related to the keyword tree (without backpointers) considered in Section 3.4. Given string S, if set $\mathcal{P}$ is defined to be the m suffixes of S, then the suffix tree for S can be obtained from the keyword tree for $\mathcal{P}$ by merging any path of nonbranching nodes into a single edge. The simple algorithm given in Section 3.4 for building keyword trees could be used to construct a suffix tree for S in $O(m^2)$ time, rather than the $O(m)$ bound we will establish.

Definition The *label of a path* from the root that ends at a *node* is the concatenation, in order, of the substrings labeling the edges of that path. The *path-label of a node* is the label of the path from the root of $\mathcal{T}$ to that node.

Definition For any node v in a suffix tree, the *string-depth* of v is the number of characters in v's label.

Definition A path that ends in the middle of an edge (u, v) splits the label on (u, v) at a designated point. Define the label of such a path as the label of u concatenated with the characters on edge (u, v) down to the designated split point.

For example, in Figure 5.1 string *xa* labels the internal node w (so node w has path-label *xa*), string *a* labels node u, and string *xabx* labels a path that ends inside edge $(w, 1)$, that is, inside the leaf edge touching leaf 1.

5.3. A motivating example

Before diving into the details of the methods to construct suffix trees, let's look at how a suffix tree for a string is used to solve the exact match problem: Given a pattern P of

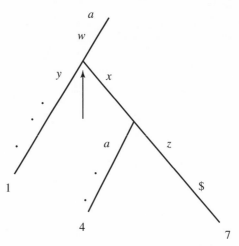

Figure 5.2: Three occurrences of *aw* in *awyawxawxz*. Their starting positions number the leaves in the subtree of the node with path-label *aw*.

length n and a text T of length m, find all occurrences of P in T in $O(n + m)$ time. We have already seen several solutions to this problem. Suffix trees provide another approach:

> Build a suffix tree $\mathcal{T}$ for text T in $O(m)$ time. Then, match the characters of P along the unique path in $\mathcal{T}$ until either P is exhausted or no more matches are possible. In the latter case, P does not appear anywhere in T. In the former case, every leaf in the subtree below the point of the last match is numbered with a starting location of P in T, and every starting location of P in T numbers such a leaf.

The key to understanding the former case (when all of P matches a path in T) is to note that P occurs in T starting at position j if and only if P occurs as a prefix of $T[j..m]$. But that happens if and only if string P labels an initial part of the path from the root to leaf j. It is the initial path that will be followed by the matching algorithm.

The matching path is unique because no two edges out of a common node can have edge-labels beginning with the same character. And, because we have assumed a finite alphabet, the work at each node takes constant time and so the time to match P to a path is proportional to the length of P.

For example, Figure 5.2 shows a fragment of the suffix tree for string $T = awyawxawxz$. Pattern $P = aw$ appears three times in T starting at locations 1, 4, and 7. Pattern P matches a path down to the point shown by an arrow, and as required, the leaves below that point are numbered 1, 4, and 7.

If P fully matches some path in the tree, the algorithm can find all the starting positions of P in T by traversing the subtree below the end of the matching path, collecting position numbers written at the leaves. All occurrences of P in T can therefore be found in $O(n+m)$ time. This is the same overall time bound achieved by several algorithms considered in Part I, but the distribution of work is different. Those earlier algorithms spend $O(n)$ time for preprocessing P and then $O(m)$ time for the search. In contrast, the suffix tree approach spends $O(m)$ preprocessing time and then $O(n + k)$ search time, where k is the number of occurrences of P in T.

To collect the k starting positions of P, traverse the subtree at the end of the matching path using any linear-time traversal (depth-first say), and note the leaf numbers encountered. Since every internal node has at least two children, the number of leaves encountered is proportional to the number of edges traversed, so the time for the traversal is $O(k)$, even though the total string-depth of those $O(k)$ edges may be arbitrarily larger than k.

If only a single occurrence of P is required, and the preprocessing is extended a bit, then the search time can be reduced from $O(n + k)$ to $O(n)$ time. The idea is to write at each node one number (say the smallest) of a leaf in its subtree. This can be achieved in $O(m)$ time in the preprocessing stage by a depth-first traversal of $\mathcal{T}$. The details are straightforward and are left to the reader. Then, in the search stage, the number written on the node at or below the end of the match gives one starting position of P in T.

In Section 7.2.1 we will again consider the relative advantages of methods that preprocess the text versus methods that preprocess the pattern(s). Later, in Section 7.8, we will also show how to use a suffix tree to solve the exact matching problem using $O(n)$ preprocessing and $O(m)$ search time, achieving the same bounds as in the algorithms presented in Part I.

5.4. A naive algorithm to build a suffix tree

To further solidify the definition of a suffix tree and develop the reader's intuition, we present a straightforward algorithm to build a suffix tree for string S. This naive method first enters a single edge for suffix $S[1..m]\$$ (the entire string) into the tree; then it successively enters suffix $S[i..m]\$$ into the growing tree, for i increasing from 2 to m. We let N_i denote the intermediate tree that encodes all the suffixes from 1 to i.

In detail, tree N_1 consists of a single edge between the root of the tree and a leaf labeled 1. The edge is labeled with the string $S\$$. Tree N_{i+1} is constructed from N_i as follows: Starting at the root of N_i find the longest path from the root whose label matches a prefix of $S[i + 1..m]\$$. This path is found by successively comparing and matching characters in suffix $S[i + 1..m]\$$ to characters along a unique path from the root, until no further matches are possible. The matching path is unique because no two edges out of a node can have labels that begin with the same character. At some point, no further matches are possible because no suffix of $S\$$ is a prefix of any other suffix of $S\$$. When that point is reached, the algorithm is either at a node, w say, or it is in the middle of an edge. If it is in the middle of an edge, (u, v) say, then it breaks edge (u, v) into two edges by inserting a new node, called w, just after the last character on the edge that matched a character in $S[i + 1..m]$ and just before the first character on the edge that mismatched. The new edge (u, w) is labeled with the part of the (u, v) label that matched with $S[i + 1..m]$, and the new edge (w, v) is labeled with the remaining part of the (u, v) label. Then (whether a new node w was created or whether one already existed at the point where the match ended), the algorithm creates a new edge $(w, i + 1)$ running from w to a new leaf labeled $i + 1$, and it labels the new edge with the unmatched part of suffix $S[i + 1..m]\$$.

The tree now contains a unique path from the root to leaf $i + 1$, and this path has the label $S[i + 1..m]\$$. Note that all edges out of the new node w have labels that begin with different first characters, and so it follows inductively that no two edges out of a node have labels with the same first character.

Assuming, as usual, a bounded-size alphabet, the above naive method takes $O(m^2)$ time to build a suffix tree for the string S of length m.

6

Linear-Time Construction of Suffix Trees

We will present two methods for constructing suffix trees in detail, Ukkonen's method and Weiner's method. Weiner was the first to show that suffix trees can be built in linear time, and his method is presented both for its historical importance and for some different technical ideas that it contains. However, Ukkonen's method is equally fast and uses far less space (i.e., memory) in practice than Weiner's method. Hence Ukkonen is the method of choice for most problems requiring the construction of a suffix tree. We also believe that Ukkonen's method is easier to understand. Therefore, it will be presented first. A reader who wishes to study only one method is advised to concentrate on it. However, our development of Weiner's method does not depend on understanding Ukkonen's algorithm, and the two algorithms can be read independently (with one small shared section noted in the description of Weiner's method).

6.1. Ukkonen's linear-time suffix tree algorithm

Esko Ukkonen [438] devised a linear-time algorithm for constructing a suffix tree that may be the conceptually easiest linear-time construction algorithm. This algorithm has a space-saving improvement over Weiner's algorithm (which was achieved first in the development of McCreight's algorithm), and it has a certain "on-line" property that may be useful in some situations. We will describe that on-line property but emphasize that the main virtue of Ukkonen's algorithm is the simplicity of its description, proof, and time analysis. The simplicity comes because the algorithm can be developed as a simple but inefficient method, followed by "common-sense" implementation tricks that establish a better worst-case running time. We believe that this less direct exposition is more understandable, as each step is simple to grasp.

6.1.1. Implicit suffix trees

Ukkonen's algorithm constructs a sequence of *implicit* suffix trees, the last of which is converted to a true suffix tree of the string S.

> **Definition**   An *implicit suffix tree* for string S is a tree obtained from the suffix tree for $S\$$ by removing every copy of the terminal symbol $\$$ from the edge labels of the tree, then removing any edge that has no label, and then removing any node that does not have at least two children.
>
>   An implicit suffix tree for a prefix $S[1..i]$ of S is similarly defined by taking the suffix tree for $S[1..i]\$$ and deleting $\$$ symbols, edges, and nodes as above.

> **Definition**   We denote the implicit suffix tree of the string $S[1..i]$ by $\mathcal{I}_i$, for i from 1 to m.

The implicit suffix tree for any string S will have fewer leaves than the suffix tree for

94

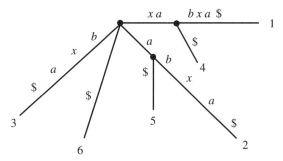

Figure 6.1: Suffix tree for string *xabxa* $.

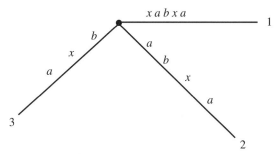

Figure 6.2: Implicit suffix tree for string *xabxa*.

string S\$ if and only if at least one of the suffixes of S is a prefix of another suffix. The terminal symbol \$ was added to the end of S precisely to avoid this situation. However, if S ends with a character that appears nowhere else in S, then the implicit suffix tree of S will have a leaf for each suffix and will hence be a true suffix tree.

As an example, consider the suffix tree for string $xabxa$\$ shown in Figure 6.1. Suffix xa is a prefix of suffix $xabxa$, and similarly the string a is a prefix of $abxa$. Therefore, in the suffix tree for $xabxa$ the edges leading to leaves 4 and 5 are labeled only with \$. Removing these edges creates two nodes with only one child each, and these are then removed as well. The resulting implicit suffix tree for $xabxa$ is shown in Figure 6.2. As another example, Figure 5.1 on page 91 shows a tree built for the string $xabxac$. Since character c appears only at the end of the string, the tree in that figure is both a suffix tree and an implicit suffix tree for the string.

Even though an implicit suffix tree may not have a leaf for each suffix, it does encode all the suffixes of S – each suffix is spelled out by the characters on some path from the root of the implicit suffix tree. However, if the path does not end at a leaf, there will be no marker to indicate the path's end. Thus implicit suffix trees, on their own, are somewhat less informative than true suffix trees. We will use them just as a tool in Ukkonen's algorithm to finally obtain the true suffix tree for S.

6.1.2. Ukkonen's algorithm at a high level

Ukkonen's algorithm constructs an implicit suffix tree $\mathcal{I}_i$ for each prefix $S[1..i]$ of S, starting from $\mathcal{I}_1$ and incrementing i by one until $\mathcal{I}_m$ is built. The true suffix tree for S is constructed from $\mathcal{I}_m$, and the time for the entire algorithm is $O(m)$. We will explain

Ukkonen's algorithm by first presenting an $O(m^3)$-time method to build all trees $\mathcal{I}_i$ and then optimizing its implementation to obtain the claimed time bound.

High-level description of Ukkonen's algorithm

Ukkonen's algorithm is divided into m *phases*. In phase $i + 1$, tree $\mathcal{I}_{i+1}$ is constructed from $\mathcal{I}_i$. Each phase $i + 1$ is further divided into $i + 1$ *extensions*, one for each of the $i + 1$ suffixes of $S[1..i + 1]$. In *extension j* of phase $i + 1$, the algorithm first finds the end of the path from the root labeled with substring $S[j..i]$. It then extends the substring by adding the character $S(i + 1)$ to its end, unless $S(i + 1)$ already appears there. So in phase $i + 1$, string $S[1..i + 1]$ is first put in the tree, followed by strings $S[2..i + 1]$, $S[3..i + 1]$, ... (in extensions 1,2, 3, ..., respectively). Extension $i + 1$ of phase $i + 1$ extends the *empty* suffix of $S[1..i]$, that is, it puts the single character string $S(i + 1)$ into the tree (unless it is already there). Tree $\mathcal{I}_1$ is just the single edge labeled by character $S(1)$. Procedurally, the algorithm is as follows:

High-level Ukkonen algorithm

Construct tree $\mathcal{I}_1$.
For i from 1 to $m - 1$ do
begin {phase $i + 1$}
    For j from 1 to $i + 1$
        begin {extension j}
        Find the end of the path from the root labeled $S[j..i]$ in the
        current tree. If needed, extend that path by adding character $S(i + 1)$,
        thus assuring that string $S[j..i + 1]$ is in the tree.
        end;
end;

Suffix extension rules

To turn this high-level description into an algorithm, we must specify exactly how to perform a *suffix extension*. Let $S[j..i] = \beta$ be a suffix of $S[1..i]$. In extension j, when the algorithm finds the end of β in the current tree, it extends β to be sure the suffix $\beta S(i + 1)$ is in the tree. It does this according to one of the following three rules:

Rule 1 In the current tree, path β ends at a leaf. That is, the path from the root labeled β extends to the end of some leaf edge. To update the tree, character $S(i + 1)$ is added to the end of the label on that leaf edge.

Rule 2 No path from the end of string β starts with character $S(i + 1)$, but at least one labeled path continues from the end of β.

In this case, a new leaf edge starting from the end of β must be created and labeled with character $S(i + 1)$. A new node will also have to be created there if β ends inside an edge. The leaf at the end of the new leaf edge is given the number j.

Rule 3 Some path from the end of string β starts with character $S(i + 1)$. In this case the string $\beta S(i + 1)$ is already in the current tree, so (remembering that in an implicit suffix tree the end of a suffix need not be explicitly marked) we do nothing.

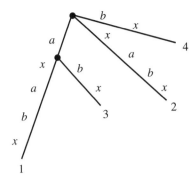

Figure 6.3: Implicit suffix tree for string *axabx* before the sixth character, *b*, is added.

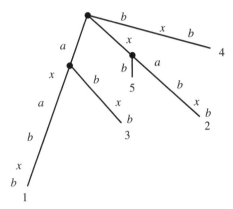

Figure 6.4: Extended implicit suffix tree after the addition of character *b*.

As an example, consider the implicit suffix tree for $S = axabx$ shown in Figure 6.3. The first four suffixes end at leaves, but the single character suffix x ends inside an edge. When a sixth character b is added to the string, the first four suffixes get extended by applications of Rule 1, the fifth suffix gets extended by rule 2, and the sixth by rule 3. The result is shown in Figure 6.4.

6.1.3. Implementation and speedup

Using the suffix extension rules given above, once the end of a suffix β of $S[1..i]$ has been found in the current tree, only constant time is needed to execute the extension rules (to ensure that suffix $\beta S(i + 1)$ is in the tree). The key issue in implementing Ukkonen's algorithm then is how to locate the ends of all the $i + 1$ suffixes of $S[1..i]$.

Naively we could find the end of any suffix β in $O(|\beta|)$ time by walking from the root of the current tree. By that approach, extension j of phase $i + 1$ would take $O(i + 1 - j)$ time, $\mathcal{I}_{i+1}$ could be created from $\mathcal{I}_i$ in $O(i^2)$ time, and $\mathcal{I}_m$ could be created in $O(m^3)$ time. This algorithm may seem rather foolish since we already know a straightforward algorithm to build a suffix tree in $O(m^2)$ time (and another is discussed in the exercises), but it is easier to describe Ukkonen's $O(m)$ algorithm as a speedup of the $O(m^3)$ method above.

We will reduce the above $O(m^3)$ time bound to $O(m)$ time with a few observations and implementation tricks. Each trick by itself looks like a sensible heuristic to accelerate the algorithm, but acting individually these tricks do not necessarily reduce the worst-case bound. However, taken together, they do achieve a linear worst-case time. The most important element of the acceleration is the use of *suffix links*.

Suffix links: first implementation speedup

Definition Let $x\alpha$ denote an arbitrary string, where x denotes a single character and α denotes a (possibly empty) substring. For an internal node v with path-label $x\alpha$, *if* there is another node $s(v)$ with path-label α, then a pointer from v to $s(v)$ is called a *suffix link*.

We will sometimes refer to a suffix link from v to $s(v)$ as the pair $(v, s(v))$. For example, in Figure 6.1 (on page 95) let v be the node with path-label xa and let $s(v)$ be the node whose path-label is the single character a. Then there exists a suffix link from node v to node $s(v)$. In this case, α is just a single character long.

As a special case, if α is empty, then the suffix link from an internal node with path-label $x\alpha$ goes to the root node. The root node itself is not considered internal and has no suffix link from it.

Although definition of suffix links does not imply that every internal node of an implicit suffix tree has a suffix link from it, it will, in fact, have one. We actually establish something stronger in the following lemmas and corollaries.

Lemma 6.1.1. *If a new internal node v with path-label $x\alpha$ is added to the current tree in extension j of some phase $i + 1$, then either the path labeled α already ends at an internal node of the current tree or an internal node at the end of string α will be created (by the extension rules) in extension $j + 1$ in the same phase $i + 1$.*

PROOF A new internal node v is created in extension j (of phase $i + 1$) only when extension rule 2 applies. That means that in extension j, the path labeled $x\alpha$ continued with some character other than $S(i + 1)$, say c. Thus, in extension $j + 1$, there is a path labeled α in the tree and it certainly has a continuation with character c (although possibly with other characters as well). There are then two cases to consider: Either the path labeled α continues only with character c or it continues with some additional character. When α is continued only by c, extension rule 2 will create a node $s(v)$ at the end of path α. When α is continued with two different characters, then there must already be a node $s(v)$ at the end of path α. The Lemma is proved in either case. $\square$

Corollary 6.1.1. In Ukkonen's algorithm, any newly created internal node will have a suffix link from it by the end of the next extension.

PROOF The proof is inductive and is true for tree $\mathcal{I}_1$ since $\mathcal{I}_1$ contains no internal nodes. Suppose the claim is true through the end of phase i, and consider a single phase $i + 1$. By Lemma 6.1.1, when a new node v is created in extension j, the correct node $s(v)$ ending the suffix link from v will be found or created in extension $j + 1$. No new internal node gets created in the last extension of a phase (the extension handling the single character suffix $S(i + 1)$), so all suffix links from internal nodes created in phase $i + 1$ are known by the end of the phase and tree $\mathcal{I}_{i+1}$ has all its suffix links. $\square$

Corollary 6.1.1 is similar to Theorem 6.2.5, which will be discussed during the treatment of Weiner's algorithm, and states an important fact about implicit suffix trees and ultimately about suffix trees. For emphasis, we restate the corollary in slightly different language.

Corollary 6.1.2. In any implicit suffix tree $\mathcal{I}_i$, if internal node v has path-label $x\alpha$, then *there is* a node $s(v)$ of $\mathcal{I}_i$ with path-label α.

Following Corollary 6.1.1, all internal nodes in the changing tree will have suffix links from them, except for the most recently added internal node, which will receive its suffix link by the end of the next extension. We now show how suffix links are used to speed up the implementation.

Following a trail of suffix links to build $\mathcal{I}_{i+1}$

Recall that in phase $i+1$ the algorithm locates suffix $S[j..i]$ of $S[1..i]$ in extension j, for j increasing from 1 to $i+1$. Naively, this is accomplished by matching the string $S[j..i]$ along a path from the root in the current tree. Suffix links can shortcut this walk and each extension. The first two extensions (for $j = 1$ and $j = 2$) in any phase $i+1$ are the easiest to describe.

The end of the full string $S[1..i]$ must end at a leaf of $\mathcal{I}_i$ since $S[1..i]$ is the longest string represented in that tree. That makes it easy to find the end of that suffix (as the trees are constructed, we can keep a pointer to the leaf corresponding to the current full string $S[1..i]$), and its suffix extension is handled by Rule 1 of the extension rules. So the first extension of any phase is special and only takes constant time since the algorithm has a pointer to the end of the current full string.

Let string $S[1..i]$ be $x\alpha$, where x is a single character and α is a (possibly empty) substring, and let $(v, 1)$ be the tree-edge that enters leaf 1. The algorithm next must find the end of string $S[2..i] = \alpha$ in the current tree derived from $\mathcal{I}_i$. The key is that node v is either the root or it is an interior node of $\mathcal{I}_i$. If it is the root, then to find the end of α the algorithm just walks down the tree following the path labeled α as in the naive algorithm. But if v is an internal node, then by Corollary 6.1.2 (since v was in $\mathcal{I}_i$) v has a suffix link out of it to node $s(v)$. Further, since $s(v)$ has a path-label that is a prefix of string α, the end of string α must end in the subtree of $s(v)$. Consequently, in searching for the end of α in the current tree, the algorithm need not walk down the entire path from the root, but can instead begin the walk from node $s(v)$. That is the main point of including suffix links in the algorithm.

To describe the second extension in more detail, let γ denote the edge-label on edge $(v, 1)$. To find the end of α, walk up from leaf 1 to node v; follow the suffix link from v to $s(v)$; and walk from $s(v)$ down the path (which may be more than a single edge) labeled γ. The end of that path is the end of α (see Figure 6.5). At the end of path α, the tree is updated following the suffix extension rules. This completely describes the first two extensions of phase $i+1$.

To extend any string $S[j..i]$ to $S[j..i+1]$ for $j > 2$, repeat the same general idea: Starting at the end of string $S[j-1..i]$ in the current tree, walk up at most one node to either the root or to a node v that has a suffix link from it; let γ be the edge-label of that edge; assuming v is not the root, traverse the suffix link from v to $s(v)$; then walk down the tree from $s(v)$, following a path labeled γ to the end of $S[j..i]$; finally, extend the suffix to $S[j..i+1]$ according to the extension rules.

There is one minor difference between extensions for $j > 2$ and the first two extensions. In general, the end of $S[j-1..i]$ may be at a node that itself has a suffix link from it, in which case the algorithm traverses that suffix link. Note that even when extension rule 2 applies in extension $j-1$ (so that the end of $S[j-1..i]$ is at a newly created internal node w), if the parent of w is not the root, then the parent of w already has a suffix link out of it, as guaranteed by Lemma 6.1.1. Thus in extension j the algorithm never walks up more than one edge.

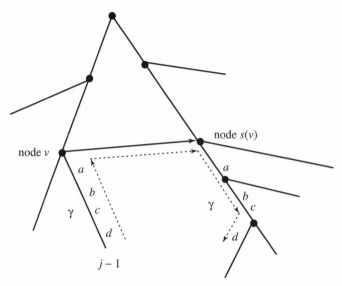

Figure 6.5: Extension $j > 1$ in phase $i + 1$. Walk up atmost one edge (labeled γ) from the end of the path labeled $S[j-1..i]$ to node v; then follow the suffix link to $s(v)$; then walk down the path specifying substring γ; then apply the appropriate extension rule to insert suffix $S[j..i + 1]$.

Single extension algorithm: SEA

Putting these pieces together, when implemented using suffix links, extension $j \geq 2$ of phase $i + 1$ is:

Single extension algorithm

Begin

1. Find the first node v at or above the end of $S[j-1..i]$ that either has a suffix link from it or is the root. This requires walking up at most one edge from the end of $S[j-1..i]$ in the current tree. Let γ (possibly empty) denote the string between v and the end of $S[j-1..i]$.

2. If v is not the root, traverse the suffix link from v to node $s(v)$ and then walk down from $s(v)$ following the path for string γ. If v is the root, then follow the path for $S[j..i]$ from the root (as in the naive algorithm).

3. Using the extension rules, ensure that the string $S[j..i]S(i + 1)$ is in the tree.

4. If a new internal node w was created in extension $j - 1$ (by extension rule 2), then by Lemma 6.1.1, string α must end at node $s(w)$, the end node for the suffix link from w. Create the suffix link $(w, s(w))$ from w to $s(w)$.

End.

Assuming the algorithm keeps a pointer to the current full string $S[1..i]$, the first extension of phase $i + 1$ need not do any up or down walking. Furthermore, the first extension of phase $i + 1$ always applies suffix extension rule 1.

What has been achieved so far?

The use of suffix links is clearly a practical improvement over walking from the root in each extension, as done in the naive algorithm. But does their use improve the worst-case running time?

The answer is that as described, the use of suffix links does not yet improve the time bound. However, here we introduce a trick that will reduce the worst-case time for the algorithm to $O(m^2)$. This trick will also be central in other algorithms to build and use suffix trees.

Trick number 1: skip/count trick

In Step 2 of extension $j + 1$ the algorithm walks *down* from node $s(v)$ along a path labeled γ. Recall that there surely must be such a γ path from $s(v)$. Directly implemented, this walk along γ takes time proportional to $|\gamma|$, the *number of characters* on that path. But a simple trick, called the *skip/count trick*, will reduce the traversal time to something proportional to the *number of nodes* on the path. It will then follow that the time for all the down walks in a phase is at most $O(m)$.

Trick 1   Let g denote the length of γ, and recall that no two labels of edges out of $s(v)$ can start with the same character, so the first character of γ must appear as the first character on exactly one edge out of $s(v)$. Let g' denote the number of characters on that edge. If g' is less than g, then the algorithm does not need to look at any more of the characters on that edge; it simply skips to the node at the end of the edge. There it sets g to $g - g'$, sets a variable h to $g' + 1$, and looks over the outgoing edges to find the correct next edge (whose first character matches character h of γ). In general, when the algorithm identifies the next edge on the path it compares the current value of g to the number of characters g' on that edge. When g is at least as large as g', the algorithm skips to the node at the end of the edge, sets g to $g - g'$, sets h to $h + g'$, and finds the edge whose first character is character h of γ and repeats. When an edge is reached where g is smaller than or equal to g', then the algorithm skips to character g on the edge and quits, assured that the γ path from $s(v)$ ends on that edge exactly g characters down its label. (See Figure 6.6).

Assuming simple and obvious implementation details (such as knowing the number of characters on each edge, and being able, in constant time, to extract from S the character at any given position) the effect of using the skip/count trick is to move from one node to the next node on the γ path in *constant* time.[1] The total time to traverse the path is then proportional to the number of *nodes* on it rather than the number of characters on it.

This is a useful heuristic, but what does it buy in terms of worst-case bounds? The next lemma leads immediately to the answer.

Definition   Define the *node-depth* of a node u to be the number of *nodes* on the path from the root to u.

Lemma 6.1.2. *Let $(v, s(v))$ be any suffix link traversed during Ukkonen's algorithm. At that moment, the node-depth of v is at most one greater than the node depth of $s(v)$.*

PROOF   When edge $(v, s(v))$ is traversed, any internal ancestor of v, which has path-label $x\beta$ say, has a suffix link to a node with path-label β. But $x\beta$ is a prefix of the path to v, so β is a prefix of the path to $s(v)$ and it follows that the suffix link from any internal ancestor of v goes to an ancestor of $s(v)$. Moreover, if β is nonempty then the node labeled by β is an internal node. And, because the node-depths of any two ancestors of v must differ, each ancestor of v has a suffix link to a distinct ancestor of $s(v)$. It follows that the node-depth of $s(v)$ is at least one (for the root) plus the number of internal ancestors of v who have path-labels more than one character long. The only extra ancestor that v can have (without a corresponding ancestor for $s(v)$) is an internal ancestor whose path-label

---

[1]   Again, we are assuming a constant-sized alphabet.

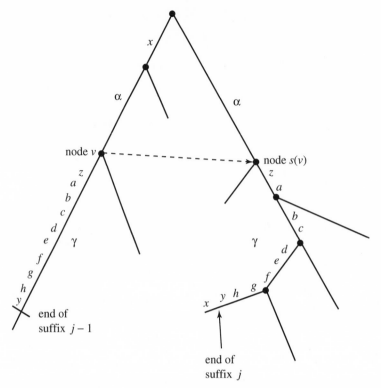

Figure 6.6: The skip/count trick. In phase $i + 1$, substring γ has length ten. There is a copy of substring γ out of node $s(v)$; it is found three characters down the last edge, after four node skips are executed.

has length one (it has label x). Therefore, v can have node-depth at most one more than $s(v)$. (See Figure 6.7). ☐

Definition   As the algorithm proceeds, the *current node-depth* of the algorithm is the node depth of the node most recently visited by the algorithm.

Theorem 6.1.1.   *Using the skip/count trick, any phase of Ukkonen's algorithm takes $O(m)$ time.*

PROOF   There are $i + 1 \leq m$ extensions in phase i. In a single extension the algorithm walks up at most one edge to find a node with a suffix link, traverses one suffix link, walks down some number of nodes, applies the suffix extension rules, and maybe adds a suffix link. We have already established that all the operations other than the down-walking take constant time per extension, so we only need to analyze the time for the down-walks. We do this by examining how the current node-depth can change over the phase.

The up-walk in any extension decreases the current node-depth by at most one (since it moves up at most one node), each suffix link traversal decreases the node-depth by at most another one (by Lemma 6.1.2), and each edge traversed in a down-walk moves to a node of greater node-depth. Thus over the entire phase the current node-depth is decremented at most $2m$ times, and since no node can have depth greater than m, the total possible increment to current node-depth is bounded by $3m$ over the entire phase. It follows that over the entire phase, the total number of edge traversals during down-walks is bounded by $3m$. Using the skip/count trick, the time per down-edge traversal is constant, so the total time in a phase for all the down-walking is $O(m)$, and the theorem is proved. ☐

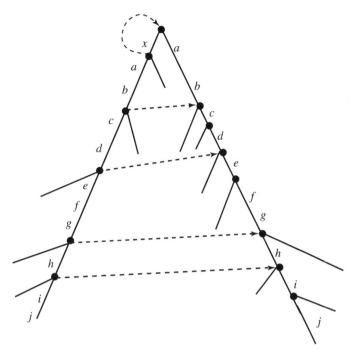

Figure 6.7: For every node v on the path $x\alpha$, the corresponding node $s(v)$ is on the path α. However, the node-depth of $s(v)$ can be one less than the node-depth of v, it can be equal, or it can be greater. For example, the node labeled *xab* has node-depth two, whereas the node-depth of *ab* is one. The node-depth of the node labeled *xabcdefg* is four, whereas the node-depth of *abcdefg* is five.

There are m phases, so the following is immediate:

Corollary 6.1.3. Ukkonen's algorithm can be implemented with suffix links to run in $O(m^2)$ time.

Note that the $O(m^2)$ time bound for the algorithm was obtained by multiplying the $O(m)$ time bound on a single phase by m (since there are m phases). This crude multiplication was necessary because the time analysis was directed to only a single phase. What is needed are some changes to the implementation allowing a time analysis that crosses phase boundaries. That will be done shortly.

At this point the reader may be a bit weary because we seem to have made no progress, since we started with a naive $O(m^2)$ method. Why all the work just to come back to the same time bound? The answer is that although we have made no progress on the time bound, we have made great conceptual progress so that with only a few more easy details, the time will fall to $O(m)$. In particular, we will need one simple implementation detail and two more little tricks.

6.1.4. A simple implementation detail

We next establish an $O(m)$ time bound for building a suffix tree. There is, however, one immediate barrier to that goal: The suffix tree may require $\Theta(m^2)$ space. As described so far, the edge-labels of a suffix tree might contain more than $\Theta(m)$ characters in total. Since the time for the algorithm is at least as large as the size of its output, that many characters makes an $O(m)$ time bound impossible. Consider the string $S = abcdefghijklmnopqrstuvwxyz$. Every suffix begins with a distinct character; hence there are 26 edges out of the root and

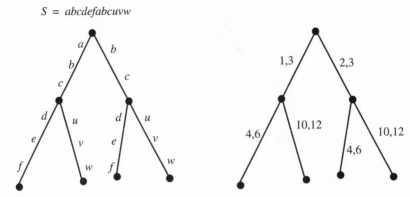

Figure 6.8: The left tree is a fragment of the suffix tree for string $S = abcdefabcuvw$, with the edge-labels written explicitly. The right tree shows the edge-labels compressed. Note that that edge with label 2, 3 could also have been labeled 8, 9.

each is labeled with a complete suffix, requiring $26 \times 27/2$ characters in all. For strings longer than the alphabet size, some characters will repeat, but still one can construct strings of arbitrary length m so that the resulting edge-labels have more than $\Theta(m)$ characters in total. Thus, an $O(m)$-time algorithm for building suffix trees requires some alternate scheme to represent the edge-labels.

Edge-label compression

A simple, alternate scheme exists for edge labeling. Instead of explicitly writing a substring on an edge of the tree, only write a *pair of indices* on the edge, specifying beginning and end positions of that substring in S (see Figure 6.8). Since the algorithm has a copy of string S, it can locate any particular character in S in constant time given its position in the string. Therefore, we may describe any particular suffix tree algorithm as if edge-labels were explicit, and yet implement that algorithm with only a constant number of symbols written on any edge (the index pair indicating the beginning and ending positions of a substring).

For example, in Ukkonen's algorithm when matching along an edge, the algorithm uses the index pair written on an edge to retrieve the needed characters from S and then performs the comparisons on those characters. The extension rules are also easily implemented with this labeling scheme. When extension rule 2 applies in a phase $i + 1$, label the newly created edge with the index pair $(i + 1, i + 1)$, and when extension rule 1 applies (on a leaf edge), change the index pair on that leaf edge from (p, q) to $(p, q + 1)$. It is easy to see inductively that q had to be i and hence the new label $(p, i + 1)$ represents the correct new substring for that leaf edge.

By using an index pair to specify an edge-label, only two numbers are written on any edge, and since the number of edges is at most $2m - 1$, the suffix tree uses only $O(m)$ symbols and requires only $O(m)$ space. This makes it more plausible that the tree can actually be built in $O(m)$ time.[2] Although the fully implemented algorithm will not explicitly write a substring on an edge, we will still find it convenient to talk about "the substring or label on an edge or path" as if the explicit substring was written there.

---

[2] We make the standard RAM model assumption that a number with up to $\log m$ bits can be read, written, or compared in constant time.

6.1.5. Two more little tricks and we're done

We present two more implementation tricks that come from two observations about the way the extension rules interact in successive extensions and phases. These tricks, plus Lemma 6.1.2, will lead immediately to the desired linear time bound.

Observation 1: Rule 3 is a show stopper   In any phase, if suffix extension rule 3 applies in extension j, it will also apply in all further extensions ($j + 1$ to $i + 1$) until the end of the phase. The reason is that when rule 3 applies, the path labeled $S[j..i]$ in the current tree must continue with character $S(i + 1)$, and so the path labeled $S[j + 1..i]$ does also, and rule 3 again applies in extensions $j + 1, j + 2, \ldots, i + 1$.

When extension rule 3 applies, no work needs to be done since the suffix of interest is already in the tree. Moreover, a new suffix link needs to be added to the tree only after an extension in which extension rule 2 applies. These facts and Observation 1 lead to the following implementation trick.

Trick 2   End any phase $i + 1$ the first time that extension rule 3 applies. If this happens in extension j, then there is no need to explicitly find the end of any string $S[k..i]$ for $k > j$.

The extensions in phase $i + 1$ that are "done" after the first execution of rule 3 are said to be done *implicitly*. This is in contrast to any extension j where the end of $S[j..i]$ is explicitly found. An extension of that kind is called an *explicit* extension.

Trick 2 is clearly a good heuristic to reduce work, but it's not clear if it leads to a better worst-case time bound. For that we need one more observation and trick.

Observation 2: Once a leaf, always a leaf   That is, if at some point in Ukkonen's algorithm a leaf is created and labeled j (for the suffix starting at position j of S), then that leaf will remain a leaf in all successive trees created during the algorithm. This is true because the algorithm has no mechanism for extending a leaf edge beyond its current leaf. In more detail, once there is a leaf labeled j, extension rule 1 will always apply to extension j in any successive phase. So once a leaf, always a leaf.

Now leaf 1 is created in phase 1, so in any phase i there is an initial sequence of consecutive extensions (starting with extension 1) where extension rule 1 or 2 applies. Let j_i denote the last extension in this sequence. Since any application of rule 2 creates a new leaf, it follows from Observation 2 that $j_i \leq j_{i+1}$. That is, the initial sequence of extensions where rule 1 or 2 applies cannot shrink in successive phases. This suggests an implementation trick that in phase $i + 1$ avoids all explicit extensions 1 through j_i. Instead, only constant time will be required to do those extensions implicitly.

To describe the trick, recall that the label on any edge in an implicit suffix tree (or a suffix tree) can be represented by two indices p, q specifying the substring $S[p..q]$. Recall also that for any leaf edge of $\mathcal{I}_i$, index q is equal to i and in phase $i + 1$ index q gets incremented to $i + 1$, reflecting the addition of character $S(i + 1)$ to the end of each suffix.

Trick 3   In phase $i + 1$, when a leaf edge is first created and would normally be labeled with substring $S[p..i + 1]$, instead of writing indices $(p, i + 1)$ on the edge, write (p, e), where e is a symbol denoting "the current end". Symbol e is a *global* index that is set to $i + 1$ once in each phase. In phase $i + 1$, since the algorithm knows that rule 1 will apply in extensions 1 through j_i at least, it need do no additional explicit work to implement

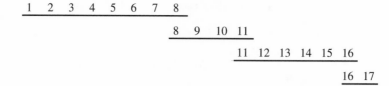

Figure 6.9: Cartoon of a possible execution of Ukkonen's algorithm. Each line represents a phase of the algorithm, and each number represents an explicit extension executed by the algorithm. In this cartoon there are four phases and seventeen explicit extensions. In any two consecutive phases, there is at most one index where the same explicit extension is executed in both phases.

those j_i extensions. Instead, it only does constant work to increment variable e, and then does explicit work for (some) extensions starting with extension $j_i + 1$.

The punch line

With Tricks 2 and 3, explicit extensions in phase $i + 1$ (using algorithm SEA) are then only required from extension $j_i + 1$ until the first extension where rule 3 applies (or until extension $i + 1$ is done). All other extensions (before and after those explicit extensions) are done implicitly. Summarizing this, phase $i + 1$ is implemented as follows:

Single phase algorithm: SPA

Begin

1. Increment index e to $i + 1$. (By Trick 3 this correctly implements all implicit extensions 1 through j_i.)

2. Explicitly compute successive extensions (using algorithm SEA) starting at $j_i + 1$ until reaching the first extension j^* where rule 3 applies or until all extensions are done in this phase. (By Trick 2, this correctly implements all the additional implicit extensions $j^* + 1$ through $i + 1$.)

3. Set j_{i+1} to $j^* - 1$, to prepare for the next phase.

End

Step 3 correctly sets j_{i+1} because the initial sequence of extensions where extension rule 1 or 2 applies must end at the point where rule 3 first applies.

The key feature of algorithm *SPA* is that phase $i + 2$ will *begin* computing explicit extensions with extension j^*, where j^* was the *last* explicit extension computed in phase $i + 1$. Therefore, two consecutive phases share *at most one* index (j^*) where an explicit extension is executed (see Figure 6.9). Moreover, phase $i + 1$ ends knowing where string $S[j^*..i + 1]$ ends, so the repeated extension of j^* in phase $i + 2$ can execute the suffix extension rule for j^* without any up-walking, suffix link traversals, or node skipping. That means the first explicit extension in any phase only takes constant time. It is now easy to prove the main result.

Theorem 6.1.2. *Using suffix links and implementation tricks* 1, 2, *and* 3, *Ukkonen's algorithm builds implicit suffix trees* $\mathcal{I}_1$ *through* $\mathcal{I}_m$ *in* $O(m)$ *total time.*

PROOF   The time for all the implicit extensions in any phase is constant and so is $O(m)$ over the entire algorithm.

As the algorithm executes explicit extensions, consider an index $\bar{j}$ corresponding to the explicit extension the algorithm is currently executing. Over the entire execution of the algorithm, $\bar{j}$ never decreases, but it does remain the same between two successive phases.

Since there are only m phases, and $\bar{j}$ is bounded by m, the algorithm therefore executes only $2m$ explicit extensions. As established earlier, the time for an explicit extension is a constant plus some time proportional to the number of node skips it does during the down-walk in that extension.

To bound the total number of node skips done during all the down-walks, we consider (similar to the proof of Theorem 6.1.1) how the current node-depth changes during successive extensions, even extensions in different phases. The key is that the first explicit extension in any phase (after phase 1) begins with extension j^*, which was the last explicit extension in the previous phase. Therefore, the current node-depth does not change between the end of one extension and the beginning of the next. But (as detailed in the proof of Theorem 6.1.1), in each explicit extension the current node-depth is first reduced by at most two (up-walking one edge and traversing one suffix link), and thereafter the down-walk in that extension increases the current node-depth by one at each node skip. Since the maximum node-depth is m, and there are only $2m$ explicit extensions, it follows (as in the proof of Theorem 6.1.1) that the maximum number of node skips done during all the down-walking (and not just in a single phase) is bounded by $O(m)$. All work has been accounted for, and the theorem is proved.   □

6.1.6. Creating the true suffix tree

The final implicit suffix tree $\mathcal{I}_m$ can be converted to a true suffix tree in $O(m)$ time. First, add a string terminal symbol $ to the end of S and let Ukkonen's algorithm continue with this character. The effect is that no suffix is now a prefix of any other suffix, so the execution of Ukkonen's algorithm results in an implicit suffix tree in which each suffix ends at a leaf and so is explicitly represented. The only other change needed is to replace each index e on every leaf edge with the number m. This is achieved by an $O(m)$-time traversal of the tree, visiting each leaf edge. When these modifications have been made, the resulting tree is a true suffix tree.

In summary,

Theorem 6.1.3. *Ukkonen's algorithm builds a true suffix tree for S, along with all its suffix links in $O(m)$ time.*

6.2. Weiner's linear-time suffix tree algorithm

Unlike Ukkonen's algorithm, Weiner's algorithm starts with the entire string S. However, like Ukkonen's algorithm, it enters one suffix at a time into a growing tree, although in a very different order. In particular, it first enters string $S(m)$$ into the tree, then string $S[m-1..m]$$, \ldots, and finally, it enters the entire string $S$$ into the tree.

Definition Suff_i denotes the suffix $S[i..m]$ of S starting in position i.

For example, Suff_1 is the entire string S, and Suff_m is the single character $S(m)$.

Definition   Define $\mathcal{T}_i$ to be the tree that has $m-i+2$ leaves numbered i through $m+1$ such that the path from the root to any leaf j ($i \le j \le m+1$) has label $\mathrm{Suff}_j$$. That is, $\mathcal{T}_i$ is a tree encoding all and only the suffixes of string $S[i..m]$$, so it is a suffix tree of string $S[i..m]$$.

Weiner's algorithm constructs trees from $\mathcal{T}_{m+1}$ *down* to $\mathcal{T}_1$ (i.e., in decreasing order of i). We will first implement the method in a straightforward inefficient way. This will

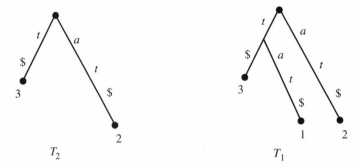

Figure 6.10: A step in the naive Weiner algorithm. The full string *tat* is added to the suffix tree for *at*. The edge labeled with the single character $ is omitted, since such an edge is part of every suffix tree.

serve to introduce and illustrate important definitions and facts. Then we will speed up the straightforward construction to obtain Weiner's linear-time algorithm.

6.2.1. A straightforward construction

The first tree $\mathcal{T}_{m+1}$ consists simply of a single edge out of the root labeled with the termination character $. Thereafter, for each i from m down to 1, the algorithm constructs each tree $\mathcal{T}_i$ from tree $\mathcal{T}_{i+1}$ and character $S(i)$. The idea of the method is essentially the same as the idea for constructing keyword trees (Section 3.4), but for a different set of strings and without putting in backpointers. As the algorithm proceeds, each tree $\mathcal{T}_i$ will have the property that for any node v in $\mathcal{T}_i$, no two edges out of v have edge-labels beginning with the same character. Since $\mathcal{T}_{m+1}$ only has one edge out of the root, this is trivially true for $\mathcal{T}_{m+1}$. We assume inductively that this property is true for tree $\mathcal{T}_{i+1}$ and will verify that it holds for tree $\mathcal{T}_i$.

In general, to create $\mathcal{T}_i$ from $\mathcal{T}_{i+1}$, start at the root of $\mathcal{T}_{i+1}$ and walk as far as possible down a path whose label matches a prefix of $\text{Suff}_i$$. Let R denote that path. In more detail, path R is found by starting at the root and explicitly matching successive characters of $\text{Suff}_i$$ with successive characters along a unique path in $\mathcal{T}_{i+1}$. The matching path is unique, since at any node v of $\mathcal{T}_{i+1}$ no two edges out of v have edge-labels beginning with the same character. Thus the matching continues on at most one edge out of v. Ultimately, because no suffix is a prefix of another, no further match will be possible. If no node exists at that point, then create a new node there. In either case, refer to the node there (old or new) as w. Finally, add an edge out of w to a new leaf node labeled i, and label the new edge (w, i) with the remaining (unmatched) part of $\text{Suff}_i$$. Since no further match had been possible, the first character on the label for edge (w, i) does not occur as the first character on any other edge out of w. Thus the claimed inductive property is maintained. Clearly, the path from the root to leaf i has label $\text{Suff}_i$$. That is, that path exactly spells out string $\text{Suff}_i$$, so tree $\mathcal{T}_i$ has been created.

For example, Figure 6.10 shows the transformation of $\mathcal{T}_2$ to $\mathcal{T}_1$ for the string *tat*.

Definition For any position i, $Head(i)$ denotes the longest prefix of $S[i..m]$ that matches a substring of $S[i+1..m]$$.

Note that $Head(i)$ could be the empty string. In fact, $Head(m)$ is always the empty string because $S[i+1..m]$ is the empty string when $i+1$ is greater than m and character $S(m) \neq $.

Since a copy of string $Head(i)$ begins at some position between $i+1$ and m, $Head(i)$ is also a prefix of $Suff_k$ for some $k > i$. It follows that $Head(i)$ is the longest prefix (possibly empty) of $Suff_i$ that is a label on some path from the root in tree $\mathcal{T}_{i+1}$.

The above straightforward algorithm to build $\mathcal{T}_i$ from $\mathcal{T}_{i+1}$ can be described as follows:

Naive Weiner algorithm

1. Find the end of the path labeled $Head(i)$ in tree $\mathcal{T}_{i+1}$.

2. If there is no node at the end of $Head(i)$ then create one, and let w denote the node (created or not) at the end of $Head(i)$. If w is created at this point, splitting an existing edge, then split its existing edge-label so that w has node-label $Head(i)$. Then, create a new leaf numbered i and a new edge (w, i) labeled with the remaining characters of $Suff_i$\$. That is, the new edge-label should be the last $m - i + 1 - |Head(i)|$ characters of $Suff_i$, followed by the termination symbol \$.

6.2.2. Toward a more efficient implementation

It should be clear that the final suffix tree $\mathcal{T} = \mathcal{T}_1$ is constructed in $O(m^2)$ time by this straightforward approach. Clearly, the difficult part of the algorithm is finding $Head(i)$, since step 2 takes only constant time for any i. So, to speed up the algorithm, we will need a more efficient way to find $Head(i)$. But, as in the discussion of Ukkonen's algorithm, a linear time bound is not possible if edge-labels are explicitly written on the tree. Instead, each edge-label is represented by two indices indicating the start and end positions of the labeling substring. The reader should review Section 6.1.4 at this point.

It is easy to implement Weiner's algorithm using an index pair to label an edge. When inserting $Suff_i$, suppose the algorithm has matched up to the kth character on an edge (u, z) labeled with interval $[s, t]$, but the next character is a mismatch. A new node w is created dividing (u, z) into two edges, (u, w) and (w, z), and a new edge is also created from w to leaf i. Edge (u, w) gets labeled $[s, s+k-1]$, edge (w, z) gets labeled $[s+k, t]$, and edge (w, i) gets labeled $[i + d(w), m]$\$, where $d(w)$ is the string-depth (number of characters) of the path from the root down to node w. These string-depths can easily be created and maintained as the tree is being built, since $d(w) = d(u) + k$. The string-depth of a leaf i is $m - i + 1$.

Finding $Head(i)$ efficiently

We now return to the central issue of how to find $Head(i)$ efficiently. The key to Weiner's algorithm are two vectors kept at each nonleaf node (including the root). The first vector is called the *indicator vector I* and the second is called the *link vector L*. Each vector is of length equal to the size of the alphabet, and each is indexed by the characters of the alphabet. For example, for the English alphabet augmented with \$, each link and indicator vector will be of length 27.

The link vector is essentially the reverse of the suffix link in Ukkonen's algorithm, and the two links are used in similar ways to accelerate traversals inside the tree.

The indicator vector is a bit vector so its entries are just 0 or 1, whereas each entry in the link vector is either null or is a pointer to a tree node. Let $I_v(x)$ specify the entry of the indicator vector at node v indexed by character x. Similarly, let $L_v(x)$ specify the entry of the link vector at node v indexed by character x.

The vectors I and L have two crucial properties that will be maintained inductively throughout the algorithm:

- For any (single) character x and any node u, $I_u(x) = 1$ in $\mathcal{T}_{i+1}$ if and only if there is a *path* from the root of $\mathcal{T}_{i+1}$ labeled $x\alpha$, where α is the path-label of node u. The path labeled $x\alpha$ need not end at a node.

- For any character x, $L_u(x)$ in $\mathcal{T}_{i+1}$ points to (internal) *node $\bar{u}$* in $\mathcal{T}_{i+1}$ if and only if $\bar{u}$ has path-label $x\alpha$, where u has path-label α. Otherwise $L_u(x)$ is null.

For example, in the tree in Figure 5.1 (page 91) consider the two internal nodes u and w with path-labels a and xa respectively. Then $I_u(x) = 1$ for the specific character x, and $L_u(x) = w$. Also, $I_w(b) = 1$, but $L_w(b)$ is null.

Clearly, for any node u and any character x, $L_u(x)$ is nonnull only if $I_u(x) = 1$, but the converse is not true. It is also immediate that if $I_u(x) = 1$ then $I_v(x) = 1$ for every ancestor node v of u.

Tree $\mathcal{T}_m$ has only one nonleaf node, namely the root r. In this tree we set $I_r(S(m))$ to one, set $I_r(x)$ to zero for every other character x, and set all the link entries for the root to null. Hence the above properties hold for $\mathcal{T}_m$. The algorithm will maintain the vectors as the tree changes, and we will prove inductively that the above properties hold for each tree.

6.2.3. The basic idea of Weiner's algorithm

Weiner's algorithm uses the indicator and link vectors to find $Head(i)$ and to construct $\mathcal{T}_i$ more efficiently. The algorithm must take care of two degenerate cases, but these are not much different than the general "good" case where no degeneracy occurs. We first discuss how to construct $\mathcal{T}_i$ from $\mathcal{T}_{i+1}$ in the good case, and then we handle the degenerate cases.

The algorithm in the good case

We assume that tree $\mathcal{T}_{i+1}$ has just been constructed and we now want to build $\mathcal{T}_i$. The algorithm starts at leaf $i + 1$ of $\mathcal{T}_{i+1}$ (the leaf for $Suff_{i+1}$) and walks toward the root looking for the first node v, if it exists, such that $I_v(S(i)) = 1$. If found, it then continues from v walking upwards toward the root searching for the first node v' it encounters (possibly v) where $L_{v'}(S(i))$ is nonnull. By definition, $L_{v'}(S(i))$ is nonnull only if $I_{v'}(S(i)) = 1$, so if found, v' will also be the first node encountered on the walk from leaf $i + 1$ such that $L_{v'}(S(i))$ is nonnull. In general, it may be that neither v nor v' exist or that v exists but v' does not. Note, however, that v or v' may be the root.

The "good case" is that *both v and v' do* exist.

Let l_i be the number of characters on the path between v' and v, and if $l_i > 0$ then let c denote the first of these l_i characters.

Assuming the good case, that both v and v' exist, we will prove below that if node v has path-label α then $Head(i)$ is precisely string $S(i)\alpha$. Further, we will prove that when $L_{v'}(S(i))$ points to node v'' in $\mathcal{T}_{i+1}$, $Head(i)$ either ends at v'', if $l_i = 0$, or else it ends *exactly l_i characters below v''* on an edge out of v''. So in either case, $Head(i)$ can be found in constant time after v' is found.

Theorem 6.2.1. *Assume that node v has been found by the algorithm and that it has path-label α. Then the string $Head(i)$ is exactly $S(i)\alpha$.*

PROOF $Head(i)$ is the longest prefix of $Suff_i$ that is also a prefix of $Suff_k$ for some $k > i$. Since v was found with $I_v(S(i)) = 1$ there is a path in $\mathcal{T}_{i+1}$ that begins with $S(i)$, so $Head(i)$ is at least one character long. Therefore, we can express $Head(i)$ as $S(i)\beta$, for some (possibly empty) string β.

Suff$_i$ and Suff$_k$ both begin with string $Head(i) = S(i)\beta$ and differ after that. For concreteness, say Suff$_i$ begins $S(i)\beta a$ and Suff$_k$ begins $S(i)\beta b$. But then Suff$_{i+1}$ begins βa and Suff$_{k+1}$ begins βb. Both $i+1$ and $k+1$ are greater than or equal to $i+1$ and less than or equal to m, so both suffixes are represented in tree $\mathcal{T}_{i+1}$. Therefore, in tree $\mathcal{T}_{i+1}$, there must be a path from the root labeled β (possibly the empty string) that extends in two ways, one continuing with character a and the other with character b. Hence there is a node u in $\mathcal{T}_{i+1}$ with path-label β, and $I_u(S(i)) = 1$ since there is a path (namely, an initial part of the path to leaf k) labeled $S(i)\beta$ in $\mathcal{T}_{i+1}$. Further, node u must be on the path to leaf $i+1$ since β is a prefix of Suff$_{i+1}$.

Now $I_v(S(i)) = 1$ and v has path-label α, so $Head(i)$ must begin with $S(i)\alpha$. That means that α is a prefix of β and so node u, with path label β, must either be v or below v on the path to leaf $i+1$. However, if $u \neq v$ then u would be a node below v on the path to leaf $i+1$, and $I_u(S(i)) = 1$. This contradicts the choice of node v, so $v = u$, $\alpha = \beta$, and the theorem is proved. That is, $Head(i)$ is exactly the string $S(i)\alpha$. □

Note that in Theorem 6.2.1 and its proof we only assume that node v exists. No assumption about v' was made. This will be useful in one of the degenerate cases examined later.

Theorem 6.2.2. *Assume both v and v' have been found and $L_{v'}(S(i))$ points to node v''. If $l_i = 0$ then $Head(i)$ ends at v''; otherwise it ends after exactly l_i characters on a single edge out of v''.*

PROOF Since v' is on the path to leaf $i+1$ and $L_{v'}(S(i))$ points to node v'', the path from the root labeled $Head(i)$ must include v''. By Theorem 6.2.1 $Head(i) = S(i)\alpha$, so $Head(i)$ must end exactly l_i characters below v''. Thus, when $l_i = 0$, $Head(i)$ ends at v''. But when $l_i > 0$, there must be an edge $e = (v'', z)$ out of v'' whose label begins with character c (the first of the l_i characters on the path from v' to v) in $\mathcal{T}_{i+1}$.

Can $Head(i)$ extend down to node z (i.e., to a node below v'')? Node z must be a branching node, for if it were a leaf then some suffix Suff$_k$, for $k > i$, would be a prefix of Suff$_i$, which is not possible. Let z have path-label $S(i)\gamma$. If $Head(i)$ extends down to branching node z, then there must be two substrings starting at or after position $i+1$ of S that both begin with string γ. Therefore, there would be a node z' with path-label γ in $\mathcal{T}_{i+1}$. Node z' would then be below v' on the path to leaf $i+1$, contradicting the selection of v'. So $Head(i)$ must not reach z and must end in the interior of edge e. In particular, it ends exactly l_i characters from v'' on edge e. □

Thus when $l_i = 0$, we know $Head(i)$ ends at v'', and when $l_i > 0$, we find $Head(i)$ from v'' by examining the edges out of v'' to identify that unique edge e whose first character is c. Then $Head(i)$ ends exactly l_i characters down e from v''. Tree $\mathcal{T}_i$ is then constructed by subdividing edge e, creating a node w at this point, and adding a new edge from w to leaf i labeled with the remainder of Suff$_i$. The search for the correct edge out of v'' takes only constant time since the alphabet is fixed.

In summary, when v and v' exist, the above method correctly creates $\mathcal{T}_i$ from $\mathcal{T}_{i+1}$, although we must still discuss how to update the vectors. Also, it may not yet be clear at this point why this method is more efficient than the naive algorithm for finding $Head(i)$. That will come later. Let us first examine how the algorithm handles the degenerate cases when v and v' do not both exist.

The two degenerate cases

The two degenerate cases are that node v (and hence node v') does not exist or that v exists but v' does not. We will see how to find $Head(i)$ efficiently in these two cases. Recall that r denotes the root node.

Case 1 $I_r(S(i)) = 0$.

In this case the walk ends at the root and no node v was found. It follows that character $S(i)$ does not appear in any position greater than i, for if it did appear, then some suffix in that range would begin with $S(i)$, some path from the root would begin with $S(i)$, and $I_r(S(i))$ would have been 1. So when $I_r(S(i)) = 0$, $Head(i)$ is the empty string and ends at the root.

Case 2 $I_v(S(i)) = 1$ for some v (possibly the root), but v' does not exist.

In this case the walk ends at the root with $L_r(S(i))$ null. Let t_i be the number of characters from the root to v. From Theorem 6.2.1 $Head(i)$ ends exactly $t_i + 1$ characters from the root. Since v exists, there is some edge $e = (r, z)$ whose edge-label begins with character $S(i)$. This is true whether $t_i = 0$ or $t_i > 0$.

If $t_i = 0$ then $Head(i)$ ends after the first character, $S(i)$, on edge e.

Similarly, if $t_i > 0$ then $Head(i)$ ends exactly $t_i + 1$ characters from the root on edge e. For suppose $Head(i)$ extends all the way to some child z (or beyond). Then exactly as in the proof of Theorem 6.2.2, z must be a branching node and there must be a node z' below the root on the path to leaf $i + 1$ such that $L_{z'}(S(i))$ is nonnull, which would be a contradiction. So when $t_i > 0$, $Head(i)$ ends exactly $t_i + 1$ characters from the root on the edge e out of the root. This edge can be found from the root in constant time since its first character is $S(i)$.

In either of these degenerate cases (as in the good case), $Head(i)$ is found in constant time after the walk reaches the root. After the end of $Head(i)$ is found and w is created or found, the algorithm proceeds exactly as in the good case.

Note that degenerate Case 2 is very similar to the "good" case when both v and v' were found, but differs in a small detail because $Head(i)$ is found $t_i + 1$ characters down on e rather than t_i characters down (the natural analogue of the good case).

6.2.4. The full algorithm for creating $\mathcal{T}_i$ from $\mathcal{T}_{i+1}$

Incorporating all the cases gives the following algorithm:

Weiner's Tree extension

1. Start at leaf $i + 1$ of $\mathcal{T}_{i+1}$ (the leaf for $Suff_{i+1}$) and walk toward the root searching for the first node v on the walk such that $I_v(S(i)) = 1$.

2. If the root is reached and $I_r(S(i)) = 0$, then $Head(i)$ ends at the root. Go to Step 4.

3. Let v be the node found (possibly the root) such that $I_v(S(i)) = 1$. Then continue walking upward searching for the first node v' (possibly v itself) such that $L_{v'}(S(i))$ is nonnull.

3a. If the root is reached and $L_r(S(i))$ is null, let t_i be the number of characters on the path between the root and v. Search for the edge e out of the root whose edge-label begins with $S(i)$. $Head(i)$ ends exactly $t_i + 1$ characters from the root on edge e.
     Else {when the condition in 3a does not hold}

3b. If v' was found such that $L_{v'}(S(i))$ is nonnull, say v'', then follow the link (for $S(i)$) to v''. Let l_i be the number of characters on the path from v' to v and let c be the first character on this path. If $l_i = 0$ then $Head(i)$ ends at v''. Otherwise, search for the edge e out of v'' whose first character is c. $Head(i)$ ends exactly l_i characters below v'' on edge e.

4. If a node already exists at the end of $Head(i)$, then let w denote that node; otherwise, create a node w at the end of $Head(i)$. Create a new leaf numbered i; create a new edge (w, i) labeled with the remaining substring of $Suff_i$ (i.e., the last $m - i + 1 - |Head(i)|$ characters of $Suff_i$), followed with the termination character $. Tree T_i has now been created.

Correctness

It should be clear from the proof of Theorems 6.2.1 and 6.2.2 and the discussion of the degenerate cases that the algorithm correctly creates tree T_i from T_{i+1}, although before it can create T_{i-1}, it must update the I and L vectors.

How to update the vectors

After finding (or creating) node w, we must update the I and L vectors so that they are correct for tree T_i. If the algorithm found a node v such that $I_v(S(i)) = 1$, then by Theorem 6.2.1 node w has path-label $S(i)\alpha$ in T_i, where node v has path-label α. In this case, $L_v(S(i))$ should be set to point to w in T_i. This is the only update needed for the link vectors since only one node can point via a link vector to any other node and only one new node was created. Furthermore, if node w is newly created, all its link entries for T_i should be null. To see this, suppose to the contrary that there is a node u in T_i with path-label $xHead(i)$, where w has path-label $Head(i)$. Node u cannot be a leaf because $Head(i)$ does not contain the character $. But then there must have been a node in T_{i+1} with path-label $Head(i)$, contradicting the fact that node w was inserted into T_{i+1} to create T_i. Consequently, there is no node in T_i with path-label $xHead(i)$ for any character x and all the L vector values for w should be null.

Now consider the updates needed to the indicator vectors for tree T_i. For every node u on the path from the root to leaf $i + 1$, $I_u(S(i))$ must be set to 1 in T_i since there is now a path for string $Suff_i$ in T_i. It is easy to establish inductively that if a node v with $I_v(S(i)) = 1$ is found during the walk from leaf $i + 1$, then every node u above v on the path to the root already has $I_u(S(i)) = 1$. Therefore, only the indicator vectors for the nodes below v on the path to leaf $i + 1$ need to be set. If no node v was found, then all nodes on the path from $i + 1$ to the root were traversed and all of these nodes must have their indicator vectors updated. The needed updates for the nodes below v can be made during the search for v (i.e., no separate pass is needed). During the walk from leaf $i + 1$, $I_u((S(i))$ is set to 1 for every node u encountered on the walk. The time to set these indicator vectors is proportional to the time for the walk.

The only remaining update is to set the I vector for a newly created node w created in the interior of an edge $e = (v'', z)$.

Theorem 6.2.3. *When a new node w is created in the interior of an edge (v'', z) the indicator vector for w should be copied from the indicator vector for z.*

PROOF It is immediate that if $I_z(x) = 1$ then $I_w(x)$ must also be 1 in T_i. But can it happen that $I_w(x)$ should be 1 and yet $I_z(x)$ is set to 0 at the moment that w is created? We will see that it cannot.

Let node z have path-label γ, and of course node w has path-label $Head(i)$, a prefix of γ. The fact that there are no nodes between u and z in T_{i+1} means that every suffix from $Suff_m$ down to $Suff_{i+1}$ that begins with string $Head(i)$ must actually begin with the longer string γ. Hence in $\mathcal{T}_{i+1}$ there can be a path labeled $xHead(i)$ only if there is also a path labeled $x\gamma$, and this holds for any character x. Therefore, if there is a path in $\mathcal{T}_i$ labeled $xHead(i)$ (the requirement for $I_w(x)$ to be 1) but no path $x\gamma$, then the hypothesized string $xHead(i)$ must begin at character i of S. That means that Suff_{i+1} must begin with the string $Head(i)$. But since w has path-label $Head(i)$, leaf $i + 1$ must be below w in $\mathcal{T}_i$ and so must be below z in $\mathcal{T}_{i+1}$. That is, z is on the root to $i + 1$ path. However, the algorithm to construct $\mathcal{T}_i$ from $\mathcal{T}_{i+1}$ starts at leaf $i + 1$ and walks toward the root, and when it finds node v or reaches the root, the indicator entry for x has been set to 1 at every node on the path from the leaf $i + 1$. The walk finishes before node w is created, and so it cannot be that $I_z(x) = 0$ at the time when w is created. So if path $xHead(i)$ exists in $\mathcal{T}_i$, then $I_z(x) = 1$ at the moment w is created, and the theorem is proved. □

6.2.5. Time analysis of Weiner's algorithm

The time to construct $\mathcal{T}_i$ from $\mathcal{T}_{i+1}$ and update the vectors is proportional to the time needed during the walk from leaf $i + 1$ (ending either at v' or the root). This walk moves from one node to its parent, and assuming the usual parent pointers, only constant time is used to move between nodes. Only constant time is used to follow a L link pointer, and only constant time is used after that to add w and edge (w, i). Hence the time to construct $\mathcal{T}_i$ is proportional to the number of nodes encountered on the walk from leaf $i + 1$.

Recall that the node-depth of a node v is the number of nodes on the path in the tree from the root to v.

For the time analysis we imagine that as the algorithm runs we keep track of what node has most recently been encountered and what its node-depth is. Call the node-depth of the most recently encountered node the *current node-depth*. For example, when the algorithm begins, the current node-depth is one and just after $\mathcal{T}_m$ is created the current node-depth is two. Clearly, when the algorithm walks up a path from a leaf the current node-depth decreases by one at each step. Also, when the algorithm is at node v'' (or at the root) and then creates a new node w below v'' (or below the root), the current node-depth increases by one. The only question remaining is how the current node-depth changes when a link pointer is traversed from a node v' to v''.

Lemma 6.2.1. *When the algorithm traverses a link pointer from a node v' to a node v'' in $\mathcal{T}_{i+1}$, the current node-depth increases by at most one.*

PROOF Let u be a nonroot node in $\mathcal{T}_{i+1}$ on the path from the root to v'', and suppose u has path-label $S(i)\alpha$ for some nonempty string α. All nodes on the root-to-v'' path are of this type, except for the single node (if it exists) with path-label $S(i)$. Now $S(i)\alpha$ is the prefix of Suff_i and of Suff_k for some $k > i$, and this string extends differently in the two cases. Since v' is on the path from the root to leaf $i + 1$, α is a prefix of Suff_{i+1}, and there must be a node (possibly the root) with path-label α on the path to v' in $\mathcal{T}_{i+1}$. Hence the path to v' has a node corresponding to every node on the path to v'', except the node (if it exists) with path-label $S(i)$. Hence the depth of v'' is at most one more than the depth of v', although it could be less. □

We can now finish the time analysis.

Theorem 6.2.4. *Assuming a finite alphabet, Weiner's algorithm constructs the suffix tree for a string of length m in $O(m)$ time.*

PROOF The current node-depth can increase by one each time a new node is created and each time a link pointer is traversed. Hence the total number of increases in the current node-depth is at most $2m$. It follows that the current node-depth can also only decrease at most $2m$ times since the current node-depth starts at zero and is never negative. The current node-depth decreases at each move up the walk, so the total number of nodes visited during all the upward walks is at most $2m$. The time for the algorithm is proportional to the total number of nodes visited during upward walks, so the theorem is proved. $\square$

6.2.6. Last comments about Weiner's algorithm

Our discussion of Weiner's algorithm establishes an important fact about suffix trees, regardless of how they are constructed:

Theorem 6.2.5. *If v is a node in the suffix tree labeled by the string $x\alpha$, where x is a single character, then there is a node in the tree labeled with the string α.*

This fact was also established as Corollary 6.1.2 during the discussion of Ukkonen's algorithm.

6.3. McCreight's suffix tree algorithm

Several years after Weiner published his linear-time algorithm to construct a suffix tree for a string S, McCreight [318] gave a different method that also runs in linear time but is more space efficient in practice. The inefficiency in Weiner's algorithm is the space it needs for the indicator and link vectors, I and L, kept at each node. For a fixed alphabet, this space is considered linear in the length of S, but the space used may be large in practice. McCreight's algorithm does not need those vectors and hence uses less space.

Ukkonen's algorithm also does not use the vectors I and L of Weiner's algorithm, and it has the same space efficiency as McCreight's algorithm.[3] In fact, the fully implemented version of Ukkonen's algorithm can be seen as a somewhat disguised version of Mc-Creight's algorithm. However, the high-level organization of Ukkonen and McCreight's algorithms are quite different, and the connection between the algorithms is not obvious. That connection was suggested by Ukkonen [438] and made explicit by Giegerich and Kurtz [178]. Since Ukkonen's algorithm has all the advantages of McCreight's, and is simpler to describe, we will only introduce McCreight's algorithm at the high level.

McCreight's algorithm at the high level

McCreight's algorithm builds the suffix tree T for m-length string S by inserting the suffixes in order, one at a time, starting from suffix one (i.e., the complete string S). (This is opposite to the order used in Weiner's algorithm, and it is superficially different from Ukkonen's algorithm.) It builds a tree encoding all the suffixes of S starting at positions 1 through $i + 1$, from the tree encoding all the suffixes of S starting at positions 1 through i.

The naive construction method is immediate and runs in $O(m^2)$ time. Using suffix links and the skip/count trick, that time can be reduced to $O(m)$. We leave this to the interested reader to work out.

---

[3] The space requirements for Ukkonen and McCreight's algorithms are determined by the need to represent and move around the tree quickly. We will be much more precise about space and practical implementation issues in Section 6.5.

6.4. Generalized suffix tree for a set of strings

We have so far seen methods to build a suffix tree for a single string in linear time. Those methods are easily extended to represent the suffixes of a set $\{S_1, S_2, \ldots, S_z\}$ of strings. Those suffixes are represented in a tree called a *generalized* suffix tree, which will be used in many applications.

A conceptually easy way to build a generalized suffix tree is to append a different end of string marker to each string in the set, then concatenate all the strings together, and build a suffix tree for the concatenated string. The end of string markers must be symbols that are not used in any of the strings. The resulting suffix tree will have one leaf for each suffix of the concatenated string and is built in time proportional to the sum of all the string lengths. The leaf numbers can easily be converted to two numbers, one identifying a string S_i and the other a starting position in S_i.

One defect with this way of constructing a generalized suffix tree is that the tree represents substrings (of the concatenated string) that span more than one of the original strings. These "synthetic" suffixes are not generally of interest. However, because each end of string marker is distinct and is not in any of the original strings, the label on any path from the root to an internal node must be a substring of one of the original strings. Hence by reducing the second index of the label on leaf edges, without changing any other parts of the tree, all the unwanted synthetic suffixes are removed.

Under closer examination, the above method can be simulated without first concatenating the strings. We describe the simulation using Ukkonen's algorithm and two strings S_1 and S_2, assumed to be distinct. First build a suffix tree for S_1 (assuming an added terminal character). Then starting at the root of this tree, match S_2 (again assuming the same terminal character has been added) against a path in the tree until a mismatch occurs. Suppose that the first i characters of S_2 match. The tree at this point encodes all the suffixes of S_1, and it implicitly encodes every suffix of the string $S_2[1..i]$. Essentially, the first i phases of Ukkonen's algorithm for S_2 have been executed on top of the tree for S_1. So, with that current tree, resume Ukkonen's algorithm on S_2 in phase $i + 1$. That is, walk up at most one node from the end of $S_2[1..i]$, etc. When S_2 is fully processed the tree will encode all the suffixes of S_1 and all the suffixes of S_2 but will have no synthetic suffixes. Repeating this approach for each of the strings in the set creates the generalized suffix tree in time proportional to the sum of the lengths of all the strings in the set.

There are two minor subtleties with the second approach. One is that the compressed labels on different edges may refer to different strings. Hence the number of symbols per edge increases from two to three, but otherwise causes no problem. The second subtlety is that suffixes from two strings may be identical, although it will still be true that no suffix is a prefix of any other. In this case, a leaf must indicate all of the strings and starting positions of the associated suffix.

As an example, if we add the string *babxba* to the tree for *xabxa* (shown in Figure 6.1), the result is the generalized suffix tree shown in Figure 6.11.

6.5. Practical implementation issues

The implementation details already discussed in this chapter turn naive, quadratic (or even cubic) time algorithms into algorithms that run in $O(m)$ worst-case time, assuming a fixed alphabet Σ. But to make suffix trees truly practical, more attention to implementation is needed, particularly as the size of the alphabet grows. There are problems nicely solved

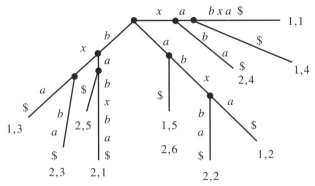

Figure 6.11: Generalized suffix tree for strings $S_1 = xabxa$ and $S_2 = babxba$. The first number at a leaf indicates the string; the second number indicates the starting position of the suffix in that string.

in theory by suffix trees, where the typical string size is in the hundreds of thousands, or even millions, and/or where the alphabet size is in the hundreds. For those problems, a "linear" time and space bound is not sufficient assurance of practicality. For large trees, paging can also be a serious problem because the trees do not have nice locality properties. Indeed, by design, suffix links allow an algorithm to move quickly from one part of the tree to a distant part of the tree. This is great for worst-case time bounds, but it is horrible for paging if the tree isn't entirely in memory. Consequently, implementing suffix trees to reduce practical space use can be a serious concern.[4] The comments made here for suffix trees apply as well to keyword trees used in the Aho–Corasick method.

The main design issues in all three algorithms are how to represent and search the branches out of the nodes of the tree and how to represent the indicator and link vectors in Weiner's algorithm. A practical design must balance the constraints of space against the need for speed, both in building the tree and in using it afterwards. We will discuss representing tree edges, since the vector issues for Weiner's algorithm are identical.

There are four basic choices possible to represent branches. The simplest is to use an *array* of size $\Theta(|\Sigma|)$ at each nonleaf node v. The array at v is indexed by single characters of the alphabet; the cell indexed by character x has a pointer to a child of v if there is an edge out of v whose edge-label begins with character x and is otherwise null. If there is such an edge, then the cell should also hold the two indices representing its edge-label. This array allows constant-time random accesses and updates and, although simple to program, it can use an impractical amount of space as $|\Sigma|$ and m get large.

An alternative to the array is to use a *linked list* at node v of characters that appear at the beginning of edge-labels out of v. When a new edge from v is added to the tree, a new character (the first character on the new edge label) is added to the list. Traversals from node v are implemented by sequentially searching the list for the appropriate character. Since the list is searched sequentially it costs no more to keep it in *sorted order*. This somewhat reduces the average time to search for a given character and thus speeds up (in practice) the construction of the tree. The key point is that it allows a faster termination of a search for a character that is not in the list. Keeping the list in sorted order will be particularly useful in some of applications of suffix trees to be discussed later.

---

[4] A very different approach to limiting space, based on changing the suffix tree into a different data structure called a *suffix array*, will be discussed in Section 7.14.

Keeping a linked list at node v works well if the number of children of v is small, but in worst-case adds time $|\Sigma|$ to every node operation. The $O(m)$ worst-case time bounds are preserved since $|\Sigma|$ is assumed to be fixed, but if the number of children of v is large then little space is saved over the array while noticeably degrading performance.

A third choice, a compromise between space and speed, is to implement the list at node v as some sort of *balanced tree* [10]. Additions and searches then take $O(\log k)$ time and $O(k)$ space, where k is the number of children of v. Due to the space and programming overhead of these methods, this alternative makes sense only when k is fairly large.

The final choice is some sort of *hashing scheme*. Again, the challenge is to find a scheme balancing space with speed, but for large trees and alphabets hashing is very attractive at least for some of the nodes. And, using perfect hashing techniques [167] the linear worst-case time bound can even be preserved.

When m and Σ are large enough to make implementation difficult, the best design is probably a mixture of the above choices. Nodes near the root of the tree tend to have the most children (the root has a child for every distinct character appearing in S), and so arrays are a sensible choice at those nodes. In addition, if the tree is dense for several levels below the root, then those levels can be condensed and eliminated from the explicit tree. For example, there are 20^5 possible amino acid substrings of length five. Every one of these substrings exists in some known protein sequence already in the databases. Therefore, when implementing a suffix tree for the protein database, one can replace the first five levels of the tree with a five-dimensional array (indexed by substrings of length five), where an entry of the array points to the place in the remaining tree that extends the five-tuple. The same idea has been applied [320] to depth seven for DNA data. Nodes in the suffix tree toward the leaves tend to have few children and lists there are attractive. At the extreme, if w is a leaf and v is its parent, then information about w may be brought up to v, removing the need for explicit representation of the edge (v, w) or the node w. Depending on the other implementation choices, this can lead to a large savings in space since roughly half the nodes in a suffix tree are leaves. A suffix tree whose leaves are deleted in this way is called a *position tree*. In a position tree, there is a one-to-one correspondence between leaves of the tree and substrings that are uniquely occurring in S.

For nodes in the middle of a suffix tree, hashing or balanced trees may be the best choice. Fortunately, most large suffix trees are used in applications where S is fixed (a dictionary or database) for some time and the suffix tree will be used repeatedly. In those applications, one has the time and motivation to experiment with different implementation choices. For a more in-depth look at suffix tree implementation issues, and other suggested variants of suffix trees, see [23].

Whatever implementation is selected, it is clear that a suffix tree for a string will take considerably more space than the representation of the string itself.[5] Later in the book we will discuss several problems involving two (or more) strings P and T, where two $O(|P| + |T|)$ time solutions exist, one using a suffix tree for P and one using a suffix tree for T. We will also have examples where equally time-efficient solutions exist, but where one uses a generalized suffix tree for two or more strings and the other uses just a suffix tree for the smaller string. In asymptotic worst-case time and space, neither approach is superior to the other, and usually the approach that builds the larger tree is conceptually simpler. However, when space is a serious practical concern (and in many problems, including

---

[5] Although, we have built suffix trees for DNA and amino acid strings more than one million characters long that can be completely contained in the main memory of a moderate-size workstation.

many in molecular biology, space is more of a constraint than is time), the size of the suffix tree for a string may dictate using the solution that builds the smaller suffix tree. So despite the added conceptual burden, we will discuss such space-reducing alternatives in some detail throughout the book.

6.5.1. Alphabet independence: all linears are equal, but some are more equal than others

The key implementation problems discussed above are all related to multiple edges (or links) at nodes. These are influenced by the size of the alphabet Σ – the larger the alphabet, the larger the problem. For that reason, some people prefer to explicitly reflect the alphabet size in the time and space bounds of keyword and suffix tree algorithms. Those people usually refer to the construction time for keyword or suffix trees as $O(m \log |\Sigma|)$, where m is the size of all the patterns in a keyword tree or the size of the string in a suffix tree. More completely, the Aho–Corasick, Weiner, Ukkonen, and McCreight algorithms all either require $\Theta(m|\Sigma|)$ space, or the $O(m)$ time bound should be replaced with the minimum of $O(m \log m)$ and $O(m \log |\Sigma|)$. Similarly, searching for a pattern P using a suffix tree can be done with $O(|P|)$ comparisons only if we use $\Theta(m|\Sigma|)$ space; otherwise we must allow the minimum of $O(|P| \log m)$ and $O(|P| \log |\Sigma|)$ comparisons during a search for P.

In contrast, the exact matching method using Z values has worst-case space and comparison requirements that are *alphabet independent* – the worst-case number of comparisons (either characters or numbers) used to compute Z values is uninfluenced by the size of the alphabet. Moreover, when two characters are compared, the method only checks whether the characters are equal or unequal, not whether one character precedes the other in some ordering. Hence no prior knowledge about the alphabet need be assumed. These properties are also true of the Knuth-Morris-Pratt and the Boyer–Moore algorithms. The alphabet independence of these algorithms makes their linear time and space bounds superior, in some people's view, to the linear time and space bounds of keyword and suffix tree algorithms: "All linears are equal but some are more equal than others". Alphabet-independent algorithms have also been developed for a number of problems other than exact matching. Two-dimensional exact matching is one such example. The method presented in Section 3.5.3 for two-dimensional matching is based on keyword trees and hence is not alphabet independent. Nevertheless, alphabet-independent solutions for that problem have been developed. Generally, alphabet-independent methods are more complex than their coarser counterparts. In this book we will not consider alphabet-independence much further, although we will discuss other approaches to reducing space that can be employed if large alphabets cause excessive space use.

6.6. Exercises

1. Construct an infinite family of strings over a fixed alphabet, where the total length of the edge-labels on their suffix trees grows faster than $\Theta(m)$ (m is the length of the string). That is, show that linear-time suffix tree algorithms would be impossible if edge-labels were written explicitly on the edges.

2. In the text, we first introduced Ukkonen's algorithm at a high level and noted that it could be implemented in $O(m^3)$ time. That time was then reduced to $O(m^2)$ with the use of suffix links and the skip/count trick. An alternative way to reduce the $O(m^3)$ time to $O(m^2)$ (without suffix links or skip/count) is to keep a pointer to the end of each suffix of $S[1..i]$.

Then Ukkonen's high-level algorithm could visit all these ends and create $\mathcal{I}_{i+1}$ from $\mathcal{I}_i$ in $O(i)$ time, so that the entire algorithm would run in $O(m^2)$ time. Explain this in detail.

3. The relationship between the suffix tree for a string S and for the reverse string S^r is not obvious. However, there is a significant relationship between the two trees. Find it, state it, and prove it.

   Hint: Suffix links help.

4. Can Ukkonen's algorithm be implemented in linear time without using suffix links? The idea is to maintain, for each index i, a pointer to the node in the current implicit suffix tree that is closest to the end of suffix i.

5. In Trick 3 of Ukkonen's algorithm, the symbol "e" is used as the second index on the label of every leaf edge, and in phase $i + 1$ the global variable e is set to $i + 1$. An alternative to using "e" is to set the second index on any leaf edge to m (the total length of S) at the point that the leaf edge is created. In that way, no work is required to update that second index. Explain in detail why this is correct, and discuss any disadvantages there may be in this approach, compared to using the symbol "e".

6. Ukkonen's algorithm builds all the implicit suffix trees I_1 through I_m in order and *on-line*, all in $O(m)$ time. Thus it can be called a linear-time on-line algorithm to construct implicit suffix trees.

   (Open question) Find an on-line algorithm running in $O(m)$ total time that creates all the *true* suffix trees. Since the time taken to explicitly store these trees is $\Theta(m^2)$, such an algorithm would (like Ukkonen's algorithm) update each tree without saving it.

7. Ukkonen's algorithm builds all the implicit suffix trees in $O(m)$ time. This sequence of implicit suffix trees may expose more information about S than does the single final suffix tree for S. Find a problem that can be solved more efficiently with the sequence of implicit suffix trees than with the single suffix tree. Note that the algorithm cannot save the implicit suffix trees and hence the problem will have to be solved in parallel with the construction of the implicit suffix trees.

8. The naive Weiner algorithm for constructing the suffix tree of S (Section 6.2.1) can be described in terms of the Aho–Corasick algorithm of Section 3.4: Given string S of length m, append \$ and let $\mathcal{P}$ be the set of patterns consisting of the $m + 1$ suffixes of string $S\$$. Then build a keyword tree for set $\mathcal{P}$ using the Aho–Corasick algorithm. Removing the backlinks gives the suffix tree for S. The time for this construction is $O(m^2)$. Yet, in our discussion of Aho–Corasick, that method was considered as a *linear* time method. Resolve this apparent contradiction.

9. Make explicit the relationship between link pointers in Weiner's algorithm and suffix links in Ukkonen's algorithm.

10. The time analyses of Ukkonen's algorithm and of Weiner's algorithm both rely on watching how the current node-depth changes, and the arguments are almost perfectly symmetric. Examine these two algorithms and arguments closely to make explicit the similarities and differences in the analysis. Is there some higher-level analysis that might establish the time bounds of both the algorithms at once?

11. Empirically evaluate different implementation choices for representing the branches out of the nodes and the vectors needed in Weiner's algorithm. Pay particular attention to the effect of alphabet size and string length, and consider both time and space issues in building the suffix tree and in using it afterwards.

12. By using implementation tricks similar to those used in Ukkonen's algorithm (particularly, suffix links and skip/count) give a linear-time implementation for McCreight's algorithm.

13. Flesh out the relationship between McCreight's algorithm and Ukkonen's algorithm, when they both are implemented in linear time.

14. Suppose one must dynamically maintain a suffix tree for a string that is growing or con-
    tracting. Discuss how to do this efficiently if the string is growing (contracting) on the left
    end, and how to do it if the string is growing (contracting) on the right end.

    Can either Weiner's algorithm or Ukkonen's algorithm efficiently handle both changes to
    the right and to the left ends of the string? What would be wrong in reversing the string so
    that a change on the left end is "simulated" by a change on the right end?

15. Consider the previous problem where the changes are in the interior of the string. If you
    cannot find an efficient solution to updating the suffix tree, explain what the technical issues
    are and why this seems like a difficult problem.

16. Consider a generalized suffix tree built for a set of k strings. Additional strings may be
    added to the set, or entire strings may be deleted from the set. This is the common case
    for maintaining a generalized suffix tree for biological sequence data [320]. Discuss the
    problem of maintaining the generalized suffix tree in this dynamic setting. Explain why this
    problem has a much easier solution than when arbitrary substrings represented in the suffix
    tree are deleted.

7

First Applications of Suffix Trees

We will see many applications of suffix trees throughout the book. Most of these applications allow surprisingly efficient, linear-time solutions to complex string problems. Some of the most impressive applications need an additional tool, the constant-time lowest common ancestor algorithm, and so are deferred until that algorithm has been discussed (in Chapter 8). Other applications arise in the context of specific problems that will be discussed in detail later. But there are many applications we can now discuss that illustrate the power and utility of suffix trees. In this chapter and in the exercises at its end, several of these applications will be explored.

Perhaps the best way to appreciate the power of suffix trees is for the reader to spend some time trying to solve the problems discussed below, without using suffix trees. Without this effort or without some historical perspective, the availability of suffix trees may make certain of the problems appear trivial, even though linear-time algorithms for those problems were unknown before the advent of suffix trees. The *longest common substring problem* discussed in Section 7.4 is one clear example, where Knuth had conjectured that a linear-time algorithm would not be possible [24, 278], but where such an algorithm is immediate with the use of suffix trees. Another classic example is the *longest prefix repeat problem* discussed in the exercises, where a linear-time solution using suffix trees is easy, but where the best prior method ran in $O(n \log n)$ time.

7.1. APL1: Exact string matching

There are three important variants of this problem depending on which string P or T is known first and held fixed. We have already discussed (in Section 5.3) the use of suffix trees in the exact string matching problem when the pattern and the text are both known to the algorithm at the same time. In that case the use of a suffix tree achieves the same worst-case bound, $O(n + m)$, as the Knuth-Morris-Pratt or Boyer–Moore algorithms.

But the exact matching problem often occurs in the situation when the text T is known first and kept fixed for some time. After the text has been preprocessed, a long sequence of patterns is input, and for each pattern P in the sequence, the search for all occurrences of P in T must be done as quickly as possible. Let n denote the length of P and k denote the number of occurrences of P in T. Using a suffix tree for T, all occurrences can be found in $O(n + k)$ time, totally independent of the size of T. That any pattern (unknown at the preprocessing stage) can be found in time proportional to its length alone, and after only spending linear time preprocessing T, is amazing and was the prime motivation for developing suffix trees. In contrast, algorithms that preprocess the pattern would take $O(n + m)$ time during the search for any single pattern P.

The reverse situation – when the pattern is first fixed and can be preprocessed before the text is known – is the classic situation handled by Knuth-Morris-Pratt or Boyer–Moore, rather than by suffix trees. Those algorithms spend $O(n)$ preprocessing time so that the search can be done in $O(m)$ time whenever a text T is specified. Can suffix trees be used in this scenario to achieve the same time bounds? Although it is not obvious, the answer is "yes". This reverse use of suffix trees will be discussed along with a more general problem in Section 7.8. Thus for the exact matching problem (single pattern), suffix trees can be used to achieve the same time and space bounds as Knuth-Morris-Pratt and Boyer–Moore when the pattern is known first or when the pattern and text are known together, but they achieve vastly superior performance in the important case that the text is known first and held fixed, while the patterns vary.

7.2. APL2: Suffix trees and the exact set matching problem

Section 3.4 discussed the *exact set matching problem*, the problem of finding all occurrences from a set of strings $\mathcal{P}$ in a text T, where the set is input all at once. There we developed a linear-time solution due to Aho and Corasick. Recall that set $\mathcal{P}$ is of total length n and that text T is of length m. The Aho–Corasick method finds all occurrences in T of any pattern from $\mathcal{P}$ in $O(n + m + k)$ time, where k is the number of occurrences. This same time bound is easily achieved using a suffix tree $\mathcal{T}$ for T. In fact, we saw in the previous section that when T is first known and fixed and the pattern P varies, all occurrences of any specific P (of length n) in T can be found in $O(n + k_P)$ time, where k_P is the number of occurrences of P. Thus the exact set matching problem is actually a simpler case because the set $\mathcal{P}$ is input at the same time the text is known. To solve it, we build suffix tree $\mathcal{T}$ for T in $O(m)$ time and then use this tree to successively search for all occurrences of each pattern in $\mathcal{P}$. The total time needed in this approach is $O(n + m + k)$.

7.2.1. Comparing suffix trees and keyword trees for exact set matching

Here we compare the relative advantages of keyword trees versus suffix trees for the exact set matching problem. Although the asymptotic time and space bounds for the two methods are the same when both the set $\mathcal{P}$ and the string T are specified together, one method may be preferable to the other depending on the relative sizes of $\mathcal{P}$ and T and on which string can be preprocessed. The Aho–Corasick method uses a keyword tree of size $O(n)$, built in $O(n)$ time, and then carries out the search in $O(m)$ time. In contrast, the suffix tree $\mathcal{T}$ is of size $O(m)$, takes $O(m)$ time to build, and is used to search in $O(n)$ time. The constant terms for the space bounds and for the search times depend on the specific way the trees are represented (see Section 6.5), but they are certainly large enough to affect practical performance.

In the case that the set of patterns is larger than the text, the suffix tree approach uses less space but takes more time to search. (As discussed in Section 3.5.1 there are applications in molecular biology where the pattern library is much larger than the typical texts presented after the library is fixed.) When the total size of the patterns is smaller than the text, the Aho–Corasick method uses less space than a suffix tree, but the suffix tree uses less search time. Hence, there is a time/space trade-off and neither method is uniformly superior to the other in time and space. Determining the relative advantages of Aho–Corasick versus suffix trees when the text is fixed and the set of patterns vary is left to the reader.

There is one way that suffix trees are better, or more robust, than keyword trees for the exact set matching problem (in addition to other problems). We will show in Section 7.8 how to use a suffix tree to solve the exact set matching problem in exactly the same time and space bounds as for the Aho–Corasick method – $O(n)$ for preprocessing and $O(m)$ for search. This is the reverse of the bounds shown above for suffix trees. The time/space trade-off remains, but a suffix tree can be used for either of the chosen time/space combinations, whereas no such choice is available for a keyword tree.

7.3. APL3: The substring problem for a database of patterns

The substring problem was introduced in Chapter 5 (page 89). In the most interesting version of this problem, a set of strings, or a database, is first known and fixed. Later, a sequence of strings will be presented and for each presented string S, the algorithm must find all the strings in the database containing S as a substring. This is the reverse of the exact set matching problem where the issue is to find which of the fixed patterns are in a substring of the input string.

In the context of databases for genomic DNA data [63, 320], the problem of finding substrings is a real one that cannot be solved by exact set matching. The DNA database contains a collection of previously sequenced DNA strings. When a new DNA string is sequenced, it could be contained in an already sequenced string, and an efficient method to check that is of value. (Of course, the opposite case is also possible, that the new string contains one of the database strings, but that is the case of exact set matching.)

One somewhat morbid application of this substring problem is a simplified version of a procedure that is in actual use to aid in identifying the remains of U.S. military personnel. Mitochondrial DNA from live military personnel is collected and a small interval of each person's DNA is sequenced. The sequenced interval has two key properties: It can be reliably isolated by the polymerase chain reaction (see the glossary page 528) and the DNA string in it is highly variable (i.e., likely differs between different people). That interval is therefore used as a "nearly unique" identifier. Later, if needed, mitochondrial DNA is extracted from the remains of personnel who have been killed. By isolating and sequencing the same interval, the string from the remains can be matched against a database of strings determined earlier (or matched against a narrower database of strings organized from missing personnel). The *substring* variant of this problem arises because the condition of the remains may not allow complete extraction or sequencing of the desired DNA interval. In that case, one looks to see if the extracted and sequenced string is a substring of one of the strings in the database. More realistically, because of errors, one might want to compute the length of the longest substring found both in the newly extracted DNA and in one of the strings in the database. That longest common substring would then narrow the possibilities for the identity of the person. The longest common substring problem will be considered in Section 7.4.

The total length of all the strings in the database, denoted by m, is assumed to be large. What constitutes a good data structure and lookup algorithm for the substring problem? The two constraints are that the database should be stored in a small amount of space and that each lookup should be fast. A third desired feature is that the preprocessing of the database should be relatively fast.

Suffix trees yield a very attractive solution to this database problem. A generalized suffix tree $\mathcal{T}$ for the strings in the database is built in $O(m)$ time and, more importantly, requires only $O(m)$ space. Any single string S of length n is found in the database, or declared not to be there, in $O(n)$ time. As usual, this is accomplished by matching the string against a path in the tree starting from the root. The full string S is in the database if and only if the matching path reaches a leaf of $\mathcal{T}$ at the point where the last character of

S is examined. Moreover, if S is a substring of strings in the database then the algorithm can find all strings in the database containing S as a substring. This takes $O(n + k)$ time, where k is the number of occurrences of the substring. As expected, this is achieved by traversing the subtree below the end of the matched path for S. If the full string S cannot be matched against a path in $\mathcal{T}$, then S is not in the database, and neither is it contained in any string there. However, the matched path does specify the longest *prefix* of S that is contained as a substring in the database.

The substring problem is one of the classic applications of suffix trees. The results obtained using a suffix tree are dramatic and not achieved using the Knuth-Morris-Pratt, Boyer–Moore, or even the Aho–Corasick algorithm.

7.4. APL4: Longest common substring of two strings

A classic problem in string analysis is to find the longest substring common to two given strings S_1 and S_2. This is the *longest common substring problem* (different from the longest common *subsequence* problem, which will be discussed in Sections 11.6.2 and 12.5 of Part III).

For example, if $S_1 = superiorcalifornialives$ and $S_2 = sealiver$, then the longest common substring of S_1 and S_2 is *alive*.

An efficient and conceptually simple way to find a longest common substring is to build a generalized suffix tree for S_1 and S_2. Each leaf of the tree represents either a suffix from one of the two strings or a suffix that occurs in both the strings. Mark each internal node v with a 1 (2) if there is a leaf in the subtree of v representing a suffix from S_1 (S_2). The path-label of any internal node marked both 1 and 2 is a substring common to both S_1 and S_2, and the longest such string is the longest common substring. So the algorithm has only to find the node with the greatest string-depth (number of characters on the path to it) that is marked both 1 and 2. Construction of the suffix tree can be done in linear time (proportional to the total length of S_1 and S_2), and the node markings and calculations of string-depth can be done by standard linear-time tree traversal methods.

In summary, we have

Theorem 7.4.1. *The longest common substring of two strings can be found in linear time using a generalized suffix tree.*

Although the longest common substring problem looks trivial now, given our knowledge of suffix trees, it is very interesting to note that in 1970 Don Knuth conjectured that a linear-time algorithm for this problem would be impossible [24, 278]. We will return to this problem in Section 7.9, giving a more space efficient solution.

Now recall the problem of identifying human remains mentioned in Section 7.3. That problem reduced to finding the longest substring in one fixed string that is also in some string in a database of strings. A solution to that problem is an immediate extension of the longest common substring problem and is left to the reader.

7.5. APL5: Recognizing DNA contamination

Often the various laboratory processes used to isolate, purify, clone, copy, maintain, probe, or sequence a DNA string will cause unwanted DNA to become inserted into the string of interest or mixed together with a collection of strings. Contamination of protein in the laboratory can also be a serious problem. During cloning, contamination is often caused by a fragment (substring) of a *vector* (DNA string) used to incorporate the desired DNA in a host organism, or the contamination is from the DNA of the host itself (for example bacteria or yeast). Contamination can also come from very small amounts of undesired foreign DNA that gets physically mixed into the desired DNA and then amplified by PCR (the polymerase chain reaction) used to make copies of the desired DNA. Without going into these and other specific ways that contamination occurs, we refer to the general phenomenon as *DNA contamination.*

Contamination is an extremely serious problem, and there have been embarrassing occurrences of large-scale DNA sequencing efforts where the use of highly contaminated clone libraries resulted in a huge amount of wasted sequencing. Similarly, the announcement a few years ago that DNA had been successfully extracted from dinosaur bone is now viewed as premature at best. The "extracted" DNA sequences were shown, through DNA database searching, to be more similar to mammal DNA (particularly human) [2] than to bird and crockodilian DNA, suggesting that much of the DNA in hand was from human contamination and not from dinosaurs. Dr. S. Blair Hedges, one of the critics of the dinosaur claims, stated: "In looking for dinosaur DNA we all sometimes find material that at first looks like dinosaur genes but later turns out to be human contamination, so we move on to other things. But this one was published." [80]

These embarrassments might have been avoided if the sequences were examined early for signs of likely contaminants, before large-scale analysis was performed or results published. Russell Doolittle [129] writes "...On a less happy note, more than a few studies have been curtailed when a preliminary search of the sequence revealed it to be a common contaminant . . . used in purification. As a rule, then, the experimentalist should search early and often".

Clearly, it is important to know whether the DNA of interest has been contaminated. Besides the general issue of the accuracy of the sequence finally obtained, contamination can greatly complicate the task of shotgun sequence assembly (discussed in Sections 16.14 and 16.15) in which short strings of sequenced DNA are assembled into long strings by looking for overlapping substrings.

Often, the DNA sequences from many of the possible contaminants are known. These include cloning vectors, PCR primers, the complete genomic sequence of the host organism (yeast, for example), and other DNA sources being worked with in the laboratory. (The dinosaur story doesn't quite fit here because there isn't yet a substantial transcript of human DNA.) A good illustration comes from the study of the nemotode *C. elegans*, one of the key model organisms of molecular biology. In discussing the need to use YACs (Yeast Artificial Chromosomes) to sequence the *C. elegans* genome, the contamination problem and its potential solution is stated as follows:

> The main difficulty is the unavoidable contamination of purified YACs by substantial amounts of DNA from the yeast host, leading to much wasted time in sequencing and assembling irrelevant yeast sequences. However, this difficulty should be eliminated (using). . . the complete (yeast) sequence. . . It will then become possible to discard instantly all sequencing reads that are recognizable as yeast DNA and focus exclusively on *C. elegans* DNA. [225]

This motivates the following computational problem:

DNA contamination problem Given a string S_1 (the newly isolated and sequenced string of DNA) and a known string S_2 (the combined sources of possible contamination), find all substrings of S_2 that occur in S_1 and that are longer than some given length l. These substrings are candidates for unwanted pieces of S_2 that have contaminated the desired DNA string.

This problem can easily be solved in linear time by extending the approach discussed above for the longest common substring of two strings. Build a generalized suffix tree for S_1 and S_2. Then mark each internal node that has in its subtree a leaf representing a suffix of S_1 and also a leaf representing a suffix of S_2. Finally, report all marked nodes that have string-depth of l or greater. If v is such a marked node, then the path-label of v is a suspicious string that may be contaminating the desired DNA string. If there are no marked nodes with string-depth above the threshold l, then one can have greater confidence (but not certainty) that the DNA has not been contaminated by the known contaminants.

More generally, one has an entire set of known DNA strings that might contaminate a desired DNA string. The problem now is to determine if the DNA string in hand has any sufficiently long substrings (say length l or more) from the known set of possible contaminants. The approach in this case is to build a generalized suffix tree for the set $\mathcal{P}$ of possible contaminants together with S_1, and then mark every internal node that has a leaf in its subtree representing a suffix from S_1 and a leaf representing a suffix from a pattern in $\mathcal{P}$. All marked nodes of string-depth l or more identify suspicious substrings.

Generalized suffix trees can be built in time proportional to the total length of the strings in the tree, and all the other marking and searching tasks described above can be performed in linear time by standard tree traversal methods. Hence suffix trees can be used to solve the contamination problem in linear time. In contrast, it is not clear if the Aho–Corasick algorithm can solve the problem in linear time, since that algorithm is designed to search for occurrences of *full* patterns from $\mathcal{P}$ in S_1, rather than for substrings of patterns.

As in the longest common substring problem, there is a more space efficient solution to the contamination problem, based on the material in Section 7.8. We leave this to the reader.

7.6. APL6: Common substrings of more than two strings

One of the most important questions asked about a set of strings is: What substrings are common to a large number of the *distinct* strings? This is in contrast to the important problem of finding substrings that occur repeatedly in a single string.

In biological strings (DNA, RNA, or protein) the problem of finding substrings common to a large number of distinct strings arises in many different contexts. We will say much more about this when we discuss database searching in Chapter 15 and multiple string comparison in Chapter 14. Most directly, the problem of finding common substrings arises because mutations that occur in DNA after two species diverge will more rapidly change those parts of the DNA or protein that are less functionally important. The parts of the DNA or protein that are critical for the correct functioning of the molecule will be more highly conserved, because mutations that occur in those regions will more likely be lethal. Therefore, finding DNA or protein substrings that occur commonly in a wide range of species helps point to regions or subpatterns that may be critical for the function or structure of the biological string.

Less directly, the problem of finding (exactly matching) common substrings in a set of distinct strings arises as a subproblem of many heuristics developed in the biological literature to *align* a set of strings. That problem, called multiple alignment, will be discussed in some detail in Section 14.10.3.

The biological applications motivate the following exact matching problem: Given a set of strings, find substrings "common" to a large number of those strings. The word "common" here means "occurring with equality". A more difficult problem is to find "similar" substrings in many given strings, where "similar" allows a small number of differences. Problems of this type will be discussed in Part III.

Formal problem statement and first method

Suppose we have K strings whose lengths sum to n.

Definition For each k between 2 and K, we define $l(k)$ to be the length of the *longest substring common to at least k of the strings.*

We want to compute a table of $K - 1$ entries, where entry k gives $l(k)$ and also points to one of the common substrings of that length. For example, consider the set of strings {*sandollar, sandlot, handler, grand, pantry*}. Then the $l(k)$ values (without pointers to the strings) are:

k	$l(k)$	one substring
2	4	sand
3	3	and
4	3	and
5	2	an

Surprisingly, the problem can be solved in linear, $O(n)$, time [236]. It really is amazing that so much information about the contents and substructure of the strings can be extracted in time proportional to the time needed just to read in the strings. The linear-time algorithm will be fully discussed in Chapter 9 after the constant-time lowest common ancestor method has been discussed.

To prepare for the $O(n)$ result, we show here how to solve the problem in $O(Kn)$ time. That time bound is also nontrivial but is achieved by a generalization of the longest common substring method for two strings. First, build a generalized suffix tree $\mathcal{T}$ for the K strings. Each leaf of the tree represents a suffix from one of the K strings and is marked with one of K unique string identifiers, 1 to K, to indicate which string the suffix is from. Each of the K strings is given a distinct termination symbol, so that identical suffixes appearing in more than one string end at distinct leaves in the generalized suffix tree. Hence, each leaf in $\mathcal{T}$ has only one string identifier.

Definition For every internal node v of $\mathcal{T}$, define $C(v)$ to be the number of *distinct* string identifiers that appear at the leaves in the subtree of v.

Once the $C(v)$ numbers are known, and the string-depth of every node is known, the desired $l(k)$ values can be easily accumulated with a linear-time traversal of the tree. That traversal builds a vector V where, for each value of k from 2 to K, $V(k)$ holds the string-depth (and location if desired) of the deepest (string-depth) node v encountered with $C(v) = k$. (When encountering a node v with $C(v) = k$, compare the string-depth of v to the current value of $V(k)$ and if v's depth is greater than $V(k)$, change $V(k)$ to the depth of v.) Essentially, $V(k)$ reports the length of the longest string that occurs *exactly* k times. Therefore, $V(k) \leq l(k)$. To find $l(k)$ simply scan V from largest to smallest index, writing into each position the maximum $V(k)$ value seen. That is, if $V(k)$ is empty or $V(k) < V(k + 1)$ then set $V(k)$ to $V(k + 1)$. The resulting vector holds the desired $l(k)$ values.

7.6.1. Computing the $C(v)$ numbers

In linear time, it is easy to compute for each internal node v the number of leaves in v's subtree. But that number may be larger than $C(v)$ since two leaves in the subtree may have the same identifier. That repetition of identifiers is what makes it hard to compute $C(v)$ in $O(n)$ time. Therefore, instead of counting the number of leaves below v, the algorithm uses $O(Kn)$ time to explicitly compute which identifiers are found below any node. For each internal node v, a K-length bit vector is created that has a 1 in bit i if there is a leaf with identifier i in the subtree of v. Then $C(v)$ is just the number of 1-bits in that vector. The vector for v is obtained by **OR**ing the vectors of the children of v. For l children, this takes lK time. Therefore over the entire tree, since there are $O(n)$ edges, the time needed to build the entire table is $O(Kn)$. We will return to this problem in Section 9.7, where an $O(n)$ time solution will be presented.

7.7. APL7: Building a smaller directed graph for exact matching

As discussed before, in many applications space is the critical constraint, and any significant reduction in space is of value. In this section we consider how to compress a suffix tree into a directed acyclic graph (DAG) that can be used to solve the exact matching problem (and others) in linear time but that uses less space than the tree. These compression techniques can also be used to build a *directed acyclic word graph* (DAWG), which is the smallest finite-state machine that can recognize suffixes of a given string. Linear-time algorithms for building DAWGs are developed in [70], [71], and [115]. Thus the method presented here to compress suffix trees can either be considered as an application of suffix trees to building DAWGs or simply as a technique to compact suffix trees.

Consider the suffix tree for a string $S = xyxaxaxa$ shown in Figure 7.1. The edge-labeled subtree below node p is *isomorphic* to the subtree below node q, except for the leaf numbers. That is, for every path from p there is a path from q with the same path-labels, and vice versa. If we only want to determine *whether* a pattern occurs in a larger text, rather than learning all the locations of the pattern occurrence(s), we could *merge* p into q by redirecting the labeled edge from p's parent to now go into q, deleting the subtree of p as shown in Figure 7.2. The resulting graph is not a tree but a directed acyclic graph.

Clearly, after merging two nodes in the suffix tree, the resulting directed graph can

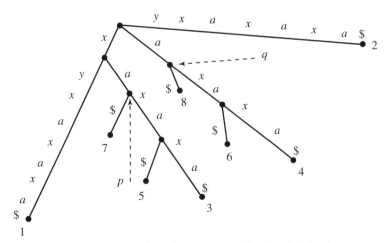

Figure 7.1: Suffix tree for string *xyxaxaxa* without suffix links shown.

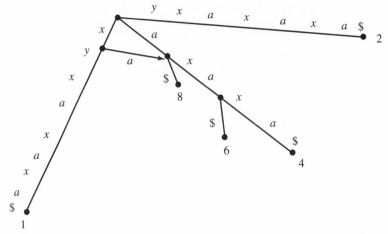

Figure 7.2: A directed acyclic graph used to recognize substrings of *xyxaxaxa*.

be used to solve the exact matching problem in the same way a suffix tree is used. The algorithm matches characters of the pattern against a unique path from the root of the graph; the pattern occurs somewhere in the text if and only if all the characters of the pattern are matched along the path. However, the leaf numbers reachable from the end of the path may no longer give the exact starting positions of the occurrences. This issue will be addressed in Exercise 10.

Since the graph is a DAG after the first merge, the algorithm must know how to merge nodes in a DAG as well as in a tree. The general merge operation for both trees and DAGs is stated in the following way:

A merge of node *p* *into* node *q* means that all edges out of *p* are removed, that the edges into *p* are directed to *q* but have their original respective edge-labels, and that any part of the graph that is now unreachable from the root is removed.

Although the merges generally occur in a DAG, the criteria used to determine which nodes to merge remain tied to the original suffix tree – node *p* can be merged into *q* if the edge-labeled subtree of *p* is isomorphic to the edge-labeled subtree of *q* in the suffix tree. Moreover, *p* can be merged into *q*, or *q* into *p*, only if the two subtrees are isomorphic. So the key algorithmic issue is how to find isomorphic subtrees in the suffix tree. There are general algorithms for subtree isomorphism but suffix trees have additional structure making isomorphism detection much simpler.

Theorem 7.7.1. *In a suffix tree T the edge-labeled subtree below a node p is isomorphic to the subtree below a node q if and only if there is a directed path of suffix links from one node to the other node, and the number of leaves in the two subtrees is equal.*

PROOF   First suppose *p* has a direct suffix link to *q* and those two nodes have the same number of leaves in their subtrees. Since there is a suffix link from *p* to *q*, node *p* has path-label $x\alpha$ while *q* has path-label α. For every leaf numbered *i* in the subtree of *p* there is a leaf numbered $i + 1$ in the subtree of *q*, since the suffix of *T* starting at *i* begins with $x\alpha$ only if the suffix of *T* starting at $i + 1$ begins with α. Therefore, for every (labeled) path from *p* to a leaf in its subtree, there is an identical path (with the same labeled edges) from *q* to a leaf in its subtree. Now the numbers of leaves in the subtrees of *p* and *q* are assumed to be equal, so *every* path out of *q* is identical to some path out of *p*, and hence the two subtrees are isomorphic.

By the same reasoning, if there is a path of suffix links from p to q going through a node v, then the number of leaves in the subtree of v must be at least as large as the number in the subtree of p and no larger than the number in the subtree of q. It follows that if p and q have the same number of leaves in their subtrees, then all the subtrees below nodes on the path have the same number of leaves, and all these subtrees are isomorphic to each other.

For the converse side, suppose that the subtrees of p and q are isomorphic. Clearly then they have the same number of leaves. We will show that there is a directed path of suffix links between p and q. Let α be the path-label of p and β be the path-label of q and assume that $|\beta| \leq |\alpha|$.

Since $\beta \neq \alpha$, if β is a suffix of α it must be a proper suffix. And, if β is a proper suffix of α, then by the properties of suffix links, there is a directed path of suffix links from p to q, and the theorem would be proved. So we will prove, by contradiction, that β must be a suffix of α.

Suppose β is not a suffix of α. Consider any occurrence of α in T and let γ be the suffix of T just to the right of that occurrence of α. That means that $\alpha\gamma$ is a suffix of T and there is a path labeled γ running from node p to a leaf in the suffix tree. Now since β is not a suffix of α, no suffix of T that starts just after an occurrence of β can have length $|\gamma|$, and therefore there is no path of length $|\gamma|$ from q to a leaf. But that implies that the subtrees rooted at p and at q are not isomorphic, which is a contradiction. $\square$

Definition   Let Q be the set of all pairs (p, q) such that a) there exists a suffix link from p to q in T, and b) p and q have the same number of leaves in their respective subtrees.

The entire procedure to compact a suffix tree can now be described.

Suffix tree compaction

begin

Identify the set Q of pairs (p, q) such that there is a suffix link from p to q and the number of leaves in their respective subtrees is equal.

While there is a pair (p, q) in Q and both p and q are in the current DAG,
Merge node p into q.

end.

The "correctness" of the resulting DAG is stated formally in the following theorem.

Theorem 7.7.2. *Let T be the suffix tree for an input string S, and let D be the DAG resulting from running the compaction algorithm on T. Any directed path in D from the root enumerates a substring of S, and every substring of S is enumerated by some such path. Therefore, the problem of determining whether a string is a substring of S can be solved in linear time using D instead of T.*

DAG D can be used to determine whether a pattern occurs in a text, but the graph seems to lose the location(s) where the pattern begins. It is possible, however, to add simple (linear-space) information to the graph so that the locations of all the occurrences can also be recovered when the graph is traversed. We address this issue in Exercise 10.

It may be surprising that, in the algorithm, pairs are merged in arbitrary order. We leave the correctness of this, a necessary part of the proof of Theorem 7.7.2, as an exercise. As a practical matter it makes sense to merge top-down, never merging two nodes that have ancestors in the suffix tree that can be merged.

DAGs versus DAWGs

DAG D created by the algorithm is not a DAWG as defined in [70], [71], and [115]. A DAWG represents a finite-state machine and, as such, each edge label is allowed to have only one character. Moreover, the main theoretical feature of the DAWG for a string S is that it is the finite-state machine with the fewest number of states (nodes) that recognizes suffixes of S. Of course, D can be converted to a finite-state machine by expanding any edge of D whose label has k characters into k edges labeled by one character each. But the resulting finite-state machine would not necessarily have the minimum number of states, and hence it would not necessarily be the DAWG for S.

Still, DAG D for string S has as few (or fewer) nodes and edges than does the associated DAWG for S, and so is as compact as the DAWG even though it may not be a finite-state machine. Therefore, construction of the DAWG for S is mostly of theoretical interest. In Exercises 16 and 17 we consider how to build the smallest finite-state machine that recognizes substrings of a string.

7.8. APL8: A reverse role for suffix trees, and major space reduction

We have previously shown how suffix trees can be used to solve the exact matching problem with $O(m)$ preprocessing time and space (building a suffix tree of size $O(m)$ for the text T) and $O(n + k)$ search time (where n is the length of the pattern and k is the number of occurrences). We have also seen how suffix trees are used to solve the exact set matching problem in the same time and space bounds (n is now the total size of all the patterns in the set). In contrast, the Knuth-Morris-Pratt (or Boyer–Moore) method preprocesses the pattern in $O(n)$ time and space, and then searches in $O(m)$ time. The Aho–Corasick method achieves similar bounds for the set matching problem.

Asymptotically, the suffix tree methods that preprocess the text are as efficient as the methods that preprocess the pattern – both run in $O(n + m)$ time and use $\Theta(n + m)$ space (they have to represent the strings). However, the practical constants on the time and space bounds for suffix trees often make their use unattractive compared to the other methods. Moreover, the situation sometimes arises that the pattern(s) will be given first and held fixed while the text varies. In those cases it is clearly superior to preprocess the pattern(s). So the question arises of whether we can solve those problems by building a suffix tree for the *pattern(s)*, not the text. This is the reverse of the normal use of suffix trees. In Sections 5.3 and 7.2.1 we mentioned that such a reverse role was possible, thereby using suffix trees to achieve exactly the same time and space bounds (preprocessing versus search time and space) as in the Knuth-Morris-Pratt or Aho–Corasick methods. To explain this, we will develop a result due to Chang and Lawler [94], who solved a somewhat more general problem, called the *matching statistics* problem.

7.8.1. Matching statistics: duplicating bounds and reducing space

Definition   Define $ms(i)$ to be the length of the longest substring of T starting at position i that matches a substring *somewhere* (but we don't know where) in P. These values are called the *matching statistics*.

For example, if $T = abcxabcdex$ and $P = wyabcwzqabcdw$ then $ms(1) = 3$ and $ms(5) = 4$.

Clearly, there is an occurrence of P starting at position i of T if and only if $ms(i) = |P|$.

Thus the problem of finding the matching statistics is a generalization of the exact matching problem.

Matching statistics lead to space reduction

Matching statistics can be used to reduce the size of the suffix tree needed in solutions to problems more complex than exact matching. This use of matching statistics will probably be more important than their use to duplicate the preprocessing/search bounds of Knuth-Morris-Pratt and Aho–Corasick. The first example of space reduction using matching statistics will be given in Section 7.9.

Matching statistics are also used in a variety of other applications described in the book. One advertisement we give here is to say that matching statistics are central to a fast approximate matching method designed for rapid database searching. This will be detailed in Section 12.3.3. Thus matching statistics provide one bridge between exact matching methods and problems of approximate string matching.

How to compute matching statistics

We want to compute $ms(i)$, for each position i in T, in $O(m)$ time using only a suffix tree for P. First, build a suffix tree $\mathcal{T}$ for P, the fixed short string, but do not remove the suffix links used during the construction of the tree. (The suffix links are either constructed by Ukkonen's algorithm or are the reverse of the link pointers in Weiner's algorithm.) This suffix tree will then be used to find $ms(i)$ for each position i in T.

The naive way to find a single $ms(i)$ value is to match, left to right, the initial characters of $T[i..m]$ against $\mathcal{T}$, by following the unique path of matches until no further matches are possible. However, repeating this for each i would not achieve the claimed linear time bound. Instead, the suffix links are used to accelerate the entire computation, similar to the way they accelerate the construction of $\mathcal{T}$ in Ukkonen's algorithm.

To learn $ms(1)$, we match characters of string T against $\mathcal{T}$, by following the unique matching path of $T[1..m]$. The length of that matching path is $ms(1)$. Now suppose in general that the algorithm has just followed a matching path to learn $ms(i)$ for $i < |m|$. That means that the algorithm has located a point b in $\mathcal{T}$ such that the path to that point exactly matches a prefix of $T[i..m]$, but no further matches are possible (possibly because a leaf has been reached).

Having learned $ms(i)$, proceed as follows to learn $ms(i+1)$. If b is an internal node v of $\mathcal{T}$ then the algorithm can follow its suffix link to a node $s(v)$. If b is not an internal node, then the algorithm can back up to the node v just above b. If v is the root, then the search for $ms(i+1)$ begins at the root. But if v is not the root, then the algorithm follows the suffix link from v to $s(v)$. The path-label of v, say $x\alpha$, is a prefix of $T[i..m]$, so α must be a prefix of $T[i+1..m]$. But $s(v)$ has path-label α, and hence the path from the root to $s(v)$ matches a prefix of $T[i+1..m]$. Therefore, the search for $ms(i+1)$ can start at node $s(v)$ rather than at the root.

Let β denote the string between node v and point b. Then $x\alpha\beta$ is the longest substring in P that matches a substring starting at position i of T. Hence $\alpha\beta$ is a string in P matching a substring starting at position $i+1$ of T. Since $s(v)$ has path-label α, there must be a path labeled β out of $s(v)$. Instead of traversing that path by examining every character on it, the algorithm uses the skip/count trick (detailed in Ukkonen's algorithm; Section 6.1.3) to traverse it in time proportional to the number of nodes on the path.

When the end of that β path is reached, the algorithm continues to match single characters from T against characters in the tree until either a leaf is reached or until no further matches are possible. In either case, $ms(i + 1)$ is the string-depth of the ending position. Note that the character comparisons done after reaching the end of the β path begin either with the same character in T that ended the search for $ms(i)$ or with the next character in T, depending on whether that search ended with a mismatch or at a leaf.

There is one special case that can arise in computing $ms(i+1)$. If $ms(i) = 1$ or $ms(i) = 0$ (so that the algorithm is at the root), and $T(i + 1)$ is not in P, then $ms(i + 1) = 0$.

7.8.2. Correctness and time analysis for matching statistics

The proof of correctness of the method is immediate since it merely simulates the naive method for finding each $ms(i)$. Now consider the time required by the algorithm. The analysis is very similar to that done for Ukkonen's algorithm.

Theorem 7.8.1. *Using only a suffix tree for P and a copy of T, all the m matching statistics can be found in $O(m)$ time.*

PROOF   The search for any $ms(i + 1)$ begins by backing up at most one edge from position b to a node v and traversing one suffix link to node $s(v)$. From $s(v)$ a β path is traversed in time proportional to the number of nodes on it, and then a certain number of additional character comparisons are done. The backup and link traversals take constant time per i and so take $O(m)$ time over the entire algorithm. To bound the total time to traverse the various β paths, recall the notion of *current node-depth* from the time analysis of Ukkonen's algorithm (page 102). There it was proved that a link traversal reduces the current depth by at most one (Lemma 6.1.2), and since each backup reduces the current depth by one, the total decrements to current depth cannot exceed $2m$. But since current depth cannot exceed m or become negative, the total increments to current depth are bounded by $3m$. Therefore, the total time used for all the β traversals is at most $3m$ since the current depth is increased at each step of any β traversal. It only remains to consider the total time used in all the character comparisons done in the "after-β" traversals. The key there is that the after-β character comparisons needed to compute $ms(i + 1)$, for $i \geq 1$, begin with the character in T that ended the computation for $ms(i)$ or with the next character in T. Hence the after-β comparisons performed when computing $ms(i)$ and $ms(i + 1)$ share at most one character in common. It follows that at most $2m$ comparisons in total are performed during all the after-β comparisons. That takes care of all the work done in finding all the matching statistics, and the theorem is proved.   □

7.8.3. A small but important extension

The number $ms(i)$ indicates the length of the longest substring starting at position i of T that matches a substring *somewhere* in P, but it does not indicate the location of any such match in P. For some applications (such as those in Section 9.1.2) we must also know, for each i, the location of at least one such matching substring. We next modify the matching statistics algorithm so that it provides that information.

Definition   For each position i in T, the number $p(i)$ specifies a starting location in P such that the substring starting at $p(i)$ matches a substring starting at position i of T for exactly $ms(i)$ places.

In order to accumulate the $p(i)$ values, first do a depth-first traversal of $\mathcal{T}$ marking each node v with the leaf number of one of the leaves in its subtree. This takes time linear in the size of $\mathcal{T}$. Then, when using $\mathcal{T}$ to find each $ms(i)$, if the search stops at a node u, the desired $p(i)$ is the suffix number written at u; otherwise (when the search stops on an edge (u, v)), $p(i)$ is the suffix number written at node v.

Back to STSs

Recall the discussion of STSs in Section 3.5.1. There it was mentioned that, because of errors, exact matching may not be an appropriate way to find STSs in new sequences. But since the number of sequencing errors is generally small, we can expect long regions of agreement between a new DNA sequence and any STS it (ideally) contains. Those regions of agreement should allow the correct identification of the STSs it contains. Using a (precomputed) generalized suffix tree for the STSs (which play the role of P), compute matching statistics for the new DNA sequence (which is T) and the set of STSs. Generally, the pointer $p(i)$ will point to the appropriate STS in the suffix tree. We leave it to the reader to flesh out the details. Note that when given a new sequence, the time for the computation is just proportional to the length of the new sequence.

7.9. APL9: Space-efficient longest common substring algorithm

In Section 7.4, we solved the problem of finding the longest common substring of S_1 and S_2 by building a generalized suffix tree for the two strings. That solution used $O(|S_1| + |S_2|)$ time and space. But because of the practical space overhead required to construct and use a suffix tree, a solution that builds a suffix tree only for the smaller of the two strings may be much more desirable, even if the worst-case space bounds remain the same. Clearly, the longest common substring has length equal to the longest matching statistic $ms(i)$. The actual substring occurs in the longer string starting at position i and in the shorter string starting at position $p(i)$. The algorithm of the previous section computes all the $ms(i)$ and $p(i)$ values using only a suffix tree for the smaller of the two strings, along with a copy of the long string. Hence, the use of matching statistics reduces the space needed to solve the longest common substring problem.

The longest common substring problem illustrates one of many space reducing applications of matching statistics to algorithms using suffix trees. Some additional applications will be mentioned in the book, but many more are possible and we will not explicitly point each one out. The reader is encouraged to examine every use of suffix trees involving more than one string, to find those places where such space reduction is possible.

7.10. APL10: All-pairs suffix-prefix matching

Here we present a more complex use of suffix trees that is interesting in its own right and that will be central in the linear-time superstring approximation algorithm to be discussed in Section 16.17.

> **Definition** Given two strings S_i and S_j, any suffix of S_i that matches a prefix of S_j is called a *suffix-prefix match* of S_i, S_j.
>
> Given a collection of strings $\mathcal{S} = S_1, S_2, \ldots, S_k$ of total length m, the **all-pairs suffix-prefix problem** is the problem of finding, for each ordered pair S_i, S_j in $\mathcal{S}$, the *longest* suffix-prefix match of S_i, S_j.

Motivation for the problem

The main motivation for the all-pairs suffix-prefix problem comes from its use in implementing fast approximation algorithms for the *shortest superstring problem* (to be discussed in Section 16.17). The superstring problem is itself motivated by sequencing and mapping problems in DNA that will be discussed in Chapter 16. Another motivation for the shortest superstring problem, and hence for the all-pairs suffix-prefix problem, arises in data compression; this connection will be discussed in the exercises for Chapter 16.

A different, direct application of the all-pairs suffix-prefix problem is suggested by computations reported in [190]. In that research, a set of around 1,400 ESTs (see Section 3.5.1) from the organism *C. elegans* (which is a worm) were analyzed for the presence of highly conserved substrings called *ancient conserved regions* (*ACRs*). One of the main objectives of the research was to estimate the number of ACRs that occur in the genes of *C. elegans*. Their approach was to extrapolate from the number of ACRs they observed in the set of ESTs. To describe the role of suffix-prefix matching in this extrapolation, we need to remember some facts about ESTs.

For the purposes here, we can think of an EST as a sequenced DNA substring of length around 300 nucleotides, originating in a gene of much greater length. If EST α originates in gene β, then the actual location of substring α in β is essentially random, and many different ESTs can be collected from the same gene β. However, in the common method used to collect ESTs, one does not learn the identity of the originating gene, and it is not easy to tell if two ESTs originate from the same gene. Moreover, ESTs are collected more frequently from some genes than others. Commonly, ESTs will more frequently be collected from genes that are more highly expressed (transcribed) than from genes that are less frequently expressed. We can thus consider ESTs as a biased sampling of the underlying gene sequences. Now we return to the extrapolation problem.

The goal is to use the ACR data observed in the ESTs to estimate the number of ACRs in the entire set of genes. A simple extrapolation would be justified if the ESTs were essentially random samples selected uniformly from the entire set of *C. elegans* genes. However, genes are not uniformly sampled, so a simple extrapolation would be wrong if the prevalence of ACRs is systematically different in ESTs from frequently or infrequently expressed genes. How can that prevalence be determined? When an EST is obtained, one doesn't know the gene it comes from, or how frequently that gene is expressed, so how can ESTs from frequently sampled genes be distinguished from the others?

The approach taken in [190] is to compute the "overlap" between each pair of ESTs. Since all the ESTs are of comparable length, the heart of that computation consists of solving the all-pairs suffix-prefix problem on the set of ESTs. An EST that has no substantial overlap with another EST was considered in the study to be from an infrequently expressed (and sampled) gene, whereas an EST that has substantial overlap with one or more of the other ESTs is considered to be from a frequently expressed gene. (Because there may be some sequencing errors, and because substring containment is possible among strings of unequal length, one should also solve the all-pairs longest common substring problem.) After categorizing the ESTs in this way, it was indeed found that ACRs occur more commonly in ESTs from frequently expressed genes (more precisely, from ESTs that overlap other ESTs). To explain this, the authors [190] conclude:

> These results suggest that moderately expressed proteins have, on average, been more highly conserved in sequence over long evolutionary periods than have rarely expressed ones and in particular are more likely to contain ACRs. This is presumably attributable in part to higher selective pressures to optimize the activities and structures of those proteins . . .

7.10.1. Solving the all-pairs suffix-prefix problem in linear time

For a single pair of strings, the preprocessing discussed in Section 2.2.4 will find the longest suffix-prefix match in time linear in the length of the two strings. However, applying the preprocessing to each of the k^2 pairs of strings separately gives a total bound of $O(km)$ time. Using suffix trees it is possible to reduce the computation time to $O(m+k^2)$, assuming (as usual) that the alphabet is fixed.

> **Definition** We call an edge a *terminal edge* if it is labeled only with a string termination symbol. Clearly, every terminal edge has a leaf at one end, but not all edges touching leaves are terminal edges.

The main data structure used to solve the all-pairs suffix-prefix problem is the generalized suffix tree $\mathcal{T}(\mathcal{S})$ for the k strings in set $\mathcal{S}$. As $\mathcal{T}(\mathcal{S})$ is constructed, the algorithm also builds a list $L(v)$ for each internal node v. List $L(v)$ contains the index i if and only if v is incident with a terminal edge whose leaf is labeled by a suffix of string S_i. That is, $L(v)$ holds index i if and only if the path label to v is a complete suffix of string S_i. For example, consider the generalized suffix tree shown in Figure 6.11 (page 117). The node with path-label ba has an L list consisting of the single index 2, the node with path-label a has a list consisting of indices 1 and 2, and the node with path-label xa has a list consisting of index 1. All the other lists in this example are empty. Clearly, the lists can be constructed in linear time during (or after) the construction of $\mathcal{T}(\mathcal{S})$.

Now consider a fixed string S_j, and focus on the path from the root of $\mathcal{T}(\mathcal{S})$ to the leaf j representing the entire string S_j. The key observation is the following: If v is a node on this path and i is in $L(v)$, then the path-label of v is a suffix of S_i that matches a prefix of S_j. So for each index i, the deepest node v on the path to leaf j such that $i \in L(v)$ identifies the longest match between a suffix of S_i and a prefix of S_j. The path-label of v is the longest suffix-prefix match of (S_i, S_j). It is easy to see that by one traversal from the root to leaf j we can find the deepest nodes for all $1 \leq i \leq k$ ($i \neq j$).

Following the above observation, the algorithm efficiently collects the needed suffix-prefix matches by traversing $\mathcal{T}(\mathcal{S})$ in a depth-first manner. As it does, it maintains k stacks, one for each string. During the depth-first traversal, when a node v is reached in a forward edge traversal, push v onto the ith stack, for each $i \in L(v)$. When a leaf j (representing the entire string S_j) is reached, scan the k stacks and record for each index i the current top of the ith stack. It is not difficult to see that the top of stack i contains the node v that defines the suffix-prefix match of (S_i, S_j). If the ith stack is empty, then there is no overlap between a suffix of string S_i and a prefix of string S_j. When the depth-first traversal backs up past a node v, we pop the top of any stack whose index is in $L(v)$.

Theorem 7.10.1. *All the k^2 longest suffix-prefix matches are found in $O(m + k^2)$ time by the algorithm. Since m is the size of the input and k^2 is the size of the output, the algorithm is time optimal.*

PROOF The total number of indices in all the lists $L(v)$ is $O(m)$. The number of edges in $\mathcal{T}(\mathcal{S})$ is also $O(m)$. Each push or pop of a stack is associated with a leaf of $\mathcal{T}(\mathcal{S})$, and each leaf is associated with at most one pop and one push; hence traversing $\mathcal{T}(\mathcal{S})$ and updating the stacks takes $O(m)$ time. Recording of each of the $O(k^2)$ answers is done in constant time per answer. □

Extensions

We note two extensions. Let $k' \leq k^2$ be the number of ordered pairs of strings that have a nonzero length suffix-prefix match. By using double links, we can maintain a linked list of the *nonempty* stacks. Then when a leaf of the tree is reached during the traversal, only the stacks on this list need be examined. In that way, all nonzero length suffix-prefix matches can be found in $O(m + k')$ time. Note that the position of the stacks in the linked list will vary, since a stack that goes from empty to nonempty must be linked at one of the ends of the list; hence we must also keep (in the stack) the name of the string associated with that stack.

At the other extreme, suppose we want to collect for every pair not just the longest suffix-prefix match, but all suffix-prefix matches no matter how long they are. We modify the above solution so that when the tops of the stacks are scanned, the entire contents of each scanned stack is read out. If the output size is k^*, then the complexity for this solution is $O(m + k^*)$.

7.11. Introduction to repetitive structures in molecular strings

Several sections of this book (Sections 7.12, 7.12.1, 9.2, 9.2.2, 9.5, 9.6, 9.6.1, 9.7, and 7.6), as well as several exercises, are devoted to discussing efficient algorithms for finding various types of *repetitive* structures in strings. (In fact, some aspects of one type of repetitive structure, *tandem repeats*, have already been discussed in the exercises of Chapter 1, and more will be discussed later in the book.) The motivation for the general topic of repetitive structures in strings comes from several sources, but our principal interest is in important repetitive structures seen in biological strings (DNA, RNA, and protein). To make this concrete, we briefly introduce some of those repetitive structures. The intent is not to write a dissertation on repetitive DNA or protein, but to motivate the algorithmic techniques we develop.

7.11.1. Repetitive structures in biological strings

One of the most striking features of DNA (and to a lesser degree, protein) is the extent to which repeated substrings occur in the genome. This is particularly true of eukaryotes (higher-order organisms whose DNA is enclosed in a cell nucleus). For example, most of the human Y chromosome consists of repeated substrings, and overall

> Families of reiterated sequences account for about one third of the human genome. [317]

There is a vast[1] literature on repetitive structures in DNA, and even in protein,

> ... reports of various kinds of repeats are too common even to list. [128]

In an analysis of 3.6 million bases of DNA from *C. elegans*, over 7,000 families of repetitive sequences were identified [5]. In contrast, prokaryotes (organisms such as bacteria whose DNA is not enclosed in a nucleus) have in total little repetitive DNA, although they still possess certain highly structured small-scale repeats.

In addition to its sheer quantity, repetitive DNA is striking for the variety of repeated structures it contains, for the various proposed mechanisms explaining the origin and maintenance of repeats, and for the biological functions that some of the repeats may play (see [394] for one aspect of gene duplication). In many texts (for example, [317], [469], and [315]) on genetics or molecular biology one can find extensive discussions of repetitive strings and their hypothesized functional and evolutionary role. For an introduction to repetitive elements in human DNA, see [253] and [255].

---

[1] It is reported in [192] that a search of the database MEDLINE using the key (*repeat OR repetitive*) AND (*protein OR sequence*) turned up over 6,000 papers published in the preceding twenty years.

$$5' \text{ TCGACCGGTCGA } 3'$$
$$3' \text{ ∀ƆƆ⊥ƆפפƆƆ∀Ɔ⊥ } ,\text{S}$$

Figure 7.3: A palindrome in the vernacular of molecular biology. The double-stranded string is the same after reflection around both the horizontal and vertical midpoints. Each strand is a *complemented palindrome* according to the definitions used in this book.

In the following discussion of repetitive structures in DNA and protein, we divide the structures into three types: local, small-scale repeated strings whose function or origin is at least partially understood; simple repeats, both local and interspersed, whose function is less clear; and more complex interspersed repeated strings whose function is even more in doubt.

Definition   A *palindrome* is a string that reads the same backwards as forwards.

For emphasis, the Random House dictionary definition of "palindrome" is: a word, sentence or verse reading the same backwards as forwards [441]. For example, the string *xyaayx* is a palindrome under this definition. Ignoring spaces, the sentence *was it a cat i saw* is another example.

Definition   A *complemented palindrome* is a DNA or RNA string that becomes a palindrome if each character in one half of the string is changed to its complement character (in DNA, $A - T$ are complements and $C - G$ are complements; in RNA $A - U$ and $C - G$ are complements). For example, *AGCTCGCGAGCT* is a complemented palindrome.[2]

Small-scale local repeats whose function or origin is partially understood include: *complemented palindromes* in both DNA and RNA, which act to regulate DNA transcription (the two parts of the complemented palindrome fold and pair to form a "hairpin loop"); *nested complemented palindromes* in tRNA (transfer RNA) that allow the molecule to fold up into a cloverleaf structure by complementary base pairing; tandem arrays of repeated RNA that flank retroviruses (viruses whose primary genetic material is RNA) and facilitate the incorporation of viral DNA (produced from the RNA sequence by reverse transcription) into the host's DNA; single copy inverted repeats that flank transposable (movable) DNA in various organisms and that facilitate that movement or the inversion of the DNA orientation; short repeated substrings (both palindromic and nonpalindromic) in DNA that may help the chromosome fold into a more compact structure; repeated substrings at the ends of viral DNA (in a linear state) that allow the concatenation of many copies of the viral DNA (a molecule of this type is called a *concatamer*); copies of genes that code for important RNAs (rRNAs and tRNAs) that must be produced in large number; clustered genes that code for important proteins (such as histone) that regulate chromosome structure and must be made in large number; families of genes that code for similar proteins (hemoglobins and myoglobins for example); similar genes that probably arose through duplication and subsequent mutation (including *pseudogenes* that have mutated

---

[2] The use of the word "palindrome" in molecular biology does not conform to the normal English dictionary definition of the word. The easiest translation of the molecular biologist's "palindrome" to normal English is: "complemented palindrome". A more molecular view is that a palindrome is a segment of double-stranded DNA or RNA such that both strands read the same when both are read in the same direction, say in the $5'$ to $3'$ direction. Alternately, a palindrome is a segment of double-stranded DNA that is symmetric (with respect to reflection) around both the horizontal axis and the midpoint of the segment. (See Figure 7.3). Since the two strands are complementary, each strand defines a complemented palindrome in the sense defined above. The term "mirror repeat" is sometimes used in the molecular biology literature to refer to a "palindrome" as defined by the dictionary.

to the point that they no longer function); common exons of eukaryotic DNA that may be basic building blocks of many genes; and common functional or structural subunits in protein (motifs and domains).

Restriction enzyme cutting sites illustrate another type of small-scale, structured, repeating substring of great importance to molecular biology. A restriction enzyme is an enzyme that recognizes a specific substring in the DNA of both prokaryotes and eukaryotes and cuts (or cleaves) the DNA every place where that pattern occurs (exactly where it cuts inside the pattern varies with the pattern). There are hundreds of known restriction enzymes and their use has been absolutely critical in almost all aspects of modern molecular biology and recombinant DNA technology. For example, the surprising discovery that eukaryotic DNA contains *introns* (DNA substrings that interrupt the DNA of protein coding regions), for which Nobel prizes were awarded in 1993, was closely coupled with the discovery and use of restriction enzymes in the late 1970s.

Restriction enzyme cutting sites are interesting examples of repeats because they tend to be *complemented palindromic substrings*. For example, the restriction enzyme *EcoRI* recognizes the complemented palindrome *GAATTC* and cuts between the *G* and the adjoining *A* (the substring *TTC* when reversed and complemented is *GAA*). Other restriction enzymes recognize *separated* (or *interrupted*) complemented palindromes. For example, restriction enzyme *BglI* recognizes *GCCNNNNNGGC*, where *N* stands for any nucleotide. The enzyme cuts between the last two *N*s. The complemented palindromic structure has been postulated to allow the two halves of the complemented palindrome (separated or not) to fold and form complementary pairs. This folding then apparently facilitates either recognition or cutting by the enzyme. Because of the palindromic structure of restriction enzyme cutting sites, people have scanned DNA databases looking for common repeats of this form in order to find additional candidates for unknown restriction enzyme cutting sites.

Simple repeats that are less well understood often arise as *tandem arrays* (consecutive repeated strings, also called "direct repeats") of repeated DNA. For example, the string *TTAGGG* appears at the ends of every human chromosome in arrays containing one to two thousand copies [332]. Some tandem arrays may originate and continue to grow by a postulated mechanism of *unequal crossing over* in meiosis, although there is serious opposition to that theory. With unequal crossing over in meiosis, the likelihood that more copies will be added in a single meiosis increases as the number of existing copies increases. A number of genetic diseases (Fragile X syndrome, Huntington's disease, Kennedy's disease, myotonic dystrophy, ataxia) are now understood to be caused by increasing numbers of tandem DNA repeats of a string three bases long. These triplet repeats somehow interfere with the proper production of particular proteins. Moreover, the number of triples in the repeat increases with successive generations, which appears to explain why the disease increases in severity with each generation. Other long tandem arrays consisting of short strings are very common and are widely distributed in the genomes of mammals. These repeats are called *satellite DNA* (further subdivided into micro and mini-satellite DNA), and their existence has been heavily exploited in genetic mapping and forensics. Highly dispersed tandem arrays of length-two strings are common. In addition to tri-nucleotide repeats, other mini-satellite repeats also play a role in human genetic diseases [286].

Repetitive DNA that is *interspersed* throughout mammalian genomes, and whose function and origin is less clear, is generally divided into SINEs (short interspersed nuclear sequences) and LINEs (long interspersed nuclear sequences). The classic example of a SINE is the *Alu* family. The Alu repeats occur about 300,000 times in the human genome and account for as much as 5% of the DNA of human and other mammalian genomes. Alu repeats are substrings of length around 300 nucleotides and occur as nearly (but not exactly) identical copies widely dispersed throughout the genome. Moreover, the interior of an Alu string itself consists of repeated substrings of length around 40, and the Alu sequence is often flanked on either side by tandem repeats of length 7–10. Those right and left flanking sequences are usually complemented palindromic copies of each other. So the Alu repeats wonderfully illustrate various kinds of phenomena that occur in repetitive DNA. For an introduction to Alu repeats see [254].

One of the most fascinating discoveries in molecular genetics is a phenomenon called *genomic* (or *gametic*) *imprinting*, whereby a particular allele of a gene is expressed only when it is inherited from one specific parent [48, 227, 391]. Sometimes the required parent is the mother and sometimes the father. The allele will be unexpressed, or expressed differently, if inherited from the "incorrect" parent. This is in contradiction to the classic Mendelian rule of *equivalence* – that chromosomes (other than the Y chromosome) have no memory of the parent they originated from, and that the same allele inherited from either parent will have the same effect on the child. In mice and humans, sixteen imprinted gene alleles have been found to date [48]. Five of these require inheritance from the mother, and the rest from the father. The DNA sequences of these sixteen imprinted genes all share the common feature that

> They contain, or are closely associated with, a region rich in direct repeats. These repeats range in size from 25 to 120 bp,[3] are unique to the respective imprinted regions, but have no obvious homology to each other or to highly repetitive mammalian sequences. The direct repeats may be an important feature of gametic imprinting, as they have been found in all imprinted genes analyzed to date, and are also evolutionarily conserved. [48]

Thus, direct repeats seem to be important in genetic imprinting, but like many other examples of repetitive DNA, the function and origin of these repeats remains a mystery.

7.11.2. Uses of repetitive structures in molecular biology

At one point, most interspersed repeated DNA was considered as a nuisance, perhaps of no functional or experimental value. But today a variety of techniques actually exploit the existence of repetitive DNA. Genetic mapping, mentioned earlier, requires the identification of features (or *markers*) in the DNA that are highly variable between individuals and that are interspersed frequently throughout the genome. Tandem repeats are just such markers. What varies between individuals is the *number* of times the substring repeats in an array. Hence the term used for this type of marker is *variable number of tandem repeats* (*VNTR*). *VNTRs* occur frequently and regularly in many genomes, including the human genome, and provide many of the markers needed for large-scale genetic mapping. These *VNTR* markers are used during the genetic-level (as opposed to the physical-level) search for specific defective genes and in forensic DNA fingerprinting (since the number of repeats is highly variable between individuals, a small set of *VNTRs* can uniquely characterize individuals in a population). Tandem repeats consisting of a very short substring, often only two characters long, are called *microsatellites* and have become the preferred marker in many genetic mapping efforts.

---

[3] A detail not contained in this quote is that the direct (tandem) repeats in the genes studied [48] have a total length of about 1,500 bases.

The existence of highly repetitive DNA, such as Alus, makes certain kinds of large-scale DNA sequencing more difficult (see Sections 16.11 and 16.16), but their existence can also facilitate certain cloning, mapping, and searching efforts. For example, one general approach to low-resolution physical mapping (finding on a true physical scale where features of interest are located in the genome) or to finding genes causing diseases involves inserting pieces of human DNA that may contain a feature of interest into the hamster genome. This technique is called *somatic cell hybridization*. Each resulting hybrid-hamster cell incorporates different parts of the human DNA, and these hybrid cells can be tested to identify a specific cell containing the human feature of interest. In this cell, one then has to identify the parts of the hamster's hybrid genome that are human. But what is a distinguishing feature between human and hamster DNA?

One approach exploits the Alu sequences. Alu sequences specific to human DNA are so common in the human genome that most fragments of human DNA longer than 20,000 bases will contain an Alu sequence [317]. Therefore, the fragments of human DNA in the hybrid can be identified by probing the hybrid for fragments of Alu. The same idea is used to isolate human *oncogenes* (modified growth-promoting genes that facilitate certain cancers) from human tumors. Fragments of human DNA from the tumor are first transferred to mouse cells. Cells that receive the fragment of human DNA containing the oncogene become transformed and replicate faster than cells that do not. This isolates the human DNA fragment containing the oncogene from the other human fragments, but then the human DNA has to be separated from the mouse DNA. The proximity of the oncogene to an Alu sequence is again used to identify the human part of the hybrid genome [471]. A related technique, again using proximity to Alu sequences, is described in [403].

Algorithmic problems on repeated structures

We consider specific problems concerning repeated structures in strings in several sections of the book.[4] Admittedly, not every repetitive string problem that we will discuss is perfectly motivated by a biological problem or phenomenon known today. A recurring objection is that the first repetitive string problems we consider concern *exact* repeats (although with complementation and inversion allowed), whereas most cases of repetitive DNA involve *nearly* identical copies. Some techniques for handling inexact palindromes (complemented or not) and inexact repeats will be considered in Sections 9.5 and 9.6. Techniques that handle more liberal errors will be considered later in the book. Another objection is that simple techniques suffice for small-length repeats. For example, if one seeks repeating DNA of length ten, it makes sense to first build a table of all the 4^{10} possible strings and then scan the target DNA with a length-ten template, hashing substring locations into the precomputed table.

Despite these objections, the fit of the computational problems we will discuss to biological phenomena is good enough to motivate sophisticated techniques for handling exact or nearly exact repetitions. Those techniques pass the "plausibility" test in that they, or the ideas that underlie them, may be of future use in computational biology. In this light, we now consider problems concerning exactly repeated substrings in a single string.

---

[4] In a sense, the longest common substring problem and the k-common substring problem (Sections 7.6 and 9.7) also concern repetitive substrings. However, the repeats in those problems occur across distinct strings, rather than inside the same string. That distinction is critical, both in the definition of the problems and for the techniques used to solve them.

7.12. APL11: Finding all maximal repetitive structures in linear time

Before developing algorithms for finding repetitive structures, we must carefully define those structures. A poor definition may lead to an avalanche of output. For example, if a string consists of n copies of the same character, an algorithm searching for all pairs of identical substrings (an initially reasonable definition of a repetitive structure) would output $\Theta(n^4)$ pairs, an undesirable result. Other poor definitions may not capture the structures of interest, or they may make reasoning about those structures difficult. Poor definitions are particularly confusing when dealing with the set of all repeats of a particular type. Accordingly, the key problem is to define repetitive structures in a way that does not generate overwhelming output and yet captures all the meaningful phenomena in a clear way. In this section, we address the issue through various notions of *maximality*. Other ways of defining and studying repetitive structures are addressed in Exercises 56, 57, and 58 in this chapter; in exercises in other chapters; and in Sections 9.5, 9.6, and 9.6.1.

> **Definition** A *maximal pair* (or a maximal repeated pair) in a string S is a pair of identical substrings α and β in S such that the character to the immediate left (right) of α is different from the character to the immediate left (right) of β. That is, extending α and β in either direction would destroy the equality of the two strings.

> **Definition** A maximal pair is represented by the triple (p_1, p_2, n'), where p_1 and p_2 give the starting positions of the two substrings and n' gives their length. For a string S, we define $\mathcal{R}(S)$ to be the set of all triples describing maximal pairs in S.

For example, consider the string $S = xabcyiizabcqabcyrxar$, where there are three occurrences of the substring abc. The first and second occurrences of abc form a maximal pair $(2, 10, 3)$, and the second and third occurrences also form a maximal pair $(10, 14, 3)$, whereas the first and third occurrences of abc do not form a maximal pair. The two occurrences of the string $abcy$ also form a maximal pair $(2, 14, 4)$. Note that the definition allows the two substrings in a maximal pair to overlap each other. For example, $cxxaxxaxxb$ contains a maximal pair whose substring is $xxaxx$.

Generally, we also want to permit a prefix or a suffix of S to be part of a maximal pair. For example, two occurrences of xa in $xabcyiizabcqabcyrxar$ should be considered as a maximal pair. To model this case, simply add a character to the start of S and one to the end of S that appear nowhere else in S. From this point on, we will assume that has been done.

It may sometimes be of interest to explicitly find and output the full set $\mathcal{R}(S)$. However, in some situations $\mathcal{R}(S)$ may be too large to be of use, and a more restricted reflection of the maximal pairs may be sufficient or even preferred.

> **Definition** Define a *maximal repeat* α as a *substring* of S that occurs in a maximal pair in S. That is, α is a *maximal repeat* in S if there is a triple $(p_1, p_2, |\alpha|) \in \mathcal{R}(S)$ and α occurs in S starting at position p_1 and p_2. Let $\mathcal{R}'(S)$ denote the set of maximal repeats in S.

For example, with S as above, both strings abc and $abcy$ are maximal repeats. Note that no matter how many times a string participates in a maximal pair in S, it is represented only once in $\mathcal{R}'(S)$. Hence $|\mathcal{R}'(S)|$ is less than or equal to $|\mathcal{R}(S)|$ and is generally much smaller. The output is more modest, and yet it gives a good reflection of the maximal pairs.

In some applications, the definition of a maximal repeat does not properly model the desired notion of a repetitive structure. For example, in $S = a\alpha bx\alpha ya\alpha b$, substring α is a maximal repeat but so is $a\alpha b$, which is a *superstring* of string α, although not every occurrence of α is contained in that superstring. It may not always be desirable to report α as a repetitive structure, since the larger substring $a\alpha b$ that sometimes contains α may be more informative.

Definition A *supermaximal repeat* is a maximal repeat that never occurs as a substring of any other maximal repeat.

Maximal pairs, maximal repeats, and supermaximal repeats are only three possible ways to define exact repetitive structures of interest. Other models of exact repeats are given in the exercises. Problems related to palindromes and tandem repeats are considered in several sections throughout the book. Inexact repeats will be considered in Sections 9.5 and 9.6.1. Certain kinds of repeats are elegantly represented in graphical form in a device called a *landscape* [104]. An efficient program to construct the landscape, based essentially on suffix trees, is also described in that paper. In the next sections we detail how to efficiently find all maximal pairs, maximal repeats, and supermaximal repeats.

7.12.1. A linear-time algorithm to find all maximal repeats

The simplest problem is that of finding all maximal repeats. Using a suffix tree, it is possible to find them in $O(n)$ time for a string of length n. Moreover, there is a *compact* representation of all the maximal repeats, and it can also be constructed in $O(n)$ time, even though the total length of all the maximal repeats may be $\Omega(n^2)$. The following lemma states a necessary condition for a substring to be a maximal repeat.

Lemma 7.12.1. *Let T be the suffix tree for string S. If a string α is a maximal repeat in S then α is the path-label of a node v in T.*

PROOF If α is a maximal repeat then there must be at least two copies of α in S where the character to the right of the first copy differs from the character to the right of the second copy. Hence α is the path-label of a node v in T. □

The key point in Lemma 7.12.1 is that path α must end at a node of T. This leads immediately to the following surprising fact:

Theorem 7.12.1. *There can be at most n maximal repeats in any string of length n.*

PROOF Since T has n leaves, and each internal node other than the root must have at least two children, T can have at most n internal nodes. Lemma 7.12.1 then implies the theorem. □

Theorem 7.12.1 would be a trivial fact if at most one substring starting at any position i could be part of a maximal pair. But that is not true. For example, in the string $S = xabcyiiizabcqabcyr$ considered earlier, both copies of substring $abcy$ participate in maximal pairs, while each copy of abc also participates in maximal pairs.

So now we know that to find maximal repeats we only need to consider strings that end at nodes in the suffix tree T. But which specific nodes correspond to maximal repeats?

Definition For each position i in string S, character $S(i-1)$ is called the *left character* of i. The *left character of a leaf* of T is the left character of the suffix position represented by that leaf.

Definition A node v of T is called *left diverse* if at least two leaves in v's subtree have different left characters. By definition, a leaf cannot be left diverse.

Note that being left diverse is a property that propagates upward. If a node v is left diverse, so are all of its ancestors in the tree.

Theorem 7.12.2. *The string α labeling the path to a node v of $\mathcal{T}$ is a maximal repeat if and only if v is left diverse.*

PROOF Suppose first that v is left diverse. That means there are substrings $x\alpha$ and $y\alpha$ in S, where x and y represent different characters. Let the first substring be followed by character p. If the second substring is followed by any character but p, then α is a maximal repeat and the theorem is proved. So suppose that the two occurrences are $x\alpha p$ and $y\alpha p$. But since v is a (branching) node there must also be a substring αq in S for some character q that is different from p. If this occurrence of αq is preceded by character x then it participates in a maximal pair with string $y\alpha p$, and if it is preceded by y then it participates in a maximal pair with $x\alpha p$. Either way, α cannot be preceded by both x and y, so α must be part of a maximal pair and hence α must be a maximal repeat.

Conversely, if α is a maximal repeat then it participates in a maximal pair and there must be occurrences of α that have distinct left characters. Hence v must be left diverse. $\square$

The maximal repeats can be compactly represented

Since the property of being left diverse propagates upward in $\mathcal{T}$, Theorem 7.12.2 implies that the maximal repeats of S are represented by some initial portion of the suffix tree for S. In detail, a node is called a "frontier" node in $\mathcal{T}$ if it is left diverse but none of its children are left diverse. The subtree of $\mathcal{T}$ from the root down to the frontier nodes precisely represents the maximal repeats in that every path from the root to a node at or above the frontier defines a maximal repeat. Conversely, every maximal repeat is defined by one such path. This subtree, whose leaves are the frontier nodes in $\mathcal{T}$, is a compact representation[5] of the set of all maximal repeats of S. Note that the total length of the maximal repeats could be as large as $\Theta(n^2)$, but since the representation is a subtree of $\mathcal{T}$ it has $O(n)$ total size (including the symbols used to represent edge labels). So if the left diverse nodes can be found in $O(n)$ time, then a tree representation for the set of maximal repeats can be constructed in $O(n)$ time, even though the total length of those maximal repeats could be $\Omega(n^2)$. We now describe an algorithm to find the left diverse nodes in $\mathcal{T}$.

Finding left diverse nodes in linear time

For each node v of $\mathcal{T}$, the algorithm either records that v is left diverse or it records the character, denoted x, that is the left character of every leaf in v's subtree. The algorithm starts by recording the left character of each leaf of the suffix tree $\mathcal{T}$ for S. Then it processes the nodes in $\mathcal{T}$ bottom up. To process a node v, it examines the children of v. If any child of v has been identified as being left diverse, then it records that v is left diverse. If none of v's children are left diverse, then it examines the characters recorded at v's children. If these recorded characters are all equal, say x, then it records character x at node v. However, if they are not all x, then it records that v is left diverse. The time to check if all children of v have the same recorded character is proportional to the number of v's children. Hence the total time for the algorithm is $O(n)$. To form the final representation of the set of maximal repeats, simply delete all nodes from $\mathcal{T}$ that are not left diverse. In summary, we have

---

[5] This kind of tree is sometimes referred to as a *compact trie*, but we will not use that terminology.

Theorem 7.12.3. *All the maximal repeats in S can be found in $O(n)$ time, and a tree representation for them can be constructed from suffix tree T in $O(n)$ time as well.*

7.12.2. Finding supermaximal repeats in linear time

Recall that a supermaximal repeat is a maximal repeat that is not a substring of any other maximal repeat. We establish here efficient criteria to find all the supermaximal repeats in a string S. To do this, we solve the more general problem of finding *near-supermaximal repeats*.

> **Definition**   A substring α of S is a *near-supermaximal repeat* if α is a maximal repeat in S that occurs *at least once* in a location where it is not contained in another maximal repeat. Such an occurrence of α is said to *witness* the near-supermaximality of α.

For example, in the string $a\alpha bx\alpha ya\alpha bx\alpha b$, substring α is a maximal repeat but not a supermaximal or a near-supermaximal repeat, whereas in $a\alpha bx\alpha ya\alpha b$, substring α is again not supermaximal, but it is near-supermaximal. The second occurrence of α witnesses that fact.

With this terminology, a supermaximal repeat α is a maximal repeat in which every occurrence of α is a witness to its near-supermaximality. Note that it is not true that the set of near-supermaximal repeats is the set of maximal repeats that are not supermaximal repeats.

The suffix tree T for S will be used to locate the near-supermaximal and the supermaximal repeats. Let v be a node corresponding to a maximal repeat α, and let w (possibly a leaf) be one of v's children. The leaves in the subtree of T rooted at w identify the locations of some (but not all) of the occurrences of substring α in S. Let $L(w)$ denote those occurrences. Do any of those occurrences of α witness the near-supermaximality of α?

Lemma 7.12.2. *If node w is an internal node in T, then none of the occurrences of α specified by $L(w)$ witness the near-supermaximality of α.*

PROOF   Let γ be the substring labeling edge (v, w). Every index in $L(w)$ specifies an occurrence of $\alpha\gamma$. But w is internal, so $|L(w)| > 1$ and $\alpha\gamma$ is the prefix of a maximal repeat. Therefore, all the occurrences of α specified by $L(w)$ are contained in a maximal repeat that begins $\alpha\gamma$, and w cannot witness the near-supermaximality of α.   □

Thus no occurrence of α in $L(w)$ can witness the near-supermaximality of α unless w is a leaf. If w is a leaf, then w specifies a single particular occurrence of substring $\beta = \alpha\gamma$. We now consider that case.

Lemma 7.12.3. *Suppose w is a leaf, and let i be the (single) occurrence of β represented by leaf w. Let x be the left character of leaf w. Then the occurrence of α at position i witnesses the near-supermaximality of α if and only if x is the left character of no other leaf below v.*

PROOF   If there is another occurrence of α with a preceding character x, then $x\alpha$ occurs twice and so is either a maximal repeat or is contained in one. In that case, the occurrence of α at i is contained in a maximal repeat.

If there is no other occurrence of α with a preceding x, then $x\alpha$ occurs only once in S. Now let y be the first character on the edge from v to w. Since w is a leaf, αy occurs only once in S. Therefore, the occurrence of α starting at i, which is preceded by x and succeeded by y, is not contained in a maximal repeat, and so witnesses the near-supermaximality of α. $\square$

In summary, we can state

Theorem 7.12.4. *A left diverse internal node v represents a near-supermaximal repeat α if and only if one of v's children is a leaf (specifying position i, say) and its left character, $S(i - 1)$, is the left character of no other leaf below v. A left diverse internal node v represents a supermaximal repeat α if and only if all of v's children are leaves, and each has a distinct left character.*

Therefore, all supermaximal and near-supermaximal repeats can be identified in linear time. Moreover, we can define the *degree* of near-supermaximality of α as the fraction of occurrences of α that witness its near-supermaximality. That degree of each near-supermaximal repeat can also be computed in linear time.

7.12.3. Finding all the maximal pairs in linear time

We now turn to the question of finding all the maximal pairs. Since there can be more than $O(n)$ of them, the running time of the algorithm will be stated in terms of the size of the output. The algorithm is an extension of the method given earlier to find all maximal repeats.

First, build a suffix tree for S. For each leaf specifying a suffix i, record its left character $S(i - 1)$. Now traverse the tree from bottom up, visiting each node in the tree. In detail, work from the leaves upward, visiting a node v only after visiting every child of v. During the visit to v, create at most σ linked lists at each node, where σ is the size of the alphabet. Each list is indexed by a left character x. The list at v indexed by x contains all the starting positions of substrings in S that match the string on the path to v and that have the left character x. That is, the list at v indexed by x is just the list of leaf numbers below v that specify suffixes in S that are immediately preceded by character x.

Letting n denote the length of S, it is easy to create (but not keep) these lists in $O(n)$ total time, working bottom up in the tree. To create the list for character x at node v, link together (but do not copy) the lists for character x that exist for each of v's children. Because the size of the alphabet is finite, the time for all linking is constant at each node. Linking without copying is required in order to achieve the $O(n)$ time bound. Linking a list created at a node v' to some other list destroys the list for v'. Fortunately, the lists created at v' will not be needed after the lists for its parent are created.

Now we show in detail how to use the lists available at v's children to find all maximal pairs containing the string that labels the path to v. At the start of the visit to node v, before v's lists have been created, the algorithm can output all maximal pairs (p_1, p_2, α), where α is the string labeling the path to v. For each character x and each child v' of v, the algorithm forms the *Cartesian* product of the list for x at v' with the union of every list for a character other than x at a child of v other than v'. Any pair in this list gives the starting positions of a maximal pair for string α. The proof of this is essentially the same as the proof of Theorem 7.12.2.

If there are k maximal pairs, then the method works in $O(n + k)$ time. The creation of the suffix tree, its bottom up traversal, and all the list linking take $O(n)$ time. Each operation used in a Cartesian product produces a maximal pair not produced anywhere else, so $O(k)$ time is used in those operations. If we only want to count the number of maximal pairs, then the algorithm can be modified to run in $O(n)$ time. If only maximal pairs of a certain minimum length are requested (this would be the typical case in many applications), then the algorithm can be modified to run in $O(n + k_m)$ time, where k_m is the number of maximal pairs of length at least the required minimum. Simply stop the bottom-up traversal at any node whose string-depth falls below that minimum.

In summary, we have the following theorem:

Theorem 7.12.5. *All the maximal pairs can be found in $O(n + k)$ time, where k is the number of maximal pairs. If there are only k_m maximal pairs of length above a given threshold, then all those can be found in $O(n + k_m)$ time.*

7.13. APL12: Circular string linearization

Recall the definition of a circular string S given in Exercise 2 of Chapter 1 (page 11). The characters of S are initially numbered sequentially from 1 to n starting at an arbitrary point in S.

Definition   Given an ordering of the characters in the alphabet, a string S_1 is *lexically* (*or lexicographically*) smaller than a string S_2 if S_1 would appear before S_2 in a normal dictionary ordering of the two strings. That is, starting from the left end of the two strings, if i is the first position where the two strings differ, then S_1 is lexically less than S_2 if and only if $S_1(i)$ precedes $S_2(i)$ in the ordering of the alphabet used in those strings.

To handle the case that S_1 is a proper prefix of S_2 (and should be considered lexically less than S_2), we follow the convention that a space is taken to be the first character of the alphabet.

The **circular string linearization problem** for a circular string S of n characters is the following: Choose a place to cut S so that the resulting linear string is the lexically smallest of all the n possible linear strings created by cutting S.

This problem arises in chemical data bases for circular molecules. Each such molecule is represented by a circular string of chemical characters; to allow faster lookup and comparisons of molecules, one wants to store each circular string by a *canonical* linear string. A single circular molecule may itself be a part of a more complex molecule, so this problem arises in the "inner loop" of more complex chemical retrieval and comparison problems.

A natural choice for canonical linear string is the one that is lexically least. With suffix trees, that string can be found in $O(n)$ time.

7.13.1. Solution via suffix trees

Arbitrarily cut the circular string S, giving a linear string L. Then, double L, creating the string LL, and build the suffix tree $\mathcal{T}$ for LL. As usual, affix the terminal symbol $\$$ at the end of LL, but interpret it to be lexically *greater* than any character in the alphabet used for S. (Intuitively, the purpose of doubling L is to allow efficient consideration of strings that begin with a suffix of L and end with a prefix of L.) Next, traverse tree $\mathcal{T}$ with the rule that, at every node, the traversal follows the edge whose first character is lexically smallest over all first characters on edges out of the node. This traversal continues until the traversed path has string-depth n. Such a depth will always be reached (with the proof left to the reader). Any leaf l in the subtree at that point can be used to cut the string. If $1 < l \leq n$, then cutting S between characters $l - 1$ and l creates a lexically smallest linearization of the circular string. If $l = 0$ or $l = n + 1$, then cut the circular string between character n and character 1. Each leaf in the subtree of this point gives a cutting point yielding the same linear string.

The correctness of this solution is easy to establish and is left as an exercise.

This method runs in linear time and is therefore time optimal. A different linear-time method with a smaller constant was given by Shiloach [404].

7.14. APL13: Suffix arrays – more space reduction

In Section 6.5.1, we saw that when alphabet size is included in the time and space bounds, the suffix tree for a string of length m either requires $\Theta(m|\Sigma|)$ space or the minimum of $O(m \log m)$ and $O(m \log |\Sigma|)$ time. Similarly, searching for a pattern P of length n using a suffix tree can be done in $O(n)$ time only if $\Theta(m|\Sigma|)$ space is used for the tree, or if we assume that up to $|\Sigma|$ character comparisons cost only constant time. Otherwise, the search takes the minimum of $O(n \log m)$ and $O(n \log |\Sigma|)$ comparisons. For these reasons, a suffix tree may require too much space to be practical in some applications. Hence a more space efficient approach is desired that still retains most of the advantages of searching with a suffix tree.

In the context of the substring problem (see Section 7.3) where a fixed string T will be searched many times, the key issues are the time needed for the search and the space used by the fixed data structure representing T. The space used during the preprocessing of T is of less concern, although it should still be "reasonable".

Manber and Myers [308] proposed a new data structure, called a *suffix array*, that is very space efficient and yet can be used to solve the exact matching problem or the substring problem almost as efficiently as with a suffix tree. Suffix arrays are likely to be an important contribution to certain string problems in computational molecular biology, where the alphabet can be large (we will discuss some of the reasons for large alphabets below). Interestingly, although the more formal notion of a suffix array and the basic algorithms for building and using it were developed in [308], many of the ideas were anticipated in the biological literature by Martinez [310].

After defining suffix arrays we show how to convert a suffix tree to a suffix array in linear time. It is important to be clear on the setting of the problem. String T will be held fixed for a long time, while P will vary. Therefore, the goal is to find a space-efficient representation for T (a suffix array) that will be held fixed and that facilitates search problems in T. However, the amount of space used during the construction of that representation is not so critical. In the exercises we consider a more space efficient way to build the representation itself.

Definition   Given an m-character string T, a *suffix array* for T, called *Pos*, is an array of the integers in the range 1 to m, specifying the lexicographic order of the m suffixes of string T.

That is, the suffix starting at position $Pos(1)$ of T is the lexically smallest suffix, and in general suffix $Pos(i)$ of T is lexically smaller than suffix $Pos(i + 1)$.

As usual, we will affix a terminal symbol \$ to the end of S, but now we interpret it to be lexically *less* than any other character in the alphabet. This is in contrast to its interpretation in the previous section. As an example of a suffix array, if T is *mississippi*, then the suffix array *Pos* is 11, 8, 5, 2, 1, 10, 9, 7, 4, 6, 3. Figure 7.4 lists the eleven suffixes in lexicographic order.

```
11:  i
 8:  ippi
 5:  issippi
 2:  ississippi
 1:  mississippi
10:  pi
 9:  ppi
 7:  sippi
 4:  sisippi
 6:  ssippi
 3:  ssissippi
```

Figure 7.4: The eleven suffixes of *mississippi* listed in lexicographic order. The starting positions of those suffixes define the suffix array *Pos*.

Notice that the suffix array holds only integers and hence contains no information about the alphabet used in string T. Therefore, the space required by suffix arrays is modest – for a string of length m, the array can be stored in exactly m computer words, assuming a word size of at least $\log m$ bits.

When augmented with an additional $2m$ values (called *Lcp* values and defined later), the suffix array can be used to find all the occurrences in T of a pattern P in $O(n + \log_2 m)$ single-character comparison and bookkeeping operations. Moreover, this bound is independent of the alphabet size. Since for most problems of interest $\log_2 m$ is $O(n)$, the substring problem is solved by using suffix arrays as efficiently as by using suffix trees.

7.14.1. Suffix tree to suffix array in linear time

We assume that sufficient space is available to build a suffix tree for T (this is done once during a preprocessing phase), but that the suffix tree cannot be kept intact to be used in the (many) subsequent searches for patterns in T. Instead, we convert the suffix tree to the more space efficient suffix array. Exercises 53, 54, and 55 develop an alternative, more space efficient (but slower) method, for *building* a suffix array.

A suffix array for T can be obtained from the suffix tree $\mathcal{T}$ for T by performing a "lexical" depth-first traversal of $\mathcal{T}$. Once the suffix array is built, the suffix tree is discarded.

Definition   Define an edge (v, u) to be *lexically less* than an edge (v, w) if and only if the first character on the (v, u) edge is lexically less than the first character on (v, w). (In this application, the end of string character $ is lexically less than any other character.)

Since no two edges out of v have labels beginning with the same character, there is a strict lexical ordering of the edges out of v. This ordering implies that the path from the root of $\mathcal{T}$ following the lexically smallest edge out of each encountered node leads to a leaf of $\mathcal{T}$ representing the lexically smallest suffix of T. More generally, a depth-first traversal of $\mathcal{T}$ that traverses the edges out of each node v in their lexical order will encounter the leaves of $\mathcal{T}$ in the lexical order of the suffixes they represent. Suffix array *Pos* is therefore just the ordered list of suffix numbers encountered at the leaves of $\mathcal{T}$ during the lexical depth-first search. The suffix tree for T is constructed in linear time, and the traversal also takes only linear time, so we have the following:

Theorem 7.14.1. *The suffix array Pos for a string T of length m can be constructed in $O(m)$ time.*

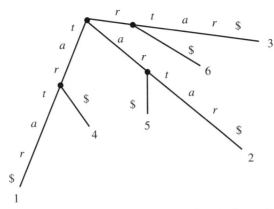

Figure 7.5: The lexical depth-first traversal of the suffix tree visits the leaves in order 5, 2, 6, 3, 4, 1.

For example, the suffix tree for $T = tartar$ is shown in Figure 7.5. The lexical depth-first traversal visits the nodes in the order 5, 2, 6, 3, 4, 1, defining the values of array *Pos*.

As an implementation detail, if the branches out of each node of the tree are organized in a *sorted* linked list (as discussed in Section 6.5, page 116) then the overhead to do a lexical depth-first search is the same as for any depth-first search. Every time the search must choose an edge out of a node v to traverse, it simply picks the next edge on v's linked list.

7.14.2. How to search for a pattern using a suffix array

The suffix array for string T allows a very simple algorithm to find all occurrences of any pattern P in T. The key is that if P occurs in T then all the locations of those occurrences will be grouped consecutively in *Pos*. For example, $P = issi$ occurs in *mississippi* starting at locations 2 and 5, which are indeed adjacent in *Pos* (see Figure 7.4). So to search for occurrences of P in T simply do binary search over the suffix array. In more detail, suppose that P is lexically less than the suffix in the middle position of *Pos* (i.e., suffix $Pos(\lceil m/2 \rceil)$). In that case, the first place in *Pos* that contains a position where P occurs in T must be in the first half of *Pos*. Similarly, if P is lexically greater than suffix $Pos(\lceil m/2 \rceil)$, then the places where P occurs in T must be in the second half of *Pos*. Using binary search, one can therefore find the smallest index i in *Pos* (if any) such that P exactly matches the first n characters of suffix $Pos(i)$. Similarly, one can find the largest index i' with that property. Then pattern P occurs in T starting at every location given by $Pos(i)$ through $Pos(i')$.

The lexical comparison of P to any suffix takes time proportional to the length of the common prefix of those two strings. That prefix has length at most n; hence

Theorem 7.14.2. *By using binary search on array Pos, all the occurrences of P in T can be found in $O(n \log m)$ time.*

Of course, the true behavior of the algorithm depends on how many long prefixes of P occur in T. If very few long prefixes of P occur in T then it will rarely happen that a specific lexical comparison actually takes $\Theta(n)$ time and generally the $O(n \log m)$ bound is quite pessimistic. In "random" strings (even on large alphabets) this method should run in $O(n + \log m)$ expected time. In cases where many long prefixes of P do occur in T, then the method can be improved with the two tricks described in the next two subsections.

7.14.3. A simple accelerant

As the binary search proceeds, let L and R denote the left and right boundaries of the "current search interval". At the start, L equals 1 and R equals m. Then in each iteration of the binary search, a query is made at location $M = \lceil (R + L)/2 \rceil$ of *Pos*. The search algorithm keeps track of the longest prefixes of *Pos*(L) and *Pos*(R) that match a prefix of P. Let l and r denote those two prefix lengths, respectively, and let $mlr = \min(l, r)$.

The value mlr can be used to accelerate the lexical comparison of P and suffix *Pos*(M). Since array *Pos* gives the lexical ordering of the suffixes of T, if i is any index between L and R, the first mlr characters of suffix *Pos*(i) must be the same as the first mlr characters of suffix *Pos*(L) and hence of P. Therefore, the lexical comparison of P and suffix *Pos*(M) can begin from position $mlr + 1$ of the two strings, rather than starting from the first position.

Maintaining mlr during the binary search adds little additional overhead to the algorithm but avoids many redundant comparisons. At the start of the search, when $L = 1$ and $R = m$, explicitly compare P to suffix *Pos*(1) and suffix *Pos*(m) to find l, r, and mlr. However, the worst-case time for this revised method is still $O(n \log m)$. Myers and Manber report that the use of mlr alone allows the search to run as fast in practice as the $O(n + \log m)$ worst-case method that we first advertised. Still, if only because of its elegance, we present the full method that guarantees that better worst-case bound.

7.14.4. A super-accelerant

Call an examination of a character in P *redundant* if that character has been examined before. The goal of the acceleration is to reduce the number of redundant character examinations to at most one per iteration of the binary search – hence $O(\log m)$ in all. The desired time bound, $O(n + \log m)$, follows immediately. The use of mlr alone does not achieve this goal. Since mlr is the minimum of l and r, whenever $l \neq r$, all characters in P from $mlr + 1$ to the maximum of l and r will have already been examined. Thus any comparisons of those characters will be redundant. What is needed is a way to begin comparisons at the *maximum* of l and r.

> **Definition** $Lcp(i, j)$ is the length of the longest common prefix of the suffixes specified in positions i and j of *Pos*. That is, $Lcp(i, j)$ is the length of the longest prefix common to suffix *Pos*(i) and suffix *Pos*(j). The term Lcp stands for *longest common prefix*.

For example, when $T = mississippi$, suffix *Pos*(3) is *issippi*, suffix *Pos*(4) is *ississippi*, and so $Lcp(3, 4)$ is four (see Figure 7.4).

To speed up the search, the algorithm uses $Lcp(L, M)$ and $Lcp(M, R)$ for each triple (L, M, R) that arises during the execution of the binary search. For now, we assume that these values can be obtained in constant time when needed and show how they help the search. Later we will show how to compute the particular Lcp values needed by the binary search during the preprocessing of T.

How to use *Lcp* values

Simplest case In any iteration of the binary search, if $l = r$, then compare P to suffix *Pos*(M) starting from position $mlr + 1 = l + 1 = r + 1$, as before.

General case When $l \neq r$, let us assume without loss of generality that $l > r$. Then there are three subcases:

$lcp(L, M)$

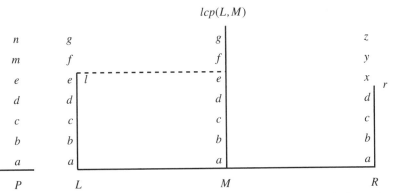

Figure 7.6: Subcase 1 of the super-accelerant. Pattern P is *abcdemn*, shown vertically running upwards from the first character. The suffixes $Pos(L)$, $Pos(M)$, and $Pos(R)$ are also shown vertically. In this case, $Lcp(L, M) > 0$ and $l > r$. Any starting location of P in T must occur in Pos to the right of M, since P agrees with suffix $Pos(M)$ only up to character l.

- If $Lcp(L, M) > l$, then the common prefix of suffix $Pos(L)$ and suffix $Pos(M)$ is longer than the common prefix of P and $Pos(L)$. Therefore, P agrees with suffix $Pos(M)$ up through character l. In other words, characters $l + 1$ of suffix $Pos(L)$ and suffix $Pos(M)$ are identical and lexically less than character $l + 1$ of P (the last fact follows since P is lexically greater than suffix $Pos(L)$). Hence all (if any) starting locations of P in T must occur to the right of position M in Pos. So in any iteration of the binary search where this case occurs, no examinations of P are needed; L just gets changed to M, and l and r remain unchanged. (See Figure 7.6.)

- If $Lcp(L, M) < l$, then the common prefix of suffix $Pos(L)$ and $Pos(M)$ is smaller than the common prefix of suffix $Pos(L)$ and P. Therefore, P agrees with suffix $Pos(M)$ up through character $Lcp(L, M)$. The $Lcp(L, M) + 1$ characters of P and suffix $Pos(L)$ are identical and lexically less than character $Lcp(L, M) + 1$ of suffix $Pos(M)$. Hence all (if any) starting locations of P in T must occur to the left of position M in Pos. So in any iteration of the binary search where this case occurs, no examinations of P are needed; r is changed to $Lcp(L, M)$, l remains unchanged, and R is changed to M.

- If $Lcp(L, M) = l$, then P agrees with suffix $Pos(M)$ up to character l. The algorithm then lexically compares P to suffix $Pos(M)$ starting from position $l + 1$. In the usual manner, the outcome of that lexical comparison determines which of L or R change, along with the corresponding change of l or r.

Theorem 7.14.3. *Using the Lcp values, the search algorithm does at most $O(n + \log m)$ comparisons and runs in that time.*

PROOF   First, by simple case analysis it is easy to verify that neither l nor r ever decrease during the binary search. Also, every iteration of the binary search terminates the search, examines no characters of P, or ends after the first mismatch occurs in that iteration.

In the two cases ($l = r$ or $Lcp(L, M) = l > r$) where the algorithm examines a character during the iteration, the comparisons start with character $\max(l, r)$ of P. Suppose there are k characters of P examined in that iteration. Then there are $k - 1$ matches during the iteration, and at the end of the iteration $\max(l, r)$ increases by $k - 1$ (either l or r is changed to that value). Hence at the start of any iteration, character $\max(l, r)$ of P may have already been examined, but the next character in P has not been. That means at most one redundant comparison per iteration is done. Thus no more than $\log_2 m$ redundant comparisons are done overall. There are at most n nonredundant comparisons of characters

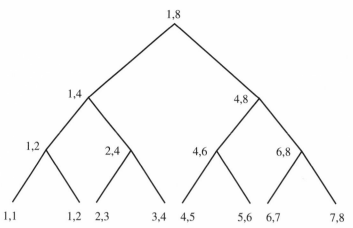

Figure 7.7: Binary tree B representing all the possible search intervals in any execution of binary search in a list of length $m = 8$.

of P, giving a total bound of $n + \log m$ comparisons. All the other work in the algorithm can clearly be done in time proportional to these comparisons. □

7.14.5. How to obtain the *Lcp* values

The *Lcp* values needed to accelerate searches are precomputed in the preprocessing phase during the creation of the suffix array. We first consider how many possible *Lcp* values are ever needed (over any possible execution of binary search). For convenience, assume m is a power of two.

> **Definition** Let B be a complete binary tree with m leaves, where each node of B is labeled with a pair of integers (i, j), $1 \leq i \leq j \leq m$. The root of B is labeled $(1, m)$. Every nonleaf node (i, j) has two children; the left one is labeled $(i, \lfloor (i + j)/2 \rfloor)$, and the right one is labeled $(\lfloor (i + j)/2 \rfloor, j)$. The leaves of B are labeled $(i, i + 1)$ (plus one labeled $(1, 1)$) and are ordered left to right in increasing order of i. (See Figure 7.7.)

Essentially, the node labels specify the endpoints (L, R) of all the possible search intervals that could arise in the binary search of an ordered list of length m. Since B is a binary tree with m leaves, B has $2m - 1$ nodes in total. So there are only $O(m)$ *Lcp* values that need be precomputed. It is therefore plausible that those values can be accumulated during the $O(m)$-time preprocessing of T; but how exactly? In the next lemma we show that the *Lcp* values at the leaves of B are easy to accumulate during the lexical depth-first traversal of $\mathcal{T}$.

Lemma 7.14.1. *In the depth-first traversal of $\mathcal{T}$, consider the internal nodes visited between the visits to leaf $Pos(i)$ and leaf $Pos(i + 1)$, that is, between the ith leaf visited and the next leaf visited. From among those internal nodes, let v denote the one that is closest to the root. Then $Lcp(i, i + 1)$ equals the string-depth of node v.*

For example, consider again the suffix tree shown in Figure 7.5 (page 151). $Lcp(5,6)$ is the string-depth of the parent of leaves 4 and 1. That string-depth is 3, since the parent of 4 and 1 is labeled with the string *tar*. The values of $Lcp(i, i + 1)$ are 2, 0, 1, 0, 3 for i from 1 to 5.

The hardest part of Lemma 7.14.1 involves parsing it. Once done, the proof is immediate from properties of suffix trees, and it is left to the reader.

If we assume that the string-depths of the nodes are known (these can be accumulated in linear time), then by the Lemma, the values $Lcp(i, i+1)$ for i from 1 to $m-1$ are easily accumulated in $O(m)$ time. The rest of the Lcp values are easy to accumulate because of the following lemma:

Lemma 7.14.2. *For any pair of positions i, j, where j is greater than $i + 1$, $Lcp(i, j)$ is the smallest value of $Lcp(k, k + 1)$, where k ranges from i to $j - 1$.*

PROOF Suffix $Pos(i)$ and Suffix $Pos(j)$ of T have a common prefix of length $lcp(i, j)$. By the properties of lexical ordering, for every k between i and j, suffix $Pos(k)$ must also have that common prefix. Therefore, $lcp(k, k + 1) \geq lcp(i, j)$ for every k between i and $j - 1$.

Now by transitivity, $Lcp(i, i+2)$ must be at least as large as the minimum of $Lcp(i, i+1)$ and $Lcp(i + 1, i + 2)$. Extending this observation, $Lcp(i, j)$ must be at least as large as the smallest $Lcp(k, k + 1)$ for k from i to $j - 1$. Combined with the observation in the first paragraph, the lemma is proved. □

Given Lemma 7.14.2, the remaining Lcp values for B can be found by working up from the leaves, setting the Lcp value at any node v to the minimum of the lcp values of its two children. This clearly takes just $O(m)$ time.

In summary, the $O(n + \log m)$-time string and substring matching algorithm using a suffix array must precompute the $2m - 1$ Lcp values associated with the nodes of binary tree B. The leaf values can be accumulated during the linear-time, lexical, depth-first traversal of T used to construct the suffix array. The remaining values are computed from the leaf values in linear time by a bottom-up traversal of B, resulting in the following:

Theorem 7.14.4. *All the needed Lcp values can be accumulated in $O(m)$ time, and all occurrences of P in T can be found using a suffix array in $O(n + \log m)$ time.*

7.14.6. Where do large alphabet problems arise?

A large part of the motivation for suffix arrays comes from problems that arise in using suffix trees when the underlying alphabet is large. So it is natural to ask where large alphabets occur.

First, there are natural languages, such as Chinese, with large "alphabets" (using some computer representation of the Chinese pictograms.) However, most large alphabets of interest to us arise because the string contains numbers, each of which is treated as a character. One simple example is a string that comes from a picture where each character in the string gives the color or gray level of a pixel.

String and substring matching problems where the alphabet contains numbers, and where P and T are large, also arise in computational problems in molecular biology. One example is the *map matching* problem. A *restriction enzyme map* for a single enzyme specifies the locations in a DNA string where copies of a certain substring (a restriction enzyme recognition site) occurs. Each such site may be separated from the next one by many thousands of bases. Hence, the restriction enzyme map for that single enzyme is represented as a string consisting of a sequence of integers specifying the distances between successive enzyme sites. Considered as a string, each integer is a character of a (huge) underlying alphabet. More generally, a map may display the sites of many different patterns of interest (whether or not they are restriction enzyme sites), so the string (map) consists of characters from a finite alphabet (representing the known patterns of interest) alternating with integers giving the distances between such sites. The alphabet is huge because the range of integers is huge, and since distances are often known with high precision, the numbers are not rounded off. Moreover, the variety of known patterns of interest is itself large (see [435]).

It often happens that a DNA substring is obtained and studied without knowing where that DNA is located in the genome or whether that substring has been previously researched. If both the new and the previously studied DNA are fully sequenced and put in a database, then the issue of previous work or locations would be solved by exact string matching. But most DNA substrings that are studied are not fully sequenced – maps are easier and cheaper to obtain than sequences. Consequently, the following matching problem on *maps* arises and translates to an matching problem on *strings* with large alphabets:

> Given an established (restriction enzyme) map for a large DNA string and a map from a smaller string, determine if the smaller string is a substring of the larger one.

Since each map is represented as an alternating string of characters and integers, the underlying alphabet is huge. This provides one motivation for using suffix arrays for matching or substring searching in place of suffix trees. Of course, the problems become more difficult in the presence of errors, when the integers in the strings may not be exact, or when sites are missing or spuriously added. That problem, called *map alignment*, is discussed in Section 16.10.

7.15. APL14: Suffix trees in genome-scale projects

Suffix trees, generalized suffix trees and suffix arrays are now being used as the central data structures in three genome-scale projects.

Arabidopsis thaliana   An *Arabidopsis thaliana* genome project,[6] at the Michigan State University and the University of Minnesota is initially creating an EST map of the *Arabidopsis* genome (see Section 3.5.1 for a discussion of ESTs and Chapter 16 for a discussion of mapping). In that project generalized suffix trees are used in several ways [63, 64, 65].

First, each sequenced fragment is checked to catch any contamination by known vector sequences. The vector sequences are kept in a generalized suffix tree, as discussed in Section 7.5.

Second, each new sequenced fragment is checked against fragments already sequenced to find duplicate sequences or regions of high similarity. The fragment sequences are kept in an expanding generalized suffix tree for this purpose. Since the project will sequence about 36,000 fragments, each of length about 400 bases, the efficiency of the searches for duplicates and for contamination is important.

Third, suffix trees are used in the search for biologically significant patterns in the obtained *Arabidopsis* sequences. Patterns of interest are often represented as regular expressions, and generalized suffix trees are used to accelerate regular expression pattern matching, where a small number of errors in a match are allowed. An approach that permits

---

[6] *Arabidopsis thaliana* is the "fruit fly" of plant genetics, i.e., the classic model organism in studying the molecular biology of plants. Its size is about 100 million base pairs.

suffix trees to speed up regular expression pattern matching (with errors) is discussed in Section 12.4.

Yeast   Suffix trees are also the central data structure in genome-scale analysis of *Saccharomyces cerevisiae* (brewer's yeast), done at the Max-Plank Institute [320]. Suffix trees are "particularly suitable for finding substring patterns in sequence databases" [320]. So in that project, highly optimized suffix trees called *hashed position trees* are used to solve problems of "clustering sequence data into evolutionary related protein families, structure prediction, and fragment assembly" [320]. (See Section 16.15 for a discussion of fragment assembly.)

Borrelia burgdorferi   *Borrelia burgdorferi* is the bacterium causing Lyme disease. Its genome is about one million bases long, and is currently being sequenced at the Brookhaven National Laboratory using a directed sequencing approach to fill in gaps after an initial shotgun sequencing phase (see Section 16.14). Chen and Skiena [100] developed methods based on suffix trees and suffix arrays to solve the fragment assembly problem for this project. In fragment assembly, one major bottleneck is overlap detection, which requires solving a variant of the suffix-prefix matching problem (allowing some errors) for all pairs of strings in a large set (see Section 16.15.1.). The Borrelia work [100] consisted of 4,612 fragments (strings) totaling 2,032,740 bases. Using suffix trees and suffix arrays, the needed overlaps were computed in about fifteen minutes. To compare the speed and accuracy of the suffix tree methods to pure dynamic programming methods for overlap detection (discussed in Section 11.6.4 and 16.15.1), Chen and Skiena closely examined cosmid-sized data. The test established that the suffix tree approach gives a 1,000 times speedup over the (slightly) more accurate dynamic programming approach, finding 99% of the significant overlaps found by using dynamic programing.

Efficiency is critical

In all three projects, the efficiency of building, maintaining, and searching the suffix trees is extremely important, and the implementation details of Section 6.5 are crucial. However, because the suffix trees are very large (approaching 20 million characters in the case of the *Arabidopsis* project) additional implementation effort is needed, particularly in organizing the suffix tree on disk, so that the number of disk accesses is reduced. All three projects have deeply explored that issue and have found somewhat different solutions. See [320], [100] and [63] for details.

7.16. APL15: A Boyer–Moore approach to exact set matching

The Boyer–Moore algorithm for exact matching (single pattern) will often make long shifts of the pattern, examining only a small percentage of all the characters in the text. In contrast, Knuth-Morris-Pratt examines all characters in the text in order to find all occurrences of the pattern.

In the case of exact *set* matching, the Aho–Corasick algorithm is analogous to Knuth-Morris-Pratt – it examines all characters of the text. Since the Boyer–Moore algorithm for a single string is far more efficient in practice than Knuth-Morris-Pratt, one would like to have a Boyer–Moore type algorithm for the exact set matching problem, that is, a method for the exact set matching problem that typically examines only a sublinear portion of T. No known simple algorithm achieves this goal and also has a linear worst-case running time. However, a synthesis of the Boyer–Moore and Aho–Corasick algorithms due to Commentz–Walter [109] solves the exact set matching problem in the spirit of the Boyer–Moore algorithm. Its shift rules allow many characters of T to go unexamined. We will not describe the Commentz–Walter algorithm but instead use suffix trees to achieve the same result more simply.

For simplicity of exposition, we will first describe a solution that uses two trees – a simple keyword tree (without back pointers) together with a suffix tree. The difficult work is done by the suffix tree. After understanding the ideas, we implement the method using only the suffix tree.

> **Definition** Let P^r denote the reverse of a pattern P, and let $\mathcal{P}^r$ be the set of strings obtained by reversing every pattern P from an input set $\mathcal{P}$.

As usual, the algorithm preprocesses the set of patterns and then uses the result of the preprocessing to accelerate the search. The following exposition interleaves the descriptions of the search method and the preprocessing that supports the search.

7.16.1. The search

Recall that in the Boyer–Moore algorithm, when the end of the pattern is placed against a position i in T, the comparison of individual characters proceeds *right to left*. However, index i is *increased* at each iteration. These high-level features of Boyer–Moore will also hold in the algorithm we present for exact set matching.

In the case of multiple patterns, the search is carried out on a simple keyword tree $\mathcal{K}^r$ (without backpointers) that encodes the patterns in $\mathcal{P}^r$. The search again chooses increasing values of index i and determines for each chosen i whether there is a pattern in set $\mathcal{P}$ ending at position i of text T. Details are given below.

The preprocessing time needed to build $\mathcal{K}^r$ is only $O(n)$, the total length of all the patterns in $\mathcal{P}$. Moreover, because no backpointers are needed, the preprocessing is particularly simple. The algorithm to build $\mathcal{K}^r$ successively inserts each pattern into the tree, following as far as possible a matching path from the root, etc. Recall that each leaf of $\mathcal{K}^r$ specifies one of the patterns in $\mathcal{P}^r$.

The test at position i

Tree $\mathcal{K}^r$ can be used to test, for any specific position i in T, whether one of the patterns in $\mathcal{P}$ ends at position i. To make this test, simply follow a path from the root of $\mathcal{K}^r$, matching characters on the path with characters in T, starting with $T(i)$ and moving right to left as in Boyer–Moore. If a leaf of $\mathcal{K}^r$ is reached before the left end of T is reached, then the pattern number written at the leaf specifies a pattern that must occur in T ending at position i. Conversely, if the matched path ends before reaching a leaf and cannot be extended, then no pattern in $\mathcal{P}$ occurs in T ending at position i.

The first test begins with position i equal to the length of the smallest pattern in $\mathcal{P}$. The entire algorithm ends when i is set larger than m, the length of T.

When the test for a specific position i is finished, the algorithm increases i and returns to the root of $\mathcal{K}^r$ to begin another test. Increasing i is analogous to shifting the single pattern in the original Boyer–Moore algorithm. But by how much should i be increased? Increasing i by one is analogous to the naive algorithm for exact matching. With a shift of only one position, no occurrences of any pattern will be missed, but the resulting computation will be inefficient. In the worst case, it will take $\Theta(nm)$ time, where n is the total size of the patterns and m is the size of the text. The more efficient algorithm will increase i by more than one whenever possible, using rules that are analogous to the bad character and good suffix rules of Boyer–Moore. Of course, no shift can be greater than the length of the shortest pattern P in $\mathcal{P}$, for such a shift could miss occurrences of P in T.

7.16.2. Bad character rule

The bad character rule from Boyer–Moore can easily be adapted to the set matching problem. Suppose the test matches some path in $\mathcal{K}^r$ against the characters from i down to $j < i$ in T but cannot extend the path to match character $T(j-1)$. A direct generalization of the bad character rule increases i to the smallest index $i_1 > i$ (if it exists) such that some pattern $\bar{P}$ from $\mathcal{P}$ has character $T(j-1)$ exactly $i_1 - j + 2$ positions from its right end. (See Figures 7.8 and 7.9.) With this rule, if i_1 exists, then when the right end of every pattern in $\mathcal{P}$ is aligned with position i_1 of T, character $j-1$ of T will be opposite a matching character in string $\bar{P}$ from $\mathcal{P}$. (There is a special case to consider if the test fails on the first comparison, i.e., at the root of $\mathcal{K}^r$. In that case, set $j = i + 1$ before applying the shift rule.)

The above generalization of the bad character rule from the two-string case is not quite correct. The problem arises because of patterns in $\mathcal{P}$ that are smaller than $\bar{P}$. It may happen that i_1 is so large that if the right ends of all the patterns are aligned with it, then the left end of the smallest pattern P_{min} in $\mathcal{P}$ would be aligned with a position greater than j in T. If that happens, it is possible that some occurrence of P_{min} (with its left end opposite a position before $j + 1$ in T) will be missed. Hence, using only the bad character information (not the suffix rules to come next), i should not be set larger than $j - 1 + |P_{min}|$. In summary, the bad character rule for a set of patterns is:

If i_1 does not exist, then increase i to $j - 1 + |P_{min}|$; otherwise increase i to the *minimum* of i_1 and $j - 1 + |P_{min}|$.

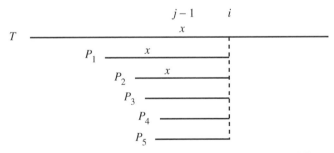

Figure 7.8: No further match is possible at position $j - 1$ of T.

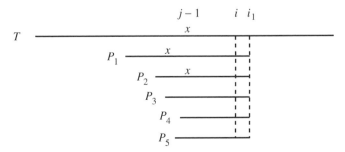

Figure 7.9: Shift when the bad character rule is applied

The preprocessing needed to implement the bad character rule is simple and is left to the reader.

The generalization of the bad character rule to set matching is easy but, unlike the case of a single pattern, use of the bad character rule alone may not be very effective. As the number of patterns grows, the typical size of $i_1 - i$ is likely to decrease, particularly if the alphabet is small. This is because *some* pattern is likely to have character $T(j - 1)$ close to, but left of, the point where the previous matches end. As noted earlier, in some applications in molecular biology the total length of the patterns in $\mathcal{P}$ is larger than the size of T, making the bad character rule almost useless. A bad character rule analogous to the simpler, unextended bad character rule for a single pattern would be even less useful. Therefore, in the set matching case, a rule analogous to the good suffix rule is crucial in making a Boyer–Moore approach effective.

7.16.3. Good suffix rule

To adapt the (weak) good suffix rule to the set matching problem we reason as follows: After a matched path in $\mathcal{K}^r$ is found (either finding an occurrence of a pattern, or not) let j be the left-most character in T that was matched along the path, and let $\alpha = T[j..i]$ be the substring of T that was matched by the traversal (but found in reverse order). A direct generalization (to set matching) of the two-string good suffix rule would shift the right ends of all the patterns in $\mathcal{P}$ to the smallest value $i_2 > i$ (if it exists) such that $T[j..i]$ matches a substring of some pattern $\bar{P}$ in $\mathcal{P}$. Pattern $\bar{P}$ must contain the substring α beginning exactly $i_2 - j + 1$ positions from its right end. This shift is analogous to the good suffix shift for two patterns, but unlike the two-pattern case, that shift may be too large. The reason is again due to patterns that are smaller than $\bar{P}$.

When there are patterns smaller than $\bar{P}$, if the right end of every pattern moves to i_2, it may happen that the left end of the smallest pattern $P_{\min}$ would be placed more than one position to the right of i. In that case, an occurrence of $P_{\min}$ in T could be missed. Even if that doesn't happen, there is another problem. Suppose that a prefix β of some pattern $P' \in \mathcal{P}$ matches a suffix of α. If P' is smaller than $\bar{P}$, then shifting the right end of P' to i_2 may shift the prefix β of P' past the substring β in T. If that happens, then an occurrence of P in T could be missed. So let i_3 be the smallest index greater than i (if i_3 exists) such that when all the patterns in $\mathcal{P}$ are aligned with position i_3 of T, a prefix of at least one pattern is aligned opposite a suffix of α in T. Notice that because $\mathcal{P}$ contains more than one pattern, that overlap might not be the largest overlap between a prefix of a pattern in $\mathcal{P}$ and a suffix of α. Then the good suffix rule is:

> Increase i to the *minimum* of i_2, i_3, or $i + |P_{\min}|$. Ignore i_2 and/or i_3 in this rule, if either or both are nonexistent.

7.16.4. How to determine i_2 and i_3

The question now is how to efficiently determine i_2 and i_3 when needed during the search. We will first discuss i_2. Recall that α denotes the substring of T that was matched in the search just ended.

To determine i_2, we need to find which pattern P in $\mathcal{P}$ contains a copy of α ending closest to its right end, but not occurring as a suffix of P. If that copy of α ends r places

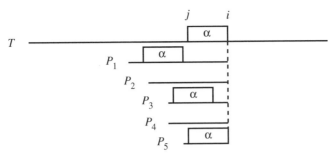

Figure 7.10: Substring α in P_5 matched from position i down to position j of T; no further match is possible to the left of position j.

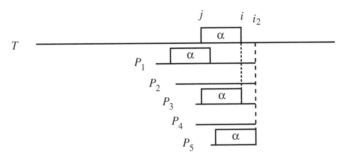

Figure 7.11: The shift when the weak good suffix rule is applied. In this figure, pattern P_3 determines the amount of the shift.

from the end of P, then i should be increased by exactly r positions, that is, i_2 should be set to $i + r$. (See Figure 7.10 and Figure 7.11.)

We will solve the problem of finding i_2, if it exists, using a suffix tree obtained by preprocessing set $\mathcal{P}^r$. The key involves using the suffix tree to search for a pattern P^r in $\mathcal{P}^r$ containing a copy of α^r starting closest to its left end but not occurring as a prefix of P^r. If that occurrence of α^r starts at position z of pattern P^r, then an occurrence of α ends $r = z - 1$ positions from the end of P.

During the preprocessing phase, build a generalized suffix tree $\mathcal{T}^r$ for the set of patterns $\mathcal{P}^r$. Recall that in a generalized suffix tree each leaf is associated with both a pattern $P^r \in \mathcal{P}^r$ and a number z specifying the starting position of a suffix of P^r.

Definition For each internal node v of $\mathcal{T}^r$, z_v denotes the smallest number z greater than 1 (if any) such that z is a suffix position number written at a leaf in the subtree of v. If no such leaf exists, then z_v is undefined.

With this suffix tree $\mathcal{T}^r$, determine the number z_v for each internal node v. These two preprocessing tasks are easily accomplished in linear time by standard methods and are left to the reader.

As an example of the preprocessing, consider the set $\mathcal{P} = \{wxa, xaqq, qxax\}$ and the generalized suffix tree for $\mathcal{P}^r$ shown in Figure 7.12. The first number on each leaf refers to a string in $\mathcal{P}^r$, and the second number refers to a suffix starting position in that string. The number z_v is the first (or only) number written at every internal node (the second number will be introduced later).

We can now describe how $\mathcal{T}^r$ is used during the search to determine value i_2, if it exists. After matching α^r along a path in $\mathcal{K}^r$, traverse the path labeled α^r from the root of $\mathcal{T}^r$. That path exists because α is a suffix of some pattern in $\mathcal{P}$ (that is what the search in $\mathcal{K}^r$

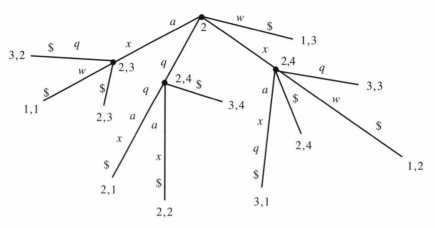

Figure 7.12: Generalized suffix tree $\mathcal{T}^r$ for the set $\mathcal{P} = \{wxa, xaqq, qxax\}$.

determined), so α^r is a prefix of some pattern in $\mathcal{P}^r$. Let v be the first node at or below the end of that path in $\mathcal{T}^r$. If z_v is defined, then i_2 can be obtained from it: The leaf defining z_v (i.e., the leaf where $z = z_v$) is associated with a string $P^r \in \mathcal{P}^r$ that contains a copy of α^r starting to the right of position one. Over all such occurrences of α^r in the strings of $\mathcal{P}^r$, P^r contains the copy of α^r starting closest to its left end. That means that P contains a copy of α that is not a suffix of P, and over all such occurrences of α, P contains the copy of α ending closest to its right end. P is then the string in $\mathcal{P}$ that should be used to set i_2. Moreover, α ends in P exactly $z_v - 1$ characters from the end of P. Hence, as argued above, i should be increased by $z_v - 1$ positions. In summary, we have

Theorem 7.16.1. *If the first node v in $\mathcal{T}^r$ at or below the end of path α^r has a defined value z_v, then i_2 equals $i + z_v - 1$.*

Using suffix tree $\mathcal{T}^r$, the determination of i_2 takes $O(|\alpha|)$ time, only doubling the time of the search used to find α. However, with proper preprocessing, the search used to find i_2 can be eliminated. The details will be given below in Section 7.16.5.

Now we turn to the computation of i_3. This is again easy assuming the proper preprocessing of $\mathcal{P}$. Again we use the generalized suffix tree $\mathcal{T}^r$ for $\mathcal{P}^r$. To get the idea of the method let $P \in \mathcal{P}$ be any pattern such that a suffix of α is a prefix of P. That means that a prefix of α^r is a suffix of P^r. Now consider the path labeled α^r in $\mathcal{T}^r$. Since some suffix of α is a prefix of P, some initial portion of the α^r path in $\mathcal{T}^r$ describes a suffix of P^r. There thus must be a leaf edge (u, z) branching off that path, where leaf z is associated with pattern P^r and the label of edge (u, z) is just the terminal character \$. Conversely, let (u, z) be any edge branching off the α^r path and labeled with the single symbol \$. Then the pattern P associated with z must have a prefix matching a suffix of α. These observations lead to the following preprocessing and search methods.

In the preprocessing phase when $\mathcal{T}^r$ is built, identify every edge (u, z) in $\mathcal{T}^r$ that is labeled only by the terminal character \$. (The number z is used both as a leaf name and as the starting position of the suffix associated with that leaf.) For each such node u, set a variable d_u to z. For example, in Figure 7.12, d_u is the second number written at each node u (d_u is not defined for the root node). In the search phase, after matching a string α in T, the value of i_3 (if needed) can be found as follows:

Theorem 7.16.2. *The value of i_3 should be set to $i + d_w - 1$, where d_w is the smallest d value at a node on the α^r path in $\mathcal{T}^r$. If no node on that path has a d value defined, then i_3 is undefined.*

The proof is immediate and is left to the reader. Clearly, i_3 can be found during the traversal of the α^r path in $\mathcal{T}^r$ used to search for i_2. If neither i_2 nor i_3 exist, then i should be increased by the length of the smallest pattern in $\mathcal{P}$.

7.16.5. An implementation eliminating redundancy

The implementation above builds two trees in time and space proportional to the total size of the patterns in $\mathcal{P}$. In addition, every time a string α^r is matched in $\mathcal{K}^r$ only $O(|\alpha|)$ additional time is used to search for i_2 and i_3. Thus the time to implement the shifts using the two trees is proportional to the time used to find the matches. From an asymptotic standpoint the two trees are as small as one, and the two traversals are as fast as one. But clearly there is superfluous work in this implementation – a single tree and a single traversal per search phase should suffice. Here's how.

Preprocessing for Boyer–Moore exact set matching

Begin

1. Build a generalized suffix tree $\mathcal{T}^r$ for the strings in $\mathcal{P}^r$. (Each leaf in the tree is numbered both by a specific pattern P^r in $\mathcal{P}^r$ and by a specific starting position z of a suffix in P^r.)

2. Identify and mark every node in $\mathcal{T}^r$, *including leaves*, that is an ancestor of a leaf numbered by suffix position one (for some pattern P^r in $\mathcal{P}^r$). Note that a node is considered to be an ancestor of itself.

3. For each marked node v, set z_v to be the smallest suffix position number z greater than one (if there is one) of any leaf in v's subtree.

4. Find every leaf edge (u, z) of $\mathcal{T}^r$ that is labeled only by the terminal character \$, and set $d_u = z$.

5. For each node v in $\mathcal{T}^r$ set d'_v equal to the smallest value of d_u for any ancestor (including v) of v.

6. Remove the subtree rooted at any unmarked node (including leaves) of $\mathcal{T}$. (Nodes were marked in step 2.)

End.

The above preprocessing tasks are easily accomplished in linear time by standard tree traversal methods.

Using $\mathcal{L}$ in the search phase

Let $\mathcal{L}$ denote the tree at the end of the preprocessing. Tree $\mathcal{L}$ is essentially the familiar keyword tree $\mathcal{K}^r$ but is more compacted: Any path of nodes with only one descendent has been replaced with a single edge. Hence, for any i, the test to see if a pattern of $\mathcal{P}$ ends at position i can be executed using tree $\mathcal{L}$ rather than $\mathcal{K}^r$. Moreover, unlike $\mathcal{K}^r$, each node v in $\mathcal{L}$ now has associated with it the values needed to compute i_2 and i_3 in *constant* time. In detail, after the algorithm matches a string α in T by following the path α^r in $\mathcal{L}$, the algorithm checks the first node v at or beneath the end of the path in $\mathcal{L}$. If z_v is defined there, then i_2 exists and equals $i + z_v - 1$. Next the algorithm checks the first node v at or above the end of the matched path. If d'_v is defined there then i_3 exists and equals $i + d'_v - 1$.

The search phase will not miss any occurrence of a pattern if either the good suffix rule or the bad character rule is used by itself. However, the two rules can be combined to increment i by the largest amount specified by either of the two rules.

Figure 7.13 shows tree $\mathcal{L}$ corresponding to the tree $\mathcal{T}^r$ shown in Figure 7.12.

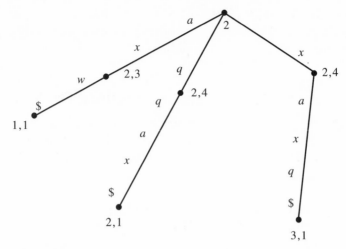

Figure 7.13: Tree $\mathcal{L}$ corresponding to tree T^r for the set $\mathcal{P} = \{wxa, xaqq, qxax\}$.

1234567890	1234567890	1234567890	1234567890
qxaxtqqpst	qxaxtqqpst	qxaxtqqpst	qxaxtqqpst
wxa	wxa	wxa	wxa
xaqq	xaqq	xaqq	xaqq
qxax	qxax	qxax	qxax

Figure 7.14: The first comparisons start at position 3 of T and match ax. The value of z_v is equal to two, so a shift of one position occurs. String $qxax$ matches; z_v is undefined, but d'_v is defined and equals 4, so a shift of three is made. The string qq matches, followed by a mismatch; z_v is undefined, but d'_v is defined to be four, so a shift of three is made, after which no further matches are found and the algorithm halts.

To see how $\mathcal{L}$ is used during the search, let T be $qxaxtqqpst$. The shifts of the $\mathcal{P}$ are shown in Figure 7.14.

7.17. APL16: Ziv–Lempel data compression

Large text or graphics files are often compressed in order to save storage space or to speed up transmission when the file is shipped. Most operating systems have compression utilities, and some file transfer programs automatically compress, ship, and uncompress the file, without user intervention. The field of text compression is itself the subject of several books (for example, see [423]) and will not be handled in depth here. However, a popular compression method due to Ziv–Lempel [487, 488] has an efficient implementation using suffix trees [382], providing another illustration of their utility.

The Ziv–Lempel compression method is widely used (it is the basis for the Unix utility *compress*), although there are actually several variants of the method that go by the same name (see [487] and [488]). In this section, we present a variant of the basic method and an efficient implementation of it using suffix trees.

Definition For any position i in a string S of length m, define the substring $Prior_i$ to be the longest prefix of $S[i..m]$ that also occurs as a substring of $S[1..i-1]$.

For example, if $S = abaxcabaxabz$ then $Prior_7$ is bax.

Definition For any position i in S, define l_i as the length of $Prior_i$. For $l_i > 0$, define s_i as the starting position of the left-most copy of $Prior_i$.

In the above example, $l_7 = 3$ and $s_7 = 2$.

Note that when $l_i > 0$, the copy of $Prior_i$ starting at s_i is totally contained in $S[1..i-1]$.

The Ziv–Lempel method uses some of the l_i and s_i values to construct a compressed representation of string S. The basic insight is that if the text $S[1..i-1]$ has been represented (perhaps in compressed form) and l_i is greater than zero, then the next l_i characters of S (substring $Prior_i$) need not be explicitly described. Rather, that substring can be described by the pair (s_i, l_i), pointing to an earlier occurrence of the substring. Following this insight, a compression method could process S left to right, outputting the pair (s_i, l_i) in place of the explicit substring $S[i..i + l_i - 1]$ when possible, and outputting the character $S(i)$ when needed. Full details are given in the algorithm below.

Compression algorithm 1

begin

$i := 1$

Repeat

    compute l_i and s_i

    if $l_i > 0$ then

    begin

        output (s_i, l_i)

        $i := i + l_i$

    end

    else

    begin

        output $S(i)$

        $i := i + 1$

    end

Until $i > n$

end.

For example, $S = abacabaxabz$ can be described as $ab(1, 1)c(1, 3)x(1, 2)z$. Of course, in this example the number of symbols used to represent S did not decrease, but rather increased! That's typical of small examples. But as the string length increases, providing more opportunity for repeating substrings, the compression improves. Moreover, the algorithm could choose to output character $S(i)$ explicitly whenever l_i is "small" (the actual rule depends on bit-level considerations determined by the size of the alphabet, etc.). For a small example where positive compression is observed, consider the contrived string $S = ababababababababababababababababab$, represented as $ab(1, 2)(1, 4)(1, 8)(1, 16)$. That representation uses 24 symbols in place of the original 32 symbols. If we extend this example to contain k repeated copies of ab, then the compressed representation contains approximately $5 \log_2 k$ symbols – a dramatic reduction in space.

To decompress a compressed string, process the compressed string left to right, so that any pair (s_i, l_i) in the representation points to a substring that has already been fully decompressed. That is, assume inductively that the first j terms (single characters or s, l pairs) of the compressed string have been processed, yielding characters 1 through $i - 1$ of the original string S. The next term in the compressed string is either character $S(i + 1)$, or it is a pair (s_i, l_i) pointing to a substring of S strictly before i. In either case, the algorithm has the information needed to decompress the jth term, and since the first term in the compressed string is the first character of S, we conclude by induction that the decompression algorithm can obtain the original string S.

7.17.1. Implementation using suffix trees

The key implementation question is how to compute l_i and s_i each time the algorithm requests those values for a position i. The algorithm compresses S left to right and does not request (s_i, l_i) for any position i already in the compressed part of S. The compressed substrings are therefore nonoverlapping, and if each requested pair (s_i, l_i) can be found in $O(l_i)$ time, then the entire algorithm would run in $O(m)$ time. Using a suffix tree for S, the $O(l_i)$ time bound is easily achieved for any request.

Before beginning the compression, the algorithm first builds a suffix tree $\mathcal{T}$ for S and then numbers each node v with the number c_v. This number equals the smallest suffix (position) number of any leaf in v's subtree, and it gives the left-most starting position in S of any copy of the substring that labels the path from r to v. The tree can be built in $O(m)$ time, and all the node numbers can be obtained in $O(m)$ time by any standard tree traversal method (or bottom-up propagation).

When the algorithm needs to compute (s_i, l_i) for some position i, it traverses the unique path in $\mathcal{T}$ that matches a prefix of $S[i..m]$. The traversal ends at point p (not necessarily a node) when i equals the string-depth of point p plus the number c_v, where v is the first node at or below p. In either case, the path from the root to p describes the longest prefix of $S[i..m]$ that also occurs in $S[1..i]$. So, s_i equals c_v and l_i equals the string-depth of p. Exploiting the fact that the alphabet is fixed, the time to find (s_i, l_i) is $O(l_i)$. Thus the entire compression algorithm runs in $O(m)$ time.

7.17.2. A one-pass version

In the implementation above we assumed S to be known ahead of time and that a suffix tree for S could be built before compression begins. That works fine in many contexts, but the method can also be modified to operate on-line as S is being input, one character at a time. Essentially, the algorithm is implemented so that the compaction of S is interwoven with the construction of $\mathcal{T}$. The easiest way to see how to do this is with Ukkonen's linear-time suffix tree algorithm.

Ukkonen's algorithm builds implicit suffix trees on-line as characters are added to the right end of the growing string. Assume that the compaction has been done for $S[1..i-1]$ and that implicit suffix tree $\mathcal{I}_{i-1}$ for string $S[1..i-1]$ has been constructed. At that point, the compaction algorithm needs to know (s_i, l_i). It can obtain that pair in exactly the same way that is done in the above implementation *if* the c_v values have been written at each node v in $\mathcal{I}_{i-1}$. However, unlike the above implementation, which establishes those c_v values in a linear time traversal of $\mathcal{T}$, the algorithm cannot traverse each of the implicit suffix trees, since that would take more than linear time overall. Instead, whenever a new internal node v is created in Ukkonen's algorithm by splitting an edge (u, w), c_v is set to c_w, and whenever a new leaf v is created, c_v is just the suffix number associated with leaf v. In this way, only constant time is needed to update the c_v values when a new node is added to the tree. In summary, we have

Theorem 7.17.1. *Compression algorithm 1 can be implemented to run in linear time as a one-pass, on-line algorithm to compress any input string S.*

7.17.3. The real Ziv–Lempel

The compression scheme given in *Compression algorithm* 1 although not the actual Ziv–Lempel method, does resemble it and capture its spirit. The real Ziv–Lempel method is a one-pass algorithm whose output differs from the output of *Compression algorithm* 1 in that, whenever it outputs a pair (s_i, l_i), it then explicitly outputs $S(i + l_i)$, the character following the substring. For example, $S = abababababababababababababababab$ would be compressed to $ab(1, 2)a(2, 4)b(1, 10)a(2, 12)b$, rather than as $ab(1, 2)(1, 4)(1, 8)$ $(1, 16)$. The one-pass version of *Compression algorithm* 1 can trivially be converted to implement Ziv–Lempel in linear time.

It is not completely clear why the Ziv–Lempel algorithm outputs the extra character. Certainly for compaction purposes, this character is not needed and seems extraneous. One suggested reason for outputting an explicit character after each (s, l) pair is that $(s_i, l_i)S(i + l_i)$ defines the shortest substring starting at position i that does not appear anywhere earlier in the string, whereas (s_i, l_i) defines the longest substring starting at i that does appear earlier. Historically, it may have been easier to reason about shortest substrings that do not appear earlier in the string than to reason about longest substrings that do appear earlier.

7.18. APL17: Minimum length encoding of DNA

Recently, several molecular biology and computer science research groups have used the Ziv–Lempel method to compress DNA strings, not for the purpose of obtaining efficient storage, but rather to compute a measure of the "complexity" or "information content" of the strings [14, 146, 325, 326, 386]. Without fully defining the central technical terms "complexity", "information", "entropy", etc., we state the basic idea, which is that substrings of greatest biological significance should be more compressable than substrings that are essentially random. One expects that random strings will have too little structure to allow high compression, since high compression is based on finding repetitive segments in the string. Therefore, by searching for substrings that are more compressable than random strings, one may be able to find strings that have a definite biological function. (On the other hand, most repetitive DNA occurs outside of exons.)

Compression has also been used to study the "relatedness"[7] of two strings S_1 and S_2 of DNA [14, 324]. Essentially, the idea is to build a suffix tree for S_1 and then compress string S_2 using only the suffix tree for S_1. That compression of S_2 takes advantage of substrings in S_2 that appear in S_1, but does not take advantage of repeated substrings in S_2 alone. Similarly, S_1 can be compressed using only a suffix tree for S_2. These compressions reflect and estimate the "relatedness" of S_1 and S_2. If the two strings are highly related, then both computations should significantly compress the string at hand.

Another biological use for Ziv–Lempel–like algorithms involves estimating the "entropy" of short strings in order to discriminate between exons and introns in eukaryotic DNA [146]. Farach et al. [146] report that the average compression of introns does not differ significantly from the average compression of exons, and hence compression by itself does not distinguish exons from introns. However, they also report the following extension of that approach to be effective in distinguishing exons from introns.

---

[7] Other, more common ways to study the relatedness or similarity of strings of two strings are extensively discussed in Part III.

Definition   For any position i in string S, let $ZL(i)$ denote the length of the longest substring beginning at i that appears somewhere in the string $S[1..i]$.

Definition   Given a DNA string S partitioned into exons and introns, the *exon-average ZL value* is the average $ZL(i)$ taken over every position i in the exons of S. Similarly, the *intron-average ZL* is the average $ZL(i)$ taken over positions in introns of S.

It should be intuitive at this point that the exon-average ZL value and the intron-average ZL value can be computed in $O(n)$ time, by using suffix trees to compute all the $ZL(i)$ values. The technique, resembles the way matching statistics are computed, but is more involved since the substring starting at i must also appear to the left of position i.

The main empirical result of [146] is that the exon-average ZL value is lower than the intron-average ZL value by a statistically significant amount. That result is contrary to the expectation stated above that biologically significant substrings (exons in this case) should be more compressable than more random substrings (which introns are believed to be). Hence, the full biological significance of string compressability remains an open question.

7.19. Additional applications

Many additional applications of suffix trees appear in the exercises below, in Chapter 9, in Sections 12.2.4, 12.3, and 12.4, and in exercises of Chapter 14.

7.20. Exercises

1. Given a set S of k strings, we want to find every string in S that is a substring of some other string in S. Assuming that the total length of all the strings is n, give an $O(n)$-time algorithm to solve this problem. This result will be needed in algorithms for the shortest superstring problem (Section 16.17).

2. For a string S of length n, show how to compute the $N(i)$, $L(i)$, $L'(i)$ and sp_i values (discussed in Sections 2.2.4 and 2.3.2) in $O(n)$ time directly from a suffix tree for S.

3. We can define the suffix tree in terms of the keyword tree used in the Aho–Corasick (AC) algorithm. The input to the AC algorithm is a set of patterns $\mathcal{P}$, and the AC tree is a compact representation of those patterns. For a single string S we can think of the n suffixes of S as a set of patterns. Then one can build a suffix tree for S by first constructing the AC tree for those n patterns, and then compressing, into a single edge, any maximal path through nodes with only a single child. If we take this approach, what is the relationship between the failure links used in the keyword tree and the suffix links used in Ukkonen's algorithm? Why aren't suffix trees built in this way?

4. A suffix tree for a string S can be viewed as a keyword tree, where the strings in the keyword tree are the suffixes of S. In this way, a suffix tree is useful in efficiently building a keyword tree when the strings for the tree are only *implicitly* specified. Now consider the following implicitly specified set of strings: Given two strings S_1 and S_2, let D be the set of all substrings of S_1 that are not contained in S_2. Assuming the two strings are of length n, show how to construct a keyword tree for set D in $O(n)$ time. Next, build a keyword tree for D together with the set of substrings of S_2 that are not in S_1.

5. Suppose one has built a generalized suffix tree for a string S along with its suffix links (or link pointers). Show how to efficiently convert the suffix tree into an Aho–Corasick keyword tree.

6. Discuss the relative advantages of the Aho–Corasick method versus the use of suffix trees for the exact set matching problem, where the text is fixed and the set of patterns is varied over time. Consider preprocessing, search time, and space use. Consider both the cases when the text is larger than the set of patterns and vice versa.

7. In what way does the suffix tree more deeply expose the structure of a string compared to the Aho–Corasick keyword tree or the preprocessing done for the Knuth-Morris-Pratt or Boyer–Moore methods? That is, the *sp* values give some information about a string, but the suffix tree gives much more information about the structure of the string. Make this precise. Answer the same question about suffix trees and Z values.

8. Give an algorithm to take in a set of k strings and to find the longest common substring of each of the $\binom{k}{2}$ pairs of strings. Assume each string is of length n. Since the longest common substring of any pair can be found in $O(n)$ time, $O(k^2 n)$ time clearly suffices. Now suppose that the string lengths are different but sum to m. Show how to find all the longest common substrings in time $O(km)$. Now try for $O(m + k^2)$ (I don't know how to achieve this last bound).

9. The problem of finding substrings common to a set of distinct strings was discussed separately from the problem of finding substrings common to a single string, and the first problem seems much harder to solve than the second. Why can't the first problem just be reduced to the second by concatenating the strings in the set to form one large string?

10. By modifying the compaction algorithm and adding a little extra (linear space) information to the resulting DAG, it is possible to use the DAG to determine not only whether a pattern occurs in the text, but to find all the occurrences of the pattern. We illustrate the idea when there is only a single merge of nodes p and q. Assume that p has larger string depth than q and that u is the parent of p before the merge. During the merge, remove the subtree of p and put a displacement number of -1 on the new u to pq edge. Now suppose we search for a pattern P in the text and determine that P is in the text. Let i be a leaf below the path labeled P (i.e., below the termination point of the search). If the search traversed the u to pq edge, then P occurs starting at position $i - 1$; otherwise it occurs starting at position i.

Generalize this idea and work out the details for any number of node merges.

11. In some applications it is desirable to know the number of times an input string P occurs in a larger string S. After the obvious linear-time preprocessing, queries of this sort can be answered in $O(|P|)$ time using a suffix tree. Show how to preprocess the DAG in linear time so that these queries can be answered in $O(|P|)$ time using a DAG.

12. Prove the correctness of the compaction algorithm for suffix trees.

13. Let S^r be the reverse of the string S. Is there a relationship between the number of nodes in the DAG for S and the DAG for S^r? Prove it. Find the relationship between the DAG for S and the DAG for S^r (this relationship is a bit more direct than for suffix trees).

14. In Theorem 7.7.1 we gave an easily computed condition to determine when two subtrees of a suffix tree for string S are isomorphic. An alternative condition that is less useful for efficient computation is as follows: Let α be the substring labeling a node p and β be the substring labeling a node q in the suffix tree for S. The subtrees of p and q are isomorphic if and only if the set of positions in S where occurrences of α end equals the set of positions in S where occurrences of β end.

Prove the correctness of this alternative condition for subtree isomorphism.

15. Does Theorem 7.7.1 still hold for a generalized suffix tree (for more than a single string)? If not, can it be easily extended to hold?

16. The DAG D for a string S can be converted to a finite-state machine by expanding each edge with more than one character on its label into a series of edges labeled by one character

each. This finite-state machine will recognize substrings of S, but it will not necessarily be the smallest such finite-state machine. Give an example of this.

We now consider how to build the smallest finite-state machine to recognize substrings of S. Again start with a suffix tree for S, merge isomorphic subtrees, and then expand each edge that it labeled with more than a single character. However, the merge operation must be done more carefully than before. Moreover, we imagine there is a suffix link from each leaf i to each leaf $i + 1$, for $i < n$. Then, there is a path of suffix links connecting all the leaves, and each leaf has zero leaves beneath it. Hence, all the leaves will get merged.

Recall that Q is the set of all pairs (p, q) such that there exists a suffix link from p to q in $\mathcal{T}$, where p and q have the same number of leaves in their respective subtrees. Suppose (p, q) is in Q. Let v be the parent of p, let γ be the label of the edge (v, p) into p, and let δ be the label of the edge into q. Explain why $|\gamma| \geq |\delta|$. Since every edge of the DAG will ultimately be expanded into a number of edges equal to the length of its edge-label, we want to make each edge-label as small as possible. Clearly, δ is a suffix of γ, and we will exploit this fact to better merge edge-labels. During a merge of p into q, remove all out edges from p as before, but the edge from v is not necessarily directed to q. Rather, if $|\delta| > 1$, then the δ edge is split into two edges by the introduction of a new node u. The first of these edges is labeled with the first character of δ and the second one, edge (u, q), is labeled with the remaining characters of δ. Then the edge from v is directed to u rather than to q. Edge (v, u) is labeled with the first $|\gamma| - |\delta| + 1$ characters of γ.

Using this modified merge, clean up the description of the entire compaction process and prove that the resulting DAG recognizes substrings of S. The finite-state machine for S is created by expanding each edge of this DAG labeled by more than a single character. Each node in the DAG is now a state in the finite-state machine.

17. Show that the finite-state machine created above has the fewest number of states of any finite-state machine that recognizes substrings of S. The key to this proof is that a deterministic finite-state machine has the fewest number of states if no state in it is equivalent to any other state. Two states are equivalent if, starting from them, exactly the same set of strings are accepted. See [228].

18. Suppose you already have the Aho–Corasick keyword tree (with backlinks). Can you use it to compute matching statistics in linear time, or if not, in some "reasonable" nonlinear time bound? Can it be used to solve the longest common substring problem in a reasonable time bound? If not, what is the difficulty?

19. In Section 7.16 we discussed how to use a suffix tree to search for all occurrences of a set of patterns in a given text. If the length of all the patterns is n and the length of the text is m, that method takes $O(n + m)$ time and $O(m)$ space. Another view of this is that the solution takes $O(m)$ preprocessing time and $O(n)$ search time. In contrast, the Aho–Corasick method solves the problem in the same total time bound but in $O(n)$ space. Also, it needs $O(n)$ preprocessing time and $O(m)$ search time.

Because there is no definite relationship between n and m, sometimes one method will use less space or preprocessing time than the other. By using a generalized suffix tree, for the set of patterns and the reverse role for suffix trees discussed in Section 7.8, it is possible to solve the problem with a suffix tree, obtaining exactly the same time and space bounds obtained by the Aho–Corasick method. Show in detail how this is done.

20. Using the reverse role for suffix trees discussed in Section 7.8, show how to solve the general DNA contamination problem of Section 7.5 using only a suffix tree for S_1, rather than a generalized suffix tree for S_1 together with all the possible contaminants.

21. In Section 7.8.1 we used a suffix tree for the small string P to compute the matching statistics $ms(i)$ for each position i in the long text string T. Now suppose we also want to

compute matching statistics $ms(j)$ for each position j in P. Number $ms(j)$ is defined as the length of the longest substring starting at position j in P that matches some substring in T. We could proceed as before, but that would require a suffix tree for the long tree T. Show how to find all the matching statistics for both T and P in $O(|T|)$ time, using only a suffix tree for P.

22. In our discussion of matching statistics, we used the suffix links created by Ukkonen's algorithm. Suffix links can also be obtained by reversing the link pointers of Weiner's algorithm, but suppose that the tree cannot be modified. Can the matching statistics be computed in linear time using the tree and link pointers as given by Weiner's algorithm?

23. In Section 7.8 we discussed the reverse use of a suffix tree to solve the exact pattern matching problem: Find all occurrences of pattern P in text T. The solution there computed the matching statistic $ms(i)$ for each position i in the text. Here is a modification of that method that solves the exact matching problem but does not compute the matching statistics: Follow the details of the matching statistic algorithm but never examine new characters in the text unless you are on the path from the root to the leaf labeled 1. That is, in each iteration, do not proceed below the string $\alpha\gamma$ in the suffix tree, until you are on the path that leads to leaf 1. When not on this path, the algorithm just follows suffix links and performs skip/count operations until it gets back on the desired path.

    Prove that this modification correctly solves the exact matching problem in linear time.

    What advantages or disadvantages are there to this modified method compared to computing the matching statistics?

24. There is a simple practical improvement to the previous method. Let v be a point on the path to leaf 1 where some search ended, and let v' be the node on that path that was next entered by the algorithm (after some number of iterations that visit nodes off that path). Then, create a direct shortcut link from v to v'. The point is that if any future iteration ends at v, then the shortcut link can be taken to avoid the longer indirect route back to v'.

    Prove that this improvement works (i.e., that the exact matching problem is correctly solved in this way).

    What is the relationship of these shortcut links to the failure function used in the Knuth-Morris-Pratt method? When the suffix tree encodes more than a single pattern, what is the relationship of these shortcut links to the backpointers used by the Aho–Corasik method?

25. We might modify the previous method even further: In each iteration, only follow the suffix link (to the end of α) and do not do any skip/count operations or character comparisons unless you are on the path to leaf 1. At that point, do all the needed skip/count computations to skip past any part of the text that has already been examined.

    Fill in the details of this idea and establish whether it correctly solves the exact matching problem in linear time.

26. Recall the discussion of STSs in Section 7.8.3, page 135. Show in more detail how matching statistics can be used to identify any STSs that a string contains, assuming there are "modest" number of errors in either the STS strings or the new string.

27. Given a set of k strings of length n each, find the longest common prefix for each pair of strings. The total time should be $O(kn + p)$, where p is the number of pairs of strings having a common prefix of length greater than zero. (This can be solved using the lowest common ancestor algorithm discussed later, but a simpler method is possible.)

28. For any pair of strings, we can compute the length of the longest prefix common to the pair in time linear in their total length. This is a simple use of a suffix tree. Now suppose we are given k strings of total length n and want to compute the minimum length of all the pairwise longest common prefixes over all of the $\binom{k}{2}$ pairs of strings, that is, the smallest

length of the pairwise pairs. The obvious way to solve this is to solve the longest common prefix problem for each of the $\binom{k}{2}$ pairs of strings in $O(k^2 + kn)$ time. Show how to solve the problem in $O(n)$ time independent of k. Consider also the problem of computing the maximum length over all the pairwise common prefixes.

29. Verify that the all-pairs suffix-prefix matching problem discussed in Section 7.10 can be solved in $O(km)$ time using any linear-time string matching method. That is, the $O(km)$ time bound does not require a suffix tree. Explain why the bound does not involve a term for k^2.

30. Consider again the all-pairs suffix-prefix matching problem. It is possible to solve the problem in the same time bound without an explicit tree traversal. First, build a generalized suffix tree $T(S)$ for the set of k strings S (as before), and set up a vector V of length k. Then successively initialize vector V to contain all zeros, and match each string in the set through the tree. The match using any string S_j ends at the leaf labeled with suffix 1 of string S_j. During this walk for S_j, if a node v is encountered containing index i in its list $L(v)$, then write the string-depth of node v into position i of vector V. When the walk reaches the leaf for suffix 1 of S_j, $V(i)$, for each i, specifies the length of the longest suffix of S_i that matches a prefix of S_j.

Establish the worst-case time analysis of this method. Compare any advantages or disadvantages (in practical space and/or time) of this method compared to the tree traversal method discussed in Section 7.10. Then propose modifications to the tree traversal method that maintain all of its advantages and also correct for its disadvantages.

31. A substring α is called a *prefix repeat* of string S if α is a prefix of S and has the form $\beta\beta$ for some string β. Give a linear-time algorithm to find the longest prefix repeat of an input string S. This problem was one of Weiner's motivations for developing suffix trees.

Very frequently in the sequence analysis literature, methods aimed at finding interesting features in a biological sequence begin by cataloging certain substrings of a long string. These methods almost always pick a *fixed-length window*, and then find all the distinct strings of that fixed length. The result of this window or *q-gram* approach is of course influenced by the choice of the window length. In the following three exercises, we show how suffix trees avoid this problem, providing a natural and more effective extension of the window approach. See also Exercise 26 of Chapter 14.

32. There are $m^2/2$ substrings of a string T whose length is m. Some of those substrings are identical and so occur more than once in the string. Since there are $\Theta(m^2)$ substrings, we cannot count the number of times each appears in T in $O(m)$ time. However, using a suffix tree we can get an *implicit* representation of these numbers in $O(m)$ time. In particular, when any string P of length n is specified, the implicit representation should allow us to compute the frequency of P in T in $O(n)$ time. Show how to construct the implicit frequency representation and how to use it.

33. Show how to count the number of distinct substrings of a string T in $O(m)$ time, where the length of T is m. Show how to enumerate one copy of each distinct substring in time proportional to the length of all those strings.

34. One way to hunt for "interesting" sequences in a DNA sequence database is to look for substrings in the database that appear much more often than they would be predicted to appear by chance alone. This is done today and will become even more attractive when huge amounts of anonymous DNA sequences are available.

Assuming one has a statistical model to determine how likely any particular substring would occur by chance, and a threshold above which a substring is "interesting", show how to efficiently find all interesting substrings in the database. If the database has total length m, then the method should take time $O(m)$ plus time proportional to the number of interesting substrings.

35. (**Smallest *k*-repeat**) Given a string S and a number k, we want to find the smallest substring of S that occurs in S exactly k times. Show how to solve this problem in linear time.

36. Theorem 7.12.1, which states that there can be at most n maximal repeats in a string of length n, was established by connecting maximal repeats with suffix trees. It seems there should be a direct, simple argument to establish this bound. Try to give such an argument. Recall that it is not true that at most one maximal repeat begins at any position in S.

37. Given two strings S_1 and S_2 we want to find all *maximal common pairs* of S_1 and S_2. A common substring C is maximal if the addition to C of any character on either the right or left of C results in a string that is not common to both S_1 and S_2. For example, if $A = aayxpt$ and $B = aqyxpw$ then the string yxp is a maximal common substring, whereas yx is not. A *maximal common pair* is a triple (p_1, p_2, n'), where p_1 and p_2 are positions in S_1 and S_2, respectively, and n' is the length of a maximal common substring starting at those positions. This is a generalization of the maximal pair in a single string.

    Letting m denote the total length of S_1 and S_2, give an $O(m + k)$-time solution to this problem, where k is the number of triples output. Give an $O(m)$-time method just to count the number of maximal common pairs and an $O(n + l)$-time algorithm to find one copy of each maximal common substring, where l is the total length of those strings. This is a generalization of the maximal repeat problem for a single string.

38. Another, equally efficient, but less concise way to identify supermaximal repeats is as follows: A maximal repeat in S represented by the left-diverse node v in the suffix tree for S is a supermaximal repeat if and only if no proper descendant of v is left diverse and no node in v's subtree (including v) is reachable via a path of suffix links from a left diverse node other than v. Prove this.

    Show how to use the above claim to find all supermaximal repeats in linear time.

39. In biological applications, we are often not only interested in repeated substrings but in occurrences of substrings where one substring is an inverted copy of the other, a complemented copy, or (almost always) both. Show how to adapt all the definitions and techniques developed for repeats (maximal repeats, maximal pairs, supermaximal repeats, near-supermaximal repeats, common substrings) to handle inversion and complementation, in the same time bounds.

40. Give a linear-time algorithm that takes in a string S and finds the longest maximal pair in which the two copies do not overlap. That is, if the two copies begin at positions $p_1 < p_2$ and are of length n', then $p_1 + n' < p_2$.

41. Techniques for handling repeats in DNA are not only motivated by repetitive structures that occur in the DNA itself but also by repeats that occur in data collected from the DNA. The paper by Leung et al. [298] gives one example. In that paper they discuss a problem of analyzing DNA sequences from *E. coli*, where the data come from more than 1,000 independently sequenced fragments stored in an *E. coli* database. Since the sequences were contributed by independent sequencing efforts, some fragments contained others, some of the fragments overlapped others, and many intervals of the *E. coli* genome were yet unsequenced. Consequently, before the desired analysis was begun, the authors wanted to "clean up" the data at hand, finding redundantly sequenced regions of the *E. coli* genome and packaging all the available sequences into a few *contigs*, i.e., strings that contain all the substrings in the data base (these contigs may or may not be the shortest possible).

    Using the techniques discussed for finding repeats, suffix-prefix overlaps, and so on, how would you go about cleaning up the data and organizing the sequences into the desired contigs?

(This application is instructive because *E. coli*, as in most prokaryotic organisms, contains little repetitive DNA. However, that does not mean that techniques for handling repetitive structures have no application to prokaryotic organisms.)

Similar clean-up problems existed in the yeast genome database, where, in addition to the kinds of problems listed above, strings from other organisms were incorrectly listed as yeast, yeast strings were incorrectly identified in the larger composite databases, and parts of cloning vectors appeared in the reported yeast strings. To further complicate the problem, more recent higher quality sequencing of yeast may yield sequences that have one tenth of the errors than sequences in the existing databases. How the new and the old sequencing data should be integrated is an unsettled issue, but clearly, any large-scale curation of the yeast database will require the kinds of computational tools discussed here.

42. ***k*-cover problem.** Given two input strings S_1 and S_2 and a parameter k, a *k-cover C* is a set of substrings of S_1, each of length k or greater, such that S_2 can be expressed as the concatenation of the substrings of C in some order. Note that the substrings contained in C may overlap in S_1, but not in S_2. That is, S_2 is a permutation of substrings of S_1 that are each of length k or greater. Give a linear-time algorithm to find a k-cover from two strings S_1 and S_2, or determine that no such cover exists.

If there is no k-cover, then find a set of substrings of S_1, each of length k or greater, that cover the most characters of S_2. Or, find the largest $k' < k$ (if any) such that there is a k'-cover. Give linear-time algorithms for these problems.

Consider now the problem of finding *nonoverlapping* substrings in S_1, each of length k or greater, to cover S_2, or cover it as much as possible. This is a harder problem. Grapple with it as best you can.

43. **exon shuffling.** In eukaryotic organisms, a gene is composed of alternating *exons*, whose concatenation specifies a single protein, and *introns*, whose function is unclear. Similar exons are often seen in a variety of genes. Proteins are often built in a modular form, being composed of distinct domains (units that have distinct functions or distinct folds that are independent of the rest of the protein), and the same domains are seen in many different proteins, although in different orders and combinations. It is natural to wonder if exons correspond to individual protein domains, and there is some evidence to support this view. Hence modular protein construction may be reflected in the DNA by modular gene construction based on the reuse and reordering of stock exons. It is estimated that all proteins sequenced to date are made up of just a few thousand exons [468]. This phenomenon of reusing exons is called *exon shuffling*, and proteins created via exon shuffling are called *mosaic proteins*. These facts suggest the following general search problem.

The problem: Given anonymous, but sequenced, strings of DNA from protein-coding regions where the exons and introns are not known, try to identify the exons by finding common regions (ideally, identical substrings) in two or more DNA strings. Clearly, many of the techniques discussed in this chapter concerning common or repeated substrings could be applied, although they would have to be tried out on real data to test their utility or limitations. No elegant analytical result should be expected. In addition to methods for repeats and common substrings, does the k-cover problem seem of use in studying exon shuffling? That question will surely require an empirical, rather than theoretical answer. Although it may not give an elegant worst-case result, it may be helpful to first find all the maximal common substrings of length k or more.

44. Prove Lemma 7.14.1.

45. Prove the correctness of the method presented in Section 7.13 for the *circular string linearization problem*.

46. Consider in detail whether a suffix array can be used to efficiently solve the more complex string problems considered in this chapter. The goal is to maintain the space-efficient properties of the suffix array while achieving the time-efficient properties of the suffix tree. Therefore, it would be cheating to first use the suffix array for a string to construct a suffix tree for that string.

47. Give the details of the preprocessing needed to implement the bad character rule in the Boyer–Moore approach to exact set matching.

48. In Section 7.16.3, we used a suffix tree to implement a *weak* good suffix rule for a Boyer–Moore set matching algorithm. With that implementation, the increment of index i was determined in constant time after any test, independent even of the alphabet size. Extend the suffix tree approach to implement a *strong* good suffix rule, where again the increment to i can be found in constant time. Can you remove the dependence on the alphabet in this case?

49. Prove Theorem 7.16.2.

50. In the Ziv–Lempel algorithm, when computing (s_i, l_i) for some position i, why should the traversal end at point p if the string-depth of p plus c_v equals i? What would be the problem with letting the match extend past character i?

51. Try to give some explanation for why the Ziv–Lempel algorithm outputs the extra character compared to *compression algorithm* 1.

52. Show how to compute all the n values $ZL(i)$, defined in Section 7.18, in $O(n)$ time. One solution is related to the computation of matching statistics (Section 7.8.1).

53. **Successive refinement methods**

Successive refinement is a general algorithmic technique that has been used for a number of string problems [114, 199, 265]. In the next several exercises, we introduce the ideas, connect successive refinement to suffix trees, and apply successive refinement to particular string problems.

Let S be a string of length n. The relation E_k is defined on pairs of suffixes of S. We say $i E_k j$ if and only if suffix i and suffix j of S agree for at least their first k characters. Note that E_k is an equivalence relation and so it partitions the elements into equivalence classes. Also, since S has n characters, every class in E_n is a singleton. Verify the following two facts:

Fact 1 For any $i \neq j$, $i E_{k+1} j$ if and only if $i E_k j$ and $i + 1 E_k j + 1$.

Fact 2 Every E_{k+1} class is a subset of an E_k class and so the E_{k+1} partition is a refinement of the E_k partition.

We use a labeled tree T, called the *refinement tree*, to represent the successive refinements of the classes of E_k as k increases from 0 to n. The root of T represents class E_0 and contains all the n suffixes of S. Each child of the root represents a class of E_1 and contains the elements in that class. In general, each node at level l represents a class of E_l and its children represent all the E_{l+1} classes that refine it.

What is the relationship of T to the keyword tree (Section 3.4) constructed from the set of n suffixes of S?

Now modify T as follows. If node v represents the same set of suffixes as its parent node v', contract v and v' to a single node. In the new refinement tree, T', each nonleaf node has at least two children. What is the relationship of T' to the suffix tree for string S? Show how to convert a suffix tree for S into tree T' in $O(n^2)$ time.

54. Several string algorithms use successive refinement without explicitly finding or representing all the classes in the refinement tree. Instead, they construct only some of the classes
or only compute the tree implicitly. The advantage is reduced use of space in practice or
an algorithm that is better suited for parallel computation [116]. The original suffix array
construction method [308] is such an algorithm. In that algorithm, the suffix array is obtained as a byproduct of a successive refinement computation where the E_k partitions are
computed only for values of k that are a power of two. We develop that method here. First
we need an extension of Fact 1:

Fact 3 For any $i \neq j$, $i E_{2k} j$ if and only if $i E_k j$ and $i + k E_k j + k$.

From Fact 2, the classes of E_{2k} refine the classes of E_k.

The algorithm of [308] starts by computing the partition E_1. Each class of E_1 simply lists all
the locations in S of one specific character in the alphabet, and the classes are arranged
in lexical order of those characters. For example, for $S = mississippi\$$, E_1 has five classes:
$\{12\}$, $\{2, 5, 8, 11\}$, $\{1\}$, $\{9, 10\}$, and $\{3, 4, 6, 7\}$. The class $\{2, 5, 8, 11\}$ lists the position of all
the i's in S and so comes before the class for the single m, which comes before the class
for the s's, etc. The end-of-string character $\$$ is considered to be lexically smaller than any
other character.

How E_1 is obtained in practice depends on the size of the alphabet and the manner that it
is represented. It certainly can be obtained with $O(n \log n)$ character comparisons.

For any $k \geq 1$, we can obtain the E_{2k} partition by refining the E_k partition, as suggested in
Fact 3. However, it is not clear how to efficiently implement a direct use of Fact 3. Instead,
we create the E_{2k} partition in $O(n)$ time, using a *reverse* approach to refinement. Rather
than examining a class C of E_k to find how C should be refined, we use C *as a refiner* to
see how it forces other E_k classes to split, or to stay together, as follows: For each number
$i > k$ in C, locate and mark number $i - k$. Then, for each E_k class A, any numbers in A
marked by C identify a complete E_{2k} class. The correctness of this follows from Fact 3.

Give a complete proof of the correctness of the reverse refinement approach to creating
the E_{2k} partition from the E_k partition.

55. Each class of E_k, for any k, holds the starting locations of a k-length substring of S. The
algorithm in [308] constructs a suffix array for S using the reverse refinement approach,
with the added detail that the classes of E_k are kept in the lexical order of the strings
associated with the classes.

More specifically, to obtain the E_2 partition of $S = mississippi\$$, process the classes of
E_1 in order, from the lexically smallest to the lexically largest class. Processing the first
class, $\{12\}$, results in the creation of the E_2 class $\{11\}$. The second E_1 class $\{2,5,8,11\}$
marks indices $\{1,4,7\}$ and $\{10\}$, and hence it creates the three E_2 classes $\{1\},\{4,7\}$ and
$\{10\}$. Class $\{9,10\}$ of E_1 creates the two classes $\{8\}$ and $\{9\}$. Class $\{3,4,6,7\}$ of E_1 creates
classes $\{2,5\}$ and $\{3,6\}$ of E_2. Each class of E_2 holds the starting locations of identical
substrings of length one or two. These classes, lexically ordered by the substrings they
represent, are: $\{12\},\{11\},\{8\},\{2,5\},\{1\},\{10\},\{9\},\{4,7\},\{3,6\}$. The classes of E_4, in lexical order
are: $\{12\},\{11\},\{8\},\{2,5\},\{1\},\{10\},\{9\},\{7\},\{4\},\{6\},\{3\}$. Note that $\{2,5\}$ remain in the same E_4
class because $\{4,7\}$ were in the same E_2 class. The E_2 classes of $\{4,7\}$ and $\{3,6\}$ are each
refined in E_4. Explain why.

Although the general idea of reverse refinement should now be clear, efficient implementation requires a number of additional details. Give complete implementation details and
analysis, proving that the E_{2k} classes can be obtained from the E_k classes in $O(n)$ time.
Be sure to detail how the classes are kept in lexical order.

Assume n is a power of two. Note that the algorithm can stop as soon as every class is a
singleton, and this must happen within $\log_2 n$ iterations. When the algorithm ends, the order

of the (singleton) classes describes a permutation of the integers 1 to n. Prove that this permutation is the suffix array for string S. Conclude that the reverse refinement method creates a suffix array in $O(n \log n)$ time. What is the space advantage of this method over the $O(n)$-time method detailed in Section 7.14.1?

56. **Primitive tandem arrays**

Recall that a string α is called a *tandem array* if α is periodic (see Section 3.2.1), i.e., it can be written as β^l for some $l \geq 2$. When $l = 2$, the tandem array can also be called a *tandem repeat*. A tandem array $\alpha = \beta^l$ contained in a string S is called *maximal* if there are no additional copies of β before or after α.

Maximal tandem arrays were initially defined in Exercise 4 in Chapter 1 (page 13) and the importance of tandem arrays and repeats was discussed in Section 7.11.1. We are interested in identifying the maximal tandem arrays contained in a string. As discussed before, it is often best to focus on a structured subset of the strings of interest in order to limit the size of the output and to identify the most informative members. We focus here on a subset of the maximal tandem arrays that succinctly and implicitly encode all the maximal tandem arrays. (In Section 9.5, 9.6, and 9.6.1 we will discuss efficient methods to find *all* the tandem repeats in a string, and we allow the repeats to contain some errors.)

We use the pair (β, l) to describe the tandem array β^l. Now consider the tandem array $\alpha = abababababababab$. It can be described by the pair $(abababab, 2)$, or by $(abab, 4)$, or by $(ab, 8)$. Which description is best? Since the first two pairs can be deduced from the last, we choose the later pair. This "choice" will now be precisely defined.

A string β is said to be *primitive* if β is not periodic. For example, the string ab is primitive, whereas $abab$ is not. The pair $(ab, 8)$ is the preferred description of $abababababababab$ because string ab is primitive. The preference for primitive strings extends naturally to the description of maximal tandem arrays that occur as substrings in larger strings. Given a string S, we use the triple (i, β, l) to mean that a tandem array (β, l) occurs in S starting at position i. A triple (i, β, l) is called a *pm-triple* if β is primitive and β^l is a maximal tandem array.

For example, the maximal tandem arrays in *mississippi* described by the pm-triples are $(2, iss, 2), (3, s, 2), (3, ssi, 2), (6, s, 2)$ and $(9, p, 2)$. Note that two or more pm-triples can have the same first number, since two different maximal tandem arrays can begin at the same position. For example, the two maximal tandem arrays ss and $ssissi$ both begin at position three of *mississippi*.

The pm-triples succinctly encode all the tandem arrays in a given string S. Crochemore [114] (with different terminology) used a successive refinement method to find all the pm-triples in $O(n \log n)$ time. This implies the very nontrivial fact that in any string of length n there can be only $O(n \log n)$ pm-triples. The method in [114] finds the E_k partition for each k. The following lemma is central:

Lemma 7.20.1. *There is a tandem repeat of a k-length substring β starting at position i of S if and only if the numbers i and $i + k$ are both contained in a single class of E_k and no numbers between i and $i + k$ are in that class.*

Prove Lemma 7.20.1. One direction is easy. The other direction is harder and it may be useful to use Lemma 3.2.1 (page 40).

57. Lemma 7.20.1 makes it easy to identify pm-triples. Assume that the indices in each class of E_k are sorted in increasing order. Lemma 7.20.1 implies that (i, β, j) is a pm-triple, where β is a k-length substring, if and only if some single class of E_k contains a maximal series of numbers $i, i + k, i + 2k, \ldots, i + jk$, such that each consecutive pair of numbers differs by k. Explain this in detail.

By using Fact 1 in place of Fact 3, and by modifying the reverse refinement method developed in Exercises 54 and 55, show how to compute all the E_k partitions for all k (not just the powers of two) in $O(n^2)$ time. Give implementation details to maintain the indices of each class sorted in increasing order. Next, extend that method, using Lemma 7.20.1, to obtain an $O(n^2)$-time algorithm to find all the pm-triples in a string S.

58. To find all the pm-triples in $O(n \log n)$ time, Crochemore [114] used one additional idea. To introduce the idea, suppose all E_k classes except one, C say, have been used as refiners to create E_{k+1} from E_k. Let p and q be two indices that are together in some E_k class. We claim that if p and q are not together in the same E_{k+1} class, then one of them (at least) has already been placed in its proper E_{k+1} class. The reason is that by Fact 1, $p+1$ and $q+1$ cannot both be in the same E_k class. So by the time C is used as a refiner, either p or q has been marked and moved by an E_k class already used as refiners.

Now suppose that each E_k class is held in a linked list and that when a refiner identifies a number, p say, then p is removed from its current linked list and placed in the linked list for the appropriate E_{k+1} class. With that detail, if the algorithm has used all the E_k classes except C as refiners, then all the E_{k+1} classes are correctly represented by the newly created linked lists plus what remains of the original linked lists for E_k. Explain this in detail. Conclude that one E_k class need not be used as a refiner.

Being able to skip one class while refining E_k is certainly desirable, but it isn't enough to produce the stated bound. To do that we have to repeat the idea on a larger scale.

Theorem 7.20.1. *When refining E_k to create E_{k+1}, suppose that for every $k > 1$, exactly one (arbitrary) child of each E_{k-1} class is skipped (i.e., not used as a refiner). Then the resulting linked lists correctly identify the E_{k+1} classes.*

Prove Theorem 7.20.1. Note that Theorem 7.20.1 allows complete freedom in choosing which child of an E_{k-1} class to skip. This leads to the following:

Theorem 7.20.2. *If, for every $k > 1$, the largest child of each E_{k-1} class is skipped, then the total size of all the classes used as refiners is at most $n \log_2 n$.*

Prove Theorem 7.20.2. Now provide all the implementation details to find all the pm-triples in S in $O(n \log n)$ time.

59. Above, we established the bound of $O(n \log n)$ pm-triples as a byproduct of the algorithm to find them. But a direct, nonalgorithmic proof is possible, still using the idea of successive refinement and Lemma 7.20.1. In fact, the bound of $3n \log_2 n$ is fairly easy to obtain in this way. Do it.

60. Folklore has it that for any position i in S, if there are two pm-triples, (i, β, l), and (i, β', l'), and if $|\beta'| > |\beta|$, then $|\beta'| \geq 2|\beta|$. That would limit the number of pm-triples with the same first number to $\log_2 n$, and the $O(n \log n)$ bound would be immediate.

Show by example that the folklore belief is false.

61. **Primer selection problem**

Let S be a set of strings over some finite alphabet Σ. Give an algorithm (using a generalized suffix tree) to find a *shortest* string S over Σ that is a substring in *none* of the strings of S. The algorithm should run in time proportional to the sum of the lengths of the strings in S.

A more useful version of the problem is to find the shortest string S that is longer than a certain minimum length and is not a substring of any string of S. Often, a string α is given along with the set S. Now the problem becomes one of finding a shortest substring of α (if any) that does not appear as a substring of any string in S. More generally, for every i, compute the shortest substring (if any) that begins at position i of α and does not appear as a substring of any string in S.

The above problems can be generalized in many different directions and solved in essentially the same way. One particular generalization is the exact matching version of the *primer selection problem*. (In Section 12.2.5 we will consider a version of this problem that allows errors.)

The primer selection problem arises frequently in molecular biology. One such situation is in "chromosome walking", a technique used in some DNA sequencing methods or gene location problems. Chromosome walking was used extensively in the location of the Cystic Fibrosis gene on human chromosome 7. We discuss here only the DNA sequencing application.

In DNA sequencing, the goal is to determine the complete nucleotide sequence of a long string of DNA. To understand the application you have to know two things about existing sequencing technology. First, current common laboratory methods can only accurately sequence a small number of nucleotides, from 300 to 500, from one end of a longer string. Second, it is possible to replicate substrings of a DNA string starting at almost any point as long as you know a small number of the nucleotides, say nine, to the left of that point. This replication is done using a technology called *polymerase chain reaction* (*PCR*), which has had a tremendous impact on experimental molecular biology. Knowing as few as nine nucleotides allows one to synthesize a string that is complementary to those nine nucleotides. This complementary string can be used to create a "primer", which finds its way to the point in the long string containing the complement of the primer. It then hybridizes with the longer string at that point. This creates the conditions that allow the replication of part of the original string to the right of the primer site. (Usually PCR is done with two primers, one for each end, but here only one "variable" primer is used. The other primer is fixed and can be ignored in this discussion.)

The above two facts suggest a method to sequence a long string of DNA, assuming we know the first nine nucleotides at the very start of the string. After sequencing the first 300 (say) nucleotides, synthesize a primer complementary to the last nine nucleotides just sequenced. Then replicate a string containing the next 300 nucleotides, sequence that substring and continue. Hence the longer string gets sequenced by successively sequencing 300 nucleotides at a time, using the end of each sequenced substring to create the primer that initiates sequencing of the next substring. Compared to the shotgun sequencing method (to be discussed in Section 16.14), this *directed* method requires much less sequencing overall, but because it is an inherently sequential process it takes longer to sequence a long DNA string. (In the Cystic Fibrosis case another idea, called *gene jumping*, was used to partially parallelize this sequential process, but chromosome walking is generally laboriously sequential.)

There is a common problem with the above chromosome walking approach. What happens if the string consisting of the last nine nucleotides appears in another place in the larger string? Then the primer may not hybridize in the correct position and any sequence determined from that point would be incorrect. Since we know the sequence to the left of our current point, we can check the known sequence to see if a string complementary to the primer exists to the left. If it does, then we want to find a nine-length substring near the end of the last determined sequence that does not appear anywhere earlier. That substring can then by used to form the primer. The result will be that the next substring sequenced will resequence some known nucleotides and so sequence somewhat fewer than 300 new nucleotides.

Problem: Formalize this primer selection problem and show how to solve it efficiently using suffix trees. More generally, for each position i in string α find the shortest substring that begins at i and that appears nowhere else in α or $\mathcal{S}$.

62. In the primer selection problem, the goal of avoiding incorrect hybridizations to the *right* of the sequenced part of the string is more difficult since we don't yet know the sequence. Still, there are some known sequences that should be avoided. As discussed in Section 7.11.1, eukaryotic DNA frequently contains regions of repeated substrings, and the most commonly occurring substrings are known. On the problem that repeated substrings cause for chromosome walking, R. Weinberg[8] writes:

They were like quicksand; anyone treading on them would be sucked in and then propelled, like Alice in Wonderland, through some vast subterranean tunnel system, only to resurface somewhere else in the genome, miles away from the starting site. The genome was riddled wih these sinkholes, called "repeated sequences." They were guaranteed to slow any chromosomal walk to a crawl.

So a more general primer problem is the following: Given a substring α of 300 nucleotides (the last substring sequenced), a string β of known sequence (the part of the long string to the left of α whose sequence is known), and a set S of strings (the common parts of known repetitive DNA strings), find the furthest right substring in α of length nine that is not a substring of β or any string in set S. If there is no such string, then we might seek a string of length larger than nine that does not appear in β or S. However, a primer much larger than nine nucleotides long may falsely hybridize for other reasons. So one must balance the constraints of keeping the primer length in a certain range, making it unique, and placing it as far right as possible.

Problem: Formalize this version of the primer selection problem and show how to apply suffix trees to it.

Probe selection

A variant of the primer selection problem is the *hybridization probe* selection problem. In DNA fingerprinting and mapping (discussed in Chapter 16) there is frequent need to see which *oligomers* (short pieces of DNA) hybridize to some target piece of DNA. The purpose of the hybridization is not to create a primer for PCR but to extract some information about the target DNA. In such mapping and fingerprinting efforts, contamination of the target DNA by vector DNA is common, in which case the oligo probe may hybridize with the vector DNA instead of the target DNA. One approach to this problem is to use specifically designed oligomers whose sequences are rarely in the genome of the vector, but are frequently found in the cloned DNA of interest. This is precisely the primer (or probe) selection problem.

In some ways, the probe selection problem is a better fit than the primer problem is to the exact matching techniques discussed in this chapter. This is because when designing probes for mapping, it is desirable and feasible to design probes so that even a single mismatch will destroy the hybrization. Such stringent probes can be created under certain conditions [134, 177].

---

[8] Racing to the Beginning of the Road; The Search for the Origin of Cancer, Harmony Books, 1996.

8

Constant-Time Lowest Common Ancestor Retrieval

8.1. Introduction

We now begin the discussion of an amazing result that greatly extends the usefulness of suffix trees (in addition to many other applications).

Definition   In a rooted tree $\mathcal{T}$, a node u is an *ancestor* of a node v if u is on the unique path from the root to v. With this definition a node is an ancestor of itself. A *proper ancestor* of v refers to an ancestor that is not v.

Definition   In a rooted tree $\mathcal{T}$, the *lowest common ancestor* (*lca*) of two nodes x and y is the deepest node in $\mathcal{T}$ that is an ancestor of both x and y.

For example, in Figure 8.1 the *lca* of nodes 6 and 10 is node 5 while the *lca* of 6 and 3 is 1.

The amazing result is that after a *linear* amount of preprocessing of a rooted tree, any two nodes can then be specified and their lowest common ancestor found in *constant* time. That is, a rooted tree with n nodes is first preprocessed in $O(n)$ time, and thereafter any lowest common ancestor query takes only constant time to solve, *independent* of n. Without preprocessing, the best worst-case time bound for a single query is $\Theta(n)$, so this is a most surprising and useful result. The *lca* result was first obtained by Harel and Tarjan [214] and later simplified by Schieber and Vishkin [393]. The exposition here is based on the later approach.

8.1.1. What do ancestors have to do with strings?

The constant-time search result is particularly useful in situations where many lowest common ancestor queries must be answered for a fixed tree. That situation often arises when applying the result to string problems. To get a feel for the connection between strings and common ancestors, note that if the paths from the root to two leaves i and j in a suffix tree are identical down to a node v, then suffix i and suffix j share a common prefix consisting of the string labeling the path to v. Hence the lowest common ancestor of leaves i and j identifies the longest common prefix of suffixes i and j. The ability to find such a longest common prefix will be an important primitive in many string problems. Some of these will be detailed in Chapter 9. That chapter can be read by taking the *lca* result as a black box, if the reader prefers to review motivating examples before diving into the technical details of the *lca* result.

The statement of the *lca* result is widely known, but the details are not. The result is usually taken as a black box in the string literature, and there is a general "folk" belief that the result is only theoretical, not practical. This is certainly not true of the Schieber–Vishkin method – it is a very practical, low-overhead method, which is simple to program and very fast in practice. It should definitely be in the standard repertoire of string packages.

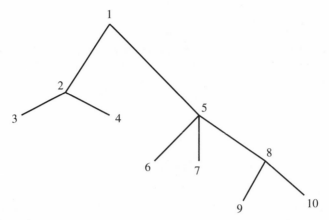

Figure 8.1: A general tree $\mathcal{T}$ with nodes named by a depth-first numbering.

However, although the method is easy to program, it is not trivial to understand at first and has been described as based on "bit magic". Nonetheless, the result has been so heavily applied in many diverse string methods, and its use is so critical in those methods, that a detailed discussion of the result is worthwhile. We hope the following exposition is a significant step toward making the method more widely understood.

8.2. The assumed machine model

Because constant retrieval time is such a demanding goal (and even linear preprocessing time requires careful attention to detail), we must be clear on what computational model is being used. Otherwise we may be accused of cheating – using an overly powerful set of primitive operations. What primitive operations are permitted in constant time? In the unit-cost RAM model when the input tree has n nodes, we certainly can allow any number with up to $O(\log n)$ bits to be written, read, or used as an address in constant time. Also, two numbers with up to $O(\log n)$ bits can be compared, added, subtracted, multiplied, or divided in constant time. Numbers that take more than $\Theta(\log n)$ bits to represent cannot be operated on in constant time. These are standard unit-cost requirements, forbidding "large" numbers from being manipulated in constant time. The *lca* result (that *lca* queries can be answered in constant time after linear-time preprocessing) can be proved in this unit-cost RAM model. However, the exposition is easier if we assume that certain additional *bit-level* operations can also be done in constant time, as long as the numbers have only $O(\log n)$ bits. In particular, we assume that the **AND**, **OR**, and **XOR** (exclusive or) of two (appropriate sized) binary numbers can be done in constant time; that a binary number can be shifted (left or right) by up to $O(\log n)$ bits in constant time; that a "mask" of consecutive 1-bits can be created in constant time; and that the position of the left-most or right-most 1-bit in a binary number can be found in constant time. On many machines, and with several high-level programming languages, these are reasonable assumptions, again assuming that the numbers involved are never more than $O(\log n)$ bits long. But for the purists, after we explain the *lca* result using these more liberal assumptions, we will explain how to achieve the same results using only the standard unit-cost RAM model.

8.3. Complete binary trees: a very simple case

We begin the discussion of the *lca* result on a particularly simple tree whose nodes are named in a very special way. The tree is a *complete binary tree* whose nodes have names

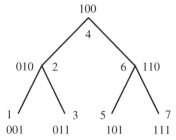

Figure 8.2: A binary tree with four leaves. The path numbers are written both in binary and in base ten.

that encode paths to them. The notation $\mathcal{B}$ will refer to this complete binary tree, and $\mathcal{T}$ will refer to an arbitrary tree.

Suppose that $\mathcal{B}$ is a *rooted complete binary tree* with p leaves ($n = 2p - 1$ nodes in total), so that every internal node has exactly two children and the number of edges on the path from the root to any leaf in $\mathcal{B}$ is $d = \log_2 p$. That is, the tree is complete and all leaves are at the same depth from the root. Each node v of $\mathcal{B}$ is assigned a $d + 1$ bit number, called its *path number*, that encodes the unique path from the root to v. Counting from the left-most bit, the ith bit of the path number for v corresponds to the ith edge on the path from the root to v: A 0 for the ith bit from the left indicates that the ith edge on the path goes to a left child, and a 1 indicates a right child.[1] For example, a path that goes left twice, right once, and then left again ends at a node whose path number begins (on the left) with 0010. The bits that describe the path are called *path bits*. Each path number is then padded out to $d + 1$ bits by adding a 1 to the right of the path bits followed by as many additional 0s as needed to make $d + 1$ bits. Thus for example, if $d = 6$, the node with path bits 0010 is named by the 7-bit number 0010100. The root node for $d = 6$ would be 1000000. In fact, the root node always has a number with left bit 1 followed by d 0s. (See Figure 8.2 for an additional example.) We will refer to nodes in $\mathcal{B}$ by their path numbers.

As the tree in Figure 8.2 suggests, path numbers have another well-known description – that of *inorder* numbers. That is, when the nodes of $\mathcal{B}$ are numbered by an inorder traversal (recursively number the left child, number the root, and then recursively number the right child), the resulting node numbers are exactly the path numbers discussed above. We leave the proof of this for the reader (it has little significance in our exposition). The path number concept is preferred since it explicitly relates the number of a node to the description of the path to it from the root.

8.4. How to solve *lca* queries in $\mathcal{B}$

Definition For any two nodes i and j, we let *lca(i,j)* denote the least common ancestor of i and j.

Given two nodes i and j, we want to find *lca(i,j)* in $\mathcal{B}$ (remembering that both i and j are path numbers). First, when *lca(i,j)* is either i or j (i.e., one of these two nodes is an ancestor of the other), then this can be detected by a very simple constant-time algorithm, discussed in Exercise 3. So assume that *lca(i,j)* is neither i nor j. The algorithm begins by taking the *exclusive or* (**XOR**) of the binary number for i and the binary number for j, denoting the result by x_{ij}. The **XOR** of two bits is 1 if and only if the two bits are different, and the **XOR** of two $d + 1$ bit numbers is obtained by independently taking the **XOR** of

---

[1] Note that normally when discussing binary numbers, the bits are numbered from right (least significant) to left (most significant). This is opposite the left-to-right ordering used for strings and for path numbers.

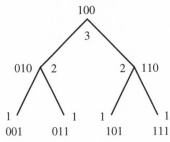

Figure 8.3: A binary tree with four leaves. The path numbers are in binary, and the position of the least-significant 1-bit is given in base ten.

each bit of the two numbers. For example, **XOR** of 00101 and 10011 is 10110. Since i and j are $O(\log n)$ bits long, **XOR** is a constant-time operation in our model.

The algorithm next finds the most significant (left-most) 1-bit in x_{ij}. If the left most 1-bit in the **XOR** of i and j is in position k (counting from the left), then the left most $k - 1$ bits of i and j are the same, and the paths to i and j must agree for the first $k - 1$ edges and then diverge. It follows that the path number for $lca(i,j)$ consists of the left most $k - 1$ bits of i (or j) followed by a 1-bit followed by $d + 1 - k$ zeros. For example, in Figure 8.2, the **XOR** of 101 and 111 (nodes 5 and 7) is 010, so their respective paths share one edge – the right edge out of the root. The **XOR** of 010 and 101 (nodes 2 and 5) is 111, so the paths to 2 and 5 have no agreement, and hence 100, the root, is their lowest common ancestor.

Therefore, to find $lca(i,j)$, the algorithm must **XOR** two numbers, find the left-most 1-bit in the result (say at position k), shift i right by $d + 1 - k$ places, set the right most bit to a 1, and shift it back left by $d + 1 - k$ places. By assumption, each of these operations can be done in constant time, and hence the lowest common ancestor of i and j can be found in constant time in $\mathcal{B}$.

In summary, we have

Theorem 8.4.1. *In a complete binary tree, after linear-time preprocessing to name nodes by their path numbers, any lowest common ancestor query can be answered in constant time.*

This simple case of a complete binary tree is very special, but it is presented both to develop intuition and because complete binary trees are used in the description of the general case. Moreover, by actually using complete binary trees, a very elegant and relatively simple algorithm can answer lca queries in constant time, if $\Theta(n \log n)$ time is allowed for preprocessing $\mathcal{T}$ and $\Theta(n \log n)$ space is available after the preprocessing. That method is explored in Exercise 12.

The lca algorithm we will present for general trees builds on the case of a complete binary tree. The idea (conceptually) is to *map* the nodes of a general tree $\mathcal{T}$ to the nodes of a complete binary tree $\mathcal{B}$ in such a way that lca retrievals on $\mathcal{B}$ will help to quickly solve lca queries on $\mathcal{T}$. We first describe the general lca algorithm assuming that the $\mathcal{T}$ to $\mathcal{B}$ mapping is explicitly used, and then we explain how explicit mapping can be avoided.

8.5. First steps in mapping $\mathcal{T}$ to $\mathcal{B}$

The first thing the preprocessing does is traverse $\mathcal{T}$ in a depth-first manner, numbering the nodes of v in the order that they are first encountered in the traversal. This is the standard *depth-first numbering* (preorder numbering) of nodes (see Figure 8.1). With this numbering scheme, the nodes in the subtree of any node v in $\mathcal{T}$ have consecutive depth-first numbers, beginning with the number for v. That is, if there are q nodes in the subtree rooted at v, and v gets numbered k, then the numbers given to the other nodes in the subtree are $k + 1$ through $k + q - 1$.

For convenience, from this point on the nodes in $\mathcal{T}$ will be referred to by their depth-first numbers. That is, when we refer to node v, v is both a node and a number. Be careful not to confuse depth-first numbers used for the general tree $\mathcal{T}$ with path numbers used only for the binary tree $\mathcal{B}$.

Definition For any number k, $h(k)$ denotes the position (counting from the right) of the least-significant 1-bit in the binary representation of k.

For example, $h(8) = 4$ since 8 in binary is 1000, and $h(5) = 1$ since 5 in binary is 101. Another way to think of this is that $h(k)$ is one plus the number of consecutive zeros at the right end of k.

Definition In a complete binary tree the *height* of a node is the number of nodes on the path from it to a leaf. The height of a leaf is one.

The following lemma states a crucial fact that is easy to prove by induction on the height of the nodes.

Lemma 8.5.1. *For any node k (node with path number k) in $\mathcal{B}$, $h(k)$ equals the height of node k in $\mathcal{B}$.*

For example, node 8 (binary 1000) is at height 4, and the path from it to a leaf has four nodes (three edges).

Definition For a node v of $\mathcal{T}$, let $I(v)$ be a node w in $\mathcal{T}$ such that $h(w)$ is maximum over all nodes in the subtree of v (including v itself).

That is, over all the nodes in the subtree of v, $I(v)$ is a node (depth-first number) whose binary representation has the largest number of consecutive zeros at its right end. Figure 8.4 shows the node numbers from Figure 8.1 in binary and base 10. Then $I(1)$, $I(5)$, and $I(8)$ are all 8, $I(2)$ and $I(4)$ are both 4, and $I(v) = v$ for every other node in the figure.

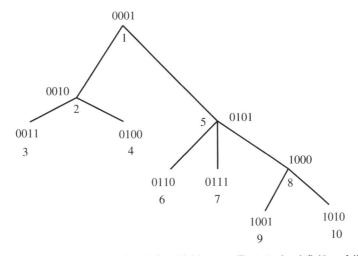

Figure 8.4: Node numbers given in four-bit binary, to illustrate the definition of $I(v)$.

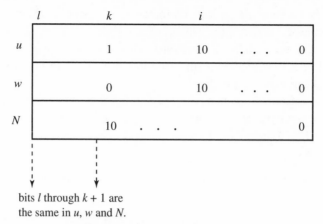

bits l through $k + 1$ are
the same in u, w and N.

Figure 8.5: Numbers u, w, and N.

Clearly, if node v is an ancestor of a node w then $h(I(v)) \geq h(I(w))$. Another way to say this is that the $h(I(v))$ values never decrease along any upward path in $\mathcal{T}$. This fact will be important in several of the proofs below.

In the tree in Figure 8.4, node $I(v)$ is uniquely determined for each node v. That is, for each node v there is exactly one w in v's subtree such that $h(w)$ is maximum. This is no accident, and it will be important in the *lca* algorithm. We now prove this fact.

Lemma 8.5.2. *For any node v in $\mathcal{T}$, there is a unique node w in the subtree of v such that $h(w)$ is maximum over all nodes in v's subtree.*

PROOF   Suppose not, and let u and w be two nodes in the subtree of v such that $h(u) = h(w) \geq h(q)$ for every node q in that subtree. Assume $h(u) = i$. By adding zeros to the left ends if needed, we can consider the two numbers u and w to have the same number of bits, say l. Since $u \neq w$, those two numbers must differ in some bit to the left of i (since by assumption bit i is 1 in both u and w, and all bits to the right of i are zero in both). Assume $u > w$, and let k be the left-most position where such a difference between u and w occurs. Consider the number N composed of the left-most $l - k$ bits of u followed by a 1 in bit k followed by $k - 1$ zeros (see Figure 8.5). Then N is strictly less than u and greater than w. Hence N must be the depth-first number given to some node in the subtree of v, because the depth-first numbers given to nodes below v form a consecutive interval. But $h(N) = k > i = h(u)$, contradicting the fact that $h(u) \geq h(q)$ for all nodes in the subtree of v. Hence the assumption that $h(u) = h(w)$ leads to a contradiction, and the Lemma is proved.   □

The uniqueness of $I(v)$ for each v can be summarized by the following corollary:

Corollary 8.5.1. The function $v \rightarrow I(v)$ is well defined.

8.6. The mapping of $\mathcal{T}$ to $\mathcal{B}$

In mapping nodes of $\mathcal{T}$ to nodes of a binary tree $\mathcal{B}$, we want to preserve enough of the ancestry relations in $\mathcal{T}$ so that *lca* relations in $\mathcal{B}$ can be used to determine *lca* queries in $\mathcal{T}$. The function $v \rightarrow I(v)$ will be central in defining that mapping. As a first step in understanding the mapping, we partition the nodes of $\mathcal{T}$ into classes of nodes whose I value is the same.

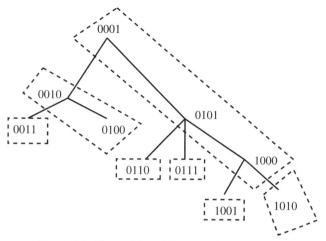

Figure 8.6: The partition of the nodes into seven runs.

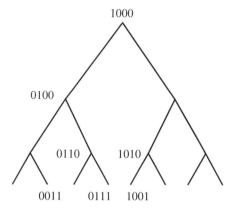

Figure 8.7: A node v in $\mathcal{B}$ is numbered if there is a node in $\mathcal{T}$ that maps to v.

Definition   A *run* in $\mathcal{T}$ is a maximal subset of nodes of $\mathcal{T}$, all of which have the same I value.

That is, two nodes u and v are in the same run if and only if $I(u) = I(v)$. Figure 8.6 shows a partition of the nodes of $\mathcal{T}$ into runs.

Algorithmically we can set $I(v)$, for all nodes, using a linear-time bottom-up traversal of $\mathcal{T}$ as follows: For every leaf v, $I(v) = v$. For every internal node v, $I(v) = v$ if $h(v)$ is greater than $h(v')$ for every child v' of v. Otherwise, $I(v)$ is set to the $I(v')$ value of the child v' whose $h(I(v'))$ value is the maximum over all children of v. The result is that each run forms an upward path of nodes in $\mathcal{T}$. And, since the $h(I())$ values never decrease along any upward path in $\mathcal{T}$, it follows that

Lemma 8.6.1. *For any node v, node $I(v)$ is the deepest node in the run containing node v.*

These facts are illustrated in Figure 8.6.

Definition   Define the *head* of a run to be the node of the run closest to the root.

For example, in Figure 8.6 node 1 (0001) is the head of a run of length three, node 2 (0010) is the head of a run of length two, and every remaining node (not in either of those two runs) is the head of a run consisting only of itself.

Finally, we can define the *tree map*.

Definition   The *tree map* is the mapping of nodes of T to nodes of a complete binary tree $\mathcal{B}$ with depth $d = \lceil \log n \rceil - 1$. In particular, node v of T maps to node $I(v)$ of $\mathcal{B}$ (recall that nodes of $\mathcal{B}$ are named by their path numbers).

The tree map is well defined because $I(v)$ is a $d + 1$ bit number, and each node of $\mathcal{B}$ is named by a distinct $d + 1$ bit number. Every node in a run of T maps to the same node in $\mathcal{B}$, but not all nodes in $\mathcal{B}$ generally have nodes in T mapping to them. Figure 8.7 shows tree $\mathcal{B}$ for tree T from Figure 8.6. A node v in $\mathcal{B}$ is numbered if there is a node in T that maps to v.

8.7. The linear-time preprocessing of T

We can now detail the linear-time preprocessing done on tree T.

Preprocessing Algorithm for T

Begin

1. Do a depth-first traversal of T to assign depth-first search numbers to the nodes. During the traversal compute $h(v)$ for each node v. For each node, set a pointer to its parent node in T.

2. Using the bottom-up algorithm described earlier, compute $I(v)$ for each v. For each number k such that $I(v) = k$ for some node v, set $L(k)$ to point to the head (or *Leader*) of the run containing node k. {Note that after this step, the head of the run containing an arbitrary node v can be retrieved in constant time: Compute $I(v)$ and then look up $L(I(v))$.}

    {This can easily be done while computing the I values. Node v is identified as the head of its run if the I value of v's parent is not $I(v)$.}

3. Let $\mathcal{B}$ be a complete binary tree with node-depth $d = \lceil \log n \rceil - 1$. Map each node v in T to node $I(v)$ in $\mathcal{B}$.

    {This mapping will be useful because it preserves enough (although not all) of the ancestry relations from T.}

    {The above three steps form the core of the preprocessing, but there is also one more technical step. For each node v in T, we want to encode *some* information about where in $\mathcal{B}$ the ancestors of v get mapped. That information is collected in the next step. Remember that $h(I(q))$ is the height in $\mathcal{B}$ of node $I(q)$ and so it is the height in $\mathcal{B}$ of the node that q gets mapped to.}

4. For each node v in T, create an $O(\log n)$ bit number A_v. Bit $A_v(i)$ is set to 1 if and only if node v has some ancestor in T that maps to height i in $\mathcal{B}$, i.e., if and only if v has an ancestor u such that $h(I(u)) = i$.

End.

   This ends the description of the preprocessing of T and the mapping of T to $\mathcal{B}$. To test your understanding of A_v, verify that the number of bits set to 1 in A_v is the number of *distinct* runs encountered on the path from the root to v. Setting the A numbers is easy by a linear-time traversal of T after all the I values are known: If v' is the parent of v then A_v is obtained by first copying $A_{v'}$ and then setting bit $A_v(i)$ to 1 if $h(I(v)) = i$ (this last step will be redundant if v and v' are on the same run, but it is always correct). As an example, consider node 3 (0011) in Figure 8.6. $A_3 = 1101(13)$ since 3 (0011) maps to height 1, 2 (0010) maps to height 3, and 1 (0001) maps to height 4 in $\mathcal{B}$.

What is this crazy mapping doing?

In the end, the programming details of this mapping (preprocessing) are very simple, and will become simpler in Section 8.9. The mapping only requires standard linear-time traversals of tree $\mathcal{T}$ (a minor programming exercise in a sophomore-level course). However, for most readers, what exactly the mapping accomplishes is quite unintuitive, because it is a many-one mapping. Certainly, ancestry relations in $\mathcal{T}$ are not perfectly preserved by the mapping into $\mathcal{B}$ [indeed, how could they be when the depth of $\mathcal{T}$ can be n while the depth of $\mathcal{B}$ is bounded by $\Theta(\log n)$], but much ancestry information is preserved, as shown in the next key lemma. Recall that a node is defined to be an ancestor of itself.

Lemma 8.7.1. *If z is an ancestor of x in $\mathcal{T}$ then $I(z)$ is an ancestor of $I(x)$ in $\mathcal{B}$. Stated differently, if z is an ancestor of x in $\mathcal{T}$ then either z and x are on the same run in $\mathcal{T}$ or node $I(z)$ is a proper ancestor of node $I(x)$ in $\mathcal{B}$.*

Figures 8.6 and 8.7 illustrate the claim in the lemma.

PROOF OF LEMMA 8.7.1   The proof is trivial if $I(z) = I(x)$, so assume that they are unequal. Since z is an ancestor of x in $\mathcal{T}$, $h(I(z)) \geq h(I(x))$ by the definition of I, but equality is only possible if $I(z) = I(x)$. So $h(I(z)) > h(I(x))$. Now $h(I(z))$ and $h(I(x))$ are the respective heights of nodes $I(z)$ and $I(x)$ in $\mathcal{B}$, so $I(z)$ is at a height greater than the height of $I(x)$ in $\mathcal{B}$.

Let $h(I(z))$ be i. We claim that $I(z)$ and $I(x)$ are identical in all bits to the left of i (recall that bits of a binary number are numbered from the right). If not, then let $k > i$ be the left-most bit where $I(z)$ and $I(x)$ differ. Without loss of generality, assume that $I(z)$ has bit 1 and $I(x)$ has bit 0 in position k. Since k is the point of left-most difference, the bits to the left of position k are equal in the two numbers, implying that $I(z) > I(x)$. Now z is an ancestor of x in $\mathcal{T}$, so nodes $I(z)$ and $I(x)$ are both in the subtree of z in $\mathcal{T}$. Furthermore, since $I(z)$ and $I(x)$ are depth-first number of nodes in the subtree of z in $\mathcal{T}$, every number between $I(x)$ and $I(z)$ occurs as a depth-first number of some node in the subtree of z. In particular, let N be the number consisting of the bits to the left of position k in $I(z)$ (or $I(x)$) followed by 1 followed by all 0s. (Figure 8.5 helps illustrate the situation, although z plays the role of u and x plays the role of w, and bit i in $I(z)$ is unknown.) Then $I(x) < N < I(z)$; therefore N is also a node in the subtree of z. But $k > i$, so $h(N) > h(I(z))$, contradicting the definition of I. It follows that $I(z)$ and $I(x)$ must be identical in the bits to the left of bit i.

Now bit i is the right most 1-bit in $I(z)$, so the bits to the left of bit i describe the complete path in $\mathcal{B}$ to node $I(z)$. Those identical bits to the left of bit i also form the initial part of the description of the path in $\mathcal{B}$ to node $I(x)$, since $I(x)$ has a 1-bit to the right of bit i. So those bits are in the path descriptions of both $I(z)$ and $I(x)$, meaning that the path to node $I(x)$ in $\mathcal{B}$ must go through node $I(z)$. Therefore, node $I(z)$ is an ancestor of node $I(x)$ in $\mathcal{B}$, and the lemma is proved. $\quad\square$

Having described the preprocessing of $\mathcal{T}$ and developed some of the properties of the tree map, we can now describe the way that *lca* queries are answered.

8.8. Answering an *lca* query in constant time

Let x and y be two nodes in $\mathcal{T}$ and let z be the *lca* of x and y in $\mathcal{T}$. Suppose we know the height in $\mathcal{B}$ of the node that z is mapped to. That is, we know $h(I(z))$. Below we show, with only that limited information about z, how z can be found in constant time.

Theorem 8.8.1. *Let z denote the lca of x and y in $\mathcal{T}$. If we know $h(I(z))$, then we can find z in $\mathcal{T}$ in constant time.*

PROOF   Consider the run containing z in $\mathcal{T}$. The path up $\mathcal{T}$ from x to z enters that run at some node $\bar{x}$ (possibly z) and then continues along that run until it reaches z. Similarly, the path up from y to z enters the run at some node $\bar{y}$ and continues along that run until z. It follows that z is either $\bar{x}$ or $\bar{y}$. In fact, z is the higher of those two nodes, and so by the numbering scheme, $z = \bar{x}$ if and only if $\bar{x} < \bar{y}$. For example, in Figure 8.6 when $x = 9$ (1001) and $y = 6$ (0110), then $\bar{x} = 8$ (1000) and $\bar{y} = z = 5$ (0101).

Given the above discussion, the approach to finding z from $h(I(z))$ is to use $h(I(z))$ to find $\bar{x}$ and $\bar{y}$, since those nodes determine z. We will explain how to find $\bar{x}$. Let $h(I(z)) = j$, so the height in $\mathcal{B}$ of $I(z)$ is j. By Lemma 8.7.1, node x (which is in the subtree of z in $\mathcal{T}$) maps to a node $I(x)$ in the subtree of node $I(z)$ in $\mathcal{B}$, so if $h(I(x)) = j$ then x must be on the same run as z (i.e., $x = \bar{x}$), and we are finished. Conversely, if $x = \bar{x}$, then $h(I(x))$ must be j. So assume from here on that $x \neq \bar{x}$.

Let w (which is possibly x) denote the node in $\mathcal{T}$ on the z-to-x path just below (off) the run containing z. Since x is not $\bar{x}$, x is not on the same run as z, and w exists. From $h(I(z))$ (which is assumed to be known) and A_x (which was computed during the preprocessing), we will deduce $h(I(w))$ and then $I(w)$, w, and $\bar{x}$.

Since w is in the subtree of z in $\mathcal{T}$ and is not on the same run as z, w maps to a node in $\mathcal{B}$ with height strictly less than the height of $I(z)$ (this follows from Lemma 8.7.1). In fact, by Lemma 8.7.1, among all nodes on the path from x to z that are not on z's run, w maps to a node of greatest height in $\mathcal{B}$. Thus, $h(I(w))$ (which is the height in $\mathcal{B}$ that w maps to) must be the largest position less than j such that A_x has a 1-bit in that position. That is, we can find $h(I(w))$ (even though we don't know w) by finding the most significant 1-bit of A_x in a position less than j. This can be done in constant time on the assumed machine (starting with all bits set to 1, shift right by $d - j + 1$ positions, **AND** this number together with A_x, and then find the left-most 1-bit in the resulting number.)

Let $h(I(w)) = k$. We will now find $I(w)$. Either w is x or w is a proper ancestor of x in $\mathcal{T}$, so either $I(w) = I(x)$ or node $I(w)$ is a proper ancestor of node $I(x)$ in $\mathcal{B}$. Moreover, by the path-encoding nature of the path numbers in $\mathcal{B}$, numbers $I(x)$ and $I(w)$ are identical in bits to the left of k, and $I(w)$ has a 1 in bit k and all 0s to the right. So $I(w)$ can be obtained from $I(x)$ (which we know) and k (which we obtained as above from $h(I(z))$ and A_x). Moreover, $I(w)$ can be found from $I(x)$ and $h(I(w))$ using constant-time bit operations.

Given $I(w)$ we can find w because $w = L(I(w))$. That is, w was just off the z run, so it must be the head of the run that it is on, and each node in $\mathcal{T}$ points to the head of its run. From w we find its parent $\bar{x}$ in constant time. $\square$

In summary, assuming we know $h(I(z))$, we can find node $\bar{x}$, which is the closest ancestor of x in $\mathcal{T}$ that is on the same run as z. Similarly, we find $\bar{y}$. Then z is either $\bar{x}$ or $\bar{y}$; in fact, z is the node among those two with minimum depth-first number in $\mathcal{T}$. Of course, we must now explain how to find $j = h(I(z))$.

How to find the height of $I(z)$

Let b be the lowest common ancestor of $I(x)$ and $I(y)$ in $\mathcal{B}$. Since $\mathcal{B}$ is a complete binary tree, b can be found in constant time as described earlier. Let $h(b) = i$. Then $h(I(z))$ can be found in constant time as follows.

Theorem 8.8.2. *Let j be the smallest position greater or equal to i such that both A_x and A_y have 1-bits in position j. Then node $I(z)$ is at height j in $\mathcal{B}$, or in other words, $h(I(z)) = j$.*

PROOF Suppose $I(z)$ is at height k in B. We will show that $k = j$. Since z is an ancestor of both x and y, both A_x and A_y have a 1-bit in position k. Furthermore, since $I(z)$ is an ancestor of both $I(x)$ and $I(y)$ in $\mathcal{B}$ (by Lemma 8.7.1), $k \geq i$, and it follows (by the selection of j) that $k \geq j$. This also establishes that a position $j \geq i$ exists where both A_x and A_y have 1-bits.

A_x has a 1-bit in position j and $j \geq i$, so x has an ancestor x' in $\mathcal{T}$ such that $I(x')$ is an ancestor of $I(x)$ in $\mathcal{B}$ and $I(x')$ is at height $j \geq i$, the height of b in $\mathcal{B}$. It follows that $I(x')$ is an ancestor of b. Similarly, there is an ancestor y' of y in $\mathcal{T}$ such that $I(y')$ is at height j and is an ancestor of b in $\mathcal{B}$. But if $I(x')$ and $I(y')$ are at the same height (j) and both are ancestors of the single node b, then it must be that $I(x') = I(y')$, meaning that x' and y' are on the same run. Being on the same run, either x' is an ancestor in $\mathcal{T}$ of y' or vice versa. Say, without loss of generality, that x' is an ancestor of y' in $\mathcal{T}$. Then x' is a common ancestor of x and y, and x' is an ancestor of z in $\mathcal{T}$. Hence x' must map to the same height or higher than z in $\mathcal{B}$. That is, $j \geq k$. But $k \geq j$ was already established, so $k = j$ as claimed, and the theorem is proved. $\square$

All the pieces for *lca* retrieval in $\mathcal{T}$ have now been described, and each takes only constant time. In summary, the lowest common ancestor z of any two nodes x and y in $\mathcal{T}$ (assuming z is neither x nor y) can be found in constant time by the following method:

Constant-time *lca* retrieval

Begin

  1. Find the lowest common ancestor b in $\mathcal{B}$ of nodes $I(x)$ and $I(y)$.

  2. Find the smallest position j greater than or equal to $h(b)$ such that both numbers A_x and A_y have 1-bits in position j. j is then $h(I(z))$.

  3. Find node $\bar{x}$, the closest node to x on the same run as z (although we don't know z) as follows:

3a. Find the position l of the right-most 1-bit in A_x.

3b. If $l = j$, then set $\bar{x} = x$ {x and z are on the same run in $\mathcal{T}$} and go to step 4. Otherwise (when $l < j$)

3c. Find the position k of the left-most 1-bit in A_x that is to the right of position j. Form the number consisting of the bits of $I(x)$ to the left of position k, followed by a 1-bit in position k, followed by all zeros. {That number will be $I(w)$, even though we don't yet know w.}
Look up node $L(I(w))$, which must be node w. Set node $\bar{x}$ to be the parent of node w in $\mathcal{T}$.

  4. Find node $\bar{y}$, the closest node to y on the same run as z, by the same approach as in step 3.

  5. If $\bar{x} < \bar{y}$ then set z to $\bar{x}$, else set z to $\bar{y}$.

End.

8.9. The binary tree is only conceptual

The astute reader will notice that the binary tree $\mathcal{B}$ can be eliminated entirely from the algorithm, although it is crucial in the exposition and the proofs. Tree $\mathcal{B}$ is certainly not used in steps 3, 4, or 5 to obtain z from $h(I(z))$. However, it is used in step 1 to find node b from $I(x)$ and $I(y)$. But all we really need from b is $h(b)$ (step 2), and that can be gotten from the right-most common 1-bit of $I(x)$ and $I(y)$. So the mapping from $\mathcal{T}$ to $\mathcal{B}$ is only conceptual, merely used for purposes of exposition.

In summary, after the preprocessing on $\mathcal{T}$, when given nodes x and y, the algorithm finds $i = h(b)$ (without first finding b) from the right-most common 1-bit in $I(x)$ and $I(y)$. Then it finds $j = h(I(z))$ from i and A_x and A_y, and from j it finds $z = lca(x,y)$. Although the logic behind this method has been difficult to convey, a program for these operations is very easy to write.

8.10. For the purists: how to avoid bit-level operations

We have assumed that the machine can do certain bit-level operations in constant time. Many of these assumptions are reasonable for existing machines, but some, such as the ability to find the right-most 1-bit, do not seem as reasonable. How can we avoid all bit-level operations, executing the *lca* algorithm on a standard RAM model? Recall that our RAM can only read, write, address, add, subtract, multiply, and divide in constant time, and only on numbers with $O(\log n)$ bits.

The idea is that during the linear-time preprocessing of $\mathcal{T}$, we also build $O(n)$-size tables that specify the results of the needed bit-level operations. Bit-level operations on a *single* $O(\log n)$ bit number are the easiest. Shifting left by $i = O(\log n)$ bits is accomplished by multiplying by 2^i, which is a permitted constant-time operation since $2^i = O(n)$. Similarly, shifting right is accomplished by division. Now consider the problem of finding the left-most 1-bit. Construct a table with n entries, one for each $\lceil \log_2 n \rceil$ bit number. The entry for binary number 1 has a value of 1, the entries for binary numbers 2 and 3 have value 2, the next 4 entries have value 3, etc. Each entry is an $O(\log n)$ bit number, so this table can easily be generated in $O(n)$ time on a RAM model. Generating the table for right-most 1-bit is a little more involved but can be done in a related manner. An even smaller table of $\lceil \log_2 n \rceil$ entries is needed for "masks". A mask of size i consists of i 0-bits on the right end of $\lceil \log_2 n \rceil$ bits and 1s in all other positions. The i mask is used in the task of finding the right-most 1-bit to the left of position i. Mask $\lceil \log_2 n \rceil$ is the binary number 0. In general, mask i is obtained from mask $i + 1$ by dividing it by two (shifting the mask to the right by one place), and then adding $2^{\lceil \log_2 n \rceil - 1}$ (which adds in the left-most 1-bit).

The tables that seem more difficult to build are the tables for *binary* operations on $O(\log n)$ bit numbers, such as **XOR, AND,** and **OR**. The full table for any of these binary operations is of size n^2 since there are n numbers with $\log n$ bits. One cannot construct an n^2-size table in $O(n)$ time. The trick is that each of the needed binary operations are done bitwise. For example, **XOR** of two $\lceil \log n \rceil$ bit numbers can be found by splitting each number into two numbers of roughly $\frac{\lceil \log n \rceil}{2}$ bits each, doing **XOR** twice, and then concatenating the answers (additional easy details are needed to do this in the RAM model). So it suffices to build an **XOR** table for numbers with only $\lceil \frac{\lceil \log n \rceil}{2} \rceil$ bits. But $2^{\frac{\lceil \log n \rceil}{2}} = \sqrt{n}$, so the **XOR** table for these numbers has only n entries, and hence it is plausible that the entire table can be constructed in $O(n)$ time. In particular, **XOR** can be implemented by **AND, OR,** and **NOT** (bit complement) operations, and tables for these can be built in $O(n)$ time (we leave this as an exercise).

8.11. Exercises

1. Using depth-first traversal, show how to construct the path numbers for the nodes of B in time proportional to n, the number of nodes in B. Be careful to observe the constraints of the RAM model.

2. Prove that the path numbers in B are the *inorder traversal* numbers.

3. The *lca* algorithm for a complete binary tree was detailed in the case that *lca(i,j)* was neither i nor j. In the case that *lca(i,j)* is one of i or j, then a very simple constant-time algorithm can determine *lca(i,j)*. The idea is first to number the nodes of the binary tree B by a depth-first numbering, and to note for each node v, the number of nodes in the subtree of v (including v). Let $I(v)$ be the *dfs* number given to node v, and let $s(v)$ be the number of nodes in the subtree of v. Then node i is an ancestor of node j if and only if $I(i) \leq I(j)$ and $I(j) < I(i) + s(i)$.

   Prove that this is correct, and fill in the details to show that the needed preprocessing can be done in $O(n)$ time.

   Show that the method extends to any tree, not just complete binary trees.

4. In the special case of a complete binary tree B, there is an alternative way to handle the situation when *lca(i,j)* is i or j. Using $h(i)$ and $h(j)$ we can determine which of the nodes i and j is higher in the tree (say i) and how many edges are on the path from the root to node i. Then we take the **XOR** of the binary for i and for j and find the left-most 1-bit as before, say in position k (counting from the left). Node i is an ancestor of j if and only if k is larger than the number of edges on the path to node i. Fill in the details of this argument and prove it is correct.

5. Explain why in the *lca* algorithm for B, it was necessary to assume that *lca(i,j)* was neither i nor j. What would go wrong in that algorithm if the issue were ignored and that case was not checked explicitly?

6. Prove that the height of any node k in B is $h(k)$.

7. Write a C program for both the preprocessing and the *lca* retrieval. Test the program on large trees and time the results.

8. Give an explicit $O(n)$-time RAM algorithm for building the table containing the right-most 1-bit in every $\log_2 n$ bit number. Remember that the entry for binary number i must be in the ith position in the table. Give details for building tables for **AND**, **OR**, and **NOT** for $\frac{\lceil \log n \rceil}{2}$ bit numbers in $O(n)$ time.

9. It may be more reasonable to assume that the RAM can shift a word left and right in constant time than to assume that it can multiply and divide in constant time. Show how to solve the *lca* problem in constant time with linear preprocessing under those assumptions.

10. In the proof of Theorem 8.8.1 we showed how to deduce $I(w)$ from $h(I(w))$ in constant time. Can we use the same technique to deduce $I(z)$ from $h(I(z))$? If so, why doesn't the method do that rather than involving nodes w, $\bar{x}$, and $\bar{y}$?

11. The constant-time *lca* algorithm is somewhat difficult to understand and the reader might wonder whether a simpler idea works. We know how to find the *lca* in constant time in a complete binary tree after $O(n)$ preprocessing time. Now suppose we drop the assumption that the binary tree is complete. So $\mathcal{T}$ is now a binary tree, but not necessarily complete. Letting d again denote the depth of $\mathcal{T}$, we can again compute $d + 1$ length path numbers that encode the paths to the nodes, and again these path numbers allow easy construction of the lowest common ancestor. Thus it might seem that even in incomplete binary trees, one can easily find the *lca* in this simple way without the need for the full *lca* algorithm. Either give the details for this or explain why it fails to find the *lca* in constant time.

If you believe that the above simple method solves the *lca* problem in constant time for any binary tree, then consider trying to use it for arbitrary trees. The idea is to use the well-known technique of converting an arbitrary tree to a binary tree by modifying every node with more than two children as follows: Suppose node v has children $v_1, v_2, \ldots, v_k$. Replace the children of v with two children v_1 and v^* and make nodes $v_2, \ldots, v_k$ children of v^*. Repeat this until each original child v_i of v has only one sibling, and place a pointer from v^* to v for every new node v^* created in this process. How is the number of nodes changed by this transformation? How does the *lca* of two nodes in this new binary tree relate to the *lca* of the two nodes in the original tree? So, assuming that the $d + 1$ length path labels can be used to solve the *lca* problem in constant-time for any binary tree, does this conversion yield a constant-time *lca* search algorithm for any tree?

12. **A simpler (but slower) *lca* algorithm.** In Section 8.4.1 we mentioned that if $\Theta(n \log n)$ preprocessing time is allowed, and $\Theta(n \log n)$ space can be allocated during both the preprocessing and the retrieval phases, then a (conceptually) simpler constant-time *lca* retrieval method is possible. In many applications, $\Theta(n \log n)$ is an acceptable bound, which is not much worse than the $O(n)$ bound we obtained in the text. Here we sketch the idea of the $\Theta(n \log n)$ method. Your problem is to flesh out the details and prove correctness.

First we reduce the general *lca* problem to a problem of finding the smallest number in an interval of a fixed list of numbers.

The reduction of *lca* to a list problem

Step 1 Execute a depth-first traversal of tree $\mathcal{T}$ to label the nodes in depth-first order and to build a multilist L of the nodes in the order that they are visited. (For any node v other than the root, the number of times v is in L equals the degree of v.) The only property of the depth-first numbering we need is that the number given to any node is smaller than the number given to any of its proper descendants. From this point on, we refer to a node only by its *dfs* number.

For example, the list for the tree in Figure 8.1 (page 182) is

$$\{1, 2, 3, 2, 4, 2, 1, 5, 6, 5, 7, 5, 8, 9, 8, 10, 8, 5, 1\}.$$

Notice that if $\mathcal{T}$ has n nodes, then L has $O(n)$ entries.

Step 2 The *lca* of any two nodes x and y can be gotten as follows: Find any occurrences of x and y in L; this defines an interval I in L between those occurrences of x and y. Then in L find the smallest number in interval I; that number is the *lca(x,y)*.

For example, if x is 6 and y is 9, then one interval I that they define is $\{6, 5, 7, 5, 8, 9\}$, implying that node 5 is *lca(6,9)*.

This is the end of the reduction. Now the first exercise.

a. Ignoring time complexity, prove that in general the *lca* of two nodes can be obtained as described in the two steps above.

Now we continue to describe the method. More exercises will follow.

With the above reduction, each *lca* query becomes the problem of finding the smallest number in a interval I of a fixed list L of $O(n)$ numbers. Let m denote the exact size of L. To be able to solve each *lca* query in constant time, we first do an $O(m \log m)$-time preprocessing of list L. For convenience assume that m is a power of 2.

Preprocessing of L

Step 1 Build a complete binary tree B with m leaves and number the leaves in left-to-right order (as given by an inorder traversal). Then for i from 1 to m, record the ith element of L at leaf i.

Step 2 For an arbitrary internal node v in B, let B_v denote the subtree of B rooted at v, and let $L_v = n_1, n_2, \ldots, n_z$ be an ordered list containing the elements of L written at the leaves of B_v, in the same left-to-right order as they appear in B. Create two lists, *Pmin(v)* and *Smin(v)*, for each internal node v. Each list will have size equal to the number of leaves in v's subtree. The kth entry of list *Pmin(v)* is the smallest number among $\{n_1, n_2, \ldots, n_k\}$. That is, the kth entry of *Pmin(v)* is the smallest number in the *prefix* of list L_v ending at position k. Similarly, the kth entry of list *Smin(v)* is the smallest number in the *suffix* of L_v starting at position k. This is the end of the preprocessing and exercises follow.

b. Prove that the total size of all the *Pmin* and *Smin* lists is $O(m \log m)$, and show how they can be constructed in that time bound.

After the $O(m \log m)$ preprocessing, the smallest number in any interval I can be found in constant time. Here's how. Let interval I in L have endpoints l and r and recall that these correspond to leaves of B. To find the smallest number in I, first find the *lca(l,r)*, say node v. Let v' and v'' be the left and right children of v in B, respectively. The smallest number in I can be found using one lookup in list *Smin(v')*, one lookup in *Pmin(v'')*, and one additional comparison.

c. Give complete details for how the smallest number in I is found, and fully explain why only constant time is used.

13. By refining the method developed in Exercise 12, the $\Theta(m \log m)$ preprocessing bound (time and space) can be reduced to only $\Theta(m \log \log m)$ while still maintaining constant retrieval time for any *lca* query. (It takes a pretty big value of m before the difference between $O(m)$ and $\Theta(m \log \log m)$ is appreciable!) The idea is to divide list L into $\frac{m}{\log m}$ blocks each of size $\log m$ and then separately preprocess each block as in Exercise 12. Also, compute the minimum number in each block, put these $\frac{m}{\log m}$ numbers in an ordered list *Lmin*, and preprocess *Lmin* as in Exercise 12.

a. Show that the above preprocessing takes $\Theta(m \log \log m)$ time and space.

Now we sketch how the retrieval is done in this faster method. Given an interval I with starting and ending positions l and r, one finds the smallest number in I as follows: If l and r are in the same block, then proceed as in Exercise 12. If they are in adjacent blocks, then find the minimum number from l to the end of l's block, find the minimum number from the start of r's block to r, and take the minimum of those two numbers. If l and r are in nonadjacent blocks, then do the above and also use *Lmin* to find the minimum number in all the blocks strictly between the block containing l and the block containing r. The smallest number in I is the minimum of those three numbers.

b. Give a detailed description of the retrieval method and justify that it takes only constant time.

14. Can the above improvement from $O(m \log m)$ preprocessing time to $O(m \log \log m)$ preprocessing time be extended to reduce the preprocessing time to $O(m \log \log \log m)$ preprocessing time? Can the improvements be continued for an arbitrary number of *logarithms*?

9

More Applications of Suffix Trees

With the ability to solve lowest common ancestor queries in constant time, suffix trees can be used to solve many additional string problems. Many of those applications move from the domain of *exact* matching to the domain of *inexact*, or approximate, matching (matching with some errors permitted). This chapter illustrates that point with several examples.

9.1. Longest common extension: a bridge to inexact matching

The *longest common extension problem* is solved as a subtask in many classic string algorithms. It is at the heart of all but the last application discussed in this chapter and is central to the *k-difference algorithm* discussed in Section 12.2.

> **Longest common extension problem** Two strings S_1 and S_2 of total length n are first specified in a preprocessing phase. Later, a *long* sequence of index pairs is specified. For each specified index pair (i, j), we must find the length of the longest substring of S_1 starting at position i that matches a substring of S_2 starting at position j. That is, we must find the length of the longest prefix of suffix i of S_1 that matches a prefix of suffix j of S_2 (see Figure 9.1).

Of course, any time an index pair is specified, the longest common extension can be found by direct search in time proportional to the length of the match. But the goal is to compute each extension in *constant* time, independent of the length of the match. Moreover, it would be cheating to allow more than linear time to preprocess S_1 and S_2.

To appreciate the power of suffix trees combined with constant-time *lca* queries, the reader should again try first to devise a solution to the longest common extension problem without those two tools.

9.1.1. Linear-time solution

The solution to the longest common extension problem first builds the generalized suffix tree $\mathcal{T}$ for S_1 and S_2, and then prepares $\mathcal{T}$ to allow constant-time *lca* queries. During this preprocessing, it also computes the string-depth of v for every node v of $\mathcal{T}$. Building and preprocessing $\mathcal{T}$ takes $O(n)$ time.

Given any specific index pair (i, j), the algorithm finds the lowest common ancestor of the leaves of $\mathcal{T}$ that correspond to suffix i in S_1 and suffix j in S_2. Let v denote that lowest common ancestor node. The key point is that the string labeling the path to v is precisely the longest substring of S_1 starting at i that matches a substring of S_2 starting at j. Consequently, the string-depth of v is the length of the *longest common extension*. Hence each longest common extension query can be answered in constant time.

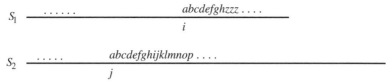

Figure 9.1: The longest common extension for pair (i, j) has length eight. The matching substring is *abcdefgh*.

9.1.2. Space-efficient longest common extension

When S_2 is much smaller than S_1, we may not wish to build the generalized suffix tree for S_1 and S_2 together. By only building the suffix tree for the smaller of the two strings, we will save considerable space, but can the longest common extension problem be efficiently solved using only this smaller tree? The answer is yes, with the aid of *matching statistics*.

Recall from Section 7.8.1 that the matching statistic $m(i)$ is the length of the longest substring of S_1 starting at position i that matches a substring starting at some position in S_2, and that $p(i)$ is one of those starting positions in S_2. In Sections 7.8.1 and 7.8.3 we showed how to compute $m(i)$ and $p(i)$ for each i in $O(|S_1|)$ total time, using only a suffix tree for S_2 and a copy of S_1.

The longest common extension query for any pair (i, j) is solved by first finding the *lca* v of leaves $p(i)$ and j in the suffix tree for S_2. The length of the longest common extension of (i, j) is then the minimum of $m(i)$ and the string-depth of node v. The proof of this is immediate and is left to the reader. Since any *lca* computation takes only constant time, we have the following:

Theorem 9.1.1. *After a preprocessing phase that takes linear time, any longest common extension query can be answered in constant time using only the space required by the suffix tree for S_2 (the smaller of the two strings) plus $2|S_1|$ words to hold the values $m(i)$ and $p(i)$.*

The space-efficient solution of the longest common extension problem usually implies a space-efficient solution to the various applications of it. This will not be explicitly mentioned in every application, and the details will be left to the reader.

9.2. Finding all maximal palindromes in linear time

Definition   An even-length substring S' of S is a *maximal palindrome of radius k* if, starting in the middle of S', S' reads the same in both directions for k characters but not for any $k' > k$ characters. An odd-length maximal palindrome S' is similarly defined after excluding the middle character of S'.

For example, if $S = aabactgaaccaat$ then both *aba* and *aaccaa* are maximal palindromes in S of radii one and three, respectively, and each occurrence of *aa* is a maximal palindrome of radius one.

Definition   A string is called a *maximal palindrome* if it is a maximal palindrome of radius k for some k.

For example, in the string *cabaabad*, both *aba* and *abaaba* are maximal palindromes.
Any palindrome is contained in some maximal palindrome with the same midpoint, so the maximal palindromes are a compact way to represent the set of all palindromes. Moreover, in most applications, the maximal palindromes are the ones of interest.

Palindrome problem: Given a string S of length n, the palindrome problem is to locate *all* maximal palindromes in S.

9.2.1. Linear-time solution

We will explain how to find all the even-length maximal palindromes in linear, $O(n)$ time – a rather severe goal. The odd-length maximal palindromes can be found by a simple modification of the even-length case.

Let S^r be the reverse of string S. Now, suppose there is an even-length maximal palindrome in S whose middle occurs just after character q of S. Let k denote the length of the palindrome. That means there is a string of length k starting at position $q + 1$ of S that is identical to a string starting at position $n - q + 1$ of S^r. Furthermore, because the palindrome is maximal, the next characters (in positions $q + k + 1$ and $n - q + k + 1$, respectively) are not identical. This implies that k is the length of the longest common extension of position $q + 1$ in S and position $n - q + 1$ in S^r. Hence for any fixed position q, the length of the maximal palindrome (if there is one) with midpoint at q can be computed in constant time.

This leads to the following simple linear-time method to find all the even length maximal palindromes in S:

1. In linear time, create the reverse string S^r from S and preprocess the two strings so that any longest common extension query can be solved in constant time.

2. For each q from 1 to $n - 1$, solve the longest common extension query for the index pair $(q + 1, n - q + 1)$ in S and S^r, respectively. If the extension has nonzero length k, then there is a maximal palindrome of radius k centered at q.

The method takes $O(n)$ time since the suffix tree can be built and preprocessed in that time, and each of the $O(n)$ extension queries is solved in constant time.

In summary, we have

Theorem 9.2.1. *All the maximal even-length palindromes in a string can be identified in linear time.*

9.2.2. Complemented and separated palindromes

Palindromes were briefly discussed in Section 7.11.1, during the general discussion of repetitive structures in biological strings. There, it was mentioned that in DNA (or RNA) the palindromes of interest are *complemented*. That means that the two halves of the substring form a palindrome (in the normal English use of the word) only if the characters in one half are converted to their complement characters, that is, A and T (or U in the case of RNA) are complements, and C and G are complements. For example, $ATTAGCTAAT$ is a complemented palindrome.

The problem of finding all complemented palindromes in a string can also be solved in linear time. Let $c(S^r)$ be the complement of string S^r (i.e., where each A is changed to T, each T to A, each C to G, and each G to C). Then proceed as in the palindrome problem, using $c(S^r)$ in place of S^r.

Another variant of the palindrome problem that comes from biological sequences is to relax the insistence that the two halves of the palindrome (complemented or not) be adjacent. When the two halves are not adjacent, the structure is called a *separated palindrome*, although in the biological literature the distinction between separated and nonseparated palindromes is sometimes blurred. The problem of finding all separated palindromes is really one of finding all inverted repeats (see Section 7.12) and hence is more complex than finding palindromes. However, if there is a fixed bound on the permitted distance of the separation, then all the separated palindromes can again be found in linear time. This is an immediate application of the longest common extension problem, the details of which are left to the reader.

Another variant of the palindrome problem, called the k-mismatch palindrome problem, will be considered below, after we discuss matching with a fixed number of mismatches.

9.3. Exact matching with wild cards

Recall the problem discussed in Sections 4.3.2 and 3.5.2 of finding all occurrences of pattern P in text T when wild cards are allowed in either string. This problem was not easily handled with Knuth-Morris-Pratt or Boyer–Moore–type methods, although the Fast Fourier transform, match-count method could be modified to handle wild cards in both strings. Using the above method to solve the longest common extension problem, when there are k wild cards distributed throughout the two strings, we can find all occurrences of P in T (allowing a wild card to match any *single* character) in $O(km)$ time, where $m \geq n$ is the length of T and n is the length of P.

At the high level, the algorithm increments i from 1 to $m - n + 1$ and checks, for each fixed i, whether P occurs in T starting at position i of T. The wild card symbol is treated as an additional character in the alphabet. The idea of the method is to align the left end of P against position i of T and then work left to right through the two strings, successively executing longest common extension queries and checking that every mismatch occurs at a position containing a wild card. After $O(n+m)$ preprocessing time (for longest common extension queries), only $O(k)$ time is used by the algorithm for any fixed i. Thus, $O(km)$ time is used overall. The following detailed algorithm works for a fixed position i of T:

Wild-card match check

Begin

1. Set j to 1 and i' to i.
2. Compute the length l of the longest common extension starting at positions j of P and i' of T.
3. If $j + l = n + 1$ then P occurs in T starting at i; stop.
4. Check if a wild card occurs in position $j + l$ of P or position $i' + l$ of T. If so then set j to $j + l + 1$, set i' to $i' + l + 1$, and go to step 2. Else, P does not occur in T starting at i; stop.

End.

The space needed by this method is $O(n + m)$, since it uses a suffix tree for the two strings. However, as detailed in Theorem 9.1.1, only a suffix tree for P plus the matching statistics for T are needed (although we must still store the original strings). Since $m > n$ we have

Theorem 9.3.1. *The exact matching problem with k wild cards distributed in the two strings can be solved in $O(km)$ time and $O(m)$ space.*

9.4. The *k*-mismatch problem

The general problem of inexact or approximate matching (matching with some errors allowed) will be considered in detail in Part III of the book (in Section 12.2), where the technique of dynamic programming will be central. But dynamic programming is not always necessary, and we have here all the tools to solve one of the classic "benchmark" problems in approximate matching: the *k-mismatch* problem.

> **Definition**   Given a pattern P, a text T, and a fixed number k that is independent of the lengths of P and T, a *k-mismatch* of P is a $|P|$-length substring of T that matches at least $|P| - k$ characters of P. That is, it matches P with at most k mismatches.

Note that the definition of a k-mismatch does not allow any insertions or deletions of characters, just matches and mismatches. Later, in Section 12.2, we will discuss bounded error problems that also allow insertions and deletions of characters.

> The **k-mismatch problem** is to find all k-mismatches of P in T.

For example, if $P = bend$, $T = abentbananaend$, and $k = 2$, then T contains three k-matches of P: P matches substring *bent* with one mismatch, substring *bana* with two mismatches, and substring *aend* with one mismatch.

Applications in molecular biology for the k-mismatch problem, along with the more general k-differences problem, will be discussed in Section 12.2.2. The k-mismatch problem is a special case of the match-count problem considered in Section 4.3, and the approaches discussed there apply. But because k is a fixed number unrelated to the lengths of P and T, faster solutions have been obtained. In particular, Landau and Vishkin [287] and Myers [341] were the first to show an $O(km)$-time solution, where P and T have lengths n and $m > n$, respectively. The value of k can never be more than n, but the motivation for the $O(km)$ result comes from applications where k is expected to be very small compared to n.

9.4.1. The solution

The idea is essentially the same as the idea for matching with wild cards (although the meaning of k in these two problems is different). For any position i in T, we determine whether a k-mismatch of P begins at position i in $O(k)$ time by simply executing up to k (constant-time) longest common extension queries. If those extensions reach the end of P, then P matches the substring starting at i with at most k mismatches. If the extensions do not reach the end of P, then more than k mismatches are needed. In either case, at most k extension queries are solved for any i, and $O(k)$ time suffices to determine whether a k-mismatch of P begins at i. Over all positions in T, the method therefore requires at most $O(km)$ time.

k-mismatch check

Begin

1. Set j to 1 and i' to i, and *count* to 0.

2. Compute the length l of the longest common extension starting at positions j of P and i' of T.

3. If $j + l = n + 1$, then a k-mismatch of P occurs in T starting at i (in fact, only *count* mismatches occur); stop.

4. If $count \leq k$, then increment $count$ by one, set j to $j + l + 1$, set i' to $i' + l + 1$, and go to step 2.

    If $count = k + 1$, then a k-mismatch of P does not occur starting at i; stop.

End.

Note that the space required for this solution is just $O(n + m)$, and that the method can be implemented using a suffix tree for the small string P alone.

We should note a different practical approach to the k-mismatch problem, based on suffix trees, that is in use in biological database search [320]. The idea is to generate every string P' that can be derived from P by changing up to k characters of P, and then to search for P' in a suffix tree for T. Using a suffix tree, the search for P' takes time just proportional to the length of P' (and can be implemented to be extremely fast), so this approach can be a winner when k and the size of the alphabet are relatively small.

9.5. Approximate palindromes and repeats

We have discussed earlier (Section 7.11.1) the importance of palindromes in molecular biology. That discussion provides most of the motivation for the palindrome problem. But in biological applications, the two parts of the "palindrome" are rarely identical. This motivates the k-mismatch palindrome problem.

Definition A *k-mismatch palindrome* is a substring that becomes a palindrome after k or fewer characters are changed. For example, *axabbcca* is a 2-mismatch palindrome.

With this definition, a palindrome is just a 0-mismatch palindrome. It is now an easy exercise to detail an $O(kn)$-time method to find all k-mismatch palindromes in a string of length n. We leave that to the reader, and we move on to the more difficult problem of finding *tandem repeats*.

Definition A *tandem repeat* α is a string that can be written as $\beta\beta$, where β is a substring.

Each tandem repeat is specified by a starting position of the repeat and the length of the substring β. This definition does not require that β be of maximal length. For example, in the string *xababababy* there are a total of six tandem repeats. Two of these begin at position two: *abab* and *abababab*. In the first case, β is *ab*, and in the second case, β is *abab*.

Using longest common extension queries, it is immediate that all tandem repeats can be found in $O(n^2)$ time – just guess a start position i and a middle position j for the tandem and do a longest common extension query from i and j. If the extension from i reaches j or beyond, then there is a tandem repeat of length $2(j - i + 1)$ starting at position i. There are $\Theta(n^2)$ choices for i and j, yielding the $O(n^2)$ time bound.

Definition A *k-mismatch tandem repeat* is a substring that becomes a tandem repeat after k or fewer characters are changed. For example, *axabaybb* is a 2-mismatch tandem repeat.

Again, all k-mismatch tandem repeats can be found in $O(kn^2)$ time, and the details are left to the reader. Below we will present a method that solves this problem in $O(kn \log(n/k))$ time. To summarize, what we have so far is

Theorem 9.5.1. *All the tandem repeats in S in which the two copies differ by at most k mismatches can be found in $O(kn^2)$ time. Typically, k is a fixed number, and the time bound is reported as $O(n^2)$.*

9.6. Faster methods for tandem repeats

The total number of tandem repeats and k-mismatch tandem repeats (even for fixed k) can grow as fast as $\Theta(n^2)$ (take the case of all n characters being the same). So, no worst-case bound better than $O(n^2)$ is possible for the problem of finding all tandem repeats. But a method whose running time depends on the *number* of tandem repeats contained in the string is possible for both the exact and the k-mismatch versions of the problem. A different approach was explored in Exercises 56, 57, and 58 of Chapter 7, where only maximal primitive tandem arrays were identified.

Landau and Schmidt [288] developed a method to find all k-mismatch tandem repeats in $O(kn \log(\frac{n}{k}) + z)$ time, where z is the number of k-mismatch tandem repeats in the string S. Now z can be as large as $\Theta(n^2)$, but in practice z is expected to be small compared to n^2, so the $O(kn \log(\frac{n}{k}) + z)$ bound is a significant improvement. Note that we will still find all tandem repeats, but the running time will depend on the actual number of repeats and not on the worst-case possible number of repeats.

We explain the method by first adapting it to find all tandem repeats (with no mismatches) in $O(n \log n + z)$ time, where z is now the total number of tandem repeats in S. That time bound for the case of no mismatches was first obtained in a paper by Main and Lorenz [307], who used a similar idea but did not use suffix trees. Their approach is explored in Exercise 8.

The Landau–Schmidt solution is a recursive, divide-and-conquer algorithm that exploits the ability to compute longest common extension queries in constant time. Let h denote $\lfloor \frac{n}{2} \rfloor$. At the highest level, the Landau–Schmidt method divides the problem of finding all tandem repeats into four subproblems:

1. Find all tandem repeats contained entirely in the first half of S (up to position h).

2. Find all tandem repeats contained entirely in the second half of S (after position h).

3. Find all tandem repeats where the first copy spans (contains) position h of S.

4. Find all tandem repeats where the second copy spans position h of S.

Clearly, no tandem repeat will be found in more than one of these four subproblems. The first two subproblems are solved by recursively applying the Landau–Schmidt solution. The second two problems are symmetric to each other, so we consider only the third subproblem. An algorithm for that subproblem therefore determines the algorithm for finding all tandem repeats.

Algorithm for problem 3

We want to find all the tandem repeats where the first copy *spans* (but does not necessarily begin at) position h. The idea of the algorithm is this: For any fixed number l, one can test in constant time whether there is a tandem repeat of length exactly $2l$ such that the first copy spans position h. Applying this test for all feasible values of l means that in $O(n)$ time we can find all the lengths of tandem repeats whose first copy spans position h. Moreover, for each such length we can enumerate all the starting points of these tandem repeats, in time proportional to the number of them. Here is how to test a number l.

Begin

1. Let $q = h + l$.

2. Compute the longest common extension (in the forward direction) from positions h and q. Let l_1 denote the length of that extension.

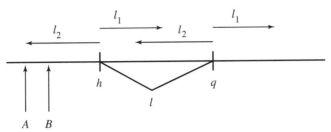

Figure 9.2: Any position between A and B inclusive is a starting point of a tandem repeat of length $2l$. As detailed in Step 4, if l_1 and l_2 are both at least one, then a subinterval of these starting points specify tandem repeats whose first copy spans h.

3. Compute the longest common extension *in the reverse direction* from positions $h - 1$ and $q - 1$. Let l_2 denote the length of that extension.

4. There is a tandem repeat of length $2l$ whose first copy spans position h if and only if $l_1 + l_2 \geq l$ and both l_1 and l_2 are at least one. Moreover, if there is such a tandem repeat of length $2l$, then it can begin at any position from $\text{Max}(h - l_2, h - l + 1)$ to $\text{Min}(h + l_1 - l, h)$ inclusive. The second copy of the repeat begins l places to the right. Output each of these starting positions along with the length $2l$. (See Figure 9.2.)

End.

To solve an instance of subproblem 3 (finding all tandem repeats whose first copy spans position h), just run the above algorithm for each l from 1 to h.

Lemma 9.6.1. *The above method correctly solves subproblem 3 for a fixed h. That is, it finds all tandem repeats whose first copy spans position h. Further, for fixed h, its running time is $O(n/2) + z_h$, where z_h is the number of such tandem repeats.*

PROOF Assume first that there is a tandem repeat whose first copy spans position h, and it has some length, say $2l$. That means that position $q = h + l$ in the second copy corresponds to position h in the first copy. Hence some substring starting at h must match a substring starting at q, in order to provide the suffix of each copy. This substring can have length at most l_1. Similarly, there must be a substring ending at $h - 1$ that matches a substring ending at $q - 1$, providing the prefix of each copy. That substring can have length at most l_2. Since all characters between h and q are contained in one of the two copies, $l_1 + l_2$ must be at least l. Conversely, by essentially the same reasoning, if $l_1 + l_2 \leq l$ and both l_1 and l_2 are at least one then one can specify a tandem repeat of length $2l$ whose first copy spans h. The necessary and sufficient condition for the existence of such a tandem is therefore proved.

The converse proof that all starting positions fall in the stated range involves similar reasoning and is left to the reader.

For the time analysis, note first that for a fixed choice of h, the method takes constant time per choice of l to execute the common extension queries, and so it takes $O(n/2)$ time for all those queries. For any fixed l, the method takes constant time per tandem that it reports, and it never reports the same tandem twice since it reports a different starting point for each repeat of length $2l$. Since each repeat is reported as a starting point and a length, it follows that over all choices of l, the algorithm never reports any tandem repeat twice. Hence the time spent to report tandem repeats is proportional to z_h, the number of tandem repeats whose first copy spans position h. □

Theorem 9.6.1. *Every tandem repeat in S is found by the execution of subproblems 1 through 4 and is reported exactly once. The time for the algorithm is $O(n \log n + z)$, where z is the total number of tandem repeats in S.*

PROOF   That all tandem repeats are found is immediate from the fact that every tandem is of a form considered by one of the subproblems 1 through 4. To show that no tandem repeat is reported twice, recall that for $h = n/2$, no tandem is of the form considered by more than one of the four subproblems. This holds recursively for subproblems 1 and 2. Further, in the proof of Lemma 9.6.1 we established that no execution of subproblem 3 (and also 4) reports the same tandem twice. Hence, over the entire execution of the four subproblems, no tandem repeat is reported twice. It also follows that the total time used to output the tandem repeats is $O(z)$.

To finish the analysis, we consider the time taken by the extension queries. This time is proportional to the number of extension queries executed. Let $T(n)$ denote the number of extension queries executed for a string of length n. Then, $T(n) = 2T(n/2) + 2n$, and $T(n) = O(n \log n)$ as claimed.   □

9.6.1. The speedup for k-mismatch tandem repeats

The idea for the $O(kn \log(\frac{n}{k}) + z)$ algorithm of Landau and Schmidt [288] is an immediate extension of the $O(n \log n + z)$ method for finding exact tandem repeats, but the implementation is a bit more involved. The method is again recursive, and again the important part is subproblem 3 (or 4), finding all k-mismatch tandem repeats whose first copy spans position h. The solution to that problem is to run k successive longest common extension queries forward from h and q and to run k successive longest common extension queries backward from $h - 1$ and $q - 1$. Now focus on the interval between h and q. To find all k-mismatch tandem repeats whose first copy spans h, find every position t (if any) in that interval where the number of mismatches from h to t (found during the forward extension) plus the number of mismatches from $t + 1$ to $q - 1$ (found during the backward extension) is at most k. Any such t provides a midpoint of the tandem repeat. We leave the correctness of that claim to the reader.

To achieve the claimed time bound, we must find all the midpoints of the k-mismatch tandem repeats whose first copy spans h in time proportional to the number of them. But unlike the case of exact tandem repeats, the set of correct midpoints need not be contiguous. How are they found? We sketch the idea and leave the details as an exercise. During the k forward extension queries, accumulate an ordered list of the positions in interval $[h, q]$ where a mismatch occurs, and do the same during the backward extension queries. Then merge (in left to right order) those two lists and calculate for each position in the list the total number of mismatches to it from h and to $q - 1$. Since each list is found in order, the time to obtain the merged list and the totals is $O(k)$. The total number of mismatches can change only at a position that is in the merged list; hence an $O(k)$ time scan of that list specifies all subintervals containing permitted midpoints of the k-mismatch tandem. In addition, every point in such a subinterval is a permitted midpoint. Thus, for a fixed h, the total query time for subproblem 3 is $O(k)$ and the total output time is kz_h. Over the entire algorithm, the total output time is $O(kz)$ and the number of queries satisfies $T(n) = 2T(n/2) + 2k$. Thus, at most $O(kn \log n)$ queries are done. In summary, we have

Theorem 9.6.2. *All k-mismatch tandem repeats in a string of length n can be found in $O(kn \log n + z)$ time.*

The bound can be sharpened to $O(kn \log(n/k) + z)$ by the observation that any $l \le k$ need not be tested in subproblems 3 and 4. We leave the details as an exercise.

We also leave it to the reader to adapt the solutions for the k-mismatch palindrome and tandem repeat problems to allow for string complementation and bounded-distance separation between copies.

9.7. A linear-time solution to the multiple common substring problem

All of the above applications are similar, exploiting the ability to solve longest common extension queries in constant time. Now we examine another use of suffix trees with constant-time *lca* that is not of this form.

The *k-common substring problem* was first discussed in Section 7.6 (the reader should review that discussion before going on). In that section, a generalized suffix tree $\mathcal{T}$ was constructed for the K strings of total length n, and the table of all the $l(k)$ values was obtained by operations on $\mathcal{T}$. That method had a running time of $O(Kn)$. In this section we reduce the time to $O(n)$. The solution was obtained by Lucas Hui [236].[1]

Recall that for any node v in $\mathcal{T}$, $C(v)$ is the number of distinct leaf string identifiers in the subtree of v, and that a table of all the $l(k)$ values can be computed in $O(n)$ time once all the $C(v)$ values are known. Recall also that $S(v)$ is the total number of leaves in the subtree of v and that $S(v)$ can easily be computed in $O(n)$ time for all nodes.

Certainly, $S(v) \geq C(v)$ for any node v, and it will be strictly greater when there are two or more leaves of the same string identifier in v's subtree. Our approach to finding $C(v)$ is to compute both $S(v)$ and a correction factor $U(v)$, which counts how many "duplicate" suffixes from the same string occur in v's subtree. Then $C(v)$ is simply $S(v) - U(v)$.

Definition $n_i(v)$ is the number of leaves with identifier i in the subtree rooted at node v. Let n_i be the total number of leaves with identifier i.

With that definition, we immediately have the following:

Lemma 9.7.1. $U(v) = \sum_{i:n_i(v)>0}(n_i(v) - 1)$ *and* $C(v) = S(v) - U(v)$.

We show below that all the correction factors for all internal nodes can be computed in $O(n)$ total time. That then gives an $O(n)$-time solution to the k-common substring problem.

9.7.1. The method

The algorithm first does a depth-first traversal of $\mathcal{T}$, numbering the leaves in the order that they are encountered. That numbering has the familiar property that for any internal node v, the numbers given to the leaves in the subtree rooted at v are consecutive (i.e., they form a consecutive interval).

For purposes of the exposition, let us focus on the single identifier i and show how to compute $n_i(v) - 1$ for each internal node v. Let L_i be the list of leaves with identifier i, in increasing order of their *dfs* numbers. For example, in Figure 9.3, the leaves with identifier i are shown boxed and the corresponding L_i is 1, 3, 6, 8, 10. By the properties of depth-first numbering, for the subtree rooted at any internal node v, all the $n_i(v)$ leaves with identifier i occur in a consecutive interval of list L_i. Call that interval $L_i(v)$. If x and

---

[1] In the introduction of an earlier unpublished manuscript [376], Pratt claims a linear-time solution to the problem but the claim doesn't specify whether the problem is for a fixed k or for all values of k. The section where the details were to be presented is not available and was apparently never finished [375].

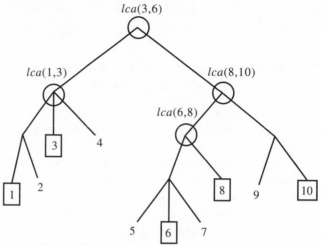

Figure 9.3: The boxed leaves have identifier i. The circled internal nodes are the lowest common ancestors of the four adjacent pairs of leaves from list L_j.

y are any two leaves in $L_i(v)$, then the *lca* of x and y is a node in the subtree of v. So if we compute the *lca* for each consecutive pair of leaves in $L_i(v)$, then all of the $n_i(v) - 1$ computed *lca*s will be found in the subtree of v. Further, if x and y are not both in the subtree of v, then the *lca* of x and y will not be a node in v's subtree. This leads to the following lemma and method.

Lemma 9.7.2. *If we compute the lca for each consecutive pair of leaves in L_i, then for any node v, exactly $n_i(v) - 1$ of the computed lcas will lie in the subtree of v.*

Lemma 9.7.2 is illustrated in Figure 9.3.

Given the lemma, we can compute $n_i(v) - 1$ for each node v as follows: Compute the *lca* of each consecutive pair of leaves in L_i, and accumulate for each node w a count of the number of times that w is the computed *lca*. Let $h(w)$ denote that count for node w. Then for any node v, $n_i(v) - 1$ is exactly $\sum[h(w) : w$ is in the subtree of $v]$. A standard $O(n)$-time bottom-up traversal of $\mathcal{T}$ can therefore be used to find $n_i(v) - 1$ for each node v.

To find $U(v)$, we don't want $n_i(v) - 1$ but rather $\sum_i[n_i(v) - 1]$. However, the algorithm must not do a separate bottom-up traversal for each identifier, since then the time bound would then be $O(Kn)$. Instead, the algorithm should defer the bottom-up traversal until each list L_i has been processed, and it should let $h(w)$ count the total number of times that w is the computed *lca* over all of the lists. Only then is a single bottom-up traversal of $\mathcal{T}$ done. At that point, $U(v) = \sum_{i:n_i>0}[n_i(v) - 1] = \sum[h(w) : w$ is in the subtree of $v]$.

We can now summarize the entire $O(n)$ method for solving the k-common substring problem.

Multiple common substring algorithm

Begin

1. Build a generalized suffix tree $\mathcal{T}$ for the K strings.

2. Number the leaves of $\mathcal{T}$ as they are encountered in a depth-first traversal of $\mathcal{T}$.

3. For each string identifier i, extract the ordered list L_i of leaves with identifier i. (The minor implementation detail needed to do this in $O(n)$ total time is left to the reader.)

4. For each node w in $\mathcal{T}$ set $h(w)$ to zero.

5. For each identifier i, compute the *lca* of each consecutive pair of leaves in L_i, and increment $h(w)$ by one each time that w is the computed *lca*.

6. With a bottom-up traversal of T, compute, for each node v, $S(v)$ and $U(v) = \sum_{i:n_i>0}[n_i(v)-1] = \sum[h(w) : w$ is in the subtree of $v]$.

7. Set $C(v) = S(v) - U(v)$ for each node v.

8. Accumulate the table of $l(k)$ values as detailed in Section 7.6.

End.

9.7.2. Time analysis

The size of the suffix tree is $O(n)$ and preprocessing of the tree for *lca* computations is done in $O(n)$ time. There are then $\sum_{i=1}^{K} |n_i - 1| < n$ *lca* computations done, each of which takes constant time, so all the *lca* computations take $O(n)$ time in total. Hence only $O(n)$ time is needed to compute all $C(v)$ values. Once these are known, only $O(n)$ additional time is needed to build the output table. That part of the algorithm is the same as in the previously discussed $O(Kn)$-time algorithm of Section 7.6. Therefore, we can state

Theorem 9.7.1. *Let S be a set of K strings of total length n, and let $l(k)$ denote the length of the longest substring that appears in at least k distinct strings of S. A table of all $l(k)$ values, for k from 2 to K, can be built in $O(n)$ time.*

That so much information about the substrings of S can be obtained in time proportional to the time needed just to read the strings is very impressive. It would be a good challenge to try to obtain this result without the use of suffix trees (or a similar data structure).

9.7.3. Related uses

The methodology developed for the k-common substring problem can be easily extended to solve related and important problems about sets of strings.

For example, suppose you are given two sets of strings S and $\mathcal{P}$, and you want to know for each string $A \in \mathcal{P}$, in how many strings of S does A appear. Let n denote the total size of all the strings in S and m denote the total size of all the strings in $\mathcal{P}$. The problem can be solved in $O(n + m)$ time, the same time bound attainable via the Aho–Corasick method. Or one could consider another problem: Given a length l, find the string of length at least l that appears in the most strings in a set given of strings. That is, find the most common substring of length at least l. That problem has applications in many multiple alignment methods. See Exercise 26 in Chapter 14.

9.8. Exercises

1. Prove Theorem 9.1.1.

2. Fill in all the details and prove the correctness of the space-efficient method solving the longest common extension problem.

3. Give the details for finding all odd-length maximal palindromes in a string in linear time.

4. Show how to solve all the palindrome problems in linear time using just a suffix tree for the string S rather than for both S and S^r.

5. Give the details for searching for complemented palindromes in a linear string.

6. Recall that a *plasmid* is a circular DNA molecule common in bacteria (and elsewhere). Some bacterial plasmids contain relatively long complemented palindromes (whose function is somewhat in question). Give a linear-time algorithm to find all maximal complemented palindromes in a *circular* string.

7. Show how to find all the k-mismatch palindromes in a string of length n in $O(kn)$ time.

8. Tandem repeats. In the recursive method discussed in Section 9.6 (page 202) for finding the tandem repeats (no mismatches), problem 3 is solved with a linear number of constant-time common extension queries, exploiting suffix trees and lowest common ancestor computations. An earlier, equally efficient, solution to problem 3 was developed by Main and Lorenz [307], without using suffix trees.

The idea is that the problem can be solved in an *amortized* linear-time bound without suffix trees. In an instance of problem 3, h is held fixed while $q = h + l - 1$ varies over all appropriate values of l. Each forward common extension query is a problem of finding *the length of the longest substring beginning at position q that matches a prefix of $S[h$ ... n]*. All those lengths must be found in linear time. But that objective can be achieved by computing Z values (again) from Chapter 1, for the appropriate substring of S. Flesh out the details of this approach and prove the linear amortized time bound.

Now show how the backward common extensions can also be solved in linear time by computing Z values on the appropriately constructed substring of S. This substring is a bit less direct than the one used for forward extensions.

9. Complete the details for the $O(kn \log n + z)$-time algorithm for the k-mismatch tandem repeat problem. Consider both correctness and time.

10. Complete the details for the $O(kn \log(n/k) + z)$ bound for the k-mismatch tandem repeat method.

11. Try to modify the Main and Lorenz method for finding all the tandem repeats (without errors) to solve the k-mismatch tandem repeat problem in $O(kn \log n + z)$ time. If you are not successful, explain what the difficulties are and how the use of suffix trees and common ancestors solves these problems.

12. The tandem repeat method detailed in Section 9.6 finds all tandem repeats even if they are not maximal. For example, it finds six tandem repeats in the string *xababababy*, even though the left-most tandem repeat *abab* is contained in the longer tandem repeat *abababab*. Depending on the application, that output may not be desirable. Give a definition of *maximality* that would reduce the size of the output and try to give efficient algorithms for the different definitions.

13. Consider the following situation: A long string S is given and remains fixed. Then a sequence of shorter strings $S_1, S_2, \ldots, S_k$ is given. After each string S_i is given (but before S_{i+1} is known), a number of longest common extension queries will be asked about S_i and S. Let r denote the total number of queries and n denote the total length of all the short strings. How can these on-line queries be answered efficiently? The most direct approach is to build a generalized suffix tree for both S and S_i when S_i is presented, preprocess it (do a depth-first traversal assigning *dfs* numbers, setting $l()$ values, etc.) for the constant-time *lca* algorithm, and then answer the queries for S_i. But that would take $\Theta(k|S| + n + r)$ time. The $k|S|$ term comes from two sources: the time to build the k generalized suffix trees and the time to preprocess each of them for *lca* queries.

Reduce that $k|S|$ term from both sources to $|S|$, obtaining an overall bound of $O(|S| + n + r)$. Reducing the time for building all the generalized suffix trees is easy. Reducing the time for the *lca* preprocessing takes a bit more thought.

Find a plausible application of the above result.

PART III

Inexact Matching, Sequence Alignment, and Dynamic Programming

At this point we shift from the general area of *exact* matching and exact pattern discovery to the general area of *inexact*, *approximate* matching, and sequence *alignment*. "Approximate" means that some errors, of various types detailed later, are acceptable in valid matches. "Alignment" will be given a precise meaning later, but generally means lining up characters of strings, allowing mismatches as well as matches, and allowing characters of one string to be placed opposite spaces made in opposing strings.

We also shift from problems primarily concerning *substrings* to problems concerning *subsequences*. A subsequence differs from a substring in that the characters in a substring *must* be contiguous, whereas the characters in a subsequence embedded in a string need not be.[1] For example, the string *xyz* is a subsequence, but not a substring, in *axayaz*. The shift from substrings to subsequences is a natural corollary of the shift from exact to inexact matching. This shift of focus to inexact matching and subsequence comparison is accompanied by a shift in *technique*. Most of the methods we will discuss in Part III, and many of the methods in Part IV, rely on the tool of *dynamic programming*, a tool that was not needed in Parts I and II.

Much of computational biology concerns sequence alignments

The area of approximate matching and sequence comparison is central in computational molecular biology both because of the presence of errors in molecular data and because of active mutational processes that (sub)sequence comparison methods seek to model and reveal. This will be elaborated in the next chapter and illustrated throughout the book. On the technical side, sequence alignment has become the central tool for sequence comparison in molecular biology. Henikoff and Henikoff [222] write:

> Among the most useful computer-based tools in modern biology are those that involve sequence alignments of proteins, since these alignments often provide important insights into gene and protein function. There are several different types of alignments: global alignments of pairs of proteins related by common ancestry throughout their lengths, local alignments involving related segments of proteins, multiple alignments of members of protein families, and alignments made during data base searches to detect homologies.

This statement provides a framework for much of Part III. We will examine in detail the four types of alignments (and several variants) mentioned above. We will also show how those different alignment models address different kinds of problems in biology. We begin, in Chapter 10, with a more detailed statement of why sequence comparison has become central to current molecular biology. But we won't forget the role of exact matching.

---

[1] It is a common and confusing practice in the biological literature to refer to a substring as a subsequence. But techniques and results for substring problems can be very different from techniques and results for the analogous subsequence problems, so it is important to maintain a clear distinction. In this book we will never use the term "subsequence" when "substring" is intended.

The role of exact matching

The centrality of *approximate* matching in molecular biology is undisputed. However, it does not follow that exact matching methods have little application there, and several biological applications of exact matching were developed in Parts I and II. As one example, recall from Section 7.15, that suffix trees are now playing a central role in several biological database efforts. Moreover, several exact matching techniques were shown earlier to directly extend or apply to approximate matching problems (the match-count problem, the wild-card problem, the k-mismatch problem, the k-mismatch palindrome problem, and the k-mismatch tandem repeat problem). In Parts III and IV we will develop additional approximate matching techniques that rely in a crucial way on efficient exact matching methods, suffix trees, etc. We will also see exact matching problems that arise as subproblems in multiple sequence comparison, in large-scale sequence comparison, in database searching, and in other biologically important applications.

10

The Importance of (Sub)sequence Comparison in Molecular Biology

Sequence comparison, particularly when combined with the systematic collection, curration, and search of databases containing biomolecular sequences, has become essential in modern molecular biology. Commenting on the (then) near-completion of the effort to sequence the entire yeast genome (now finished), Stephen Oliver says

> In a short time it will be hard to realize how we managed without the sequence data. Biology will never be the same again. [478]

One fact explains the importance of molecular sequence data and sequence comparison in biology.

The first fact of biological sequence analysis

The *first fact* of biological sequence analysis   In biomolecular sequences (DNA, RNA, or amino acid sequences), high sequence similarity usually implies significant functional or structural similarity.

Evolution reuses, builds on, duplicates, and modifies "successful" structures (proteins, exons, DNA regulatory sequences, morphological features, enzymatic pathways, etc.). Life is based on a repertoire of structured and interrelated molecular building blocks that are shared and passed around. The same and related molecular structures and mechanisms show up repeatedly in the genome of a single species and across a very wide spectrum of divergent species. "Duplication with modification" [127, 128, 129, 130] is the central paradigm of protein evolution, wherein new proteins and/or new biological functions are fashioned from earlier ones. Doolittle emphasizes this point as follows:

> The vast majority of extant proteins are the result of a continuous series of genetic duplications and subsequent modifications. As a result, redundancy is a built-in characteristic of protein sequences, and we should not be surprised that so many new sequences resemble already known sequences. [129]

He adds that

> . . . all of biology is based on an enormous redundancy . . . .[130]

The following quotes reinforce this view and suggest the utility of the "enormous redundancy" in the practice of molecular biology. The first quote is from Eric Wieschaus, cowinner of the 1995 Nobel prize in medicine for work on the genetics of *Drosophila* development. The quote is taken from an Associated Press article of October 9, 1995. Describing the work done years earlier, Wieschaus says

> We didn't know it at the time, but we found out everything in life is so similar, that the same genes that work in flies are the ones that work in humans.

212

And fruit flies aren't special. The following is from a book review on DNA repair [424]:

> Throughout the present work we see the insights gained through our ability to look for sequence homologies by comparison of the DNA of different species. Studies on yeast are remarkable predictors of the human system!

So "redundancy", and "similarity" are central phenomena in biology. But similarity has its limits – humans and flies do differ in some respects. These differences make *conserved* similarities even more significant, which in turn makes *comparison* and *analogy* very powerful tools in biology. Lesk [297] writes:

> It is characteristic of biological systems that objects that we observe to have a certain form arose by evolution from related objects with similar but not identical from. They must, therefore, be robust, in that they retain the freedom to tolerate some variation. We can take advantage of this robustness in our analysis: By identifying and comparing related objects, we can distinguish variable and conserved features, and thereby determine what is crucial to structure and function.

The important "related objects" to compare include much more than sequence data, because biological universality occurs at many levels of detail. However, it is usually easier to acquire and examine sequences than it is to examine fine details of genetics or cellular biochemistry or morphology. For example, there are vastly more protein sequences known (deduced from underlying DNA sequences) than there are known three-dimensional protein structures. And it isn't just a matter of convenience that makes sequences important. Rather, the biological sequences *encode* and reflect the more complex common molecular structures and mechanisms that appear as features at the cellular or biochemical levels. Moreover, "nowhere in the biological world is the Darwinian notion of 'descent with modification' more apparent than in the sequences of genes and gene products" [130]. Hence a tractable, though partly heuristic, way to search for functional or structural universality in biological systems is to search for similarity and conservation at the *sequence* level. The power of this approach is made clear in the following quotes:

> Today, the most powerful method for inferring the biological function of a gene (or the protein that it encodes) is by sequence similarity searching on protein and DNA sequence databases. With the development of rapid methods for sequence comparison, both with heuristic algorithms and powerful parallel computers, discoveries based solely on sequence homology have become routine. [360]

> Determining function for a sequence is a matter of tremendous complexity, requiring biological experiments of the highest order of creativity. Nevertheless, with only DNA sequence it is possible to execute a computer-based algorithm comparing the sequence to a database of previously characterized genes. In about 50% of the cases, such a mechanical comparison will indicate a sufficient degree of similarity to suggest a putative enzymatic or structural function that might be possessed by the unknown gene. [91]

Thus large-scale sequence comparison, usually organized as database search, is a very powerful tool for biological inference in modern molecular biology. And that tool is almost universally used by molecular biologists. It is now standard practice, whenever a new gene is cloned and sequenced, to translate its DNA sequence into an amino acid sequence and then search for similarities between it and members of the protein databases. No one today would even think of publishing the sequence of a newly cloned gene without doing such database searches.

The final quote reflects the potential total impact on biology of the *first fact* and its exploitation in the form of sequence database searching. It is from an article [179] by Walter Gilbert, Nobel prize winner for the coinvention of a practical DNA sequencing method. Gilbert writes:

> The new paradigm now emerging, is that all the 'genes' will be known (in the sense of being resident in databases available electronically), and that the starting point of biological investigation will be theoretical. An individual scientist will begin with a theoretical conjecture, only then turning to experiment to follow or test that hypothesis.

Already, hundreds (if not thousands) of journal publications appear each year that report biological research where sequence comparison and/or database search is an integral part of the work. Many such examples that support and illustrate the *first fact* are distributed throughout the book. In particular, several in-depth examples are concentrated in Chapters 14 and 15 where multiple string comparison and database search are discussed. But before discussing those examples, we must first develop, in the next several chapters, the techniques used for approximate matching and (sub)sequence comparison.

Caveat

The *first fact of biological sequence analysis* is extremely powerful, and its importance will be further illustrated throughout the book. However, there is not a one-to-one correspondence between sequence and structure or sequence and function, because the converse of the *first fact* is not true. That is, high sequence similarity usually implies significant structural or functional similarity (the first fact), but structural or functional similarity does not necessarily imply sequence similarity. On the topic of protein structure, F. Cohen [106] writes "... similar sequences yield similar structures, but quite distinct sequences can produce remarkably similar structures". This *converse* issue is discussed in greater depth in Chapter 14, which focuses on multiple sequence comparison.

11

Core String Edits, Alignments, and Dynamic Programming

11.1. Introduction

In this chapter we consider the inexact matching and alignment problems that form the core of the field of inexact matching and others that illustrate the most general techniques. Some of those problems and techniques will be further refined and extended in the next chapters. We start with a detailed examination of the most classic inexact matching problem solved by dynamic programming, the *edit distance* problem. The motivation for inexact matching (and, more generally, sequence comparison) in molecular biology will be a recurring theme explored throughout the rest of the book. We will discuss many specific examples of how string comparison and inexact matching are used in current molecular biology. However, to begin, we concentrate on the purely formal and technical aspects of defining and computing inexact matching.

11.2. The edit distance between two strings

Frequently, one wants a measure of the difference or *distance* between two strings (for example, in evolutionary, structural, or functional studies of biological strings; in textual database retrieval; or in spelling correction methods). There are several ways to formalize the notion of distance between strings. One common, and simple, formalization [389, 299], called *edit distance*, focuses on *transforming* (or editing) one string into the other by a series of edit operations on individual characters. The permitted edit operations are *insertion* of a character into the first string, the *deletion* of a character from the first string, or the *substitution* (or *replacement*) of a character in the first string with a character in the second string. For example, letting I denote the insert operation, D denote the delete operation, R the substitute (or replace) operation, and M the nonoperation of "match," then the string "*vintner*" can be edited to become "*writers*" as follows:

```
RIMDMDMMI
v intner
wri t ers
```

That is, v is replaced by w, r is inserted, i matches and is unchanged since it occurs in both strings, n is deleted, t is unchanged, n is deleted, er match and are unchanged, and finally s is inserted. We now more formally define edit transcripts and string transformations.

Definition   A string over the alphabet I, D, R, M that describes a transformation of one string to another is called an *edit transcript*, or transcript for short, of the two strings.

In general, given the two input strings S_1 and S_2, and given an edit transcript for S_1 and S_2, the transformation is accomplished by successively applying the specified operation in the transcript to the next character(s) in the appropriate string(s). In particular, let $next_1$ and

215

$next_2$ be pointers into S_1 and S_2. Both pointers begin with value one. The edit transcript is read and applied left to right. When symbol "I" is encountered, character $next_2$ is inserted before character $next_1$ in S_1, and pointer $next_2$ is incremented one character. When "D" is encountered, character $next_1$ is deleted from S_1 and $next_1$ is incremented by one character. When either symbol "R" or "M" is encountered, character $next_1$ in S_1 is replaced or matched by character $next_2$ from S_2, and then both pointers are incremented by one.

Definition   The *edit distance* between two strings is defined as the minimum number of edit operations – insertions, deletions, and substitutions – needed to transform the first string into the second. For emphasis, note that matches are not counted.

Edit distance is sometimes referred to as *Levenshtein distance* in recognition of the paper [299] by V. Levenshtein where edit distance was probably first discussed.

We will sometimes refer to an edit transcript that uses the minimum number of edit operations as an *optimal transcript*. Note that there may be more than one optimal transcript. These will be called "cooptimal" transcripts when we want to emphasize the fact that there is more than one optimal.

The **edit distance problem** is to compute the edit distance between two given strings, along with an optimal edit transcript that describes the transformation.

The definition of edit distance implies that all operations are done to one string only. But edit distance is sometimes thought of as the minimum number of operations done on either of the two strings to transform both of them into a common third string. This view is equivalent to the above definition, since an insertion in one string can be viewed as a deletion in the other and vice versa.

11.2.1. String alignment

An edit transcript is a way to *represent* a particular transformation of one string to another. An alternate (and often preferred) way is by displaying an explicit *alignment* of the two strings.

Definition   A (global) *alignment* of two strings S_1 and S_2 is obtained by first inserting chosen spaces (or dashes), either into or at the ends of S_1 and S_2, and then placing the two resulting strings one above the other so that every character or space in either string is opposite a unique character or a unique space in the other string.

The term "global" emphasizes the fact that for each string, the entire string is involved in the alignment. This will be contrasted with local alignment to be discussed later. Notice that our use of the word "alignment" is now much more precise than its use in Parts I and II. There, alignment was used in the colloquial sense to indicate how one string is placed relative to the other, and spaces were not then allowed in either string.

As an example of a global alignment, consider the alignment of the strings *qacdbd* and *qawxb* shown below:

$$
\begin{array}{ccccccc}
q & a & c & \_ & d & b & d \\
q & a & w & x & \_ & b & \_
\end{array}
$$

In this alignment, character c is mismatched with w, both the ds and the x are opposite spaces, and all other characters match their counterparts in the opposite string.

Another example of an alignment is shown on page 215 where *vintner* and *writers* are aligned with each other below their edit transcript. That example also suggests a duality between alignment and edit transcript that will be developed below.

Alignment versus edit transcript

From the mathematical standpoint, an alignment and an edit transcript are equivalent ways to describe a relationship between two strings. An alignment can be easily converted to the equivalent edit transcript and vice versa, as suggested by the *vintner–writers* example. Specifically, two opposing characters that mismatch in an alignment correspond to a substitution in the equivalent edit transcript; a space in an alignment contained in the first string corresponds in the transcript to an insertion of the opposing character into the first string; and a space in the second string corresponds to a deletion of the opposing character from the first string. Thus the edit distance of two strings is given by the alignment minimizing the number of opposing characters that mismatch plus the number of characters opposite spaces.

Although an alignment and an edit transcript are mathematically equivalent, from a modeling standpoint, an edit transcript is quite different from an alignment. An edit transcript emphasizes the putative *mutational events* (point mutations in the model so far) that transform one string to another, whereas an alignment only displays a relationship between the two strings. The distinction is one of *process* versus *product*. Different evolutionary models are formalized via different permitted string operations, and yet these can result in the same alignment. So an alignment alone blurs the mutational model. This is often a pedantic point but proves helpful in some discussions of evolutionary modeling.

We will switch between the language of edit transcript and alignment as is convenient. However, the language of alignment will often be preferred since it is more neutral, making no statement about process. And, the language of alignment will be more natural in the area of multiple sequence comparison.

11.3. Dynamic programming calculation of edit distance

We now turn to the algorithmic question of how to compute, via dynamic programming, the edit distance of two strings along with the accompanying edit transcript or alignment. The general paradigm of dynamic programming is probably well known to the readers of this book. However, because it is such a crucial tool and is used in so many string algorithms, it is worthwhile to explain in detail both the general dynamic programming approach and its specific application to the edit distance problem.

> **Definition**   For two strings S_1 and S_2, $D(i, j)$ is defined to be the edit distance of $S_1[1..i]$ and $S_2[1..j]$.

That is, $D(i, j)$ denotes the minimum number of edit operations needed to transform the first i characters of S_1 into the first j characters of S_2. Using this notation, if S_1 has n letters and S_2 has m letters, then the edit distance of S_1 and S_2 is *precisely* the value $D(n, m)$.

We will compute $D(n, m)$ by solving the more general problem of computing $D(i, j)$ for all combinations of i and j, where i ranges from zero to n and j ranges from zero to m. This is the standard *dynamic programming* approach used in a vast number of computational problems. The dynamic programming approach has three essential components – the *recurrence relation*, the *tabular computation*, and the *traceback*. We will explain each one in turn.

11.3.1. The recurrence relation

The recurrence relation establishes a *recursive* relationship between the value of $D(i, j)$, for i and j both positive, and values of D with index pairs smaller than i, j. When there are no smaller indices, the value of $D(i, j)$ must be stated explicitly in what are called the *base conditions* for $D(i, j)$.

For the edit distance problem, the base conditions are

$$D(i, 0) = i$$

and

$$D(0, j) = j.$$

The base condition $D(i, 0) = i$ is clearly correct (that is, it gives the number required by the definition of $D(i, 0)$) because the only way to transform the first i characters of S_1 to zero characters of S_2 is to delete all the i characters of S_1. Similarly, the condition $D(0, j) = j$ is correct because j characters must be inserted to convert zero characters of S_1 to j characters of S_2.

The recurrence relation for $D(i, j)$ when both i and j are strictly positive is

$$D(i, j) = \min[D(i - 1, j) + 1, D(i, j - 1) + 1, D(i - 1, j - 1) + t(i, j)],$$

where $t(i, j)$ is defined to have value 1 if $S_1(i) \neq S_2(j)$, and $t(i, j)$ has value 0 if $S_1(i) = S_2(j)$.

Correctness of the general recurrence

We establish correctness in the next two lemmas using the concept of an edit transcript.

Lemma 11.3.1. *The value of $D(i, j)$ must be $D(i, j - 1) + 1$, $D(i - 1, j) + 1$, or $D(i - 1, j - 1) + t(i, j)$. There are no other possibilities.*

PROOF   Consider an edit transcript for the transformation of $S_1[1..i]$ to $S_2[1..j]$ using the minimum number of edit operations, and focus on the last symbol in that transcript. That last symbol must either be I, D, R, or M. If the last symbol is an I then the last edit operation is the insertion of character $S_2(j)$ onto the end of the (transformed) first string. It follows that the symbols in the transcript before that I must specify the minimum number of edit operations to transform $S_1[1..i]$ to $S_2[1..j - 1]$ (if they didn't, then the specified transformation of $S_1[1..i]$ to $S_2[1..j]$ would use more than the minimum number of operations). By definition, that latter transformation takes $D(i, j - 1)$ edit operations. Hence if the last symbol in the transcript is I, then $D(i, j) = D(i, j - 1) + 1$.

Similarly, if the last symbol in the transcript is a D, then the last edit operation is the deletion of $S_1(i)$, and the symbols in the transcript to the left of that D must specify the minimum number of edit operations to transform $S_1[1..i - 1]$ to $S_2[1..j]$. By definition, that latter transformation takes $D(i - 1, j)$ edit operations. So if the last symbol in the transcript is D, then $D(i, j) = D(i - 1, j) + 1$.

If the last symbol in the transcript is an R, then the last edit operation replaces $S_1(i)$ with $S_2(j)$, and the symbols to the left of R specify the minimum number of edit operations to transform $S_1[1..i - 1]$ to $S_2[1..j - 1]$. In that case $D(i, j) = D(i - 1, j - 1) + 1$. Finally, and by similar reasoning, if the last symbol in the transcript is an M, then $S_1(i) = S_2(j)$ and $D(i, j) = D(i - 1, j - 1)$. Using the variable $t(i, j)$ introduced earlier [i.e., that $t(i, j) = 0$ if $S_1(i) = S_2(j)$; otherwise $t(i, j) = 1$] we can combine these last two cases as one: If the last transcript symbol is R or M, then $D(i, j) = D(i - 1, j - 1) + t(i, j)$.

Since the last transcript symbol must either be I, D, R, or M, we have covered all cases and established the lemma.   □

Now we look at the other side.

Lemma 11.3.2. $D(i, j) \leq \min[D(i-1, j)+1, D(i, j-1)+1, D(i-1, j-1)+t(i, j)]$.

PROOF   The reasoning is very similar to that used in the previous lemma, but it achieves a somewhat different goal. The objective here is to demonstrate constructively the existence of transformations achieving each of the three values specified in the inequality. Then since all three values are feasible, their minimum is certainly feasible.

First, it *is* possible to transform $S_1[1..i]$ into $S_2[1..j]$ with exactly $D(i, j-1)+1$ edit operations. Simply transform $S_1[1..i]$ to $S_2[1..j-1]$ with the minimum number of edit operations, and then use one more to insert character $S_2(j)$ at the end. By definition, the number of edit operations in that particular way to transform S_1 to S_2 is exactly $D(i, j-1)+1$. Second, it *is* possible to transform $S_1[1..i]$ to $S_2[1..j]$ with exactly $D(i-1, j)+1$ edit operations. Transform $S_1[1..i-1]$ to $S_2[1..j]$ with the fewest operations, and then delete character $S_1(i)$. The number of edit operations in that particular transformation is exactly $D(i-1, j)+1$. Third, it *is* possible to do the transformation with exactly $D(i-1, j-1)+t(i, j)$ edit operations, using the same argument.   □

Lemmas 11.3.1 and 11.3.2 immediately imply the correctness of the general recurrence relation for $D(i, j)$.

Theorem 11.3.1. *When both i and j are strictly positive,* $D(i, j) = \min[D(i-1, j)+1, D(i, j-1)+1, D(i-1, j-1)+t(i, j)]$.

PROOF   Lemma 11.3.1 says that $D(i, j)$ must be equal to one of the three values $D(i-1, j)+1$, $D(i, j-1)+1$, or $D(i-1, j-1)+t(i, j)$. Lemma 11.3.2 says that $D(i, j)$ must be *less than or equal* to the smallest of those three values. It follows that $D(i, j)$ must therefore be equal to the smallest of those three values, and we have proven the theorem.   □

This completes the first component of the dynamic programming method for edit distance, the recurrence relation.

11.3.2. Tabular computation of edit distance

The second essential component of any dynamic program is to use the recurrence relations to efficiently compute the value $D(n, m)$. We could easily code the recurrence relations and base conditions for $D(i, j)$ as a recursive computer procedure using any programming language that allows recursion. Then we could call that procedure with input m, n and sit back and wait for the answer.[1] This *top-down* recursive approach to evaluating $D(n, m)$ is simple to program but extremely inefficient for large values of n and m.

The problem is that the number of recursive calls grows exponentially with n and m (an easy exercise to establish). But there are only $(n + 1) \times (m + 1)$ combinations of i and j, so there are only $(n + 1) \times (m + 1)$ *distinct* recursive calls possible. Hence the inefficiency of the top-down approach is due to a massive number of redundant recursive calls to the procedure. A nice discussion of this phenomenon is contained in [112]. The key to a (vastly) more efficient computation of $D(n, m)$ is to abandon the simplicity of top-down computation and instead compute *bottom-up*.

---

[1] and wait, and wait, . . .

$D(i,j)$				w	r	i	t	e	r	s
			0	1	2	3	4	5	6	7
		0	0	1	2	3	4	5	6	7
v	1	1								
i	2	2								
n	3	3								
t	4	4								
n	5	5								
e	6	6								
r	7	7								

Figure 11.1: Table to be used to compute the edit distance between *vintner* and *writers*. The values in row zero and column zero are already included. They are given directly by the base conditions.

Bottom-up computation

In the bottom-up approach, we first compute $D(i, j)$ for the smallest possible values for i and j, and then compute values of $D(i, j)$ for increasing values of i and j. Typically, this bottom-up computation is organized with a dynamic programming table of size $(n + 1) \times (m+1)$. The table holds the values of $D(i, j)$ for all the choices of i and j (see Figure 11.1). Note that string S_1 corresponds to the vertical axis of the table, while string S_2 corresponds to the horizontal axis. Because the ranges of i and j begin at zero, the table has a zero row and a zero column. The values in row zero and column zero are filled in directly from the base conditions for $D(i, j)$. After that, the remaining $n \times m$ subtable is filled in one row at a time, in order of increasing i. Within each row, the cells are filled in order of increasing j.

To see how to fill in the subtable, note that by the general recurrence relation for $D(i, j)$, all the values needed for the computation of $D(1, 1)$ are known once $D(0, 0)$, $D(1, 0)$, and $D(0, 1)$ have been computed. Hence $D(1, 1)$ can be computed after the zero row and zero column have been filled in. Then, again by the recurrence relations, after $D(1, 1)$ has been computed, all the values needed for the computation of $D(1, 2)$ are known. Following this idea, we see that the values for row one can be computed in order of increasing index j. After that, all the values needed to compute the values in row two are known, and that row can be filled in, in order of increasing j. By extension, the entire table can be filled in one row at a time, in order of increasing i, and in each row the values can be computed in order of increasing j (see Figure 11.2).

Time analysis

How much work is done by this approach? When computing the value for a specific cell (i, j), only cells $(i - 1, j - 1)$, $(i, j - 1)$, and $(i - 1, j)$ are examined, along with the two characters $S_1(i)$ and $S_2(j)$. Hence, to fill in one cell takes a constant number of cell examinations, arithmetic operations, and comparisons. There are $O(nm)$ cells in the table, so we obtain the following theorem.

Theorem 11.3.2. *The dynamic programming table for computing the edit distance between a string of length n and a string of length m can be filled in with $O(nm)$ work. Hence, using dynamic programming, the edit distance $D(n, m)$ can be computed in $O(nm)$ time.*

$D(i,j)$			w	r	i	t	e	r	s
		0	1	2	3	4	5	6	7
	0	0	1	2	3	4	5	6	7
v	1	1	1	2	3	4	5	6	7
i	2	2	2	2	2	3	4	5	6
n	3	3	3	3	3	3	4	5	6
t	4	4	4	4	4	4	*		
n	5	5	5						
e	6	6	6						
r	7	7	7						

Figure 11.2: Edit distances are filled in one row at a time, and in each row they are filled in from left to right. The example shows the edit distances $D(i, j)$ to column 3 of row 4. The next value to be computed is $D(4, 4)$, where an asterisk appears. The value for cell $(4, 4)$ is 3, since $S_1(4) = S_2(4) = t$ and $D(3, 3) = 3$.

The reader should be able to establish that the table could also be filled in *columnwise* instead of rowwise, after row zero and column zero have been computed. That is, column one could be first filled in, followed by column two, etc. Similarly, it is possible to fill in the table by filling in successive anti-diagonals. We leave the details as an exercise.

11.3.3. The traceback

Once the value of the edit distance has been computed, how is the associated optimal edit transcript extracted? The easiest way (conceptually) is to establish *pointers* in the table as the table values are computed.

In particular, when the value of cell (i, j) is computed, set a pointer from cell (i, j) to cell $(i, j - 1)$ if $D(i, j) = D(i, j - 1) + 1$; set a pointer from (i, j) to $(i - 1, j)$ if $D(i, j) = D(i - 1, j) + 1$; and set a pointer from (i, j) to $(i - 1, j - 1)$ if $D(i, j) = D(i - 1, j - 1) + t(i, j)$. This rule applies to cells in row zero and column zero as well. Hence, for most objective functions, each cell in row zero points to the cell to its left, and each cell in column zero points to the cell just above it. For other cells, it is possible (and common) that more than one pointer is set from (i, j). Figure 11.3 shows an example.

The pointers allow easy recovery of an optimal edit transcript: Simply follow *any* path of pointers from cell (n, m) to cell $(0, 0)$. The edit transcript is recovered from the path by interpreting each *horizontal* edge in the path, from cell (i, j) to cell $(i, j-1)$, as an *insertion* (I) of character $S_2(j)$ into S_1; interpreting each *vertical* edge, from (i, j) to $(i - 1, j)$, as a *deletion* (D) of $S_1(i)$ from S_1; and interpreting each *diagonal* edge, from (i, j) to $(i-1, j-1)$, as a *match* (M) if $S_1(i) = S_2(j)$ and as a *substitution* (R) if $S_1(i) \neq S_2(j)$. That this traceback path specifies an optimal edit transcript can be proved in a manner similar to the way that the recurrences for edit distances were established. We leave this as an exercise.

Alternatively, in terms of *aligning* S_1 and S_2, each horizontal edge in the path specifies a space inserted into S_1, each vertical edge specifies a space inserted into S_2, and each diagonal edge specifies either a match or a mismatch, depending on the specific characters.

For example, there are three traceback paths from cell $(7, 7)$ to cell $(0, 0)$ in the example given in Figure 11.3. The paths are identical from cell $(7, 7)$ to cell $(3, 3)$, at which point

$D(i,j)$			w	r	i	t	e	r	s
		0	1	2	3	4	5	6	7
	0	0	← 1	← 2	← 3	← 4	← 5	← 6	← 7
v	1	↑ 1	↖ 1	↖ ← 2	↖ ← 3	↖ ← 4	↖ ← 5	↖ ← 6	↖ ← 7
i	2	↑ 2	↖ ↑ 2	↖ 2	↖ 2	← 3	← 4	← 5	← 6
n	3	↑ 3	↖ ↑ 3	↖ ↑ 3	↖ ↑ 3	↖ 3	↖ ← 4	↖ ← 5	↖ ← 6
t	4	↑ 4	↖ ↑ 4	↖ ↑ 4	↖ ↑ 4	↖ 3	↖ ← 4	↖ ← 5	↖ ← 6
n	5	↑ 5	↖ ↑ 5	↖ ↑ 5	↖ ↑ 5	↑ 4	↖ 4	↖ ← 5	↖ ← 6
e	6	↑ 6	↖ ↑ 6	↖ ↑ 6	↖ ↑ 6	↑ 5	↖ 4	↖ ← 5	↖ ← 6
r	7	↑ 7	↖ ↑ 7	↖ 6	↖←↑ 7	↑ 6	↑ 5	↖ 4	← 5

Figure 11.3: The complete dynamic programming table with pointers included. The arrow ← in cell (i, j) points to cell $(i, j - 1)$, the arrow ↑ points to cell $(i - 1, j)$, and the arrow ↖ points to cell $(i - 1, j - 1)$.

it is possible to either go up or to go diagonally. The three optimal alignments are:

$$
\begin{array}{ccccccc}
w & r & i & t & \_ & e & r & s \\
v & i & n & t & n & e & r & \_
\end{array}
$$

$$
\begin{array}{cccccccc}
w & r & i & \_ & t & \_ & e & r & s \\
v & \_ & i & n & t & n & e & r & \_
\end{array}
$$

and

$$
\begin{array}{cccccccc}
w & r & i & \_ & t & \_ & e & r & s \\
\_ & v & i & n & t & n & e & r & \_
\end{array}
$$

If there is more than one pointer from cell (n, m), then a path from (n, m) to $(0, 0)$ can start with either of those pointers. Each of them is on a path from (n, m) to $(0, 0)$. This property is repeated from any cell encountered. Hence a traceback path from (n, m) to $(0, 0)$ can start simply by following any pointer out of (n, m); it can then be extended by following any pointer out of any cell encountered. Moreover, every cell except $(0, 0)$ has a pointer out of it, so no path from (n, m) can get stuck. Since any path of pointers from (n, m) to $(0, 0)$ specifies an optimal edit transcript or alignment, we have the following:

Theorem 11.3.3. *Once the dynamic programming table with pointers has been computed, an optimal edit transcript can be found in* $O(n + m)$ *time.*

We have now completely described the three crucial components of the general dynamic programming paradigm, as illustrated by the edit distance problem. We will later consider ways to increase the speed of the solution and decrease its needed space.

The pointers represent all optimal edit transcripts

The pointers that are built up while computing the values of the table do more than allow one optimal transcript (or optimal alignment) to be retrieved. They allow *all* optimal transcripts to be retrieved.

Theorem 11.3.4. *Any path from* (n, m) *to* $(0, 0)$ *following pointers established during the computation of* $D(i, j)$ *specifies an edit transcript with the minimum number of edit*

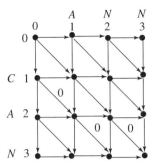

Figure 11.4: Edit graph for the strings *CAN* and *ANN*. The weight on each edge is one, except for the three zero-weight edges marked in the figure.

operations. *Conversely, any optimal edit transcript is specified by such a path. Moreover, since a path describes only one transcript, the correspondence between paths and optimal transcripts is one-to-one.*

The theorem can be proven by essentially the same reasoning that established the correctness of the recurrence relations for $D(i, j)$, and this is left to the reader. An alternative way to find the optimal edit transcript(s), without using pointers, is discussed in Exercise 9. Once the pointers have been established, all the cooptimal edit transcripts can be enumerated in $O(n + m)$ time per transcript. That is the focus of Exercise 12.

11.4. Edit graphs

It is often useful to represent dynamic programming solutions of string problems in terms of a *weighted edit graph*.

> **Definition**   Given two strings S_1 and S_2 of lengths n and m, respectively, a *weighted edit graph* has $(n + 1) \times (m + 1)$ nodes, each labeled with a distinct pair (i, j) $(0 \le i \le n, 0 \le j \le m)$. The specific edges and their edge weights depend on the specific string problem.

In the case of the edit distance problem, the edit graph contains a directed edge from each node (i, j) to each of the nodes $(i, j + 1)$, $(i + 1, j)$, and $(i + 1, j + 1)$, provided those nodes exist. The weight on the first two of these edges is one; the weight on the third (diagonal) edge is $t(i + 1, j + 1)$. Figure 11.4 shows the edit graph for strings *CAN* and *ANN*.

The central property of an edit graph is that any *shortest path* (one whose total weight is minimum) from start node $(0, 0)$ to destination node (n, m) specifies an edit transcript with the minimum number of edit operations. Equivalently, any shortest path specifies a global alignment of minimum total weight. Moreover, the following theorem and corollary can be stated.

Theorem 11.4.1. *An edit transcript for S_1, S_2 has the minimum number of edit operations if and only if it corresponds to a shortest path from $(0, 0)$ to (n, m) in the edit graph.*

Corollary 11.4.1. The set of all shortest paths from $(0, 0)$ to (n, m) in the edit graph exactly specifies the set of all optimal edit transcripts of S_1 to S_2. Equivalently, it specifies all the optimal (minimum weight) alignments of S_1 and S_2.

Viewing dynamic programming as a shortest path problem is often useful because there are many tools for investigating and compactly representing shortest paths in graphs. This view will be exploited in Section 13.2 when suboptimal solutions are discussed.

11.5. Weighted edit distance

11.5.1. Operation weights

An easy, yet crucial, generalization of edit distance is to allow an arbitrary *weight* or *cost* or *score*[2] to be associated with every edit operation, as well as with a match. Thus, any insertion or deletion has a weight denoted d, a substitution has a weight r, and a match has a weight e (which is usually small compared to the other weights and is often zero). Equivalently, an *operation-weight alignment* is one where each mismatch costs r, each match costs e, and each space costs d.

> **Definition**   With arbitrary operation weights, the *operation-weight edit distance problem* is to find an edit transcript that transforms string S_1 into S_2 with the minimum total operation weight.

In these terms, the edit distance problem we have considered so far is just the problem of finding the minimum operation-weight edit transcript when $d = 1, r = 1$, and $e = 0$. But, for example, if each mismatch has a weight of 2, each space has a weight of 4, and each match a weight of 1, then the alignment

$$
\begin{array}{cccccccc}
w & r & i & t & \_ & e & r & s \\
v & i & n & t & n & e & r & \_
\end{array}
$$

has a total weight of 17 and is an optimal alignment.

Because the objective function is to minimize total weight and because a substitution can be achieved by a deletion followed by an insertion, if substitutions are to be allowed then a substitution weight should be less than the sum of the weights for a deletion plus an insertion.

Computing operation-weight edit distance

The operation-weight edit distance problem for two strings of length n and m can be solved in $O(nm)$ time by a minor extension of the recurrences for edit distance. $D(i, j)$ now denotes the minimum total weight for edit operations transforming $S_1[1..i]$ to $S_2[1..j]$. We again use $t(i, j)$ to handle both substitution and equality, where now $t(i, j) = e$ if $S_1(i) = S_2(j)$; otherwise $t(i, j) = r$. Then the base conditions are

$$D(i, 0) = i \times d$$

and

$$D(0, j) = j \times d.$$

The general recurrence is

$$D(i, j) = \min[D(i, j - 1) + d, D(i - 1, j) + d, D(i - 1, j - 1) + t(i, j)].$$

---

[2] The terms "weight" or "cost" are heavily used in the computer science literature, while the term "score" is used in the biological literature. We will use these terms more or less interchangeably in discussing algorithms, but the term "score" will be used when talking about specific biological applications.

The operation-weight edit distance problem can also be represented and solved as a shortest path problem on a weighted edit graph, where the edge weights correspond in the natural way to the weights of the edit operations. The details are straightforward and are thus left to the reader.

11.5.2. Alphabet-weight edit distance

Another critical, yet simple, generalization of edit distance is to allow the weight or score of a substitution to depend on exactly which character in the alphabet is being removed and which is being added. For example, it may be more costly to replace an *A* with a *T* than with a *G*. Similarly, we may want the weight of a deletion or insertion to depend on exactly which character in the alphabet is being deleted or inserted. We call this form of edit distance the *alphabet-weight edit distance* to distinguish it from the operation-weight edit distance problem.

The operation-weight edit distance problem is a special case of the alphabet-weight problem, and it is trivial to modify the previous recurrence relations (for operation-weight edit distance) to compute alphabet-weight edit distance. We leave that as an exercise. We will usually use the simple term *weighted edit distance* when we mean the alphabet-weight version. Notice that in weighted edit distance, the weight of an operation depends on what characters are involved in an operation but not on *where* those characters appear in the string.

When comparing proteins, "the edit distance" almost always means the alphabet-weight edit distance over the alphabet of amino acids. There is an extensive literature (and continuing research) on what scores should be used for operations on amino acid characters and how they should be determined. The dominant amino acid scoring schemes are now the PAM matrices of Dayhoff [122] and the newer BLOSUM scoring matrices of the Henikoffs [222], although these matrices are actually defined in terms of a maximization problem (similarity) rather than edit distance.[3] Recently, a mathematical theory has been developed [16, 262] concerning the way scores should be interpreted and how a scoring scheme should relate both to the data it is obtained from and to the types of searches it is designed for. We will briefly discuss this issue again in Section 15.11.2.

When comparing DNA strings, unweighted or operation-weight edit distance is more often computed. For example, the popular database searching program, BLAST, scores identities as +5 and mismatches as −4. However, alphabet-weighted edit distance is also of interest and alphabet-based scoring schemes for DNA have been suggested (for example see [252]).

11.6. String similarity

Edit distance is one of the ways that the relatedness of two strings has been formalized. An alternate, and often preferred, way of formalizing the relatedness of two strings is to measure their *similarity* rather than their distance. This approach is chosen in most biological applications for technical reasons that should be clear later. When focusing on

---

[3] In a pure computer science or mathematical discussion of alphabet-weight edit distance, we would prefer to use the general term "weight matrix" for the matrix holding the alphabet-dependent substitution scores. However, molecular biologists use the terms "amino acid substitution matrix" or "nucleotide substitution matrix" for those matrices, and they use the term "weight matrix" for a very different object (See Section 14.3.1). Therefore, to maintain generality, and yet to keep in some harmony with the molecular biology literature, we will use the general term "scoring matrix".

similarity, the language of alignment is usually more convenient than the language of edit transcript. We now begin to develop a precise definition of similarity.

Definition   Let Σ be the alphabet used for strings S_1 and S_2, and let Σ' be Σ with the added character "_" denoting a space. Then, for any two characters x, y in Σ', $s(x, y)$ denotes the value (or *score*) obtained by aligning character x against character y.

Definition   For a given alignment $\mathcal{A}$ of S_1 and S_2, let S_1' and S_2' denote the strings after the chosen insertion of spaces, and let l denote the (equal) length of the two strings S_1' and S_2' in $\mathcal{A}$. The *value* of alignment $\mathcal{A}$ is defined as $\sum_{i=1}^{l} s(S_1'(i), S_2'(i))$.

That is, every position i in $\mathcal{A}$ specifies a pair of opposing characters in the alphabet Σ', and the value of $\mathcal{A}$ is obtained by summing the value contributed by each pair.

For example, let $\Sigma = \{a, b, c, d\}$ and let the pairwise scores be defined in the following matrix:

s	a	b	c	d	_
a	1	−1	−2	0	−1
b		3	−2	−1	0
c			0	−4	−2
d				3	−1
_					0

Then the alignment

c	a	c	_	d	b	d
c	a	b	b	d	b	_

has a total value of $0 + 1 - 2 + 0 + 3 + 3 - 1 = 4$.

In string similarity problems, scoring matrices usually set $s(x, y)$ to be greater than or equal to zero if characters x, y of Σ' match and less than zero if they mismatch. With such a scoring scheme, one seeks an alignment with as *large* a value as possible. That alignment will emphasize matches (or similarities) between the two strings while penalizing mismatches or inserted spaces. Of course, the meaningfulness of the resulting alignment may depend heavily on the scoring scheme used and how match scores compare to mismatch and space scores. Numerous character-pair scoring matrices have been suggested for proteins and for DNA [81, 122, 127, 222, 252, 400], and no single scheme is right for all applications. We will return to this issue in Sections 13.1, 15.7, and 15.10.

Definition   Given a pairwise scoring matrix over the alphabet Σ', the *similarity* of two strings S_1 and S_2 is defined as the value of the alignment $\mathcal{A}$ of S_1 and S_2 that maximizes total alignment value. This is also called the *optimal alignment value* of S_1 and S_2.

String similarity is clearly related to alphabet-weight edit distance, and depending on the specific scoring matrix involved, one can often transform one problem into the other. An important difference between similarity and weighted edit distance will become clear in Section 11.7, after we discuss local alignment.

11.6.1. Computing similarity

The similarity of two strings S_1 and S_2, and the associated optimal alignment, can be computed by dynamic programming with recurrences that should by now be very intuitive.

Definition $V(i, j)$ is defined as the *value* of the optimal alignment of prefixes $S_1[1..i]$ and $S_2[1..j]$.

Recall that a dash ("–") is used to represent a space inserted into a string. The base conditions are

$$V(0, j) = \sum_{1 \le k \le j} s(\_, S_2(k))$$

and

$$V(i, 0) = \sum_{1 \le k \le i} s(S_1(k), \_).$$

For i and j both strictly positive, the general recurrence is

$$V(i, j) = \max[V(i - 1, j - 1) + s(S_1(i), S_2(j)), V(i - 1, j) + s(S_1(i), \_),$$
$$V(i, j - 1) + s(\_, S_2(j))].$$

The correctness of this recurrence is established by arguments similar to those used for edit distance. In particular, in any alignment $\mathcal{A}$, there are three possibilities: characters $S_1(i)$ and $S_2(j)$ are in the same position (opposite each other), $S_1(i)$ is in a position after $S_2(j)$, or $S_1(i)$ is in a position before $S_2(j)$. The correctness of the recurrence is based on that case analysis. Details are left to the reader.

If S_1 and S_2 are of length n and m, respectively, then the value of their optimal alignment is given by $V(n, m)$. That value, and the entire dynamic programming table, can be obtained in $O(nm)$ time, since only three comparisons and arithmetic operations are needed per cell. By leaving pointers while filling in the table, as was done with edit distance, an optimal alignment can be constructed by following any path of pointers from cell (n, m) to cell $(0, 0)$. So the optimal (global) alignment problem can be solved in $O(nm)$ time, the same time as for edit distance.

11.6.2. Special cases of similarity

By choosing an appropriate scoring scheme, many problems can be modeled as special cases of optimal alignment or similarity. One important example is the *longest common subsequence problem*.

Definition In a string S, a *subsequence* is defined as a subset of the characters of S arranged in their original "relative" order. More formally, a subsequence of a string S of length n is specified by a list of indices $i_1 < i_2 < i_3 < \ldots < i_k$, for some $k \le n$. The subsequence specified by this list of indices is the string $S(i_1)S(i_2)S(i_3)\ldots S(i_k)$.

To emphasize again, a *subsequence* need not consist of contiguous characters in S, whereas the characters of a *substring* must be contiguous.[4] Of course, a substring satisfies the definition for a subsequence. For example, "its" is a subsequence of "winters" but not a substring, whereas "inter" is both a substring and a subsequence.

Definition Given two strings S_1 and S_2, a *common subsequence* is a subsequence that appears both in S_1 and S_2. The *longest common subsequence problem* is to find a longest common subsequence (*lcs*) of S_1 and S_2.

---

[4] The distinction between subsequence and substring is often lost in the biological literature. But algorithms for substrings are usually quite different in spirit and efficiency than algorithms for subsequences, so the distinction is an important one.

The *lcs* problem is important in its own right, and we will discuss some of its uses and some ideas for improving its computation in Section 12.5. For now we show that it can be modeled and solved as an optimal alignment problem.

Theorem 11.6.1. *With a scoring scheme that scores a one for each match and a zero for each mismatch or space, the matched characters in an alignment of maximum value form a longest common subsequence.*

The proof is immediate and is left to the reader. It follows that the longest common subsequence of strings of lengths n and m, respectively, can be computed in $O(nm)$ time.

At this point we see the first of many differences between substring and subsequence problems and why it is important to clearly distinguish between them. In Section 7.4 we established that the longest common substring could be found in $O(n + m)$ time, whereas here the bound established for finding longest common subsequence is $O(n \times m)$ (although this bound can be reduced somewhat). This is typical – substring and subsequence problems are generally solved by different methods and have different time and space complexities.

11.6.3. Alignment graphs for similarity

As was the case for edit distance, the computation of similarity can be viewed as a path problem on a directed acyclic graph called an *alignment graph*. The graph is the same as the edit graph considered earlier, but the weights on the edges are the specific values for aligning a specific pair of characters or a character against a space. The start node of the alignment graph is again the node associated with cell $(0, 0)$, and the destination node is associated with cell (n, m) of the dynamic programming table, but the optimal alignment comes from the longest start to destination path rather than from the shortest path. It is again true that the longest paths in the alignment graph are in one-to-one correspondence with the optimal (maximum value) alignments. In general, computing longest paths in graphs is difficult, but for directed acyclic graphs the longest path is found in time proportional to the number of edges in the graph, using a variant of dynamic programming (which should come as no surprise). Hence for alignment graphs, the longest path can be found in $O(nm)$ time.

11.6.4. End-space free variant

There is a commonly used variant of string alignment called *end-space free alignment*. In this variant, any spaces at the end or the beginning of the alignment contribute a weight of zero, no matter what weight other spaces contribute. For example, in the alignment

$$
\begin{array}{ccccccccc}
\_ & \_ & c & a & c & \_ & d & b & d \\
l & t & c & a & b & b & d & b & \_
\end{array}
$$

the two spaces at the left end of the alignment are free, as is the single space at the right end.

Making end spaces free in the objective function encourages one string to align in the interior of the other, or the suffix of one string to align with a prefix of the other. This is desirable when one believes that those kinds of alignments reflect the "true" relationship of the two strings. Without a mechanism to encourage such alignments, the optimal alignment might have quite a different shape and not capture the desired relationship.

One example where end-spaces should be free is in the *shotgun sequence assembly* (see Sections 16.14 and 16.15). In this problem, one has a large set of partially overlapping substrings that come from many copies of one original but unknown string; the problem is to use comparisons of pairs of substrings to infer the correct original string. Two random substrings from the set are unlikely to be neighbors in the original string, and this is reflected by a low end-space free alignment score for those two substrings. But if two substrings do overlap in the original string, then a "good-sized" suffix of one should align to a "good-sized" prefix of the other with only a small number of spaces and mismatches (reflecting a small percentage of sequencing errors). This overlap is detected by an end-space free weighted alignment with high score. Similarly the case when one substring contains another can be detected in this way. The procedure for deducing candidate neighbor pairs is thus to compute the end-space free alignment between every pair of substrings; those pairs with high scores are then the best candidates. We will return to shotgun sequencing and extend this discussion in Part IV, Section 16.14.

To implement free end spaces in computing similarity, use the recurrences for global alignment (where all spaces count) detailed on page 227, but change the base conditions to $V(i, 0) = V(0, j) = 0$, for every i and j. That takes care of any spaces on the left end of the alignment. Then fill in the table as in the case of global alignment. However, unlike global alignment, the value of the optimal alignment is not necessarily found in cell (n, m). Rather, the value of the optimal alignment with free ends is the maximum value over all cells in row n or column m. Cells in row n correspond to alignments where the last character of string S_1 contributes to the value of the alignment, but characters of S_2 to its right do not. Those characters are opposite end spaces, which are free. Cells in column m have a similar characterization. Clearly, optimal alignment with free end spaces is solved in $O(nm)$ time, the same time as for global alignment.

11.6.5. Approximate occurrences of P in T

We now examine another important variant of global alignment.

Definition Given a parameter δ, a substring T' of T is said to be an *approximate occurrence* of P if and only if the optimal alignment of P to T' has value at least δ.

The problem of determining if there is an approximate occurrence of P in T is an important and natural generalization of the exact matching problem. It can be solved as follows: Use the same recurrences (given on page 227) as for global alignment between P and T and change only the base condition for $V(0, j)$ to $V(0, j) = 0$ for all j. Then fill in the table (leaving the standard backpointers). Using this variant of global alignment, the following theorem can be proved.

Theorem 11.6.2. *There is an approximate occurrence of P in T ending at position j of T if and only if $V(n, j) \geq \delta$. Moreover, $T[k..j]$ is an approximate occurrence of P in T if and only if $V(n, j) \geq \delta$ and there is a path of backpointers from cell (n, j) to cell $(0, k)$.*

Clearly, the table can be filled in using $O(nm)$ time, but if all approximate occurrence of P in T are to be explicitly output, then $\Theta(nm)$ time may not be sufficient. A sensible compromise is to identify every position j in T such that $V(n, j) \geq \delta$, and then for each such j, explicitly output only the *shortest* approximate occurrence of P that ends at position j. That substring T' is found by traversing the backpointers from (n, j) until a cell in row zero is reached, breaking ties by choosing a vertical pointer over a diagonal one and a diagonal one over a horizontal one.

11.7. Local alignment: finding substrings of high similarity

In many applications, two strings may not be highly similar in their entirety but may contain regions that are highly similar. The task is to find and extract a pair of regions, one from each of the two given strings, that exhibit high similarity. This is called the *local alignment* or *local similarity problem* and is defined formally below.

> **Local alignment problem**   Given two strings S_1 and S_2, find substrings α and β of S_1 and S_2, respectively, whose similarity (optimal global alignment value) is maximum over all pairs of substrings from S_1 and S_2. We use v^* to denote the value of an optimal solution to the local alignment problem.

For example, consider the strings $S_1 = pqraxabcstvq$ and $S_2 = xyaxbacsll$. If we give each match a value of 2, each mismatch a value of -2, and each space a value of -1, then the two substrings $\alpha = axabcs$ and $\beta = axbacs$ of S_1 and S_2, respectively, have the following optimal (global) alignment

$$
\begin{array}{ccccccc}
a & x & a & b & \_ & c & s \\
a & x & \_ & b & a & c & s
\end{array}
$$

which has a value of 8. Furthermore, over all choices of pairs of substrings, one from each of the two strings, those two substrings have maximum similarity (for the chosen scoring scheme). Hence, for that scoring scheme, the optimal local alignment of S_1 and S_2 has value 8 and is defined by substrings $axabcs$ and $axbacs$.

It should be clear why local alignment is defined in terms of similarity, which maximizes an objective function, rather than in terms of edit distance, which minimizes an objective. When one seeks a pair of substrings to minimize distance, the optimal pairs would be exactly matching substrings under most natural scoring schemes. But the matching substrings might be just a single character long and would not identify a region of high similarity. A formulation such as local alignment, where matches contribute positively and mismatches and spaces contribute negatively, is more likely to find more meaningful regions of high similarity.

Why local alignment?

Global alignment of protein sequences is often meaningful when the two strings are members of the same protein family. For example, the protein *cytochrome c* has almost the same length in most organisms that produce it, and one expects to see a relationship between two cytochromes from any two different species over the entire length of the two strings. The same is true of proteins in the *globin* family, such as *myoglobin* and *hemoglobin*. In these cases, global alignment is meaningful. When trying to deduce evolutionary history by examining protein sequence similarities and differences, one usually compares proteins in the same sequence family, and so global alignment is typically meaningful and effective in those applications.

However, in many biological applications, local similarity (local alignment) is far more meaningful than global similarity (global alignment). This is particularly true when long stretches of anonymous DNA are compared, since only *some* internal sections of those strings may be related. When comparing protein sequences, local alignment is also critical because proteins from very different families are often made up of the same structural or functional subunits (motifs or domains), and local alignment is appropriate in searching for these (unknown) subunits. Similarly, different proteins are often made from related motifs that form the inner core of the protein, but the motifs are separated by outside surface looping regions that can be quite different in different proteins.

A very interesting example of conserved domains comes from the proteins encoded by *homeobox* genes. Homeobox genes [319, 381] show up in a wide variety of species, from fruit flies to frogs to humans. These genes regulate high-level embryonic development, and a single mutation in these genes can transform one body part into another (one of the original mutation experiments causes fruit fly antenna to develop as legs, but it doesn't seem to bother the fly very much). The protein sequences that these genes encode are very different in each species, except in one region called the *homeodomain*. The homeodomain consists of about sixty amino acids that form the part of the regulatory protein that binds to DNA. Oddly, homeodomains made by certain insect and mammalian genes are particularly similar, showing about 50 to 95% identity in alignments without spaces. Protein-to-DNA binding is central in how those proteins regulate embryo development and cell differentiation. So the amino acid sequence in the most biologically critical part of those proteins is highly conserved, whereas the other parts of the protein sequences show very little similarity. In cases such as these, local alignment is certainly a more appropriate way to compare protein sequences than is global alignment.

Local alignment in protein is additionally important because particular isolated characters of related proteins may be more highly conserved than the rest of the protein (for example, the amino acids at the *active site* of an enzyme or the amino acids in the *hydrophobic core* of a globular protein are the most highly conserved). Local alignment will more likely detect these conserved characters than will global alignment. A good example is the family of *serine proteases* where a few isolated, conserved amino acids characterize the family. Another example comes from the Helix-Turn-Helix motif, which occurs frequently in proteins that regulate DNA transcription by binding to DNA. The tenth position of the Helix-Turn-Helix motif is very frequently occupied by the amino acid glycine, but the rest of the motif is more variable.

The following quote from C. Chothia [101] further emphasizes the biological importance of protein domains and hence of local string comparison.

> Extant proteins have been produced from the original set not just by point mutations, insertions and deletions but also by combinations of genes to give chimeric proteins. This is particularly true of the very large proteins produced in the recent stages of evolution. Many of these are built of different combinations of protein domains that have been selected from a relatively small repertoire.

Doolittle [129] summarizes the point: "The underlying message is that one must be alert to regions of similarity even when they occur embedded in an overall background of dissimilarity."

Thus, the dominant viewpoint today is that local alignment is the most appropriate type of alignment for comparing proteins from different protein families. However, it has also been pointed out [359, 360] that one often sees extensive global similarity in pairs of protein strings that are first recognized as being related by strong local similarity. There are also suggestions [316] that in some situations global alignment is more effective than local alignment in exposing important biological commonalities.

11.7.1. Computing local alignment

Why not look for regions of high similarity in two strings by first globaly aligning those strings? A global alignment between two long strings will certainly be influenced by regions of high similarity, and an optimal global alignment might well align those corresponding regions with each other. But more often, local regions of high local similarity would get lost in the overall optimal global alignment. Therefore, to identify high local similarity it is more effective to search explicitly for local similarity.

We will show that if the lengths of strings S_1 and S_2 are n and m, respectively, then the local alignment problem can be solved in $O(nm)$ time, the same time as for global alignment. This efficiency is surprising because there are $\Theta(n^2 m^2)$ pairs of substrings, so even if a global alignment could be computed in constant time for each chosen pair, the time bound would be $\Theta(n^2 m^2)$. In fact, if we naively use $O(kl)$ for the bound on the time to align strings of lengths k and l, then the resulting time bound for the local alignment problem would be $O(n^3 m^3)$, instead of the $O(nm)$ bound that we will establish. The $O(nm)$ time bound was obtained by Temple Smith and Michael Waterman [411] using the algorithm we will describe below.

In the definition of local alignment given earlier, any scoring scheme was permitted for the global alignment of two chosen substrings. One slight restriction will help in computing local alignment. We assume that the global alignment of two empty strings has value zero. That assumption is used to allow the local alignment algorithm to choose two empty substrings for α and β. Before describing the solution to the local alignment problem, it will be helpful to consider first a more restricted version of the problem.

Definition   Given a pair of indices $i \leq n$ and $j \leq m$, the *local suffix alignment problem* is to find a (possibly empty) suffix α of $S_1[1..i]$ and a (possibly empty) suffix β of $S_2[1..j]$ such that $V(\alpha, \beta)$ is the maximum over all pairs of suffixes of $S_1[1..i]$ and $S_2[1..j]$. We use $v(i, j)$ to denote the value of the optimal local suffix alignment for the given index pair i, j.

For example, suppose the objective function counts 2 for each match and -1 for each mismatch or space. If $S_1 = abcxdex$ and $S_2 = xxxcde$, then $v(3, 4) = 2$ (the two cs match), $v(4, 5) = 1$ (cx aligns with cd), $v(5, 5) = 3$ (x_d aligns with xcd), and $v(6, 6) = 5$ (x_de aligns with $xcde$).

Since the definition allows either or both of the suffixes to be empty, $v(i, j)$ is always greater than or equal to zero.

The following theorem shows the relationship between the local alignment problem and the local suffix alignment problem. Recall that v^* is the value of the optimal local alignment for two strings of length n and m.

Theorem 11.7.1. $v^* = \max[v(i, j) : i \leq n, j \leq m]$.

PROOF   Certainly $v^* \geq \max[v(i, j) : i \leq n, j \leq m]$, because the optimal solution to the local suffix alignment problem for any i, j is a feasible solution to the local alignment problem. Conversely, let α, β be the substrings in an optimal solution to the local alignment problem and suppose α ends at position i^* and β ends at j^*. Then α, β also defines a local suffix alignment for index pair i^*, j^*, and so $v^* \leq v(i^*, j^*) \leq \max[v(i, j) : i \leq n, j \leq m]$, and both directions of the lemma are established.   □

Theorem 11.7.1 only specifies the value v^*, but its proof makes clear how to find substrings whose alignment have that value. In particular,

Theorem 11.7.2. *If i', j' is an index pair maximizing $v(i, j)$ over all i, j pairs, then a pair of substrings solving the local suffix alignment problem for i', j' also solves the local alignment problem.*

Thus a solution to the local suffix alignment problem solves the local alignment problem. We now turn our attention to the problem of finding $\max[v(i, j) : i \leq n, j \leq m]$ and a pair of strings whose alignment has maximum value.

11.7.2. How to solve the local suffix alignment problem

First, $v(i, 0) = 0$ and $v(0, j) = 0$ for all i, j, since we can always choose an empty suffix.

Theorem 11.7.3. *For $i > 0$ and $j > 0$, the proper recurrence for $v(i, j)$ is*

$$v(i, j) = \max[0, v(i - 1, j - 1) + s(S_1(i), S_2(j)),$$

$$v(i - 1, j) + s(S_1(i), \_), v(i, j - 1) + s(\_, S_2(j))].$$

PROOF   The argument is similar to the justifications of previous recurrence relations. Let α and β be the substrings of S_1 and S_2 whose global alignment establishes the optimal local alignment. Since α and β are permitted to be empty suffixes of $S_1[1..i]$ and $S_2[1..j]$, it is correct to include 0 as a *candidate* value for $v(i, j)$. However, if the optimal α is not empty, then character $S_1(i)$ must either be aligned with a space or with character $S_2(j)$. Similarly, if the optimal β is not empty, then $S_2(j)$ is aligned with a space or with $S_1(i)$. So we justify the recurrence based on the way characters $S_1(i)$ and $S_2(j)$ may be aligned in the optimal local suffix alignment for i, j.

If $S_1(i)$ is aligned with $S_2(j)$ in the optimal local i, j suffix alignment, then those two characters contribute $s(S_1(i), S_2(j))$ to $v(i, j)$, and the remainder of $v(i, j)$ is determined by the local suffix alignment for indices $i - 1, j - 1$. That local suffix alignment must be optimal and so has value $v(i - 1, j - 1)$. Therefore, if $S_1(i)$ and $S_2(j)$ are aligned with each other, $v(i, j) = v(i - 1, j - 1) + s(S_1(i), S_2(j))$.

If $S_1(i)$ is aligned with a space, then by similar reasoning $v(i, j) = v(i - 1, j) + s(S_1(i), \_)$, and if $S_2(j)$ is aligned with a space then $v(i, j) = v(i, j - 1) + s(\_, S_2(j))$. Since all cases are exhausted, we have proven that $v(i, j)$ must either be zero or be equal to one of the three other terms in the recurrence.

On the other hand, for each of the four terms in the recurrence, there *is* a way to choose suffixes of $S_1[1..i]$ and $S_2[1..j]$ so that an alignment of those two suffixes has the value given by the associated term. Hence the optimal suffix alignment value is at least the maximum of the four terms in the recurrence. Having proved that $v(i, j)$ must be one of the four terms, and that it must be greater than or equal to the maximum of the four terms, it follows that $v(i, j)$ must be equal to the maximum which proves the theorem. $\square$

The recurrences for local suffix alignment are almost identical to those for global alignment. The only difference is the inclusion of zero in the case of local suffix alignment. This makes intuitive sense. In both global alignment and local suffix alignment of prefixes $S_1[1..i]$ and $S_2[1..j]$ the end characters of any alignment are specified, but in the case of local suffix alignment, any number of initial characters can be ignored. The zero in the recurrence implements this, acting to "restart" the recurrence.

Given Theorem 11.7.2, the method to compute v^* is to compute the dynamic programming table for $v(i, j)$ and then find the largest value in *any* cell in the table, say in cell (i^*, j^*). As usual, pointers are created while filling in the values of the table. After cell

(i^*, j^*) is found, the substrings α and β giving the optimal local alignment of S_1 and S_2 are found by tracing back the pointers from cell (i^*, j^*) until an entry (i', j') is reached that has value zero. Then the optimal local alignment substrings are $\alpha = S_1[i'..i^*]$ and $\beta = S_2[j'..j^*]$.

Time analysis

Since it takes only four comparisons and three arithmetic operations per cell to compute $v(i, j)$, it takes only $O(nm)$ time to fill in the entire table. The search for v^* and the traceback clearly require only $O(nm)$ time as well, so we have established the following desired theorem:

Theorem 11.7.4. *For two strings S_1 and S_2 of lengths n and m, the local alignment problem can be solved in $O(nm)$ time, the same time as for global alignment.*

Recall that the pointers in the dynamic programming table for edit distance, global alignment, and similarity encode all the optimal alignments. Similarly, the pointers in the dynamic programming table for local alignment encode the optimal local alignments as follows.

Theorem 11.7.5. *All optimal local alignments of two strings are represented in the dynamic programming table for $v(i, j)$ and can be found by tracing any pointers back from any cell with value v^*.*

We leave the proof as an exercise.

11.7.3. Three final comments on local alignment

Terminology for local and global alignment

In the biological literature, global alignment (similarity) is often referred to as a Needleman–Wunsch [347] alignment after the authors who first discussed global similarity. Local alignment is often referred to as a Smith–Waterman [411] alignment after the authors who introduced local alignment. There is, however, some confusion in the literature between Needleman–Wunsch and Smith–Waterman as *problem statements* and as *solution methods*. The original solution given by Needleman–Wunsch runs in cubic time and is rarely used. Hence "Needleman–Wunsch" usually refers to the global alignment *problem*. The Smith–Waterman method runs in quadratic time and is commonly used, so "Smith–Waterman" often refers to their specific solution as well as to the problem statement. But there are solution methods to the (Smith–Waterman) local alignment problem that differ from the Smith–Waterman solution and yet are sometimes also referred to as "Smith–Waterman".

Using Smith–Waterman to find several regions of high similarity

Very often in biological applications it is not sufficient to find just a single pair of substrings of input strings of S_1 and S_2 with the optimal local alignment. Rather, what is required is to find all or "many" pairs of substrings that have similarity above some threshold. A specific application of this kind will be discussed in Section 18.2, and the general problem will be studied much more deeply in Section 13.2. Here we simply point out that, in practice, the dynamic programming table used to solve the local suffix alignment problem is often used to find additional pairs of substrings with "high" similarity. The key observation is that for any cell (i, j) in the table, one can find a pair of substrings of S_1 and S_2 (by traceback)

```
c   t   t   t   a   a   c   _   _   a   _   a   c
c   _   _   _   c   a   c   c   c   a   t   _   c
```

Figure 11.5: An alignment with seven spaces distributed into four gaps.

with similarity (global alignment value) of $v(i, j)$. Thus, an easy way to look for a set of highly similar substrings is to find a set of cells in the table with a value above some set threshold. Not all similar substrings will be identified in this way, but this approach is common in practice.

The need for good scoring schemes

The utility of optimal local alignment is affected by the scoring scheme used. For example, if matches are scored as one, and mismatches and spaces as zero, then the optimal local alignment will be determined by the longest common *subsequence*. Conversely, if mismatches and spaces are given large negative scores, and each match is given a score of one, then the optimal local alignment will be the longest common *substring*. In most cases, neither of these is the local alignment of interest and some care is required to find an application-dependent scoring scheme that yields meaningful local alignments. For local alignment, the entries in the scoring matrix must have an average score that is negative. Otherwise the resulting "local" optimal alignment tends to be a global alignment. Recently, several authors have developed a rather elegant theory of what scoring schemes for local alignment mean in the context of database search and how they should be derived. We will briefly discuss this theory in Section 15.11.2.

11.8. Gaps

11.8.1. Introduction to Gaps

Until now the central constructs used to measure the value of an alignment (and to define similarity) have been *matches, mismatches,* and *spaces*. Now we introduce another important construct, *gaps*. Gaps help create alignments that better conform to underlying biological models and more closely fit patterns that one expects to find in meaningful alignments.

Definition   A gap is any *maximal, consecutive run* of spaces in a *single* string of a given alignment.[5]

A gap may begin before the start of S, in which case it is bordered on the right by the first character of S, or it may begin after the end of S, in which case it is bordered on the left by the last character of S. Otherwise, a gap must be bordered on both sides by characters of S. A gap may be as small as a single space. As an example of gaps, consider the alignment in Figure 11.5, which has four gaps containing a total of seven spaces. That alignment would be described as having five matches, one mismatch, four gaps, and seven spaces. Notice that the last space in the first string is followed by a space in the second string, but those two spaces are in two gaps and do not form a single gap.

By including a term in the objective function that reflects the gaps in the alignment one has some influence on the *distribution* of spaces in an alignment and hence on the overall shape of the alignment. In the simplest objective function that includes gaps,

---

[5] Sometimes in the biology literature the term "space" (as we use it) is not used. Rather, the term "gap" is used both for "space" and for "gap" (as we have defined it here). This can cause much confusion, and in this book the terms "gap" and "space" have distinct meanings.

each gap contributes a constant weight W_g, independent of how long the gap is. That is, each individual space is free, so that $s(x, \_) = s(\_, x) = 0$ for every character x. Using the notation established in Section 11.6, (page 226), we write the value of an alignment containing k gaps as

$$\sum_{i=1}^{l} s(S_1'(i), S_2'(i)) - kW_g.$$

Changing the value of W_g relative to the other weights in the objective function can change how spaces are distributed in the optimal alignment. A large W_g encourages the alignment to have few gaps, and the aligned portions of the two strings will fall into a few substrings. A smaller W_g allows more fragmented alignments. The influence of W_g on the alignment will be discussed more deeply in Section 13.1.

11.8.2. Why gaps?

Most of the biological justifications given for the importance of local alignment (see Section 11.7) apply as well to justify the gap as an explicit concept in string alignment.

Just as a space in an alignment corresponds to an insertion or deletion of a single character in the edit transcript, a gap in string S_1 opposite substring α in string S_2 corresponds to either a deletion of α from S_1 or to an insertion of α into S_2. The concept of a gap in an alignment is therefore important in many biological applications because the insertion or deletion of an entire substring (particularly in DNA) often occurs as single mutational event. Moreover, many of these single mutational events can create gaps of quite varying sizes with almost equal likelihood (within a wide, but bounded, range of sizes). Much of the repetitive DNA discussed in Section 7.11.1 is caused by single mutational events that copy and insert long pieces of DNA. Other mutational mechanisms that make long insertions or deletions in DNA include: *unequal crossing-over* in meiosis (causing an insertion in one string and a reciprocal deletion in the other); DNA *slippage* during replication (where a portion of the DNA is repeated on the replicated copy because the replication machinery loses its place on the template, slipping backwards and repeating a section); insertion of *transposable* elements (jumping genes) into a DNA string; insertions of DNA by *retroviruses*; and *translocations* of DNA between chromosomes [301, 317]. See Figure 11.6 for an example of gaps in genomic sequence data.

When computing alignments for the purpose of deducing evolutionary history over a long period of time, it is often the gaps that are the most informative part of the alignments. In DNA strings, single character substitutions due to point mutations occur continuously and usually at a much faster rate than (nonfatal) mutational events causing gaps. The analogous gene (specifying the "same" protein) in two species can thus be very different at the DNA sequence level, making it difficult to sort out evolutionary relationships on the basis of string similarity (without gaps). But large insertions and deletions in molecules that show up as gaps in alignments occur less frequently than substitutions. Therefore, common gaps in pairs of aligned strings can sometimes be the key features used to deduce the overall evolutionary history of a set of strings [45, 405]. Later, in Section 17.3.2, we will see that such gaps can be considered as evolutionary characters in certain approaches to building evolutionary trees.

At the protein level, recall that many proteins are "built of different combinations of protein domains that have been selected from a relatively small repertoire"[101]. Hence two protein strings might be relatively similar over several intervals but differ in intervals where one contains a protein domain that the other does not. Such an interval most naturally

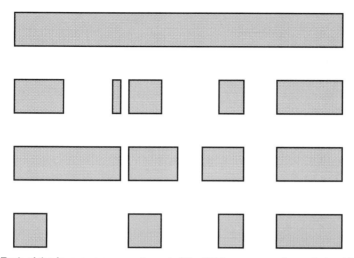

Figure 11.6: Each of the four rows represents part of the RNA sequence of one strain of the HIV-1 virus. The HIV virus mutates rapidly, so that mutations can be observed and traced. The bottom three rows are from virus strains that have each mutated from an ancestral strain represented in the top row. Each of the bottom sequences is shown aligned to the top sequence. A dark box represents a substring that matches the corresponding substring in the top sequence, while each white space represents a gap resulting from a known sequence deletion. This figure is adapted from one in [123].

long string

pieces of shorter string interspersed with gaps

Figure 11.7: In cDNA matching, one expects the alignment of the smaller string with the longer string to consist of a few regions of very high similarity, interspersed with relatively long gaps.

shows up as a gap when two proteins are aligned. In some contexts, many biologists consider the proper identification of the major (long) gaps as *the* essential problem of protein alignment. If the long (major) gaps have been selected correctly, the rest of the alignment – reflecting point mutations – is then relatively easy to obtain.

An alignment of two strings is intended to reflect the cost (or likelihood) of mutational events needed to transform one string to another. Since a gap of more than one space can be created by a single mutational event, the alignment model should reflect the true distribution of spaces into gaps, not merely the number of spaces in the alignment. It follows that the model must specify how to weight gaps so as to reflect their biological meaning. In this chapter we will discuss different proposed schemes for weighting gaps, and in later chapters we will discuss additional issues in scoring gaps. First we consider a concrete example illustrating the utility of the gap concept.

11.8.3. cDNA matching: a concrete illustration

One concrete illustration of the use of gaps in the alignment model comes from the problem of *cDNA matching*. In this problem, one string is much longer than the other, and the alignment best reflecting their relationship should consist of a few regions of very high similarity interspersed with "long" gaps in the shorter string (see Figure 11.7). Note that the matching regions can have mismatches and spaces, but these should be a small percentage of the region.

Biological setting of the problem

In eukaryotes, a gene that codes for a protein is typically made up of alternating *exons* (expressed sequences), which contribute to the code for the protein, and *introns* (intervening sequences), which do not. The number of exons (and hence also introns) is generally modest (four to twenty say), but the lengths of the introns can be huge compared to the lengths of the exons.

At a very coarse level, the protein specified by a eukaryotic gene is made in the following steps. First, an RNA molecule is transcribed from the DNA of the gene. That RNA transcript is a complement of the DNA in the gene in that each A in the gene is replaced by U (uracil) in the RNA, each T is replaced by A, each C by G, and each G by C. Moreover, the RNA transcript covers the entire gene, introns as well as exons. Then, in a process that is not completely understood, each intron-exon boundary in the transcript is located, the RNA corresponding to the introns is spliced out (or *snurp*ed out by a molecular complex called a *snrp* [420]), and the RNA regions corresponding to exons are concatenated. Additional processing occurs that we will not describe. The resulting RNA molecule is called the *messenger RNA (mRNA)*; it leaves the cell nucleus and is used to create the protein it encodes.

Each cell (usually) contains a copy of all the chromosomes and hence of all the genes of the entire individual, yet in each specialized cell (a liver cell for example) only a small fraction of the genes are expressed. That is, only a small fraction of the proteins encoded in the genome are actually produced in that specialized cell. A standard method to determine which proteins are expressed in the specialized cell line, and to hunt for the location of the encoding genes, involves capturing the mRNA in that cell after it leaves the cell nucleus. That mRNA is then used to create a DNA string complementary to it. This string is called *cDNA* (complementary DNA). Compared to the original gene, the cDNA string consists only of the concatenation of exons in the gene.

It is routine to capture mRNA and make cDNA libraries (complete collections of a cell's mRNA) for specific cell lines of interest. As more libraries are built up, one collects a reflection of all the genes in the genome and a taxonomy of the cells that the genes are expressed in. In fact, a major component of the Human Genome Project [111], [399] is to obtain cDNAs reflecting most of the genes in the human genome. This effort is also being conducted by several private companies and has led to some interesting disputes over patenting cDNA sequences.

After cDNA is obtained, the problem is to determine where the gene associated with that cDNA resides. Presently, this problem is most often addressed with laboratory methods. However, if the cDNA is sequenced or partially sequenced (and in the Human Genome Project, for example, the intent is to sequence parts of each of the obtained cDNAs), and if one has sequenced the part of the genome containing the gene associated with that cDNA (as, for example, one would have after sequencing the entire genome), then the problem of finding the gene site given a cDNA sequence becomes a string problem. It becomes one of aligning the cDNA string against the longer string of sequenced DNA in a way that reveals the exons. It becomes the cDNA matching problem discussed above.

Why gaps are needed in the objective function

If the objective function includes terms only for matches, mismatches, and spaces, there seems no way to encourage the optimal alignment to be of the desired form. It's worth a moment's effort to explain why.

Certainly, you don't want to set a large penalty for spaces, since that would align all the cDNA string close together, rather than allowing gaps in the alignment corresponding to the long introns. You would also want a rather high penalty for mismatches. Although there may be a few sequencing errors in the data, so that some mismatches will occur even when the cDNA is properly cut up to match the exons, there should not be a large percentage of mismatches. In summary, you want small penalties for spaces, relatively large penalties for mismatches, and positive values for matches.

What kind of alignment would likely result using an objective function that has low space penalty, high mismatch penalty, positive match value of course, and no term for gaps? Remember that the long string contains more than one gene, that the exons are separated by long introns, and that DNA has an alphabet of only four letters present in roughly equal amounts. Under these conditions, the optimal alignment would probably be the *longest common subsequence* between the short cDNA string and the long anonymous DNA string. And because the introns are long and DNA has only four characters, that common subsequence would likely match *all* of the characters in the cDNA. Moreover, because of small but real sequencing errors, the true alignment of the cDNA to its exons would not match all the characters. Hence the longest common subsequence would likely have a higher score than the correct alignment of the cDNA to exons. But the longest common subsequence would fragment the cDNA string over the longer DNA and not give an alignment of the desired form – it would not pick out its exons.

Putting a term for gaps in the objective function rectifies the problem. By adding a constant gap weight W_g for each gap in the alignment, and setting W_g appropriately (by experimenting with different values of W_g), the optimal alignment can be induced to cut up the cDNA to match its exons in the longer string.[6] As before, the space penalty is set to zero, the match value is positive, and the mismatch penalty is set high.

Processed pseudogenes

A more difficult version of cDNA matching arises in searching anonymous DNA for *processed pseudogenes*. A pseudogene is a near copy of a working gene that has mutated sufficiently from the original copy so that it can no longer function. Pseudogenes are very common in eukaryotic organisms and may play an important evolutionary role, providing a ready pool of diverse "near genes". Following the view that new genes are created by the process of *duplication with modification* of existing genes [127, 128, 130], pseudogenes either represent trial genes that failed or future genes that will function after additional mutations.

A pseudogene may be located very far from the gene it corresponds to, even on a different chromosome entirely, but it will usually contain both the introns and the exons derived from its working relative. The problem of finding pseudogenes in anonymous sequenced DNA is therefore related to that of finding repeated substrings in a very long string.

A more interesting type of pseudogene, the *processed pseudogene*, contains only the exon substrings from its originating gene. Like cDNA, the introns have been removed and the exons concatenated. It is thought that a processed pseudogene originates as an mRNA that is retranscribed back into DNA (by the enzyme Reverse Transcriptase) and inserted into the genome at a random location.

Now, given a long string of anonymous DNA that might contain both a processed pseudogene and its working ancestor, how could the processed pseudogenes be located?

---

[6] This really works, and it is a very instructive exercise to try it out empirically.

The problem is similar to cDNA matching but more difficult because one does not have the cDNA in hand. We leave it to the reader to explore the use of repeat finding methods, local alignment, and gap weight selection in tackling this problem.

Caveat

The problems of cDNA and pseudogene matching illustrate the utility of including gaps in the alignment objective function and the importance of weighting the gaps appropriately. It should be noted, however, that in practice one can approach these matching problems by a judicious use of local alignment without gaps. The idea is that in computing local alignment, one can find not only the most similar pair of substrings but many other highly similar pairs of substrings (see Sections 13.2.4, and 11.7.3). In the context of cDNA or pseudogene matching, these pairs will likely be the exons, and so the needed match of cDNA to exons can be pieced together from a number of nonoverlapping local alignments. This is the more typical approach in practice.

11.8.4. Choices for gap weights

As illustrated by the example of cDNA matching, the appropriate use of gaps in the objective function aids in the discovery of alignments that satisfy an expected shape. But clearly, the way gaps are weighted critically influences the effectiveness of the gap concept. We will examine in detail four general types of gap weights: *constant, affine, convex,* and *arbitrary*.

The simplest choice is the *constant* gap weight introduced earlier, where each individual space is free, and each gap is given a weight of W_g independent of the number of spaces in the gap. Letting W_m and W_{ms} denote weights for matches and mismatches, respectively, the *operator-weight* version of the problem is:

Find an alignment $\mathcal{A}$ to maximize $[W_m(\#\ matches) - W_{ms}(\#\ mismatches) - W_g\ (\#\ gaps)]$.

More generally, if we adopt the alphabet-dependent weights for matches and mismatches, the objective in the constant gap weight model is:

Find an alignment $\mathcal{A}$ to maximize $\left(\sum_{i=1}^{l}[s(S_1'(i), S_2'(i))] - W_g(\#\ gaps) \right)$,

where $s(x, \_) = s(\_, x) = 0$ for every character x, and S_1' and S_2' represent the strings S_1 and S_2 after insertion of spaces.

A generalization of the constant gap weight model is to add a weight W_s for each space in the gap. In this case, W_g is called the *gap initiation weight* because it can represent the cost of starting a gap, and W_s is called the *gap extension weight* because it can represent the cost of extending the gap by one space. Then the operator-weight version of the problem is:

Find an alignment to maximize $[W_m(\#\ matches) - W_{ms}(\#\ mismatches) - W_g(\#\ gaps) - W_s(\#\ spaces)]$.

This is called the *affine gap weight* model[7] because the weight contributed by a single gap of length q is given by the affine function $W_g + qW_s$. The constant gap weight model is simply the affine model with $W_s = 0$.

---

[7] The affine gap model is sometimes called the *linear* weight model, and I prefer that term. However, "affine" has become the dominant term in the biological literature, and "linear" there usually refers to an affine function with $W_g = 0$.

The alphabet-weight version of the affine gap weight model again sets $s(x, \_) = s(\_, x) = 0$ and has the objective of finding an alignment to

$$\text{maximize} \left(\sum_{i=1}^{l} [s(S_1'(i), S_2'(i))] - W_g(\# \, gaps) - W_s(\# \, spaces) \right).$$

The affine gap weight model is probably the most commonly used gap model in the molecular biology literature, although there is considerable disagreement about what W_g and W_s should be [161] (in addition to questions about W_m and W_{ms}). For aligning amino acid strings, the widely used search program FASTA [359] has chosen the default settings of $W_g = 10$ and $W_s = 2$. We will return to the question of the choice of these settings in Section 13.1.

It has been suggested [57, 183, 466] that some biological phenomena are better modeled by a gap weight function where each additional space in a gap contributes less to the gap weight than the preceding space (a function with negative second derivative). In other words, a gap weight that is a *convex*,[8] but not affine, function of its length. An example is the function $W_g + \log_e q$, where q is the length of the gap. Some biologists have suggested that a gap function that initially increases to a maximum value and then decreases to near zero would reflect a *combination* of different biological phenomena that insert or delete DNA.

Finally, the most general gap weight we will consider is the *arbitrary gap weight*, where the weight of a gap is an arbitrary function $w(q)$ of its length q. The constant, affine, and convex weight models are of course subcases of the arbitrary weight model.

Time bounds for gap choices

As might be expected, the time needed to optimally solve the alignment problem with arbitrary gap weights is greater than for the other models. In the case that $w(q)$ is a totally arbitrary function of gap length, the optimal alignment can be found in $O(nm^2 + n^2m)$ time, where n and $m \geq n$ are the lengths of the two strings. In the case that $w(q)$ is convex, we will show that the time can be reduced to $O(nm \log m)$ (a further reduction is possible, but the algorithm is much too complex for our interests). In the affine (and hence constant) case the time bound is $O(nm)$, which is the same time bound established for the alignment model without the concept of gaps. In the next sections we will first discuss alignment for arbitrary gap weights and then show how to reduce the running time for the case of affine weight functions. The $O(nm \log m)$-time algorithm for convex weights is more complex than the others and is deferred until Chapter 13.

11.8.5. Arbitrary gap weights

This case was first introduced and solved in the classic paper of Needleman and Wunsch [347], although with somewhat different detail and terminology than used here.

For arbitrary gap weights, we will develop recurrences that are similar to (but more detailed than) the ones used in Section 11.6.1 for optimal alignment without gaps. There is, however, a subtle question about whether these recurrences correctly model the biologist's view of gaps. We will examine that issue in Exercise 45.

To align strings S_1 and S_2, consider, as usual, the prefixes $S_1[1..i]$ of S_1 and $S_2[1..j]$ of S_2. Any alignment of those two prefixes is one of the following three types (see Figure 11.8):

---

[8] Some call this *concave*.

Figure 11.8: The recurrences for alignment with gaps are divided into three types of alignments: 1. those that align $S_1(i)$ to the left of $S_2(j)$, 2. those that align $S_1(i)$ to the right of $S_2(j)$, and 3. those that align them opposite each other.

1. Alignments of $S_1[1..i]$ and $S_2[1..j]$ where character $S_1(i)$ is aligned to a character strictly to the left of character $S_2(j)$. Therefore, the alignment ends with a gap in S_1.

2. Alignments of the two prefixes where $S_1(i)$ is aligned strictly to the right of $S_2(j)$. Therefore, the alignment ends with a gap in S_2.

3. Alignments of the two prefixes where characters $S_1(i)$ and $S_2(j)$ are aligned opposite each other. This includes both the case that $S_1(i) = S_2(j)$ and that $S_1(i) \neq S_2(j)$.

Clearly, these three types of alignments cover all the possibilities.

Definition Define $E(i, j)$ as the maximum value of any alignment of type 1; define $F(i, j)$ as the maximum value of any alignment of type 2; define $G(i, j)$ as the maximum value of any alignment of type 3; and finally define $V(i, j)$ as the maximum value of the three terms $E(i, j)$, $F(i, j)$, $G(i, j)$.

Recurrences for the case of arbitrary gap weights

By dividing the types of alignments into three cases, as above, we can write the following recurrences that establish $V(i, j)$:

$$V(i, j) = \max[E(i, j), F(i, j), G(i, j)],$$

$$G(i, j) = V(i - 1, j - 1) + s(S_1(i), S_2(j)),$$

$$E(i, j) = \max_{0 \leq k \leq j-1} [V(i, k) - w(j - k)],$$

$$F(i, j) = \max_{0 \leq l \leq i-1} [V(l, j) - w(i - l)].$$

To complete the recurrences, we need to specify the base cases and where the optimal alignment value is found. If all spaces are included in the objective function, even spaces that begin or end an alignment, then the optimal value for the alignment is found in cell (n, m), and the base case is

$$V(i, 0) = -w(i),$$

$$V(0, j) = -w(j),$$

$$E(i, 0) = -w(i),$$

$$F(0, j) = -w(j),$$

where $G(0, 0) = 0$, but $G(i, j)$ is undefined when exactly one of i or j is zero. Note that $V(0, 0) = w(0)$, which will most naturally be assigned to be zero.

When end spaces, and hence end gaps, are free, then the optimal alignment value is the maximum value over any cell in row n or column m, and the base cases are

$$V(i, 0) = 0,$$

$$V(0, j) = 0.$$

Time analysis

Theorem 11.8.1. *Assuming that $|S_1| = n$ and $|S_2| = m$, the recurrences can be evaluated in $O(nm^2 + n^2m)$ time.*

PROOF   We evaluate the recurrences by the usual approach of filling in an $(n+1) \times (m+1)$ size table one row at time, where each row is filled from left to right. For any cell (i, j), the algorithm examines one other cell to evaluate $G(i, j)$, j cells of row i to evaluate $E(i, j)$, and i cells of column j to evaluate $F(i, j)$. Therefore, for any fixed row, $m(m + 1)/2 = \Theta(m^2)$ cells are examined to evaluate all the E values in that row, and for any fixed column, $\Theta(n^2)$ cells are examined to evaluate all the F values of that column. The theorem then follows since there are n rows and m columns. $\square$

The increase in running time over the previous case ($O(nm)$ time when gaps are not in the model) is caused by the need to look j cells to the left and i cells above to determine $V(i, j)$. Before gaps were included in the model, $V(i, j)$ depended only on the three cells adjacent to (i, j), and so each $V(i, j)$ value was computed in constant time. We will show next how to reduce the number of cell examinations for the case of affine gap weights; later we will show a more complex reduction for the case of convex gap weights.

11.8.6. Affine (and constant) gap weights

Here we examine in detail the simplest affine gap weight model and show that optimal alignments in that model can be computed in $O(nm)$ time. That bound is the same as for the alignment model without a gap term in the objective function. So although an explicit gap term in the objective function makes the alignment model much richer, it does not increase the running time used (in an asymptotic, worst-case sense) to find an optimal alignment. This important result was derived by several different authors (e.g., [18], [166], [186]). The same result then holds immediately for constant gap weights.

Recall that the objective is to find an alignment to

$$\text{maximize}[W_m(\# matches) - W_{ms}(\# mismatches) - W_g(\# gaps) - W_s(\# spaces)].$$

We will use the same variables $V(i, j)$, $E(i, j)$, $F(i, j)$, and $G(i, j)$ used in the recurrences for arbitrary gap weights. The definition and meanings of these variables remain unchanged, but the recurrence relations will be modified for the case of affine gap weights.

The key insight leading to greater efficiency in the affine gap case is that the increase in the total weight of a gap contributed by each additional space is a constant W_s independent of the size of the gap to that point. In other words, in the affine gap weight model $w(q+1) - w(q) = W_s$ for any gap length q greater than zero. This is in contrast to the arbitrary weight case where there is no predictable relationship between $w(q)$ and $w(q+1)$. Because the gap weight increases by the same W_s for each space after the first one, when evaluating $E(i, j)$ or $F(i, j)$ we need not be concerned with exactly where a gap begins, but only whether it has already begun or whether a new gap is being started (either opposite character i of S_1 or opposite character j of S_2). This insight, as usual, is formalized in a set of recurrences.

The recurrences

For the case where end gaps are included in the alignment value, the base case is easily seen to be

$$V(i, 0) = E(i, 0) = -W_g - iW_s,$$

$$V(0, j) = F(0, j) = -W_g - jW_s,$$

so that the zero row and columns of the table for V can be filled in easily. When end gaps are free, then $V(i, 0) = V(0, j) = 0$.

The general recurrences are

$$V(i, j) = \max[E(i, j), F(i, j), G(i, j)],$$

$$G(i, j) = \begin{cases} V(i - 1, j - 1) + W_m, \text{if } S_1(i) = S_2(j) \\ V(i - 1, j - 1) - W_{ms}, \text{if } S_1(i) \neq S_2(j), \end{cases}$$

$$E(i, j) = \max[E(i, j - 1), V(i, j - 1) - W_g] - W_s,$$

$$F(i, j) = \max[F(i - 1, j), V(i - 1, j) - W_g] - W_s.$$

To better understand these recurrences, consider the recurrence for $E(i, j)$. By definition, $S_1(i)$ will be aligned to the left of $S_2(j)$. The recurrence says that either 1. $S_1(i)$ is exactly one place to the left of $S_2(j)$, in which case a gap begins in S_1 opposite character $S_2(j)$, and $E(i, j) = V(i, j-1) - W_g - W_s$ or 2. $S_1(i)$ is to the left of $S_2(j-1)$, in which case the same gap in S_1 is opposite both $S_2(j - 1)$ and $S_2(j)$, and $E(i, j) = E(i, j - 1) - W_s$. An explanation for $F(i, j)$ is similar, and $G(i, j)$ is the simple case of aligning $S_1(i)$ opposite $S_2(j)$.

As before, the value of the optimal alignment is found in cell (n, m) if right end spaces contribute to the objective function. Otherwise the value of the optimal alignment is the maximum value in the nth row or mth column.

The reader should be able to verify that these recurrences are correct but might wonder why $V(i, j - 1)$ and not $G(i, j - 1)$ is used in the recurrence for $E(i, j)$. That is, why is $E(i, j)$ not $\max[E(i, j - 1), G(i, j - 1) - W_g] - W_s$? This recurrence would be incorrect because it would not consider alignments that have a gap in S_2 bordered on the left by character $j - 1$ of S_2 and ending opposite character i of S_1, followed immediately by a gap in S_1. The expanded recurrence $E(i, j) = \max[E(i, j - 1), G(i, j - 1) - W_g, V(i, j - 1) - W_g] - W_s$ would allow for all alignments and would be correct, but the inclusion of the middle term ($G(i, j - 1) - W_g$) is redundant because the last term ($V(i, j - 1) - W_g$) includes it.

Time analysis

Theorem 11.8.2. *The optimal alignment with affine gap weights can be computed in $O(nm)$ time, the same time as for optimal alignment without a gap term.*

PROOF   Examination of the recurrences shows that for any pair (i, j), each of the terms $V(i, j)$, $E(i, j)$, $F(i, j)$, and $G(i, j)$ is evaluated by a constant number of references to previously computed values, arithmetic operations, and comparisons. Hence $O(nm)$ time suffices to fill in all the $(n + 1) \times (m + 1)$ cells in the dynamic programming table.   □

11.9. Exercises

1. Write down the edit transcript for the alignment example on page 226.

2. The definition given in this book for string transformation and edit distance allows at most one operation per position in each string. But part of the motivation for string transformation and edit distance comes from an attempt to model evolution, where there is no restriction on the number of mutations that could occur at the same position. A deletion followed by an insertion and then a replacement could all happen at the same position. However, even though multiple operations at the same position are allowed, they will not occur in the transformation that uses the fewest number of operations. Prove this.

3. In the discussion of edit distance, all transforming operations were assumed to be done to one string only, and a "hand-waiving" argument was given to show that no greater generality is gained by allowing operations on both strings. Explain in detail why there is no loss in generality in restricting operations to one string only.

4. Give the details for how the dynamic programming table for edit distance or alignment can be filled in columnwise or by successive antidiagonals. The antidiagonal case is useful in the context of practical parallel computation. Explain this.

5. In Section 11.3.3, we described how to create an edit transcript from the traceback path through the dynamic programming table for edit distance. Prove that the edit transcript created in this way is an optimal edit transcript.

6. In Part I we discussed the exact matching problem when don't-care symbols are allowed. Formalize the edit distance problem when don't-care symbols are allowed in both strings, and show how to handle them in the dynamic programming solution.

7. Prove Theorem 11.3.4 showing that the pointers in the dynamic programming table completely capture all the optimal alignments.

8. Show how to use the optimal (global) alignment value to compute the edit distance of two strings and vice versa. Discuss in general the formal relationship between edit distance and string similarity. Under what circumstances are these concepts essentially equivalent, and when are they different?

9. The method discussed in this chapter to construct an optimal alignment left back-pointers while filling in the dynamic programming (DP) table, and then used those pointers to trace back a path from cell (n, m) to cell $(0, 0)$. However, there is an alternate approach that works even if no pointers are available. If given the full DP table without pointers, one can construct an alignment with an algorithm that "works through" the table in a single pass from cell (n, m) to cell $(0, 0)$. Make this precise and show it can be done as fast as the algorithm that fills in the table.

10. For most kinds of alignments (for example, global alignment without arbitrary gap weights), the traceback using pointers (as detailed in Section 11.3.3) runs in $O(n + m)$ time, which is less than the time needed to fill in the table. Determine which kinds of alignments allow this speedup.

11. Since the traceback paths in a dynamic programming table correspond one-to-one with the optimal alignments, the number of distinct cooptimal alignments can be obtained by computing the number of distinct traceback paths. Give an algorithm to compute this number in $O(nm)$ time.

    Hint: Use dynamic programming.

12. As discussed in the previous problem, the cooptimal alignments can be found by enumerating all the traceback paths in the dynamic programming table. Give a backtracking method to find each path, and each cooptimal alignment, in $O(n + m)$ time per path.

13. In a dynamic programming table for edit distance, must the entries along a row be

nondecreasing? What about down a column or down a diagonal of the table? Now discuss the same questions for optimal global alignment.

14. Give a complete argument that the formula in Theorem 11.6.1 is correct. Then provide the details for how to find the longest common subsequence, not just its length, using the algorithm for weighted edit distance.

15. As shown in the text, the longest common subsequence problem can be solved as an optimal alignment or similarity problem. It can also be solved as an operation-weight edit distance problem.

    Let u represent the length of the longest common subsequence of two strings of lengths n and m. Using the operation weights of $d = 1, r = 2$, and $e = 0$, we claim that $D(n, m) = m+n-2u$ or $u = (m+n-D(n, m))/2$. So, $D(n, m)$ is minimized by maximizing u. Prove this claim and explain in detail how to find a longest common subsequence using a program for operation-weight edit distance.

16. Write recurrences for the longest common subsequence problem that do not use weights. That is, solve the *lcs* problem more directly, rather than expressing it as a special case of similarity or operation-weighted edit distance.

17. Explain the correctness of the recurrences for similarity given in Section 11.6.1.

18. Explain how to compute edit distance (as opposed to similarity) when end spaces are free.

19. Prove the one-to-one correspondence between shortest paths in the edit graph and minimum weight global alignments.

20. Show in detail that the end-space free variant of the similarity problem is correctly solved using the method suggested in Section 11.6.4.

21. Prove Theorem 11.6.2, and show in detail the correctness of the method presented for finding the shortest approximate occurrence of P in T ending at position j.

22. Explain how to use the dynamic programming table and traceback to find all the optimal solutions (pairs of substrings) to the local alignment problem for two strings S_1 and S_2.

23. In Section 11.7.3, we mentioned that the dynamic programming table is often used to identify pairs of substrings of high similarity, which may not be optimal solutions to the local alignment problem. Given similarity threshold t, that method seeks to find pairs of substrings with similarity value t or greater. Give an example showing that the method might miss some qualifying pairs of substrings.

24. Show how to solve the alphabet-weight alignment problem with affine gap weights in $O(nm)$ time.

25. The discussions for alignment with gap weights focused on how to compute the values in the dynamic programming table and did not detail how to construct an optimal alignment. Show how to augment the algorithm so that it constructs an optimal alignment. Try to limit the amount of additional space required.

26. Explain in detail why the recurrence $E(i, j) = \max[E(i, j-1), G(i, j-1) - W_g, V(i, j-1) - W_g] - W_s$ is correct for the affine gap model, but is redundant, and that the middle term $(G(i, j-1) - W_g)$ can be removed.

27. The recurrences relations we developed for the affine gap model follow the logic of paying $W_g + W_s$ when a gap is "initiated" and then paying W_s for each additional space used in that gap. An alternative logic is to pay $W_g + W_s$ at the point when the gap is "completed." Write recurrences relations for the affine gap model that follow that logic. The recurrences should compute the alignment in $O(nm)$ time. Recurrences of this type are developed in [166].

28. In the *end-gap free* version of alignment, spaces and gaps at either end of the alignment

do not contribute to the cost of the alignment. Show how to use the affine gap recurrences developed in the text to solve the end-gap free version of the affine gap model of alignment. Then consider using the alternate recurrences developed in the previous exercise. Both should run in $O(nm)$ time. Is there any advantage to using one over the other of these recurrences?

29. Show how to extend the *agrep* method of Section 4.2.3 to allow character insertions and deletions.

30. Give a *simple* algorithm to solve the local alignment problem in $O(nm)$ time if no spaces are allowed in the local alignment.

31. **Repeated substrings.** Local alignment between two different strings finds pairs of substrings from the two strings that have high similarity. It is also important to find substrings of a single string that have high similarity. Those substrings represent *inexact repeated substrings*. This suggests that to find inexact repeats in a single string one should locally align of a string against itself. But there is a problem with this approach. If we do local alignment of a string against itself, the best substring will be the entire string. Even using all the values in the table, the best path to a cell (i, j) for $i \neq j$ may be strongly influenced by the main diagonal. There is a simple fix to this problem. Find it. Can your method produce two substrings that overlap? Is that desirable? Later in Exercise 17 of Chapter 13, we will examine the problem of finding the most similar *nonoverlapping* substrings in a single string.

32. **Tandem repeats.** Let P be a pattern of length n and T a text of length m. Let P^m be the concatenation of P with itself m times, so P^m has length mn. We want to compute a local alignment between P^m and T. That will find an interval in T that has the best global alignment (according to standard alignment criteria) with some tandem repeat of P. This problem differs from the problem considered in Exercise 4 of Chapter 1, because errors (mismatches and insertions and deletions) are now allowed. The particular problem arises in studying the secondary structure of proteins that form what is called a *coiled-coil* [158]. In that context, P represents a *motif* or *domain* (a pattern for our purposes) that can repeat in the protein an unknown number of times, and T represents the protein. Local alignment between P^m and T picks out an interval of T that "optimally" consists of tandem repeats of the motif (with errors allowed). If P^m is explicitly created, then standard local alignment will solve the problem in $O(nm^2)$ time. But because P^m consists of identical copies of P, an $O(nm)$-time solution is possible. The method essentially simulates what the dynamic programming algorithm for local alignment would do if it were executed with P^m and T explicitly. Below we outline the method.

The dynamic programming algorithm will fill in an $m + 1$ by $n + 1$ table V, whose rows are numbered 0 to n, and whose columns are numbered 0 to m. Row 0 and column 0 are initialized to all 0 entries. Then in each row i, from 1 to m, the algorithm does the following: It executes the standard local alignment recurrences in row i; it sets $V(i, 0)$ to $V(i, n)$; and then it executes the standard local alignment recurrences in row i again. After completely filling in each row, the algorithm selects the cell with largest V value, as in the standard solution to the local alignment problem.

Clearly, this algorithm only takes $O(nm)$ time. Prove that it correctly finds the value of the optimal local alignment between P^m and T. Then give the details of the traceback to construct the optimal local alignment. Discuss why P was (conceptually) expanded to P^m and not a longer or shorter string.

33. **a.** Given two strings S_1 and S_2 (of lengths n and m) and a parameter δ, show how to construct the following matrix in $O(nm)$ time: $\mathcal{M}(i, j) = 1$ if and only if there is an alignment of S_1 and S_2 in which characters $S_1(i)$ and $S_2(j)$ are aligned with each other and the value of the

alignment is within δ of the maximum value alignment of S_1 and S_2. That is, if $V(S_1, S_2)$ is the value of the optimal alignment, then the best alignment that puts $S_1(i)$ opposite $S_2(j)$ should have value at least $V(S_1, S_2) - \delta$. This matrix $\mathcal{M}$ is used [490] to provide *some* information, such as common or uncommon features, about the set of *suboptimal* alignments of S_1 and S_2. Since the biological significance of the *optimal* alignment is sometimes uncertain, and optimality depends on the choice of (often disputed) weights, it is useful to efficiently produce or study a set of suboptimal (but close) alignments in addition to the optimal one. How can the matrix $\mathcal{M}$ be used to produce or study these alignments?

b. Show how to modify matrix $\mathcal{M}$ so that $\mathcal{M}(i, j) = 1$ if and only if $S_1(i)$ and $S_2(j)$ are aligned in every alignment of S_1 and S_2 that has value at least $V(S_1, S_2) - \delta$. How efficiently can this matrix be computed? The motivation for this matrix is essentially the same as for the matrix described in the preceding problem and is used in [443] and [445].

34. Implement the dynamic programming solution for alignment with a gap term in the objective function, and then experiment with the program to find the right weights to solve the cDNA matching problem.

35. The process by which intron-exon boundaries (called *splice sites*) are found in mRNA is not well understood. The simplest hope – that splice sites are marked by patterns that always occur there and never occur elsewhere – is false. However, it is true that certain short patterns very frequently occur at the splice sites of introns. In particular, most introns start with the dinucleotide *GT* and end with *AG*. Modify the dynamic programming recurrences used in the cDNA matching problem to enforce this fact.

There are additional pattern features that are known about introns. Search a library to find information about those conserved features – you'll find a lot of interesting things while doing the search.

36. Sequence to structure deduction via alignment

An important application for aligning protein strings is to deduce unknown secondary structure of one protein from known secondary structure of another protein. From that secondary structure, one can then try to determine the three-dimensional structure of the protein by model building methods. Before describing the alignment exercise, we need some background on protein structure.

A string of a typical globular protein (a typical enzyme) consists of substrings that form the tightly wrapped *core* of the protein, interspersed by substrings that form *loops* on the exterior of the protein. There are essentially three types of secondary structures that appear in globular proteins: α-helixes and β-sheets, which make up the core of the protein, and loops on the exterior of the protein. There are also turns, which are smaller than loops. The structure of the core of the protein is highly conserved over time, so that any large insertions or deletions are much more likely to occur in the loops than in the core.

Now suppose one knows the secondary (or three-dimensional) structure of a protein from one species, and one has the *sequence* of the homologous protein from another species, but not its two- or three-dimensional structure. Let S_1 denote the string for the first protein and S_2 the second. Determining two- or three-dimensional structure by crystallography or NMR is very complex and expensive. Instead, one would like to use sequence alignment of S_1 and S_2 to identify the α and β structures in S_2. The hope is that with the proper alignment model, scoring matrix, and gap penalties, the substrings of the α and β structures in the two strings will align with each other. Since the locations of the α and β regions are known in S_1, a "successful" alignment will identify the α and β regions in S_2. Now, insertions and deletions in core regions are rare, so an alignment that successfully identifies the α and β regions in S_2 should not have large gaps in the α and β regions in S_1. Similarly, the alignment should not have large gaps in the substrings of S_2 that align to the known α and β regions of S_1.

Usually a scoring matrix is used to score matches and mismatches, and a affine (or linear) gap penalty model is also used. Experiments [51, 447] have shown that the success of this approach is very sensitive to the exact choice of the scoring matrix and penalties. Moreover, it has been suggested that the gap penalty must be made higher in the substrings forming the α and β regions than in the rest of the string (for example, see [51] and [296]). That is, no *fixed* choice for gap penalty and space penalty (gap initiation and gap extension penalties in the vernacular of computational biology) will work. Or at least, having a higher gap penalty in the secondary regions will more likely result in a better alignment. High gap penalties tend to keep the α and β regions unbroken. However, since insertions and deletions do definitely occur in the loops, gaps in the alignment of regions outside the core should be allowed.

This leads to the following alignment problem: How do you modify the alignment model and penalty structure to achieve the requirements outlined above? And, how do you find the optimal alignment within those new constraints?

Technically, this problem is not very hard. However, the application to deducing secondary structure is very important. Orders of magnitude more protein sequence data are available than are protein structure data. Much of what is "known" about protein structure is actually obtained by deductions from protein sequence data. Consequently, deducing structure from sequence is a central goal.

A multiple alignment version of this structure prediction problem is discussed in the first part of Section 14.10.2.

37. Given two strings S_1 and S_2 and a text T, you want to find whether there is an occurrence of S_1 and S_2 *interwoven* (without spaces) in T. For example, the strings *abac* and *bbc* occur interwoven in *cabbabccdw*. Give an efficient algorithm for this problem. (It may have a relationship to the longest common subsequence problem.)

38. As discussed earlier in the exercises of Chapter 1, bacterial DNA is often organized into circular molecules. This motivates the following problem: Given two linear strings of lengths n and m, there are n circular shifts of the first string and m circular shifts of the second string, and so there are nm pairs of circular shifts. We want to compute the global alignment for each of these nm pairs of strings. Can that be done more efficiently than by solving the alignment problem from scratch for each pair? Consider both worst-case analysis and "typical" running time for "naturally occurring" input.

Examine the same problem for local alignment.

39. The stuttering subsequence problem [328]. Let P and T be strings of n and m characters each. Give an $O(m)$-time algorithm to determine if P occurs as a *subsequence* of T.

Now let P^i denote the string P where each character is repeated i times. For example, if $P = abc$ then P^3 is *aaabbbccc*. Certainly, for any fixed i, one can test in $O(m)$ time whether P^i occurs as a subsequence of T. Give an algorithm that runs in $O(m \log m)$ time to determine the largest i such that P^i is a subsequence of T. Let $Maxi(P, T)$ denote the value of that largest i.

Now we will outline an approach to this problem that reduces the running time from $O(m \log m)$ to $O(m)$. You will fill in the details.

For a string T, let d be the number of distinct characters that occur in T. For string T and character x in T, define $odd(x)$ to be the positions of the odd occurrences of x in T, that is, the positions of the first, third, fifth, etc. occurrence of x in T. Since there are d distinct characters in T, there are d such odd sets. For example, if $T = 012000211202222201100001$ then $odd(1)$ is 2,9,18. Now define $half(T)$ as the subsequence of T that remains after *removing* all the characters in positions specified by the d *odd* sets. For example, $half(T)$

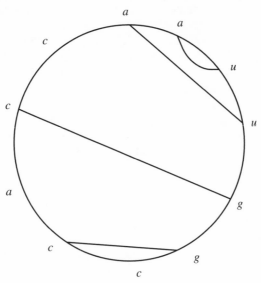

Figure 11.9: A nested pairing, not necessarily of maximum cardinality.

above is 0021220101. Assuming that the number of distinct symbols, d, is fixed ahead of time, give an $O(m)$-time algorithm to find $half(T)$. Now argue that the length of $half(T)$ is at most $m/2$. This will be used later in the time analysis.

Now prove that $|Maxi(P, T) - 2Maxi(P, half(T))| \leq 1$.

This fact is the critical one in the method.

The above facts allow us to find $Maxi(P, T)$ in $O(m)$ time by a divide-and-conquer recursion. Give the details of the method: Specify the termination conditions of the divide and conquer, prove correctness of the method, set up a recurrence relation to analyze the running time, and then solve the relation to obtain an $O(m)$ time bound.

Harder problem: What is a realistic application for the stuttering subsequence problem?

40. As seen in the previous problem, it is easy to determine if a single pattern P occurs as a subsequence in a text T. This takes $O(m)$ time. Now consider the problem of determining if *any* pattern in a *set* of patterns occurs in a text. If n is the length of all the patterns in the set, then $O(nm)$ time is obtained by solving the problem for each pattern separately. Try for a time bound that is significantly better than $O(nm)$. Recall that the analogous substring set problem can be solved in $O(n + m)$ time by Aho–Corasik or suffix tree methods.

41. **The tRNA folding problem.** The following is an extremely crude version of a problem that arises in predicting the secondary (planar) structure of transfer RNA molecules. Let S be a string of n characters over the RNA alphabet a, c, u, g. We define a *pairing* as set of disjoint pairs of characters in S. A pairing is called *proper* if it only contains (a, u) pairs or (c, g) pairs. This constraint arises because in RNA a and u are complementary nucleotides, as are c and g. If we draw S as a circular string, we define a *nested pairing* as a proper pairing where each pair in the pairing is connected by a line inside the circle, and where the lines do not cross each other. (See Figure 11.9). The problem is to find a nested pairing of largest cardinality. Often one has the additional constraint that a character may not be in a pair with either of its two immediate neighbors. Show how to solve this version of the tRNA folding problem in $O(n^3)$ time using dynamic programming.

Now modify the problem by adding weights to the objective function so that the weight of an $a-u$ pair is different than the weight of a $c-g$ pair. The goal now is to find a nested pairing of maximum total weight. Give an efficient algorithm for this weighted problem.

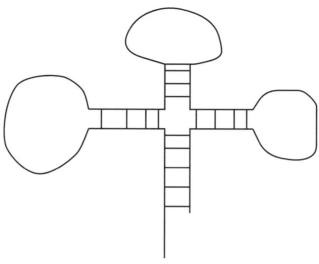

Figure 11.10: A rough drawing of a cloverleaf structure. Each of the small horizontal or vertical lines inside a stem represents a base pairing of $a-u$ or $c-g$.

42. Transfer RNA (tRNA) molecules have a distinctive planar secondary structure called the *cloverleaf* structure. In a cloverleaf, the string is divided into alternating *stems* and *loops* (see Figure 11.10). Each stem consists of two parallel substrings that have the property that any pair of opposing characters in the stem must be complements (a with u; c with g). Chemically, each complementary stem pair forms a bond that contributes to the overall stability of the molecule. A $c-g$ bond is stronger than an $a-u$ bond.

    Relate this (very superficial) description of tRNA secondary structure to the weighted nested pairing problem discussed above.

43. The true bonding pattern of complementary bases (in the stems) of tRNA molecules mostly conforms to the noncrossing condition in the definition of a nested pairing. However, there are exceptions, so that when the secondary structure of known tRNA molecules is represented by lines through the circle, a few lines may cross. These violations of the noncrossing condition are called *psuedoknots*.

    Consider the problem of finding a maximum cardinality proper pairing where a fixed number of psuedoknots are allowed. Give an efficient algorithm for this problem, where the complexity is a function of the permitted number of crossings.

44. **RNA sequence and structure alignment.** Because of the nested pairing structure of RNA, it is easy to incorporate some structural considerations when aligning RNA strings. Here we examine alignments of this kind.

    Let P be an RNA pattern string with a known pairing structure, and let T be a larger RNA text string with a known pairing structure. To represent pairing structure in P, let $O_P(i)$ be the *offset* (positive or negative) of the mate of the character at position i, if any. For example, if the character at position 17 is mated to the character at position 46, then $O_P(17) = 29$ and $O(46) = -29$. If the character at position i has no mate, then $O_P(i)$ is zero. The structure of T is similarly represented by an offset vector O_T. Then P exactly occurs in T starting at position j if and only if $P(i) = T(j+i-1)$ and $O_P(i) = O_T(j+i-1)$, for each position i in P.

    a. Assuming the lengths of P and T are n and m, respectively, give an $O(n+m)$-time algorithm to find every place that P exactly occurs in T.

    b. Now consider a more liberal criteria for deciding that P occurs in T starting at position j. We again require that $P(i) = T(j+i-1)$ for each position i in P, but now only require that $O_P(i) = O_T(j+i-1)$ when $O_P(i)$ is not zero.

Give an efficient algorithm to find all locations where P occurs in T under the more liberal definition of occurrence. The naive, $O(nm)$-time solution of explicitly aligning P to every starting position j and then checking for a match is not efficient. An efficient solution can be obtained using only methods in Part I and II of the book.

c. Discuss when the more liberal definition is reasonable and when it may not be.

45. A gap modeling question

The recurrences given in Section 11.8.5 for the case of arbitrary gap weights raise a subtle question about the proper gap model when the gap penalty w is arbitrary. With those recurrences, any single gap can be considered as two or more gaps that just happen to be adjacent. Suppose, for example, $w(q) = 1$ for $q \leq 5$, and $w(q) = 10^6$ for $i > 5$. Then, a gap of length 10 would have weight 10^6 if considered as a single gap, but would only have weight 2 if considered as two adjacent gaps of length five each. The recurrences from Section 11.8.5 would treat those ten spaces as two adjacent gaps with total weight 2. Is this the proper gap model?

There are two viewpoints on this question. In one view, the goal is to model the most likely set of mutation events transforming one string into another, and the alignment is just an aid in *displaying* this transformation. The primitive mutational events allowed are the transformation of single characters (mismatches in the alignment) and insertion and deletion of blocks of characters of arbitrary lengths (each of which causes a gap in the alignment). With this view, it is perfectly proper to have two adjacent gaps on the same string. These are just two block insertions or deletions that happen to have occurred next to each other. If the gap weights correctly model the costs of such block operations, and the cost is a concave increasing function of length as in the above example, then it is much more likely that a long gap will be created by several insertion or deletion events than by a single such event. With this view, one should insist that the dynamic program allow adjacent gaps when they are advantageous.

In the other view, one is just interested in how "similar" two strings are, and there may be no explicit mutational model. Then, a given alignment of two strings is simply one way to demonstrate the similarity of the two strings. In that view, a gap is a maximal set of adjacent spaces and so should not be broken into smaller gaps.

With arbitrary gap weights, the dynamic programming recurrences presented correctly model the first view, but not the second. Also, in the case of convex (and hence affine or constant) gap weights, the given recurrences correctly model both views, since there is no incentive to break up a gap into shorter gaps. However, if gap weights with concave increasing sections are thought proper, then different recurrences are required to correctly model the second view. The recurrences below correctly implement the second view:

$$V(i, j) = \max[E(i, j), F(i, j), G(i, j)],$$

$$G(i, j) = V(i - 1, j - 1) + s(S_1(i), S_2(j)),$$

$$E(i, j) = \max[G(i, k) - w(j - k)] \text{ (over } 0 \leq k \leq j - 1),$$

$$F(i, j) = \max[G(l, j) - w(i - l)] \text{ (over } 0 \leq l \leq i - 1).$$

These equations differ from the recurrences of Section 11.8.5 by the change of $V(i, k)$ and $V(l, j)$ to $G(i, k)$ and $G(l, j)$ in the equations for $E(i, j)$ and $F(i, j)$, respectively. The effect is that every gap except the left-most one must be preceded by two aligned characters; hence there cannot be two adjacent gaps in the same string. However, this also prohibits two adjacent gaps where each is in a different string. For example, the alignment

$$
\begin{array}{cccccccccc}
x & x & a & b & c & \_ & \_ & \_ & y & y \\
x & x & \_ & \_ & \_ & i & d & e & y & y
\end{array}
$$

would never be found by these modified recurrences.

There seems no modeling justification to prohibit adjacent gaps in opposite strings. In fact some mutations, such as *substring inversions* (which are common in DNA), would be best represented in an alignment as adjacent gaps of this type, unless the model of alignment has an explicit notion of inversion (we will consider such a model in Chapter 19). Another example where adjacent spaces would be natural occurs when comparing two mRNA strings that arise from alternative intron splicing. In eukaryotes, genes are often comprised of alternating regions of exons and introns. In the normal mode of transcription, every intron is eventually spliced out, so that the mRNA molecule reflects a concatenation of the exons. But it can also happen, in what is called *alternative splicing*, that exons can be spliced out as well as introns. Consider then the situation where all the introns plus exon i are spliced out, and the situation where all the introns plus exon $i + 1$ are spliced out. When these two mRNA strings are compared, the best alignment may very well put exon i against a gap in the second string, and then put exon $i + 1$ against a gap in the first string. In other words, the informative alignment *would* have two adjacent gaps in alternate strings. In that case, the recurrences above do not correctly implement the second viewpoint.

Write recurrences for arbitrary gap weights to allow adjacent gaps in the two opposite strings and yet prohibit adjacent gaps in a single string.

12

Refining Core String Edits and Alignments

In this chapter we look at a number of important refinements that have been developed for certain core string edit and alignment problems. These refinements either speed up a dynamic programming solution, reduce its space requirements, or extend its utility.

12.1. Computing alignments in only linear space

One of the defects of dynamic programming for all the problems we have discussed is that the dynamic programming tables use $\Theta(nm)$ space when the input strings have length n and m. (When we talk about the space used by a method, we refer to the maximum space ever in use simultaneously. Reused space does not add to the count of space use.) It is quite common that the limiting resource in string alignment problems is not time but space. That limit makes it difficult to handle large strings, no matter how long we may be willing to wait for the computation to finish. Therefore, it is very valuable to have methods that reduce the use of space without dramatically increasing the time requirements.

Hirschberg [224] developed an elegant and practical space-reduction method that works for many dynamic programming problems. For several string alignment problems, this method reduces the required space from $\Theta(nm)$ to $O(n)$ (for $n < m$) while only doubling the worst-case time bound. Miller and Myers expanded on the idea and brought it to the attention of the computational biology community [344]. The method has since been extended and applied to many more problems [97]. We illustrate the method using the dynamic programming solution to the problem of computing the optimal weighted global alignment of two strings.

12.1.1. Space reduction for computing similarity

Recall that the *similarity* of two strings is a *number*, and that under the similarity objective function there is an optimal alignment whose value equals that number. Now *if* we only require the similarity $V(n, m)$, and not an actual alignment with that value, then the maximum space needed (in addition to the space for the strings) can be reduced to $2m$. The idea is that when computing V values for row i, the only values needed from previous rows are from row $i - 1$; any rows before $i - 1$ can be discarded. This observation is clear from the recurrences for similarity. Thus, we can implement the dynamic programming solution using only two rows, one called row C for *current*, and one called row P for *previous*. In each iteration, row C is computed using row P, the recurrences, and the two strings. When that row C is completely filled in, the values in row P are no longer needed and C gets copied to P to prepare for the next iteration. After n iterations, row C holds the values for row n of the full table and hence $V(n, m)$ is located in the last cell of that row. In this way, $V(n, m)$ can be computed in $O(m)$ space and $O(nm)$ time. In fact, any

254

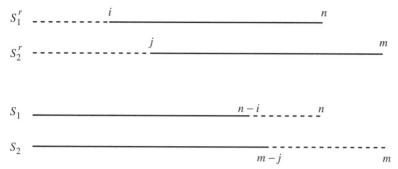

Figure 12.1: The similarity of the first i characters of S_1^r and the first j characters of S_2^r equals the similarity of the last i characters of S_1 and the last j characters of S_2. (The dotted lines denote the substrings being aligned.)

single row of the full table can be found and stored in those same time and space bounds. This ability will be critical in the method to come.

As a further refinement of this idea, the space needed can be reduced to one row plus one additional cell (in addition to the space for the strings). Thus $m + 1$ space is all that is needed. And, if $n < m$ then space use can be further reduced to $n + 1$. We leave the details as an exercise.

12.1.2. How to find the optimal alignment in linear space

The above idea is fine *if* we only want the similarity $V(n, m)$ or just want to store one preselected row of the dynamic programming table. But what can we do if we actually want an *alignment* that achieves value $V(n, m)$? In most cases it is such an alignment that is sought, not just its value. In the basic algorithm, the alignment would be found by traversing the pointers set while computing the full dynamic programming table for similarity. However, the above linear space method does not store the whole table and linear space is insufficient to store the pointers.

Hirschberg's high-level scheme for finding the optimal alignment in only linear space performs several smaller alignment computations, each using only linear space and each determining a bit more about an actual optimal alignment. The net result of these computations is a full description of an optimal alignment. We first describe how the initial piece of the full alignment is found using only linear space.

Definition   For any string α, let α^r denote the reverse of string α.

Definition   Given strings S_1 and S_2, define $V^r(i, j)$ as the similarity of the string consisting of the first i characters of S_1^r, and the string consisting of the first j characters of S_2^r. Equivalently, $V^r(i, j)$ is the similarity of the last i characters of S_1 and the last j characters of S_2 (see Figure 12.1).

Clearly, the table of $V^r(i, j)$ values can be computed in $O(nm)$ time, and any single preselected row of that table can be computed and stored in $O(nm)$ time using only $O(m)$ space.

The initial piece of the full alignment is computed in linear space by computing $V(n, m)$ in two parts. The first part uses the original strings; the second part uses the reverse strings. The details of this two-part computation are suggested in the following lemma.

Lemma 12.1.1. $V(n, m) = \max_{0 \le k \le m}[V(n/2, k) + V^r(n/2, m - k)]$.

PROOF   This result is almost obvious, and yet it requires a proof. Recall that $S_1[1..i]$ is the prefix of string S_1 consisting of the first i characters and that $S_1^r[1..i]$ is the reverse of the suffix of S_1 consisting of the last i characters of S_1. Similar definitions hold for S_2 and S_2^r.

For any fixed position k' in S_2, there is an alignment of S_1 and S_2 consisting of an alignment of $S_1[1..n/2]$ and $S_2[1..k']$ followed by a disjoint alignment of $S_1[n/2 + 1..n]$ and $S_2[k' + 1..m]$. By definition of V and V^r, the best alignment of the first type has value $V(n/2, k')$ and the best alignment of the second type has value $V^r(n/2, m - k')$, so the combined alignment has value $V(n/2, k') + V^r(n/2, m - k') \leq \max_k[V(n/2, k) + V^r(n/2, m - k)] \leq V(n, m)$.

Conversely, consider an optimal alignment of S_1 and S_2. Let k' be the right-most position in S_2 that is aligned with a character at or before position $n/2$ in S_1. Then the optimal alignment of S_1 and S_2 consists of an alignment of $S_1[1..n/2]$ and $S_2[1..k']$ followed by an alignment of $S_1[n/2 + 1..n]$ and $S_2[k' + 1..m]$. Let the value of the first alignment be denoted p and the value of the second alignment be denoted q. Then p must be equal to $V(n/2, k')$, for if $p < V(n/2, k')$ we could replace the alignment of $S_1[1..n/2]$ and $S_2[1..k']$ with the alignment of $S_1[1..n/2]$ and $S_2[1..k']$ that has value $V(n/2, k')$. That would create an alignment of S_1 and S_2 whose value is larger than the claimed optimal. Hence $p = V(n/2, k')$. By similar reasoning, $q = V^r(n/2, m - k')$. So $V(n, m) = V(n/2, k') + V^r(n/2, m - k') \leq \max_k[V(n/2, k) + V^r(n/2, m - k)]$.

Having shown both sides of the inequality, we conclude that $V(n, m) = \max_k[V(n/2, k) + V^r(n/2, m - k)]$.   □

Definition   Let k^* be a position k that maximizes $[V(n/2, k) + V^r(n/2, m - k)]$.

By Lemma 12.1.1, there is an optimal alignment whose traceback path in the full dynamic programming table (if one had filled in the full n by m table) goes through cell $(n/2, k^*)$. Another way to say this is that there is an optimal (longest) path L from node $(0, 0)$ to node (n, m) in the alignment graph that goes through node $(n/2, k^*)$. That is the key feature of k^*.

Definition   Let $L_{n/2}$ be the subpath of L that starts with the last node of L in row $n/2 - 1$ and ends with the first node of L in row $n/2 + 1$.

Lemma 12.1.2.   *A position k^* in row $n/2$ can be found in $O(nm)$ time and $O(m)$ space. Moreover, a subpath $L_{n/2}$ can be found and stored in those time and space bounds.*

PROOF   First, execute dynamic programming to compute the optimal alignment of S_1 and S_2, but stop after iteration $n/2$ (i.e., after the values in row $n/2$ have been computed). Moreover, when filling in row $n/2$, establish and save the normal traceback pointers for the cells in that row. At this point, $V(n/2, k)$ is known for every $0 \leq k \leq m$. Following the earlier discussion, only $O(m)$ space is needed to obtain the values and pointers in rows $n/2$. Second, begin computing the optimal alignment of S_1^r and S_2^r but stop after iteration $n/2$. Save both the values for cells in row $n/2$ along with the traceback pointers for those cells. Again, $O(m)$ space suffices and value $V^r(n/2, m - k)$ is known for every k. Now, for each k, add $V(n/2, k)$ to $V^r(n/2, m - k)$, and let k^* be an index k that gives the largest sum. These additions and comparisons take $O(m)$ time.

Using the first set of saved pointers, follow any traceback path from cell $(n/2, k^*)$ to a cell k_1 in row $n/2 - 1$. This identifies a subpath that is on an optimal path from cell $(0, 0)$ to cell $(n/2, k^*)$. Similarly, using the second set of traceback pointers, follow any traceback

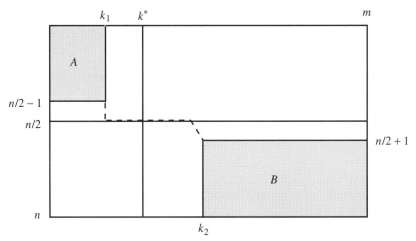

k_1 k^* m

A

$n/2 - 1$

$n/2$

$n/2 + 1$

B

n

k_2

Figure 12.2: After finding k^*, the alignment problem reduces to finding an optimal alignment in section A of the table and another optimal alignment in section B of the table. The total area of subtables A and B is at most $cnm/2$. The subpath $L_{n/2}$ through cell $(n/2, k^*)$ is represented by a dashed path.

path from cell $(n/2, k^*)$ to a cell k_2 in row $n/2 + 1$. That path identifies a subpath of an optimal path from $(n/2, k^*)$ to (n, m). These two subpaths taken together form the subpath $L_{n/2}$ that is part of an optimal path L from $(0, 0)$ to (n, m). Moreover, that optimal path goes through cell $(n/2, k^*)$. Overall, $O(nm)$ time and $O(m)$ space is used to find k^*, k_1, k_2, and $L_{n/2}$. □

To analyze the full method to come, we will express the time needed to fill in the dynamic programming table of size p by q as cpq, for some unspecified constant c, rather than as $O(pq)$. In that view, the $n/2$ row of the first dynamic program computation is found in $cnm/2$ time, as is the $n/2$ row of the second computation. Thus, a total of cnm time is needed to obtain and store both rows.

The key point to note is that with a cnm-time and $O(m)$-space computation, the algorithm learns k^*, k_1, k_2, and $L_{n/2}$. This specifies part of an optimal *alignment* of S_1 and S_2, and not just the value $V(n, m)$. By Lemma 12.1.1 it learns that there is an optimal alignment of S_1 and S_2 consisting of an optimal alignment of the first $n/2$ characters of S_1 with the first k^* characters of S_2, followed by an optimal alignment of the last $n/2$ characters of S_1 with the last $m - k^*$ characters of S_2. In fact, since the algorithm has also learned the subpath (subalignment) $L_{n/2}$, the problem of aligning S_1 and S_2 reduces to two smaller alignment problems, one for the strings $S_1[1..n/2 - 1]$ and $S_2[1..k_1]$, and one for the strings $S_1[n/2 + 1..n]$ and $S_2[k_2..m]$. We call the first of the two problems the *top* problem and the second the *bottom* problem. Note that the top problem is an alignment problem on strings of lengths at most $n/2$ and k^*, while the bottom problem is on strings of lengths at most $n/2$ and $m - k^*$.

In terms of the dynamic programming table, the top problem is computed in section A of the original n by m table shown in Figure 12.2, and the bottom problem is computed in section B of the table. The rest of the table can be ignored. Again, we can determine the values in the middle row of A (or B) in time proportional to the total size of A (or B). Hence the middle row of the top problem can be determined at most $ck^*n/2$ time, and the middle row in the bottom problem can be determined in at most $c(m - k^*)n/2$ time. These two times add to $cnm/2$. This leads to the full idea for computing the optimal alignment of S_1 and S_2.

12.1.3. The full idea: use recursion

Having reduced the original n by m alignment problem (for S_1 and S_2) to two smaller alignment problems (the top and bottom problems) using $O(nm)$ time and $O(m)$ space, we now solve the top and bottom problems by a recursive application of this reduction. (For now, we ignore the space needed to save the subpaths of L.) Applying exactly the same idea as was used to find k^* in the n by m problem, the algorithm uses $O(m)$ space to find the best column in row $n/4$ to break up the top $n/2$ by k_1 alignment problem. Then it reuses $O(m)$ space to find the best column to break up the bottom $n/2$ by $m - k_2$ alignment problem. Stated another way, we have two alignment problems, one on a table of size at most $n/2$ by k^* and another on a table of size at most $n/2$ by $m - k^*$. We can therefore find the best column in the middle row of each of the two subproblems in at most $cnk^*/2 + cn(m - k^*)/2 = cnm/2$ time, and recurse from there with four subproblems.

Continuing in this recursive way, we can find an optimal alignment of the two original strings with $\log_2 n$ levels of recursion, and at no time do we ever use more than $O(m)$ space. For convenience, assume that n is a power of two so that each successive halving gives a whole number. At each recursive call, we also find and store a subpath of an optimal path L, but these subpaths are edge disjoint, and so their total length is $O(n + m)$. In summary, the recursive algorithm we need is:

Hirschberg's linear-space optimal alignment algorithm

Procedure OPTA(l, l', r, r');
> begin
> $h := (l' - l)/2$;
> In $O(l' - l) = O(m)$ space, find an index k^* between l and l', inclusively, such that there is an optimal alignment of $S_1[l..l']$ and $S_2[r..r']$ consisting of an optimal alignment of $S_1[l..h]$ and $S_2[r..k^*]$ followed by an optimal alignment of $S_1[h + 1..l']$ and $S_2[k^* + 1..r']$. Also find and store the subpath L_h that is part of an optimal (longest) path L' from cell (l, r) to cell (l', r') and that begins with the last cell k_1 on L' in row $h - 1$ and ends with the first cell k_2 on L' in row $h + 1$. This is done as described earlier.
> Call *OPTA($l, h - 1, r, k_1$)*; {new top problem}
> Output subpath L_h;
> Call *OPTA($h + 1, l', k_2, r'$)*; {new bottom problem}
> end.

The call that begins the computation is to *OPTA*$(1, n, 1, m)$. Note that the subpath L_h is output between the two *OPTA* calls and that the top problem is called before the bottom problem. The effect is that the subpaths are output in order of increasing h value, so that their concatenation describes an optimal path L from $(0, 0)$ to (n, m), and hence an optimal alignment of S_1 and S_2.

12.1.4. Time analysis

We have seen that the first level of recursion uses cnm time and the second level uses at most $cnm/2$ time. At the ith level of recursion, we have 2^{i-1} subproblems, each of which has $n/2^{i-1}$ rows but a variable number of columns. However, the columns in these subproblems are distinct so the total size of all the problems is at most the total number of columns, m, times $n/2^{i-1}$. Hence the total time used at the ith level of recursion is at most $cnm/2^{i-1}$. The final dynamic programming pass to describe the optimal alignment takes cnm time. Therefore, we have the following theorem:

Theorem 12.1.1. *Using Hirschberg's procedure OPTA, an optimal alignment of two strings of length n and m can be found in $\sum_{i=1}^{\log n} cnm/2^{i-1} \le 2cnm$ time and $O(m)$ space.*

For comparison, recall that cnm time is used by the original method of filling in the full n by m dynamic programming table. Hirschberg's method reduces the space use from $\Theta(nm)$ to $\Theta(m)$ while only doubling the worst-case time needed for the computation.

12.1.5. Extension to local alignment

It is easy to apply Hirschberg's linear-space method for (global) alignment to solve the local alignment problem for strings S_1 and S_2. Recall that the optimal local alignment of S_1 and S_2 identifies substrings α and β whose global alignment has maximum value over all pairs of substrings. Hence, if substrings α and β can be found using only linear space, then their actual alignment can be found in linear space, using Hirschberg's method for global alignment.

From Theorem 11.7.1, the value of the optimal local alignment is found in the cell (i^*, j^*) containing the maximum v value. The indices i^* and j^* specify the *ends* of strings α and β whose global alignment has a maximum similarity value. The v values can be computed rowwise, and the algorithm must store values for only two rows at a time. Hence the end positions i^* and j^* can be found in linear space. To find the starting positions of the two strings, the algorithm can execute a reverse dynamic program using linear space (we leave this to the reader to detail). Alternatively, the dynamic programming algorithm for v can be extended to set a pointer $h(i, j)$ for each cell (i, j), as follows: If $v(i, j)$ is set to zero, then set the pointer $h(i, j)$ to (i, j); if $v(i, j)$ is set greater than zero, and if the normal traceback pointer would point to cell (p, q), then set $h(i, j)$ to $h(p, q)$. In this way, $h(i^*, j^*)$ specifies the starting positions of substrings α and β, respectively. Since α and β can be found in linear space, the local alignment problem can be solved in $O(nm)$ time and $O(m)$ space. More on this topic can be found in [232] and [97].

12.2. Faster algorithms when the number of differences is bounded

In Sections 9.4 and 9.5 we considered several alignment and matching problems where the number of allowed mismatches was bounded by a parameter k, and we obtained algorithms that run faster than without the imposed bound. One particular problem was the *k-mismatch problem*, finding all places in a text T where a pattern P occurs with at most k mismatches. A direct dynamic programming solution to this problem runs in $O(nm)$ time for a pattern of length n and a text of length m. But in Section 9.4 we developed an $O(km)$-time solution based on the use of a suffix tree, without any need for dynamic programming.

The $O(km)$-time result for the k-mismatch problem is useful because many applications seek only exact or nearly exact occurrences of P in T. Motivated by the same kinds of applications (and additional ones to be discussed in Section 12.2.1), we now extend the k-mismatch result to allow both mismatches and spaces (insertions and deletions from the viewpoint of edit distance). We use the term "differences" to refer to both mismatches and spaces.

Two specific bounded difference problems

We study two specific problems: *the k-difference global alignment problem* and the more involved *k-difference inexact matching problem*. This material was developed originally in the papers of Ukkonen [439], Fickett [155], Myers [341], and Landau and Vishkin [289]. The latter paper was expanded and illustrated with biological applications by Landau, Vishkin, and Nussinov [290]. There is much additional algorithmic work exploiting the assumption that the number of differences may be small [341, 345, 342, 337, 483, 94, 93, 95, 373, 440, 482, 413, 414, 415]. A related topic, algorithms whose expected running time is fast, is studied in Section 12.3.

> **Definition** Given strings S_1 and S_2 and a fixed number k, the *k-difference global alignment problem* is to find the best global alignment of S_1 and S_2 containing at most k mismatches and spaces (if one exists).

The k-difference global alignment problem is a special case of edit distance and is useful when S_1 and S_2 are believed to be fairly similar. It also arises as a subproblem in more complex string processing problems, such as the approximate PCR primer problem considered in Section 12.2.5. The solution to the k-difference global alignment problem will also be used to speed up global alignment when no bound k is specified.

> **Definition** Given strings P and T, the *k-difference inexact matching problem* is to find all ways (if any) to match P in T using at most k character substitutions, insertions, and deletions. That is, find all occurrences of P in T using at most k mismatches and spaces. (End spaces in T but not P are free.)

The inclusion of spaces, in addition to mismatches, allows a more robust version of the k-mismatch problem discussed in Section 9.4, but it complicates the problem. Unlike our solution to the k-mismatch problem, the k-differences problem seems to require the use of dynamic programming. The approach we take is to speed up the basic $O(nm)$-time dynamic programming solution, making use of the assumption that only alignments with at most k differences are of interest.

12.2.1. Where do bounded difference problems arise?

There is a large (and growing) computer science literature on algorithms whose efficiency is based on assuming a bounded number of differences. (See [93] for a survey and comparison of some of these, along with an additional method.) It is therefore appropriate, before discussing specific algorithmic results, to ask whether bounded difference problems arise frequently enough to justify the extensive research effort.

Bounded difference problems arise naturally in situations where a text is repeatedly modified (edited). Alignment of the text before and after modification can highlight the places where changes were made. A related application [345] concerns updating a graphics screen after incremental changes have been made to the displayed text. The assumption behind incremental screen update is that the text has changed by only a small amount, and that changing the text on the screen is slow enough to be seen by the user. The alignment of the old and new text then specifies the fewest changes to the existing screen needed to display the new text. Graphic displays with random access can exploit this information to very rapidly update the screen. This approach has been taken by a number of text editors. The effects of the speedup are easily seen and are often quite dramatic.

12.2.2. Illustrations from molecular biology

In biological applications of alignment, it may be less apparent that a bound on the number of allowed (or expected) differences between strings is ever justified. It has been explicitly stated by some computer scientists that bounded difference alignment methods have no relevance in biology. Certainly, the *major* open problems in aligning and comparing biological sequences arise from strings (usually protein) that have very *little* overall similarity. There is no argument on that point. Still, there are many sequence problems in molecular biology (particularly problems that come from genomics and handling DNA sequences rather than proteins) where it is appropriate to restrict the number of allowed (or expected) differences. A few hours of skimming biology journals will turn up many such examples.[1] We have already discussed one application, that of searching for STSs and ESTs in newly sequenced DNA (see Section 7.8.3). We have also mentioned the approximate PCR primer problem, which will be discussed in detail in Section 12.2.5. We mention here a few additional examples of alignment problems in biology where setting a bound on the number of differences is appropriate.

Chang and Lawler [94] point out that present DNA sequence assembly methods (see Sections 16.14 and 16.15.1) solve a massive number of instances of the approximate suffix-prefix matching problem. These methods compute, for every pair of strings S_1, S_2 in a large set of strings, the best match of a suffix of S_1 with a prefix of S_2, where the match is permitted to contain a "modest" percentage of differences. Using standard dynamic programming methods, those suffix-prefix computations have accounted for over 90% of the computation time used in past sequence assembly projects [363]. But in this application, the only suffix-prefix matches of interest are those with a modest number of differences. Accordingly, it is appropriate to use a faster algorithm that explicitly exploits that assumption. A related problem occurs in the "BAC-PAC" sequencing method involving hundreds of thousands of sequence alignments (see Section 16.13.1).

Another example arises in approaches to locating genes whose mutation causes or contributes to certain genetic diseases. The basic idea is to first identify (through genetic linkage analysis, functional analysis, or other means) a gene, or a region containing a gene, that is believed to cause or contribute to the disease of interest. Copies of that gene or region are then obtained and sequenced from people who are affected by the disease and people (usually relatives) who are not. The sequenced DNA from the affected and unaffected individuals is compared to find any consistent differences. Since many genetic diseases are caused by very small changes in a gene (possibly a single base change, deletion, or inversion), the problem involves comparing strings that have a very small number of differences. Systematic investigation of gene *polymorphisms* (differences) is an active area of research, and there are databases holding all the different sequences that have been found for certain specific genes. These sequences generally will be very similar to one another, so alignment and string manipulation tools that assume a bounded number of differences between strings are useful in handling those sequences.

A similar situation arises in the emerging field of "molecular epidemiology" where one tries to trace the transmission history of a pathogen (usually a virus) whose genome is mutating rapidly. This fine-scale analysis of the changing viral DNA or RNA gives rise to string comparisons between very similar strings. Aligning pairs of these strings to reveal

---

[1] I recently attended a meeting concerning the Human Genome Project, where numerous examples were presented in talks. I stopped taking notes after the tenth one.

their similarities and differences is a first step in sorting out their history and the constraints on how they can mutate. The history of their mutations is then represented in the form of an evolutionary tree (see Chapter 17). Collections of HIV viruses have been studied in this way. Another good example of molecular epidemiology [348] arises in tracing the history of *Hantavirus* infections in the southwest United States that appeared during the early 1990s.

The final two examples come from the milestone paper [162] reporting the first complete DNA sequencing of a free-living organism, the bacteria *Haemophilus influenzae Rd*. The genome of this bacteria consists of 1,830,137 base pairs and its full sequence was determined by pure shotgun sequencing without initial mapping (see Section 16.14). Before the large-scale sequencing project, many small, disparate pieces of the bacterial genome had been sequenced by different groups, and these sequences were in the DNA databases. One of the ways the sequencers checked the quality of their large-scale sequencing was to compare, when possible, their newly obtained sequence to the previously determined sequence. If they could not match the appropriate new sequences to the old ones with only a small number of differences, then additional steps were taken to assure that the new sequences were correct. Quoting from [162], "The results of such a comparison show that our sequence is 99.67 percent identical overall to those GenBank sequences annotated as *H. influenzae Rd*".

From the standpoint of alignment, the problem discussed above is to determine whether or not the new sequences match the old ones with few differences. This application illustrates both kinds of bounded difference alignment problems introduced earlier. When the location in the genome of the database sequence is known, the corresponding string in the full sequence can be extracted for comparison. The resulting comparison problem is then an instance of the *k-difference global alignment problem* that will be discussed next, in Section 12.2.3. When the genome location of the database sequence *P* is *not* known (and this is common), the comparison problem is to find all the places in the full sequence where *P* occurs with a very small number of allowed differences. That is then an instance of the *k*-difference inexact matching problem, which will be considered in Section 12.2.4.

The above story of *H. influenzae* sequencing will be repeated frequently as systematic large-scale DNA sequencing of various organisms becomes more common. Each full sequence will be checked against the shorter sequences for that organism already in the databases. This will be done not only for quality control of the large-scale sequencing, but also to correct entries in the databases, since it is generally believed that large-scale sequencing is more accurate.

The second application from [162] concerns building a *nonredundant* database of bacterial proteins (NRBP). For a number of reasons (for example, to speed up the search or to better evaluate the statistical significance of matches that are found), it is helpful to reduce the number of entries in a sequence database (in this case, bacterial protein sequences) by culling out, or combining in some way, highly similar, "redundant" sequences. This was done in the work presented in [162], and a "nonredundant" version of GenBank is regularly compiled at The National Center for Biotechnology Information. Fleischmann et al. [162] write:

Redundancy was removed from NRBP at two stages. All DNA coding sequences were extracted from GenBank . . . and sequences from the same species were searched against each other. Sequences having more than 97 percent identity over regions longer than 100 nucleotides were combined. In addition, the sequences were translated and used in protein

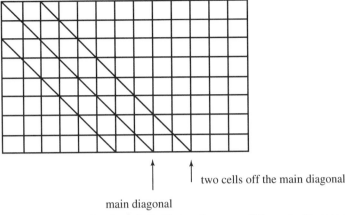

two cells off the main diagonal

main diagonal

Figure 12.3: The main diagonal and a strip that is $k = 2$ spaces off the main diagonal on each side.

comparisons with all sequences in SwissProt ... Sequences belonging to the same species and having more than 98 percent similarity over 33 amino acids were combined.

A similar example is discussed in [399] where roughly 170,000 DNA sequences "were subjected to an optimal alignment procedure to identify sequence pairs with at least 97% identity". In these alignment problems, one can impose a bound on the number of allowed differences. Alignments that exceed that bound are not of interest – the computation only needs to determine whether two sequences are "sufficiently similar" or not. Moreover, because these applications involve a large number of alignments (all database entries against themselves), efficiency of the method is important.

Admittedly, not every bounded-difference alignment problem in biology requires a sophisticated algorithm. But applications are so common, the sizes of some of the applications are so large, and the speedups so great, that it seems unproductive to completely dismiss the potential utility to molecular biology of bounded-difference and bounded-mismatch methods. With this motivation, we now discuss specific techniques that efficiently solve bounded-difference alignment problems.

12.2.3. k-difference global alignment

The problem is to find the best global alignment subject to the added condition that the alignment contains at most k mismatches and spaces, for a given value k. The goal is to reduce the time bound for the solution from $O(nm)$ (based on standard dynamic programming) to $O(km)$. The basic approach is to compute the *edit distance* of S_1 and S_2 using dynamic programming but fill in only an $O(km)$-size portion of the full table.

The key observation is the following: If we define the *main diagonal* of the dynamic programming table as the cells (i, i) for $i \le n \le m$, then any path in the dynamic programming table that defines a k-difference global alignment must not contain any cell $(i, i + l)$ or $(i, i - l)$ where l is greater than k (see Figure 12.3). To understand this, note that any path specifying a global alignment begins on the main diagonal (in cell $(0, 0)$) and ends on, or to the right of, the main diagonal (in cell (n, m)). Therefore, the path must introduce one space in the alignment for every horizontal move that the path makes off the main diagonal. Thus, only those paths that are never more than k horizontal cells from the main diagonal are candidates for specifying a k-difference global alignment. (Note that this implies that $m - n \leq k$ is a necessary condition for there to be any solution.) Therefore, to find any k-difference global alignment, it suffices to fill in the dynamic programming table in a strip consisting of $2k + 1$ cells in each row, centered on the main diagonal. When assigning values to cells in that strip, the algorithm follows the established recurrence relations for edit distance except for cells on the upper and lower border of the strip. Any cell on the upper border of the strip ignores the term in the recurrence relation for the cell above it (since it is out of the strip); similarly, any cell on the lower border ignores the term in the recurrence relation for the cell to its left. If $m = n$, the size of the strip can be reduced by half (Exercise 4).

If there is no global alignment of S_1 and S_2 with k or fewer differences, then the value obtained for cell (n, m) will be greater than k. That value, greater than k, is not necessarily the correct edit distance of S_1 and S_2, but it will indicate that the correct value for (n, m) is greater than k. Conversely, if there is a global alignment with $d \leq k$ differences, then the corresponding path is contained inside the strip and so the value in cell (n, m) will be correctly set to d. The total area of the strip is $O(kn)$ which is $O(km)$, because n and m can differ by at most k. In summary, we have

Theorem 12.2.1. *There is a global alignment of S_1 and S_2 with at most k differences if and only if the above algorithm assigns a value of k or less to cell (n, m). Hence the k-difference global alignment problem can be solved in $O(km)$ time and $O(km)$ space.*

What if k is not specified?

The solution presented above can be used in somewhat different context. Suppose the edit distance of S_1 and S_2 is k^*, but we don't know k^* or any bound on it ahead of time. The straightforward dynamic programming solution to compute the edit distance, k^*, takes $\Theta(nm)$ time and space. We will reduce those bounds to $\Theta(k^*m)$. So when the edit distance is small, the method runs fast and uses little space. When the edit distance is large, the method only uses $O(nm)$-time and space, the same as for the standard dynamic programming solution.

The idea is to successively guess a bound k on k^* and use Theorem 12.2.1 to determine if the guessed bound is big enough. In detail, the method starts with $k = 1$ and checks if there is a global alignment with at most one difference. If so, then the best global alignment (with zero or one difference) has been found. If not, then the method doubles k and again checks if there is a k-difference global alignment. At each successive iteration the method doubles k and checks whether the current k is sufficient. The process continues until a global alignment is found that has at most k differences, for the current value of k. When the method stops, the best alignment in the present strip (of width k on either side of the main diagonal) must have value k^*. The reason is that the alignment paths are divided into two types: those contained entirely in the present strip and those that go out of the strip. The alignment in hand is the best alignment of the first type, and any path that goes out of the strip specifies an alignment with more than k spaces. It follows that the current value of cell (n, m) must be k^*.

Theorem 12.2.2. *By successively doubling k until there is a k-difference global alignment, the edit distance k^* and its associated alignment are computed in $O(k^*m)$ time and space.*

PROOF Let k' be the largest value of k used in the method. Clearly, $k' \leq 2k^*$. So the total work in the method is $O(k'm + k'm/2 + k'm/4 + \cdots + m) = O(k'm) = O(k^*m)$. □

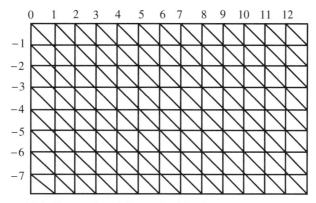

Figure 12.4: The numbered diagonals of the dynamic programming table.

12.2.4. The return of the suffix tree: k-difference inexact matching

We now consider the problem of inexactly matching a pattern P to a text T, when the number of differences is required to be at most k. This is an extension of the k-mismatch problem but is more difficult because it allows spaces in addition to mismatches. The k-mismatch problem was solved using suffix trees alone, but suffix trees are not well structured to handle insertion and deletion errors. The k-difference inexact matching problem is also more difficult than the k-difference global alignment problem because we seek an alignment of P and T in which the end spaces occurring in T are not counted. Therefore, the sizes of P and T can be very different, and we cannot restrict attention to paths that stay within k cells of the main diagonal.

Even so, we will again obtain an $O(km)$ time and space method, combining dynamic programming with the ability to solve longest common extension queries in constant time (see Section 9.1). The resulting solution will be the first of several examples of *hybrid dynamic programming*, where suffix trees are used to solve subproblems within the framework of a dynamic programming computation. The $O(km)$-time result was first obtained by Landau and Vishkin [287] and Myers [341] and extended in a number of papers. Good surveys of many methods for this problem appear in [93] and [421].

Definition   As before, the *main diagonal* of the n by m dynamic programming table consists of cells (i, i) for $0 \leq i \leq n \leq m$. The diagonals above the main diagonal are numbered 1 through m; the diagonal starting in cell $(0, i)$ is diagonal i. The diagonals below the main diagonal are numbered -1 through $-n$; the diagonal starting in cell $(i, 0)$ is diagonal $-i$. (See Figure 12.4.)

Since end spaces in the text T are free, row zero of the dynamic programming table is initialized with all zero entries. That allows a left end of T to be opposite a gap without incurring any penalty.

Definition   A d-path in the dynamic programming table is a path that starts in row zero and specifies a total of exactly d mismatches and spaces.

Definition   A d-path is *farthest-reaching in diagonal i* if it is a d-path that ends in diagonal i, and the index of it's ending column c (along diagonal i) is greater than or equal to the ending column of any other d-path ending in diagonal i.

Graphically, a d-path is farthest reaching in diagonal i if no other d-path reaches a cell further along diagonal i.

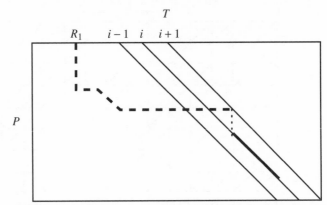

Figure 12.5: Path R_1 consists of a farthest-reaching $(d-1)$-path on diagonal $i+1$ (shown with dashes), followed by a vertical edge (dots), which adds the dth difference to the alignment, followed by a maximal path (solid line) on diagonal i that corresponds to (maximal) identical substrings in P and T.

Hybrid dynamic programming: the high-level idea

At the high level, the $O(km)$ method will run in k iterations, each taking $O(m)$ time. In every iteration $d \leq k$, the method finds the end of the farthest-reaching d-path on diagonal i, for each i from $-n$ to m. The farthest-reaching d-path on diagonal i is found from the farthest-reaching $(d-1)$-paths on diagonals $i-1$, i, and $i+1$. This will be explained in detail below. Any farthest-reaching d-path that reaches row n specifies the end location (in T) of an occurrence of P with exactly d differences. We will implement each iteration in $O(n+m)$ time, yielding the desired $O(km)$-time bound. Space will be similarly bounded.

Details

To begin, when $d = 0$, the farthest-reaching 0-path ending on diagonal i corresponds to the longest common extension of $T[i..m]$ and $P[1..n]$, since a 0-path allows no mismatches or spaces. Therefore, the farthest-reaching 0-path ending on diagonal i can be found in constant time, as detailed in Section 9.1.

For $d > 0$, the farthest-reaching d-path on diagonal i can be found by considering the following three particular paths that end on diagonal i.

- Path R_1 consists of the farthest-reaching $(d-1)$-path on diagonal $i+1$, followed by a vertical edge (a space in text T) to diagonal i, followed by the maximal extension along diagonal i that corresponds to identical substrings in P and T. (See Figure 12.5). Since R_1 begins with a $(d-1)$-path and adds one more space for the vertical edge, R_1 is a d-path.

- Path R_2 consists of the farthest-reaching $(d-1)$-path on diagonal $i-1$, followed by a horizontal edge (a space in pattern P) to diagonal i, followed by the maximal extension along diagonal i that corresponds to identical substrings in P and T. Path R_2 is a d-path.

- Path R_3 consists of the farthest-reaching $(d-1)$-path on diagonal i, followed by a diagonal edge corresponding to a mismatch between a character of P and a character of T, followed by a maximal extension along diagonal i that corresponds to identical substrings from P and T. Path R_3 is a d-path. (See Figure 12.6.)

Each of the paths R_1, R_2, and R_3 ends with a maximal extension corresponding to identical substrings of P and T. In the case of R_1 (or R_2), the starting positions of the two substrings are given by the last entry point of R_1 (or R_2) into diagonal i. In the case of R_3, the starting position is the position just past the last mismatch on R_3.

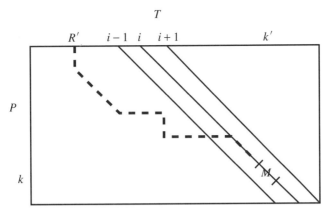

Figure 12.6: The dashed line shows path R', the farthest-reaching $(d-1)$-path ending on diagonal i. The edge M on diagonal i just past the end of R' must correspond to a mismatch between P and T (the characters involved are denoted $P(k)$ and $T(k')$ in the figure).

Theorem 12.2.3. *Each of the three paths R_1, R_2, and R_3 are d-paths ending on diagonal i. The farthest-reaching d-path on diagonal i is the path R_1, R_2, or R_3 that extends the farthest along diagonal i.*

PROOF   Each of the three paths is an extension of a $(d-1)$-path, and each extension adds either one more space or one more mismatch. Hence each is a d-path, and each ends on diagonal i by definition. So the farthest-reaching d-path on diagonal i must either be the farthest-reaching of R_1, R_2, and R_3, or it must reach farther on diagonal i than any of those three paths.

Let R' be the farthest-reaching $(d-1)$-path on diagonal i. The edge of the alignment graph along diagonal i that immediately follows R' must correspond to a mismatch, otherwise R' would not be the farthest-reaching $(d-1)$-path on i. Let M denote that edge (see Figure 12.6).

Let R^* denote the farthest-reaching d-path on diagonal i. Since R^* ends on diagonal i, there is a point where R^* enters diagonal i for the last time and then never leaves diagonal i. If R^* enters diagonal i for the last time above edge M, then R^* must traverse edge M, otherwise R^* would not reach as far as R_3. When R^* reaches M (which marks the end of R'), it must also have $(d-1)$ differences; if that portion of R^* had less than a total of $(d-1)$ differences, then it could traverse M creating a $(d-1)$-path on diagonal i that reached farther on diagonal i than R', contradicting the definition of R'. It follows that if R^* enters diagonal i above M, then it will have d differences after it traverses M, and so it will end exactly where R_3 ends. So if R^* is not R_3, then R^* must enter diagonal i *below* edge M.

Suppose R^* enters diagonal i for the last time below edge M. Then R^* must have d differences, at that point of entry; if it had fewer differences then R' would again fail to be the farthest-reaching $(d-1)$-path on diagonal i. Now R^* enters diagonal i for the last time either from diagonal $i-1$ or diagonal $i+1$, say $i+1$ (the case of $i-1$ is symmetric). So R^* traverses a vertical edge from diagonal $i+1$ to diagonal i, which adds a space to R^*. That means that the point where R^* ends on diagonal $i+1$ defines a $(d-1)$-path on diagonal $i+1$. Hence R^* leaves diagonal $i+1$ at or above the point where the path R_1 does. Then R_1 and R^* each have d spaces or mismatches at the points where they enter diagonal i for the last time, and then they each run along diagonal i until reaching an edge corresponding to a mismatch. It follows that R^* cannot reach farther along diagonal i then R_1 does. So in this case, R^* ends exactly where R_1 ends.

The case that R^* enters diagonal i for the last time from diagonal $i - 1$ is symmetric, and R^* ends exactly where R_2 ends. In each case we have shown that R^*, the assumed farthest-reaching d-path on diagonal i, ends at the ending point of either R_1, R_2, or R_3. Hence the farthest-reaching d-path on diagonal i is the farthest-reaching of R_1, R_2, and R_3.   □

Theorem 12.2.3 is the key to the $O(km)$-time method.

Hybrid dynamic programming: k-differences algorithm

begin
$d := 0$
for $i := 0$ to m do
find the longest common extension between $P[1..n]$ and $T[i..m]$. This specifies the end column of the farthest-reaching 0-path on diagonal i.

For $d = 0$ to k do
    begin

    For $i = -n$ to m do
        begin
        using the farthest-reaching $(d - 1)$-paths on diagonals $i, i - 1$, and $i + 1$,
        find the end, on diagonal i, of paths R_1, R_2, and R_3. The farthest-reaching
        of these three paths is the farthest-reaching d-path on diagonal i;
        end;
    end;
    Any path that reaches row n in column c say, defines an inexact match of P in
    T that ends at character c of T and that contains at most k differences.
end.

Implementation and time analysis

For each value of d and each diagonal i, we record the column in diagonal i where the farthest-reaching d-path ends. Since d ranges from 0 to k and there are only $O(n + m)$ diagonals, all of these values can be stored in $O(km)$ space. In iteration d, the algorithm only needs to retrieve the values computed in iteration $(d - 1)$. The entire set of stored values can be used to reconstruct any alignment of P in T with at most k differences. We leave the details of that reconstruction as an exercise.

Now we proceed with the time analysis. For each d and each i, the end of three particular $(d - 1)$-paths must be retrieved. For a fixed d and i, this takes constant time, so these retrievals take $O(km)$-time over the entire algorithm. There are also $O(km)$ path extensions, each along a diagonal, that must be computed. But each path extension corresponds to a maximal identical substring in P and T starting at particular known positions in P and T. Hence each path extension requires finding the longest substring starting at a given location in T that matches a substring starting at a given location of P. In other words, each path extension requires a *longest common extension* computation. In Section 9.1 on page 196 we showed that any longest common extension computation can be done in constant time, after linear preprocessing of the strings. Hence the $O(km)$ extensions can all be computed in $O(n + m + km) = O(km)$ total time. Furthermore, as shown in Section 9.1.2, these extensions can be implemented using only a copy of the two strings and a suffix tree for the smaller of the two strings. In summary, we have

Theorem 12.2.4. *All locations in T where pattern P occurs with at most k differences can be found in $O(km)$-time and $O(km)$ space. Moreover, the actual alignment of P and T for each of these locations can be reconstructed in $O(km)$ total time.*

Sometimes this k differences result is reported in a somewhat simpler but less useful form, requiring less space. If one is only interested in the *end locations* in T where P inexactly matches in T with at most k differences, then the $O(km)$ space bound can be reduced to $O(n + m)$. The idea is that the ends of the farthest-reaching $(d - 1)$-paths in each diagonal would then not be needed after iteration d and could be discarded. Thus only $O(n + m)$ space is needed to solve the simpler problem.

Theorem 12.2.5. *In $O(km)$-time and $O(n + m)$ space, the algorithm can find all the end locations in T where P matches T with at most k differences.*

12.2.5. The primer (and probe) selection problem revisited – An application of bounded difference matching

In Exercise 61 of Chapter 7, we introduced an exact matching version of the *primer (and probe) selection problem*. The simplest version of that problem starts with two strings α and β. The exact matching version is:

Exact matching primer (and probe) problem   For each index j past some starting point, find the shortest substring γ of α (if any) that begins at position j and that does *not* appear as a substring of β.

That problem can be solved in time proportional to the sum of the lengths, of α and β.

The exact matching version of the primer selection problem may not fully model the real primer selection problem (although as noted earlier, the exact matching version may be realistic for probe selection). Recall that primers are short substrings of DNA that *hybridize* to the desired part of string α and that ideally should not hybridize to any parts of another string β. Exact matching is not an adequate model of practical hybridization because a substring of DNA can hybridize, under the right conditions, to another string of DNA even without exact matching; inexact matching of the right type may be enough to allow hybridization. A more realistic version of the primer selection problem moves from exact matching to inexact matching as follows:

Inexact matching primer problem   Given a parameter p, find for each index j (past some starting point), the shortest substring γ of α (if any) that begins at position j and that has *edit distance* at least $|\gamma|/p$ from any substring in β.

We solve the above problem efficiently by solving the following-k-difference problem:

k-difference primer problem   Given a parameter k, find for each index j (past some starting point), the shortest substring γ of α (if any) that begins at position j and that has edit distance at least k from any substring in β.

Changing $|\gamma|/p$ to k in the problem statement (converting the Inexact matching primer problem to the k-difference primer problem) makes the solution easier but does not reduce the utility of the solution. The reason is that the length of a practical primer must be within a fixed and fairly narrow range, so for fixed p, $|\gamma|/p$ also falls in a small range. Hence for a specified p, the k-difference primer problem can be solved for a small range of choices for k and still be expected to pick out useful primer candidates.

How to solve the k-difference primer problem

We follow the approach introduced in [243]. The method examines each position j in α separately. For any position j, the k-difference primer problem becomes:

> Find the shortest prefix of string $\alpha[j..n]$ (if it exists) that has edit distance at least k from *every* substring in β.

The problem for a fixed j is essentially the "reverse" of the k-differences inexact matching problem. In the k-difference inexact matching problem we want to find the substrings of T that P matches, with *at most* k differences. But now, we want to *reject* any prefix of $\alpha[j..n]$ that matches a substring of β with less than k differences. The viewpoint is reversed, but the same machinery works.

The solution is to run the k-differences algorithm with string $\alpha[j..n]$ playing the role of P and β playing the role of T. The algorithm computes the farthest-reaching d-paths, for $d = k$, in each diagonal. If row n is reached by any d-path for $d \leq k - 1$, then the entire string $\alpha[j..n]$ matches a substring of β with less than k differences, so no acceptable primer can start at j. But, if none of the farthest-reaching $(k - 1)$-paths reach row n, then there is an acceptable primer starting at position j. In detail, if none of the farthest-reaching of the d-paths for $d = k - 1$ reach row $r < n$, then the substring $\gamma = \alpha[j..r]$ has edit distance at least k from every substring in β. Moreover, if r is the smallest row with that property, then $\alpha[j..r]$ is the shortest substring starting at j that has edit distance at least k from every substring in β.

The above algorithm is applied to each potential starting position j in α, yielding the following theorem:

Theorem 12.2.6. *If α has length n and β has length m, then the k-differences primer selection problem can be solved in $O(knm)$ total time.*

12.3. Exclusion methods: fast expected running time

The k-mismatch and k-difference methods we have presented so far all have worst-case running times of $\Theta(km)$. For $k \ll n$, these speedups are significant improvements over the $\Theta(nm)$ bound for straight dynamic programming. Still, even greater efficiency is desired when m (the size of the text T) is large. The typical situation is that T represents a large database of sequences, and the problem is to find an approximate match of a pattern P in T. The goal is to obtain methods that are significantly faster than $\Theta(km)$ not in worst case, but in *expected* running time. This is reminiscent of the way that the Boyer–Moore method, which typically skips over a large fraction of the text, has an expected running time that is sublinear in the size of the text.

Several methods have been devised for approximate matching problems whose expected running times are faster than $\Theta(km)$. In fact, some of the methods have an expected running time that is *sublinear* in m, for a reasonable range of k. These methods artfully mix *exact* matching with dynamic programming and explicitly use many of the ideas in Parts I and II of the book. Although the details differ considerably, all the methods we will discuss have a similar high-level flavor. We focus on methods due to Baeza-Yates and Perleberg [36], Chang and Lawler [94], and Myers [342], although only the first method will be explained and analyzed in full detail. Two other methods (Wu-Manber [482] and Pevzner-Waterman [373]) will also be mentioned. These methods do not completely achieve the goal of *provable* linear and sublinear expected running times for all practical ranges of errors (and this remains a superb open problem), but they do achieve the goal when the error rate k/n is "modest".

Let σ be the size of the alphabet used in P and T. As usual, n is the length of P and m is the length of T. For the general discussion, an occurrence of P in T with at most k errors (mismatches or differences depending on the particular problem) will be called an *approximate occurrence* of P. The high-level outline of most of the methods is the following:

Partition approach to approximate matching

a. **Partition** T or P into consecutive regions of a given length r (to be specified later).

b. **Search phase** Using various exact matching methods, search T to find length-r intervals of T (or regions, if T was partitioned) that could be contained in an approximate occurrence of P. These are called *surviving* intervals. The nonsurviving intervals are definitely not contained in any approximate occurrence of P, and the goal of this phase is to eliminate as many intervals as possible.

c. **Check phase** For each surviving interval R of T, use some approximate matching method to explicitly check if there is an approximate occurrence of P in a larger interval around R.

The methods differ primarily in the choice of r, in the choice of string to partition, and in the exact matching methods used in the search phase. The methods also differ in the definition of a region but are not generally affected by the specific choice of checking algorithm. The point of the partition approach is to exclude a large amount of T, using only (sub)linear expected time in the search phase, so that only (sub)linear expected time is needed to check the few surviving intervals. A balance is needed between searching and checking because a reduction in the time used in one phase causes an increase in the time used in the other phase.

12.3.1. The BYP method

The first specific method we will look at is due to R. Baeza-Yates and C. Perleberg [36]. Its expected running time is $O(m)$ for modest error rates (made precise below).

Let $r = \lfloor \frac{n}{k+1} \rfloor$, and partition P into consecutive r-length regions (the last region may be of length less than r). By the choice of r, there are $k + 1$ regions that have the full length r. The utility of this partition is suggested in the following lemma.

Lemma 12.3.1. *Suppose P matches a substring T' of T with at most k differences. Then T' must contain at least one interval of length r that exactly matches one of the r-length regions of the partition of P.*

PROOF In the alignment of P to T', each region of P aligns to some part of T' (see Figure 12.7), defining $k + 1$ subalignments. If each of those $k + 1$ subalignments were to contain at least one error (mismatch or space), then there would be more than k differences in total, a contradiction. Therefore, one of the first $k + 1$ regions of P must be aligned to an interval of T' without any errors. □

Note that the lemma also holds even for the k-mismatch problem (i.e., when no space

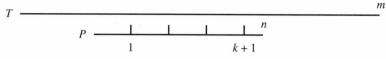

Figure 12.7: The first $k + 1$ regions of P are each of length $r = \lfloor \frac{n}{k+1} \rfloor$.

insertions are allowed). Lemma 12.3.1 leads to the following approximate matching algorithm:

Algorithm BYP

a. Let $\mathcal{P}$ be the set of $k + 1$ substrings of P taken from the first $k + 1$ regions of P's partition.

b. Build a keyword tree (Section 3.4) for the set of "patterns" $\mathcal{P}$.

c. Using the Aho–Corasik algorithm (Section 3.4), find $\mathcal{I}$, the set of all starting locations in T where any pattern in $\mathcal{P}$ occurs exactly.

d. For each index $i \in \mathcal{I}$ use an approximate matching algorithm (usually based on dynamic programming) to locate the end points of all approximate occurrences of P in the substring $T[i - n - k..i + n + k]$ (i.e., in an appropriate-length interval around i).

By Lemma 12.3.1, it is easy to establish that the algorithm correctly finds all approximate occurrences of P in T. The point is that the interval around each i is "large enough" to align with any approximate occurrence of P that spans i, and there can be no approximate occurrence of P outside such an interval. A formal proof is left as an exercise. Now we focus on specific implementation details and time analysis.

Building the keyword tree takes $O(n)$ time, and the Aho-Corasik algorithm takes $O(m)$ (worst-case) time (Section 3.4). So steps b and c take $O(n + m)$ time. There are a number of alternate implementations for steps b and c. One is to build a suffix tree for T, and then use it to find every occurrence in T of a pattern in $\mathcal{P}$ (see Section 7.1). However, that would be very space intensive. A space-efficient version of this approach is to construct a generalized suffix tree for only $\mathcal{P}$, and then match T to it (in the way that matching statistics are computed in Section 7.8.1). Both approaches take $\Theta(n + m)$ worst-case time, but are no faster in expected time because every character in T is examined. A faster approach in practice is to use the Boyer–Moore *set matching* method based on suffix trees, which was developed in Section 7.16. That algorithm will skip over parts of T, and hence it breaks the $\Theta(m)$ bottleneck. A different variation was developed by Wu and Manber [482] who implement steps b and c using the *Shift-And* method (Section 4.2) on a set of patterns. Another approach, found in the paper of Pevzner and Waterman [373] and elsewhere, uses *hashing* to identify long exact matching substrings of P and T. Of course, one can use suffix trees to find long common substrings, and one could develop a Karp–Rabin type method as well. Hashing, or approaches based on suffix trees, that look directly for long common substrings between P and T, seem a bit more robust than BYP because there is no string partition involved. But the only stated time bounds in [373] are the same as those for BYP.

In the checking phase, step d, the algorithm executes some approximate matching algorithm between P and an interval of T of length $O(n)$, for each index in $\mathcal{I}$. Naively, each of these checks can be done in $O(n^2)$ time by dynamic programming (global alignment). Even this time bound will be adequate to establish an expected $O(m)$ overall running time for the range of error rates that will be detailed below. Alternately, the Landau–Vishkin method (Section 12.2) based on suffix trees could be used, so that each check takes only $O(kn)$ worst-case time. If no spaces are allowed in the alignment of P to T' (only matches and mismatches) then the simpler $O(kn)$-time approach based on longest common extension (Section 9.1) can be used, or if attention is paid to exactly where in P any match is found, then $O(n)$ time suffices for each check.

12.3.2. Expected time analysis of algorithm BYP

Since steps b and c run in $O(m)$ worst-case time, we only need to analyze step d. The key is to estimate the expected size of set $\mathcal{I}$.

In the following analysis, we assume that each character of T is drawn uniformly (i.e., with equal probability) from an alphabet of size σ. However, P can be an arbitrary string. Consider any pattern $p \in \mathcal{P}$. Since p has length r, and T contains roughly m substrings of length r, the expected number of exact occurrences of p in T is m/σ^r. Therefore, the expected total number of occurrences in T of patterns from $\mathcal{P}$ (i.e., the expected size of $\mathcal{I}$) is $m(k+1)/\sigma^r$.

For each $i \in \mathcal{I}$, the algorithm spends $O(n^2)$ time (or less if faster methods are used) in the checking phase. So the expected checking time is $mn^2(k+1)/\sigma^r$. The goal is to make the expected checking time linear in m for modest k, so we must determine what values of k make

$$\frac{mn^2(k+1)}{\sigma^r} < cm,$$

for some constant c.

To simplify the analysis, replace k by $n-1$, and solve for r in

$$\frac{mn^3}{\sigma^r} = cm.$$

This gives $\sigma^r = \frac{n^3}{c}$, so $r = \log_\sigma n^3 - \log_\sigma c$. But $r = \lfloor \frac{n}{k+1} \rfloor$, so

Theorem 12.3.1. *Algorithm BYP runs in $O(m)$ time for $k = O\left(\frac{n}{\log n}\right)$.*

Stated another way, as long as the error rate is less than one in $\log_\sigma n$ characters, algorithm BYP will run in linear time as a function of m.

The bottleneck in the BYP method is the $\Theta(m)$ time required to run the Aho–Corasik algorithm. Using the Boyer–Moore set matching method should reduce that time in practice, but we cannot present a time analysis for that approach. However, the Chang–Lawler method has an expected time bound that is provably sublinear for $k = O\left(\frac{n}{\log n}\right)$.

12.3.3. The Chang–Lawler method

For ease of exposition, we will explain the Chang–Lawler (CL) method [94] for the k-mismatches problem; we leave the extension to k-differences as an exercise.

In CL, it is string T, not P, that is partitioned into consecutive fixed regions of length $r = n/2$. These regions are large compared to the regions in BYP. The purpose of the length $n/2$ is to assure that no matter how P is aligned to T (without inserted spaces), at least one of the fixed regions in T's partition is completely contained in the interval spanned by P (see Figure 12.8). Therefore, if P occurs in T with at most k mismatches, there must be one region of T that is spanned by that occurrence of P and, of course, that region matches its counterpart in P with at most k mismatches. Based on this observation, the search phase of CL examines each region in the partition of T to find regions that cannot match

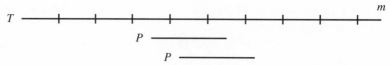

Figure 12.8: Each full region in T has length $r = n/2$. This assures that no matter how P is aligned with T, P spans one full region.

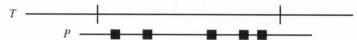

Figure 12.9: Blowup of one region in T aligned with one copy of P. Each black box shows a mismatch between a character in P and its counterpart in T.

any substring of P with at most k mismatches. These regions are excluded, and then an interval around each surviving region is checked using an approximate matching method, as in BYP. The search phase of CL relies heavily on the *matching statistics* discussed in Section 7.8.1.

Recall that the value of matching statistic $ms(i)$ is the length of the longest substring starting at position i of T that matches a substring *somewhere* (an unspecified location) in P. Recall also, that for any string S, all the matching statistics for the positions in S can be computed in $O(|S|)$ total time. This is true even when S is a substring of a larger string T.

Now let T' be the substring of one of the regions of T's partition that matches a substring P' of P with at most k mismatches (see Figure 12.9). The alignment of P' and T' can be divided into at most $k + 1$ intervals where no mismatches occur, alternating with intervals containing only mismatches. Let i be the starting position of any one of those matching intervals, and let l be its length. Then clearly, $ms(i) \geq l$. The CL search phase exploits this observation. It executes the following algorithm for each region R in the partition of T:

The CL search in region R

    Set j to the starting position j^* of region R in T.
    $cn := 0$;
    Repeat
    compute $ms(j)$;
    $j := j + ms(j) + 1$;
    $cn := cn + 1$;
    Until $cn = k$ or $j - j^* > n/2$.
    If $j - j^* > n/2$ then region R survives, otherwise it is excluded.

If R is a surviving region, then in the checking phase CL executes an approximate matching algorithm for P against a neighborhood of T that starts $n/2$ positions to the left of R and ends $n/2$ positions to its right. This neighborhood is of size $3n/2$, and so each check can be executed in $O(kn)$ time.

The correctness of the CL method comes from the following lemma, and the fact that the neighborhoods are "large enough".

Lemma 12.3.2. *When the CL search declares a region R excluded, then there is no occurrence of P in T with at most k mismatches that completely contains region R.*

The proof is easy and is left to the reader, as is its use in a formal proof of the correctness of CL. Now we consider the time analysis.

The CL search is executed on $2m/n$ regions of T. For any region R let j' be the last value of j (i.e., the value of j when cn reaches k or when $j - j^*$ exceeds $n/2$). Thus, in R, matching statistics are computed for the interval of length $j' - j^* \leq n/2$. With the matching statistics algorithm in Section 7.8.1, the time used to compute those matching statistics is $O(j' - j^*)$. Now the expected value of $j' - j^*$ is less than or equal to k times the expected value of $ms(i)$, for any i. Let $E(M)$ denote the expected value of a matching statistic, and let e denote the expected number of regions that survive the search phase. Then the expected time for the search phase is $O(2mkE(M)/n)$, and the expected time for the checking phase is $O(kne)$.

In the following analysis, we assume that P is a random string where each character is chosen uniformly from an alphabet of size σ.

Lemma 12.3.3. $E(M)$, the expected value of a matching statistic, is $O(\log_\sigma n)$.

PROOF   For fixed length d, there are roughly n substrings of length d in P, and there are σ^d substrings of length d that can be constructed. So, for any specific string α of length d, the probability that α is found somewhere in P is *less than* n/σ^d. This is true for any d, but vacuously true until $\sigma^d = n$ (i.e., when $d = \log_\sigma n$).

Let X be the random variable that has value $\log_\sigma n$ for $ms(i) \leq \log_\sigma n$; otherwise it has value $ms(i)$. Then

$$E(M) < E(X) < \log_\sigma n + \sum_{l=\log_\sigma n}^{\infty} \frac{l}{\sigma^l} = \log_\sigma n + 2.$$

□

Corollary 12.3.1.   The expected time that CL spends in the search phase is $O(2mk \log_\sigma n/n)$, which is sublinear in m for $k < n/\log_\sigma n$.

The analysis for e, the expected number of surviving regions is too difficult to present here. It is shown in [94] that when $k = O(n/\log_\sigma n)$, then $e = m/n^4$, so the expected time that CL spends in the checking phase is $O(km/n^3) = o(m)$. The search phase of CL is so effective in excluding regions of T that the checking phase has very small expected running time.

12.3.4. Multiple filtration for *k*-mismatches

Both the BYP and the CL methods use fairly simple combinatorial criteria in their search phases to exclude intervals of T. One can devise more stringent conditions that are *necessary* for an interval of T to be contained in an approximate occurrence of P. In the context of the k-mismatches problem, conditions of this type (called filtration conditions) were developed and studied by Pevzner and Waterman [373]. These conditions are used together with substring hashing to obtain another linear expected-time method for the k-mismatch problem. Empirical results are given in [373] that show faster running times in practice than other methods for the k-mismatch problem.

12.3.5. Myers's sublinear-time method

Gene Myers [342, 337] developed an exclusion method that is more sophisticated than the ones we have discussed so far and that runs in sublinear time for a wider range of error rates. The method handles approximate matching with insertions and deletions as well as mismatches. The full algorithm and its analysis are too complex for detailed discussion here, but we can introduce some of the ideas it uses to address deficiencies in the other exclusion methods.

There are two basic problems with the Baeza-Yates-Perlberg and the Chang–Lawler methods (and the other exclusion methods we have mentioned). First, the exclusion criteria they use permit a large expected number of surviving regions compared to the expected number of true approximate matches. That is, not every initial surviving region is actually contained in an approximate match, and the ratio of expected survivors to expected matches is fairly high (for random patterns and text). Further, the higher the permitted error rate, the more severe is the problem. Second, when a surviving region is first located, the methods move directly to full dynamic programming computations (or some other relatively expensive operations) to check for an approximate match in a large interval around the surviving region. Hence the methods are required to do a large amount of computation for a large number of intervals that don't contain any approximate match.

Compared to the other exclusion methods, Myers's method contains two different ideas to make it both more selective (finding fewer initial surviving regions) and less expensive to test the ones that are found. Myers's algorithm begins in a manner similar to the other exclusion methods. It partitions P into short substrings (to be specified later) and then finds all locations in T where these substrings appear with a small number of allowed differences. The details of the search are quite different from the other methods, but the intent (to exclude a large portion of T from further consideration) is the same. Each of these initial alignments of a substring of P that is found (approximately) in T is called a *surviving match*. A surviving match roughly plays the role of a surviving *region* in the other exclusion methods, but it specifies two substrings (one in P and one in T) rather than just a single substring, as a surviving region does. Another way to think of a surviving region is as a roughly diagonal subpath in the alignment graph for P and T.

Having found the initial surviving matches (or surviving regions), all the other exclusion methods we have mentioned would next check a full interval of length roughly $2n$ around each surviving region in T to see if it contains an approximate match to P. In contrast, Myers's method will *incrementally extend* and check a growing interval around each initial surviving match to create longer surviving matches or to exclude a surviving match from further consideration. This is done in about $O(\log n)$ iterations. (Recall that n is the length of the pattern and m is the length of the text.)

Definition For a given error rate ϵ, a string S ϵ-matches a substring of T if S matches the substring using at most $\epsilon|S|$ insertions, deletions, and mismatches.

For example, let $S = aba$ and $\epsilon = 2/3$. Then ac ϵ-matches S using one mismatch and one deletion operation.

In the first iteration, the pattern P is partitioned into consecutive, nonoverlapping subpatterns of length $\log_\sigma m$ (assumed to be an integer), and the algorithm finds all substrings in T that ϵ-match one of these short subpatterns (discussed in more detail below). The length of these subpatterns is short enough that all the ϵ-matches can be found in sublinear expected time for a wide range of ϵ values. These ϵ-matches are the initial surviving matches.

The algorithm next tries to extend each initial surviving match to become an ϵ-match between substrings (in P and T) that are roughly twice as long as those in the current surviving match. This is done by dynamic programming in an appropriate interval around the surviving match. In each successive iteration, the method applies a more selective and expensive filter, trying to double the length of the ϵ-match around each surviving match.

Since the intervals of interest double in length, the time used per interval grows four fold in each successive iteration. However, the number of surviving matches is expected to fall hyper-exponentially in each successive iteration, more than offsetting the increase in computation time per interval.

With this iterative expansion, the effort expended to check any initial surviving match is doled out incrementally throughout the $O\left(\log\frac{n}{\log m}\right)$ iterations, and is not continued for any surviving match past an iteration where it is excluded. We now describe in a bit more detail how the initial surviving matches are found and how they are incrementally extended in successive iterations.

The first iteration

Definition   For a string S and value of ϵ, let $d = \epsilon|S|$. The *d-neighborhood* of S is the set of all strings that ϵ-match S.

For example, over the two-letter alphabet $\{a,b\}$, if $S = aba$ and $d = 1$, then the *1-neighborhood* of S is *{bba, aaa, abb, aaba, abaa, baba, abba, abab, ba, aa, ab}*. It is created from S by the operations of mismatch, insertion and deletion respectively. The *condensed d-neighborhood* of S is created from the d-neighborhood of S by removing any substring that is a *prefix* of another string in the d-neighborhood. The condensed 1-neighborhood S is *{bba, aaa, aaba, abaa, baba, abba, abab}*.

Recall that pattern P is initially partitioned into subpatterns of length $\log_\sigma m$ (assumed to be an integer). Let $\mathcal{P}$ be the set of these subpatterns. In the first iteration, the algorithm (conceptually) constructs the condensed d-neighborhood for each subpattern in $\mathcal{P}$, and then finds all locations of substrings in text T that *exactly* match one of the substrings in one of the condensed d-neighborhoods. In this way, the method finds all substrings of T that ϵ-match one of the subpatterns in $\mathcal{P}$. These ϵ-matches form the initial surviving matches.

In actuality, the tasks of generating the substrings in the condensed d-neighborhoods and of searching for their exact occurrences in T are intertwined and require text T to have been preprocessed into some index structure. This structure could be a suffix tree, a suffix array or a hash table holding short substrings of T. Details are found in [342].

Myers [342] shows that when the length of the subpatterns is $O(\log_\sigma m)$, then the first iteration can be implemented to run in $O(km^{p(\epsilon)}\log m)$ expected time. The function $p(\epsilon)$ is complicated, but it is convex (negative second derivative) increasing, and increases more slowly as the alphabet size grows. For DNA, it has value less than one for $\epsilon \leq \frac{1}{3}$, and for proteins it has value less than one for $\epsilon \leq 0.56$.

Successive iterations

To explain the central idea, let $\alpha = \alpha_0\alpha_1$, where $|\alpha_0|$ is assumed equal to $|\alpha_1|$.

Lemma 12.3.4. *Suppose α ϵ-matches β. Then β can be divided into two substrings β_0 and β_1 such that $\beta = \beta_0\beta_1$, and either α_0 ϵ-matches β_0 or α_1 ϵ-matches β_1.*

This lemma (used in reverse) is the key to determining how to expand the intervals around the surviving matches in each iteration. For simplicity, assume that n is a power of two and that $\log_\sigma m$ is also a power of two. Let B be a binary tree representing successive divisions of P into two equal size parts, until each part has length $\log_\sigma m$ (see Figure 12.10). The substrings written at the leaves are the subpatterns used in the first iteration of Myers's algorithm. Iteration i of the algorithm examines substrings of P that label (some) nodes of B i levels above the leaves (counting the leaves as level 1).

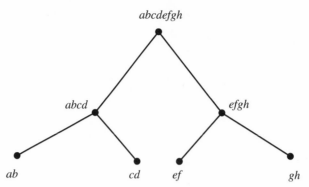

Figure 12.10: Binary tree B defining the successive divisions of P and its partition into regions of length $\log_\sigma m$ (equal to two in this figure).

Suppose at iteration $i - 1$ that substrings P' and T' in the query and text, respectively, form a surviving match (i.e., are found to align to form an ϵ-match). Let P'' be the parent of P' in tree B. If P' is a left child of P'', then in iteration i, the algorithm tries to ϵ-match P'' to a substring of T in an interval that extends T' to the right. Conversely, if P' is a right child of P'', then the algorithm tries to ϵ-match P'' with a substring in an interval that extends T' to its left. By Lemma 12.3.4, if the ϵ-match of P' to T' is part of an ϵ-match of P to a substring of T, then P'' will ϵ-match the appropriate substring of T. Moreover, the specified interval in T that must be compared against P'' is just twice as long as the interval for T'. The end result, as detailed in [342], is that all of the checking, and hence the entire algorithm, runs in $O(km^{p(\epsilon)} \log m)$ expected time.

Final comments on Myers's method

There are several points to emphasize. First, the exposition given above is only intended to be an outline of Myers's method, without any analysis. The full details of the algorithm and analysis are found in [342]; [337] provides an overview, in relation to other exclusion methods. Second, unlike the BYP and CL methods, the error rates that establish sublinear (or linear) running times do not depend on the length of P. In BYP and CL, the permitted error rate *decreases* as the length of P increases. In Myers's method, the permitted error rate depends only on the alphabet size. Third, although the expected running times for both CL and for Myers's method are sublinear (for the proper range of error rates), there is an important difference in the nature of these sublinearities. In the CL method, the sublinearity is due to a multiplicative factor that is less than one. But in Myers's method, the sublinearity is due to an *exponent* that is less than one. So as a function of m, the CL bound increases linearly (although for any fixed value of m the expected running time is less than m), while the bound for Myers's method increases sublinearly in m. This is an important distinction since many databases are rapidly increasing in size.

However, Myers's method assumes that the text T has already been preprocessed into some index structure, and the time for that preprocessing (while linear in m) is not included in the above time bounds. In contrast, the running times of the BYP and CL methods include all the work needed for those methods. Finally, Myers has shown that in experiments on problems of meaningful size in molecular biology (patterns of length 80 on texts of length 3 million), the k-difference algorithms of Sections 12.2.4 and 12.2.3 run 100 to 500 times slower than his expected sublinear method.

12.3.6. Final comment on exclusion methods

The fast expected-time exclusion methods have all been developed with the motivation of searching large DNA and protein databases for approximate occurrences of query strings. But the proven results are a bit weak for the case of protein database search, because error rates as high as 85% (the so-called twilight zone) are of great interest when comparing protein sequences [127, 360]. In the twilight zone, evidence of common ancestry may still remain, but it takes some skill to determine if a given match is meaningful or not. Another problem with the exclusion methods presented here is that not all of the methods or analyses extend nicely to the case of weighted or local alignment.

Nonetheless, these results are promising, and the open problem of finding sublinear expected-time algorithms for higher error rates is very inviting. Moreover, we will see in Chapter 15 on database searching that the most effective practical database search methods in use today (BLAST, FASTA, and variants) can be considered as exclusion methods and are based on ideas similar to some of the more formal methods presented here.

12.4. Yet more suffix trees and more hybrid dynamic programming

Although the suffix tree was initially designed and employed to handle complex problems of exact matching, it can be used to great advantage in various problems of *inexact matching*. This has already been demonstrated in Sections 9.4 and 12.2 where the k-mismatch and k-difference problems were discussed. The suffix tree in the latter application was used in combination with dynamic programming to produce a *hybrid dynamic programming* method that is faster than dynamic programming alone. One deficiency of that approach is that it does not generalize nicely to problems of *weighted* alignment. In this section, we introduce a different way to combine suffix trees with dynamic programming for problems of weighted alignment. These ideas have been claimed to be very effective in practice, particularly for large computational projects. However, the methods do not always lend themselves to greatly improved *provable, worst-case* time bounds. The ideas presented here loosely follow the published work of Ukkonen [437] and an unpublished note of Gonnet and Baeza-Yates [34]. The thesis by Bieganski [63] discusses a related idea for using suffix trees in regular expression pattern matching (with errors) and its large-scale application in managing genomic databases. The method of Gonnet and Baeza-Yates has been implemented and extensively used for large-scale protein comparisons [57], [183].

Two problems

We assume the existence of a scoring matrix used to compute the value of any alignment, and hence "edit distance" here refers to *weighted* edit distance. We will discuss two problems in the text and introduce two more related problems in the exercises.

1. **The *P*-against-all problem** Given strings P and T, compute the edit distance between P and every substring T' of T.

2. **The threshold all-against-all problem** Given strings P and T and a threshold d, find every pair of substrings P' of P and T' of T such that the edit distance between P' and T' is less than d.

The threshold all-against-all problem is similar to problems mentioned in Section 12.2.1 concerning the construction of nonredundant sequence databases. However, the threshold all-against-all problem is harder, because it asks for the alignment of all pairs of *substrings*, not just the alignment of all pairs of strings. This critical distinction has been the source of some confusion in the literature [50], [56].

12.4.1. The P-against-all problem

The P-against-all problem is an example of a *large-scale alignment* problem that asks for a great amount of related alignment information. If not done carefully, its solution will involve a large amount of redundant computation.

Assume that P has length n and T has length $m > n$. The most naive solution to the P-against-all problem is to enumerate all $\binom{m}{2}$ substrings of T, and then separately compute the edit distance between P and each substring of T. This takes $\Theta(nm^3)$ total time. A moment's thought leads to an improvement. Instead of choosing all substrings of T, we need only choose each *suffix* S of T and compute the dynamic programming edit distance table for strings P and S. If S begins at position i of T, then the last row of that table gives the edit distance between P and every substring of T that begins at position i. That is, the edit distance between P and $T[i..j]$ is found in cell $(n, j - i + 1)$ of the table. This approach takes $\Theta(nm^2)$ total time.

We are interested in the P-against-all problem when T is very long. In that case, the introduction of a suffix tree may greatly speed up the dynamic programming computation, depending on how much repetition is contained in string T.[2] (See also Section 7.11.1.) To get the basic idea of the method, consider two substrings T' and T'' of T that are identical for their first n' characters. In the dynamic programming approach above, the edit distances between P and T' and between P and T'' would be computed separately. But if we compute edit distance *columnwise* (instead of in the usual rowwise manner), then we can combine the two edit distance computations for the first n' columns, since the first n' characters of T' and T'' are the same (see Figure 12.11). It would be redundant to compute the first n by n' subtable separately for the two edit distances. This idea of using the commonality of T' and T'' can be formalized and fully exploited through the use of a suffix tree for string T.

Consider a suffix tree $\mathcal{T}$ for string T and recall that any path from the root of $\mathcal{T}$ specifies some substring S of T. If we traverse a path from the root of $\mathcal{T}$, and we let S denote the growing substring corresponding to that path, then during the traversal we can build up (columnwise) the dynamic programming table for the edit distance between P and the growing substring S of T. The full idea then is to traverse $\mathcal{T}$ in a depth-first manner, computing the appropriate dynamic programming column (from the column to its left) for every substring S specified by the current path. When the traversal reaches a node v of $\mathcal{T}$, it stores there the last (most recently generated) column and last subrow of the current subtable (the last row will always be row n). That is, if S is the substring specified by the path to a node v, then what will be stored at v is the last row and column of the dynamic programming table for the edit distance between P and S. When the depth-first traversal visits a child v' of v, it adds columns (one for each character on the (v, v') edge) to this table to correspond to the extension of substring S. When the depth-first traversal reaches a leaf of $\mathcal{T}$ corresponding to the suffix starting at a position i (say) of T, it can then output the values in the last row of the current table. Those values specify the edit distances

---

[2] Recent estimates put the amount of repeated human DNA at 50 to 60%. That is, 50 to 60% of all human DNA is contained in *nontrivial length*, structured substrings that show up repeatedly throughout the genome. Similar levels of redundancy appear in many other organisms.

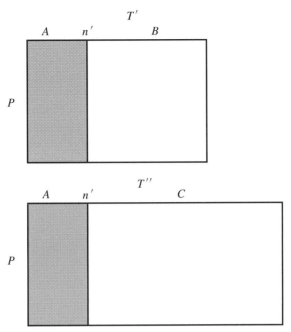

Figure 12.11: A cartoon of the dynamic programming tables for computing the edit distance between P and substring T' (top) and between P and substring T'' (bottom). The two tables share the subtable for P and substring A (shown as a shaded rectangle). This shaded subtable only needs to be computed once.

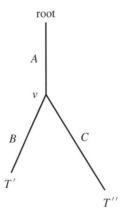

Figure 12.12: A piece of the suffix tree for T. The traversal from the root to node v is accompanied by the computation of subtable A (from the previous figure). At that point, the last row and column of subtable A are stored at node v. Computing the subtable B corresponds to the traversal from v to the leaf representing substring T'. After the traversal reaches the leaf for T', it backs up to node v, retrieves the row and column stored there, and uses them to compute the subtable C needed to compute the edit distance between P and T''.

between P and every substring beginning at position i of T. When the depth-first traversal backs up to a node v, and v has an unvisited child v', the row and column stored at v are retrieved and extended as the traversal follows a new (v, v') edge (see Figure 12.12).

It should be clear that this suffix-tree approach does correctly compute the edit distance between P and every substring of T, and it does exploit repeated substrings (small or large) that may occur in T. But how effective is it compared to the $\Theta(nm^2)$-time dynamic programming approach?

Definition   The *string-length* of an edge label in a suffix tree is the length of the string labeling that edge (even though the label is compactly represented by a constant number of characters). The *length of a suffix tree* is the sum of the string-lengths for all of its edges.

The length for a suffix tree $\mathcal{T}$ for a string T of length m can be anywhere between $\Theta(m)$ and $\Theta(m^2)$, depending on how much repetition exists in T. In computational experiments using long substrings of mammalian DNA (length around one million), the string-lengths of the resulting suffix trees have been around $m^2/10$. Now the number of dynamic programming columns that are generated during the depth-first traversal of $\mathcal{T}$ is exactly the length of $\mathcal{T}$. Each column takes $\Theta(n)$ time to generate, and so we can state

Lemma 12.4.1.   *The time used to generate the needed columns in the depth-first traversal is $\Theta(n \times (length\ of\ \mathcal{T}))$.*

We must also account for the time and space used to write the rows and columns stored at each node of $\mathcal{T}$. In a suffix tree with m leaves there are $\Theta(m)$ internal nodes and a single row and column take at most $O(m + n)$ time and space to write. Therefore, the time and space needed for the row and column stores is $\Theta(m^2 + nm) = \Theta(m^2)$. Hence, we have

Theorem 12.4.1.   *The total time for the suffix-tree approach is $\Theta(n \times (length\ of\ \mathcal{T}) + m^2)$, and the maximum space used is $\Theta(m^2)$.*

Reducing space

The size of the required output is $\Theta(m^2)$, since the problem calls for the edit distance between P and *each* of $\Theta(m^2)$ substrings of T, making the $\Theta(m^2)$ term in the time bound acceptable. On the other hand, the space used seems excessive since the space needed by the dynamic programming solution without using a suffix tree is just $\Theta(nm)$ and can be reduced to $O(m)$. We now modify the suffix-tree approach to also use only $O(n + m)$ space and the same time bounds as before.

First, there is no need to store the current column at each node v. When backing up from a child v' of v, we can use the current column at v' and the string labeling edge (v, v') to recompute the column for node v. This does, however, double the total time for computing the columns. There is also no need to keep the current row n at each node v. Instead, only $O(n)$ space is needed for row entries. The key idea is that the current table is expanded columnwise, so if the string-depth of v' is j and the string-depth of v' is $j + d$, then the row n stored at v and v' would be identical for the first j entries. We leave it as an exercise to work out the details. In summary, we have

Theorem 12.4.2.   *The hybrid suffix-tree/dynamic programming approach to the P-against-all problem can be implemented to run in $\Theta[n(length\ of\ \mathcal{T}) + m^2]$ time and $O(n + m)$ space.*

The above time and space bounds should be compared to the $\Theta(nm^2)$ time and $O(n+m)$ space bounds that result from a straightforward application of dynamic programming. The effectiveness in practice of this method depends on the length of $\mathcal{T}$ for realistic strings. It is known that for random strings, the length of $\mathcal{T}$ is $\Theta(m^2)$, making the method unattractive. (For random strings, the suffix tree is bushy for string-depths of $\log_\sigma m$ or less, where σ is the size of the alphabet. But beyond that depth, the suffix tree becomes very sparse, since the probability is very low that a substring of length greater than $\log_\sigma m$ occurs more than once in the string.) However, strings with more structured repetitions (as occur in DNA) should give rise to suffix trees with lengths that are small enough to make this method useful. We examined this question empirically for DNA strings up to one million characters, and the lengths of the resulting suffix trees were around $m^2/10$.

12.4.2. The (threshold) all-against-all problem

Now we consider a more ambitious problem: Given strings P and T, find every pair of substrings where the edit distance is below a fixed threshold d. Computations of this type have been conducted when P and T are both equal to the combined set of protein strings in the database Swiss-Prot [183]. The importance of this kind of large-scale computation and the way in which its results are used are discussed in [57]. The way suffix trees are used to accelerate the computation is discussed in [34].

Since P and T have respective lengths of n and m, the full all-against-all problem (with threshold ∞) calls for the computation of n^2m^2 pieces of output. Hence no method for this problem can run faster than $\Theta(n^2m^2)$ time. Moreover, that time bound is easily achieved: Pick a pair of starting positions in P and T (in nm possible ways), and for each choice of starting positions i, j fill in the dynamic programming table for the edit distance of $P[i..n]$ and $T[j..m]$ (in $O(nm)$-time). For any choice of i and j, the entries in the corresponding table give the edit distance for every pair of substrings that begin at position i in P and at position j in T. Thus, achieving the $O(n^2m^2)$ bound for the full all-against-all problem does not require suffix trees.

But the full all-against-all problem calls for an amount of output that is often excessive, and the output can be reduced by choosing a meaningful threshold. Or the criteria for reporting a substring pair might be a function of both length and edit distance. Whatever the specific reporting criteria, if it is no longer necessary to report the edit distance of every pair, it is no longer certain that $\Theta(n^2m^2)$ time is required. Here we develop a method whose worst-case running time is expressed as $O(C + R)$, where C is a computation time that may be less than $\Theta(n^2m^2)$ and R is the output size (i.e., the number of reported pairs of substrings). In this setting, the use of suffix trees may be quite valuable depending on the size of the output and the amount of repetition in the two strings.

An $O(C + R)$-time method

The method uses a suffix tree $\mathcal{T}_P$ for string P and a suffix tree $\mathcal{T}_T$ for string T. The worst-case time for the method will be shown to be $O(C + R)$, where C is the length of $\mathcal{T}_P$ times the length of $\mathcal{T}_T$ *independent of whatever the output criteria are*, and R is the size of the output. (The definition of the length of a suffix tree is found in Section 12.4.1.) That is, the method will compute certain dynamic programming cell values, which will be the same no matter what the output criteria are, and then when a cell value satisfies the particular output criteria, the algorithm will collect the relevant substrings associated with that cell. Hence our description of the method holds for the full all-against-all problem, the threshold version of the problem, or any other version with different reporting criteria.

To start, recall that each node in $\mathcal{T}_P$ represents a substring of P and that every substring of P is a prefix of a substring represented by a node of $\mathcal{T}_P$. In particular, each suffix of P is represented by a leaf of $\mathcal{T}_P$. The same is true of T and $\mathcal{T}_T$.

Definition   The dynamic programming table for a pair of nodes (u, v), from $\mathcal{T}_P$ and $\mathcal{T}_T$, respectively, is defined as the dynamic programming table for the edit distance between the string represented by node u and the string represented by node v.

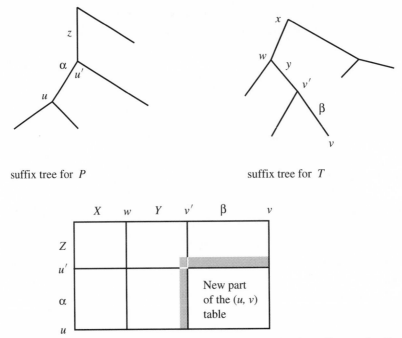

suffix tree for *P*   suffix tree for *T*

Figure 12.13: The dynamic programming table for (u,v) is shown below the suffix trees for P and T. The string on the path to node u is $Z\alpha$ and the string to node v is $XY\beta$. Every cell in the (u,v) table, except any in the lower right rectangle, is also in the (u,v'), (u',v), or (u',v') tables. The new part of the (u,v) table can be computed from the shaded entries and substrings α and β. The shaded entries contain exactly one entry from the (u',v') table; $|\alpha|$ entries from the last column in the (u,v') table; and $|\beta|$ entries from the last row in the (u',v) table.

The threshold all-against-all problem could be solved (ignoring time) by computing the dynamic programming table for each pair of leaves, one from each tree, and then examining every entry in each of those tables. Hence it certainly would be solved by computing the dynamic programming table for each pair of nodes and then examining each entry in those tables. This is essentially what we will do, but we proceed in a way that avoids redundant computation and examination. The following lemma gives the key observation.

Lemma 12.4.2. *Let u' be the parent of node u in $\mathcal{T}_P$ and let α be the string labeling the edge between them. Similarly, let v' be the parent of v in $\mathcal{T}_T$ and let β be the string labeling the edge between them. Then, all but the bottom right $|\alpha||\beta|$ entries in the dynamic programming table for the pair (u, v) appear in one of the tables for (u', v'), (u', v), or (u, v'). Moreover, that bottom right part of the (u, v) table can be obtained from the other three tables in $O(|\alpha||\beta|)$ time. (See Figure 12.13.)*

The proof of this lemma is immediate from the definitions and the edit distance recurrences.

The computation for the new part of the (u, v) table produces an $|\alpha|$ by $|\beta|$ rectangular subtable that forms the lower right section of the (u, v) table. In the algorithm to be developed below, we will store and associate with each node pair (u, v) the last column and the last row of this $|\alpha|$ by $|\beta|$ subtable.

We can now fully describe the algorithm.

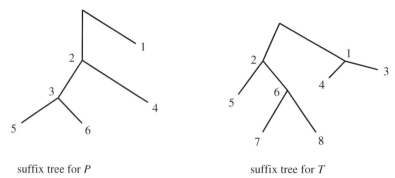

suffix tree for P suffix tree for T

Figure 12.14: The suffix trees for P and T with nodes numbered by string-depth. Note that these numbers are not the standard suffix position numbers that label the leaves. The ordered list of node pairs begins (1,1),(1,2),(1,3)... and ends with (6,8).

Details of the algorithm

First, number the nonroot nodes of $\mathcal{T}_P$ according to string-depth, with smaller string-depth first.[3] Separately, number the nodes of $\mathcal{T}_T$ according to string-depth. Then form a list L of all pairs of node numbers, one from each tree, in lexicographic order. Hence, pair (u, v) appears before pair (p, q) in the list if and only if u is less than p, or if u is equal to p and v is less than q. (See Figure 12.14). It follows that if u' is the parent of u in $\mathcal{T}_P$ and v' is the parent of v in $\mathcal{T}_T$, then (u', v') appears before (u, v).

Next, process each pair of nodes (u, v) in the order that it appears in L. Assume again that u' is the parent of u, that v' is the parent of v, and that the labels on the respective edges are α and β. To process a node pair (u, v), retrieve the value in the single lower right cell from the stored part of the (u', v') table; retrieve the column stored with the pair (u, v'), and retrieve the row stored with the pair (u', v). These three pairs of nodes have already been processed, due to the lexicographic ordering of the list. From those retrieved values, and from the substrings α and β, compute the new $|\alpha|$ by $|\beta|$ subtable completing the (u, v) table. Store with pair (u, v) the last row and column of newly computed subtable.

Now suppose cell (i, j) is in the new $|\alpha|$ by $|\beta|$ subtable, and its value satisfies the output criteria. The algorithm must find and output all locations of the two substrings specified by (i, j). As usual, a depth-first traversal to the leaves below u and v will then find all the starting positions of those strings. The length of the strings is determined by i and j. Hence, when it is required to output pairs of substrings that satisfy the reporting criteria, the time to collect the pairs is just proportional to the number of them.

Correctness and time analysis

The correctness of the method follows from the fact that at the highest level of description, the method computes the edit distance for every pair of substrings, one from each string. It does this by generating and examining every cell in the dynamic programming table for every pair of substrings (although it avoids redundant examinations). The only subtle point is that the method generates and examines the cells in each table in an incremental manner to exploit the commonalities between substrings, and hence it avoids regenerating and reexamining any cell that is part of more than one table. Further, when the method finds a cell satisfying the reporting criteria (a function of value and length), it can find all

---

[3] Actually, any topological numbering will do, but string-depth has some advantages when heuristic accelerations are added.

the pairs of substrings specified by that cell using a traversal to a subset of leaves in the two suffix trees. A formal proof of correctness is left to the reader as an exercise.

For the time analysis, recall that the length of $\mathcal{T}_P$ is the sum of lengths of all the edge labels in $\mathcal{T}_P$. If P has length n, then the length of $\mathcal{T}_P$ ranges between n and $n^2/2$, depending on how repetitive P is. The length of $\mathcal{T}_T$ is similarly defined and ranges between m and $m^2/2$, where m is the length of T.

Lemma 12.4.3. *The time used by the algorithm for all the needed dynamic programming computations and cell examinations is proportional to the product of the length of $\mathcal{T}_P$ and the length of $\mathcal{T}_T$. Hence that time, defined as C, ranges between nm and $n^2 m^2$.*

PROOF   In the algorithm, each pair of nodes is processed exactly once. At the point a pair (u, v) is processed, the algorithm spends $O(|\alpha||\beta|)$ time to compute a subtable and examine it, where α and β are the labels on the edges into u and v, respectively. Each edge-label in $\mathcal{T}_P$ therefore forms exactly one dynamic programming table with each of the edge-labels in $\mathcal{T}_T$. The time to build those tables is $|\alpha|$(length of $\mathcal{T}_T$). Summing over all edges in $\mathcal{T}_P$ gives the claimed time bound.   □

The above lemma counts all the time used in the algorithm except the time used to collect and report pairs of substrings (by their starting position, length, and edit distance). But since the algorithm collects substrings when it sees a cell value that satisfies the reporting criteria, the time devoted to output is just the time needed to traverse the tree to collect output pairs. We have already seen that this time is proportional to the number of pairs collected, R. Hence, we have

Theorem 12.4.3. *The complete time for the algorithm is $O(C + R)$.*

How effective is the suffix tree approach?

As in the P-against-all problem, the effectiveness of this method in practice depends on the lengths of $\mathcal{T}_P$ and $\mathcal{T}_T$. Clearly, the product of those lengths, C, falls as P and T increase in repetitiveness. We have built a suffix tree for DNA strings of total length around one million bases and have observed that the tree length is around one tenth of the maximum possible. In that case, C is around $n^2 m^2/100$, so all else being equal (which is unrealistic), standard dynamic programming for the all-against-all problem should run about one hundred times slower than the hybrid dynamic programming approach.

A vastly larger "all-against-all" computation on amino acid strings was reported in [183]. Although their description is very vague, they essentially used the suffix tree approach described here, computing similarity instead of edit distance. But, rather than a hundred-fold speedup, they claim to have achieved nearly a million-fold speedup over standard dynamic programming.[4] That level of speedup is not supported by theoretical considerations (recall that for a random string S of length m, a substring of length greater than $\log_\sigma m$ is very unlikely to occur in S more than once). Nor is it supported by the experiments we have done. The explanation may be the incorporation of an early stopping rule described in [183] only by the vague statement "Time is saved because the matching of patricia[5] subtrees is aborted when the score falls below a liberally chosen similarity limit". That rule is apparently very effective in reducing running time, but without a

---

[4] They finish a computation in 405 cpu days that they claim would otherwise have taken more than a million cpu years without the use of suffix trees.

[5] A patricia tree is a variant of a suffix tree.

clearer description of it we cannot define precisely what specific all-against-all problem was solved.

12.5. A faster (combinatorial) algorithm for longest common subsequence

The longest common subsequence problem (*lcs*) is a special case of general weighted alignment or edit distance, and it can be solved in $\Theta(nm)$ time either by applying those general methods or with more direct recurrences (Exercise 16 of Chapter 11). However, the *lcs* problem plays a special role in the field of string algorithms and merits additional discussion. This is partly for historical reasons (many string and alignment ideas were first worked out for the special case of *lcs*) and partly because *lcs* often seems to capture the desired relationship between the strings of interest.

In this section we present an alternative (combinatorial) method for *lcs* that is *not* based on dynamic programming. For two strings of lengths n and $m > n$, the method runs in $O(r \log n)$ worst-case time, where r is a parameter that is typically small enough to make this bound attractive compared to $\Theta(nm)$. The main idea is to reduce the *lcs* problem to a simpler sounding problem, the *longest increasing subsequence problem* (*lis*). The method can also be adapted to compute the length of the *lcs* in $O(r \log n)$ time, using only linear space, without the need for Hirschberg's method. That will be considered in Exercise 23.

12.5.1. Longest increasing subsequence

Definition   Let Π be a list of n integers, not necessarily distinct. An *increasing subsequence* of Π is a subsequence of Π whose values *strictly* increase from left to right.

For example, if $\Pi = 5, 3, 4, 9, 6, 2, 1, 8, 7, 10$ then $\{3, 4, 6, 8, 10\}$ and $\{5, 9, 10\}$ are both increasing subsequences in Π. (Recall the distinction between subsequences and substrings.) We are interested in the problem of computing a *longest increasing subsequence* in Π. The method we develop here will later be used to solve the problem of finding the *longest common subsequence* of two (or more) strings.

Definition   A *decreasing subsequence* of Π is a subsequence of Π where the numbers are *nonincreasing* from left to right.

For example, under this definition, $\{8, 5, 5, 3, 1, 1\}$ is a decreasing subsequence in the sequence $4, 8, 3, 9, 5, 2, 5, 3, 10, 1, 9, 1, 6$. Note the asymmetry in the definitions of *increasing* and *decreasing* subsequences. The term "decreasing" is slightly misleading. Although "nonincreasing" is more precise, it is too clumsy a term to use in high repetition.

Definition   A *cover* of Π is a set of decreasing subsequences of Π that contain all the numbers of Π.

For example, $\{5, 3, 2, 1\}$; $\{4\}$; $\{9, 6\}$; $\{8, 7\}$; $\{10\}$ is a cover of $\Pi = 5, 3, 4, 9, 6, 2, 1, 8, 7, 10$. It consists of five decreasing subsequences, two of which contain only a single number.

Definition   The *size* of the cover is the number of decreasing subsequences in it, and a *smallest* cover is a cover with minimum size among all covers.

We will develop an $O(n \log n)$-time method that simultaneously constructs a longest increasing subsequence (*lis*) and a smallest cover of Π. The following lemma is the key.

5	4	9	8	10
3		6	7	
2				
1				

Figure 12.15: Decreasing cover of {5, 3, 4, 9, 6, 2, 1, 8, 7, 10}

Lemma 12.5.1. *If I is an increasing subsequence of Π with length equal to the size of a cover of Π, call it C, then I is a longest increasing subsequence of Π and C is a smallest cover of Π.*

PROOF   No increasing subsequence of Π can contain more than one number contained in any decreasing subsequence of Π, since the numbers in an increasing subsequence strictly increase left to right, whereas the numbers in a decreasing subsequence are nonincreasing left to right. Hence no increasing subsequence of Π can have length greater than the size of any cover of Π.

Now assume that the length of I is equal to the size of C. This implies that I is a longest increasing subsequence of Π because no other increasing subsequence can be longer than the size of C. Conversely, C must be a smallest cover of Π, for if there were a smaller cover C' then I would be longer than the size of C', which is impossible. Hence, if the length of I equals the size of C, then I is a longest increasing subsequence and C is a smallest cover.   □

Lemma 12.5.1 is the basis of a method to find a longest increasing subsequence and a smallest cover of Π. The idea is to decompose Π into a cover C such that there is an increasing subsequence I containing exactly one number from each decreasing subsequence in C. Without concern for efficiency, a cover of Π can be built in the following straightforward way:

Naive cover algorithm   Starting from the left of Π, examine each successive number in Π and place it at the end of the first (left-most) decreasing subsequence that it can extend. If there are no decreasing subsequences it can extend, then start a new (decreasing) subsequence to the right of all the existing decreasing subsequences.

To elaborate, if x denotes the current number from Π being examined, then x extends a subsequence i if x is smaller than or equal to the current number at the end of subsequence i, and if x is strictly larger than the last number of each subsequence to the left of i.

For example, with Π as before the first two numbers examined are put into a decreasing subsequence {5, 3}. Then the number 4 is examined, which is in position 3 of Π. Number 4 cannot be placed at the end of the first subsequence because 4 is larger than 3. So 4 begins a new subsequence of its own to the right of the first subsequence. Next, the number 9 is considered and since it cannot be added to the end of either subsequence {5,3} or 4, it begins a third subsequence. Next, 6 is considered; it can be added to 9 but not to the end of any of the two subsequences to the left of 9. The final cover of Π produced by the algorithm is shown in Figure 12.15, where each subsequence runs vertically.

Clearly, this algorithm produces a cover of Π, which we call the *greedy cover*. To see whether a number x can be added to any particular decreasing subsequence, we only have to compare x to the number, say y, currently at the end of the subsequence – x can be added if and only if $x \leq y$. Hence if there are k subsequences at the time x is considered, then the time to add x to the correct subsequence is $O(k)$. Since $k \leq n$, we have the following:

Lemma 12.5.2. *The greedy cover of Π can be built in $O(n^2)$ time.*

We will shortly see how to reduce the time needed to find the greedy cover to $O(n \log n)$, but we first show that the greedy cover is a smallest cover of Π and that a longest increasing subsequence can easily be extracted from it.

Lemma 12.5.3. *There is an increasing subsequence I of Π containing exactly one number from each decreasing subsequence in the greedy cover C. Hence I is the longest possible, and C is the smallest possible.*

PROOF   Let x be an arbitrary number placed into decreasing subsequence $i > 1$ (counting from the left) by the greedy algorithm. At the time x was considered, the last number y of subsequence $i - 1$ must have been smaller than x. Also, since y was placed before x was, y appears before x in Π, and $\{y, x\}$ forms an increasing subsequence in Π. Since x was arbitrary, the same argument applies to y, and if $i - 1 > 1$ then there must be a number z in subsequence $i - 2$ such that $z < y$ and z appears before y in Π. Repeating this argument until the first subsequence is reached, we conclude that there is an increasing subsequence in Π containing one number from each of the first i subsequences in the greedy cover and ending with x. Choosing x to be any number in the last decreasing subsequence proves the lemma.   □

Algorithmically, we can find a longest increasing subsequence *given* the greedy cover as follows:

Longest increasing subsequence algorithm

begin

0. Set i to be the number of subsequences in the greedy cover. Set I to the empty list; pick any number x in subsequence i and place it on the front of list I.

1. While $i > 1$ do
        begin

2. Scanning down from the *top* of subsequence $i - 1$, find the first number y that is smaller than x.

3. Set x to y and i to $i - 1$.

4. Place x on the front of list I.
        end

end.

Since no number is examined twice during this algorithm, a longest increasing subsequence can be found in $O(n)$ time given the greedy cover.

An alternate approach is to use pointers. As the greedy cover is being constructed, whenever a number x is added to subsequence i, connect a pointer from x to the number at the current end of subsequence $i - 1$. After the greedy algorithm finishes, pick any number in the last decreasing subsequence and follow the unique path of pointers starting from it and ending at the first subsequence.

Faster construction of the greedy cover

Now we reduce the time to construct a greedy cover to $O(n \log n)$, reducing the overall running time to find a longest increasing subsequence to $O(n \log n)$ as well.

At any point during the running of the greedy cover algorithm, let L be the ordered list containing the last number of each of the decreasing subsequences built so far. That is, the last number from any subsequence $i - 1$ appears in L before the last number from subsequence i.

Lemma 12.5.4. *At any point in the execution of the algorithm, the list L is sorted in increasing order.*

PROOF   Assume inductively that the lemma holds through iteration $k-1$. When examining the kth number in Π, call it x, suppose x is to be placed at the end of subsequence i. Let w be the current number at the end of subsequence $i - 1$, let y be the current number at the end of subsequence i (if any), and let z be the number at the end of subsequence $i + 1$ (if it exists). Then $w < x \le y$ by the workings of the algorithm, and since $y < z$ by the inductive assumption, $x < z$ also. In summary, $w < x < z$, so the new subsequence L remains sorted. $\square$

Note that L itself need not be (and generally will not be) an increasing subsequence of Π. Although $x < z$, x appears to the right of z in Π. Despite this, the fact that L is in sorted order means that we can use *binary search* to implement each iteration of the algorithm building the greedy cover. Each iteration k considers the kth number x in Π and the current list L to find the left-most number in L larger than x. Since L is in sorted order, this can be done in $O(\log n)$ time by binary search. The list Π has n numbers, so we have

Theorem 12.5.1. *The greedy cover can be constructed in $O(n \log n)$ time. A longest increasing subsequence and a smallest cover of Π can therefore be found in $O(n \log n)$ time.*

In fact, if p is the length of the *lis*, then it can be found in $O(n \log p)$ time.

12.5.2. Longest common subsequence reduces to longest increasing subsequence

We will now solve the *longest common subsequence problem* for a pair of strings, using the method for finding a longest increasing subsequence in a list of integers.

Definition   Given strings S_1 and S_2 (of length m and n, respectively) over an alphabet Σ, let $r(i)$ be the number of times that the ith character of string S_1 appears in string S_2.

Definition   Let r denote the sum $\sum_{i=1}^{m} r(i)$.

For example, suppose we are using the normal English alphabet; when $S_1 = abacx$ and $S_2 = baabca$ then $r(1) = 3, r(2) = 2, r(3) = 3, r(4) = 1$, and $r(5) = 0$, so $r = 9$. Clearly, for any two strings, r will fall in the range 0 to nm. We will solve the *lcs* problem in $O(r \log n)$ time (where $n \le m$), which is inferior to $O(nm)$ when the r is large. However, r is often substantially smaller than nm, depending on the alphabet Σ. We will discuss this more fully later.

The reduction

For each alphabet character x that occurs at least once in S_1, create a list of the positions where character x occurs in string S_2; write this list in *decreasing* order. Two distinct alphabet characters will have totally disjoint lists. In the above example ($S_1 = abacx$ and $S_2 = baabca$) the list for character a is 6, 3, 2 and the list for b is 4, 1.

Now create a list called $\Pi(S_1, S_2)$ of length r, in which each character *instance* in S_1 is replaced with the associated list for that character. That is, for each position i in S_1, insert the list associated with the character $S_1(i)$. For example, list $\Pi(S_1, S_2)$ for the above two strings is 6, 3, 2, 4, 1, 6, 3, 2, 5.

To understand the importance of $\Pi(S_1, S_2)$, we examine what an increasing subsequence in that list means in terms of the original strings.

Theorem 12.5.2. *Every increasing subsequence I in $\Pi(S_1, S_2)$ specifies an equal length common subsequence of S_1 and S_2 and vice versa. Thus a longest common subsequence of S_1 and S_2 corresponds to a longest increasing subsequence in the list $\Pi(S_1, S_2)$.*

PROOF First, given an increasing subsequence I of $\Pi(S_1, S_2)$, we can create a string S and show that S is a subsequence of both S_1 and S_2. String S is successively built up during a left-to-right scan of I. During this scan, also construct two lists of indices specifying a subsequence of S_1 and a subsequence of S_2. In detail, if number j is encountered in I during the scan, and number j is contained in the sublist contributed by character i of S_1, then add character $S_1(i)$ to the right end of S, add number i to the right end of the first index list, and add j to the right end of the other index list.

For example, consider $I = 3, 4, 5$ in the running example. The number 3 comes from the sublist for character 1 of S_1, the number 4 comes from the sublist for character 2, and the number 5 comes from the sublist for character 4. So the string S is *abc*. That string is a subsequence of S_1 found in positions 1, 2, 4 and is a subsequence of S_2 found in positions 3, 4, 5.

The list $\Pi(S_1, S_2)$ contains one sublist for every position in S_1, and each such sublist in $\Pi(S_1, S_2)$ is in decreasing order. So at most one number from any sublist is in I and any position in S_1 contributes at most one character to S. Further, the m lists are arranged left to right corresponding to the order of the characters in S_1, so S is certainly a subsequence of S_1. The numbers in I strictly increase and correspond to positions in S_2, so S is also a subsequence of S_2.

In summary, we have proven that every increasing subsequence in $\Pi(S_1, S_2)$ can be used to create an equal length common subsequence in S_1 and S_2. The converse argument, that a common subsequence yields an increasing subsequence, is very similar and is left as an exercise. $\square$

$\Pi(S_1, S_2)$ is a list of r integers, and the longest increasing subsequence problem can be solved in $O(r \log l)$ time on an r-length list when the longest increasing subsequence is of length l. If $n \leq m$ then $l \leq n$, yielding the following theorem:

Theorem 12.5.3. *The longest common subsequence problem can be solved in $O(r \log n)$ time.*

The $O(r \log n)$ result for *lcs* was first obtained by Hunt and Szymanski [238]. Their algorithm is superficially very different than the one above, but in retrospect one can see similar ideas embodied in it. The relationship between the *lcs* and *lis* problems was partly identified by Apostolico and Guerra [25, 27] and made explicit by Jacobson and Vo [244] and independently by Pevzner and Waterman [370].

The *lcs* method based on *lis* is an example of what is called *sparse dynamic programming*, where the input is a relatively sparse set of pairs that are permitted to align. This approach, and in fact the solution technique discussed here, has been very extensively generalized by a number of people and appears in detail in [137] and [138].

12.5.3. How good is the method

How good is the *lcs* method based on the *lis* compared to the original $\Theta(nm)$-time dynamic programming approach? It depends on the size of r. Let σ denote the size of the alphabet Σ. A very naive analysis would say that r can be expected to be about nm/σ. This assumes that each character in Σ appears with equal probability and hence is expected to appear n/σ times in the short string. That means that $r_i = n/\sigma$ for each i. The long string has length m, so r is expected to be nm/σ. But of course, equal distribution of characters is not really typical, and the value of r is then highly dependent on the specific strings.

For the Roman alphabet with capital letters, digits, and punctuation marks added, σ is around 100, but the assumption of equal distribution is clearly flawed. Still, one can ask whether $(nm/100)\log n$ looks attractive compared to nm. For such alphabets, the speedup doesn't look so compelling, although the method retains its simplicity and space efficiency. Thus for typical English text, the *lis*-based approach may not be much superior to the dynamic programming approach. However, in many applications, the "alphabet" size is quite large and grows with the size of the text.[6] This is true, for example, in the unix utility *diff* where each line in the text is considered as a character in the "alphabet" used for the *lcs* computation. In certain applications in molecular biology the alphabet consists of patterns or substrings, rather than the four-character alphabet of DNA or the twenty-character alphabet of protein. These substrings might be genes, exons, or restriction enzyme recognition sequences. In those cases, the alphabet size is large compared to the string size, so r is small and $r\log n$ is quite attractive compared to nm.

Constrained *lcs*

The *lcs* method based on *lis* has another advantage over the standard dynamic programming approach. In some applications there are additional constraints imposed on which pairs of positions are permitted to align in the *lcs*. That is, in addition to the constraint that position i in S_1 can align with position j in S_2 only if $S_1(i) = S_2(j)$, some additional constraints may apply. The reduction of *lcs* to *lis* can be easily modified to incorporate these additional constraints, and we leave the details to the reader. The effect is to reduce the size of r and consequently to speed up the entire *lcs* computation. This is another example and variant of sparse dynamic programming.

12.5.4. The *lcs* of more than two strings

One of the nice features of the *lcs* method based on *lis* is that it easily generalizes to the *lcs* problem for more than two strings. That problem is a special case of *multiple sequence alignment*, a crucial problem in computational molecular biology that we will more fully discuss in Chapter 14. The generalization from two to many strings will be presented here for three strings, S_1, S_2, and S_3.

The idea is to again reduce the *lcs* problem to the *lis* problem. As before, we start by creating a list for each character x in S_1. In particular, the list for x will contain pairs of integers, each pair containing a position in S_2 where x occurs and a position in S_3 where x occurs. Further, the list for character x will be ordered so that the pairs in the list are in *lexically decreasing* order. That is, if pair (i, j) appears before pair (i', j') in the list for x, then either $i > i'$ or $i = i'$ and $j > j'$. For example, if $S_1 =$

---

[6] This is one of the few places in the book where we deviate from the standard assumption that the alphabet is fixed.

$abacx$ and $S_2 = baabca$ (as above) and $S_3 = babbac$, then the list for character a is $(6, 5), (6, 2), (3, 5), (3, 2), (2, 5), (2, 2)$.

The lists for each character are again concatenated in the order that the characters appear in string S_1, forming the sequence of pairs $\Pi(S_1, S_2, S_3)$. We define an increasing subsequence in $\Pi(S_1, S_2, S_3)$ to be a subsequence of pairs such that the first numbers in each pair form an increasing subsequence of integers, and the second numbers in each pair also form an increasing subsequence of integers. We can easily modify the greedy cover algorithm to find a longest increasing subsequence of pairs under this definition. This increasing subsequence is used as follows.

Theorem 12.5.4. *Every increasing subsequence in $\Pi(S_1, S_2, S_3)$ specifies an equal length common subsequence of S_1, S_2, S_3 and vice versa. Therefore, a longest common subsequence of S_1, S_2, S_3 corresponds to a longest increasing subsequence in $\Pi(S_1, S_2, S_3)$.*

The proof of this theorem is similar to the case of two strings and is left as an exercise. Adaptation of the greedy cover algorithm and its time analysis for the case of three strings is also left to the reader. Extension to more than three strings is immediate. The combinatorial approach to computing *lcs* also has a nice space-efficiency feature that we will explore in the exercises.

12.6. Convex gap weights

Overwhelmingly, the affine gap weight model is the model most commonly used by molecular biologists today. This is particularly true for aligning amino acid sequences. However, a richer gap model, the *convex* gap weight, was proposed and studied by Waterman in 1984 [466], and has been more extensively examined since then. In discussing the common use of the affine gap weight, Benner, Cohen and Gonnet state "There is no justification either theoretical or empirical for this treatment" [183] and forcefully argue that "a non-linear gap penalty is the only one that is grounded in empirical data" [57]. They propose [57] that to align two protein sequences that are d PAM units diverged (see Section 15.7.2), a gap of length q should be given the weight:

$$35.03 - 6.88 \log_{10} d + 17.02 \log_{10} q$$

Under this weighting model, the cost to initiate a gap is at most 35.03, and declines with increasing evolutionary (PAM) distance between the two sequences. In addition to this initiation weight, the function adds $17.02 \log_{10} q$ for the actual length, q, of the gap.

It is hard to believe that a function this precise could be correct, but the key point is that, for a fixed PAM distance, the proposed gap weight is a *convex* function of its length.[7]

The alignment problem with convex gap weights is more difficult to solve than with affine gap weights, but it is not as difficult as the problem with arbitrary gap weights. In this section we develop a practical algorithm to optimally align two strings of lengths n and $m > n$, when the gap weights are specified by a *convex* function of the gap length. The algorithm runs in $O(nm \log m)$ time, in contrast to the $O(nm)$-time bound for affine gap weights and the $O(nm^2)$ time for arbitrary gap weights. The speedup for the convex case was established by Miller and Myers [322] and independently by Galil and Giancarlo

---

[7] Unfortunately, there is no standard agreement on terminology, and some of the papers refer to the model as the "convex" gap weight model, while others call it the "concave" gap model. In this book, a convex function is one with a negative or zero second derivative, and a concave function is one with a positive second derivative.

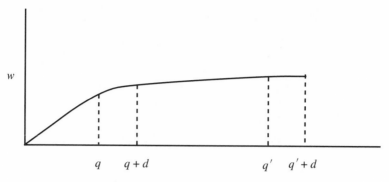

Figure 12.16: A convex function w.

[170]. However, the solution in the second paper is given in terms of edit distance rather than similarity. Similarity is often more useful than edit distance because it can be used to handle the extremely important case of local comparison. Hence we will discuss convex gap weights in terms of similarity (maximum weighted alignment) and leave it to the reader to derive the analogous algorithms for computing edit distance with convex gap weights. More advanced results on alignment with convex or concave gap weights appear in [136], [138], and [276].

Recall from the discussion of arbitrary gap weights that $w(q)$ is the weight given to a gap of length q. That gap then contributes a penalty of $-w(q)$ to the total weight of the alignment.

> **Definition**   Assume that $w(q)$ is a nonnegative function of q. Then $w(q)$ is *convex* if and only if $w(q + 1) - w(q) \leq w(q) - w(q - 1)$ for every q.

That is, as a gap length increases, the additional penalty contributed by the gap decreases for each additional unit of the gap. It follows that $w(q + d) - w(q) \geq w(q' + d) - w(q')$ for $q < q'$ and any fixed d (see Figure 12.16). Note that the function w can have regions of both positive and negative slope, although any region of positive slope must be to the left of the region of negative slope. Note that the definition allows $w(q)$ to become negative for large enough n and m. At that point, $-w(q)$ becomes positive, which is probably not desirable. Hence, gap weight functions with negative slope must be used with care.

The convex gap weight was introduced in [466] with the suggestion that mutational events that insert or delete varying length blocks of DNA can be more meaningfully modeled by convex gap weights, compared to affine or constant gap weights. A convex gap penalty allows the modeler more specificity in reflecting the cost or probability of different gap lengths, and yet it can be more efficiently handled than arbitrary gap weights. One particular convex function that is appealing in this context is the *log* function, although it is not clear which base of the logarithm might be most meaningful.

The argument for or against convex gap weights is still open, and the affine gap model remains dominant in practice. Still, even if the convex gap model never becomes popular in molecular biology it could well find application elsewhere. Furthermore, the algorithm for alignment with convex gaps is of interest in itself, as a representative of a number of related algorithms in the general area of "sparse dynamic programming".

Speeding up the general recurrences

To solve the convex gap weight case we use the same dynamic programming recurrences developed for arbitrary gap weights (page 242), but reduce the time needed to evaluate those recurrences. For convenience, we restate the general recurrences for arbitrary gap weights.

$$V(i, j) = \max[E(i, j), F(i, j), G(i, j)],$$

$$G(i, j) = V(i - 1, j - 1) + s(S_1(i), S_2(j)),$$

$$E(i, j) = \max_{0 \le k \le j-1} [V(i, k) - w(j - k)],$$

$$F(i, j) = \max_{0 \le l \le i-1} [V(l, j) - w(i - l)],$$

$$V(i, 0) = -w(i),$$

$$V(0, j) = -w(j),$$

$$E(i, 0) = -w(i),$$

$$F(0, j) = -w(j).$$

$G(i, j)$ is undefined when i or j is zero.

Even with arbitrary gap weights, the work required by the first and second recurrences is $O(m)$ per row, which is within our desired time bound. It is the recurrences for $E(i, j)$ and $F(i, j)$ that respectively require $\Theta(m^2)$ time per *row* and $\Theta(n^2)$ time per *column* when the function w is arbitrary. Hence, it is the evaluation of E and F for any given row or column that will be improved in the case where w is convex. We will focus on the computation of E for a single row. The computation of F and the associated time analysis for a single column is symmetric, with one caveat to be discussed later.

Simplifying notation

The value $E(i, j)$ depends on i only through the values $V(i, k)$ for $k < i$. Hence, in any fixed row, we can drop the reference to the row index i, simplifying the recurrence for E. That is, in any fixed row we define

$$E(j) = \max_{0 \le k \le j-1} [V(k) - w(j - k)].$$

Further, we introduce the following notation to simplify the recurrence:

$$Cand(k, j) = V(k) - w(j - k);$$

therefore,

$$E(j) = \max_{0 \le k \le j-1} Cand(k, j).$$

The term *Cand* stands for "candidate"; the meaning of this will become clear later.

12.6.1. Forward dynamic programming

It will be useful in the exposition to change the way we normally implement dynamic programming. Normally when setting the value $E(j)$, we would look *backwards* in the row to compare all the $Cand(k, j)$ values for $k < j$, taking the largest one to be the value $E(j)$. But an alternative *forward-looking* implementation is also possible and is more helpful in this exposition.[8]

---

[8] Gene Lawler pointed out that in some circles forward and backward implementations are referred to as "*push you – pull me*" dynamic programming. The reader may determine which term denotes forwards and which denotes backwards.

In the forward implementation, we first initialize a variable $\overline{E}(j')$ to $Cand(0, j')$ for each cell $j' > 0$ in the row. The E values are set left to right in the row, as in backward dynamic programming. However, to set the value of $E(j)$ (for any $j > 0$) the algorithm merely sets $E(j)$ to the current value of $\overline{E}(j)$, since every cell to the left of j will have contributed a candidate value to cell j. Then, before setting the value of $E(j + 1)$, the algorithm traverses *forwards* in the row to set $\overline{E}(j')$ (for each $j' > j$) to be the maximum of the current $\overline{E}(j')$ and $Cand(j, j')$. To summarize, the forward implementation for a fixed row is:

Forward dynamic programming for a fixed row

For $j := 1$ to m do
begin
$\overline{E}(j) := Cand(0, j)$;
$b(j) := 0$
end;

For $j := 1$ to m do
begin
$E(j) := \overline{E}(j)$;
$V(j) := \max[G(j), E(j), F(j)]$;
{We assume, but do not show that $F(j)$ and $G(j)$
have been computed for cell j in the row.}

For $j' := j + 1$ to m do {Loop 1}
    if $\overline{E}(j') < Cand(j, j')$ then
        begin
        $\overline{E}(j') := Cand(j, j')$;
        $b(j') := j$; {This sets a pointer from j' to j to be explained later.}
        end
end;

An alternative way to think about forward dynamic programming is to consider the weighted edit graph for the alignment problem (see Section 11.4). In that (acyclic) graph, the optimal path (shortest or longest distance, depending on the type of alignment being computed) from cell $(0, 0)$ to cell (n, m) specifies an optimal alignment. Hence algorithms that compute optimal distances in (acyclic) graphs can be used to compute optimal alignments, and distance algorithms (such as Dijkstra's algorithm for shortest distance) can be described as forward looking. When the correct distance $d(v)$ to a node v has been computed, and there is an edge from v to a node w whose correct distance is still unknown, the algorithm adds $d(v)$ to the distance on the edge (v, w) to obtain a candidate value for the correct distance to w. When the correct distances have been computed to all nodes with a direct edge to w, and each has contributed a candidate value for v, the correct distance to v is the best of those candidate values.

It should be clear that exactly the same arithmetic operations and comparisons are done in both backward and forward dynamic programming – the only difference is the order in which the operations take place. It follows that the forward algorithm correctly sets all the E values in a fixed row and still requires $\Theta(m^2)$ time per row. Thus forward dynamic programming is no faster than backwards dynamic programming, but the concept will help explain the speedup to come.

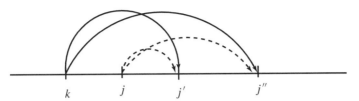

Figure 12.17: Graphical illustration of the *key observation*. Winning candidates are shown with a solid curve and losers with a dashed curve. If the candidate from j loses to the candidate from k at cell j', then the candidate from j will lose to the candidate from k at every cell j'' to the right of j'.

12.6.2. The basis of the speedup

At the point when $E(j)$ is set, call cell j the *current cell*. We interpret $Cand(j, j')$ as the "candidate value" for $E(j')$ that cell j "sends forward" to cell j'. When j is the current cell, it "sends forward" $m - j$ candidate values, one to each cell $j' > j$. Each such $Cand(j, j')$ value is compared to the current $\overline{E}(j')$; it either *wins* (when $Cand(j, j')$ is greater than $\overline{E}(j')$) or *loses* the comparison. The speedup works by identifying and eliminating large numbers of candidate values that have no chance of winning any comparison. In this way, the algorithm avoids a large number of useless comparisons. This approach is sometimes called a *candidate list* approach. The following is the key observation used to identify "loser" candidates:

> **Key observation**   Let j be the current cell. If $Cand(j, j') \le \overline{E}(j')$ for *some* $j' > j$, then $Cand(j, j'') \le \overline{E}(j'')$ for *every* $j'' > j'$. That is, "one strike and you're out".

Hence the current cell j need not send forward any candidate values to the right of the first cell $j' > j$ where $Cand(j, j')$ is less than or equal to $\overline{E}(j')$. This suggests the obvious practical speedup of stopping the loop labeled {Loop 1} in the Forward dynamic programming algorithm as soon as j's candidate loses. But this improvement does not lead directly to a better (worst-case) time bound. For that, we will have to use one more trick. But first, we prove the key observation with the following more precise lemma.

Lemma 12.6.1. *Let $k < j < j' < j''$ be any four cells in the same row. If $Cand(j, j') \le Cand(k, j')$ then $Cand(j, j'') \le Cand(k, j'')$. See Figure 12.17 for reference.*

PROOF $Cand(k, j') \ge Cand(j, j')$ implies that $V(k) - w(j' - k) \ge V(j) - w(j' - j)$, so $V(k) - V(j) \ge w(j' - k) - w(j' - j)$.

Trivially, $(j' - k) = (j' - j) + (j - k)$. Similarly, $(j'' - k) = (j'' - j) + (j - k)$. For future use, note that $(j' - k) < (j'' - k)$.

Now let q denote $(j' - j)$, let q' denote $(j'' - j)$, and let d denote $(j - k)$. Since $j' < j''$, then $q < q'$. By convexity, $w(q + d) - w(q) \ge w(q' + d) - w(q')$ (see Figure 12.16). Translating back, we have $w(j' - k) - w(j' - j) \ge w(j'' - k) - w(j'' - j)$. Combining this with the result in the first paragraph gives $V(k) - V(j) \ge w(j'' - k) - w(j'' - j)$, and rewriting gives $V(k) - w(j'' - k) \ge V(j) - w(j'' - j)$, i.e., $Cand(k, j'') \ge Cand(j, j'')$, as claimed.   □

Lemma 12.6.1 immediately implies the key observation.

12.6.3. Cell pointers and row partition

Recall from the details of the forward dynamic programming algorithm that the algorithm maintains a variable $b(j')$ for each cell j'. This variable is a pointer to the left-most cell

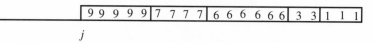

j

Figure 12.18: Partition of the cells $j + 1$ through m into maximal blocks of consecutive cells such that all the cells in any block have the same b value. The common b value in any block is less than the common b value in the preceding block.

$k < j'$ that has contributed the best candidate yet seen for cell j'. Pointer $b(j')$ is updated every time the value of $\overline{E}(j')$ changes. The use of these pointers combined with the next lemma leads ultimately to the desired speedup.

Lemma 12.6.2. *Consider the point when j is the current cell, but before j sends forward any candidate values. At that point, $b(j') \geq b(j' + 1)$ for every cell j' from $j + 1$ to $m - 1$.*

PROOF For notational simplicity, let $b(j') = k$ and $b(j' + 1) = k'$. Then, by the selection of k, $Cand(k, j') \geq Cand(k', j')$. Now suppose $k < k'$. Then, by Lemma 12.6.1, $Cand(k, j' + 1) \geq Cand(k', j' + 1)$, in which case $b(j' + 1)$ should be set to k, not k'. Hence $k \geq k'$ and the lemma is proved. □

The following corollary restates Lemma 12.6.2 in a more useful way.

Corollary 12.6.1. At the point that j is the current cell but before j sends forward any candidates, the values of the b pointers form a nonincreasing sequence from left to right. Therefore, cells $j, j + 1, j + 2, \ldots, m$ are partitioned into maximal blocks of consecutive cells such that all b pointers in the block have the same value, and the pointer values decline in successive blocks.

> **Definition** The partition of cells j through m referred to in Corollary 12.6.1 is called the current *block-partition*. See Figure 12.18.

Given Corollary 12.6.1, the algorithm doesn't need to explicitly maintain a b pointer for every cell but only record the common b pointer for each block. This fact will next be exploited to achieve the desired speedup.

Preparation for the speedup

Our goal is to reduce the time per row used in computing the E values from $\Theta(m^2)$ to $O(m \log m)$. The main work done in a row is to update the $\overline{E}$ values and to update the current block-partition with its associated pointers. We first focus on updating the block-partition and the b pointers; after that, the treatment of the $\overline{E}$ values will be easy. So for now, assume that all the $\overline{E}$ values are maintained for free.

Consider the point where j is the current cell, but before it sends forward any candidate values. After $E(j)$ (and $F(j)$ and then $V(j)$) have been computed, the algorithm must update the block-partition and the needed b pointers. To see the new idea, take the case of $j = 1$. At this point, there is only one block (containing cells 1 through m), with common b pointer set to cell zero (i.e., $b(j') = 0$ for each cell j' in the block). After $E(1)$ is set to $\overline{E}(1) = Cand(0, 1)$, any $\overline{E}(j')$ value that then changes will cause the block-partition to change as well. In particular, if $\overline{E}(j')$ changes, then $b(j')$ changes from zero to one. But since the b values in the new block-partition must be nonincreasing from left to right, there are only three possibilities for the new block-partition:[9]

- Cells 2 through m might remain in a single block with common pointer $b = 0$. By Lemma 12.6.1, this happens if and only if $Cand(1, 2) \leq \overline{E}(2)$.

---

[9] The $\overline{E}$ values in these three cases are the values before any $\overline{E}$ changes.

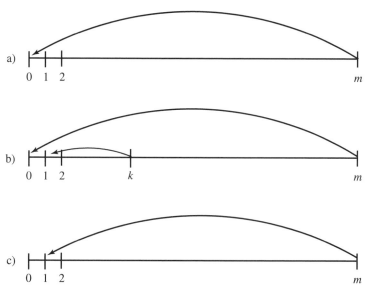

Figure 12.19: The three possible ways that the block partition changes after $E(1)$ is set. The curves with arrows represent the common pointer for the block and leave from the last entry in the block.

- Cells 2 through m might get divided into two blocks, where the common pointer for the first block is $b = 1$, and the common pointer for the second is $b = 0$. This happens (again by Lemma 12.6.1) if and only if for some $k < m$ $Cand(1, j') > \overline{E}(j')$ for j' from 2 to k and $Cand(1, j') \leq \overline{E}(j')$ for j' from $k + 1$ to m.

- Cells 2 through m might remain in a single block, but now the common pointer b is set to 1. This happens if and only if $Cand(1, j') > \overline{E}(j')$ for j' from 2 to m.

Figure 12.19 illustrates the three possibilities.

Therefore, before making any changes to the $\overline{E}$ values, the new partition of the cells from 2 to m can be efficiently computed as follows: The algorithm first compares $\overline{E}(2)$ and $Cand(1, 2)$. If $\overline{E}(2) \geq Cand(1, 2)$ then all the cells to the right of 2 remain in a single block with common b pointer set to zero. However, if $\overline{E}(2) < Cand(1, 2)$ then the algorithm searches for the left-most cell $j' > 2$ such that $\overline{E}(j') \geq Cand(1, j')$. If j' is found, then cells 2 through $j' - 1$ form a new block with common pointer to cell one, and the remaining cells form another block with common pointer to cell zero. If no j' is found, then all cells 2 through m remain in a single block, but the common pointer is changed to one.

Now for the punch line: By Corollary 12.6.1, this search for j' can be done by binary search. Hence only $O(\log m)$ *comparisons* are used in searching for j'. And, since we only record one b pointer per block, at most one pointer update is needed.

Now consider the general case of $j > 1$. Suppose that $E(j)$ has just been set and that the cells $j + 1, \ldots, m$ are presently partitioned into r maximal blocks ending at cells $p_1 < p_2 < \cdots < p_r = m$. The block ending at p_i will be called the ith block. We use b_i to denote the common pointer for cells in block i. We assume that the algorithm has a list of the *end-of-block* positions $p_1 < p_2 < \cdots < p_r$ and a parallel list of common pointers $b_1 > b_2 > \cdots > b_r$.

After $E(j)$ is set, the new partition of cells $j + 1$ through m is found in the following way: First, if $\overline{E}(j + 1) \geq Cand(j, j + 1)$ then, by Lemma 12.6.1, $\overline{E}(j') \geq Cand(j, j')$ for all $j' > j$, so the partition of cells greater than j remains unchanged. Otherwise (if $\overline{E}(j + 1) < Cand(j, j + 1)$), the algorithm successively compares $\overline{E}(p_i)$ to $Cand(j, p_i)$

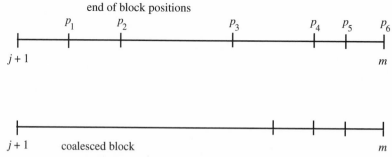

Figure 12.20: To update the block-partition the algorithm successively examines cell p_i to find the first index s where $\bar{E}(p_s) \geq Cand(j, p_s)$. In this figure, s is 4. Blocks 1 through $s - 1 = 3$ coalesce into a single block with some initial part of block $s = 4$. Blocks to the right of s remain unchanged.

for i from 1 to r, until either the end-of-block list is exhausted, or until it finds the first index s with $\bar{E}(p_s) \geq Cand(j, p_s)$. In the first case, the cells $j + 1, \ldots, m$ fall into a single block with common pointer to cell j. In the second case, the blocks $s + 1$ through r remain unchanged, but all the blocks 1 through $s - 1$ *coalesce* with some initial part (possibly all) of block s, forming one block with common pointer to cell j (see Figure 12.20). Note that every comparison but the last one results in two neighboring blocks coalescing into one.

Having found block s, the algorithm finds the proper place to split block s by doing binary search over the cells in the block. This is exactly as in the case already discussed for $j = 1$.

12.6.4. Final implementation details and time analysis

We have described above how to update the block-partition and the common b pointers, but that exposition uses $\bar{E}$ values that we assumed could be maintained for free. We now deal with that problem.

The key observation is that the algorithm retrieves $\bar{E}(j)$ only when j is the current cell and retrieves $\bar{E}(j')$ only when examining cell j' in the process of updating the block-partition. But the current cell j is always in the first block of the current block-partition (whose endpoint is denoted p_1), so $b(j) = b_1$, and $\bar{E}(j)$ equals $Cand(b_1, j)$, which can be computed in constant time when needed. In addition, when examining a cell j' in the process of updating the block-partition, the algorithm knows the block that j' falls into, say block i, and hence it knows b_i. Therefore, it can compute $\bar{E}(j')$ in constant time by computing $Cand(b_i, j')$. The result is that *no* explicit $\bar{E}$ values ever need to be stored. They are simply computed when needed. In a sense, they are only an expositional device. Moreover, the number of $\bar{E}$ values that need to be computed on the fly is proportional to the number of comparisons that the algorithm does to maintain the block-partition. These observations are summarized in the following:

Revised forward dynamic programming for a fixed row

Initialize the end-of-block list to contain the single number m.
Initialize the associated pointer list to contain the single number 0.

For $j := 1$ to m do
begin
      Set k to be the first pointer on the b-pointer list.
      $E(j) := Cand(k, j)$;

$V(j) := \max[G(j), E(j), F(j)];$
{As before we assume that the needed F and G values have been computed.}

{Now see how j's candidates change the block-partition.}
Set j' equal to the first entry on the end-of-block list.

{look for the first index s in the end-of-block list where j loses}
If $Cand(b(j'), j+1) < Cand(j, j+1)$ then {j's candidate wins one}
begin
    While
    The end-of-block list is not empty and $Cand(b(j'), j') < Cand(j, j')$ do
        begin
        remove the first entry on the end-of-block list,
        and remove the corresponding b-pointer
        If the end-of-block list is not empty then
        set j' to the new first entry on the end-of-block list.
        end;
    end {while};
If the end-of-block list is empty then
place m at the head of that list;
Else {when the end-of-block list is not empty}
    begin
    Let p_s denote the first end-of-block entry.
    Using binary search over the cells in block s, find the
    right-most point p in that block such that $Cand(j, p) > Cand(b_s, p)$.
    Add p to the head of the end-of-block list;
    end;

    Add j to the head of the b pointer list.

    end;
end.

Time analysis

An $\overline{E}$ value is computed for the current cell, or when the algorithm does a comparison involved in maintaining the current block-partition. Hence the total time for the algorithm is proportional to the number of those comparisons. In iteration j, when j is the current cell, the comparisons are divided into those used to find block s and those used in the binary search to split block s. If the algorithm does $l > 2$ comparisons to find s in iteration j, then at least $l - 1$ full blocks coalesce into a single block. The binary search then splits at most one block into two. Hence if, in iteration j, the algorithm does $l > 2$ comparisons to find s, then the total number of blocks decreases by at least $l - 2$. If it does one or two comparisons, then the total number of blocks at most increases by one. Since the algorithm begins with a single block and there are m iterations, it follows that over the entire algorithm there can be at most $O(m)$ comparisons done to find every s, excluding the comparisons done during the binary searches. Clearly, the total number of comparisons used in the m binary searches is $O(m \log m)$. Hence we have

Theorem 12.6.1. *For any fixed row, all the $E(j)$ values can be computed in $O(m \log m)$ total time.*

The case of F values is essentially symmetric

A similar algorithm and analysis is used to compute the F values, except that for $F(i, j)$ the lists partition column j from cell i through n. There is, however, one point that might cause confusion: Although the analysis for F focuses on the work in a single column and is symmetric to the analysis for E in a single row, the computations of E and F are actually *interleaved* since, by the recurrences, each $V(i, j)$ value depends on both $E(i, j)$ and $F(i, j)$. Even though both the E values and the F values are computed rowwise (since V is computed rowwise), one row after another, $E(i, j)$ is computed just prior to the computation of $E(i, j + 1)$, while between the computation of $F(i, j)$ and $F(i + 1, j)$, $m - 1$ other F values will be computed ($m - j$ in row i and $j - 1$ in row $i + 1$). So although the analysis treats the work in a column as if it is done in one contiguous time interval, the algorithm actually breaks up the work in any given column.

Only $O(nm)$ total time is needed to compute the G values and to compute every $V(i, j)$ once $E(i, j)$ and $F(i, j)$ is known. In summary we have

Theorem 12.6.2. *When the gap weight w is a convex function of the gap length, an optimal alignment can be computed in $O(nm \log m)$ time, where $m > n$ are the lengths of the two strings.*

12.7. The Four-Russians speedup

In this section we will discuss an approach that leads both to a theoretical and to a practical speedup of many dynamic programming algorithms. The idea, comes from a paper [28] by four authors, Arlazarov, Dinic, Kronrod, and Faradzev, concerning boolean matrix multiplication. The general idea taken from this paper has come to be known in the West as the Four-Russians technique, even though only one of the authors is Russian.[10] The applications in the string domain are quite different from matrix multiplication, but the general idea suggested in [28] applies. We illustrate the idea with the specific problem of computing (unweighted) *edit distance*. This application was first worked out by Masek and Paterson [313] and was further discussed by those authors in [312]; many additional applications of the Four-Russians idea have been developed since then (for example [340]).

12.7.1. t-blocks

Definition   A t-block is a t by t square in the dynamic programming table.

The rough idea of the Four-Russians method is to partition the dynamic programming table into t-blocks and compute the essential values in the table one t-block at a time, rather than one cell at a time. The goal is to spend only $O(t)$ time per block (rather than $\Theta(t^2)$ time), achieving a factor of t speedup over the standard dynamic programming solution. In the exposition given below, the partition will not be exactly achieved, since neighboring t-blocks will overlap somewhat. Still, the rough idea given here does capture the basic flavor and advantage of the method presented below. That method will compute the edit distance in $O(n^2 / \log n)$ time, for two strings of length n (again assuming a fixed alphabet).

---

[10] This reflects our general level of ignorance about ethnicities in the then Soviet Union.

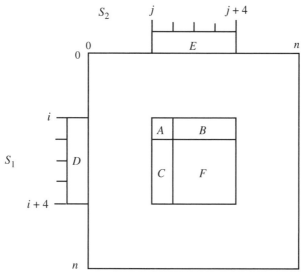

Figure 12.21: A single block with $t = 4$ drawn inside the full dynamic programming table. The distance values in the part of the block labeled F are determined by the values in the parts labeled A, B, and C together with the substrings of S_1 and S_2 in D and E. Note that A is the intersection of the first row and column of the block.

Consider the standard dynamic programming approach to computing the edit distance of two strings S_1 and S_2. The value $D(i, j)$ given to any cell (i, j), when i and j are both greater than 0, is determined by the values in its three neighboring cells, $(i - 1, j - 1)$, $(i - 1, j)$, and $(i, j - 1)$, and by the characters in positions i and j of the two strings. By extension, the values given to the cells in an entire t-block, with upper left-hand corner at position (i, j) say, are determined by the values in the first row and column of the t-block together with the substrings $S_1[i..i + t - 1]$ and $S_2[j..j + t - 1]$ (see Figure 12.21). Another way to state this observation is the following:

Lemma 12.7.1. *The distance values in a t-block starting in position (i, j) are a function of the values in its first row and column and the substrings $S_1[i..i + t - 1]$ and $S_2[j..j + t - 1]$.*

Definition Given Lemma 12.7.1, and using the notation shown in Figure 12.21, we define the *block function* as the function from the five inputs (A, B, C, D, E) to the output F.

It follows that the values in the last row and column of a t-block are also a function of the inputs (A, B, C, D, E). We call the function from those inputs to the values in the last row and column of a t-block, the *restricted block function*.

Notice that the total size of the input and the size of the output of the restricted block function is $O(t)$.

Computing edit distance with the restricted block function

By Lemma 12.7.1, the edit distance between S_1 and S_2 can be computed using the restricted block function. For simplicity, suppose that S_1 and S_2 are both of length $n = k(t - 1)$, for some k.

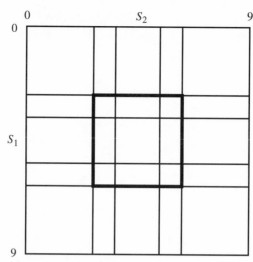

Figure 12.22: An edit distance table for $n = 9$. With $t = 4$, the table is covered by nine overlapping blocks. The center block is outlined with darker lines for clarity. In general, if $n = k(t-1)$ then the $(n+1)$ by $(n+1)$ table will be covered by k^2 overlapping t-blocks.

Block edit distance algorithm

Begin

1. Cover the $(n + 1)$ by $(n + 1)$ dynamic programming table with t-blocks, where the last column of every t-block is shared with the first column of the t-block to its right (if any), and the last row of every t-block is shared with the first row of the t-block below it (if any). (See Figure 12.22). In this way, and since $n = k(t - 1)$, the table will consist of k rows and k columns of partially overlapping t-blocks.

2. Initialize the values in the first row and column of the full table according to the base conditions of the recurrence.

3. In a rowwise manner, use the *restricted* block function to successively determine the values in the last row and last column of each block. By the overlapping nature of the blocks, the values in the last column (or row) of a block are the values in the first column (or row) of the block to its right (or below it).

4. The value in cell (n, n) is the edit distance of S_1 and S_2.

end.

Of course, the heart of the algorithm is step 3, where specific instances of the restricted block function must be computed. Any instance of the restricted block function can be computed $O(t^2)$ time, but that gains us nothing. So how is the restricted block function computed?

12.7.2. The Four-Russians idea for the restricted block function

The general Four-Russians observation is that a speedup can often be obtained by *precomputing* and storing information about all possible instances of a subproblem that might arise in solving a problem. Then, when solving an instance of the full problem and specific subproblems are encountered, the computation can be accelerated by looking up the answers to precomputed subproblems, instead of recomputing those answers. If the subproblems are chosen correctly, the total time taken by this method (including the time for the precomputations) will be less than the time taken by the standard computation.

In the case of edit distance, the precomputation suggested by the Four-Russians idea is to enumerate all possible inputs to the restricted block function (the proper size of the block will be determined later), compute the resulting output values (a t-length row and a t-length column) for each input, and store the outputs indexed by the inputs. Every time a specific restricted block function must be computed in step 3 of, the *block edit distance algorithm*, the value of the function is then retrieved from the precomputed values and need not be computed. This clearly works to compute the edit distance $D(n, n)$, but is it any faster than the original $O(n^2)$ method? Astute readers should be skeptical, so please suspend disbelief for now.

Accounting detail

Assume first that all the precomputation has been done. What time is needed to execute the *block edit distance algorithm*? Recall that the sizes of the input and the output of the restricted block function are both $O(t)$. It is not difficult to organize the input-output values of the (precomputed) restricted block function so that the correct output for any specific input can be retrieved in $O(t)$ time. Details are left to the reader. There are $\Theta(n^2/t^2)$ blocks, hence the total time used by the *block edit distance algorithm* is $O(n^2/t)$. Setting t to $\Theta(\log n)$, the time is $O(n^2/\log n)$. However, in the unit-cost RAM model of computation, each output value can be retrieved in constant time since $t = O(\log n)$. In that case, the time for the method is reduced to $O(n^2/(\log n)^2)$.

But what about the precomputation time? The key issue involves the number of input choices to the restricted block function. By definition, every cell has an integer from zero to n, so there are $(n + 1)^t$ possible values for any t-length row or column. If the alphabet has size σ, then there are σ^t possible substrings of length t. Hence the number of distinct input combinations to the restricted block function is $(n + 1)^{2t}\sigma^{2t}$. For each input, it takes $\Theta(t^2)$ time to evaluate the last row and column of the resulting t-block (by running the standard dynamic program). Thus the overall time used in this way to precompute the function outputs to all possible input choices is $\Theta((n + 1)^{2t}\sigma^{2t}t^2)$. But t must be at least one, so $\Omega(n^2)$ time is used in this way. No progress yet! The idea is right, but we need another trick to make it work.

12.7.3. The trick: offset encoding

The dominant term in the precomputation time is $(n + 1)^{2t}$, since σ is assumed to be fixed. That term comes from the number of distinct choices there are for two t-length subrows and subcolumns. But $(n + 1)^t$ overcounts the number of different t-length subrows (or subcolumns) that could appear in a real table, since the value in a cell is not independent of the values of its neighbors. We next make this precise.

Lemma 12.7.2. *In any row, column, or diagonal of the dynamic programming table for edit distance, two adjacent cells can have a value that differs by at most one.*

PROOF Certainly, $D(i, j) \leq D(i, j - 1) + 1$. Conversely, if the optimal alignment of $S_1[1..i]$ and $S_2[1..j]$ matches $S_2(j)$ to some character of S_1, then by simply omitting $S_2(j)$ and aligning its mate against a space, the distance increases by at most one. If $S_2(j)$ is not matched then its omission reduces the distance by one. Hence $D(i, j - 1) \leq D(i, j) + 1$, and the lemma is proved for adjacent row cells. Similar reasoning holds along a column.

In the case of adjacent cells in a diagonal, it is easy to see that $D(i, j) \leq D(i - 1, j - 1) + 1$. Conversely, if the optimal alignment of $S_1[1..i]$ and $S_2[1..j]$ aligns i against j, then $D(i-1, j-1) \le D(i, j)+1$. If the optimal alignment doesn't align i against j, then at least one of the characters, $S_1(i)$ or $S_2(j)$, must align against a space, and $D(i-1, j-1) \le D(i, j)$. $\square$

Given Lemma 12.7.2, we can *encode* the values in a row of a t-block by a t-length vector specifying the value of the first entry in the row, and then specifying the difference (offset) of each successive cell value to its left neighbor: A zero indicates equality, a one indicates an increase by one, and a minus one indicates a decrease by one. For example, the row of distances 5, 4, 4, 5 would be encoded by the row of offsets 5, -1, 0, $+1$. Similarly, we can encode the values in any column by such offset encoding. Since there are only $(n + 1)3^{t-1}$ distinct vectors of this type, a change to offset encoding is surely a move in the right direction. We can, however, reduce the number of possible vectors even further.

Definition   The *offset vector* is a t-length vector of values from $\{-1, 0, 1\}$, where the first entry must be zero.

The key to making the Four-Russians method efficient is to compute edit distance using only offset vectors rather than actual distance values. Because the number of possible offset vectors is much less than the number of possible vectors of distance values, much less precomputation will be needed. We next show that edit distance can be computed using offset vectors.

Theorem 12.7.1. *Consider a t-block with upper left corner in position (i, j). The two offset vectors for the last row and last column of the block can be determined from the two offset vectors for the first row and column of the block and from substrings $S_1[1..i]$ and $S_2[1..j]$. That is, no D value is needed in the input in order to determine the offset vectors in the last row and column of the block.*

PROOF   The proof is essentially a close examination of the dynamic programming recurrences for edit distance. Denote the unknown value of $D(i, j)$ by C. Then for column q in the block, $D(i, q)$ equals C plus the total of the offset values in row i from column $j + 1$ to column q. Hence even if the algorithm doesn't know the value of C, it can express $D(i, q)$ as C plus an integer that it can determine. Each $D(q, j)$ can be similarly expressed. Let $D(i, j + 1)$ be $C + J$ and let $D(i + 1, j)$ be $C + I$, where the algorithm can know I and J. Now consider cell $(i + 1, j + 1)$. $D(i + 1, j + 1)$ is equal to $D(i, j) = C$ if character $S_1(i)$ matches $S_2(j)$. Otherwise $D(i + 1, j + 1)$ equals the minimum of $D(i, j + 1) + 1$, $D(i + 1, j) + 1$, and $D(i, j) + 1$, i.e., the minimum of $C + I + 1, C + J + 1$, and $C + 1$. The algorithm can make this comparison by comparing I and J (which it knows) to the number zero. So the algorithm can correctly express $D(i + 1, j + 1)$ as $C, C + I + 1$, $C + J + 1$, or $C + 1$. Continuing in this way, the algorithm can correctly express each D value in the block as an unknown C plus some integer that it can determine. Since every term involves the same unknown constant C, the offset vectors can be correctly determined by the algorithm. $\square$

Definition   The function that determines the two offset vectors for the last row and last column from the two offset vectors for the first row and column of a block together with substrings $S_1[1..i]$ and $S_2[1..j]$ is called the *offset function*.

We now have all the pieces of the Four-Russians–type algorithm to compute edit distance. We again assume, for simplicity, that each string has length $n = k(t - 1)$ for some k.

Four-Russians edit distance algorithm

1. Cover the n by n dynamic programming table with t-blocks, where the last column of every t-block is shared with the first column of the t-block to its right (if any), and the last row of every t-block is shared with the first row of the t-block below it (if any).

2. Initialize the values in the first row and column of the full table according to the base conditions of the recurrence. Compute the offset values in the first row and column.

3. In a rowwise manner, use the *offset* block function to successively determine the offset vectors of the last row and column of each block. By the overlapping nature of the blocks, the offset vector in the last column (or row) of a block provides the next offset vector in the first column (or row) of the block to its right (or below it). Simply change the first entry in the next vector to zero.

4. Let Q be the total of the offset values computed for cells in row n. $D(n, n) = D(n, 0) + Q = n + Q$.

Time analysis

As in the analysis of the *block edit distance algorithm*, the execution of the *four-Russians edit distance algorithm* takes $O(n^2 / \log n)$ time (or $O[n^2/(\log n)^2]$ time in the unit-cost RAM model) by setting t to $\Theta(\log n)$. So again, the key issue is the time needed to precompute the block offset function. Recall that the first entry of an offset vector must be zero, so there are $3^{2(t-1)}$ possible offset vectors. There are σ^t ways to specify a substring over an alphabet with σ characters, and so there are $3^{2(t-1)}\sigma^{2t}$ ways to specify the input to the offset function. For any specific input choice, the output is computed in $O(t^2)$ time (via dynamic programming), hence the entire precomputation takes $O(3^{2t}\sigma^{2t}t^2)$ time. Setting t equal to $(\log_{3\sigma} n)/2$, the precomputation time is just $O(n(\log n)^2)$. In summary, we have

Theorem 12.7.2. *The edit distance of two strings of length n can be computed in* $O\left(\frac{n^2}{\log n}\right)$ *time or* $O\left(\frac{n^2}{(\log n)^2}\right)$ *time in the unit-cost RAM model.*

Extension to strings of unequal lengths is easy and is left as an exercise.

12.7.4. Practical approaches

The theoretical result that edit distance can be computed in $O\left(\frac{n^2}{\log n}\right)$ time has been extended and applied to a number of different alignment problems. For truly large strings, these theoretical results are worth using. But the Four-Russians method is primarily a theoretical contribution and is not used in its full detail. Instead, the basic idea of precomputing either the restricted block function or the offset function is used, but only for *fixed* size blocks. Generally, t is set to a fixed value independent of n and often a rectangular 2 by t block is used in place of a square block. The point is to pick t so that the restricted block or offset function can be determined in constant time on practical machines. For example, t could be picked so that the offset vector fits into a single computer word. Or, depending on the alphabet and the amount of space available, one might hash the input choices for rapid function retrieval. This should lead to a computing time of $O\left(\frac{n^2}{t}\right)$, although practical programming issues become important at this level of detail. A detailed experimental analysis of these ideas [339] has shown that this approach is one of the most effective ways to speed up the practical computation of edit distance, providing a factor of t speedup over the standard dynamic programming solution.

12.8. Exercises

1. Show how to compute the *value* $V(n,m)$ of the optimal alignment using only $\min(n,m) + 1$ space in addition to the space needed to represent the two input strings.

2. Modify Hirschberg's method to work for alignment with a gap penalty (affine and general) in the objective function. It may be helpful to use both the affine gap recurrences developed in the text, and the alternative recurrences that pay for a gap when terminated. The latter recurrences were developed in the exercise 27 of Chapter 11.

3. Hirschberg's method computes one optimal alignment. Try to find ways to modify the method to produce more (all?) optimal alignments while still achieving substantial space reduction and maintaining a good time bound compared to the $O(nm)$-time and space method? I believe this is an open area.

4. Show how to reduce the size of the strip needed in the method of Section 12.2.3, when $|m - n| < k$.

5. Fill in the details of how to find the actual alignments of P in T that occur with at most k differences. The method uses the $O(km)$ values stored during the k differences algorithm. The solution is somewhat simpler if the k differences algorithm also stores a sparse set of pointers recording how each farthest-reaching d-path extends a farthest-reaching $(d-1)$-path. These pointers only take $O(km)$ space and are a sparse version of the standard dynamic programming pointers. Fill in the details for this approach as well.

6. The k differences problem is an unweighted (or unit weighted) alignment problem defined in terms of the number of mismatches and spaces. Can the $O(km)$ result be extended to operator- or alphabet-weighted versions of alignment? The answer is: not completely. Explain why not. Then find special cases of weighted alignment, and plausible uses for these cases, where the result does extend.

7. Prove Lemma 12.3.2 from page 274.

8. Prove Lemma 12.3.4 from page 277.

9. Prove Theorem 12.4.2 that concerns space use in the P-against-all problem.

10. **The threshold P-against-all problem**

    The P-against-all problem was introduced first because it most directly illustrates one general approach to using suffix trees to speed up dynamic programming computations. And, it has been proposed that such a massive study of how P relates to substrings of T can be important in certain problems [183]. Nonetheless, for most applications the output of the P-against-all problem is excessive and a more focused computation is desirable. The *threshold P-against-all problem* is of this type: Given strings P and T and a threshold d, find every substring T' of T such that the edit distance between P and T' is less than d. Of course, it would be cheating to first solve the P-against-all problem and then filter out the substrings of T whose edit distance to P is d or greater. We want a method whose speed is related to d. The computation should increase in speed as d falls.

    The idea is to follow the solution to the P-against-all problem, doing a depth-first traversal of suffix tree $\mathcal{T}$, but recognize subtrees that need not be traversed. The following lemma is the key.

    Lemma 12.8.1. *In the P-against-all problem, suppose that the current path in the suffix tree specifies a substring S of T and that the current dynamic programming column (including the zero row) contains no values below d. Then the column representing an extension of S will also contain no values below d. Hence no columns need be computed for any extensions of S.*

Prove the lemma and then show how to exploit it in the solution to the threshold P-against-all problem. Try to estimate how effective the lemma is in practice. Be sure to consider how the output is efficiently collected when the dynamic programming ends high in the tree, before a leaf is reached.

11. Give a complete proof of the correctness of the all-against-all suffix tree algorithm.

12. Another, faster, alternative to the P-against-all problem is to change the problem slightly as follows: For each position i in T such that there is a substring starting at i with edit distance less than d from P, report only the *smallest* such substring starting at position i. This is the (P-against-all) *starting location problem*, and it can be solved by modifying the approach discussed for the threshold P-against-all problem. The starting location problem (actually the equivalent ending location problem) is the subject of a paper by Ukkonen [437]. In that paper, Ukkonen develops three hybrid dynamic programming methods in the same spirit as those presented in this chapter, but with additional technical observations. The main result of that paper was later improved by Cobbs [105].

    Detail a solution to the starting location problem, using a hybrid dynamic programming approach.

13. Show that the suffix tree methods and time bounds for the P-against-all and the all-against-all problems extend to the problem of computing similarity instead of edit distance.

14. Let R be a regular expression. Show how to modify the P-against-all method to solve the R-against-all problem. That is, show how to use a suffix tree to efficiently search for a substring in a large text T that matches the regular expression R. (This problem is from [63].)

    Now extend the method to allow for a bounded number of errors in the match.

15. Finish the proof of Theorem 12.5.2.

16. Show that in any permutation of n integers from 1 to n, there is either an increasing subsequence of length at least $\sqrt{n}$ or a decreasing subsequence of length at least $\sqrt{n}$. Show that, averaged over all the $n!$ permutations, the average length of the longest increasing subsequence is at least $\sqrt{n}/2$. Show that the lower bound of $\sqrt{n}/2$ cannot be tight.

17. What do the results from the previous problem imply for the *lcs* problem?

18. If S is a subsequence of another string S', then S' is said to be a *supersequence* of S. If two strings S_1 and S_2 are subsequences of S', then S' is a *common supersequence* of S_1 and S_2. That leads to the following natural question: Given two strings S_1 and S_2, what is the *shortest* supersequence common to both S_1 and S_2. This problem is clearly related to the longest common subsequence problem. Develop an explicit relationship between the two problems, and the lengths of their solutions. Then develop efficient methods to find a shortest common supersequence of two strings. For additional results on subsequences and supersequences see [240] and [241].

19. Can the results in the previous problem be generalized to the case of more than two strings? For instance, is there a natural relationship between the longest common subsequence and the shortest common supersequence of three strings?

20. Let T be a string whose characters come from an alphabet Σ with σ characters. A subsequence S of T is *nondecreasing* if each successive character in S is lexically greater than or equal to the preceding character. For example, using the English alphabet let $T = characterstring$; then $S = aacrst$ is a nondecreasing subsequence of T. Give an algorithm that finds the longest nondecreasing subsequence of a string T in time $O(n\sigma)$, where n is the length of T. How does this bound compare to the $O(n \log n)$ bound given for the longest increasing subsequence problem over integers.

21. Recall the definition of r given for two strings in Section 12.5.2 on page 290. Extend the

definition for r to the longest common subsequence problem for more than two strings, and use r to express the time for finding an *lcs* in this case.

22. Show how to model and solve the *lis* problem as a shortest path problem in a directed, acyclic graph. Are there any advantages to viewing the problem in this way?

23. Suppose we only want to learn the length of the *lcs* of two strings S_1 and S_2. That can be done, as before, in $O(r \log n)$ time, but now only using linear space. The key is to keep only the last element in each list of the cover (when computing the *lis*), and not to generate all of $\Pi(S_1, S_2)$ at once, but to generate (in linear space) parts of $\Pi(S_1, S_2)$ on the fly. Fill in the details of these ideas and show that the length of the *lcs* can be computed as quickly as before in only linear space.

    Open problem: Extend the above combinatorial ideas, to show how to compute the actual *lcs* of two strings using only linear space, without increasing the needed time. Then extend to more than two strings.

24. (This problem requires a knowledge of systolic arrays.) Show how to implement the longest increasing subsequence algorithm to run in $O(n)$ time on an $O(n)$-element systolic array (remember that each array element has only constant memory). To make the problem simpler, first consider how to compute the length of the *lis*, and then work out how to compute the actual increasing subsequence.

25. Work out how to compute the *lcs* in $O(n)$ time on an $O(n)$-element systolic array.

26. We have reduced the *lcs* problem to the *lis* problem. Show how to do the reduction in the opposite direction.

27. Suppose each character in S_1 and S_2 is given an individual weight. Give an algorithm to find an increasing subsequence of maximum total weight.

28. Derive an $O(nm \log m)$-time method to compute edit distance for the convex gap weight model.

29. The idea of forward dynamic programming can be used to speed up (in practice) the (global) alignment of two strings, even when gaps are not included in the objective function. We will explain this in terms of computing unweighted edit distance between strings S_1 and S_2 (of lengths n and m respectively), but the basic idea works for computing similarity as well. Suppose a cell (i, j) is reached during the (forward) dynamic programming computation of edit distance and the value there is $D(i, j)$. Suppose also that there is a fast way to compute a *lower bound*, $L(i, j)$, on the distance between substrings $S_1[i + 1, \ldots, n]$ and $S_2[j + 1, \ldots, m]$. If $D(i, j) + L(i, j)$ is greater than or equal to a known distance between S_1 and S_2 obtained from some particular alignment, then there is no need to propogate candidate values forward from cell (i, j). The question now is to find efficient methods to compute "effective" values of $L(i, j)$. One simple one is $|n - m + j - i|$. Explain this. Try it out in practice to see how effective it is. Come up with other simple lower bounds that are much more effective.

    Hint: Use the count of the number of times each character appears in each string.

30. As detailed in the text, the Four-Russians method precomputes the offset function for $3^{2(t-1)}\sigma^{2t}$ specifications of input values. However, the problem statement and time bound allow the precomputation of the offset function to be done *after* strings S_1 and S_2 are known. Can that observation be used to reduce the running time?

    An alternative encoding of strings allows the σ^{2t} term to be changed to $(t + 2)^t$ even in problem settings where S_1 and S_2 are not known when the precomputation is done. Discover and explain the encoding and how edit distance is computed when using it.

31. Consider the situation when the edit distance must be computed for each pair of strings from a large set of strings. In that situation, the precomputation needed by the Four-Russians

method seems more justified. In fact, why not pick a "reasonable" value for t, do the pre-computation of the offset function once for that t, and then embed the offset function in an edit distance algorithm to be used for all future edit distance computations. Discuss the merits and demerits of this proposal.

32. The Four-Russians method presented in the text only computes the edit distance. How can it be modified to compute the edit transcript as well?

33. Show how to apply the Four-Russians method to strings of unequal length.

34. What problems arise in trying to extend the Four-Russians method and the improved time bound to the *weighted* edit distance problem? Are there restrictions on weights (other than equality) that make the extension easier?

35. Following the lines of the previous question, show in detail how the Four-Russians approach can be used to solve the longest common subsequence problem between two strings of length n, in $O(n^2 / \log n)$ time.

13

Extending the Core Problems

In this chapter we look in detail at alignment problems in the more complex contexts typical of string problems that currently arise in computational molecular biology. These more complex problems require techniques that extend (rather than refine) the core alignment methods.

13.1. Parametric sequence alignment

13.1.1. Introduction

When using sequence alignment methods to study DNA or amino acid sequences, there is often considerable disagreement about how to weight matches, mismatches, insertions and deletions (indels), and gaps. The most commonly used alignment software packages require the user to specify fixed values for those parameters, and it is widely observed that the biological significance of the resulting alignment can be greatly affected by the choice of parameter settings. The following relates to alignments of proteins from the globin family and is representative of frequently seen comments in the biological literature:

> ... one must be able to vary the gap and gap size penalties independently and in a query dependent fashion in order to obtain the maximal sensitivity of the search. [81]

A similar comment appears in [432]:

> Sequence alignment is sensitive to the choices of gap penalty and the form of the relatedness matrix, and it is often desirable to vary these ...

Finally, from [446],

> One of the most prominent problems is the choice of parametric values, especially gap penalties. When very similar sequences are compared, the choice is not critical; but when the conservation is low, the resulting alignment is strongly affected.

Parametric sequence alignment is a tool that efficiently explores such penalty variation. It avoids the problem of choosing fixed parameter settings by computing the optimal alignment as a *function* of variable parameters for weights and penalties. The goal is to partition the parameter space into regions (which we will show are necessarily convex) such that in each region one alignment is optimal throughout and such that each region is maximal for this property. Thus parametric alignment allows one to see explicitly, and completely, the effect of parameter choices on the optimal alignment. Parametric sequence alignment was first used in a paper by Fitch and Smith [161] to demonstrate that an alignment objective function lacking an explicit term for gaps did not produce a biologically correct alignment (for sequences from hemoglobin chains) no matter what combination of parameter settings were used.

13.1.2. Definitions and first results

Parametric alignment problems arise both with the use of character-specific scoring matrices (amino acid substitution matrices such as PAM or BLOSUM matrices) and without their use. The treatment of these two cases differs somewhat, and each will be discussed separately. We first consider the case that no scoring matrix is used.

Definition For any alignment $\mathcal{A}$ of two strings, let mt_A, ms_A, id_A, and gp_A, respectively, denote the *number* of matches, mismatches, indels, and gaps contained in $\mathcal{A}$.

Without the use of character-specific scoring matrices, the value of $\mathcal{A}$ is

$$v_A(\alpha, \beta, \gamma, \delta) \equiv \alpha \times mt_A - \beta \times ms_A - \gamma \times id_A - \delta \times gp_A,$$

where α, β, γ, and δ are parameters that can be varied to adjust the relative contributions of the matches, mismatches, indels, and gaps. Note that the value of the alignment is a linear function, of the four parameters. When these four parameters have fixed values α_0, β_0, γ_0, and δ_0, then the resulting problem is the standard *fixed-parameter* problem of finding an alignment $\mathcal{A}$ maximizing the objective function: $\alpha_0 \times mt_A - \beta_0 \times ms_A - \gamma_0 \times id_A - \delta_0 \times gp_A$.

Parametric alignment studies how the optimal alignment changes as a function of the variable parameters α, β, γ, and δ. It is not hard to see that there are actually only three independent parameters, rather than four. However, because it is difficult to make three-dimensional displays, we will restrict attention to parametric problems where only two of the four parameters are varied. The remaining parameters are given fixed values. We leave it to the reader to extend the results to three variable parameters.

For illustration, let γ and δ be the variable parameters and let α and β be fixed at one. This is the typical choice when studying protein alignment (although a scoring matrix is also usually employed). If you imagine γ and δ to define two axes while a third (perpendicular) axis defines alignment value, then the value of alignment $\mathcal{A}$ as a function of γ and δ describes a *plane* in this three-dimensional space. Two distinct planes that are not parallel intersect at a line, which gives us

Lemma 13.1.1. *If the planes for alignments $\mathcal{A}$ and $\mathcal{A}'$ intersect and are distinct, then there is a line L in the γ, δ space along which $\mathcal{A}$ and $\mathcal{A}'$ have equal value; $\mathcal{A}$ has larger value than $\mathcal{A}'$ on one of the half-planes defined by L, and $\mathcal{A}$ has smaller value on the other half-plane. If the planes for $\mathcal{A}$ and $\mathcal{A}'$ do not intersect, then one of the alignments has larger value than the other at every γ, δ point.*

Consider now the subset of the γ, δ space where a fixed alignment $\mathcal{A}$ is optimal. If $\mathcal{A}$ is optimal at a point p, then p must be in the correct half-plane with respect to each alignment $\mathcal{A}'$ discussed above. That is, $\mathcal{A}$ is optimal in a subset of the space defined by the intersection of half-planes. Hence, we have

Corollary 13.1.1. *If $\mathcal{A}$ is optimal for at least one point p in the γ, δ space, then it is optimal only for point p, or it is optimal only for a line segment that contains p, or it is optimal only for a convex polygon that contains p.*

It then follows that

Theorem 13.1.1. *Given two strings S_1 and S_2, the γ, δ parameter space decomposes into convex polygons such that any alignment that is optimal for some γ, δ point in the interior of a polygon $\mathcal{P}$ is optimal for all points in $\mathcal{P}$ and nowhere else.*

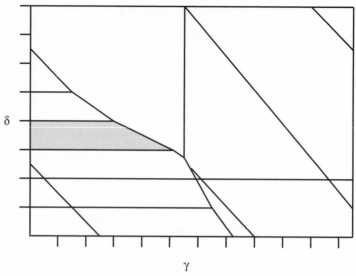

Figure 13.1: A polygonal decomposition of the γ, δ space for two strings not shown. In the program XPARAL the user selects a polygon (shown shaded) and the program displays the alignments that are optimal for that entire region.

Figure 13.1 illustrates one such polygonal decomposition of the γ, δ parameter space. Parametric alignment programs, such as XPARAL [203, 205] or the program developed by Waterman et al. [447, 463], or the program discussed in [486], take in two strings and a choice of two variable parameters and then compute and display the resulting polygonal decomposition. Further detailed examination of alignments and their corresponding polygons is then possible in an interactive mode.

13.1.3. Parametric alignment with the use of scoring matrices

When a character-specific scoring matrix is used (such as a PAM or BLOSUM matrix) that gives scores for specific matches and mismatches, then the parametric objective function changes somewhat.

Definition   For any alignment $\mathcal{A}$ of two strings, let $smt_\mathcal{A}$ and $sms_\mathcal{A}$, respectively, denote the total score (obtained from the scoring matrix) for the specific matches in $\mathcal{A}$ and the total score for the specific mismatches in $\mathcal{A}$. As before, $id_\mathcal{A}$ and $gp_\mathcal{A}$ denote the number of indels and gaps contained in $\mathcal{A}$.

Using scoring matrices, the parametric value of alignment $\mathcal{A}$ is $\alpha \times smt_\mathcal{A} + \beta \times sms_\mathcal{A} - \gamma \times id_\mathcal{A} - \delta \times gp_\mathcal{A}$. The term $\beta \times sms_\mathcal{A}$ is added to the expression because the scoring matrix incorporates the appropriate sign.

All of the results established in Section 13.1.2 still hold when scoring matrices are employed. We leave it to the reader to verify this claim.

13.1.4. Efficient algorithms for computing a polygonal decomposition

We now develop an efficient algorithm to compute polygonal decompositions given two strings, an objective function, and a choice of which two parameters are variable. The algorithm will be described at a high level and will work whether or not a scoring matrix is

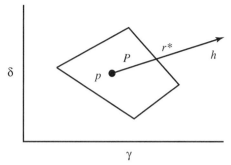

Figure 13.2: Alignment $\mathcal{A}$ is optimal in polygon P. Starting from point p, $\mathcal{A}$ remains optimal along ray h through point r^*.

employed. For illustration, we again assume that γ and δ are the two variable parameters, but the same method applies to any choice of parameters. The following problem, called the *ray-search problem*, is solved repeatedly in the inner loop of the decomposition algorithm:

Ray-search problem   Given an alignment $\mathcal{A}$, a point p where $\mathcal{A}$ is optimal, and a ray h in γ, δ space starting at p, find the furthest point (call it r^*) from p on ray h where $\mathcal{A}$ remains optimal. If $\mathcal{A}$ remains optimal until h reaches a border of the parameter space, then r^* is that border point on h. It is also possible that $r^* = p$. (See Figure 13.2.)

The ray-search problem is solved as follows:

Newton's ray-search algorithm

Set r to the (γ, δ) point where h intersects a border of the parameter space.

While $\mathcal{A}$ is not an optimal alignment at point r do
begin
Find an optimal alignment $\mathcal{A}^*$ at point r.
Set r to be the unique point on h where the value of $\mathcal{A}$ equals the value of $\mathcal{A}^*$.
end;
Set r^* to r.

This algorithm is Newton's classic zero-finding method specialized to a piecewise-linear function. The following three facts (Newton's laws) will be needed in the analysis. They are easy to establish and are left to the reader.

Lemma 13.1.2. 1) *Newton's ray-search algorithm finds r^* exactly.* 2) *Unless $\mathcal{A}$ is optimal at the initial setting of r, the last computed alignment $\mathcal{A}^*$ is cooptimal with $\mathcal{A}$ at r^* and yet is also optimal on h for some nonzero distance beyond r^*.* 3) *When Netwon's ray-search algorithm computes an alignment at a point r on h, none of the alignments computed previously (in this execution of Newton's algorithm) are optimal at r.*

It follows that if $r^* = p$, then Newton's method discovers this and returns an alignment $\mathcal{A}^*$ that is optimal at p and also optimal for some nonzero distance along h. It also follows that for any polygon P intersected by h, a single ray-search computes alignments at no more than two points of P.

Finding a polygon of the decomposition

Let $\mathcal{A}$ be an alignment that is optimal in the interior of an (unknown) polygon $\mathcal{P}(\mathcal{A})$, and let p be a known point where $\mathcal{A}$ is optimal. We now explain how to find the polygon $\mathcal{P}(\mathcal{A})$.

First pick any ray h from p and solve the ray-search problem along h. There are two degenerate cases that can occur: one is that the resulting r^* lies on a border of the parameter space; the other is that r^* is a vertex of the decomposition. We will consider those degenerate cases later and assume for now that they do not occur. Therefore, the ray search along h will find a point r^* that lies on an edge e of polygon $\mathcal{P}(\mathcal{A})$. By Newton's second law, the ray search will also return an alignment $\mathcal{A}^*$ that is optimal in the interior of the polygon bordering edge e. The intersection of the two planes for $\mathcal{A}$ and $\mathcal{A}^*$ describes a line l^* that contains edge e; then the full extent of e can be determined by solving two more ray-search problems using $\mathcal{A}$. In one problem, ray h is the half-line of l^* starting at r^* and running in one direction along l^*, and in the other problem ray h is the remaining half-line of l^* in the other direction. These two ray searches find the opposite endpoints of edge e. Once edge e is fully described, we look for another edge of $\mathcal{P}(\mathcal{A})$ by selecting another ray h from p that does not intersect edge e. By linking identical endpoints of edges that are found in this way, it is easy to continue selecting rays from p that do not intersect previously discovered edges or vertices of $\mathcal{P}(\mathcal{A})$. This method continues until all the discovered edges link together in a closed cycle, which then exactly describes the edges and vertices of $\mathcal{P}(\mathcal{A})$.

Consider now the two degenerate cases that may occur when trying to find an edge of $\mathcal{P}(\mathcal{A})$. In the case that r^* is on a border of the parameter space, then one of the edges of $\mathcal{P}(\mathcal{A})$ runs along that border and can be found by using the borderline as l^*. In the other case, when r^* is a vertex, the algorithm will realize this when it does the ray searches along l^* – alignment $\mathcal{A}$ will not be optimal past r^* on at least one of the two rays on l^* from r^*. When this occurs, the algorithm stops its current search for an edge and begins a new ray search from p using a ray that avoids r^* and all other previously discovered vertices and edges.

Filling in the parameter space

To compute a full polygonal decomposition, one first finds an alignment that is sure to be optimal in the interior of some (unknown) polygon. This is easy to do with a constant number of ray searches. For example, we could pick a point p, find an optimal alignment $\mathcal{A}$ at p, and then do a ray search along any ray h from p. If the resulting point r^* is not p, then $\mathcal{A}$ is optimal for some nonzero distance along h. If $r^* = p$, then the ray search returns a different alignment $\mathcal{A}^*$ that is optimal for some nonzero distance along h. Assume the first case occurs. We next need to determine if $\mathcal{A}$ is optimal in the interior of a polygon or just on an edge of it that runs along h. So pick a point p^* on the interior of the segment of h where $\mathcal{A}$ is known to be optimal, and do a ray search from p^* along a ray perpendicular to h. That ray search either confirms that $\mathcal{A}$ is optimal in the interior of a polygon or it returns an alignment that is optimal in the interior of a polygon. Either way, we find a first alignment to use to construct the first polygon of the decomposition.

Now we explain how the algorithm finds successive polygons to fill in the polygonal decomposition. When finding the first polygon $\mathcal{P}(\mathcal{A})$ (and for each additional polygon it finds), the algorithm inserts into a list, L, one vector $(smt_{\mathcal{A}^*}, sms_{\mathcal{A}^*}, id_{\mathcal{A}^*}, gp_{\mathcal{A}^*})$ (if the vector isn't already there) for each alignment $\mathcal{A}^*$ found to be optimal in the interior of a polygon bordering $\mathcal{P}(\mathcal{A})$. When $\mathcal{P}(\mathcal{A})$ is finished, the algorithm finds and marks one of the unmarked vectors from L, say for $\mathcal{A}'$, and then finds the polygon $\mathcal{P}(\mathcal{A}')$ where $\mathcal{A}'$ is optimal. Continuing in this way, the parameter space will be fully decomposed when all vectors in L are marked. Since the algorithm never chooses a marked vector, nor inserts two equal vectors, and during its construction of any polygon, it finds one alignment that is optimal in the interior of each of its neighboring polygons, it follows that each polygon in the full polygonal decomposition will be found exactly once by this approach.

13.1.5. Time analysis and the next idea

The above details lead to the following time analysis: Let R, E, and V be the number of polygons, edges, and vertices, respectively, in a decomposition, and let $O(nm)$ be the time to compute a single fixed-parameter alignment for sequence of lengths n and $m > n$. How many ray searches are executed to find a polygon $\mathcal{P}(\mathcal{A})$, given $\mathcal{A}$ and p? Let d be the number of edges of $\mathcal{P}(\mathcal{A})$. Then $3d$ ray searches are done to find the edges of $\mathcal{P}(\mathcal{A})$, and in the highly degenerate case that the selected rays from p intersect all the vertices of $\mathcal{P}(\mathcal{A})$, then another $3d$ (wasted) ray searches may be done as well. Hence at most $6d$ ray searches suffice to describe $\mathcal{P}(\mathcal{A})$. Each edge lies on at most two polygons, so the algorithm does at most $12E$ ray searches to find the complete decomposition. Further, from Newton's third law each ray search requires at most R fixed-parameter alignment computations, so the complete decomposition requires at most $12RE$ fixed-parameter alignments, which can be done in $O(ERnm)$ time. This unsatisfactory bound will be improved with one additional idea.

The new idea is to modify Newton's algorithm (given $\mathcal{A}$) to pick the initial point r far enough on h to be at or beyond the (unknown) point r^*, yet as close to r^* as present information allows. Consider any alignment $\mathcal{A}'$ computed before the present execution of Newton's method. Compute the intersection of the planes for $\mathcal{A}$ and $\mathcal{A}'$ and project that line onto the γ-δ plane. If the projection intersects h, then the initial r need not be any further from p than r', since $\mathcal{A}'$ has greater value than $\mathcal{A}$ beyond r'. If the projection misses h, then $\mathcal{A}$ has greater value than $\mathcal{A}'$ at every point on h. Any intersection and projection can be done in constant time. Repeating this for each previously computed alignment $\mathcal{A}'$, we set the initial r to the point on h closest to p among all the computed r' points.

The modified Newton's method clearly reduces the number of needed alignments in practice, but by how much, and how much added time is needed to implement it? We consider the second question first. During its entire execution, the decomposition algorithm keeps a list L' that is a superset of list L. Whenever an alignment is computed, the vector for that alignment is placed into L' (if it isn't already there), whether or not that alignment is optimal for a polygon, for an edge or only for a vertex. When we begin any execution of Newton's ray-search method, we use L' as explained above to find the initial point r. Since each point r' is computed in constant time, the added time needed to use L' in a single ray search is proportional to the size of L'. We claim that the size of L' is at most $V + E + R$, and over the entire execution of the algorithm, only $V + E + R$ distinct vectors are computed. The reason is that dynamic programming is deterministic, so no matter how many times an optimal alignment is computed at a fixed vertex, the same alignment (and the same vector) is computed each time. Similarly, even if alignments are computed at many points along a polygon edge, deterministic dynamic programming code will return the same alignment each time, since any alignment that is optimal for some point on the edge is optimal for all points on the edge. Finally, the same vector is returned from each alignment inside a polygon, since all optimal alignments inside a polygon have the same vector. Thus, the algorithm only returns one vector for each vertex, edge, or polygon of the decomposition, or $V + E + R$ vectors in total. The added bookkeeping time for using L' is just $O(V + E + R)$ per ray search or $O(12E(V + E + R))$ overall.

For the analysis of the number of fixed-parameter alignments computed using the modified algorithm, call an alignment computation *redundant* if it returns an alignment with a vector that is already in L'. We claim that in any single ray search (using Newton's modified method with $\mathcal{A}$) only the last alignment computation in that ray search can be redundant. Consider the two cases: Either $\mathcal{A}$ is optimal at the initial point r or it is not. If $\mathcal{A}$ is optimal at the initial r, then only one alignment is computed in the ray search and the claim holds vacuously. Otherwise, when $\mathcal{A}$ is not optimal at the initial r, any redundant alignment, $\mathcal{A}'$ say, is computed at a point closer to p than the initial r. That means that the intersection of $\mathcal{A}$ and $\mathcal{A}'$ projects to a point on h closer to p than the initial r. Since, by Newton's third law, each vector computed during the ray search is distinct, $\mathcal{A}'$ must have been in L' before the present ray search began. But that would contradict the choice of the initial point r.

We can now analyze the time to find a complete polygonal decomposition. Use of the modified Newton's method does not change the fact that at most $12E$ ray searches are done. Only one alignment computation per ray search is redundant, so each of the other fixed-parameter alignment computations must find a new vector to add to L'. Since the size of L' is bounded by $V + E + R$, it follows that the complete polygonal decomposition is computed using at most $V + 13E + R$ fixed-parameter alignments. This leads to an overall time bound of $O(12E(V + E + R) + (V + 13E + R)nm)$. Now a polygonal decomposition can be viewed as a connected planar graph, and when each vertex is incident with at least three edges (as in the case of a polygonal decomposition), then $V \le E \le 3R$. This is easy to show using Euler's theorem on planar graphs, although it is not true in general for planar graphs. Thus the terms $12E$, $V + E + R$, and $V + 13E + R$ are each $O(R)$. Hence, the above time bound becomes $O(R^2 + Rnm)$, which is $O(R + nm)$ per polygon.

The bound of $O(R + nm)$ per polygon holds no matter what parametric choices are made. Moreover, if we use some form of alignment where the fixed-parameter computation takes $O(C)$ time, rather than $O(nm)$ time, then the time bound is $O(R + C)$ per polygon. However, we will show below that when no character-specific scoring matrices are used, then $R = O(nm)$. In fact, when global alignment is computed and no scoring matrices are used, then $R < n^{2/3}$. When scoring matrices are used, but γ and δ are the chosen variable parameters, then again $R = O(nm)$. Putting all the pieces together will establish

Theorem 13.1.2. *For most of the (important) parameter choices, a full polygonal decomposition can be found in $O(nm)$ time per polygon (i.e., proportional to the time needed to compute just a single fixed-parameter alignment).*

13.1.6. Bounding the number of polygons in the decomposition

In this section we consider how many polygons there can be in various types of polygonal decompositions. Although this question is interesting in its own right, it has two practical consequences. First, it addresses the concern that there may be so many polygons that the decomposition will give no interesting information. Second, because the polygonal decomposition method discussed in the previous section runs in $O(R + nm)$ time per polygon, we must bound R to establish Theorem 13.1.2. We will discuss two results that are detailed in [203].

Theorem 13.1.3. *Consider the following scoring scheme that does not use scoring matrices: $\alpha \times mt - \beta \times ms - \gamma \times id - \delta \times gp$. No matter which two of the four parameters are chosen to be variable, and no matter what type of alignment (global, local, etc.) is computed, the polygonal decomposition can contain at most $O(nm)$ polygons.*

PROOF   For illustration, suppose γ and δ are variable and that α and β are fixed at α_0 and β_0. Then for any alignment $\mathcal{A}$, $\alpha_0 \times mt_A - \beta_0 \times ms_A$ is a constant, say C_A, and the value of $\mathcal{A}$ is $C_A - \gamma \times id_A - \delta \times gp_A$. We associate $\mathcal{A}$ with the triple (C_A, id_A, gp_A). Now if $\mathcal{A}'$ is another alignment that has id_A spaces and gp_A gaps, and $C_A < C_{A'}$, then $\mathcal{A}$ can never be optimal at any γ, δ point. Therefore, among all triples whose last two terms are id_A, gp_A, at most one of those triples is associated with an alignment that is optimal at some point. If $n \le m$ are the lengths of the two strings, then there can be at most $n+1$ gaps and $m + n$ spaces in an alignment of the two strings, so there are at most $O(nm)$ triples associated with an optimal alignment. The theorem follows because any two alignments associated with the same triple are optimal at exactly the same points.   □

It should also be clear that when a scoring matrix is used to score the contribution of the matches and mismatches, then there are still only $O(nm)$ polygons possible as long as γ and δ are the variable parameters. However, when scoring matrices are used and α and β are the variable parameters, then the $O(nm)$ bound does not follow from the above reasoning. In fact, we do not know any reasonable bound on the number of polygons in that case, and empirically the number seen in the decompositions is much larger than when γ and δ are the variable parameters.

The bound of $R = O(nm)$ is enough to complete the proof of Theorem 13.1.2. However, when no scoring matrices are used and the alignments are global, then R is much smaller than nm, as we will show next.

The special case of global alignment

We want to bound the number of polygons that can be obtained in the parametric decomposition when no scoring matrices are used and when the alignment is global. Without loss of generality, consider the following two-parameter objective function: Maximize $mt_A - \beta \times ms_A - \gamma \times id_A$. The next Lemma is the key.

Lemma 13.1.3. *For any alignment $\mathcal{A}$ with corresponding vector (mt, ms, id): $2mt + 2ms + id = N$, where $N = n + m$ is the sum of the two sequence lengths. Hence $mt + ms + id/2 = N/2$ for any global alignment.*

PROOF   A match or a mismatch involves two characters. Thus the total number of characters that form part of a match or mismatch is $2(mt + ms)$. An indel involves only one character from one sequence. All spaces are counted in a global alignment, so the number of characters involved in indels is id. Each of the N characters is counted once as part of a match, a mismatch, or an indel.   □

Corollary 13.1.2. *Every global alignment has the same value (N) at the point $\beta = -1$, $\gamma = -1/2$.*

PROOF   Plugging into the objective function, we see that at point $(-1, -1/2)$ every global alignment CA has value $mt_A - ms_A - id_A/2$, which equals N by Lemma 13.1.3.   □

Since every alignment has the same value at $(-1, -1/2)$, any boundary between two adjacent polygons in the decomposition must go through point $(-1, -1/2)$. This says that each polygon in the decomposition is a semi-infinite (pie-shaped) polygon radiating from the point $(-1, -1/2)$. We now establish a bound on the number of polygons.

Theorem 13.1.4. *In the case of global alignment with no scoring matrices, there can be at most $O(n)$ polygons in the parametric decomposition, where $n \le m$.*

PROOF   Since each polygon boundary radiates from the point $(-1, -1/2)$, each polygon boundary in the positive β, γ quadrant (the area of interest) must intersect either the horizontal or the vertical axis of the space. We will show that the number of intersections of the vertical axis cannot exceed n.

Along the vertical (γ) axis, β is zero, so the parametric problem along that axis is to maximize $mt_A - \gamma \times id_A$ as a function of the single parameter γ. Clearly, as γ increases, mt must decrease whenever the optimal changes (i.e., at each breakpoint along the γ axis). But since the number of matches can only vary from zero to n, there can be at most n polygon boundaries that intersect the γ axis.

The same upper bound of n boundaries holds (by the same reasoning) along the horizontal axis. This proves the theorem.   □

Actually, one can prove stronger statements. First, it is easy to show that no polygon boundaries actually intersect the horizontal (β) axis. We leave that as an exercise. Moreover, it was shown [203] that the number of polygons is bounded by $O(n^{\frac{2}{3}})$. So in the case of global alignment without scoring matrices, the number of polygons in the parametric decomposition is quite small.

13.1.7. Uses for parametric alignment

We will discuss two uses for parametric alignment.

Sensitivity analysis

The first use is simple *sensitivity analysis* to determine the robustness of an alignment. When an optimal alignment is computed using fixed parameter settings (a point p in the parameter space), one should check to see how sensitive the alignment is to changes in the parameters. Pearson [359] puts it this way:

> One should be very cautious when evaluating "significant" alignments that are seen only with one or two combinations of scoring matrices and gap penalties.

Parametric alignment does not directly change scoring matrices (although it can change the weighted contribution of the scores), but gap initiation and extension penalties can be directly varied. The sensitivity of an alignment at point p can be examined by determining how far p is from a polygon boundary of the parametric decomposition. If p is contained in a polygon $\mathcal{P}$, then additional information can be obtaining from the size of $\mathcal{P}$ and from how much the alignment value changes over $\mathcal{P}$ and over the neighboring polygons of $\mathcal{P}$.

Efficient computation of all cooptimals

A second example draws on a study done by Barton and Sternberg [51]. This example illustrates that it is often important to compute all the cooptimal alignments, rather than just a single optimal alignment, and it shows how parametric alignment can make this a more tractable task.

Without going into the full details, Barton and Sternberg aligned certain pairs of proteins using (essentially) the parametric objective function: $V(\mathcal{A}, \gamma, \delta) = smt_A + sms_A - \gamma \times id_A - \delta \times gp_A$. They computed one optimal alignment for each of 121 γ, δ combinations, where γ and δ each took on integer values in the range zero to ten. They were interested in whether there are optimal alignments, for any of the 121 parameter settings, having a particular property (which we won't detail). None of the 121 alignments they computed had that property. They concluded that the standard alignment model (as encoded by the above objective function) was not sufficient to find alignments with the desired property, and that a different alignment model was necessary.

However, it was later shown [205] that there are integral choices for γ and δ in the proper range for which an optimal alignment does have the property of interest. Moreover, there are nonintegral choices for the parameters where *most* of the optimal alignments have that property. The point is that generally, for any point p in the parameter space, there will be many cooptimal alignments at p. All the cooptimal alignments will have equal value at point p, but other properties of the cooptimal alignments can differ significantly. If one is interested in whether there is an optimal alignment $\mathcal{A}$ (optimal with respect to a given objective function) with a certain additional property, it is not sufficient to compute only a single optimal alignment at each query point. One has to look at all the cooptimals. Otherwise, as occurred in [51], one may be unlucky and miss the cooptimal alignments of interest. Similarly, it is noted in [486], that a large empirical study on how different weights, penalties and matrices affect the quality of the alignment [448], often missed significant alignments that a full parametric search later found.

The advice – to compute all the cooptimals – is well accepted but not easily followed because it appears very demanding to compute all cooptimals at all points of interest in a parameter space. This is where parametric alignment can be very helpful.

To find every alignment that is optimal at one or more (unknown) point(s) in a given parameter space, first compute the parametric decomposition of the parameter space. Then compute every cooptimal alignment at each *vertex* of that decomposition. In this way, every alignment that is optimal for some point(s) in the parameter space will be found. Thereafter, the cooptimals can be computed in $O(n + m)$ time per alignment (see Exercise 12 on page 245). For greater efficiency and robustness it is preferable to select one point in the interior of each polygon and compute the cooptimals only at those points. By this method, all the (robust) alignments will be found that are optimal for some nonzero area of the parameter space, and each such alignment will be computed only once.

This approach to computing robust cooptimal alignments is empirically observed to be very efficient. In the case of the study in [51], there are only 17 polygons in the parametric decomposition of the entire γ, δ space (partly displayed in Figure 13.1). Moreover, only 120 alignments are optimal for some interior point of those polygons. Parametric alignment makes it tractable to produce each of those 120 robust cooptimal alignments, once and only once, by first identifying a set a 17 interior points where the cooptimals should be enumerated. This is in contrast to the obvious approach of computing all the cooptimal alignments at the 121 integral points that were selected in [51]. In that case, 1,105 alignments would be computed. That nearly tenfold redundancy is due to many integral points that fall into the same polygon (since there are only 17 polygons and 121 integral points). Worse, many of the 121 integral points in Figure 13.1 lie on vertices or edges of the decomposition (for reasons explained in [205]), where every alignment that is optimal in a neighboring polygon will be produced when all the cooptimals are computed at that point. This large redundant generation of alignments makes enumeration of cooptimal alignments seem unattractive. But parametric decomposition pinpoints places to query so that all, and only, robust cooptimal alignments will be produced and so that each will be produced exactly once.

13.2. Computing suboptimal alignments

Optimal alignment, even with a wide range of models and parameter choices, does not always identify the biological phenomena that it is intended to reflect. The problem is fivefold: The available objective functions might not reflect the full range of biological forces that cause differences between strings; the objective functions might not induce the optimal alignment to form the desired shape (recall the discussion of cDNA matching); the data might contain errors that confound the algorithms; there may be ties for the optimal alignment; or there may be many nearly optimal alignments that are biologically more significant than any optimal one.[1]

One response to this problem is to accept that optimal alignment is only a crude reflection of biological significance and should not be relied on exclusively. Instead, it is desirable to generate a larger, yet still manageable, set of alignments that may contain something of more biological interest. Yet, even though optimality may not correspond to "true biological significance", if the objective function has been constructed to model the biology, then a set of alignments *close* to optimal should be preferred to a random collection of alignments. Following this philosophy, optimal alignment is used as part of a filtering mechanism to find a candidate set of alignments of manageable size that can then be evaluated by some additional criteria or by human expertise. In the previous sections we saw one way to do this, namely by varying parameters, finding the resulting polygons, and obtaining one or more optimal alignments per polygon. Another concrete approach is to generate "near-optimal" alignments for some fixed parameter setting(s). This approach was initially developed in [490], [443], [445], and [460]. Our discussion follows the paper of Naor and Brutlag [346]; a more space efficient version of their approach appears in [96].

13.2.1. First definitions and first results

In this discussion we will rely heavily on the concept of alignment graphs introduced in Section 11.6.3 (page 228), and on the correspondence between string alignments and paths in an alignment graph. The results we will develop hold for the model of *maximum-value global alignment*, and we therefore do not specify any specific objective function. Instead, we simply let $s(u, v)$ denote the value of the edge (u, v) in the alignment graph, and let $V(S_1, S_2)$ denote the value of the optimal alignment of strings S_1 and S_2, for some fixed but unspecified maximizing objective function. Extending these results to other models of alignment or to edit distance is left as an exercise. We begin with some needed definitions about optimal and near-optimal paths in an alignment graph.

> **Definition**   Let R be a path in the alignment graph from the start node $s \equiv (0, 0)$ to the destination node $t \equiv (n, m)$. Path R corresponds to some global alignment (not necessarily optimal) of strings S_1 and S_2.

> **Definition**   For any pair of nodes x, y in the alignment graph, let $l(x, y)$ be the length of the longest path from x to y. Let R^* be the path corresponding to an optimal global alignment, so that the length of R^* is equal to $l(s, t) = V(S_1, S_2)$.

> **Definition**   For path R, let $\delta(R)$ be the length of R^* minus the length of R. $\delta(R)$ is called the "deviation" of R (from the optimal), and R is called a $\delta(R)$-*near-optimal* path. A path R is called δ-*near-optimal* if $\delta(R) = \delta$.

> **Definition**   For an edge $e = (u, v)$, let $\delta(e) = l(s, t) - [l(s, u) + s(u, v) + l(v, t)]$. That is, $\delta(e)$ is the difference between the length of R^* and the length of the longest s-to-t path that is *forced* to go through edge e.

---

[1] It is crucial to emphasize here that the designation "optimal" is only with respect to a given *objective function*. It should not be taken as synonymous with "correct" or "desirable", although the literature is often very unclear on this issue. The whole point of this section is that "optimal" is not always "most biologically significant", or even "adequate for biological interest".

Computing the optimal alignment of S_1 and S_2 gives $l(s, u)$ for each node u, and computing the optimal alignment of the reverse of those two strings gives $l(v, t)$ for each node v. Hence we have

Lemma 13.2.1. $\delta(e)$ *can be computed for all edges in the time used to compute two optimal alignments plus time proportional to the number of edges.*

Δ near-optimal alignments

One way to study near-optimal paths (or alignments) is to specify a cutoff value Δ and then compute information about the set of all paths whose deviation from the optimal is at most Δ, that is, the set of all paths that are δ-near-optimal for some $\delta \leq \Delta$. We will see how to efficiently *count* and *enumerate* this set of paths. A less explicit alternative is explored in Exercise 11 at the end of this chapter.

13.2.2. A useful reweighting

The following ideas facilitate both counting and enumerating near-optimal paths.

Definition For an edge $e = (u, v)$, let $\epsilon(e) = l(u, t) - [s(u, v) + l(v, t)]$.

The interpretation of $\epsilon(e)$ is that it is the "additional penalty for using e on the path from u to t, rather than following the optimal (longest) path from u to t directly" [346]. The following theorem suggests the importance of these values.

Theorem 13.2.1. *For any s-to-t path R, $\delta(R) = \sum_{e \in R} \epsilon(e)$.*

PROOF Let $R = v_0, v_1, \ldots, v_{k-1}, v_k$, where $v_0 = s$ and $v_k = t$. Also, let $|R|$ denote the length of R. Then

$$\sum_{e \in R} \epsilon(e) = \sum_{i=0}^{k-1} \epsilon(v_i, v_{i+1})$$

$$= \sum_{i=0}^{k-1} [l(v_i, t) - s(v_i, v_{i+1}) - l(v_{i+1}, t)]$$

$$= \sum_{i=0}^{k-1} [l(v_i, t) - l(v_{i+1}, t)] - \sum_{i=0}^{k-1} s(v_i, v_{i+1})$$

$$= \left(\sum_{i=1}^{k-1} [l(v_i, t) - l(v_i, t)] + l(v_0, t) - l(v_k, t) \right) - |R|$$

$$= l(s, t) - |R| = |R^*| - |R| = \delta(R). \qquad \square$$

Corollary 13.2.1. Consider a path R' from s to u and let δ denote $\sum_{e \in R'} \epsilon(e)$. Then the s-to-t path R consisting of path R' followed by the longest u-to-t path is a δ-near-optimal path.

PROOF By definition of $\epsilon(e)$, $\epsilon(e) = 0$ for any edge e on the longest u-to-t path. Hence $\delta(R) = \delta$ by Theorem 13.2.1. $\square$

This corollary will be the key to efficient counting and enumeration of near-optimal paths. It allows one to determine whether an initial part of an s-to-t path will be on a full δ-near-optimal path from s to t, without looking at the full path.

13.2.3. Counting and enumerating near-optimal paths

Why count?

One of the main problems in the use of alignment models in biology is that an alignment with a large similarity score or large statistical significance (compared to alignments of random strings) may not be an alignment of greatest biological significance. Consequently, a number of researchers have sought effective "rules-of-thumb" to try to identify alignments with useful biological information (see [73, 127]). One approach, discussed in [346], uses the *number* of near-optimal alignments as a rule-of-thumb. From empirical studies of specific molecular sequences, they report that alignments with a large number of suboptimals close to the optimal tend not to correctly highlight important biological information. However, alignments with a small number of suboptimals close to the optimal do tend to correctly reflect important biological information in the sequences. Therefore, the number of suboptimal solutions "near" the optimal may be an effective rule-of-thumb for separating biologically informative alignments from uninformative ones. Thus the ability to efficiently count the number of near-optimal paths may be the most important result we develop on the subject of near-optimal alignments.

How to count

Definition Let $N(v, \delta)$ be the *number* of δ-near-optimal s-to-t paths that go through node v.

For a given value Δ, the number of s-to-t paths whose deviation from R^* is at most Δ is

$$\sum_{\delta \leq \Delta} N(t, \delta).$$

We compute that sum by evaluating the following recurrence for each node v and for each "needed" value of δ:

$$N(v, \delta) = \sum_{(u,v) \in E} [N(u, \delta - \epsilon(u, v))].$$

That is, for node v, we will compute and store $N(v, \delta)$ for each number $\delta = \delta' + \epsilon(u, v)$ such that $N(u, \delta')$ has been computed for some predecessor node u of node v. In practice that means scanning the N values computed for each predecessor of v, and using those values to update the appropriate N value for v.

Note that, by Corollary 13.2.1, there are actually $N(v, \delta)$ paths through v whose deviation from R^* is exactly δ, since each of the $N(v, \delta)$ paths to v can be completed using the longest v-to-t path. Without the use of $\epsilon(e)$ values (or some similar idea), one would have to count all paths to v, even those whose deviation would ultimately be greater than Δ. Without $\epsilon(e)$ values (or some similar device), the algorithm would not know whether a path to v could be extended to a full path to t with deviation at most Δ. We return to this issue in Exercise 13.

Enumeration

The δ-near-optimal paths can be enumerated in order of increasing δ, and the enumeration can be terminated when $\delta = \Delta$ or when some fixed number of paths have been found. The general idea is one that is used in almost all enumeration methods. A tree enumerating partial paths is maintained. Each internal node x in the tree corresponds to a path R' from

s to a node u in the alignment graph. The value $d(x) = \sum_{e \in R'} \epsilon(e)$ is stored at node x, and by Corollary 13.2.1 there is a path from s to t of exactly that deviation from R^*. Each internal node x in the tree is expanded once for each edge out of u in the alignment graph. An expansion adds the edge out of u onto path R' and creates a new internal node in the tree. To enumerate δ-near-optimal paths in order of increasing δ, the algorithm keeps an ordered list (or priority queue) of known deviations for paths already discovered, and it chooses to expand the node x of smallest $d(x)$ value; at that point it outputs the path whose deviation is exactly $d(x)$. This method is simpler if all the paths with deviation at most Δ need not be enumerated in order of deviation. An additional improvement for enumerating the paths in order is developed in [346].

13.2.4. An alternative approach to suboptimal alignment

Enumerating all the suboptimal alignments within a fixed deviation of the optimal can be very useful, but it may also produce an overwhelming number of "similar-looking" alignments, and it may be very time consuming. Waterman and Eggert [462] suggested another, more efficient way to enumerate a broad range of suboptimal alignments. Their method will miss many of the alignments generated by the enumeration method presented above, but for many purposes it may generate a sufficient range of alignments for the biologist to examine and choose from.

To explain the idea in [462], it is useful to again consider the alignment graph. An optimal alignment corresponds to a longest path from the upper left corner to the lower right corner in the alignment graph. The dynamic program that finds one optimal alignment also finds all edges in the graph that are contained in some longest path. The method of Waterman seeks to find a next best alignment (possibly also optimal) that differs in a "significant" way from the first optimal alignment (i.e., is not just a minor deviation from the first alignment). After it finds the first longest path, it removes all the edges that appear on that path along with any edges that touch those edges. It then finds a longest path in the remaining graph and iterates in this way. The paths it finds specify alignments that are fairly different from each other. This method works both for local and for global alignment.

13.3. Chaining diverse local alignments

Having found many optimal or suboptimal pairs of highly similar substrings in two strings S and S', how should a "good" subset of these pairs be selected to display a relationship between strings S and S' over their entire lengths? This problem of *chaining* together regions of high similarity given a set of diverse local alignments is a general problem of importance that often arises in computational molecular biology. From the perspective of biological modeling, it is essentially unsolved because one rarely has compelling criteria to evaluate the goodness of a chosen subset. Nevertheless, the most common model presently used (to be explained below) does allow an efficient algorithm for selecting a best subset of pairs (under that model). We begin with a one-dimensional version of this problem.

A one-dimensional chaining problem

Consider a set of r (possibly) overlapping intervals drawn on the line R, where each interval j has some associated value $v(j)$. The problem is to select a subset of *nonoverlapping* intervals whose values sum to as large a number as possible. See Figure 13.3.

One place this problem arises is in the *gene assembly* problem (see Section 18.2). In

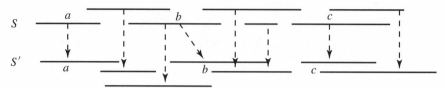

Figure 13.3: Line *R* and the *r* smaller intervals in *R*. Each interval has an associated value, and the problem is to pick a subset of nonoverlapping intervals with largest total value.

Figure 13.4: Pairs of highly similar substrings from *S* and *S'*. Each pair is connected by a directed line. The pairs denoted *a,b,* and *c* form a chain. The values of the pairs are not shown.

that problem, *R* is a DNA sequence of a eukaryotic gene, and each interval is a putative exon in that gene. The value $v(j)$ reflects the "goodness" or "likelihood" that the interval is in fact an exon. Generally (with the exception of alternative splicing), real exons do not overlap in a gene, but the putative exons do. Therefore, the gene assembly problem is to select a nonoverlapping subset of putative exons to assemble into a single deduced gene.

One-dimensional Algorithm

Let *I* be a list of all the $2r$ numbers representing the locations of the endpoints of the intervals in *L*. Sort the numbers in *I*, and annotate each entry in *I* with the name of the interval it is part of and whether it is a left or a right endpoint. For convenience, let *I* be a one-dimensional array.

Set *max* to zero.
For *i* from 1 to $2r$ do
begin
If $I[i]$ represents the *left* end of an interval, say interval *j*, then set $V[j]$ to $v(j)+ max$.
If $I[i]$ represents the *right* end of interval *j*, then set *max* to the maximum of *max* and $V[j]$.
end.

The value of *max* at the end of the algorithm is the value of the optimal solution. The time for the algorithm is $O(r \log r)$ and is dominated by the time to sort the numbers in *I*. We leave as an exercise the problem of outputting the set of intervals for the optimal solution.

The two-dimensional chain problem

Now we return to the original problem of selecting pairs of highly similar substrings. Given two strings *S* and *S'*, one may know several pairs of substrings of *S* and *S'* that have high similarity. These pairs might be identified using the techniques of Sections 13.2.4 or 11.7.3. However, the identified substrings in *S* (or *S'*) may overlap each other, and the problem is to select a "good" subset of substring pairs so that none of the selected substrings from *S* overlap each other and none of the selected substrings from *S'* overlap each other. (See Figure 13.4.) One example of this problem involves the alignment of DNA strings to protein strings, which will be discussed in Section 18.1.

Abstractly, the input to the problem can be represented as a set of rectangles drawn in

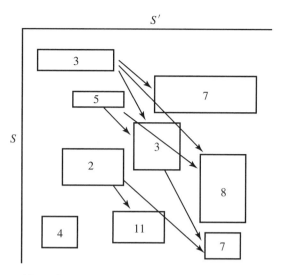

Figure 13.5: Each directed line points from a rectangle to a rectangle that can succeed it in a chain. There are two lines missing that are implied by transitivity. The value of each rectangle is written inside it. The optimal chain contains rectangles of values 5, 3, and 7 and has total value 15.

the plane, where the vertical sides of each rectangle are parallel to the y axis and horizontal sides are parallel to the x axis. If S is written on the y axis and S' is written on the x axis, then the vertical side of a rectangle represents a substring in S, and the horizontal side of a rectangle represents a substring in S'. Each rectangle i has an associated value $v(i)$, which, for example, could be the similarity value of the two substrings.

> **Definition**   A subset of the rectangles is called a *chain* if no horizontal or vertical line intersects more than one rectangle in the subset and if the rectangles can be ordered so that each one is below and to the right of its predecessor (see Figure 13.5). The value of a chain is the sum of the values of the rectangles in the chain.

The Chain Problem   Find a chain with maximum value over all chains.

A conceptually simple way to solve the chain problem is to construct a directed acyclic graph G with one node for each rectangle and a directed edge from node i to node j if and only if rectangle j could follow rectangle i in some chain. Graph G also contains a start node s with an out-edge directed to every node in G and a terminal node t with an in-edge directed from every node in G. The length of any edge into a node i other than t is set to $v(i)$. The maximum value chain of rectangles corresponds to the longest path from s to t. If there are r rectangles and e edges, then the longest path can be found in $O(e) = O(r^2)$ time.

A faster solution

We solve the chain problem with an algorithm that runs in $O(r \log r)$ time and that generalizes the solution to the one-dimensional problem. The method, first explicitly detailed for the problem of finding nonoverlapping pairs of substrings with high similarity [251], is typical of many methods in the general area of *sparse dynamic programming*.

To begin, consider the line segments making up the horizontal boundaries of the r rectangles, and let I be the sorted list of their endpoints. That is, each entry in I is the x coordinate of either the left most or the right most point of some rectangle, and I is sorted in left to right order. Annotate each entry in I with the rectangle it comes from and whether the point is a left end or a right end.

The chain algorithm processes the entries in I in order (left to right), as in the algorithm for the one-dimensional problem. But because rectangles are two-dimensional objects, the algorithm must also consider the y coordinates of each rectangle. For each rectangle j, let h_j be the y coordinate of its highest point, and let l_j be the y coordinate of its lowest point. The algorithm will keep a list L of certain triples $(l_j, V(j), j)$. The third entry, j, specifies a particular rectangle, and $V(j)$ is the maximum value of any chain that ends with rectangle j. The triples in L will be kept in sorted order according to the value of the l entry. The triple with the largest l entry will be first in L, and the triple with the smallest l entry will be last. Roughly then, rectangles whose bottoms are higher in the plane are represented in L before rectangles whose bottoms are lower in the plane.

List L will be updated as the algorithm processes points in I. Intuitively, after the algorithm processes point $\bar{x}$ in I, the entry $(l_j, V(j), j)$ in L indicates that the best chain completely contained in the upper left quadrant bounded by x coordinate $\bar{x}$ and y coordinate l_j (i.e., to the left of $\overline{X}$ and above l_j) ends with rectangle j and has value $V(j)$.

The complete algorithm is

Two-dimensional chain algorithm

List L begins empty.
For i from to $2r$ do
begin

If $I[i]$ is the left end of a rectangle, say rectangle k, then
    begin
    search L for the last triple where l_j is greater than h_k. That is,
    find the closest (in the y dimension) rectangle j with a triple in L
    whose lowest point is strictly above the highest point of rectangle k.
    Set $V(k)$ to $v(k) + V(j)$.
    end

Else
If $I[i]$ is the right end of rectangle k, then
    begin
    Search L for the first triple where l_j is less than or equal to l_k.

    If $l_j < l_k$, or $l_j = l_k$ and $V(k) > V(j)$, then insert the triple
    $(l_k, V(k), k)$ into L, in the proper location to keep the triples
    sorted by their l values.

    Delete from L the triple for every rectangle j' where $l_{j'} \leq l_k$
    and $V(k) > V(j')$.
    end

end.

At the end of the algorithm, the last entry in L gives the value of the optimal chain and specifies the last rectangle in the chain. We leave it as an exercise to extend the algorithm so that it actually constructs an optimal chain.

Correctness and implementation

We leave the correctness as an exercise. Before doing the time analysis, note that after every iteration, if a triple for rectangle j appears in L after a triple for rectangle j', then

$V(j) \geq V(j')$. That is, the triples in L appear in sorted (nondecreasing) order of their V values, even though the algorithm never explicitly sorts by V value.

For the time analysis, consider first the time used during a search of L (either for the last triple where l_j is greater than h_k or for the first triple where l_j is less than or equal to l_k). By keeping list L in a balanced binary search tree, such as an AVL tree [10], each search can be done in $O(\log r)$ time. Moreover, any element can be inserted or deleted in $O(\log r)$ time as well. Now consider the deletions after a new triple $(l_k, V(k), k)$ is inserted into L. Since the triples in L appear in nondecreasing order of V value, all the triples to be deleted appear in a contiguous block starting just after the new triple $(l_k, V(k), k)$. So the deletions can be implemented by successively examining the list after the triple for k until the first triple is found where $V(j') > V(k)$. In that way, all but the last triple examined is deleted. The time for the deletions in that iteration is therefore proportional to the number of triples deleted times $\log r$. Further, each deleted triple is associated with a rectangle j', and once deleted, no additional triple for j' will be reinserted into L. Therefore, over the entire algorithm, there are at most r deletions. In summary, we have

Theorem 13.3.1. *An optimal chain can be found in $O(r \log r)$ time.*

13.4. Exercises

1. Establish Newton's three laws stated in Lemma 13.1.2.

2. Show that in the case of parametric global alignment as defined in Section 13.1.6, no polygon boundaries touch the horizontal (β) of the positive quadrant.

3. Show that all the results in Section 13.1.2 hold when character-specific scoring matrices are used in the alignment objective function.

4. Analyze the number and shape of the polygons for global parametric alignment without scoring matrices when the objective function is $\alpha \times mt_A - \beta \times ms_A - id_A$, i.e., indels cost 1 and the gap penalty is set to zero.

5. Analyze the situation when a term for gaps is included in the objective function, and two (of the four) parameters are chosen to be variable. Consider again the case of global alignment without scoring matrices.

6. Answer the same question as in Exercise 5, but now allow three parameters to be variable.

7. Consider the following single-parameter objective function for alignment $\mathcal{A}$: $v_A(\lambda) \equiv mt_A - \lambda \times [ms_A + id_A]$. For any particular fixed value λ_0, there may be many different alignments that are optimal at λ_0. If there are more than one, then only two of these can be optimal for points other than λ_0. Explain this. Then show how to find, in $O(nm)$ time, the alignment that is optimal at λ_0 and also optimal for some value of λ greater than λ_0.

8. (This problem is for the graph theoretically inclined) Using Euler's famous theorem on planar graphs, show that when each node of the graph has degree at least three, then the number of regions (faces) of the graph is proportional to the number of vertices of the graph. (Recall that the number of vertices and edges are proportional to each other in any planar graph.)

9. Taylor [432] has suggested the usefulness of the following kind of parametric analysis: Instead of using a single scoring matrix, two scoring matrices are given, denoted say MD and ID, along with a multipliers x and y. To score an alignment $\mathcal{A}$ at a choice of x and y, multiply the score of the alignment given by matrix MD by x, multiply the score given by matrix ID by y, and add these two scores. By varying the values of x and y, the relative importance of MD versus ID can be changed. Show that this kind of parametric analysis can be handled with the kind of tools developed in the text.

10. It is common in the biological literature to add a constant to each of the values of an amino acid scoring matrix. However, what the specific constant should be is a subject of some debate. Although this is a problem of *adding* to each entry of a scoring matrix, rather than *multiplying* each entry by some amount, the problem of studying how the alignment depends on the choice of additive constant can be handled with the kind of parametric tools developed in this chapter. Show this in detail.

11. For a given value Δ, let G_Δ be the union of the edges in all δ-near-optimal paths, for all $\delta \leq \Delta$. Graph G_Δ gives an indirect reflection of the set of paths that are δ-near-optimal for all $\delta \leq \Delta$. Prove the following theorem:

Theorem 13.4.1. For any given value Δ, let $E_\Delta = \{e|\delta(e) \leq \Delta\}$. Then $E_\Delta = G_\Delta$.

12. Show, by example, a path in some G_Δ that is not δ-near-optimal for every $\delta \leq \Delta$. Given this, what uses are there for G_Δ? There are some.

13. Compare in detail the efficiency of the methods based on $\epsilon(e)$ values that count and enumerate near-optimal paths to similar methods based on the original edge values. Devise a way to use the original edge distances in the same efficiency as with $\epsilon(e)$ values.

14. Show how to simplify the enumeration method for near-optimal paths, if the Δ-near-optimal solutions need not be enumerated in order of increasing deviation from optimal.

15. Prove the correctness of the $O(r \log r)$-time method for the two-dimensional chain problem in Section 13.3.

16. Consider the problem of computing the optimal chain if each edge is given a weight, in addition to the weights given the rectangles. Can the $O(r \log r)$-time result still be obtained if the weights are arbitrary? What if the weights are monitonically increasing with the Euclidian distance between the rectangles?

17. Finding nonoverlapping approximate repeats

Exercise 31 in Chapter 11 considered the problem of finding two substrings in a string S that have maximum similarity over all pairs of substrings of S. In that problem, the two substrings are allowed to overlap. In some situations we impose the extra requirement that the substrings not overlap. One example is in searching DNA to find pairs of substrings where one substring in the pair was duplicated from the other, reinserted into the full DNA, and then modified separately from the first. In this process, the two substrings are physically distinct. So the formal string problem is: Given string S, find two nonoverlapping substrings in S whose similarity is as great as any pair of nonoverlapping substrings. This problem has been studied in a series of papers [60, 257, 321, 396] and can be solved in $O(n^2 \log n)$ time [396], where n is the length of S. That result, is too complex to develop here. Instead, we outline the first approach, by Miller [321]. You must flesh out the details.

The basic approach is to fill in a dynamic programming table for a local alignment computation of S against itself, but with some additions detailed below. The nonoverlap requirement means that when the algorithm considers a substring ending at a position i and another ending at position $j > i$, the second string must begin *after* position i. How do we incorporate this into the dynamic program computation for local alignment? The obvious way is to keep, for each cell (i, j) in the table, the value of the best path starting at each cell to its upper left. Then the best nonoverlapping path that ends at (i, j) is the best of these upper-left paths that begins at a column $k > i$. That means that each cell has associated with it $\Theta(n^2)$ values, and the time required is $\Theta(n^4)$. Explain the method and its correctness in detail.

The $\Theta(n^4)$ time bound can be reduced. A little thought shows that cell (i, j) needs to keep only the value of the best path starting in each *column* to its left. So only $\Theta(n)$ values are maintained at each cell, and the time is reduced to $\Theta(n^3)$. Explain this in detail.

Miller's method is a refinement of the above method and also has worst-case running time of $O(n^3)$. However, the typical time is more like $\Theta(n^2)$. To see the idea, consider the best paths to cell (i, j) starting at a column k and at a column $k' > k$. If the path from k' is better than the path from k, then the path from k cannot be extended to become the best nonoverlapping path to any cell to the lower right of (i, j). Hence column k is no longer a candidate for those cells, and hence, no cell to the lower right of (i, j) needs to compute a value for a path starting in column k. This reduces the size of the lists kept at those cells. Fully explain and justify this idea.

For efficient implementation, each cell (i, j) keeps an ordered list of candidate columns (columns that could possibly contribute the best path to (i, j)). The candidates are ordered left to right by column numbers and have the property that the values of the associated paths *decrease*. Using such lists, the algorithm can be implemented in time proportional to the total size of all the lists. Give the details for this implementation. Explain why it is reasonable to expect that the total size of the lists will be smaller than $\Theta(n^3)$. In fact, $\Theta(n^2)$ has been observed in practice.

18. Explain the high-level similarities of Miller's algorithm, the algorithm for convex gap weights discussed in Section 12.6, and the algorithm to chain diverse local alignments discussed in Section 13.3.

19. Miller's method is guaranteed to find the most similar nonoverlapping pair of substrings, but it is somewhat complex. A simpler approach is to use the algorithms from Section 13.2.4 that find many (but not all) good local alignments in order of decreasing similarity value, stopping the first time that a nonoverlapping pair is output. Under what circumstances will this approach likely miss the best nonoverlapping pair?

14

Multiple String Comparison – The Holy Grail

In this chapter we begin the discussion of *multiple string comparison*, one of the most important methodological issues and most active research areas in current biological sequence analysis. We first discuss some of the reasons for the importance of multiple string comparison in molecular biology. Then we will examine multiple string *alignment*, one common way that multiple string comparison has been formalized. We will precisely define three variants of the multiple alignment problem and consider in depth algorithms for attacking those problems. Other variants will be sketched in this chapter; additional multiple alignment issues will be discussed in Part IV.

14.1. Why multiple string comparison?

For a computer scientist, the multiple string comparison problem may at first seem like a generalization for generalization's sake – "two strings good, four strings better". But in the context of molecular biology, multiple string comparison (of DNA, RNA, or protein strings) is much more than a technical exercise. It is the most critical cutting-edge tool for *extracting and representing* biologically important, yet faint or widely dispersed, commonalities from a set of strings. These (faint) commonalities may reveal evolutionary history, critical conserved motifs or conserved characters in DNA or protein, common two- and three-dimensional molecular structure, or clues about the common biological function of the strings. Such commonalities are also used to characterize families or superfamilies of proteins. These characterizations are then used in database searches to identify other potential members of a family. Because many important commonalities are faint or widely dispersed, they might not be apparent when comparing two strings alone but may become clear, or even obvious, when comparing a set of related strings. As an example, see Figure 14.1.

One central technique for multiple string comparison involves *multiple alignment*. Although the main algorithmic issues involved in multiple alignment will be discussed later, we need here a clear definition in order to introduce some of its uses.

Definition A global multiple alignment of $k > 2$ strings $\mathcal{S} = \{S_1, S_2, \ldots, S_k\}$ is a natural generalization of alignment for two strings. Chosen spaces are inserted into (or at either end of) each of the k strings so that the resulting strings have the same length, defined to be l. Then the strings are arrayed in k rows of l columns each, so that each character and space of each string is in a unique column.

For example, see Figure 14.1 and 14.2.

Multiple comparison can be inverse to two-string comparison

In one sense, multiple string comparison is a natural extension of two-string comparison (via alignment), when one is looking for weakly conserved patterns rather than strongly

```
                                   *vvvvv*

HUMA      VLSPADKTNVKAAWGKVGAHAGEYGAEALERMFLSFPTTKTYFPHF  DLSH      GS
HAOR      MLTDAEKKEVTALWGKAAGHGEEYGAEALERLFQAFPTTKTYFSHF  DLSH      GS
HADK      VLSAADKTNVKGVFSKIGGHAEEYGAETLERMFIAYPQTKTYFPHF  DLSH      GS
HBHU      VHLTPEEKSAVTALWGKV  NVDEVGGEALGRLLVVYPWTQRFFESFGDLSTPDAVMGN
HBOR      VHLSGGEKSAVTNLWGKV  NINELGGEALGRLLVVYPWTQRFFEAFGDLSSAGAVMGN
HBDK      VHWTAEEKQLITGLWGKV  NVADCGAEALARLLIVYPWTQRFFASFGNLSSPTAILGN
MYHU       GLSDGEWQLVLNVWGKVEADIPGHGQEVLIRLFKGHPETLEKFDKFKHLKSEDEMKAS
MYOR       GLSDGEWQLVLKVWGKVEGDLPGHGQEVLIRLFKTHPETLEKFDKFKGLKTEDEMKAS
IGLOB     SPLTADEASLVQSSWK    AVSHNEVEILAAVFAAYPDIQNKFSQFA1GKDLASIKDT
GPUGNI    ALTEKQEALLKQSWEVLKQNIPAHSLRLFALIIEAAPESKYVFSFLKDSNEIPE   NN
GPYL      GVLTDVQVALVKSSFEEFNANIPKNTHRFFTLVLEIAPGAKDLFSFLKGSSEVPQ  NN
GGZLB     MLDQQTINIIKATVPVLKEHGVTITTTFYKNLFAKHPEVRPLF     DMGRQE    SL


          vvvvv                        vvvv*

HUMA      AQVKGHGKKVADALTNAV      AHVDDM    PNALSALSDLHAHKLRVDPVNFKLLS
HAOR      AQIKAHGKKVADALSTAA      GHFDDM    DSALSALSDLHAHKLRVDPVNFKLLA
HADK      AQIKAHGKKVAAALVEAV      NHVDDI    AGALSKLSDLHAQKLRVDPVNFKFLG
HBHU      PKVKAHGKKVLGAFSDGL      AHLDNL    KGTFATLSELHCDKLHVDPENFRLLG
HBOR      PKVKAHGAKVLTSFGDAL      KNLDDL    KGTFAKLSELHCDKLHVDPENFNRLG
HBDK      PMVRAHGKKVLTSFGDAV      KNLDNI    KNTFAQLSELHCDKLHVDPENFRLLG
MYHU      EDLKKHGATVLTALGGIL      KKKGHH    EAEIKPLAQSHATKHKIPVKYLEFIS
MYOR      ADLKKHGGTVLTALGNIL      KKKGQH    EAELKPLAQSHATKHKISIKFLEYIS
IGLOB     GAFATHATRIVSFLSEVIAL1SGNTSNAAAV   NSLVSKLGDDHKARGVSAAQ1FGEFR
GPUGNI    PKLKAHAAVIFKTICESA  TELRQKGHAVWDNNTLKRLGSIH LKNKITDPHFEVMK
GPYL      PDLQAHAGKVFKLTYEAA  IQLEVNGAVASDATLKSLGSVHVSKGVVDA HFPVVK
GGZLB     EQPKALAMTVLAAAQNI      ENLPAI    LPAVKKIAVKHC QAGVAAAHYPIVG


          vvvvv               vvv

HUMA      HCLLVTLAAHLPAEFTPAVHASLDKFLASVSTVLTSKYR
HAOR      HCILVVLARHCPGEFTPSAHAAMDKFLSKVATVLTSKYR
HADK      HCFLVVVAIHHPAALTPEVHASLDKFMCAVGAVLTAKYR
HBHU      NVLVCVLAHHFGKEFTPPVQAAYQKVVAGVANALAHKYH
HBOR      NVLIVVLARHFSKDFSPEVQAAWQKLVSGVAHALGHKYH
HBDK      DILIIVLAAHFTKDFTPECQAAWQKLVRVVAHALARKYH
MYHU      ECIIQVLQSKHPGDFGADAQGAMNKALELFRKDMASNYKELGFQG
MYOR      EAIIHVLQSKHSADFGADAQAAMGKALELFRNDMAAKYKEFGFQG
IGLOB     TALVAYLQANVS    WGDNVAAAWNKAL1DNTFAIVVPRL
GPUGNI    GALLGTIKEAIKENWSDEMGQAWTEAYNQLVATIKAEMKE
GPYL      EAILKTIKEVVGDKWSEELNTAWTIAYDELAIIIKKEMKDAA
GGZLB     QELLGAIKEVLGDAATDDILDAWGKAYGVIADVFIQVEADLYAQAVE
```

Figure 14.1: A multiple alignment of several amino acid sequences of globin proteins modified from the paper of McClure, Vasi, and Fitch [316]. The abbreviations on the left indicate the organisms that the globin sequences are from. Because of the length of the sequences, the multiple alignment is displayed using three lines for each sequence. Columns in the alignment containing a high concentration of similar residues in regions of known secondary structure are marked with a "v", and columns with identical residues are marked with a star. Two residues are considered similar if they belong to the same class of the following partition: (F,Y), (M,L,I,V), (A,G), (T,S), (Q,N), (K,R), and (E,D).

conserved patterns. Quoting from Arthur Lesk [233]: "One or two homologous sequences whisper ... a full multiple alignment shouts out loud".

But in another sense, multiple string comparison is used in a fundamentally different way than two-string comparison. It can be used for biological inference problems that are *inverse* to problems addressed by two-string comparison. In the context of database searching, two-string alignment finds strings that have common subpatterns (substrings or subsequences) but may not have been known to be biologically related. Indeed, the greatest value of database searching comes from using *apparent* string similarities to identify unsuspected biological relationships. The inverse problem is to move from known biological relationships to unknown substring or subsequence similarities. This direction uses multiple string comparison to deduce unknown conserved subpatterns from a set of strings already known to be biologically related.

14.1.1. Biological basis for multiple string comparison

Recall the *first fact of biological sequence comparison* from Chapter 10 on page 212. This first fact underlies the effectiveness of two-sequence comparison and of biological database searching. But as powerful as the *first fact* is, it does not capture all the important biology forming the basis for biological sequence comparison. There is an additional fact that compels *multiple* string comparison.

The second fact of biological sequence comparison   Evolutionarily and functionally related molecular strings can *differ significantly* throughout much of the string and yet preserve the same three-dimensional structure(s), or the same two-dimensional substructure(s) (motifs, domains), or the same active sites, or the same or related dispersed residues (DNA or amino acid).

This *second fact* was implicit in the earlier discussion of local alignment (Section 11.7). What is added here is a matter of degree. Two strings specifying the "same" protein in different species may be so different that the few observed similarities may just be due to chance. Typically, secondary or three-dimensional structure is the most well preserved feature of a large set of evolutionarily related proteins. Conservation of function is less common, and conservation of amino-acid sequence is the least common. "One of the most intriguing observations to arise from recent analyses of protein structures is that similar folds recur, even in the absence of sequence similarity" [428]. (This point was also made by the quote from Cohen on page 214.)

For example, hemoglobin is a nearly universal protein containing four chains of about 140 amino acids each. It is contained in organisms as diverse as mammals and insects, and it functions in essentially the same way (it binds and transports oxygen). However, in the time since insects and invertebrates diverged (some 600 million years ago), multiple amino acid substitutions have occurred in insect and invertebrate hemoglobin sequences, so that on average there have been about 100 amino acid mutations in each chain of the two sequences[1][11]. A pairwise alignment of hemoglobin chains from two mammals will suggest a functional similarity of the protein (human and chimpanzee sequences are in fact identical), but a two-string alignment from a mammal and an insect may reveal very little. Moreover, many proteins mutate faster than hemoglobin, compounding the problem.

As another, more recent example, the three-dimensional structure of part of a *cell adhesion molecule* called *E-Cadherin* was worked out [354, 451]. The deduced structure was unexpectedly found to show "remarkable similarity to the Ig fold (although there is no sequence homology)" [451]. The Ig fold is a common structure first seen in the immunoglobulin (antibody) superfamily but later seen in many other proteins as well. The term "homology" here should be interpreted as "similarity", and "no homology" should be interpreted as "no more similarity than expected between random strings". So the

---

[1] This doesn't translate immediately into the number of positions where amino acid differences occur between mammal and insect hemoglobin because more than one mutation can occur at a given position.

fold found is a conserved physical structure relating the Ig proteins with the E-Cadherin protein, despite the lack of sequence similarity. A related discussion in [451] about a different protein (the growth hormone receptor) illustrates the same point: "Previously, the growth hormone receptor structure was found to be built of domains related to this (Ig) evolutionarily successful motif; in this case too, no sequence homology exists".

The ability of many proteins to preserve structure and function in the face of massive amino acid substitution is simply an empirical fact, although it may seem counterintuitive. One might intuitively expect[2] that every detail of an amino acid string or its encoding DNA is essential, so that even small changes to the string would destroy the structure or function of the protein. Indeed, many gene-related diseases are caused by such small mutations. Often even a single change at a *particular* amino acid position, which might be caused by a single DNA nucleotide change, leads to a serious disease such as sickle-cell anemia, cancer (the *ras* oncogene, for example, is activated by a single nucleotide point mutation [403, 110]), or a prion disease such as *Mad Cow Disease* [377]. But on whole, many amino acid positions in a protein string are noncritical and over evolutionary history have mutated to a wide selection of other amino acids without a destructive effect on the protein. So, although the *first fact* of biological sequence analysis, that sequence similarity implies structural or functional similarity, is very powerful and widely exploited, the converse of that fact simply is not true.

The extent of permissive mutations in structurally or functionally conserved molecules may be such that comparing two strings at a time reveals little of the critically conserved patterns or of the critical amino acids. There may be so few conserved amino acids or they may be so widely dispersed that the best alignment between two related protein strings is statistically indistinguishable from the best alignment of two random amino acid strings. The problem occurs at the other extreme as well. Comparison of two proteins from highly related species might not reveal biologically important conserved patterns because the critical similarities are lost in the large number of similarities due only to recent shared evolution. Thus when doing two-string comparison to find critical common patterns, the challenge is to pick species whose level of divergence is "most informative" (a vague and difficult task).

Multiple string comparison (often via multiple alignment) is a natural response to the problems introduced above. With multiple string comparison, it is not as crucial (although it is still helpful) to pick species whose level of divergence is "most informative". Often, biologically important patterns that cannot be revealed by comparison of two strings alone become clear when many related strings are simultaneously compared. Moreover, with multiple alignments, it is sometimes possible to arrange a set of related strings (often in a tree) to demonstrate *continuous* changes along a path connecting two extreme strings that by themselves show little pairwise similarity. In particular, multiple alignments of this kind are very useful in deducing evolutionary history.

14.2. Three "big-picture" biological uses for multiple string comparison

Multiple sequence comparison is a critical task in three kinds of inquiries that address "big-picture" biological questions: the representation of protein families and superfamilies, the identification and representation of conserved sequence features of DNA or protein that

---

[2] And some early papers on protein evolution even state this expectation as a "fact". For example, see [121].

correlate with structure and/or function, and the deduction of evolutionary history (usually in the form of a tree) from DNA or protein sequences.[3]

Although useful, this taxonomy is misleading because these three inquiries are often quite interrelated. The unifying idea is that common structure or function of the strings usually results from common evolutionary history. So even if one has no explicit interest in deducing history and one defines a protein family just on the basis of common function or structure, or one looks for conserved sites to elucidate the mechanics of structure or function, evolution still resides in the background. Knowledge of evolutionary history therefore can aid in the first two kinds of inquiries, and conversely, results from the first two inquiries contribute to understanding evolutionary history.

In the next sections, we will discuss the first two of the three "big-picture" uses of multiple string comparison. The third use, deducing evolutionary history, will be discussed in Sections 14.8 and 14.10.2 of this chapter and again in Chapter 17 of Part IV.

14.3. Family and superfamily representation

Often a set of strings (a family) is defined by biological similarity, and one wants to find subsequence commonalities that characterize or represent the family. The conserved sequence commonalities in a family may give clues to better understand the function or structure of family members. Moreover, the representation of the family may be useful in identifying potential new members of the family while excluding strings that are not in the family. Protein families (or super-families) are typical examples, where a family is a set of proteins that have similar biological function, or have similar two- or three-dimensional structure, or have a known shared evolutionary history.[4] Specific examples of protein families include the famous *globins* (hemoglobins, myoglobins) and the *immunoglobulin* (*antibody*) proteins. The following quote from C. Chothia [101] emphasizes the centrality of studying families in molecular biology.

> Proteins are clustered into families whose members have diverged from a common ancestor and so have similar folds; they also usually have similar sequences and functions. This classification is central to our understanding of how life has evolved, and makes the elucidation and definition of such families one of the principal concerns of molecular biology.

The word "folds" means "two and three dimensional (sub)structures".

---

[3] In addition to these three major applications, multiple string comparison (and particularly alignment) often arises in specific laboratory protocols, such as PCR primer design, gene cloning, and shotgun sequence assembly. The third of those three applications will be discussed in Sections 16.14 and 16.15.

[4] There is serious ambiguity in the biological literature in the use of the words "family" and "superfamily". To the average working biologist a family (superfamily) means a set of proteins that are related by structure, function, or known evolutionary history. The problem then is to find sufficient sequence similarity to characterize the family. This meaning is implicit in the quote by Chothia to follow, and it is the meaning we intend in this book. However, sometimes "family" and "superfamily" are used in almost the exact opposite way. Dayhoff established a classification of proteins into families and superfamilies based exclusively on the degree of their *sequence* similarity. In that scheme, no known functional or structural relationship between family members is needed, and the problem is to determine what, if any, biological relationship the family members have. Of course, shared evolutionary history is postulated to be the reason for sequence similarity, and recent evolutionary history also roughly implies common structure and function. Under that view, both notions of family lead to the same insights and to the dual use of the term. But that theoretical connection between the two uses of "family" is not perfect in practice, so when reading the biological literature, one still has to carefully determine which meaning is intended. Is a biological commonality for the family members actually known, or is only a sequence similarity known?

$$
\begin{array}{ccccc}
a & b & c & \_ & a \\
a & b & a & b & a \\
a & c & c & b & \_ \\
c & b & \_ & b & c
\end{array}
$$

Figure 14.2: A multiple alignment that will be used to generate a profile.

Family representations are very useful

By identifying the family or super-family that a newly identified protein belongs to, the researcher gains tremendous clues about the physical structure or biological function of the protein. This approach is very powerful, because it has been estimated that the 100,000 or so proteins in humans might be organized into as few as one thousand meaningful protein families [101]. Other estimates put the number as low as a few hundred.

When considering a protein for membership in an established family, or when searching a protein database for new candidate members of a protein family, it is usually much more effective to align the candidate string to a *representation* of the known family members, rather than aligning to single members of the family.[5] That leads to the questions of which kinds of representations are most useful for identifying additional members (while excluding non-members), and how are good representations obtained? This is a wide-open and active research area, but it often involves some kind of multiple alignment of the protein strings in the family; the family representation is then extracted from that multiple string alignment. The paper by McClure, Vasi, and Fitch [316] compares different multiple alignment methods, motivated by that very question: How well do the differing methods find and define known protein-family patterns?

Three common representations

There are three common kinds of family representations that come from multiple string comparison: *profile* representations, *consensus sequence* representations, and *signature* representations. The first two representations are based on multiple alignment and will be discussed in Sections 14.3.1 and 14.7. Signature representations are not all based on multiple alignment and will be discussed in Section 14.3.2.

14.3.1. Family representations and alignments with profiles

Definition   Given a multiple alignment of a set of strings, a *profile* for that multiple alignment specifies for each column the *frequency* that each character appears in the column. A profile is sometimes also called a *weight matrix* in the biological literature. Profiles were initially developed and exploited in molecular biology by Gribskov, Eisenberg, and coworkers [194, 193, 192].

For example, consider the multiple alignment of four strings shown in Figure 14.2. The profile derived from it has five columns as shown in Figure 14.3.

Often the values in the profile are converted to *log-odds ratios*. That is, if $p(y, j)$ is the frequency that character y appears in column j, and $p(y)$ is the frequency that character y appears anywhere in the multiply aligned sequences, then $\log p(y, j)/p(y)$ is commonly used as the y, j profile (or weight matrix) entry. The meaningful part is the

---

[5] The representation seems particularly effective when it is developed in an iterative process. One starts with a representation based on a few known members of a family, and that representation is used to find additional members in the database. The expanded set is then used to create a new representation which is used to search the database, etc.

	$C1$	$C2$	$C3$	$C4$	$C5$
a	.75		.25		.50
b		.75		.75	
c	.25	.25	.50		.25
$-$			.25	.25	.25

Figure 14.3: The profile extracted from the previous multiple alignment.

a	a	b		b	c
1		2	3	4	5

Figure 14.4: An alignment of string *aabbc* to the column positions of the previous alignment.

ratio. The logorithm is primarily for convenience and any particular base can be used. For our purposes, when we use the term "profile" we will not need to specify whether the value is a raw frequency or a log-odds ratio.

Aligning a string to a profile

Given a profile $\mathcal{P}$ and a new string S, one often asks how well S, or some substring of S, fits the profile $\mathcal{P}$. Since a space is a legal character of a profile, a fit of S to $\mathcal{P}$ should also allow the insertion of spaces into S, and hence the question of how well S fits $\mathcal{P}$ is naturally formalized as an easy generalization of pure string alignment. Consider a string C of profile column positions, and then align S to C by inserting spaces into S and C etc., as in pure string alignment. For example, an alignment of the string *aabbc* to the previous profile is shown in Figure 14.4. The key issues are how to score an alignment of a string to a profile, and given a scoring scheme, how to compute an optimal alignment.

How to score a string/profile alignment

Assume an alphabet-weight scoring scheme for pure string alignment is known. Using that pure string-scoring scheme, the common approach to scoring an alignment of a character in S to a column in C is to consider the string character aligned to every character in the column and then compute a weighted sum of scores based on the frequency of the characters in the column. The score for the full alignment is the sum of the column scores. For example, suppose the alphabet-weight scheme gives a score of 2 to an *a,a* alignment, a score of -1 to an *a,b* or an alignment of *a* with a space, and a score of -3 to an *a,c* alignment. Then, the first column of the previous alignment of string *aabbc* to C contributes a score of $0.75 \times 2 - 0.25 \times 3$ to the overall alignment score. The second column contributes a score of -1, since it aligns character *a* to a space.

How to optimally align a string to a profile

We view alignment in terms of maximizing similarity and compute the optimal alignment by dynamic programming. The details are very similar to the way pure string similarity is computed (see Section 11.6.1). However, the notation is a little more involved. Recall that for two characters x and y, $s(x, y)$ denotes the alphabet-weight value assigned to aligning x with y in the pure string alignment problem.

Definition For a character y and column j, let $p(y, j)$ be the frequency that character y appears in column j of the profile, and let $S(x, j)$ denote $\sum_y [s(x, y) \times p(y, j)]$, the score for aligning x with column j.

Definition Let $V(i, j)$ denote the value of the optimal alignment of substring $S[1..i]$ with the first j columns of C.

With that notation, the recurrences for computing optimal string to profile alignment are

$$V(0, j) = \sum_{k \leq j} S(\_, k)$$

and

$$V(i, 0) = \sum_{k \leq i} s(S_1(k), \_).$$

For i and j both strictly positive, the general recurrence is

$$V(i, j) = \max[V(i - 1, j - 1) + S(S_1(i), j), V(i - 1, j) + s(S_1(i), \_), V(i, j - 1) + S(\_, j)].$$

The correctness of these recurrences is immediate and is left to the reader. As usual in dynamic programming, the actual alignment is obtained during a traceback phase, after the recurrences have been evaluated. The time to evaluate the recurrences is $O(\sigma nm)$, where n is the length of S and σ is the size of the alphabet. Verification of the time bound is also left to the reader. We see from this that aligning a string to a profile costs little more than aligning two strings. A specific use for aligning strings to profiles is discussed in the first part of Section 14.10.2. See [193] for a more in-depth discussion of profiles in computational biology.

Profile to profile alignment

Another way that profiles are used is to compare one protein set to another. In that case, the profile for one set is compared to the profile of the other. Alignments of this kind are sometimes used as a subtask in certain multiple alignment methods (see Section 14.10.1). It is straightforward to formalize optimal profile to profile alignment and to obtain the recurrences to compute it. That is again left as an exercise.

14.3.2. Signature representations of families

We now turn to the second major family representation, the *signature* or *motif signature*. The major collections of signatures in protein are the PROSITE database and the BLOCKS database derived from it. These are discussed in Sections 15.8 and 15.9. Here, we will illustrate the signature approach and its typical use with a study done by Gorbalenya and Koonin [185] in the paper "Helicases: Amino Acid Sequence Comparisons and Structure-Function Relationships".

Signatures for helicase proteins

DNA is a double-stranded molecule that takes the shape of a double-stranded helix – two (sugar) backbones that wind around each other and are connected on the inside by (complementary) pairs of nucleotides. Since the backbone is identical along its length, the information stored by DNA (for example, in a gene coding for a protein) is contained in the specific nucleotide sequence in the interior of the double helix. To extract that stored information, the helix must be slightly "unwound" and "opened". Helicases are proteins that help unwind double-stranded DNA so that the DNA can be read for duplication, transcription, recombination, or repair. The recently discovered gene for Werner's syndrome is believed to code for a helicase protein [365]. An individual afflicted with Werner's syndrome ages at a vastly faster rate than normal. There are several hundred helicases from different organisms whose amino acid sequences have been deduced, and

A large fraction of the available information on the structure and possible functions of the helicases has been obtained by computer-assisted comparative analysis of their amino acid sequences. This approach has led to the delineation of motifs and patterns that are conserved in different subsets of the helicases. [185]

Such motifs have been used to search the protein and DNA databases (translated to the protein alphabet) to find previously unknown helicases. As stated in Gorbalenya and Koonin [185], "These motifs have been widely used for prediction of new helicases. . . .".

In our terminology, the helicases form a family of proteins related by common biological function – they all act to help unwind DNA. By extracting and representing sequence commonalities, greater understanding of the mechanics of helicases may be gained, and unknown helicases may be identified. At the sequence level, all the known helicases contain certain conserved "sequence signatures" or motifs,[6] but these ubiquitous motifs may not be unique features of helicases. They therefore do not discriminate perfectly between helicases and other nonhelicase proteins, although they may identify the functionally most important amino acids.

The goal in [185] was to evaluate the claim that "additional motifs *are* unique identifiers of distinct groups of (putative) helicases". Gorbalenya and Koonin found that "subsets of helicases possess additional conserved motifs. . . ." Some of these motifs can be used as reliable identifiers of the respective groups of helicases in database searches. The two largest groups[7] "share similar patterns of seven conserved sequence motifs, some of which are separated by long poorly conserved spacers". An example of one of the motifs is

$$[\&H][\&A]D[DE]x_n[TSN]x_4[QK]Gx_7[\&A],$$

where "&" stands for any amino acid from the group (I, L, V, M, F, Y, W), "x" stands for any amino acid, and alternative amino acids are bracketed together. The subscript on x gives the length of a permitted substring; "n" indicates that any length substring is possible. Clearly, a motif in this context is not an identically conserved substring, but rather a *regular expression*. Hence one can search for an instance of the pattern in a string by using regular expression pattern matching without allowing errors (see Section 3.6).

Helicase motifs may be functional units

Motifs may be more than good determinants of family membership. They may actually be important functional units of the protein: ". . . the recognition of motifs that are conserved in vast groups of helicases . . . has led to the obvious hypothesis that these motifs contain the amino acid residues most important for the helicase functions" [185]. These hypotheses can be checked by doing *site-directed mutagenesis*, where a selected position in the gene coding for the helicase is changed so that only a single selected amino acid in the helicase is affected. After the modified DNA is reintroduced into the organism and the modified helicase protein is expressed, the experimenter can see whether the selected sequence change results in a functional change. This kind of experiment can determine if the amino acids in the conserved motifs are the critical ones for helicase function.

Another good illustration of the signature approach, both the identification and the use of signatures, is contained in the paper by Posfai et al. [374].

---

[6] There is a subfamily of helicases known as the DEAD helicases because they all contain that four letter substring.

[7] Unfortunately for the lay reader, they often refer to groups related only by sequence similarity as "superfamilies" and "families", following the tradition of Dayhoff, but often refer to them simply as "groups" or "subsets", consistent with our terminology.

14.4. Multiple sequence comparison for structural inference

Quoting from a recent paper on multiple alignment, "One of the most successful applications of multiple sequence alignment has been to improve the accuracy of secondary structure prediction" [306].

We now illustrate the second "big-picture" use for multiple alignment, that of deducing biomolecular structure. We focus on a specific problem of secondary structure in tRNA. Another illustration of the use of multiple alignment in structure prediction is discussed in Section 14.10.2 (page 361). Another use of sequence alignment in structural inference was developed in Exercise 36 of Chapter 11.

The problem of computing the secondary (planar) structure of a *tRNA* molecule from its nucleotide sequence was introduced in Exercises 41, 42, 43, and 44 at the end of Chapter 11.[8] There, a dynamic programming approach was explored to find a maximum cardinality *nested pairing* of nucleotides. Recall that a nested pairing does not allow any crossing lines in its (plane) representation.

More recently, Cary and Stormo [89] explored an approach to computing the secondary structure of *tRNA* that is based on weighted matching in general (nonbipartite) graphs. That approach does not explicitly incorporate the noncrossing constraints that were so integral to the nested pairing approach (and generally in agreement with known *tRNA* structures).

Matching allowing crosses

In the method of [89], each nucleotide of the *tRNA* molecule is represented as a node in a graph, with an edge between each pair of nodes and a weight on each edge representing the "goodness" (we will not be too technical here) of selecting the pair of nucleotides connected by that edge. The method then finds a maximum weight *matching* in the graph. A matching in a graph is a subset E' of edges from the graph such that no node in the graph is touched by more than one edge in E'. The weight of a matching is the sum of the weights of all the edges in E'. Clearly, a matching in the graph specifies a pairing of nucleotides (as defined in Exercise 41 of Chapter 11).

Finding a maximum weight matching in a graph is a classic combinatorial problem that has a relatively efficient solution [294], and so the above approach to pairing nucleotides was examined [89] to see how well it recovers the secondary structure of *tRNA* molecules. Since this approach does not rule out crossing pairs (called pseudoknots), it may do better than the nested pairing approach in recovering true *tRNA* secondary structure. However, since nested pairing is the general rule and pseudoknots are the exception, one might think that a method based on general matching would allow too many crosses to determine correct secondary structure. So the justification for this approach has to rely on what structural information is *encoded* in the edge weights. Here is where *multiple string alignment* enters the picture.

The edge weights

Without being too technical (or correct), the edge weights are obtained as follows. A large number of related *tRNA* sequences, including the one of interest, are multiply aligned (exactly how will be left vague), so that each nucleotide in each sequence is assigned to a specific column of the alignment. Then, each pair of columns (i, j) is examined to find

---

[8] The reader should review the text material in Exercises 41 and 42 of Chapter 11 for basic definitions and biological facts.

the extent that character changes in column i correlate with character changes in column j (we will again be vague in the precise definition of this correlation). For example, if half the sequences have an A in column 16 and a U in column 92, while the other half have a C in 16 and a G in 92, then columns 16 and 92 would be considered highly correlated. The explanation for high correlation in this example is that the two nucleotides occupying positions 16 and 92 (defined by the alignment) in the *tRNA* are base-paired with each other, even though the specific nucleotides in those positions may mutate and be different in different sequences. The point is that in order to preserve structural integrity, point mutations in paired nucleotides must generally be correlated.[9] At the other extreme, if changes in column 16 were totally independent of changes in column 92, then one could deduce that the nucleotides in those two positions were probably not base-paired. In the first situation, the weight of edge (16, 92) in the graph should be set high, whereas in the second situation, the edge weight should be set low. Of course, these two extreme cases are for illustration only; most correlation data will be more ambiguous, which is why a weighted matching computation is needed to decide on the overall nucleotide pairing.

We have not been precise on how pairwise column correlation is translated into edge weights [89], but we have presented the essential intuition. In summary, multiple string comparison (alignment) of *tRNA* sequences seems to expose enough information about *tRNA* secondary structure so that nucleotide pairing based on general weighted matching is "fairly" successful in recovering the true secondary structure from sequence information alone.

14.5. Introduction to computing multiple string alignments

Having motivated multiple string comparison with the preceding discussions, we now move to the purely technical issues of how to compute *multiple string alignment*. The definition of *global* multiple alignment was given on page 332, and examples are shown in Figures 14.2 and 14.6b. It is also natural to define *local* multiple alignment. Two-string local alignment was defined as a global alignment of substrings, and multiple local alignment can be similarly defined.

Definition   Given a set of $k > 2$ strings $S = \{S_1, S_2, \ldots, S_k\}$, a *local* multiple alignment of S is obtained by selecting one substring S_i' from each string $S_i \in S$ and then globally aligning those substrings.

All of the biological justifications for preferring local alignment to global alignment of two strings (see Section 11.7) also apply to local versus global multiple alignment. These justifications are further strengthened by the *second fact* of biological sequence analysis, discussed above. However, the best theoretical results (from a computer science perspective) on multiple alignment have been obtained for global alignment, and so our emphasis will be on global alignment. We will briefly discuss local multiple alignment in Section 14.10.3.

14.5.1. How to score multiple alignments

Although the notion of a multiple alignment is easily extended from two strings to many strings, the *score* or goodness of a multiple alignment is not as easily generalized. To date,

---

[9]   However, this explanation does not cover all the cases of high correlation, for there are pairs of positions with high correlation that do not correspond to base-paired nucleotides in the *tRNA*. Also, positions at which the nucleotide does not mutate are not well handled by this method.

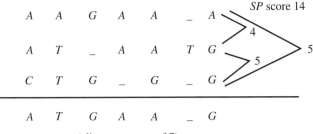

Figure 14.5: Multiple alignment $\mathcal{M}$ of three strings shown above the horizontal line. Using the pairwise scoring scheme of *ms* + *id*, i.e., #(mismatches) + #(spaces opposite a nonspace), the three induced pairwise alignments have scores of 4, 5, and 5 for a total *SP* score of 14. Note that a space opposite a space contributes a zero. Using the plurality rule, the consensus string $S_{\mathcal{M}}$ (defined in Section 14.7.2) for alignment $\mathcal{M}$ is shown below the horizontal line. It has an alignment error of seven.

there is no objective function that has been as well accepted for multiple alignment as (weighted) edit distance or similarity has been for two-string alignment. In fact, as we will see later, some popular methods to find multiple alignments do so without using any explicit objective function. The goodness of those methods is judged by the biological meaning of the alignments that they produce, and so the biological insight of the evaluator is of critical importance.

Not being biologists, we will emphasize multiple alignment methods that *are* guided by explicit objective functions, although we will sketch other methods as well. We will discuss three types of objective functions in this chapter: *sum-of-pairs functions*, *consensus functions*, and *tree functions*. Although these objective functions are quite different, we will explore approximation algorithms for these problems that are highly related. In Part IV of the book we will discuss how multiple alignment problems interact with problems of deducing trees representing evolutionary history.

Definition   Given a multiple alignment $\mathcal{M}$, the *induced pairwise alignment* of two strings S_i and S_j is obtained from $\mathcal{M}$ by removing all rows except the two rows for S_i and S_j. That is, the induced alignment is the multiple alignment $\mathcal{M}$ restricted to S_i and S_j. Any two opposing spaces in that induced alignment can be removed if desired.

Definition   The *score* of an induced pairwise alignment is determined using any chosen scoring scheme for two-string alignment in the standard manner.

For an example, see Figure 14.5.

Notice that the definition of "score" does not specify whether a score is a weighted distance or a similarity. As before, similarity is more natural for treating local alignment. Most of this chapter concerns global alignment, where the algorithms and consequent theorems all require the score to be a weighted distance.

14.6. Multiple alignment with the sum-of-pairs (*SP*) objective function

The sum-of-pairs (*SP*) score

Definition   The *sum of pairs* (*SP*) score of a multiple alignment $\mathcal{M}$ is the sum of the scores of pairwise global alignments induced by $\mathcal{M}$. See Figure 14.5.

Although one can give "handwaving" arguments for the significance of the *SP* score, it is difficult to give a theoretical justification for it (or any other multiple alignment scoring scheme). However, the *SP* score is easy to work with and has been used by many people studying multiple alignment. The *SP* score was first introduced in [88], and was subsequently used in [33, 334], and [187]. A similar score is used in [153]. The *SP* score is also used in a subtask in the multiple alignment package MACAW [398] developed at the National Institutes of Health (NIH), National Center for Biotechnology Information.

The SP alignment problem   Compute a global multiple alignment $\mathcal{M}$ with *minimum* sum-of-pairs score.

14.6.1. An exact solution to the *SP* alignment problem

As might be expected, the *SP* problem can be solved exactly (optimally) via dynamic programming [334]. Unfortunately, if there are k strings and each is of length n say, dynamic programming takes $\Theta(n^k)$ time and hence is practical for only a small number of strings. Moreover, the exact *SP* alignment problem has been proven to be NP-complete [454]. We will therefore develop the dynamic programming recurrences only for the case of three strings. Extension to any larger number of strings is straightforward, although it proves to be impractical for even five strings whose lengths are in the low hundreds (typical of proteins). We will also develop an accelerant to the basic dynamic programming solution that somewhat increases the number of strings that can be optimally aligned.

Definition   Let S_1, S_2, and S_3 denote three strings of lengths n_1, n_2, and n_3, respectively, and let $D(i, j, k)$ be the optimal *SP* score for aligning $S_1[1..i]$, $S_2[1..j]$, and $S_3[1..k]$. The score for a match, mismatch, or space is specified by the variables *smatch, smis*, and *sspace*, respectively.

The dynamic programming table D used to align three strings forms a three-dimensional cube. Each cell (i, j, k) that is not on a boundary of the table (i.e., that has no index equal to zero) has seven neighbors that must be consulted to determine $D(i, j, k)$. The general recurrences for computing the cost of a nonboundary cell are similar in spirit to the recurrences for two strings but are a bit more involved. The recurrences are encoded in the following pseudocode:

Recurrences for a nonboundary cell (i, j)

```
for i := 1 to n₁ do
    for j := 1 to n₂ do
        for k := 1 to n₃ do
            begin
            if (S₁(i) = S₂(j)) then cij := smatch
            else cij := smis;
            if (S₁(i) = S₃(k)) then cik := smatch
            else cik := smis;
            if (S₂(j) = S₃(k)) then cjk := smatch
            else cjk := smis;

            d1 := D(i−1, j−1, k−1) + cij + cik + cjk;
            d2 := D(i−1, j−1 ,k) + cij + 2*sspace;
            d3 := D(i−1, j, k−1) + cik + 2*sspace;
```

$$d4 := D(i, j{-}1, k{-}1) + cjk + 2*\text{sspace};$$
$$d5 := D(i{-}1, j, k) + 2*\text{sspace};$$
$$d6 := D(i, j{-}1, k) + 2*\text{sspace};$$
$$d7 := D(i, j, k{-}1) + 2*\text{sspace};$$

$$D(i, j, k) := \text{Min}[d1, d2, d3, d4, d5, d6, d7];$$
$$\text{end};$$

What remains is to specify how to compute the D values for the boundary cells on the three initial faces of the table (i.e., when $i = 0$, $j = 0$, or $k = 0$). To do this, let $D_{1,2}(i, j)$ denote the familiar pairwise distance between substrings $S_1[1..i]$ and $S_2[1..j]$, and let $D_{1,3}(i, k)$ and $D_{2,3}(j, k)$ denote the analogous pairwise distances involving pairs S_1, S_3 and S_2, S_3. These distances are computed in the standard way. Then,

$$D(i, j, 0) = D_{1,2}(i, j) + (i + j) * sspace,$$

$$D(i, 0, k) = D_{1,3}(i, k) + (i + k) * sspace,$$

$$D(0, j, k) = D_{2,3}(j, k) + (j + k) * sspace,$$

and

$$D(0, 0, 0) = 0.$$

The correctness of the recurrences and the fact that they can be evaluated in $O(n_1 n_2 n_3)$ time is left to the reader. The recurrences can be generalized to incorporate alphabet-weighted scores, and this is also left to the reader.

A speedup for the exact solution

Carillo and Lipman [88] suggested a way to reduce some of the work needed in computing the optimal *SP* multiple alignment in practice. An extension of their idea was implemented in the program called MSA [303], but that extension was not described in the paper. Here we will explain the idea in the MSA version of the speedup. Additional refinements (particularly space reductions) to MSA are reported in [197]. We again restrict attention to aligning three strings.

The program for multiple alignment shown in the previous section is an example of using recurrences in a *backward* direction. At the time that the algorithm sets $D(i, j, k)$, the program looks backward to retrieve the D values of the seven (at most) cells that influence the value of $D(i, j, k)$.

In the alternative approach of forward dynamic programming (see also Section 12.6.1), when $D(i, j, k)$ is set, $D(i, j, k)$ is then sent *forward* to the seven (at most) cells whose D value can be influenced by cell (i, j, k). Another way to say this is to view optimal alignment as a shortest path problem in the weighted edit graph corresponding to a multiple alignment table. In that view, when the shortest path from the source s (cell $(0, 0, 0)$) to a node v (cell (i, j, k)) has been computed, the best-yet distance from s to any out-neighbor of v is then updated. In more detail, let $D(v)$ be the shortest distance from s to v, let (v, w) be a directed edge of weight $p(v, w)$, and let $p(w)$ be the shortest distance yet found from s to w. After $D(v)$ is computed, $p(w)$ is immediately updated to be $\min[p(w), D(v){+}p(v, w)]$. Moreover, the true shortest distance from s to w, $D(w)$, must be $p(w)$ after $p(w)$ has been updated from every node v having an edge pointing into w. One more detail is needed. The forward dynamic programming implementation will keep a *queue* of nodes (cells) whose final D value has not yet been set. The algorithm will set the D value of the node v at the head of the queue and remove that node. When it does, it updates $p(w)$ for each out-neighbor of v, and if w is not yet in the queue, it adds w to the end of the queue. It is easy to verify that when a node comes to the head of the queue, all the nodes that point to it in the graph have already been removed from the queue.

For the problem of aligning three strings of n characters each, the edit graph has roughly n^3 nodes and $7n^3$ edges, and the backward dynamic programming computation will do some work for each of these edges. However, if one can identify a large number of nodes as not possibly being on any optimal (shortest) path from $(0, 0, 0)$ to (n, n, n), then work can be avoided for each edge out of those excluded nodes. The Carillo and Lipman speedup excludes some nodes before the main dynamic programming computation is begun, and forwards or backwards dynamic programming is equally convenient. In the extension used in MSA, nodes are excluded *during* the main algorithm, and forward dynamic programming is critical.

Definition Let $d_{1,2}(i, j)$ be the edit distance between suffixes $S_1[i..n]$ and $S_2[j..n]$ of strings S_1 and S_2. Define $d_{1,3}(i, k)$ and $d_{2,3}(j, k)$ analogously.

Certainly, all these d values can be computed in $O(n^2)$ time by reversing the strings and computing three pairwise distances. Also, the shortest path from node (i, j, k) to node (n, n, n) in the edit graph for S_1, S_2, and S_3 must have distance at least $d_{1,2}(i, j) + d_{1,3}(i, k) + d_{2,3}(j, k)$.

Now suppose that some multiple alignment of S_1, S_2, and S_3 is known (perhaps from a bounded-error heuristic of the type to be discussed later) and that the alignment has SP score of z. The idea of the heuristic speedup is:

Key idea Recall that $D(i, j, k)$ is the optimal SP score for aligning $S_1[1..i]$, $S_2[1..j]$, and $S_3[1..k]$. If $D(i, j, k) + d_{1,2}(i, j) + d_{1,3}(i, k) + d_{2,3}(j, k)$ is greater than z, then node (i, j, k) cannot be on any optimal path and so (in a forward computation) $D(i, j, k)$ need not be sent forward to any cell.

When this heuristic idea applies to a cell (i, j, k), it saves the work of not sending a D value forward to the (seven) out-neighbors of cell (i, j, k). But more important than the speedup observed at cell (i, j, k) is that when the heuristic applies at many cells, additional cells downstream may be automatically excluded from consideration because they never get placed in the queue. In this way, the heuristic may prune off a large part of the alignment graph and dramatically reduce the number of cells whose D value is set when determining $D(n, n, n)$. This is a standard heuristic in shortest path computations. Notice that the computation remains exact and will find the optimal alignment.

If no initial z value is known, then the heuristic can be implemented as an A^* heuristic [230, 379] so that the most "promising" alignments are computed first. In that approach, during the running of the algorithm the best alignment computed so far provides the (falling) z value. It can be proven [230] that this approach prunes out as much of the alignment graph as the Carillo and Lipman method does, but usually it will exclude much more. It is reported in [303] that MSA can align six strings of lengths around 200 characters in a "practical" amount of time. It doesn't seem likely, however, that this approach will make it practical to optimaly align tens or hundreds of strings, unless one begins with an extremely good value for z. Even then, it is not clear that MSA would be practical.

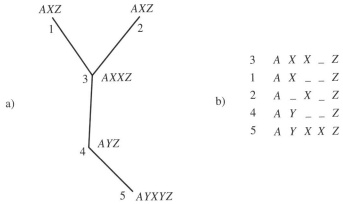

Figure 14.6: a. A tree with its nodes labeled by a (multi)set of strings. b. A multiple alignment of those strings that is consistent with the tree. The pairwise scoring scheme scores a zero for each match and a one for each mismatch or space opposite a character. The reader can verify that each of the four induced alignments specified by an edge of the tree has a score equal to its respective optimal distance. However, the induced alignment of two strings which do not label adjacent nodes may have a score greater than their optimal pairwise distance.

14.6.2. A bounded-error approximation method for *SP* alignment

Because the exact solution to the *SP* alignment problem is feasible for only a small number of strings, most practical (heuristic) multiple alignment methods do not insist on finding the optimal *SP* alignment. However, little is usually known about how much the alignments produced by those heuristics can deviate from the optimal *SP* alignment. In contrast, in this section we discuss one of the few methods where such error analysis has been possible. We develop here a *bounded-error approximation* method, from [201], for *SP* alignment. That is, the method is provably fast (runs in polynomial worst-case time) and yet produces alignments whose *SP* score is guaranteed to be less than twice the score of the optimal *SP* alignment. This will be the first of several bounded-error approximation methods presented in this book, which will include additional methods for different multiple alignment problems.

An initial key idea: alignments *consistent* with a tree

The *SP* approximation we present, the improvements of it, and several other methods for other problems all use a key idea that relates multiple alignments to trees.[10] We develop that idea in general before continuing with *SP* alignment in particular. Recall that for two strings, $D(S_i, S_j)$ is the (optimal) weighted edit distance between S_i and S_j.

> **Definition** Let S be a set of strings, and let T be a tree where each node is labeled with a distinct string from S. Then, a multiple alignment $\mathcal{M}$ of S is called *consistent* with T if the induced pairwise alignment of S_i and S_j has score $D(S_i, S_j)$ for each pair of strings (S_i, S_j) that label adjacent nodes in T. For an example, see Figure 14.6.

Theorem 14.6.1. *For any set of strings S and for any tree T whose nodes are labeled by distinct strings of S, we can efficiently find a multiple alignment $\mathcal{M}(T)$ of S that is consistent with T.*

---

[10] This connection between trees and multiple alignments should not be confused with the connection between *evolutionary* trees and multiple alignments, to be discussed in Sections 14.8 and 14.10.2.

Note that the role of T in this theorem is very special and the induced alignment of two strings S_i and S_j that do not label adjacent nodes will generally have a score greater than $D(S_i, S_j)$.

PROOF OF THEOREM 14.6.1   We will construct the multiple alignment $\mathcal{M}(T)$ one string at a time and show inductively that Theorem 14.6.1 holds after each new string is added to the alignment. To start, choose any two strings S_i and S_j that label adjacent nodes in T and form a two-string alignment of S_i and S_j with distance $D(S_i, S_j)$. The theorem trivially holds at this point. Assume that the theorem holds after some arbitrary number of strings have been added to the multiple alignment. To continue, select any string S' not yet included in the alignment such that S' labels a node adjacent to a node whose label, S_i say, is already in the alignment. In that existing multiple alignment, some spaces may have been inserted into S_i, and we use $\overline{S}_i$ to denote the string S_i with those spaces included. Note that for every string in the existing multiple alignment, each character of that string is in a distinct column with exactly one character of $\overline{S}_i$.

Next, optimally align string S' with $\overline{S}_i$ using a scoring scheme for two-string alignment, with the added rule that two opposing spaces have zero cost (they are considered a match). It is immediately apparent that the score of that resulting pairwise alignment is exactly $D(S_i, S')$. Now we will add S' to the existing multiple alignment so that the induced alignment of S_i and S' has score $D(S_i, S')$ and so that all the induced scores of the strings already in the alignment remain unchanged. Let $\overline{S}'$ be the string S' with any spaces inserted by the alignment of S' and $\overline{S}_i$. If the pairwise alignment of S' and $\overline{S}_i$ does not insert any new spaces into $\overline{S}_i$, then append $\overline{S}'$ to the existing multiple alignment. The result is a multiple alignment with one more string, where the induced score of S_i and S' is $D(S_i, S')$ and where all the induced scores from the previous multiple alignment remain unchanged.

However, if the pairwise alignment does insert a new space in $\overline{S}_i$ between characters l and $l+1$, say, then insert a space between characters l and $l+1$ in every string in the *existing* multiple alignment. This will create new column(s) in the existing multiple alignment in which every character is a space, but otherwise retains the columns of the existing multiple alignment. (For example, suppose that the first four strings from Figure 14.6 have been multiply aligned and that $\overline{S}_4$ at that point is AY_Z. The alignment of S_5 to $\overline{S}_4$ inserts an additional space into $\overline{S}_4$ at the fourth position. The first four rows of Figure 14.6 show the resulting multiple alignment after that space is replicated in strings S_1, S_2, and S_3.) The result of adding a column containing only spaces is a multiple alignment in which the score of all pairwise induced alignments is the same as before any spaces were inserted. Then appending $\overline{S}'$ to that multiple alignment creates a multiple alignment of one more string, where the statement of the theorem holds. The existence of $\mathcal{M}(T)$ then follows by induction.

The time needed to compute $\mathcal{M}(T)$ is dominated by the time to compute $k-1$ pairwise alignments. If each string has length n, then each pairwise alignment takes time $O(n^2)$ and the time to construct $\mathcal{M}(T)$ is $O(kn^2)$. $\square$

Although Theorem 14.6.1 has become a part of "folklore", it may have been first explicitly stated in [153]. We can now return to the approximation method for the *SP* alignment problem.

The center star method for *SP* alignment

We will describe the method in terms of an alphabet-weighted scoring scheme for two-string alignment, and let $s(x, y)$ be the score contributed when a character x (possibly a space) is aligned opposite a character y (possibly a space).

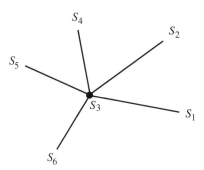

Figure 14.7: A generic center star for six strings, where the center string S_c is S_3.

Definition A scoring scheme satisfies the *triangle inequality* if for any three characters x, y, and z, $s(x, z) \leq s(x, y) + s(y, z)$.

Triangle inequality makes good intuitive sense in the context of edit distance. It says that the "cost" of transforming x to z directly is no more than the cost of first transforming x to an intermediate character y, and then transforming y to z. It should be noted, however, that not all scoring matrices in use in computational biology satisfy the triangle inequality.

Definition Given a set of k strings S, define a *center* string $S_c \in S$ as a string in S that minimizes $\sum_{S_j \in S} D(S_c, S_j)$, and let M denote that minimum sum. Define the *center star* to be a star tree of k nodes, with the center node labeled S_c and with each of the $k - 1$ remaining nodes labeled by a distinct string in $S - S_c$. For an example, see Figure 14.7.

Definition Define the multiple alignment $\mathcal{M}_c$ of the set of strings S to be the multiple alignment *consistent* with the center star.

Definition Define $d(S_i, S_j)$ as the score of the pairwise alignment of strings S_i and S_j *induced* by $\mathcal{M}_c$. Denote the score of an alignment $\mathcal{M}$ as $d(\mathcal{M})$.

Clearly, $d(S_i, S_j) \geq D(S_i, S_j)$, and $d(\mathcal{M}_c) = \sum_{i<j} d(S_i, S_j)$. Also, since $\mathcal{M}_c$ is consistent with the star tree, and string S_c is at the center of the star, $d(S_i, S_c) = D(S_i, S_c)$ for each string S_i. We will show that $d(\mathcal{M}_c)$ is at most twice the score of the optimal *SP* multiple alignment of S, provided that the scoring scheme used for pairwise alignment satisfies the triangle inequality.

Lemma 14.6.1. *Assume that the two-string scoring scheme satisfies the triangle inequality. Then, for any strings S_i and S_j in S, $d(S_i, S_j) \leq d(S_i, S_c) + d(S_c, S_j) = D(S_i, S_c) + D(S_c, S_j)$.*

PROOF Consider any single column in the multiple alignment $\mathcal{M}_c$, and let x, y, and z be the three characters in this column from the three strings S_i, S_c, and S_j, respectively. By the triangle inequality, $s(x, z) \leq s(x, y) + s(y, z)$, and so the claimed inequality follows by the definition of d. The claimed equality follows because the pairwise alignment of S_i and S_c induced by $\mathcal{M}_c$ is an optimal alignment of S_i and S_c, and this is true also for the alignment of S_c and S_j. □

Definition Let $\mathcal{M}^*$ be the optimal multiple alignment of the k strings of S. Let $d^*(S_i, S_j)$ be the score of the pairwise alignment of strings S_i and S_j induced by $\mathcal{M}^*$. Then $d(\mathcal{M}^*) = \sum_{i<j} d^*(S_i, S_j)$.

We can now state and prove the main theorem of this section.

Theorem 14.6.2. $d(\mathcal{M}_c)/d(\mathcal{M}^*) \leq 2(k - 1)/k < 2$.

PROOF   First, define $v(\mathcal{M}_c) \equiv \sum_{(i,j)} d(S_i, S_j)$ and $v(\mathcal{M}^*) \equiv \sum_{(i,j)} d^*(S_i, S_j)$, where the pair (i, j) is an *ordered* pair in each case. Clearly, $v(\mathcal{M}_c) = 2d(\mathcal{M}_c)$ and $v(\mathcal{M}^*) = 2d(\mathcal{M}^*)$, and so the ratios $d(\mathcal{M}_c)/d(\mathcal{M}^*)$ and $v(\mathcal{M}_c)/v(\mathcal{M}^*)$ are equal. It is more convenient to work with the second ratio. Recall that the minimum sum of distances, M, is defined as $\sum_j D(S_c, S_j)$. Now $v(\mathcal{M}_c) = \sum_{(i,j)} d(S_i, S_j) \le \sum_{(i,j)}[D(S_i, S_c) + D(S_c, S_j)]$, by Lemma 14.6.1. For any fixed j, $D(S_c, S_j)$ (which equals $D(S_j, S_c)$) shows up in this expression exactly $2(k-1)$ times. So $v(\mathcal{M}_c) \le 2(k-1) \times \sum_j D(S_c, S_j) = 2(k-1)M$.

From the other side, $v(\mathcal{M}^*) = \sum_{(i,j)} d^*(S_i, S_j) \ge \sum_{(i,j)} D(S_i, S_j) = \sum_i \sum_j D(S_i, S_j) \ge k \times \sum_j D(S_c, S_j) = kM$ (by the choice of S_c). So $d(\mathcal{M}_c)/d(\mathcal{M}^*) = v(\mathcal{M}_c)/v(\mathcal{M}^*) \le 2(k-1)M/kM = 2(k-1)/k = 2 - 2/k < 2$.   □

Note that for $k = 3$ the guaranteed upper bound is $4/3$. That is, for three strings, the multiple alignment produced by the center star method will never be more that 34% more than the optimal *SP* score. Translated into lower bounds this says that for $k = 3$, $d(\mathcal{M}^*) \ge .75d(\mathcal{M}_c)$. For $k = 4$ the upper bound is only 1.5, and for $k = 6$ (a problem size considered to be too large for efficient exact solution with strings of length 200) the bound is still only 1.67.

Corollary 14.6.1.
$kM \le \sum_{i<j} D(S_i, S_j) \le d(\mathcal{M}^*) \le d(\mathcal{M}_c) \le [2(k-1)/k] \sum_{i<j} D(S_i, S_j)$.

In practice one can better measure the goodness of $\mathcal{M}_c$ by the ratio $d(\mathcal{M}_c)/\sum_{i<j} D(S_i, S_j)$. By Corollary 14.6.1 this ratio is always less than two, but the analysis used there is worst case, so one can expect the ratio to often be considerably less than two. Similarly, one should expect that $d(\mathcal{M}_c)/d(\mathcal{M}^*)$ will often be considerably less than two, since typically $\sum_{(i,j)} D(S_i, S_j)$ will be considerably larger than kM; that $d(\mathcal{M}^*)$ will not generally be close to $\sum_{i<j} D(S_i, S_j)$ for any strings except those that are very similar; and that $D(S_i, S_j)$ will be less than $D(S_i, S_c) + D(S_c, S_j)$ for most typical strings.

Corollary 14.6.1 is also useful in the MSA speedup discussed in Section 14.6.1 for the exact solution to *SP* alignment, since that method requires knowing an efficiently computed upper bound z on the optimal *SP* alignment score.

14.6.3. Weighted *SP* alignment

A generalization of *SP* alignment was introduced by Altschul and Lipman [20], in which the induced pairwise score of each pair (S_i, S_j) is multiplied by a weight $w(i, j)$. Then the score of a multiple alignment $\mathcal{M}$ is $\sum_{i<j} w(i, j)d(S_i, S_j)$, where $d(S_i, S_j)$ is the score of the pairwise alignment induced by $\mathcal{M}$. The weights change the importance given to different pairs of strings and are often intended to reflect known evolutionary distance between the organisms from which the strings are obtained. Using weights, one can try to induce the multiple alignment to more accurately reflect known evolutionary history.

The optimal *weighted SP* alignment can again be computed in exponential time by dynamic programming, but little is known about approximating weighted *SP* alignment. However, the weights are primarily desired as a reflection of evolutionary distance, and there is a different multiple alignment objective function that addresses the same goal of using evolutionary history to influence the resulting multiple alignment. That objective function is contained in the *phylogenetic alignment* problem, for which there is an approximation method quite related to the center star method. We will discuss those topics in Section 14.8. But before doing so, we discuss the consensus objective function and an approximation algorithm even more related to the center star method.

14.7. Multiple alignment with consensus objective functions

Many multiple string alignment or comparison methods used in biology have as their goal to derive a *consensus representation* of the critical common features of a *set* of strings. Although there is no universal consensus on how to define "consensus", we examine three definitions that capture the spirit of most of the methods. We will then show that the three definitions lead to the same string and the same multiple alignment, which can again be approximated by the center star method. For clarity of exposition, we will use some notions of consensus that do not appear in the computational biology literature, and consequently we will define some new terminology as well.

14.7.1. Steiner consensus strings

The first notion of a consensus sequence is stated without any explicit connection to multiple alignment. As before, we assume the existence of a two-string scoring scheme that satisfies the triangle inequality and recall that $D(S_i, S_j)$ denotes the weighted edit distance of strings S_i and S_j.

Definition   Given a set of strings S, and given another string $\overline{S}$, the *consensus error* of a string $\overline{S}$ *relative to* S is $E(\overline{S}) = \sum_{S_i \in S} D(\overline{S}, S_i)$. Note that $\overline{S}$ need not be from S.

Definition   Given a set of strings S, an optimal *Steiner string* S^* for S is a string that minimizes the consensus error $E(S^*)$ over all possible strings.

Note that S^* need not be from S and generally will not be. We will usually refer to a "Steiner consensus string" as a "Steiner string".

The Steiner string S^* attempts to capture and reflect, in a single string, the common characteristics of the set of strings S. There is no known efficient method to find S^*, but there is an approximation method that finds a string $\overline{S}$ such that the ratio $E(\overline{S})/E(S^*)$ is never more than two.

Lemma 14.7.1. *Let S have k strings, and assume that the two-string scoring scheme satisfies the triangle inequality. Then there exists a string $\overline{S} \in S$ such that $E(\overline{S})/E(S^*) \leq 2 - 2/k < 2$.*

PROOF   Let $\overline{S}$ be any string in S. By the triangle inequality of the pairwise scoring scheme, $D(\overline{S}, S_i) \leq D(\overline{S}, S^*) + D(S^*, S_i)$ for any S_i. So $E(\overline{S}) = \sum_{S_i \in S} D(\overline{S}, S_i) \leq \sum_{S_i \neq \overline{S}} [D(\overline{S}, S^*) + D(S^*, S_i)] = (k-2)D(\overline{S}, S^*) + E(S^*)$.

Now pick $\overline{S}$ to be a string in S that is closest to the optimal Steiner string S^*. That is, choose $\overline{S}$ so that $D(\overline{S}, S^*)$ is less than or equal to $D(S_i, S^*)$, for any $S_i \in S$. (Of course, $\overline{S}$ is not known constructively since S^* is not known, but $\overline{S}$ does exist.) Then $E(S^*) = \sum_{S_i \in S} D(S^*, S_i) \geq kD(\overline{S}, S^*)$. Therefore, $E(\overline{S})/E(S^*) \leq (k-2)D(\overline{S}, S^*)/kD(\overline{S}, S^*) + 1 = (k-2)/k + 1 = 2 - 2/k < 2$. $\square$

Recall that the *center string* S_c is a string in S that minimizes $\sum_{S_i \in S}(S_c, S_i)$ over all strings in S.

Theorem 14.7.1. *Assuming that the scoring scheme satisfies the triangle inequality, $E(S_c)/E(S^*) \leq 2 - 2/k$.*

PROOF   The theorem follows immediately from Lemma 14.7.1 and the fact that $E(S_c) \leq E(\overline{S})$ by definition. $\square$

Thus the center string has a consensus error that is at most $2 - 2/k$ times the error of the optimal Steiner consensus string. This result is a specialization of an approximation method used in the phylogenetic alignment problem, which will be discussed in Section 14.8.

14.7.2. Consensus strings from multiple alignment

Although the definition of the optimal Steiner string makes no mention of multiple alignment, the problem of finding S^* is equivalent to more traditional consensus problems that *are* defined in terms of multiple alignments. We now make those consensus problems precise.

> **Definition**   Given a multiple alignment $\mathcal{M}$ of a set of strings $\mathcal{S}$, the *consensus character* of column i of $\mathcal{M}$ is the character that minimizes the summed distance to it from all the characters in column i. Let $d(i)$ denote that minimum sum in column i.

Since the alphabet is finite, a consensus character for each column of $\mathcal{M}$ exists and can be found by enumeration. As one simple special case, if the pairwise scoring scheme scores a match with a zero and a mismatch or a space opposite a character with a one, then the consensus character in column i is the *plurality* character (i.e., the character occurring the most often in column i). Note that the plurality character can be a space. See Figure 14.5 for an example.

> **Definition**   The *consensus string* $S_\mathcal{M}$ *derived from alignment* $\mathcal{M}$ is the concatenation of the consensus characters for each column of $\mathcal{M}$.

It is common in the computational biology literature to compute a multiple alignment $\mathcal{M}$ of a set of strings and then represent those strings by the consensus string $S_\mathcal{M}$ derived from $\mathcal{M}$. It is then natural to use the goodness of string $S_\mathcal{M}$ as a way to evaluate the goodness of the multiple alignment $\mathcal{M}$. One way to do this is to define the goodness of $\mathcal{M}$ as $\sum_i D(S_\mathcal{M}, S_i)$. That is, the goodness of $\mathcal{M}$ is reflected by how well $S_\mathcal{M}$ acts as a *Steiner string* for the set $\mathcal{S}$. Another (seemingly different) way to evaluate the goodness of multiple alignment $\mathcal{M}$ through its consensus string $S_\mathcal{M}$ is more closely tied to $\mathcal{M}$.

> **Definition**   Let $\mathcal{M}$ be a multiple alignment of the set of strings $\mathcal{S}$, and let $S_\mathcal{M}$ be its consensus string containing q characters. Then the *alignment error* of $S_\mathcal{M}$ equals $\sum_{i=1}^{i=q} d(i)$, and the alignment error of $\mathcal{M}$ is defined as the alignment error of $S_\mathcal{M}$. (See Figure 14.5.)

> **Definition**   The *optimal consensus multiple alignment* is a multiple alignment $\mathcal{M}$ for input set $\mathcal{S}$ whose consensus string $S_\mathcal{M}$ has smallest alignment error over all possible multiple alignments of $\mathcal{S}$.

This gives us three notions of how to define a consensus of a set of strings $\mathcal{S}$ and two related notions of how to evaluate a multiple alignment $\mathcal{M}$. The first notion of consensus is the *Steiner string* S^* defined from $\mathcal{S}$ without any mention of multiple alignment. The second notion is the consensus string $S_\mathcal{M}$, which is derived from a multiple alignment $\mathcal{M}$, but whose goodness is evaluated by how well it acts as a Steiner string for $\mathcal{S}$. In this case, the goodness of the multiple alignment $\mathcal{M}$ is defined as the consensus error of its consensus string $S_\mathcal{M}$. The third notion is the most closely tied to multiple alignment. In that notion, string $S_\mathcal{M}$ is both derived from $\mathcal{M}$ and evaluated by how well it reflects the *columnwise* properties of $\mathcal{M}$. In this case, the goodness of $\mathcal{M}$ is again based on the goodness of $S_\mathcal{M}$, in particular, on the alignment error of $S_\mathcal{M}$. These three seemingly different notions are equivalent in that they lead to the same multiple alignment.

Definition   Given set S of k strings, let T be the star tree with Steiner string S^* at the root and each of the k strings of S at distinct leaves of T. Then the multiple alignment of $S \cup S^*$ *consistent* with T is said to be consistent with S^*.

Theorem 14.7.2. *Let S denote the consensus string of the optimal consensus multiple alignment. Then, removal of the spaces from S creates the optimal Steiner string S^*. Conversely, removal of the row for S^* from the multiple alignment consistent with S^* creates the optimal consensus multiple alignment of S.*

PROOF   Let S have k strings, and let $\mathcal{M}$ be any multiple alignment of S. By definition, each character of the consensus string $S_\mathcal{M}$ derived from $\mathcal{M}$ is associated with a distinct column of $\mathcal{M}$, and this association induces a particular pairwise alignment between $S_\mathcal{M}$ and each S_i in S. Clearly, the score of that alignment is at least $D(S_i, S_\mathcal{M})$. Now the alignment error of $S_\mathcal{M}$ is exactly the sum of the scores of those k pairwise induced alignments, and so the alignment error of $S_\mathcal{M}$ is at least $\sum_i D(S_i, S_\mathcal{M})$, which is the consensus error of $S_\mathcal{M}$ for S. But by definition, S^* has the minimum consensus error for S, so the alignment error of $S_\mathcal{M}$ is at least the consensus error of S^*.

Now consider the multiple alignment $\mathcal{M}^*$ of $S \cup S^*$ consistent with S^*, and for any string α in $S \cup S^*$, let $\bar{\alpha}$ denote the string in the row of $\mathcal{M}^*$ corresponding to α. By consistency, the score of the induced alignment in $\mathcal{M}^*$ of $\bar{S}^*$ and $\bar{S}_i$ is $D(S^*, S_i)$ for any $S_i \in S$. Let $\mathcal{M}'$ be $\mathcal{M}^*$ after removing the row for S^*. Then, the alignment error of $\bar{S}^*$ with $\mathcal{M}'$ is exactly $\sum_i D(S^*, S_i)$, which is the consensus error of S^* for S. Hence, using the conclusion from the first paragraph, the alignment error of any other consensus string $S_\mathcal{M}$ for any other multiple alignment $\mathcal{M}$ must be at least as large as the alignment error of $\bar{S}^*$ for $\mathcal{M}'$. It follows that $\mathcal{M}'$ is the optimal consensus multiple alignment for S and that $\bar{S}^*$ is its consensus string. Therefore, since S^* is obtained from $\bar{S}^*$ by removing the spaces in $\bar{S}^*$, the theorem is proved.   □

Hence the optimal consensus multiple alignment specifies the optimal Steiner string S^*, and conversely, S^* can be used to construct the optimal consensus multiple alignment. A Steiner string and the consensus string for the optimal consensus multiple alignment both capture the same commonalities in S, and the optimal consensus multiple alignment allows us to explicitly display those commonalities in a columnwise manner.

14.7.3. Approximating the optimal consensus multiple alignment

From the proof of Theorem 14.7.2 and the statement of Theorem 14.7.1, it should now be clear how to find a multiple alignment $\mathcal{M}$ whose *alignment error* is never more than $2 - 2/k$ times the alignment error of the optimal consensus multiple alignment (assuming the triangle inequality of the two-string scoring scheme). Just find the center string S_c as before, place it at the center of a k node star, label each leaf with one of the remaining strings in S, and construct the multiple alignment $\mathcal{M}$ consistent with this tree.

The multiple alignment $\mathcal{M}$ just described is exactly the alignment $\mathcal{M}_c$ used to approximate the *SP* objective function. We have thus established the following, perhaps surprising result:

Theorem 14.7.3. *Assuming the triangle inequality, the multiple alignment $\mathcal{M}_c$ created by the center star method has an SP score that is never more than $2 - 2/k$ times the SP score of the optimal SP alignment, and it has a (consensus) alignment error that is never more than $2 - 2/k$ times the alignment error of the optimal consensus multiple alignment.*

14.8. Multiple alignment to a (phylogenetic) tree

As mentioned in the discussion of weighted *SP* alignment, it is often desirable to use known evolutionary history to influence the multiple alignment computed for a set of strings. The method should be encouraged to align more closely those sequences from closely related organisms. Evolutionary history is most often represented by an *evolutionary tree* where known sequences of (extant) organisms are represented at the leaves of the tree, and their unknown ancestors are represented at internal nodes of the tree. When the tree is known (from previous data and deductions) the problem is to deduce sequences for the internal nodes to optimize an objective function defined below. Once these sequences are in hand, one can find a multiple alignment consistent with the labeled tree and then remove the deduced sequences from that multiple alignment. Alternatively, the labeled tree itself gives a deduced picture of the evolutionary history that led to the known (leaf) sequences. The latter use for phylogenetic alignment will be discussed in Section 17.6.

> **Definition** Given an input tree T with a distinct string (from a set of strings S) written at each leaf, a *phylogenetic alignment* for T is an assignment of one string to each internal node of T. Note that the strings assigned to internal nodes need not be distinct and need not be from the input strings S.

The phylogenetic tree, T, is meant to represent the "established" evolutionary history of a set of taxa (read "objects of interest"), with the convention that each extant taxon ("object") is represented at a unique leaf of T. Each edge (u, v) represents some mutational history that transforms the string at u (assuming u is the parent of v) to the string at v. The cost of that transformation is given by the edit distance between those two strings, and so the cost of the phylogenetic alignment is the sum of all those edge costs. This is now stated more formally.

Recall that $D(S, S')$ denotes the edit distance between string S and string S'. All that is assumed about the function D is that it obeys the usual triangle inequality conditions.

> **Definition** If strings S and S' are assigned to the endpoints of an edge (i, j), then (i, j) has *edge distance* $D(S, S')$. The distance along a path is the sum of the distances on the edges in the path. The distance of a phylogenetic alignment is the total of all the edge distances in the tree.

> **The phylogenetic alignment problem for** T Find an assignment of strings to internal nodes of T (one string to each node) that minimizes the distance of the alignment.

The phylogenetic alignment problem, also called the *tree alignment problem* was developed principally by Sankoff [387, 388]. We use the term "phylogenetic alignment" instead of "tree alignment" because the latter term is also used for a very different problem in the string literature [117].

Note that the consensus string problem studied in Section 14.7 is just a special case of the phylogenetic alignment problem (i.e., when tree T is a star). The connection between phylogenetic alignment and the consensus problem was first observed in [20]. In a similar manner, the algorithm to be presented here for the phylogenetic alignment problem will be a natural generalization of the algorithm developed for the *SP* and consensus objective functions.

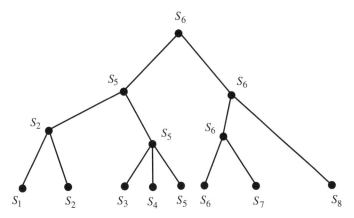

Figure 14.8: An abstract lifted alignment. Each node has the abstract name of a string written by it.

14.8.1. A heuristic for phylogenetic alignment

The general phylogenetic alignment problem is *NP-complete* [454], but it has an exponential-time solution via dynamic programming [387]. A limited version of the problem is approached heuristically in [457]. Here we discuss an efficient heuristic method that was developed and analyzed by Wang, Jiang, and Lawler [249]. Their method produces phylogenetic alignments whose distance is never more than twice the distance of the minimum alignment. However, our analysis follows a simplification of the original proof, suggested by Mike Paterson. For this result, we only assume that the edit distance function D satisfies the triangle inequality. The factor of two result also leads to a polynomial approximation scheme allowing a trade-off between computation time and guaranteed accuracy [249]. A newer result along those lines leads to a practical approximation scheme where a deviation of at most 1.58 can be guaranteed for protein-length strings [453].

Definition   A phylogenetic alignment is called a *lifted alignment* if for every internal node v, the string assigned to v is also assigned to one of v's children.

For example, the solution in Figure 14.8 is a lifted alignment. Clearly, each node v in a lifted alignment is assigned a string that labels a leaf in the subtree rooted at v.

We will show that the best lifted alignment in T has a total distance less than twice that of the optimal phylogenetic alignment. We do this by constructing a particular lifted alignment T^L with that property. T^L is found by transforming the optimal phylogenetic alignment to the lifted alignment T^L. Later we show that the best lifted alignment can be computed efficiently by dynamic programming. Because the best lifted alignment has total distance at most that of T^L, one can efficiently find a lifted alignment with total distance less than twice that of the optimal phylogenetic alignment.

The transformation creating T^L

Let T^* be the optimal phylogenetic alignment for tree T. We will transform T^* into a lifted alignment T^L by a series of string replacements at internal nodes of T. This transformation is only conceptual since we do not know T^*, but it serves to define the lifted alignment T^L.

We say that a node has been *lifted* after it has been labeled by a string in the leaf set S. By definition, each leaf begins lifted. The lifting process will successively "lift" each of the internal nodes of T in any order, provided that a node is lifted only after all of

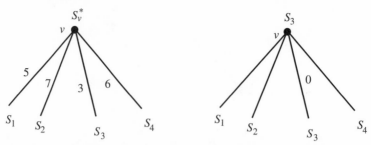

Figure 14.9: The lifting operation at node v. The numbers on the edges are the distances from S_v^* to the lifted strings labeling its children. Note that after the lift, one edge will have zero distance.

its children have been lifted. The tree that results at the end of the process is the lifted alignment T^L.

We lift a node v as follows: Let S_v^* be the string labeling internal node v in T^*. Without loss of generality, assume that v's children have been labeled with the lifted strings $S_1, S_2, \ldots, S_k$ from S, and we refer to v's children by their string labels. Let S_j be the string from among v's children that is closest to S_v^*. That is, $D(S_v^*, S_j) \leq D(S_v^*, S_i)$ for any i from 1 to k. To lift v, replace string S_v^* by S_j (see Figure 14.9). This changes the distance on each edge out of v to $D(S_j, S_i)$. In particular, the distance becomes zero on the edge between v and its child labeled S_j.

The error analysis

Theorem 14.8.1. *The lifted alignment T^L has total distance less or equal to twice that of the optimal phylogenetic alignment T^* of T.*

PROOF Let $e = (v, w)$ be any edge in T, where v is the parent of w. Suppose that in T^L, node v is labeled with string $S_j \in S$ and node w is labeled with string $S_i \in S$. If $S_j = S_i$, then edge e has a zero-length edge in T^L and so we will not be concerned with it. But if $S_j \neq S_i$, then the distance of edge e in T^L is $D(S_j, S_i) \leq D(S_j, S_v^*) + D(S_v^*, S_i) \leq 2 \times D(S_v^*, S_i)$. The first inequality is due to the triangle inequality. To see the second inequality, note that at the point in the lifting process that v becomes labeled, S_i labeled a child (w) of v and hence was a candidate to label v. So $D(S_v^*, S_j)$ must be no larger than $D(S_v^*, S_i)$, and so $D(S_j, S_i) \leq 2 \times D(S_v^*, S_i)$. Now consider the path in T^* from v to the leaf labeled S_i. Denote that path by P_e. By the triangle inequality, $D(S_v^*, S_i)$ is at most the *total* of the edge distances on P_e in T^*. Hence the distance of edge e in T^L is at most twice the total path length of P_e. Note that path P_e is only defined for an edge e that has a nonzero distance in T^L.

For a nonzero-length edge $e = (v, w)$, path P_e is the path along which the leaf string S_i was lifted to node w in T^L (see Figure 14.10). Hence, every node on P_e, except v, is labeled by S_i, and no node outside of P_e is labeled with S_i. So if $e' = (v', w')$ is any other nonzero-length edge, and $P_{e'}$ is the path along which a leaf string was lifted to w', then P_e and $P_{e'}$ have no edges in common (again, see Figure 14.10). This defines a mapping from each nonzero-length edge e in T^L to a path P_e in T^* such that: a. the distance in T^L of edge e is at most twice the total distance in T^* of the edges on P_e and b. no edge in T^* is mapped to by more than one edge in T^L. Further, if the root is labeled in T^L with string $S \in S$, then no edge in T^* on the path from the root to leaf S is mapped to by any edge in T^L. Therefore, the total distance of the lifted alignment T^L is less than or equal to twice the total distance of the optimal phylogenetic alignment T^*. $\square$

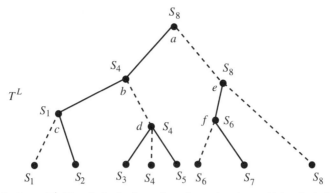

Figure 14.10: Lifted tree T^L. The dashed edges show the paths along which a leaf string was lifted to some internal node. In T^L each of these dashed edges has zero length. Path $P_{(a,b)}$ (as an example) is the path b, d, S_4 along which the string labeling b was lifted. Edge (a, b) has distance in T^L at most twice that of path $P_{(a,b)}$ in T^*.

The proof can be extended to show strict inequality. We can now state the main result.

Corollary 14.8.1. The best lifted alignment has total distance less than twice that of the optimal phylogenetic alignment T^*.

Computing the minimum distance lifted alignment

The best lifted alignment is computed by dynamic programming.

Definition   Let T_v be the subtree of T rooted at node v. Let $d(v, S)$ denote the distance of the best lifted alignment of T_v under the requirement that string S is assigned to node v (assuming of course that S is a string at a leaf of T_v).

After the computation is finished, if r represents the root node, then the value of the best lifted alignment is the minimum of $d(r, S)$, where S ranges over all strings written at the leaves of T. The dynamic programming algorithm computes the values of $d(v, S)$ from the leaves up. A node will be processed only after all of its children have been processed. The algorithm starts with the assumption that all the leaves have already been processed. If v is an internal node all of whose children are leaves, then the algorithm sets $d(v, S) = \sum_{S'} D(S, S')$, where S' is the string written at child of v.

For a general internal node v, the dynamic programming recurrence is

$$d(v, S) = \sum_{v'} \min_{S'} [D(S, S') + d(v', S')],$$

where v' is a child of v and S' is a string at a leaf in tree $T_{v'}$.

It should be apparent that these recurrences are correct and that the algorithm correctly finds the distance of the best lifted phylogenetic alignment. To actually assign the strings at the internal node, the algorithm must do a traceback through the recurrences (or pointers) in the usual dynamic programming way.

For the time analysis, let k be the number of leaves of T and assume that all the $\binom{k}{2}$ distances between pairs of strings are computed in a preprocessing stage. That preprocessing takes $O(N^2)$ time, where N is the total length of all the k strings. Then the work at any internal node is $O(k^2)$, and the overall work of the algorithm is $O(N^2 + k^3)$. In the exercises, we suggest an improvement that reduces the running time to $O(N^2 + k^2)$. In summary, we have

Theorem 14.8.2. *The optimal lifted alignment can be computed in polynomial time as a function of size of the tree and the lengths of the input strings.*

Hence a phylogenetic alignment whose distance is within a factor of two of the optimal can be computed in polynomial time. This phylogenetic alignment will be the center star alignment discussed earlier for the *SP* and consensus objective functions.

14.9. Comments on bounded-error approximations

Three bounded-error approximation methods were presented in the previous sections. Additional bounded-error methods will be presented in Part IV. For the reader who is not familiar with bounded-error approximations, it may be worthwhile emphasizing a few points.

First, Theorem 14.6.2 says that the alignment $\mathcal{M}_c$ will have a score that is never more than twice that of the optimal *SP* score. It does *not* say that the score of $\mathcal{M}_c$ *will* be twice the optimal (either always or typically) or even that it *could* be twice that of the optimal score. In fact, in limited tests of the center star method, the score of $\mathcal{M}_c$ deviated from the bound $\sum_{i<j} D(S_i, S_j)$ (and hence from the optimal score) by amounts between 2 and 16% [201].

Second, bounded-error approximation methods are not always the most effective or practical methods, nor are they necessarily intended to be used as "stand-alone, off-the-shelf" methods. Rather, they are often theoretical devices used to establish *provable* claims about the behavior of algorithms on hard problems. However, bounded-error methods can lead to, or be a part of, very effective methods whose behavior cannot be easily proven. One way is in combination with local improvement methods, which take a solution and iteratively look for small changes that improve it. Another way is in combination with branch-and-bound methods, where the result of the approximation method can be used both as an upper bound and to create a lower bound on the score of the optimal solution.[11] Similarly, approximation methods can sometimes be used to establish guaranteed bounds in practice for methods that lack guaranteed bounds in general but are claimed to be superior. Suppose an approximation method truly is "bad", meaning that its results are often close to its guaranteed worst-case bound, and some other heuristic is truly "good", meaning that its results are often close to the optimal. Then, on specific problem instances, the ratio of the bad value to the good value will be close to the guaranteed approximation bound, *proving* that the good solution truly deviates from the optimal by a small amount. In this manner, an approximation method with a guaranteed error bound will always be a winner – if it generally produces solutions better than other methods, then it wins the upper bound race; if it generally produces solutions much worse than other methods, then it wins too by providing a *proven* bound on the goodness of the better solution. An example, in the case of phylogenetic alignment, is developed in the exercises and is studied more deeply in [206].

Third, bounded-error methods with *fixed* error bounds are often improvable to methods that have a *provable trade-off* between bounded-error and running time. With such methods, called *polynomial time approximation schemes*, the user knows the amount of

---

[11] Some bounded-error methods have the desirable feature that for any specific instance of the problem, the algorithm computes a nonobvious *lower bound* (as well as an upper bound) on the value of the optimal solution. This is not true of the center-star method, where the lower bound used is immediate and unrelated to the upper bound computation.

running time that will be sufficient to achieve any particular level of confidence (i.e., any chosen limit on the maximum possible deviation from the optimal). Such a method has been developed for the *SP* alignment problem by Bafna, Lawler, and Pevzner [37]. That method aligns k strings and produces a multiple alignment whose *SP* score can never exceed $2 - q/k$ times the optimal *SP* score, for any chosen q. For any fixed value of q, the running time of the method is polynomial in n (the assumed string length of each string) and k. As q increases, so does the *provable accuracy* of the result, but the worst-case running time also increases. Approximation schemes have also been developed for the phylogenetic alignment problem [249, 453].

Finally, bounded-error approximation results sometimes evoke an "ignorance-is-bliss" reaction. That reaction is characterized by the attitude that a proven bound on the maximum possible error is a strike against an algorithm (rather than a point in its favor), unless the proven bound is very small. In its extreme form, the attitude is that a heuristic method whose maximum possible deviation from the optimal is unknown is better than one whose maximum possible deviation is known, unless the proven bound is very small. This attitude may again be a result of confusing a limit on the maximum possible error for a claim that the *typical* error is at that limit. Stated in this extreme cartoon form, the ignorance-is-bliss attitude seems unlikely to occur, but in fact it does (even in the computer science community). In contrast, the viewpoint here is that an algorithm with a proven error bound is a good thing to obtain, even though the algorithms with the best proven bounds are not always the most effective algorithms in practice.

14.10. Common multiple alignment methods

The center-star alignment algorithm was developed as a bounded-error heuristic and is not widely used by practitioners. In this section we focus on ideas embodied in methods that are widely used in practice for multiple alignment.

There are numerous multiple alignment methods (and programs) that have been published in the computational biology literature, and hundreds (if not thousands) of papers report the results of multiple alignments computed from molecular data of interest. Two papers, [92] and [316], give a starting point for examining some of the specific methods in use. We will not attempt a survey here. But virtually all of the methods used in practice are variants of one of two ideas (or a mixture of them), and so it is worth discussing those two ideas in general terms, with some details provided by specific illustrations. After that, we will also briefly mention two different, more recent ideas.

14.10.1. Iterative pairwise alignment

Most of the efforts in global multiple alignment follow a general strategy of iteratively merging two multiple alignments of two subsets of strings into a single multiple alignment of the union of those subsets. In the simplest form, this approach uses pairwise alignment scores to iteratively add one additional string to a growing multiple alignment. What differs between the methods are the criteria for selecting the pair of alignments to merge (and hence the order in which the merges are done) and the style of alignment used to create the new merged alignment.

The simplest example of this approach starts by aligning the two strings whose edit distance is minimum over all pairs of strings. Then the method successively merges in the string with smallest edit distance from any of the strings already in the multiple alignment.

The new string is added to the growing multiple alignment exactly as described in the proof of Theorem 14.6.1. Clearly, this method can be viewed as finding the multiple alignment consistent with a *minimum spanning tree* (see [112]) formed from the pairwise edit distance data. When similarity scores are used in place of edit distance, the methods can be seen as finding multiple alignments consistent with *maximum spanning trees* formed from similarity data.

Since there are several, superficially different, ways to construct a minimum (or maximum) spanning tree [112], there are also several multiple alignment methods that seem different but in fact produce the equivalent multiple alignment. The method described in the paragraph above follows Prim's minimum spanning tree algorithm, whereas a method based on Kruskal's algorithm would generally merge multiple alignments of subsets of strings. Nonetheless, even that method would determine which pair of multiple alignments to merge on the basis of pairwise edit distances (or similarity scores) of original strings, and it would use two original strings to merge the two multiple alignments as in the proof of Theorem 14.6.1.

A more realistic variant of the above approach might choose two multiple alignments to be merged based on the *average* edit distance (or similarity) between the pairs of strings in the two subsets. This is called the *average linkage method*, and is also known as UPGMA, for "Unweighted Pair-Group Method using arithmetic Averages". At each merge step, the new multiple alignment could be created by aligning some representation of the two smaller alignments (for example, by aligning profiles or consensus sequences), rather than using only one specific string from each subset. In fact, if more computation time is permitted, at each iteration one might first compute all pairwise alignments of existing multiple alignments and then choose to merge the pair of alignments whose merged alignment is best. (Another variant of the maximum spanning tree approach will be discussed in the first part of Section 14.10.2).

Iterative alignment and clustering

It should also be clear that multiple alignment methods based on pairwise iterative alignment are direct applications of *clustering algorithms* [217]. Many other global multiple alignment methods can likewise be seen as translations of clustering methods applied to edit distance data. Essentially, any specific clustering method defines a history of merges, building larger clusters from smaller ones. Hence almost any method that appears in the pure clustering literature can be adapted to become a multiple alignment method.

The issue of which, if any, of these clustering variants leads to the most effective multiple alignment method is still wide open and is probably very dependent on the particular biological application. Consequently, many additional variants of this type of alignment will no doubt be devised and used in practice.

14.10.2. Two specific illustrations of iterative pairwise alignment

The above discussion is intended to give the flavor of iterative pairwise alignment methods and, as the reader should be able to imagine, there are many specific variants of the idea that have been proposed and studied. Here we briefly discuss two variants that appear in the literature. One method is intended to identify secondary structure in proteins; the other method is intended to partially reconstruct evolutionary history.

Iterative multiple alignment to identify protein secondary structure

Barton and Sternberg [52] developed and tested a multiple alignment method that is a variant of the maximum spanning tree approach discussed in the last section. The goal of their method is to align related protein sequences to reveal conserved β-*strands* (see Exercise 36 in Chapter 11) in the proteins.

They tested their method on a set of immunoglobulin sequences where the true locations of the β-strands were known from three-dimensional x-ray crystallography studies. The x-ray data allow one to easily match up corresponding regions of secondary structure. In this application, a biologically informative multiple alignment method should, without knowing the true locations of the β-strands, produce a multiple alignment that lines up all the corresponding β-strands in the same way that the x-ray data do. That is, the multiple alignment method should be "tuned" so that the most critical feature of the alignment is how it treats (unknown) regions of secondary structure.

The multiple alignment method in [52] first computes pairwise similarity scores between each pair of strings, but it does not directly use those scores for the multiple alignment. Instead, it uses scores that are believed to be more biologically informative and that are derived from the similarity scores. For each pair of strings S_i, S_j, it randomly permutes the characters of the two strings and computes the similarity score of the permuted strings. This is done one hundred times per pair, and for each pair, the mean and standard deviation of those one hundred similarity scores is also computed. (This is called "jumbling" in [127].) Then for each pair of strings S_i, S_j, the score $sd(i, j)$ is defined as the similarity score of the original strings S_i, S_j divided by the standard deviation computed from the permuted S_1, S_2 strings.

The intuition for the $sd(i, j)$ score is straightforward. If S_i and S_j contain nonrandom structures (presumably the desired secondary protein structure) interleaved with more random sequences, then the alignment score for S_1, S_2 should be much larger than the mean alignment score from permuted copies of the two strings. In permuted strings, the original nonrandom structure is destroyed. It is reported in [52] that there is a very good correlation between the $sd(i, j)$ scores and the extent to which the optimal alignment of S_i and S_j correctly aligns the known secondary structures in the strings. It is reported that for the strings studied, when the optimal pairwise alignment has an sd score of greater than five, that alignment has a better than 70% agreement with the reference alignment determined by x-ray data. So $sd(i, j)$ values can be used to give some confidence that the optimal alignment is biologically informative, even when the alignment is obtained from proteins where the secondary structure is not known.

Having defined sd scores derived from similarity scores, we can now describe the iterative multiple alignment method in [52]. The method begins by first finding the pair of strings with maximum sd score and optimally aligning those two strings. Then, successively, it finds the largest $sd(i, j)$ score among all pairs of strings S_i, S_j, where S_i is already in the alignment and S_j is not, and merges S_j into the growing multiple alignment. The merge of S_j is accomplished by aligning S_j to the profile (Section 14.3.1) of the existing alignment.

Clearly, the specific sd scores selected by the above method are exactly the scores that would appear on every maximum spanning tree formed from the sd data. This method is thus a variant of a maximum spanning tree clustering method. However, we cannot simply describe the resulting multiple alignment as the multiple alignment consistent with some node labeling of the maximum spanning tree, because string S_j is aligned with the profile of the previous alignment, rather than with the single string S_i.

The multiple alignments computed in [52] were evaluated by looking at induced pairwise alignments (induced from a multiple alignment) compared to the optimal pairwise alignment, to see how well the secondary structures (known from the x-ray data) lined up. The results were somewhat contradictory. The induced pairwise alignment was much better at aligning secondary structure than was the optimal pairwise alignment, when the optimal alignment only poorly aligned the secondary structure. However, when the optimal pairwise alignment did very well at aligning secondary structure, then the induced pairwise alignment did somewhat worse. So compared to optimal pairwise alignments, the induced alignments greatly improved the quality of the poor alignments, while slightly reducing the quality of the better ones.

Iterative multiple alignment to build evolutionary trees

We have illustrated multiple string comparison and alignment for the purpose of characterizing protein families and for identifying important molecular structure, but as Doolittle [127] states, for some purposes

> What we're really interested in is a *historical* alignment. The historical alignment ought to reflect, as accurately as possible, the series of divergences that led to the contemporary sequences.

This is the third of the three "big-picture" uses for multiple alignment.

Iterative alignment methods determine a sequence of merges of disjoint subsets of strings. Hence the history of those merges can be described by a binary tree T. Each leaf of T represents a single string from the input set, and each node of T specifies a merge of the strings found at the leaves of its subtree (or equivalently, a merge of the two subsets found at its two children). Each node also represents a multiple alignment created by the merge done at that node.

If the purpose of the algorithm is merely to find a multiple alignment of the input strings, then tree T is a byproduct of the iterative alignment method that may be of no independent interest. But, if the criteria for selecting merges are believed to reflect evolutionary history (for example, using the idea that more similar strings are more recently related), then T also represents a deduced evolutionary tree of the taxa whose strings are the input to the multiple alignment problem. Therefore, iterative (cluster-based) multiple alignment methods play a role not only in producing family representations and in identifying important molecular structure, but in the deduction of evolutionary history. We will return to this connection between tree building and multiple alignment in Section 17.7.

Progressive alignment

Perhaps the cleanest argument for the interrelationship between iterative multiple alignment and the deduction of evolutionary trees appears in the classic paper by Feng and Doolittle [153] (see also [127]) entitled "Progressive Sequence Alignment as a Prerequisite to Correct Phylogenetic Trees".

The key idea is that the pair of strings with minimum edit distance (or greatest similarity) is likely obtained from the pair of taxa that has most recently diverged and that the pairwise alignment of those two specific strings provides the most "reliable" information that can be extracted from the input strings. Therefore any spaces (gaps) that appear in the optimal pairwise alignment of those two strings should be preserved in the overall multiple alignment – "once a gap, always a gap" [153]. The idea of never removing gaps is repeatedly applied throughout the entire sequence of successive merges. As in the proof of Theorem

14.6.1, it is easy to successively merge alignments without removing any previously inserted gaps. In fact, if the merged alignment is determined by aligning the profiles of the two existing alignments, then existing pairwise gaps are automatically preserved.

The progressive alignment method of Feng and Doolittle is explicitly aimed at building an evolutionary tree from molecular data while simultaneously constructing an evolutionarily informative multiple alignment. It is a clear example of iterative pairwise alignment. However, the criteria used in [153] for choosing which two multiple alignments to merge at each step are a bit involved and the interested reader is referred to the original paper for details. We will return to the interrelationship between evolutionary trees and multiple alignment in Part IV of the book.

14.10.3. Repeated-motif methods

We now turn to the second major approach commonly used in multiple alignment methods (both local and global). That approach relies on first finding a substring or a small similar subsequence that is common to many of the strings in the set. This common subsequence (often required to be a substring) is given various names such as a *motif, anchor, core, block, core block, region, identity segment, model, leg, q-gram, prime window, consensus pattern*, etc. We will use the term "motif"; its "width" refers to the length of the common subsequence (or substring), and "multiplicity" refers to the number of strings that it appears in.

Once a first "good" motif (wide and with high multiplicity) is found, the strings containing it are shifted so that the occurrences of the motif are aligned with each other. The problem of completing the multiple alignment of those strings then decomposes into two smaller subproblems, one for substrings on each side of the motif. This recursion continues until no sufficiently wide or high motif is found; the remaining subproblems may be tackled by iterative alignment methods. Strings that did not contain the first good motif are aligned separately, and the two alignments are merged afterwards. In the case of local multiple alignment the output of the algorithm may just consist of the best motifs found in this process, as these are the substrings of interest[12].

The above outline is almost an exact description of the multiple alignment method developed in [374]. Another good representative of this repeated motif approach was developed by Waterman and coauthors [455, 456, 467]. The commercial-quality program MACAW [398] developed at the National Library for Medicine at the NIH is a widely available program that uses the repeated motif approach at the top level, although it switches to the *SP* objective function to align the strings between the anchors.

There are a variety of ways that existing methods try to find good motifs, but they are mostly based on similar ideas. Usually all the (overlapping) substrings of a fixed size (window) are collected and examined to find a substring that appears in many of the strings being aligned. Sometimes this comparison is accomplished by hashing techniques, sometimes by more straightforward comparisons. It certainly can be done by using suffix trees (see Exercises 32, 33 in Chapter 7) or sorting methods, etc. Sometimes a small number of characters in each window string are changed, generating a set of similar substrings for each original substring. Those substrings are then treated as before, and in this way

---

[12] We note, however, that McClure et al. [316] report that good motifs are aligned (and therefore easily identified) better by existing *global* alignment methods, which do not explicitly look for motifs, than by existing repeated motif methods. The conclusion is that a local alignment might be better computed by extracting ordered motifs from a good global alignment. This conclusion is contested.

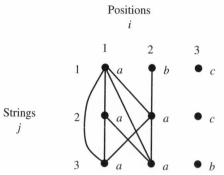

Figure 14.11: Graph D for three strings of length three each. In this example, two nodes are connected if they represent the starting positions of length-two substrings, having at least one matching character.

a motif can be formed from *similar* substrings rather than requiring identical substrings. Once a substring is found that is contained in many different strings, typical methods try to extend the candidate motif by looking at additional characters on either side of the fixed length substring.

There are, as in the case of iterative pairwise methods, a vast number of ways that this general approach can be turned into a concrete method, and it surely will continue to be the basis for many methods to come. There seems to be no theoretical way to devise a *best* variant, and so empirical evaluation remains the basis for comparing these methods. However, there is one theoretical improvement that can be incorporated into many published methods that follow the repeated-motif approach. That improvement will be explored in Exercise 26. Now we discuss a variant of the repeated-motif approach that has a more precise, combinatorial flavor than many other repeated-motif methods.

The Vingron/Argos method

Martin Vingron and Pat Argos [446] developed a repeated-motif multiple alignment method that, for three strings, can be succinctly described in graph theoretic terms. In language different from that used in [446], the method is the following.

First, create $3n$ nodes in a graph D, one node for each distinct pair (i, j), where i represents a position in string j (see Figure 14.11). Then, for a fixed length l chosen by the modeler, determine for each pair of nodes (i, j) and (i', j') with $j \neq j'$, whether the length-l substring starting at i in string j is "sufficiently similar" to the length-l substring starting at i' of j'. The criteria for sufficient similarity are also in the hands of the modeler. Connect node (i, j) to node (i', j') if there is sufficient similarity between those two substrings. Next, find and remove every edge in the graph that is not part of any clique of size three. A clique of size three represents three substrings in the three strings that are mutually "sufficiently similar". These three substring form a *motif*. Now, represent the clique (motif) formed from nodes $(i, 1)$, $(i', 2)$, and $(i'', 3)$ by (i, i', i''). Clique (i, i', i'') is said to be *to the left* of clique (z, z', z'') if and only if $i < z$, $i' < z'$, and $i'' < z''$. Two cliques are *noncrossing* if one is to the left of the other. At this point, the algorithm tries to find a set of noncrossing cliques that are "nicely spaced" in a manner to be next made precise.

If clique (i, i', i'') is to the left of clique (z, z', z''), then the algorithm assigns that pair of cliques a *spacing weight* based on how well the differences $i' - i$ and $i'' - i'$ match to $z' - z$ and $z'' - z'$. This again is in the hands of the modeler. For example, it is good if $i' - i = z' - z$ and $i'' - i' = z'' - z'$, for then a multiple alignment that aligns position i in string 1 with positions i' and i'' in strings 2 and 3, and aligns z with z' and z'', will have consistent spacing between the two columns of aligned motifs. The motifs in that case are essentially "parallel", and a high weight should be given to that pair of cliques. A pair of badly spaced cliques (representing motifs that are not parallel) should be given a small weight.

To find a good set of *well-spaced, noncrossing cliques*, build a directed, weighted graph G. Each node of G represents a clique in D. Add a directed edge from a node v to a node w of G if and only if v represents a clique in graph D to the left of the clique represented by w. The spacing weight for those two cliques is assigned to edge (v, w). Each node (representing a clique/motif) is also given a (large) weight, reflecting the goodness of the motif. These node weights are again determined by the modeler, but the general scheme allows the use of any desired scoring matrix and choice of multiple alignment objective function. Now find a *maximum weight* directed path in G, and suppose there are k nodes on that path. Those nodes represent k noncrossing cliques in graph D, and hence they specify a set of k noncrossing motifs that form "anchors" in the three strings to be aligned. This use of longest path in multiple alignment was first developed in [444]. The substrings between those k anchors can then be aligned by any method, even this one. By altering the relative weights of edges versus nodes of G, the modeler can change the emphasis from finding a multiple alignment with many good, noncrossing motifs to finding one with a few well-spaced motifs. How can one study the trade-offs between these two objectives? With parametric optimization of course.

The importance of shape

The ability to explicitly incorporate notions of shape or spacing into a multiple alignment objective function is important because a "good" shape is one indicator that the alignment is "biologically meaningful". When using multiple alignment to deduce structural information, a multiple alignment displaying a few short regions of high similarity, interspersed by longer regions of low similarity, is more believable than a multiple alignment whose matches are widely dispersed throughout the sequences [433]. For evolutionary deduction, when related sequences are "well aligned", the rows can often be partitioned into a few classes such that in each class the spaces that appear in one row are similar to the spaces that appear in any other row of that class. An example is the alignment in Figure 14.1. A multiple alignment with many spaces, where the spaces in each row are wildly different from the spaces in any other row, will generally be considered suspect in biological applications. This expectation is exploited in the work reported in [45] (discussed in Section 17.3.2).

The extension of the Vingron–Argos method to $r > 3$ sequences should ideally be based on finding cliques of size r in graph D, now containing rn nodes. However, such a search is computationally difficult; instead, a superset of the edges contained in the size-r cliques of D is identified. An efficient, graph theoretic algorithm for use in identifying the superset is developed in [369]. Once the superset of edges is in hand, dense subsets of edges play the role of cliques in finding good noncrossing motifs to align. The details are too complex for inclusion here.

14.10.4. Two newer approaches to multiple string comparison

There are two newer, well-publicized approaches to multiple string comparison that have a very different flavor than the methods discussed so far. Both of the methods come out of a more statistical or stochastic approach to sequence analysis. The first method, called the *Gibbs sampling method* [295], is a local alignment or local comparison method. The interested reader is directed to [295].

The second method (initially brought to computational biology through work at the University of California at Santa Cruz by David Haussler) is based on *hidden Markov models* (HMM) [285]. Essentially, the HMM method tries to build a Markov model (network) that could generate a family of molecular sequences of interest (for example, globins) using undetermined edge transition probabilities and node output probabilities. It then tries to determine good edge and node probabilities by finding the best fit of the chosen sequence data to the Markov model. That subproblem is a difficult problem in statistical optimization and is totally outside the scope of this book. An optimized HMM can then be used to determine if additional sequences are good candidates for family membership. A candidate string is evaluated by computing the probability that the HMM would generate it. This computation is an easy exercise in dynamic programming.

Once the HMM is built and optimized, it can also be used to construct a multiple alignment of a set of strings. First, for each string, the most probable path in the HMM that generates that string is computed. That path places spaces in each sequence. Then, the sequences with their spaces are simply lined up to obtain the multiple alignment.

14.11. Exercises

1. Prove the correctness of the recurrences for profile alignment given in Section 14.3.1.

2. Formalize an objective function for profile to profile alignment, and then give recurrences to optimally compute that alignment.

3. John Kececioglu [267] has observed that many specialized string problems can be cast as multiple alignment problems by selecting the appropriate way to score the columns of the multiple alignment. For example, a *shortest common supersequence* of a set of strings S is a shortest string that contains each string in S as a subsequence. If each column of a multiple alignment of S is scored by the number of distinct characters contained in that column, then a multiple alignment minimizing the total column scores identifies a shortest common supersequence of S. Explain this in detail.

4. Discuss the appropriateness of the *SP* score for evaluating multiple alignment. That is, why and/or when should one expect that multiple alignments produced under the *SP* objective function will highlight sequence information that biologists want to capture? Discuss the appropriateness of producing consensus strings from multiple alignments and of evaluating the goodness of a multiple alignment by evaluating their consensus string.

5. Prove the correctness of the dynamic programming recurrences for multiple alignment given in Section 14.6.1.

6. Explain in more detail why forward dynamic programming is important when implementing the speedup presented in Section 14.6.1.

7. In the center-star method, the multiple alignment is constructed by successively aligning each new string to the center string S_c. However, the order that strings are aligned to S_c was not specified. Show that the same multiple alignment is built, no matter what order is used.

8. In the proof of Theorem 14.6.1 (constructing a multiple alignment $\mathcal{M}(T)$ that is consistent with labeled tree T), the order that the edges of T are processed can vary. It may at first seem that different orders will result in different multiple alignments. Prove that the same multiple alignment results, regardless of the order that the edges are processed.

9. Intuitively, in the center-star method, it seems that a better *SP* alignment would be produced by aligning each successive string to the profile of the existing alignment. Certainly, the first time this is done, the resulting alignment is better. But if this is done for every new

string, would the resulting final alignment necessarily be better than the alignment gotten by the original center-star method? Does the factor of two approximation result still hold if the merges are done this way? How does the order that the strings are merged now affect the quality of the resulting alignment?

10. Recall the *sd* scores discussed in Section 14.10.2. We would like to use *sd* scores in the bounded approximation methods developed in this chapter. However, *sd* scores do not necessarily satisfy the triangle inequality needed for the bounded error guarantees. Will *sd* scores likely satisfy the triangle inequality in practice? Perform an empirical investigation of that question.

11. Recall the weighted *SP* value defined in Section 14.6.3, in which the induced pairwise score of each pair (S_i, S_j) is multiplied by a weight $w(i, j)$. Try to extend the center string approximation method to this weighted version of the problem.

12. If we weaken the statement of Lemma 14.7.1 to say that there exists a string $\bar{S} \in \mathcal{S}$ such that $E(\bar{S}) \leq 2$ (rather than $2 - 2/k$), then the proof can be simplified. Do it.

13. The proof of Lemma 14.7.1 can be modified to establish that if $\bar{S}$ is either the closest, second closest, or third closest string in $\mathcal{S}$ to S^*, then $E(\bar{S})/E(S^*) \leq 2$. Prove this. (This problem was suggested by Mike Paterson)

14. Suppose, for the *SP* problem, that instead of using M_c as the center string, one chooses a string S_i from the input set at random to be placed at the center of the star. Show that, averaging over all choices for S_i, the expected *SP* value of the resulting multiple alignment is at most $2M$. Next, use that average and the fact that the minimum value is M to show that the median value is at most $3M$. Finally, argue that by using a random string S_i at the center of the star, the resulting multiple alignment is within a factor of three of the optimal *SP* value, with probability of at least one half. Do a similar analysis for the consensus objective function.

15. The optimal *SP* multiple alignment must have value at least $\sum_{i<j} D(S_i, S_j)$. This lower bound on the optimal value is useful not only to establish a guaranteed bound of $2 - 2/k$ for the center string approximation, but also to establish a better empirical error bound when the method is run on any *particular* set of strings. In the discussion of the *consensus criteria* for multiple alignment, no similar lower bounds were established, and no method was given to compute empirical error bounds. Below we give one. Prove that it gives a lower bound on the cost of the optimal Steiner string.

    For simplicity, assume that, $\mathcal{S}$ has an even number of strings. Construct the complete, weighted graph G_D, where each node represents one string in $\mathcal{S}$. Assign the weight $D(S_i, S_j)$ to the edge between the nodes representing S_i and S_j. A *complete matching* C in G_D is a set of edges such that each node touches exactly one of the edges in C. The weight of C is the sum of the weights on the edges in C. The weight of a maximum weight complete matching in G_D can be found efficiently. That weight is a lower bound on the cost of the optimal Steiner string.

16. Give a detailed justification that the optimal lifted alignment can be correctly computed in $O(N^2 + k^3)$ time.

17. The time to compute an optimal lifted alignment can be reduced to $O(N^2 + k^2)$ time with the following definition and observation: An ordered pair of sequences (S, S') is called *legal* for an edge (v, v') if S is a string at a leaf in T_v and S' is a string at a leaf of $T_{v'}$. Suppose v is the parent of v'. When computing the recurrence $d(v, S) = \sum_{v'} \min_{S'} [D(S, S') + d(v', S')]$, only legal pairs for each (v, v') need be considered.

    Show that an ordered pair of strings can be a legal pair for *at most* one edge in the tree. Give an $O(k^2)$-time algorithm to find the legal pairs for every edge. Then, flesh out the details of the $O(N^2 + k^2)$-time method.

18. In the phylogenetic alignment problem, we would like to have an efficient algorithm that can take in any problem instance and compute a *lower bound* on the value of the optimal solution. Prove that, for any problem instance, the cost of the optimal phylogenetic alignment is at least half the weight of the minimum spanning T tree formed from the pairwise distance data, as discussed in Section 17.5.2.

    Show, by example, that the full cost of the minimum spanning tree is not a lower bound, and explain why this is so.

19. **Uniform lifted alignments [453].** In this problem we consider another way to transform the optimal phylogenetic alignment T^* to a lifted alignment, called T^u. First, partition the nodes of T into levels according to the number of edges from the root. We will lift one level at a time, bottom up. For simplicity, assume that T is a binary tree, so each internal node has two children, but T need not be a complete binary tree. For each internal node v, arbitrarily label one of its children as the *red* child $r(v)$ and the other as the *blue* child $b(v)$. Then, a lifted alignment is called a *uniform lifted* alignment if, at each level, either all internal nodes receive their lifted label from their red child or all internal nodes receive their lifted label from their blue child.

    The transformation of T^* to T^u is performed one level at a time, bottom up. Let $V(i)$ be the internal nodes at level i. To lift labels to level i (from level $i + 1$) consider two sums: $\sum_{v \in V(i)} D(S_v^*, S_{r(v)})$ and $\sum_{v \in V(i)} D(S_v^*, S_{b(v)})$, where $S_{v(r)}$ and $S_{v(b)}$ are, respectively, the labels that have been lifted to the red and blue children of v. If the first sum is smaller than or equal to the second, then assign each internal node in level i the lifted label of its red child; otherwise assign each internal node in level i the lifted label of its right child. When all levels have been done, the resulting uniform lifted alignment is T^u.

    Problem: Prove that the uniform lifted alignment T^u has total distance less than twice that of the optimal phylogenetic alignment T^*. The proof can be modeled on the analysis of the lifted alignment T^L presented in the text.

20. Assuming again a binary tree, give a polynomial time algorithm to find the uniform lifted alignment with smallest total distance. It is possible to find it in $O(n^2)$ time rather than the $O(n^3)$ time bound established in the text for the optimal lifted alignment.

21. Generalize the definitions, methods, and analyses for a uniform lifted alignment to any tree, not necessarily a binary one.

22. Clearly, the best uniform lifted alignment will have total distance no less than the best lifted alignment. Ignoring time efficiency, what value would there be in finding a uniform lifted alignment with an error bound of two?

    Hint: Note that the assignment of sibling nodes to red or blue is arbitrary, and recall the comments in Section 14.9 on the utility of (bad) bounded-error approximations.

23. Assume again that T is a binary tree, which is not necessarily balanced. A *layout* of T is defined by choosing for each internal node v which of its two children to label the left child and which to label the right child. How many distinct layouts are there of T? How many distinct lifted alignments are there for T?

24. Clearly, every lifted alignment is a uniform lifted alignment for some layout(s) of T. Let d be the depth of T (the maximum number of edges on a path from the root to a leaf) of T. Show that every lifted alignment is a uniform lifted alignment for exactly 2^d layouts of T.

25. Using the answers to the previous two problems, establish that in any collection of 2^d distinct lifted alignments, at least one of them has a distance that is no more than twice the distance of the optimal phylogenetic alignment T^*. Next, improve this to two distinct lifted alignments. Hence, in a balanced binary tree with n leaves, at least $2/n$ of all lifted alignments are within a factor of two of the optimal.

26. The multiple substring problem, considered in Sections 7.6 and 9.7, allows a clean formal-

ization of the exact-matching phase used in several motif detection and multiple alignment methods. The linear-time solution to the multiple substring problem (and variants of it) thus provides a powerful tool that can be used in the heart of many biological sequence analysis programs. To be concrete we consider one specific program, that due to Leung et al. [298], to identify multiple matches with errors in a set of long strings.

The program of Leung et al. finds substrings that are similar and that appear in a "sufficient" number of the input strings. In an exact-matching phase it first identifies certain "core blocks". A core block of *multiplicity m* is a set of m substrings, each from a different string (in the most interesting case), that are *identical* and that are maximal for that property. That is, any extension of the strings by one character on either side would cause at least two of the m strings to differ. In the program, the user sets m and a parameter c_m, which is the minimal acceptable length of the strings in any core block to be reported. Intuitively, if m and c_m are reasonably large (as determined by the user), then a core block is a nontrivial conserved exact pattern that appears in a significant number of the strings. A local multiple alignment is built in the program by flanking the core block with additional *identity blocks* (blocks of identical substrings that are not as long as the core block), which are within a preset distance from the previous identity block. The output of the program is significantly influenced by the choice of m and c_m. The choices suggested in [298] for m and c_m are based on statistical results for random strings and are not determined by the actual input strings.

Describe the general task of finding core blocks in terms of the multiple common substring problem.

Recall that the linear-time algorithm in Section 9.7 constructs a table showing, for each possible value of k, the length of the longest substring common to at least k of the input strings. This table lays out *all* the trade-offs between m and c_m for the given input strings. Explain this in detail, and explain the advantages of using the table to choose m and c_m over using preselected values of m and c_m. Show that once a choice of m and c_m is made, the strings in that core block can be located efficiently using the suffix tree that was used to construct the table.

Show that by using the multiple common substring algorithm, one can select the values for m and c_m on any given input, by concentrating either on substring length or on multiplicity. That is, one can first select the desired string length c and then find the largest m such that there is an identical substring of length c in m of the q strings or one can go in the opposite direction from m to c.

For another effort to formalize the definitions and computations involved in repeated motif methods, see [385].

15

Sequence Databases and Their Uses – The Mother Lode

We now turn to the dominant, most mature and most successful application of string algorithms in computational biology: the building and searching of databases holding molecular sequence data.[1] We start by illustrating some uses of sequence databases and by describing a bit about existing databases. The impact of database searching (and expectations of greater future impact), explains a large part of the interest among biologists for algorithms that search, manipulate and compare strings. In turn, the biologists' activities have stimulated additional interest in string algorithms among computer scientists.

After describing the "why" and the "what" of sequence databases, we will discuss some "how" issues in string algorithms that are particular to database organization and search.

Why sequence databases?

Comprehensive databases/archives holding DNA and protein sequences are firmly established as central tools in current molecular biology – "electronic databases are fast becoming the lifeblood of the field" [452]. The fundamental reason is the power of biomolecular *sequence comparison*. This was made explicit in the *first fact of biological sequence analysis* (Chapter 10, page 212). Given the effectiveness of sequence comparison in molecular biology, it is natural to *stockpile* and *systematically organize* the biosequences to be compared; this has naturally led to the growth of sequence databases. We start this chapter with a few illustrations of the power of sequence comparison in the form of sequence database search.

15.1. Success stories of database search

15.1.1. The first success story

Database organization and searching have become industries, and "discoveries based solely on sequence homology have become routine" [360]. Yet, it was only fourteen years ago that the first major success of this approach was reported [132].

It was well established in the early 1970s that infection by certain viruses could cause particular cells in culture (*in vitro*) to grow without bound. This cancerlike *transformation* of cultured cells by viruses suggested that viral infection could be a cause of cancer in animals, but the mechanisms were unknown. It was hypothesized that certain genes in the infecting viruses (termed *oncogenes*) encode cellular *growth factors*, which are particular proteins needed to stimulate or continue growth of a cell colony. The virus-infected cells would thus produce uncontrolled quantities of the growth factor, allowing the cell colony to grow beyond its normal limits. That hypothesis is now generally accepted. "However,

---

[1] Some people argue that there are no true molecular sequence *databases* in biology, only sequence *libraries* or *archives*. They argue that a true database must integrate the data in a more productive way and allow more complex queries than is presently done. However, here we will use the term "database" and "archive" interchangeably.

the link between oncogenes and growth factors did not come from a direct test of this hypothesis – instead it was an unanticipated result of merging two independent sets of data via a computer search" [110].

Simian sarcoma virus is a *retrovirus* (meaning that its genome consists of double-stranded RNA that must be converted in the infected cell to DNA before the virus can replicate) that was known by the early 1970s to cause cancer in a specific species of monkeys. The oncogene responsible, named *v-sis*, was isolated and sequenced in 1983. About that same time, a partial amino acid sequence of an important growth factor, the *platelet-derived growth factor* (*PDGF*) was worked out and published. R. F. Doolittle had been keeping a home-grown database of published amino acid sequences (entered by hand with the partial aid of family-supplied labor) and had previously entered the translated amino acid sequence of the *v-sis* oncogene. When the partial amino acid sequence of PDGF was published, he compared that sequence to those in his database and found one region of 31 amino acid residues with 26 exact matches between the PDGF sequence and the *v-sis* protein sequence. In another region of 39 residues, he found 35 exact matches [132]. This search, and a similar computer search by another group, "provided for the first time a known physiological context in which to view the action of oncogenes – neoplastic transformation was the apparent consequence of inappropriate expression of a protein whose function was to stimulate the proliferation of normal cells" [110]. For our purposes, "neoplastic transformation" can be translated as "cancer".

That first-established connection between an oncogene and a normal protein affected the way oncogenesis has been seen and understood since then. Many additional oncogenes have now been shown to be highly similar (but not identical) in their sequence to genes that encode growth-regulating proteins in normal cells. Several of these oncogene-to-protein connections were also established by similarity searches in protein databases. The normal genes for those growth-related proteins have been termed *proto-oncogenes*, and the theory is that a previously harmless virus becomes oncogenic by incorporating the proto-oncogene of its host into its own genome. In the viral genome, the proto-oncogene is then mutated, or moved close to a strong enhancer, or away from a strong repressor, so that now an excessive amount of the proto-oncogene product (a growth factor for example) is produced when the virus (with its oncogene) infects a normal cell.

15.1.2. A more recent example of successful database search

The first complete DNA sequence of a free-living organism was reported in the summer of 1995 [162]. A total of 1,743 putative coding regions (putative genes, essentially) were identified in the sequence by means we won't discuss here (one approach is briefly discussed in Section 18.2). Each of those 1,743 strings was then translated into one or more amino acid sequence(s) (if the reading frame was in question) and used to search for "sufficiently similar" sequences in the protein sequence database Swiss-Prot. In this way, 1,007 of the putative genes not only matched entries in the database, but matched in such an unambiguous manner that the specific biochemical function could be deduced (or conjectured) for each one. These deductions were made possible by linking Swiss-Prot to a different database (the Riley database), where known, sequenced proteins have been divided up into 102 very detailed biochemical roles. In this way, when a putative protein sequence encoded by one of the 1,743 genes closely matched a known protein sequence, the biochemical function of the new protein could be very specifically conjectured. And such specific conjectures were possible in 1,007 of the cases. A full list of the

102 biochemical roles appears in [162], and it is worth looking at just to get a sense of how much detailed information about the putative genes is being deduced (or conjectured) via a database search using the sequence alone.

15.1.3. Indirect applications of database search

The previous two examples illustrate the most direct use of biological sequence databases, to search for similar genes, regulatory sequences, or related proteins when new sequences are obtained. This is now routine. Whenever a new gene (or a new protein) is cloned and sequenced, visiting the appropriate databases is the next step. But sequence databases are used in other ways as well. There is a great deal of database search done for the purpose of *clustering* similar sequences into sequence families. Such families may reveal important conserved biological phenomena that had not been observed by laboratory work and that would be hard to recognize by looking at two sequences alone. There are also many clever ways that people have devised to use sequence databases to tackle both biological and biotechnical problems. We will illustrate this with two examples. The first is an example of a technical problem, and the second is an example of a problem whose solution has revealed important biology. Other illustrations of an indirect use of database search were presented in Sections 7.3 and 7.5. Additional illustrations will appear in Sections 16.9 and 18.2.1.

Sequence assembly in bacteria

Some aspects of sequence assembly were discussed in previous sections of the book, and sequence assembly will be discussed in detail in Section 16.15. Here, we need only consider the following abstract problem that arises in assembly. Suppose that a very long DNA substring from a specific (target) bacteria has been cut up into many shorter substrings (called *fragments*), and you are presented with the fragments completely out of order.[2] That is, the fragments have been permuted so that their original order in the long substring has been lost. The sequence assembly problem is to recover the original fragment order.

Two key facts are needed. In most bacteria, most DNA (about 85%) is contained in genes, and bacterial genes are rarely broken up by introns (intervening sequences inside the gene sequence). Thus a random stretch of bacterial DNA is likely to be involved in specifying a protein made by the bacteria. This fact, along with the power of comparative analysis of molecular sequences, has been exploited by Fleischmann et al. [162] to facilitate the bacterial sequence assembly.

One *element* of their assembly method is to sequence each of the fragments; translate each fragment into an amino acid sequence (actually three sequences because the reading frame is unknown); and then, for each fragment, search in protein databases for substrings of known proteins that are highly similar to the fragment. The goal is to identify pairs of DNA fragments that are either adjacent in the target bacterial sequence or are close to one another. Database search helps to find some of these pairs. Two fragments found to be highly similar to two intervals in a single protein sequence (in the database) are assumed to come from an interval in the target bacterial genome that codes for a single protein. That protein in the target bacterium is assumed to be homologous to the protein in the database to which the two fragments matched. In addition, if the two fragments match to adjacent regions in the database(d) protein, then they are assumed to be adjacent in the

---

[2] In reality, many copies of the long substring are cut up, and the cuts are overlapping. That detail is important in general, but not to the specific use of databases being illustrated here.

bacterial genome as well. In this way, some pieces (or contigs) of the correct fragment order can be established. Of course, these database matches are not enough to assemble the full order, but their use is an integral part of the assembly method in [162].

Multiple Sclerosis and database search

A clear example of how database search can be combined with laboratory molecular biology was mentioned in the New York Times [79]. Multiple Sclerosis (MS) is a debilitating neurological disease that is not well understood. However, it is understood that MS is an *autoimmune disease*, meaning that the immune system incorrectly identifies native cells as foreign invaders. In MS, the *myelin sheath* encasing nerve cells is attacked by the immune system, disrupting the normal transmission of signals along the nerve. The first line of attack in the immune system are the T-cells, which identify foreign matter. Once identified, other elements of the immune system attack and destroy the identified matter. The body develops specific T-cells in reaction to exposure to different foreign (nonself) matter (antigens).

Recently, specific T-cells were found that identify proteins or protein segments (polypeptides) that appear on the surface of myelin cells. It was then natural to conjecture that those T-cells (that identify proteins on the myelin surface as foreign, incorrectly leading to their destruction) had previously been generated by the immune system to (correctly) identify highly similar proteins on the surface of bacteria or viruses. That is, the immune system attacks the myelin sheath because it confuses certain proteins on its surface for proteins on the outer surface of certain bacteria and viruses that had previously infected the individual. But how could this conjecture be tested? Which bacteria and which viruses are involved?

Using the *sequences* of myelin surface proteins, a search was conducted in the protein databases for highly similar proteins in bacteria and viruses. About one hundred proteins were found. Laboratory work then verified that the specific T-cells that attack myelin sheath also attack particular proteins found by the database search. This combined database/laboratory approach not only confirmed the general conjecture stated above, but it identified the particular bacterial and viral proteins that are confused with proteins on the myelin surface. The hope now is that by examining the similarities among those bacterial and viral protein sequences (an example of multiple sequence comparison), one might better understand what features of the myelin surface proteins are used by the T-cells to mistakenly identify myelin cells as foreign.

15.2. The database industry

The major DNA archives hold sequences whose lengths total more than 500,000,000 base pairs, obtained from parts of 300,000 different genes in many diverse organisms. The major protein sequence archives hold sequences totaling around 25,000,000 amino acids, obtained from around 100,000 different protein sequences. These numbers have been growing exponentially, and the standard ballpark figure is that the number of base pairs held in DNA archives increases by about 75% each year [479]. And, of course, when the various genome projects get into full-swing DNA sequencing, the present numbers will look like antiques.

Not only are the databases growing in size, but their number is proliferating as well [1]. The granddaddy of the comprehensive DNA archives is GenBank, which is charged with storing and facilitating retrieval of all DNA sequences ever made public.[3] For most of its life, GenBank was maintained at the Los Alamos National Labs (LANL), but it is now maintained by the National Center for Biotechnology Information (NCBI) at the National Library of Medicine (NLM), which is part of the National Institutes of Health (NIH). The Europeans maintain an essentially equivalent DNA archive, the EMBL data library, and another one, the DNA DataBase of Japan (DDBJ), exists in Japan. There is also a new comprehensive DNA database, the Genome Sequence DataBase (GSDB), which is a spin-off of GenBank, and a project of the National Center for Genome Resources (NCGR), which is supported by the Department of Energy (DOE). These four complete DNA databases share information between them, and a submission to one is effectively a submission to all. On the protein side, the major protein sequence archives are the Protein Information Resource (PIR) in the United States and Swiss-Prot in Europe.

Although this may seem like many acronyms already, there are many more. Over one hundred specialized databases (holding sequence data as well as other data) have been created in the past several years (see [1] for an overview of several of these). Some of these specialized databases and their uses were mentioned earlier in Section 3.5.1. Databases with identifiers such as dbEST, dbSTS, FLYBASE, HAEMA, HAEMB, HLA, p53, RE-BASE, REPBASE, TRANSFAC, SGD, YPD, EcoCyc, XREFdb, CGSC, ECDC, NRBP, EGAD, HCD, SST, ALU, BERLIN, EPD, KABAT, LISTA, SBASE, YEAST, GOBASE are popping up all the time.[4] The specialized databases differ along several lines. Some specialize in a particular organism or cell type, some concentrate on particular biological functions, some follow specialized terminology and taxonomic style that are particular to a subfield of biology, some try to record all mutations and differences (polymorphisms) that have been discovered in a given gene or set of genes, and some differ in the way that the sequence data are stored and integrated with other biological information, as well as the kinds of retrieval and query processing services that are offered. (GenBank has traditionally been kept as a flat file, whereas some newer databases are relational or object-oriented, and many try to integrate different types of biological data.)

All the databases function as sequence repositories, allowing the retrieval of any sequence by a title or accession number. But a more important aspect than this data-storage and data-transfer function is that the databases act to "generate new knowledge" [130], by allowing sequence retrievals based on *sequence similarity searching* of one type or another. Moreover, many of the databases are accessible over the Internet (and the World Wide Web) [76, 215, 452], and there is a large array of software packages that allow the user to search, view, and manipulate the sequence data from the archives (the most widely used package is GCG (Genetics Computer Group), originally developed at the University of Wisconsin).

The use of sequence databases is growing as rapidly as the databases themselves. Swiss-Prot clocks about 400,000 query requests a month. Similarly, GenBank currently services about 10,000 retrieval requests a day, but the full use of GenBank (or EMBL) can't be properly estimated since multiple copies of these databases are distributed to servers throughout the world, and anyone can download the current versions and search them

---

[3] In the early days, which weren't so long ago, GenBank staff would search the journals for published DNA sequences and enter them into the database by hand. Now most submissions are electronic, whether the sequences are published or not, and timely submission is often required by funding agencies.

[4] For someone not immersed in the world of biological databases or acronyms, it is quite a challenge to keep the various archives straight, let alone to try to understand the differences in biological communities and political histories that have contributed to this explosion of database species.

on their own machines. That ability is in fact important to biotech and pharmaceutical companies who want to keep their searches private so that no one can infer from their queries what they are working on.

With all the different sequence databases (not to mention all the *DNA map* databases), servers, and software packages for accessing, searching, viewing, comparing, and downloading biological sequence data, it is difficult for users to know the best way to take advantage of the stored information (keeping the names of the databases straight is hard enough). Consequently, a major effort is underway to link and integrate, or *federate*, various databases in order to share information and automatically process complex queries that require interaction between several databases [76, 452, 479, 480]. Multidatabase searchers now exist (*Entrez* and its network-based relative *nEntrez* from NCBI are the current standard-bearers [397]), and other database *browsers* and *crawlers* are being created. People are even thinking (perhaps fancifully) about "knowbots" [55], programs that will automatically "explore and mine" the databases, using unexpected matches and correlations to formulate and report interesting biological hypotheses.

In summary, sequence databases and the software to search them have become vital tools in modern molecular biology. "It would be hard to find a recent DNA-based discovery that *didn't* use these tools" [55].

15.3. Algorithmic issues in database search

15.3.1. Should there be any?

Sequence database search is the dominant application of string algorithms in molecular biology, and there has been a huge investment of research directed for that application. But why should special effort be required? With all the tools available to compare two strings at a time, when one has a query string and wants to find similar strings in a database, why not just align the query string in turn to each of the strings in the database and then report the "best" alignments? Furthermore, one can use the various advanced methods discussed in Sections 12.1–12.4 to speed up the computations and reduce the space consumption. Of course, there remain important details (local vs. global alignment, gap model, parameter choices, scoring matrices, thresholds for reporting a similarity, statistical significance estimates, etc.), but those are modeling and statistical issues, not algorithmic issues. So after fourteen chapters on matching and alignment, is there anything left to say about *how* to search in sequence databases?

In fact, some people claim that the day will come when there will be little additional algorithmic detail to discuss and that the approach outlined above will work just fine. The key is that although databases are getting larger, proteins are not. The size of the query and database strings (at least for protein database search) will rarely be larger than five hundred and will usually be less. Therefore, the computation time will only grow *linearly* as the number of strings in the database increases. Given the increase in computer speed, the decline in processor cost, and the ability to trivially divide up a database search among many processors, one can see the day when it may be practical to search the databases by repeatedly optimizing a precise objective function using some version of dynamic programming. Even today, services are available to search sequence databases using a 4,000-processor MASPAR computer, and there are specialized chips and other hardware to implement sequence alignment very rapidly. Some of the specialized hardware is sold as commercial products (for example, machines from Time Logic, Inc. and from

Compugen Ltd. in Israel), and other hardware is being developed in academic institutions (for example, the Kestral project at the University of California, Santa Cruz [235]).

The day may well come when sequence database search will just involve the repeated application of precise two-string alignments. But that day is not here yet. Moreover, issues other than computer efficiency also dictate the use of strategies other than dynamic programming. When looking for known motifs, or semicharacterized features of importance in a query string, a more sensitive, selective, and time-efficient search can be achieved by using a specialized strategy or (refined) database, rather than relying on general similarity search in a raw sequence database. Hence, we discuss next what is presently done in practice when a one has a new sequence and uses a database to search for important features of it or for sequences highly similar to it. As usual, the emphasis is on the algorithmic ideas behind the specific programs and databases in common use today.

15.4. Real sequence database search

Doug Brutlag has suggested what the professional database searcher does when a new protein sequence is determined.[5] First, one compares the new sequence with the PROSITE database and the BLOCKS database for well-characterized *sequence motifs*. Then one searches the DNA and protein sequence databases (GenBank, Swiss-Prot, etc.) for sequences highly similar to the new sequence (usually a local similarity). When looking for these highly similar sequences, one does not initially attempt to compute optimal similarity (as defined in Section 11.6) between the new sequence and each database entry. Rather, one runs the approximate (heuristic) methods BLAST and FASTA. Only then, if needed, does one commit the resources required to compute optimal similarity based on dynamic programming. As a further refinement, when amino acid substitution matrices are employed, one usually tries both some variant of a Dayhoff PAM matrix and of a BLOSUM matrix.

Given this summary, we now briefly discuss the algorithmic and modeling ideas behind each of the capitalized words in the previous paragraph: FASTA, BLAST, PAM, PROSITE, BLOCKS, and BLOSUM. These six terms (along with "dynamic programming", "similarity", "GenBank", and "Swiss-Prot") make up the essential vocabulary of any serious sequence-database searcher.

FASTA and BLAST

FASTA (short for "fast-all" and pronounced "fast-AY" not "FAST-uh") and BLAST (short for "basic local alignment search tool" and pronounced "BLAST") are two related programs that are nearly universally used to approximate local alignment and local similarity. In this context, the term "approximate" has the normal colloquial meaning, rather than the mathematical meaning of a bounded-error approximation algorithm. Both of these programs are actually suites of programs containing tuned variants for use in different problem domains (searching DNA vs. searching protein, searching a database vs. aligning two strings, etc.). We will not discuss the mechanics of using these programs, only the algorithmic ideas behind them.

Both FASTA and BLAST are based on the same kinds of ideas used in the *exclusion* algorithms discussed in Section 12.3. However, unlike those algorithms, FASTA and BLAST are more complex and more heuristic (and more effective); they therefore do

---

[5] Others have suggested a different order, but still using the same or similar set of tools.

not permit precise analyses of their speed or accuracy. Our treatment of both of these methods will be sketchy but should give the essential flavor and allow the reader to see the connection "in spirit" of FASTA and BLAST to the exclusion methods discussed earlier. FASTA was developed by D. Lipman and W. Pearson [304, 361], and BLAST was developed later by S. Altschul, W. Gish, W. Miller, E. Myers, and D. Lipman [19].

15.5. FASTA

FASTA is a heuristic, exclusion method that we will describe in terms of the standard dynamic programming table for local (weighted) alignment of a query string against a single text string. It is applied in turn to each text string in the database being searched. There are several variants of FASTA, and it continues to evolve. The following description is "generic" and is based on the original FASTA papers [304, 361], along with some more recent comments on the current status of FASTA [360]. For additional commentary and analysis of the effectiveness of FASTA see also [356], [357], and [358].

To start, the user selects a value for a parameter called *ktup*, and in the first step of FASTA, *ktup*-length "hot-spots" in the dynamic programming table are identified as follows. FASTA finds pairs (i, j) such that the *ktup*-length substring (called here a k-tuple) starting at position i in the query string *exactly* matches the k-tuple starting at position j in the text. Such a pair is called a *hot-spot*. The standard recommended values for *ktup* are six for DNA searches and two for protein. For such small values of *ktup*, the hot-spot matches can be efficiently found by hashing the k-tuples of the query and/or the text string. For example, the k-tuples in the database can be hashed to a hash table before every release of the database, if there is sufficient space. In that case, when a query is presented, the method looks up in the table every k-tuple of the query. If that approach requires too much space, then the k-tuples in a presented query can be hashed to their own table and each k-tuple in the text then found in that table. The latter approach is the one most identified with FASTA.

Each hot-spot (i, j) found in the first step should be considered as a *ktup*-length interval in diagonal $(i - j)$ of the full dynamic programming table. (The main diagonal is numbered zero; the diagonals above it are identified by positive numbers, while the diagonals below it are identified by negative numbers.) Next, FASTA locates the ten best *diagonal runs* of hot-spots in the table. A diagonal run is a sequence of consecutive hot-spots in a single diagonal. A run need not contain all the hot-spots in that diagonal, and a diagonal can contain more than one of the ten selected runs. To evaluate diagonal runs, FASTA gives each hot-spot a positive score, and it gives the space between consecutive hot-spots a negative score that decreases with increasing distance. The score of a diagonal run is the sum of the hot-spot scores and the interspot scores. FASTA finds the ten highest scoring diagonal runs under this evaluation scheme.

With a query of length n and a database of length m, it may seem that finding the best diagonal runs would take $\Theta(nm)$ time, destroying the claimed efficiency of FASTA. But the diagonal search only needs time proportional to the number of hot-spots found during the hash table lookups. Suppose the hash table holds the k-tuples of the *query* string. Then the method should look up the k-tuples of text in left-to-right order. In this way, a linked list of hot-spots can be kept in each diagonal and extended in constant time. Moreover, as the lists are updated, the method can maintain candidates for good diagonal runs, with the rule of continuing a candidate run as long as the score of the run remains positive.

Each of the ten diagonal runs selected above specifies a pair of aligned substrings. Each such alignment may now contain both matches (from the hot-spots) and mismatches (from the interspot regions), but it does not contain spaces because it is derived from a single diagonal. In step two, each of these ten alignments is examined for a pair of aligned substrings (a subalignment) with maximum score, where the score now is determined by using an amino acid substitution matrix (in the case of aligning protein sequences). Originally, the recommended substitution matrix was a Dayhoff PAM matrix, but currently it is the BLOSUM50 matrix (see Sections 15.7 and 15.10 for discussion of PAM and BLOSUM). FASTA reports the single best subalignment found at this step and calls it *init1*.

In step three, FASTA attempts to combine "good" subalignments (subalignments from step two whose score is above some specified *cutoff*) into a single larger high-scoring alignment that does allow some spaces. To explain the general strategy, let each of the ten subalignments with score above the cutoff be represented by a weighted node in a directed graph, where the weight of the node equals the score (from step two) of the subalignment it represents. Let u represent one of the chosen subalignments, starting at position (i, j) and ending at position $(i+d, j+d)$ in the table. Let v be another of the subalignments starting at position (i', j'). Then extend an edge from node u to node v in the graph if and only if $i' > i + d$. That is, v should start at a row lower in the table than where u ends. Apply a weight to that edge to penalize any gaps that would be created in an alignment where subalignment u is followed by subalignment v. So a large penalty (negative weight) should be applied to the edge (u, v) if i' is much greater than $i + d$, or if there is a great distance between the two diagonals containing u and v. Alternately [360], one can use a constant penalty for a gap no matter how long it is. Essentially, FASTA then finds a maximum-weight path in this graph. The selected path specifies a single local alignment between the two strings. The output from this step is called *initn*. This local alignment may not be the optimal local alignment that a full dynamic programming algorithm (e.g., Smith–Waterman) would pick out, but the intuition behind FASTA (supported by experience) is that it will score well in comparison to the dynamic programming optimal.

Step three is a small example of a more general problem – combining conflicting subalignments into one larger alignment – that was discussed in Section 13.3.

Finally, in step four, FASTA computes an alternative local alignment score, in addition to *initn*. It returns to *init1*, the output of step two, and forms a band in the table around the diagonal containing *init1*. For protein, if the value of *ktup* is set to two, then the band consists of the 16 diagonals around the diagonal containing *init1*. For *ktup* set to one, the band contains 32 diagonals. Then FASTA uses the Smith–Waterman algorithm to compute the optimal *local* alignment in the subtable restricted to those 16 or 32 diagonals. The output of this step is reported as *opt*.

In the current version of FASTA, when a query sequence is compared to the sequences in a database, the default setting is to rank the database sequences according to the resulting *opt* scores. Previously, the default setting had been to rank by *initn* scores. Whichever score is used to identify a (few) promising sequences in the database, FASTA finally aligns those sequences to the query sequence using a full dynamic programming computation (i.e., the Smith–Waterman local alignment algorithm).

As FASTA is applied to successive text strings in the database (with a fixed query string), it collects statistics on the scores *initn*, *init1*, and *opt*, although the current version only reports statistics on the score used to rank the database sequences. These statistics are used to evaluate the statistical significance of the best local alignments found between the query string and all the text strings in the database.

15.6. BLAST

Almost immediately upon its release in 1990, BLAST became the dominant searching engine for biological sequence databases. The initial reasons for its success were speed, the fact that it outputs a range of solutions, and that each match is accompanied by an estimate of statistical significance (essentially the probability that a match of that value or better would be found in random strings). Since the introduction of BLAST, FASTA has evolved and now also outputs multiple solutions and estimates of statistical significance. Nonetheless, BLAST remains dominant.

BLAST originated from three convergent efforts. The first was a general effort by David Lipman, Warren Gish, and others at NCBI to increase the speed of FASTA by introducing more stringent rules to locate (fewer and better) alignment hot-spots. The second was the sublinear searching work of Myers [337, 342] (discussed in Section 12.3.5), which introduced the idea of substring neighborhoods and finite-state machines to locate initial alignment hot-spots. The third was the work of Karlin, Altschul, and Dembo [124, 262], who derived the probability results used in BLAST to evaluate the statistical significance of the reported matches.

BLAST concentrates on finding regions of high *local similarity* in alignments without gaps, evaluated by an alphabet-weight scoring matrix. Alignments with some gaps can be created by chaining together several locally similar regions that BLAST finds. The fundamental objects that concern BLAST are *segment pairs*, *locally maximal segment pairs*, and *maximal segment pairs*.

Definition   Given two strings S_1 and S_2, a *segment pair* is a pair of equal- length substrings of S_1 and S_2, aligned without spaces. A *locally maximal segment pair* is a segment pair whose alignment score (without spaces) would fall either by expanding or shortening the segments on either side. A *maximal segment pair* (MSP) in S_1, S_2 is a segment pair with the *maximum* score over all segment pairs in S_1, S_2.

When comparing all the sequences in a database with a fixed query sequence P, BLAST attempts to find all those database sequences that, together with P, contain a MSP above some set cutoff score C. The choice of C is guided by the theorems of [262] and [124], based on the scoring matrix and on characteristics of P and of the database sequences. Those theorems identify the lowest value of C for which a MSP with score above C is unlikely to occur by chance in any of the database sequences. Hence any sequence with a MSP score above C is considered "significant" and is reported. BLAST also reports sequences that do not have an MSP above C but that have several segment pairs that in combination are statistically significant. The statistical method used to identify such multiple segment pairs is described in [263].

15.6.1. The hit (hot-spot) strategy of BLAST

BLAST is actually a collection of programs, each tuned to a different problem domain. We will separately discuss BLASTP, the version of BLAST used to search protein databases, and BLAST applied to DNA.

To find sequences with a MSP above C in a protein database, BLASTP uses the following strategy: With a fixed length w and a fixed threshold t, BLAST finds all length-w substrings of S (called "words" in BLAST) that align to some length-w substring of P with an alignment score above t. Each such hot-spot (called a "hit" in BLAST) is then extended to see if it is contained in a segment pair with score above C.

To find the length-w hits, BLASTP follows an idea used in Myers's sublinear matching method (see Section 12.3.5). For each length-w substring α in P, BLAST constructs all the length-w words whose similarity to α is at least value t. Altschul et al. [19] state that "If a little care is taken in programming, the list of words can be generated in time essentially proportional to the length of the list". During the generation, these words are essentially placed into a *keyword tree* (actually a deterministic finite state automaton) for use in an algorithm similar, but not identical, to the *Aho–Corasick* algorithm (see Section 3.4) that locates every exact occurrence in the database of one of those words. So in essentially linear time in the number of words generated, and then in linear time in the length of the database, BLASTP finds every length-w substring in the database that has an alignment score at least t with some length-w substring in P.

To find all sequences with a MSP above C in a DNA database, BLAST follows a much simpler strategy. It uses a word list consisting of all the length-w substrings in the query string P. In practice, w is around twelve.

In either case (protein or DNA), once a hit is located, it is ideally extended to see if it is contained in a locally maximal segment pair with alignment score at least C. In practice, these extensions are truncated early if the running alignment score falls too far below the best score found previously for a shorter extension. This shortcut, and the use of length-w hits, means that BLAST is not guaranteed to find all segment pairs with score at least C.

Note that BLAST for DNA is essentially a standard *exclusion-type* method. It locates initial survivors (hot-spots, hits) of a fixed length by looking for *exact* matches to fixed-length substrings of P. A region around each hit is then examined. However, in contrast to the particular exclusion methods discussed in Section 12.3, BLAST does not use dynamic programming to examine those regions, because it disallows spaces in the alignments.

15.6.2. The effectiveness of BLAST

Clearly, the choices of scoring matrix, of w and of t, in relation to C, are critical to the efficiency and effectiveness of BLAST. Lowering t reduces the chance that a sequence with an MSP score above C will be missed but increases the amount of computation required. These choices have been extensively studied empirically, and the defaults have changed over time. We will not attempt to describe or justify the current recommendations or defaults except to note that for protein searches, w is in the range of three to five amino acid residues, and for DNA searches it is around twelve nucleotides.

Initially, BLASTP was reported [19] to run an order of magnitude faster than FASTA and many times faster than computing an optimal local alignment using dynamic programming. However, comparisons of BLASTP with current versions of FASTA [359] show a much smaller difference in speed, although both remain significantly faster than running dynamic programming.

The issue of biological effectiveness – the sensitivity and selectivity of the methods[6] – also remains an active area of empirical investigation. These empirical results will continue to vary as the methods, data, and biological concerns vary, but the current message seems to be this: BLAST may be a bit less effective than FASTA in identifying important sequence matches, particularly when the most biologically significant alignments contain spaces, but generally, BLAST is competitive with FASTA. In some limited circumstances,

---

[6] In an experiment where one knows which sequences are biologically similar to the query and which are not, "sensitivity" is the (false negative) rate at which the method fails to identify similar sequences, and "selectivity" is the (false positive) rate at which the method identifies sequences that are in fact not biologically similar to the query.

both FASTA and BLAST are notably less effective than optimal local alignment, but for general use FASTA and BLAST perform well compared to optimal local alignment. The recommendation to use both BLAST and FASTA makes good sense, as does the recommendation to compute a Smith–Waterman local alignment on the entire length of any two sequences found to be locally similar by the heuristic methods. Pearson [359] states "while BLASTP is effective in identifying distant relationships, a Smith–Waterman alignment should always be used when BLASTP matches are analyzed and displayed."

15.7. PAM: the first major amino acid substitution matrices

The sequence alignments computed for a protein database search are almost always *weighted* alignments, and the choice of the *scoring matrix* (amino acid substitution matrix) used can have a large effect on the search results. It is sometimes suggested that the proper scoring matrix is the most critical technical element in a successful search of protein databases. Ideally, the scores in the matrix should reflect the biological phenomena that the alignment seeks to expose. In the case of sequence divergence due to evolutionary mutations, the numbers in the scoring matrix should ideally be derived from empirical observation of ancestral sequences and their present-day descendants. In the case of conserved motifs or well-defined sequence-to-structure correlations the numbers should be derived from collections of sequences containing those motifs or exhibiting those structural features.

15.7.1. PAM units and PAM matrices

The term "PAM", which is an acronym for "point accepted mutation" or "percent accepted mutations" has two related uses. First, it is used as a *unit* to measure of the amount of evolutionary divergence (or evolutionary distance) between two amino acid sequences. In that context, one might say that sequence S_1 is 5 PAMs diverged (or 5 PAMs distant) from sequence S_2. Second, the term "PAM" is used to refer to certain amino acid substitution matrices (scoring matrices) whose scores have a relationship to PAM units. The methodology of PAM units and the first specific PAM matrices were developed by Margaret Dayhoff and coworkers [122, 400].

To discuss PAM units and PAM matrices, it is useful to distinguish the *ideal* PAM objects, which are defined in terms of data that are unavailable, from the *real* PAM objects that are computed using less than ideal data.

15.7.2. PAM units

Definition Ideally, two sequences S_1 and S_2 are defined as being *one PAM unit diverged* if a series of *accepted* point mutations (and no insertions or deletions) has converted S_1 to S_2 with an average of one accepted point-mutation event per one-hundred amino acids.

The term "accepted" here means a mutation that was incorporated into the protein and passed on to its progeny. Therefore, either the mutation did not change the function of the protein or the change in the protein was beneficial to the organism (or was at least nonlethal).

The above definition of one PAM unit sounds like "one PAM unit of divergence between S_1 and S_2 implies a one percent sequence difference between those sequences." It does not. The reason is that a single position can undergo more than a single mutation, so that (for example) if there have been eight point mutations in an 800-length amino acid sequence, it could happen that the same position mutates twice. It could even happen that the position mutates back to its original amino acid. Therefore, if we align S_1 and S_2 (recalling that there have been no insertions or deletions), the expected number of positions where an amino acid difference is observed is (somewhat) less than eight.

The difference between the correct definition of a PAM unit and the (common) mis-statement of its meaning is slight for sequences that are only one PAM unit diverged. But the difference grows as the number of units does. For example, it is not true that two sequences that are 100 PAM units diverged are expected to be different in every position. In fact, even amino acid sequences that have diverged by 200 PAM units are expected to be identical in about 25% of their positions, and sequences that are 250 PAM units diverged can generally be distinguished from a pair of random sequences.

There are a number of criticisms of the entire concept of a PAM unit. For now let's ignore these larger issues and ask: How does one determine the number of PAM units separating two sequences S_i and S_j? In the ideal case, when one extant sequence has diverged from an ancestral sequence due only to point mutations, one simply counts the number of observed positions where the two sequences differ. Then a simple stochastic formula is applied that relates the expected number of observed differences to the expected number of point mutation events.[7]

In practice, there are two problems in implementing this ideal scenario. The first is that all the sequences we know today are from *extant* organisms. We don't know any protein sequences where one is actually derived from the other. The second problem is that insertions and deletions do occur in protein evolution, so that one sometimes cannot be certain of the correct correspondence between sequence positions. Additional criticisms of the PAM unit concept include the fact that not all positions mutate with equal frequency.

The first problem, the lack of ancestral protein sequences, is handled by appeal to the molecular clock theory (see Section 17.1.4). The two sequences S_i and S_j have each diverged from some common ancestor S_{ij}, and the molecular clock theory implies that the expected PAM distance (number of PAM units of divergence) between S_{ij} and S_i equals the expected PAM distance between S_{ij} and S_j. So one uses half the number of differences in the alignment of S_i to S_j to calculate the PAM distance between S_{ij} and its two derived sequences S_i and S_j. For this to be correct, one needs also to assume that amino acid mutations are reversible and equally likely in either direction.

The second problem, of insertions and deletions, is more difficult. To determine the proper correspondence between positions in the two sequences, one must unambiguously identify the true historical gaps, or at least identify large intervals in the two sequences where the correspondence is correct. That cannot always be done with certainty, especially when the sequences are diverged by a large number of PAM units.

---

[7] This may be additionally complicated by the fact that the mutations occur at the DNA level, but the observations occur at the amino acid level. The degeneracy of the genetic code, plus the fact of "silent mutations", means that some assumptions must also be made about codon frequencies in order to translate the number of *observed* amino acid changes into the *expected* number of true amino acid mutations. For example, TCT is one of the codons for the amino acid serine, and TAT and AGC are two codons for tyrosine. TCT mutates to TAT with a single (DNA) point mutation, whereas it takes three point mutations to transform to AGC. So if one observes a change of serine to tyrosine, the question of how many mutations occurred must involve details at the DNA level. There is some controversy about the correct way to connect the DNA level to the amino acid level [477].

15.7.3. PAM matrices

PAM matrices are amino acid substitution matrices (scoring matrices) that encode and summarize expected evolutionary change at the amino acid level. Each PAM matrix is designed to be used to compare pairs of sequences that are a specific number of PAM units diverged. For example, the PAM 120 matrix is ideally used to compare two sequences known to be 120 PAM units diverged. For any specific pair of amino acid characters, denoted A_i, A_j, the (i, j) entry in the PAM n matrix reflects the frequency that A_i is expected to replace A_j in two sequences that are n PAM units diverged.

The PAM n matrix can be obtained *ideally* (but not practically) as follows: Collect many distinct pairs of homologous sequences that are known to be n PAM units diverged (by point mutations only). Align each pair of sequences, and for each amino acid pair A_i, A_j, count the number of times that A_i aligns opposite A_j, and divide that number by the total number of pairs in all the aligned data. Let $f(i, j)$ denote the resulting frequency. Let f_i and f_j, respectively, be the frequencies that amino acids A_i and A_j appear in the sequences. That is, f_i is the number of times A_i appears in all the sequences, divided by the total lengths of the sequences. Then, the (i, j) entry for the ideal PAM n matrix is

$$\log \frac{f(i, j)}{f(i)f(j)}.$$

This is another example of a *log-odds* ratio . The reason for dividing $f(i, j)$ by $f(i)f(j)$ is to normalize the true (historical) replacement frequency by the replacement frequency one expects due to chance alone. The logarithm is used because addition is easier than multiplication.

15.7.4. How are PAM matrices actually derived?

The above description of the PAM n matrix is the ideal case. But when insertions and deletions have occurred, the proper correspondence between positions can't be known unless the gaps are deduced correctly. In other words, one needs to align the sequences in a way that correctly recovers the history of their divergence before collecting the needed statistics to construct a PAM matrix. However, a scoring matrix is usually used to obtain the alignment. That creates a circular bind. How did Dayhoff resolve this problem?

Dayhoff took *highly similar* sequences that were each believed to have diverged from a common ancestor by only a low number of PAM units. "In order to minimize the occurrence of changes caused by successive accepted mutations at one site, the sequences . . . were less than 15% different from one another" [122]. Because the divergence was small, the sequences in each pair were essentially the same length, and the few insertions and deletions that might have occurred were easily spotted and accounted for. Therefore, establishing the correct correspondence between characters in those sequences was not a problem, and accumulation of the statistics followed essentially the ideal case.[8] This solves the problem for a low number of PAM units, but not for a high number.

To construct the PAM matrix for high number of PAM units, Dayhoff (roughly, but not exactly) proceeded as follows: Assume that the data consist of aligned pairs of sequences that are one PAM unit diverged. From those data, calculate the frequency, denoted $M(i, j)$, that amino acid A_i mutates to amino acid A_j in one PAM unit. That is, given that a position

---

[8] Actually, there is an additional detail in Dayhoff's method, involving the use of phylogenetic trees. We will discuss that detail in Section 17.6.2 of Part IV, after phylogenetic trees have been more fully introduced.

contains amino acid A_i (no matter how frequently or infrequently it does) $M(i, j)$ is the frequency that the position mutates to A_j in one PAM unit. Let M denote the 20 by 20 matrix of these frequencies. Then matrix M^n (the result of multiplying M by itself n times) gives the probability that any particular amino acid mutates to another particular one in n PAM units. The (i, j) entry for the PAM n matrix is therefore

$$\log \frac{f(i)M^n(i, j)}{f(i)f(j)} = \log \frac{M^n(i, j)}{f(j)},$$

where $f(i)$ and $f(j)$ are the observed frequencies of amino acids A_i and A_j. This approach assumes that the frequencies of the amino acids remain constant over time and that the mutational processes causing replacements in an interval of one PAM unit operate the same for longer periods. These assumptions, which are made for practicality, are further deviations from the ideal description of PAM matrices.

The values in the PAM matrices are usually multiplied by 10, rounded to integers, and often people add a constant (between 2 to 8) to each entry in the matrix, other than an amino acid character against a space. This has the effect of muting some of the differences between the entries, and the use of this muted matrix has been observed to be beneficial in obtaining "better" alignments in some contexts.

15.7.5. The use of the PAM matrix

Although the PAM matrices were defined and derived from aligned pairs of sequences containing few mismatches (point mutations) and few insertions and deletions, their major use is in weighted alignment with affine gap weights where there may be many mismatches and indels. The alignment of two sequences that are 250 PAM units diverged will generally not have a large number of matches, even after an optimal insertion of gaps. Moreover, when aligning two sequences, one should ideally use the particular PAM matrix corresponding to their PAM distance (the number of PAM units that the two sequences have diverged), but the PAM distance of the two sequences is not generally known. Indeed, one purpose of computing the alignment is to insert gaps to reveal ungapped segments that can be used to compute a PAM distance. Because of these two problems, one cannot know from first principles how effective a PAM matrix will be in finding "biologically informative" alignments.

Yet, it seems that the general experience of practitioners is that PAM matrices are very effective in finding alignments that highlight important biological phenomena. In particular, the PAM 250 matrix is reported to be very effective for aligning sequences used in evolutionary studies, and until fairly recently it had often been regarded as the "canonical" protein scoring matrix.[9] However, the BLOSUM62 matrix is increasingly displacing PAM250 as the default matrix.

---

[9] I once attended a workshop that was designed for computer scientists, mathematicians, and biologists to develop joint activities in computational biology. It was assumed that most of the nonbiologists attending would have little background in computational biology, and most of the talks were introductory. However, one speaker (whose discipline I have forgotten) launched into a rather advanced talk that concerned fine points of PAM matrices. He assumed that everyone knew the basics of PAM matrices and so did not introduce them, or say anything about their function, origin, or construction. Finally, about halfway into the talk, after the speaker had mentioned "the PAM matrix" perhaps thirty times, a frustrated listener (who had never previously heard of PAM matrices) raised his hand and said something like "you've mentioned the PAM matrix repeatedly, but you've never told us what it is. Just *what is* this PAM matrix you are talking about?" The speaker rapidly responded "it's 250" and went on with his talk without hesitation.

15.8. PROSITE

PROSITE is a "dictionary of sites and patterns in proteins" [41, 42] that is linked to the protein sequence database Swiss-Prot. Some aspects of PROSITE were mentioned earlier in Section 3.6 and Exercise 32 of Chapter 3. The goal of the PROSITE developer, Amos Bairoch, is to identify and represent *biologically significant* patterns in protein families (particularly those believed to be important to the function of the protein) that allow new protein sequences to be reliably assigned to the proper family. The emphasis on biologically significant patterns distinguishes PROSITE from other efforts to find good *discriminators*, which only seek to reliably identify known family members while excluding known nonmembers. A PROSITE pattern is represented either as a signature motif (see Section 14.3.2) or as a profile (see Section 14.3.1) derived from a multiple alignment of the family members. PROSITE currently contains around one thousand patterns. Each one comes with cross references to the entries in Swiss-Prot where the pattern is found, along with known false positive and false negative matches in Swiss-Prot.

The *signature* patterns in PROSITE are written as *regular expressions* (see Section 3.6). For example, G-$[GN]$-$[SGA]$-G-x-R-x-$[SGA]$-C-$x(2)$-$[IV]$ is a PROSITE pattern. Each capital letter specifies a specific amino acid, capital letters inside brackets indicate that any single one of the enclosed amino acids is permitted, x indicates any amino acid is permitted, and $x(2)$ indicates any pair of amino acids is permitted. When searching a new sequence for an occurrence of a PROSITE pattern, one can use regular expression pattern matching methods (Section 3.6). If the total length of all the regular expressions is n and the length of the new sequence is m, then this approach takes $\Theta(nm)$ time. However, a PROSITE pattern only recognizes finite-length substrings. Moreover, as noted in Exercise 5 (page 84) of Chapter 4, wild cards cause no problems for methods such as *Shift-And* or *agrep*. Thus, as addressed in Exercise 6 of Chapter 4, PROSITE patterns that do not contain a large number of repetition-range specifiers can be more efficiently handled using *Shift-And* or *agrep* style methods.

PROSITE patterns represented as *profiles* are each derived from a multiple alignment of family members and are used when the family similarity is insufficient to derive effective motif signatures. Additional ability to recognize new family members is obtained by "extrapolating" the profile, that is, by using knowledge of amino acid substitutability to give scores for amino acids in certain positions where those amino acids have not been observed. For example, if the amino acid Leucine is seen frequently in a given position of the multiple alignment, but Isoleucine is not, Isoleucine may still be given a large score for that position because Isoleucine and Leucine have similar chemical properties and are frequently substituted for each other.

15.9. BLOCKS and BLOSUM

BLOCKS is a database of protein motifs that is derived from the PROSITE library, and BLOSUM is an amino acid substitution matrix that is derived from BLOCKS. Both BLOCKS and BLOSUM were developed by Steven and Jorja Henikoff [220, 221, 222].

The BLOCKS database attempts to represent the most highly conserved substrings (motifs) in amino acid sequences of related proteins. At this level of description, that goal applies equally well to the PROSITE library of motifs. However, the PROSITE motifs were collected with particular attention paid to known functions or known structures of the proteins. Each motif in PROSITE is expected to have a known biological meaning. In contrast, the motifs in BLOCKS are based on conserved sequence similarity, even if no function is known for the motifs.

> **Definition**   A *block* is a short contiguous interval in a multiple alignment of amino acid sequences.

The BLOCKS database holds roughly 3,000 blocks of short, highly conserved sequences derived from roughly 800 groups of related proteins. The system that builds BLOCKS, called PROTOMAT, separately processes each of the groups of related proteins in the PROSITE library. For each PROSITE group, PROTOMAT multiply aligns the sequences in that group and then searches for intervals of up to sixty positions where the aligned amino acids are highly similar in a "critical number" (at least 50%) of the sequences. The details are intentionally left vague and only the underlying ideas are described, since the methods and specific parameters used are highly heuristic and have been obtained through extensive experimentation.

After PROTOMAT finds individual blocks in the multiple alignment, it selects a subset of the blocks to form a "path" (i.e., an ordered set of nonoverlapping blocks that occur in a "critical number" of sequences). A path identifies regions of high local similarity in the set of proteins without concern for the length of the gaps between its constituent blocks. Given a score for each individual block (based on its width, the level of similarity, the number of sequences it contains, etc.), one can find a path whose block scores sum to the largest number. This is exactly the *chaining problem* discussed in Section 13.3. The resulting path might identify an effective set of motifs to characterize that protein group. However, the criteria in [221] for evaluating a path are more complex than the sum of block scores, and the "best path" in BLOCKS is found by using brute force enumeration to examine all possible paths. The BLOCKS database holds the best path found for each PROSITE protein group.

Using BLOCKS to classify new sequences

The blocks (or paths) in the BLOCKS database are used to identify potential new members of the protein group that the block (or the path) is derived from. A *profile* (see Section 14.3.1) is created for each block, and a new sequence is aligned to each block profile. Sequences that align well to the profiles from many blocks in a single path are considered highly likely to be members of the underlying protein group.

15.10. The BLOSUM substitution matrices

The BLOSUM[10] amino acid substitution matrices are derived from the BLOCKS database [222] and are currently the main competition for the Dayhoff PAM matrices. The basic BLOSUM amino acid substitution score for a pair of amino acids i, j is roughly computed as follows: Define P as the set of all pairs of positions in the blocks contained in BLOCKS, such that the two positions in any pair in P are contained in the same column of the same block. For example, if there are n columns contained in all the blocks in BLOCKS, and each block contains exactly k rows, then P consists of $n \times \binom{k}{2}$ pairs of positions. Now let $n(i, j)$ be the number of pairs in P such that one position of the pair contains amino acid i and the other contains amino acid j. Intuitively, $n(i, j)$ increases if amino acids

---

[10] BLOSUM stands for *blocks substitution matrix*.

i and j tend to appear together in the same columns of the blocks, and it decreases if they tend not to appear together. To normalize $n(i, j)$, define $f(i)$ and $f(j)$ as the fraction of all positions in BLOCKS occupied by amino acids i and j, respectively, and define $e(i, j)$ as $|P| f(i) f(j)$. Next, define $s(i, j)$ as $\log_2 \frac{n(i,j)}{e(i,j)}$. Finally, $s(i, j)$ is multiplied by two and rounded to the nearest integer to form the i, j entry in the basic BLOSUM matrix. Intuitively, $s(i, j)$ is the (log of the) ratio of the number of times that amino acids i and j appear together in the same column, divided by the number of times one would (roughly) expect to see i, j pairs in the same column if the placement of amino acids i and j were random throughout BLOCKS. Like the values in a PAM matrix, a BLOSUM value is an example of a log-odds ratio.

An important additional refinement to the basic BLOSUM methodology is the deemphasis of pairs of rows that are "too" highly similar in BLOCKS. For example, if two rows that appear in a given block are identical, then one is removed to reduce the influence of that pair of rows in the resulting BLOSUM scores. Extending this idea, the BLOSUM x matrix (for x generally between 50 and 80) is created after successively removing one row from any pair of rows in a block that are more than x percent identical. The result is that any two remaining rows in a given block will be less than x percent identical.

The purpose of reducing the influence of closely related sequences is to better capture more distant, yet conserved, sequence motifs from a set of widely diverged sequences. Ideally, the sequences in the set should uniformly sample the full evolutionary range of the species being studied. But often the sequences do not provide a uniform sample and are clustered into subsets of more highly related sequences. Iteratively removing a row of a block that is "too similar" to some remaining row is an attempt to create a block reflecting a less biased data set. The effectiveness of this approach is a purely empirical question. Current opinion is that the BLOSUM 62 matrix has been found to be particularly effective as a less biased reflection of important amino acid conservation.

How BLOSUM matrices differ from PAM matrices

The scores in Dayhoff's PAM matrices are extrapolated from data obtained from very similar sequences. In fact, as explained earlier, the sequences used were intentionally selected to be highly similar. However, "the most common task involving substitution matrices is the detection of much more distant relationships" [222], and the BLOSUM matrices were developed as a way to explicitly represent those important distant relationships. The belief is that highly conserved sequence segments from otherwise highly diverged protein sequences (of proteins in a given family) lead to substitution scores that more effectively encourage local alignment algorithms to produce alignments highlighting biologically important similarities. The general view is that the BLOSUM matrices have been successful in that goal.

15.11. Additional considerations for database searching

This chapter has focused on the use of string algorithms in database search and the search for biologically important substrings (motifs). However, we have also discussed the related, but nonalgorithmic, topics of the PAM and BLOSUM scoring matrices, since their use affects the biological utility of searching algorithms. In this section we briefly discuss additional nonalgorithmic issues impacting the practical effectiveness of database searching.

15.11.1. Statistical significance

We begin with a quote from William Pearson.

> The major technical improvement in protein sequence comparison over the past five years has been the incorporation of statistical estimates in widely used similarity searching programs. [359]

Better estimates of statistical significance have led to improved sensitivity and selectivity, and therefore, to more effective search programs. These estimates are due to theorems on the statistics of MSPs [124, 262] initially incorporated into BLAST and to more recent work of Karlin and Altschul [263] and Waterman and Vingron [459] on alignments that allow gaps. Any discussion of these statistical results is well beyond the scope of this book. We refer the reader to the original papers, to the text by Waterman [461], and to a very readable, general-audience paper by Altschul et al. [17].

Since the introduction of BLAST, FASTA has also incorporated the theorems of [262] and [124] into its significance estimates. However, since FASTA finds alignments with gaps, the statistical results on MSPs are not as immediately applicable to FASTA output. There is some theoretical and empirical work [21, 263, 459] suggesting that the results of [262] and [124] extend in a natural way to local alignments with gaps, but the statistical significance of an alignment with gaps is often estimated by some form of *jumbling* or *shuffling* computation [127, 359]. In that computation, one aligns the query string to random strings that have the same average length and character composition as the strings in the database. Statistics based on those alignment scores are used to evaluate the significance of any high-scoring alignment found between the query and an original string in the database.

Using probability: a cautionary tale concerning BRCA1

It is desirable to make database searching as much of a "push-button exercise" as possible, without requiring great expertise from the user. But today, effective database searching often still requires a judicious mix of biological and statistical insights. Recent experience with the BRCA1 gene illustrates this point.

Familial breast and ovarian cancer has been linked to mutations in two genes, BRCA1 and BRCA2, which have both been precisely located, cloned, and sequenced. Finding and cloning these genes was a major milestone in cancer research, but therapies or preventions for these cancers require an understanding of the proteins that these genes code for. In the case of BRCA1, there was some early hope that such an understanding had begun, after a database search found a PROSITE motif (see Section 15.8) in the amino acid sequence derived from BRCA1 gene [245, 417]. A substring of the BRCA1 sequence matched the consensus sequence for a family of proteins called *granins* (its consensus sequence is displayed in Section 3.6). Granins had not previously been connected to breast cancer, but based on the database match, several experiments were performed that supported a role for granins in beast cancer. However, other experiments contradicted those conclusions. What was exciting and caused a great deal of attention is that granins are proteins that are secreted outside of the cells where they are made. This is in contrast to most known cancer-related proteins, which work inside the cell (or inside the nucleus of the cell), where it is hard to get at them or to supplement them. If the BRCA1 protein does in fact function like the known granins and is secreted outside the cell, successful strategies against oncogenic mutants of the BRCA1 protein seemed much more likely.

Unfortunately, a deeper examination of the original PROSITE search concluded that "the granin motif has no statistical support" [282]. The first paper [245] had reported that "the probability of the granin consensus occurring by chance in BRCA1 is approximately . . . 0.00175", and hence the authors concluded that the granin match reflected an important biological phenomenon. However, without prior evidence that granins are involved in breast cancer, the computed number (which is a reasonable approximation of the probability that the granin sequence occurs in BRCA1 by chance.) does not answer the right question. The more informative question is: What is the probability that *some* PROSITE motif (as complex as the granin motif) would occur by chance in BRCA1. That probability was reported in [282] to be 0.87. This high number is due to three factors: PROSITE contains well over 1,000 motifs, the granin motif is only ten positions long and three positions are wild cards, and BRCA1 is a relatively long protein (1,863 residues). With such a high probability of a random match to some PROSITE motif, additional work is needed to make an effective use of the sequence and motif database. For example, one should look for more extensive similarities between the BRCA1 protein sequence and the sequence containing the matched motif, *beyond* the motif region itself. Or one should first identify short stretches of the BRCA1 protein sequence where a matched motif would more likely reflect a real biological phenomenon, and where probability computations would be more informative.

The above approach was followed by Koonin, Altschul and Bork [282]. They first identified six regions of the BRCA1 sequence with a predicted globular structure, since globular regions are more likely to contain conserved functional motifs. Matching each of these six regions against a sequence database, they found a moderate match between a 202-residue-long globular region and an analogously located region (at the C-end of the proteins) in a human protein, 53BP1, known to bind to the universal tumor suppressor p53. Moreover, the match consisted of two distinct segments that were separated by almost the same number of residues in each protein (100 and 97). The similar position and shape of the matching substrings suggest that the match has some biological importance. The use of such evidence is sometimes important in augmenting pure probability computations. The biological plausibility of a connection between BRCA1 and p53 (which is associated with about half of all cancers) is strengthened by the fact that cancer-causing mutations in the BRCA1 gene frequently occur in the 3′ end of the gene, which corresponds to the C-end of the BRCA1 protein. This line of reasoning, and additional evidence, lead the authors to suggest that the BRCA1 gene is associated with a well-established cancer agent and not a granin.[11]

15.11.2. A theory of log-odds scores

We have examined two amino acid scoring schemes in detail (PAM and BLOSUM matrices) and noted that the values in both matrices are log-odds ratios. Other substitution matrices, such as those presented in [57, 183] also contain log-odds values. Similarly, profiles characterizing significant motifs in biological sequences are often converted to log-odds ratios as well (see Section 14.3.1). Recently, a theory has been developed [16, 262] that justifies the prevalence of the log-odds ratio in a few specific contexts. The theory is most naturally applied to the problem of characterizing selected (ungapped) substrings

---

[11] They did offer one surprising suggestion. Searching the database with a profile constructed from the most conserved parts of BRCA1 (from human and mouse) and 53BP1, they found a protein in yeast that might be structurally related to p53.

of some family and to the problem of finding ungapped local alignments that produce a family of selected alignments. A very readable overview of this theory, incorporating other nonalgorithmic issues in database search, appears in [17].

To explain the simplest form of the theory, consider the problem of searching for a substring of a fixed length t that has some given property (for example, the property of being a promoter sequence in *E. coli*). Initially, one is given a set $\mathcal{M}$ of selected substrings that have the desired property. Later, one will be given a set of sequences where each sequence may or may not contain a substring with the desired property. The problem of biological interest is to determine which sequences are likely to contain a t-length substring with the desired property, and which are not. The theory of [16] and [262] concerns methods that look for these substrings in the following way: Given a sequence S, each t-length substring in S is explicitly examined and scored by the sum of scores given to each character in the substring. Individual character scores can depend both on the specific character and on its position in the substring.

If the scoring scheme is effective, substrings with the desired property should have a high score, while substrings without the property should have a low score. That vague goal is made precise by the requirement that substrings from $\mathcal{M}$ have a statistically significant higher score than random substrings with the same overall distribution of characters. Under this scenario, what scoring scheme is best for distinguishing members of $\mathcal{M}$ from random substrings? That is the precise question that has been answered.

Let $p(x, j)$ be the frequency with which character x occurs in position j of a string in $\mathcal{M}$, and let $p(x)$ be the frequency with which character x occurs anywhere in the input set. It has been proved [16, 262] that the scores that optimally distinguish the substrings in $\mathcal{M}$ from random substrings (using the frequency $p(x)$ for each character x) are of the form $\log[p(x, j)/p(x)]$ where the base of the log is arbitrary. This theorem is consistent with the way profile values are often created (Section 14.3.1).

In the context of ungapped local alignment, let $\mathcal{M}$ now be a set of local alignments of interest, let $p(x, y)$ be the frequency that characters x and y are aligned with each other in $\mathcal{M}$, and let $p(x)$ be the frequency with which character x occurs in the sequences to be locally aligned. Then the scores that best distinguish the local alignments of interest from random local alignments are of the form $\log\{p(x, y)/[p(x)p(y)]\}$. Again, a log-odds score.

15.11.3. Importance of searching protein with protein

Overwhelmingly today, new protein sequences are obtained by sequencing the underlying DNA in the gene (or the mRNA) that codes for the protein. Therefore, most of the entries in the "protein" databases are actually "derived amino acid sequences", and their originating DNA sequences are contained in the DNA databases. So, when a searcher wants to search for a similar protein sequence, the searcher, can often use either the DNA sequence of the query protein to search the DNA databases or the translated sequence to search the protein databases. There are many reasons to maintain and explore the underlying DNA sequence, but to detect similarities in highly diverged sequences, the standard advice (from Doolittle [127] in this case) is to

> Translate those DNA Sequences!
> Some beginning sequence comparers are under the impression there is more to be gained by searching the actual DNA sequence rather than the amino acid sequence derived from it. Such a course is greatly mistaken . . .

The reason to translate is that more sensitive and informative comparisons are possible between amino acid sequences than are possible between DNA sequences, *using the common alignment methods and objective functions of* the type detailed in this book.[12] All (and more) of the "information" contained in a derived amino acid sequence is contained in its underlying DNA sequence. However, that DNA information is not in a form that directly allows common alignment methods and objective functions to be the most effective.

Derived amino acid strings allow more meaningful alignment by using scoring matrices that reflect the evolutionary, biological, or chemical similarities of specific amino acid pairs. For example, replacement of leucine by isoleucine is generally considered a minor change, and that fact is reflected in the PAM and BLOSUM matrices. But alignment scores derived by summing the scores from matches and mismatches of the underlying nucleotide pairs would not capture the close relationship of leucine and isoleucine. Similarly, most amino acids are coded for by more than one codon, and two codons differing only in their third position often code for the same amino acid. This kind of correlated, third-position effect would be hard to correctly model in alignments of DNA using alignment methods discussed in this book. The mismatch score of a pair of nucleotides must reflect the appearance of those nucleotides in the codes for many different amino acids. And since there are only sixteen different nucleotide pairs, the sum of pair scores over a codon would rarely have the specificity of an amino acid substitution score from a matrix for 400 pairs. The derived amino acid sequence encodes these position dependancies, allowing standard algorithms (along with substitution matrices) to be more effective. There are, however, some efforts to get the best of both representations, by using derived amino acid sequences but remembering the specific codons in the underlying DNA. This ideally requires a codon substitution matrix of size 64 by 64.

Although it is standard to search protein sequences with translated DNA sequences, there is a major technical problem with this approach, discussed in Section 18.1.

15.12. Exercises

1. One feature that has been implemented in some databases is to record every search request (query string) made of the database and the address of the searcher. Then when new sequences are added to the database, the new sequences are compared to the prior query strings, and a report is sent to the searcher if a significant similarity is found. Before such services were available, there were several significant cases where the searcher missed an important finding because they looked in a copy of a database that didn't contain the most updated information (GenBank now releases the current version six times a year), while their competitors used the most current version and made the important match.

   What data structures and search algorithms would be best to use to implement the database feature mentioned above? The answer might depend on parameters such as how many queries there will be and how many new sequences are added to the database at each update.

---

[12] Translation is often mistakenly justified by statements like "amino acid similarities are much better preserved through evolution than similarities in the underlying DNA". In fact, that is the statement I had in a draft of this book before Stephen Altschul pointed out that the critical issue is not what information is conserved in the sequences, but how well common alignment and statistical methods can use that form of information. And, in fact, there are cases where the DNA sequences are claimed to be more highly conserved than the translated amino acid sequences [314].

2. Related to the previous problem, some databases use the collection of prior queries strings to screen future queries. That is, whenever a query string comes in, it is checked against the prior query strings to see if that query, or a highly related one, has been made previously. Each query in the collection points to strings in the database that are highly similar to it, so that when a repeat query is received, a full search can be avoided. In some databases, it is common that the same query is made repeatedly.

Discuss choices of data structures and search algorithms to implement this database feature.

3. When a (log) odds-score is used as the basis for amino acid substitution values, it is possible for the value of an identical match to be smaller than the value for some particular mismatch. Explain how this can happen. Is this an undesirable feature of a substitution matrix?

PART IV

Currents, Cousins, and Cameos

In the previous three parts of the book we developed general techniques and specific string algorithms whose importance is either already well established or is likely to be established. We expect that the material of those three parts will be relevant to the field of string algorithms and molecular sequence analysis for many years to come. In this final part of the book we branch out from well established techniques and from problems strictly defined on strings. We do this in three ways.

First, we discuss techniques that are very current but may not stand the test of time although they may lead to more powerful and effective methods. Similarly, we discuss string problems that are tied to current technology in molecular biology but may become less important as that technology changes.

Second, we discuss problems, such as physical mapping, fragment assembly, and building phylogenetic (evolutionary) trees, that, although related to string problems, are not themselves string problems. These cousins of string problems either motivate specific *pure* string problems or motivate string problems generally by providing a more complete picture of how biological sequence data are obtained, or they *use the output* of pure string algorithms.

Third, we introduce a few important cameo topics without giving as much depth and detail as has generally been given to other topics in the book.

Of course, some topics to be presented in this final part of the book cross the three categories and are simultaneously currents, cousins, *and* cameos.

16

Maps, Mapping, Sequencing, and Superstrings

16.1. A look at some DNA mapping and sequencing problems

In this chapter we consider a number of theoretical and practical issues in creating and using genome maps and in large-scale (genomic) DNA sequencing. These areas are considered in this book for two reasons: First, we want to more completely explain the origin of molecular sequence data, since string problems on such data provide a large part of the motivation for studying string algorithms in general. Second, we need to more completely explain specific problems on strings that arise in obtaining molecular sequence data.

We start with a discussion of mapping in general and the distinction between *physical* maps and *genetic* maps. This leads to the discussion of several physical mapping techniques such as *STS-content* mapping and *radiation-hybrid* mapping. Our discussion emphasizes the combinatorial and computational aspects common to those techniques. We follow with a discussion of the tightest layout problem, and a short introduction to *map comparison* and *map alignment*. Then we move to *large-scale sequencing* and its relation to physical mapping. We emphasize *shotgun sequencing* and the string problems involved in *sequence assembly* under the shotgun strategy. Shotgun sequencing leads naturally to a beautiful *pure* string problem, the *shortest common superstring problem*. This pure, exact string problem is motivated by the practical problem of shotgun sequence assembly and deserves attention if only for the elegance of the results that have been obtained. We next return to DNA sequencing, discussing *sequencing by hybridization*, a highly structured special case of the shortest superstring problem.

16.2. Mapping and the genome project

The ultimate goal of the Human Genome Project is to sequence the entire human genome, producing the complete DNA transcript – the full three-billion-nucleotide-long string. Although this goal is the best publicized aspect of the project, in fact, production-scale sequencing of human DNA has only just begun.[1] A lesser known initial goal of the Human Genome Project and other genome projects is to produce *maps* of the genome showing the locations of "landmarks" (such as STSs and ESTs discussed in Section 3.5.1) at regular intervals over the genome. Genome maps are critical in hunting for specific genes of interest; they are used in the first stages of most, but not all, large-scale DNA sequencing

---

[1] However, the genomes of several small bacteria with lengths approaching 2 million base pairs have been fully sequenced. Additionally, the full sequence of the (brewer's) yeast genome *Saccharamyces cerevisiae* (length around 12 million base pairs) was released in spring 1996, the genome of *E. coli* (length 4.7 million base pairs) was finished on January 16, 1997, and a large percentage of the genome of the worm *C. elegans* (length 100 million base pairs) has been completed. Additionally, there is now a consensus to undertake large-scale sequencing of the human genome faster than previously planned [309, 352, 176].

projects (see Section 16.13), and they are useful for the further physical examination of DNA that is required for other parts of the genome project.

We will present a few topics in mapping and map use that are intertwined with string problems, or relate to string problems through their use in DNA sequencing strategies discussed later. We emphasize combinatorial aspects of these problems as this best fits the style of this book. However, we will not attempt a comprehensive survey of the various maps and mapping strategies or a detailed discussion of how maps are produced or used, as that would take us far from the focus of this book.

16.3. Physical versus genetic maps

At the high level, there are two kinds of genome maps: *genetic maps* and *physical maps*. Problems with acquiring and using physical maps are closer to string problems than are problems involved with genetic maps. Therefore, the emphasis here will be on physical mapping problems.

The term "physical mapping" refers to establishing the true physical location of landmarks (markers) or of features of interest (often genes or known patterns) in the genome. The metric used is based on observable or countable physical features, usually the number of nucleotides between two points. The earliest physical maps consisted of the *banding patterns* that appear on chromosomes when various stains are applied. These bands are reproducible and visible under a microscope and so provide a low-resolution physical map. This map has been used to locate the rough position of genes whose deletion or expansion causes disease, when those chromosome changes are large enough to change the observable banding patterns. For example, the tumor suppressor gene for the cancer *retinoblastoma* was initially localized to a band on chromosome 13 by observing a deletion in that region for some people with the disease [472]. Conversely, isolated fragments of a chromosome can often be located on the full chromosome because the banding pattern is irregular, making regions distinctive. In a human chromosome, the number of bands produced with the widely used stains is generally under one hundred, and so observable bands provide only a low-resolution physical map. In other organisms, such as *Drosophila*, the number of bands is much greater (around 5,000), making their utility greater.[2]

In current high-resolution physical mapping, the goal is to locate features of interest in terms of their actual base-pair location on a chromosome. That is, locations are measured and described in terms of the number of nucleotides (base pairs) between the feature and some other established reference point. The sequence tagged site (STS) (see Section 3.5.1) is one such feature of particular importance, and one of the initial goals of the Human Genome Project was to establish a map of STSs spaced roughly every 100,000 bases throughout the entire genome [234, 399]. Compared to genetic mapping, high-resolution physical mapping on a large scale is relatively new, is very dependent on recombinant DNA technology, and has become a central component of the human and other genome projects. A good introduction to physical mapping (and other techniques) in the Human Genome Project can be found in [111].

---

[2] In the salivary glands of *Drosophila* larva, chromosomes align and replicate but then do not separate as they normally should. Instead, one thousand or more copies of a chromosome stack up lengthwise, in register, creating a very thick *polytene chromosome* whose bands are easily seen under a light microscope. These polytene chromosomes allowed physical mapping and physical study of *Drosophila* genetics well before the advent of recombinant DNA technology. For example, the famous mutation in *Drosophila* that creates *curly* wings was seen early on to be due to an *inversion* of an interval of one of the fly's chromosomes.

In contrast to physical mapping, *genetic* (or *linkage*) *mapping* establishes the relative order of features of interest on a chromosome or establishes that two features are likely on different chromosomes. Genetic mapping uses a unit called a *centimorgan* that is only monotonic with physical distance.[3] Genetic mapping is based on observing how frequently the states (alleles) of two loci on the DNA (possibly genes) are inherited together or are changed by an odd number of "cross-over" (or recombination) events in meiosis. The higher the frequency of coinheritance, the closer the two loci are deduced to be. To establish a linkage map, one therefore wants to find a set of uniformly dispersed loci in the DNA where the alleles of those loci are highly variable (or "polymorphic"). Originally, the loci were genes responsible for observable features at the organism (or "phenotypic") level, but in the past decade, features observed directly in the DNA have become the critical markers. The first markers of this type were *restriction-fragment-length polymorphisms* (*RLFPs*) [476], but increasingly microsatellites have become the markers of choice (see Section 7.11.1). A good introduction to genetic mapping through linkage analysis can be found in [426], and [353] provides an in-depth treatment.

Genetic mapping is much older and more established than high-resolution physical mapping, and yet it is still of great importance in providing the rough location of genes of interest and in developing tests to find carriers of certain defective genes. It is also very important in applications such as plant and animal breeding, where one tries to establish that a gene of interest is located close to a gene whose effect on the organism is more easily observed. The breeder then tries to breed individuals that have an allele (state) of the easily observed gene that is highly correlated with a state of the more hidden gene of interest.

Genetic mapping also aids in physical mapping by finding a region of interest in which the more demanding physical mapping can then be carried out. This combination of genetic and physical mapping has had an important impact, and will have a growing impact, on efforts to locate genes for inherited diseases even when the defective protein associated with the disease is unknown. This approach to finding defective genes is called *positional cloning*, and it is the reverse of the older approach where the protein involved in the disease is first identified and studied. Positional cloning is responsible for locating close to one hundred genes in the past several years (the best known cases are cystic fibrosis, Huntington's disease, and inherited forms of breast cancer). The rough idea is to use a genetic map to analyze patterns of disease inheritance in several generations of an affected family. By seeing which loci have alleles that are frequently *coinherited* with the disease, one can pin down the disease-causing gene between two locations on the genetic map. Then, physical maps of that region allow one to pull out substrings of DNA for closer examination and/or sequencing. Further study is done on those substrings that have the "look and feel" of a gene (for example, an open reading frame, a large concentration of Gs and Cs, or a codon usage that is typical of genes in that organism). See Section 18.2 for more detail. The goal is to find a small substring of DNA (often only a single nucleotide) from a putative gene that exhibits a systematic difference between disease-afflicted and unafflicted individuals. The gene containing that substring is then the

---

[3] It is widely reported in elementary texts that one centimorgan corresponds roughly to one million DNA base pairs. But despite the repetition of this claim, it is known that no simple linear correspondence exists between physical and genetic distances. For example, the limited correspondence that does exist differs in different species (the correlation is very poor in tomatoes but relatively good in *Arabadopsis*), differs at different locations on the genome, and differs in males and females. Moreover, the correlation between physical and genetic distance becomes increasingly poor as the physical distance increases.

prime candidate for the disease-causing gene. As genetic and physical maps become more refined, the positional cloning approach to gene location will become faster, cheaper, and more routine.

The creation of high-resolution genetic maps forms a central part of genome projects. However, we will not discuss genetic mapping further because the computational problems involved in genetic mapping are less "stringlike" and are also less combinatorial than problems in physical mapping.

16.4. Physical mapping

Although large-scale physical mapping is relatively new, there are a variety of distinct methods in use and many additional promising proposals. This contrasts with DNA sequencing, where there are only a small number of widely used methods, and is unlike genetic mapping, where there is one major approach (based on crossing-over in meiosis) along with related methods that work in specialized contexts.

We will discuss a few computational, combinatorial issues that arise in obtaining physical maps and one computational issue involved in comparing maps (physical or genetic). These problems are not string problems in the strictest sense but are close relatives to string problems, and their solution often shares techniques also used to solve certain pure string problems. Moreover, some specific string problems arise in the context of mapping, and conversely, large-scale DNA sequencing strategies usually start by obtaining a physical map of the DNA.

16.5. Physical mapping: STS-content mapping and ordered clone libraries

We introduce physical mapping by discussing the specific case of one important mapping strategy: *STS-content mapping* [189, 234].

Recall from Section 3.5.1 (page 61) that an STS is essentially, but not exactly, a short substring of DNA that occurs in only one place in the genome. An EST is an STS obtained from a transcribed sequence (essentially a gene). The most common process of obtaining an STS does not determine where the STS is located in the genome – it merely determines that the STS does occur uniquely [111, 317]. To be of greatest value, the physical location of that STS must be determined on a physical STS map.

A *clone-library* comprises a set of short DNA fragments (called *clones*) that can be maintained under laboratory conditions and that originate in a stretch of DNA of interest. It is important to note that the clones in the library can overlap; thus any particular site on the DNA may be covered by two or more clones. In the case of physical mapping, the clones in the library should ideally cover the DNA of interest. However, the original location of the clones in the underlying DNA, or even their relative order, may be unknown. The problem of building an *ordered clone-library* involves deducing the order and physical position of the clones on the underlying DNA string.

STS-content mapping is an approach that simultaneously determines the ordering of the STSs and builds a physical map for the ordered clone-library.[4] The procedure first

---

[4] Some researchers care more about ordering the STSs than about laying out the clones. The reason is that many existing clone-libraries have errors, which make the clones themselves less of interest, although they can be used to establish the STS order.

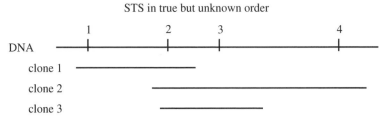

Figure 16.1: A schematic of three clones and four STSs. The figure shows the true locations of the STSs and clones, although their locations are unknown and must be deduced by the algorithm.

$$
\begin{array}{c|cccc}
 & \multicolumn{4}{c}{\text{STS}} \\
 & 2 & 4 & 1 & 3 \\
\hline
1 & 1 & 0 & 1 & 0 \\
\text{CLONE} \quad 2 & 1 & 1 & 0 & 1 \\
3 & 1 & 0 & 0 & 1 \\
\end{array}
$$

Figure 16.2: A data matrix for the clones in the previous figure. The STSs are written in permuted order since the correct order of the STSs is generally unknown.

determines which clones in the library contain which STSs. (See Figure 16.1.) This can be done by attempting to hybridize each STS to each clone. Alternately, PCR[5] can be used to determine whether any specific STS is contained in any specific library clone. The ends of each STS contain known and unique PCR primer templates. Therefore, to determine if a particular STS is contained in a particular clone in the library, one can use the known primers for that STS to determine if PCR generates large amounts of new DNA from the DNA in the clone. The particular STS is located in the clone if and only if the PCR generates new DNA using those primers. Running this experiment for each STS/clone pair gives data that can be summarized in a binary matrix. Each row corresponds to a clone, and each column corresponds to one STS. A column labeled j has a 1 in row i if clone i contains STS j. (See Figure 16.2.)

Of course, the true experimental protocol is more complex than described here, and the computational problem is always complicated by errors (which we will consider later). But our treatment captures all the features relevant to the combinatorial aspects of the problem. We now turn to the question of how to use the STS-content data to reconstruct the order of the STSs and place the DNA clones on a physical map. There are two related ideas behind all approaches to deducing STS order. First, the distance between STSs i and i' is inversely related to the number of clones that contain both i and i'. That is, if i and i' appear together on "many" clones, then one deduces that they are "close" to each other on the DNA string. Second, if STS i appears in both clones j and j', then those clones must overlap at the site where STS i is located.

16.5.1. Reconstruction of STS order

The reconstruction problem is greatly simplified by the assumption that the data are error free. That means that the data correctly reflect the assumption that each STS occurs at only

---

[5] See the glossary for a short introduction to PCR.

STS

		1	2	3	4
	1	1	1	0	0
CLONE	2	0	1	1	1
	3	0	1	1	0

Figure 16.3: When the STS-content data are error free, then the columns of the matrix can be permuted so that in every row all the 1 s appear in consecutive order. The above matrix is the matrix from the previous figure, with the column permuted to exhibit the consecutive ones property in each row.

one site in the DNA, and that if two clones overlap at that site then the PCR procedure will report that both clones contain that STS. Under these assumptions, the problem of reconstructing the STS order is an application of the classic *consecutive ones problem* [74]. The input to the consecutive ones problem is a binary matrix, and the goal is to permute the columns of the matrix so that, in each row, all ones become located in a block of consecutive entries (see Figure 16.3). By the assumption that the data are error free, there is such a permutation corresponding to the true, but unknown, order of the STSs on the DNA.

A beautiful linear-time algorithm due to Booth and Lueker [74] finds a permutation creating consecutive ones in each row or declares that no such permutation exits. In case more than one permutation creates consecutive ones in each row, the method of [74] characterizes them all in a compact yet revealing way. Clearly, any consecutive ones ordering specifies a permutation that establishes an ordering of the STSs consistent with the data. That STS order fixes the location of each clone relative to the ordering. If the lengths of the clones are known, those lengths can be used to determine a physical layout of the clones and a physical map of the clone locations. Of course, the precise physical layout of the clones might not be determined from the data, even if the order of the STSs has been correctly deduced.

In summary, the STS reconstruction problem could be efficiently solved by the consecutive ones approach *if* the data were error free. Unfortunately, such is rarely the case. We will next consider several types of error that arise in STS-content mapping.

Errors in STS-content data

There are three common systematic errors (in addition to errors caused by contamination): *false positives, false negatives*, and errors due to *chimeric clones*. A false positive is the report that a clone contains a particular STS when in fact it does not, and a false negative is the report that a clone does not contain an STS when it actually does. Chimeric clones are a more complex type of error. In the process of obtaining and maintaining clones, it sometimes happens that two different DNA fragments join and then behave as a single clone. This new "single" clone is called a *chimeric* clone, and its two fragments can come from widely differing locations in the DNA. Chimeric clones clearly complicate the process of determining a correct STS order from PCR data. In some clone libraries, more than 50% of the clones are chimeric, and the tendency to form *chimers* seems to increase as the size of the fragments increases. It is a particularly serious problem for fragments inserted into *yeast artificial chromosomes* (YACs), which typically take inserts of around 500,000 bases and megaYAC clones with inserts of up to 2,000,000 bases. Chimers consisting of more than two fragments can also occur, but this seem to happen much less frequently than chimers of two fragments.

There are many proposed approaches to handling errors in STS-content mapping, but in this glimpse of mapping we cannot attempt a comprehensive discussion of these approaches. Instead, we will discuss one combinatorial approach based on a "traveling salesman" formulation. That formulation is a generalization of the consecutive ones problem, and the resulting algorithm is also used to reconstruct the STS order in a different mapping method called *radiation- hybrid mapping*. We will see that the problem of handling errors in STS-content mapping can be viewed as a special case of the problem of ordering STSs in radiation-hybrid mapping. We will therefore postpone the discussion of how to handle errors in STS-content mapping until after the following discussion of radiation-hybrid mapping.

16.6. Physical mapping: radiation-hybrid mapping

A mapping technique called *radiation-hybrid mapping* [113, 317, 408] has become popular as a means to map large clones that may comprise 10 to 25% of a single human chromosome (hence each clone can be tens of millions of bases long). The method is actually a hybrid between genetic and physical mapping and is being used in several genome efforts [234, 399]. As we will see, certain computational problems involved with radiation-hybrid mapping have been modeled as a traveling salesman problem.

Radiation-hybrid mapping exploits the ability of human DNA to become incorporated into rodent cells. To simplify the discussion, we will concentrate on its application to a single human chromosome, although it can also be used at the whole genome level. The outline of the method is the following: A single chromosome of human DNA is isolated and irradiated, breaking the chromosome at random locations into a small number of fragments. These fragments (from one copy of a single chromosome) are then merged into a hamster cell, which replicates the human DNA along with hamster DNA. In successive generations, each hamster cell replicates less of the human DNA, so that eventually a cell will contain only a fraction of the single human chromosome. The result is that, in the last generation, any single hamster cell contains roughly five to ten disconnected nonoverlapping fragments from a single human chromosome, and the total length of those fragments is about 15 to 20% of the human chromosome (see Figure 16.4). This process of obtaining a set of fragments that partially covers a single human chromosome is repeated with different copies of the same chromosome; it is sufficiently random so that two hamster cells containing fragments from copies of the same chromosome will contain a different set of fragments. The two sets generally will cover different but possibly overlapping portions of the chromosome (Figure 16.4).

At this point one can determine, for each cell, which of the set of unordered STSs are located in the human fragments contained in the cell. As in the case of STS-content mapping, this determination can be done by hybridization or by PCR: For each STS/cell pair, just test if the primers for that STS allow the generation of large amounts of new

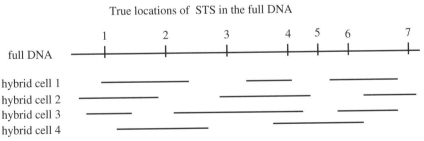

Figure 16.4: A schematic of human DNA fragments that are contained in four hybrid hamster cells. The first line represents a full human DNA chromosome and shows the true (but unknown) locations of seven STSs. The four succeeding rows show the true (but unknown) locations of the fragments. For simplicity, fewer than five fragments are shown per cell.

		STS						
		2	4	5	3	1	7	6
	1	1	1	0	0	1	0	1
Fragment	2	0	1	0	1	1	1	0
	3	0	1	0	1	1	0	1
	4	1	1	1	0	0	0	1

Figure 16.5: Example of radiation-hybrid data from the previous figure. A one in a column labeled j of row i indicates that STS j occurs in a fragment located in hybrid cell i. Because the true order of the STSs is unknown and must be deduced, the column labels are shown in permuted order.

DNA from the DNA in the hybrid cell. Although this resembles STS-content mapping, there is one big difference: The test determines whether the STS is contained in one of the human fragments in the cell, but it can't tell *which* fragment it is in. In fact, even the number of fragments is unknown and must be deduced. The data obtained by these tests are again summarized in a binary matrix. Each row corresponds to a hybrid cell, and each column corresponds to one STS (see Figure 16.5). A column labeled j has a one in row i if the fragments in hybrid cell i contain STS j.

The problem now is to use the binary data to deduce the correct order of the STSs. The intuition behind all approaches to deducing the order is that the frequency with which two STSs are present together or are absent together in a hybrid cell depends on the physical distance between the two STSs on the chromosome. That follows because the closer they are, the more likely it is that both are contained in a fragment of the cell (in particular in the same fragment) whenever either one is contained in a fragment. The experimental protocol essentially takes many random samples of the chromosome, where each sample reports the STSs found in a random set of nonoverlapping intervals covering only a portion of the chromosome.

Again, the true experimental protocol is more complex than described above and is additionally complicated by errors. However, the combinatorial elements are clear and we will now describe one approach to reconstructing the STS order from hybridization data.

16.6.1. Reconstruction of STS order in radiation hybrids

The STS ordering problem from error-free radiation-hybrid data would be the same as the reconstruction problem from error-free STS-content data, *if* each hybrid cell contained only a single human fragment. But since there is more than one fragment per hybrid cell, one suggested way [13] to establish the correct STS order is to find a permutation of the columns that minimizes the total number of blocks of consecutive ones. Each block is then interpreted as a fragment in the cell where the data were obtained. Another proposal is to find a permutation to minimize the maximum number of blocks that appear in any single row. These are both generalizations of the consecutive ones problem. However, unlike the consecutive ones problem, no fast (worst-case) exact method is known for either of these computational problems. Still, we will see that the first problem can be cast as a traveling salesman problem on an undirected graph, and the second problem can be cast as a "bottleneck" version of the traveling salesman. We will discuss the first variant in some detail.

16.6.2. Traveling salesman formulation of STS ordering

Given a binary matrix of radiation-hybrid data, we create an undirected graph with one node for each STS (column of the matrix), one extra node labeled s, and one undirected edge between each pair of nodes. A traveling salesman tour through the graph is a path

	STS			
	2	4	3	1
1	1	1	0	1
Fragment 2	0	1	1	1
3	0	1	1	1
4	1	1	0	0

Figure 16.6: An extract of radiation-hybrid data from the previous example. The undirected graph for these data is shown in Figure 16.7.

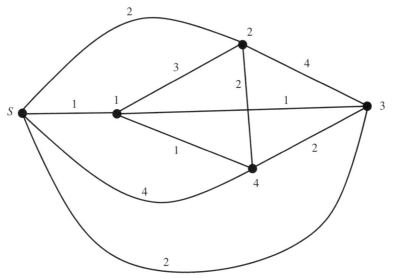

Figure 16.7: The graph used in the traveling salesman approach to the STS ordering problem for the data in the previous figure. Unfortunately, in this example the optimal tour is s, 3, 1, 4, 2, s, which minimizes the number of blocks created (four blocks) but does not correctly reconstruct the original ordering of the STSs.

that starts and ends at node s and that visits each other node exactly once. Other than node s, a traveling salesman tour visits each node of the graph exactly once, so it establishes a permutation (ordering) of the matrix columns. Hence, each traveling salesman tour establishes an STS ordering.

We next assign a weight to each edge in the graph. The weight of edge (s, v), for any node v, is the total number of 1 entries in column v. The weight of any other edge (u, v) is the *Hamming distance* between the columns for STS u and STS v. That is, the weight of edge (u, v) is the number of rows where the entries in the column for u and the column for v differ. The *traveling salesman problem* is to find a traveling salesman tour in the graph whose total edge weight is as small as possible. See Figures 16.6 and 16.7.

What is the intuition for this weighting? Suppose first that column u has a 1 in row i, that column v has a 0 in row i, and that edge (u, v) is traversed in a traveling salesman tour. The column ordering given by the tour creates a block of 1s in row i that ends with column u. By the matrix entries, STS u appears in hybrid cell i but STS v does not. Therefore, the STS ordering corresponding to the column ordering implies that a fragment in hybrid cell i must end after STS u but before STS v. The weight on edge (u, v) thus counts the number of blocks of 1s that will end with column u (if edge (u, v) is traversed), and hence it also counts the number of fragments that end after STS u in the implied fragment reconstruction. Now suppose the column for u has a 0 while the column for v has a 1 in row i. Then a traversal of edge (u, v) implies that a fragment in hybrid cell i *begins* with STS v and also that a new block of 1s in row i begins with the column for v.

If a tour traverses an edge from start node s to any other node v, then the edge weight on (s, v) counts the number of blocks of 1s that start in the left-most column of the permuted matrix. Similarly, if a tour traverses an edge from v to s, the edge weight on (s, v) counts the number of blocks that end at the right-most column of the permuted matrix. Thus, with the given edge weighting, each block (or implied fragment) created by the tour is counted exactly twice. In summary, we have

Theorem 16.6.1. *A traveling salesman tour that has total edge weight of w specifies a permutation of columns that creates exactly $w/2$ consecutive blocks of ones. Hence, the traveling salesman tour of minimum total weight produces a permutation of the columns that minimizes the number of blocks of consecutive ones. That permutation produces an ordering of the STSs that minimizes the total number of fragments deduced to be in the radiation-hybrid cells.*

Therefore, *if* the proper STS ordering is actually the one that minimizes the total number of blocks of consecutive ones (as has been proposed), then that ordering can be found by solving an instance of a traveling salesman problem. Unfortunately, the optimal solution to the salesman problem does not always reconstruct the original ordering, as illustrated in Figure 16.7.

The traveling salesman problem is NP-hard [171], but some very impressive efforts have shown that it can be solved exactly in reasonable time for graphs with hundreds of nodes. Moreover, in those methods, an exact solution is found rather quickly, and most of the time is used to prove that the exact solution has been obtained. If no proof is required, then the methods are particularly fast. At present, the number of STSs used in radiation-hybrid mapping (for a single chromosome) is small enough that the associated traveling salesman problems can be solved in this way. Alternatively, very effective heuristic methods are known. These methods, based on local search or techniques such as simulated annealing, can very quickly obtain a near-optimal solution, although without any guarantee on the closeness of the solution [13]. To obtain a guarantee, one can use a polynomial-time approximation method that finds a permutation where the number of resulting blocks of consecutive ones is never greater than 50% more than in the best permutation. That is, if the best permutation produces q blocks of ones, then the permutation given by the approximation algorithm produces at most $1.5q$ blocks. This follows from the method due to Christofides [102, 250], which finds a traveling salesman tour guaranteed to be at most 1.5 times the shortest traveling salesman tour when the edge distances satisfy the triangle inequality. It is easy to establish that Hamming distance satisfies the triangle inequality, and so the Christofides method applies. As discussed earlier (see Section 14.9), guaranteed approximation bounds are generally overly pessimistic, and the true performance of the methods in practice is much better than the guarantee.

Generally, the traveling salesman model of the STS-ordering problem is not expected to establish the STS order entirely on its own. Rather, it establishes a first-pass ordering that can be validated or corrected by other data. For example, a proposed STS order may be used in conjunction with some STS-content information from single fragments, with various types of *fingerprinting information* (discussed below), with an existing genetic map, or with other physical maps of the same DNA. These additional data can either be incorporated into the traveling salesman model as side constraints or used in an ad hoc manner to modify the STS order given by the salesman tour.

16.6.3. Back to STS-content mapping: the case of errors

We deferred the discussion of how to handle errors in STS-content mapping until after the discussion of radiation-hybrid mapping. It should now be clear why this was done: The problem of STS-content mapping in the presence of errors (false positives, false negatives, and chimers) has the same combinatorial structure as the problem of STS ordering from radiation-hybrid data without errors. Accordingly, the two problems can be handled in very similar ways.

STS-content data from chimeric clones have all the features of data from radiation-hybrid mapping, except that the number of fragments (clones) per cell is smaller. In radiation-hybrid mapping the number of fragments per hybrid cell is often between five and ten, whereas a chimeric clone rarely consists of more than two fragments. Nonetheless, chimeric STS-content data can be used to construct an undirected weighted graph, exactly as was done for radiation-hybrid data. A salesman tour, visiting every node in the graph exactly once, again determines an ordering of the STSs and induces blocks of consecutive ones in each row of the data matrix. Since the set of clones contains chimers, we expect that there will be some rows with more than one block of consecutive ones. The question now becomes: What objective function do we use in selecting a salesman tour?

In the case of chimeric clones, no more than two blocks of consecutive ones per row are expected, so it may no longer be appropriate to look for the salesman tour with the minimum *total* cost. The tour with minimum total cost induces a column ordering of the matrix that minimizes the total number of blocks of consecutive ones, but it does not provide a constraint on the number of blocks per row. A better approach would be to minimize the total tour cost *subject to* the added constraint that the induced permuted matrix never contains more than two or maybe three blocks of ones per row. An alternative approach is to find an ordering (a permutation of columns) that minimizes the maximum number of resulting blocks of consecutive ones in any row. This problem is also NP-hard, but an approximation method, along with software based on that approach, has been developed by Istrail et al. [191, 242].

Consider now the case of false positive or false negative data without the complication of chimeric clones. Without any errors, the columns of the matrix of STS-content data can be permuted so that every row has the consecutive ones property. Relative to that correct ordering, the effect of a false positive test for a given STS/clone pair is either to extend the block for that clone by one column (on either side) or to create a separate block consisting of a single entry. If false positives occur with equal probability for any STS/clone pair, then the probability that a false positive will create another block is much greater than the probability that it will extend a block. Similarly, the effect of a false negative is to shrink a block by one column or to break up a block into two blocks separated by a single zero. Again, without systematic bias, the probability that a false negative will break a block is greater than the probability that it will shrink it. The result is that we can focus on those false positive and negative errors that create additional blocks rather than changing the size of existing blocks. With this view, the problem of false positives and negatives can again be modeled as the problem of finding an ordering of the columns that either minimizes the total number of resulting blocks or that minimizes the maximum number of blocks in any row. Which of the two is the more appropriate objective function depends on the frequency of errors. We leave it to the reader to consider problems where chimeric clones occur together with false positive and negative tests.

16.7. Physical mapping: fingerprinting for general map construction

STS-content mapping and radiation-hybrid mapping using STSs exploit the convenient property that each STS occurs at a unique location in the DNA. This simplifies the problem of ordering the STSs and the problem of laying out a map showing the positions of the clones or fragments. As difficult as those problems are, the uniqueness of STSs clearly is an advantage. There are, however, other physical mapping methods where each probe does not have a unique location in the DNA. (See Figure 16.8) Nonunique probes are often needed to obtain a finer-scale map than an STS map. The reconstruction problem with nonunique probes is usually based on *fingerprints*.

To define fingerprints, consider a clone-library and a set of probes. A typical example of a probe in this setting is either an oligonucleotide (a short string of DNA) that hybridizes to clones containing a substring complementary to the oligo or a restriction-enzyme that cuts either at every occurrence of a specific short substring in a clone or at every occurrence of a class of short substrings. A less informative form of a "probe" gives only the length of a DNA substring between two restriction-enzyme cutting sites. Whatever the specifics of the probe, the set of probes that test positive for a specific clone constitutes the *fingerprint* of the clone. So abstractly, fingerprint data are again represented by a binary matrix where each row is a clone and each column is a probe. This resembles STS-content mapping, but a probe is now expected to occur at more than a single location in the DNA. To further complicate the problem, it is not always possible to determine the number of places in the full DNA or in a clone where a given probe tests positive.

The problem is to lay out, or assemble, the overlapping pattern of clones by examining commonalities in the clone fingerprints. Since a probe does not in general have a unique location in the DNA, approaches based on the consecutive ones problem are no longer appropriate. Instead, most approaches use the idea that the more two clones overlap, the more their fingerprints are expected to share or to exclude the same probes. Many of the suggested assembly methods first examine the clones pairwise, using the fingerprints of the two clones to estimate the length over which those clones overlap. This estimate involves the number of shared probes, the number of coexcluded probes, and the known or estimated frequency of the particular probes. After these pairwise estimates are made, clones are assembled in a fairly greedy manner; the two clones with greatest estimated overlap are put together, with copies of their shared probes located in the overlapping

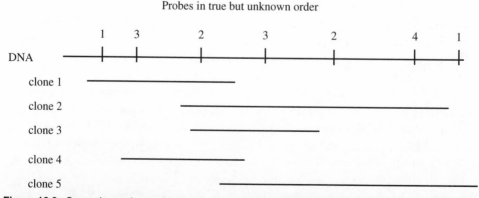

Figure 16.8: General mapping problem using fingerprints from nonunique probes. The probe/clone data from this example would be represented in a five by four binary matrix. In this example, it would be very difficult to correctly reconstruct the clone layout from that binary matrix.

region of the two clones. This merged clone (called a *contig*) is then added to the set of clones and the greedy step is repeated, although the problem is now constrained because of the placement of probe copies in the contig. A full discussion of the method and its many variants is not possible here. The key paper setting out the analytical and statistical issues in fingerprinting (estimating the extent of clone overlap from shared fingerprints and determining the number of clones needed etc.) is by Lander and Waterman [292]. A textbook discussion of these issues appears in [461]. However, the above sketch is sufficient to motivate an assembly subproblem, discussed in the next section, that has a particularly clean combinatorial solution.

16.8. Computing the tightest layout

This section explores a clone layout subproblem that is solved in an approach to physical mapping developed by R. Karp et al. [12]. The particular subproblem arises as a very refined piece of the larger mapping problem but is interesting in its own right.

Definition   Let $\mathcal{P}$ be a set of probes (or names) and C be a set of n clones. For each clone $c \in C$, let P_c be the set of probes known to occur in clone c.

In this problem, a probe may occur at more than one location, and the number of times a probe occurs in the "true" layout is not known.

Definition   Given the input data P_c for each $c \in C$, a *feasible layout* is a mapping of clones (as intervals) and of probe occurrences (as points) onto the real line, such that the interval for each clone $c \in C$ spans at least one copy of each probe in P_c and spans no copies of probes outside of P_c.

Because there is no restriction on the number of locations that any probe might appear at, a feasible layout is always possible – simply lay out each clone c totally isolated from the others and then place one copy of each required probe in that clone interval. A more useful task is to find a "tightest" feasible layout (one that occupies the least span on the real line and/or that requires the fewest probe copies), from which all other feasible layouts can be derived.

This problem is addressed in [12], but it is restricted in two ways. First, each clone is assumed to be of equal length. That assumption approximately models today's mapping protocols where clone fragments are "size selected" to be fairly uniform. Being of equal length, if the left end of one clone appears to the left of the left end of a second clone in some layout, then the right end of the first clone also appears to the left of the second right end. Therefore, the left-to-right order of the clones in any fixed layout is fully specified by a *permutation* of the clone names, although an infinite number of layouts give rise to the same permutation. The second restriction made in [12] is that a predetermined permutation of the clone names is given as input to the problem.

In summary, the problem solved in [12] is to find the tightest feasible layout (of the clones and probes), given the probe/clone data and a permutation of the clone names as input and assuming that the clone intervals are of equal length. Two overlapping clones in a feasible layout may be placed arbitrarily close to each other on the line, so long as some nonzero distance separates their left ends. In all that follows, we assume that a fixed permutation has been given as input. Assume (by renaming if necessary) that in the fixed permutation, the clones are numbered left to right by the increasing integers $1, 2, \ldots, n$.

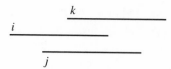

Figure 16.9: If clones i and k overlap, then clone j (for $i < j < k$) must be completely contained in their span. The reason is that the left end of clone j is to the right of the left end of clone i, and the right end of clone j is to the left of the right end of clone k. Hence if clones i and k overlap, then no probe $p \in P_j$ can be placed to hit the interval for j and yet avoid the intervals for both clones i and k.

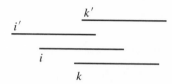

Figure 16.10: If clones i' and k' overlap, then clones i and k must also, for $i' \le i < k \le k'$.

Finding the tightest layout

The key is a simple condition that excludes the overlap of two clones i and k in any feasible layout. Let $i < j < k$ be three clones numbered according to the input permutation. If the probe/clone data specify that some probe p occurs in clone j but in neither clone i nor k, then clones i and k cannot overlap in any feasible layout (see Figure 16.9). Moreover, if i, j, k are as assumed, and $i' \le i < k \le k'$, then clone i' can not overlap clone k' in any feasible layout. To see this, note that since all clones have the same length and the order of left ends is fixed, an overlap of clones i' and k' would force an overlap of i and k as well (see Figure 16.10).

> **Definition**   A pair of clones i', k' is called an *excluded pair* if $i' \le i < j < k \le k'$ for some clones i, j, k, where P_j contains a probe p that is contained in neither P_i nor P_k. A pair is called *permitted* if it is not excluded.

Note that the definitions of excluded and permitted pair are relative to a fixed permutation of the clones. A pair can be excluded in one permutation but permitted in others. The next lemma follows immediately from the definitions and will be used later.

Lemma 16.8.1. *If i', k is an excluded pair and $i < i' < k$, then i, k is also an excluded pair. Similarly, if i, k is a permitted pair and $i < i' < k$, then i, i' and i', k are also permitted pairs.*

Being "permitted" is a necessary condition for a pair to overlap in some feasible layout, but we will prove something much stronger. Guided by Lemma 16.8.1, we will construct a feasible layout where *every* permitted pair i, k overlaps (and of course no excluded pair does). Moreover, that layout can be made to have the smallest span of any feasible layout and to use the minimum number of probe copies. Hence it is the "tightest" feasible layout.

The greedy clone layout

Place clone 1 at some left starting point of the layout.
For k from 2 to n do
begin
Find the smallest index $i < k$ such that i, k is a permitted pair. (Note that i exists

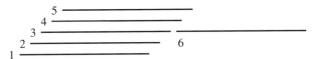

Figure 16.11: Each clone 2 through 5 forms a permitted pair with clone 1. Clone 6 forms an excluded pair with clones 1 through 3, but pair (4,6) is permitted.

because $(k-1), k$ is always a permitted pair.) Place clone k so that it overlaps with clone i, it does not overlap with clone $i-1$, and its left end is to the right of the left end of clone $k-1$ (see Figure 16.11).

end

Theorem 16.8.1. *In the greedy clone layout, two clones overlap if and only if they are a permitted pair.*

PROOF   The proof is by induction on k. Consider the algorithm up until the start of iteration k. Assume for every $\bar{k} < k$, and every $j < \bar{k}$, that clone $\bar{k}$ overlaps the right end of clone j if and only if $j, \bar{k}$ is a permitted pair. In iteration k, if i is the smallest index such that i, k is a permitted pair, then $(i-1), k$ is an excluded pair. Then by Lemma 16.8.1, pair i', k is also excluded for every $i' \leq i-1$. By construction, clone k is placed to avoid overlap with clone $i-1$, and hence it also avoids overlap with any clone less than $i-1$. Therefore, the induction step is established for every pair i', k with $i' \leq i-1$.

Now consider each index i' between i and $k-1$. By Lemma 16.8.1, pair i, i' is also permitted for every index i' between $i+1$ and $k-1$. So (by the induction hypothesis), every clone i' between $i+1$ and $k-1$ has been placed to overlap the right end of clone i. By construction, since clone k is placed to overlap the right end of clone i, k will also overlap every clone i' between i and $k-1$, and so the induction step is proven for all pairs of clones through index k.   □

Below we will show (by placing the probes) that the greedy clone layout can be made a feasible layout, but we first show that the greedy clone layout can be made to have smaller span than any other feasible layout. In the greedy clone layout, let i_2 be the smallest index greater than 1 such that $(1, i_2)$ is an excluded pair. By construction, the clones strictly between 1 and i_2 all overlap clone 1, and the successive displacements of their left ends can be made arbitrarily small. Therefore, the gap between the right end of clone 1 and the left end of clone i_2 can be made arbitrarily small. Repeating this analysis, let $1 < i_2 < i_3 < \cdots < i_r$ be a sequence of indices where, for every $l < r$, i_{l+1} is the smallest index greater than i_l such that i_l, i_{l+1} is an excluded pair. Then in the greedy clone layout, the gap between the right end of clone i_l and the left end of clone i_{l+1} can be made arbitrarily small. Therefore, the span of the greedy clone layout can be made arbitrarily close to the length of r clones. Now each successive pair of clones in this sequence is excluded; consequently, by Lemma 16.8.1, every pair in the sequence is excluded and *no* feasible layout can have length less than r clones. In summary, we have

Theorem 16.8.2. *The greedy clone layout can be made to have a span on the real line that is less than the span of any other feasible layout.*

Given the greedy clone layout, we now place probe copies to make the greedy clone layout a feasible layout. This layout will use the smallest number of probe copies of any feasible layout. We assume an artificial clone $n+1$ that contains no probes. Each probe p will be placed separately with the following algorithm:

Greedy probe placement for probe p

Let i^* be the smallest index such that p is in P_{i^*}.

repeat

Set j to the smallest index greater than i^* such that either (i^*, j) is an excluded pair or $p \notin P_j$. (The clones from i^* to $j - 1$ are called a *run*.)

Place a copy of probe p inside the interval for clone i^* as far right as possible such that the interval for clone j does not span that copy of p.

If $p \in P_k$ for some $k \geq j$, then set i^* to the smallest such index k.

until there is no index $k \geq j$ with $p \in P_k$.

Theorem 16.8.3. *The greedy probe placement makes the greedy clone layout feasible. Moreover, this placement uses as few copies of probes as any feasible layout.*

PROOF   For any fixed probe p, consider an index that is assigned to i^*, and let $j > i^*$ be as defined in the algorithm. In the case that (i^*, j) is an excluded pair, clones i^* and j have no overlap, so the copy of p is placed at the extreme right end of clone i^*. Therefore, that copy of p is only spanned by i^* and the clones that overlap the right end of i^*. Since all clones have the same length, those clones are precisely the ones in the run defined by i^*. But by the choice of j, probe p is in P_l for each clone l in that run.

In the case that (i^*, j) is a permitted pair (and so $p \notin P_j$), clone j will overlap clone i^*, but there could also be a clone $l < i^*$ that overlaps the left end of clone i^*. Clone l does not constrain the placement of p in the interval for i^* unless p is not in P_l. But if $p \notin P_l$, then pair l, j is excluded (due to clone i^*); thus clones l and j do not overlap in the greedy clone layout. Therefore, by placing the copy of p as far right in i^* as possible, but to the left of j, only the clones in the run defined by i^* span the placed copy of p.

By the probe placement algorithm, if p is in P_l for some clone l, then l is in one of the runs defined by the algorithm. Hence clone l spans at least one copy of probe p.

Next we consider the number of copies of probe p used. Let I^* be the set of indices that variable i^* is assigned during the probe placement algorithm. Note that the number of copies of probe p placed by the algorithm equals $|I^*|$. But in any feasible layout, no copy of p can be spanned by more than one clone in I^*. To see this, let i and i' be two consecutive indices in I^*. By construction, either i, i' is an excluded pair or there is a index j between i and i' such that $p \notin P_j$. In the first case, clones i and i' can't overlap in any feasible layout. In the second case, they may overlap but that overlapping region must be spanned by the interval for clone j, and hence no copy of probe p can be placed in that region. Hence any feasible layout must use at least $|I^*|$ copies of probe p, and the theorem is proved. $\square$

The set of all feasible layouts

With the kind of reasoning used in the previous section, the following theorem is easy to establish.

Theorem 16.8.4. *Assume a fixed permutation of the clones. Any clone layout whose left-to-right order corresponds to the permutation can be made into a feasible layout (by placing probes) if and only if no excluded pair of clones overlap in the clone layout.*

Theorem 16.8.4 shows that the tightest feasible layout has one additional desirable feature: It can be used to derive every other feasible clone layout. Starting from the tightest layout, just pull clones apart in any way desired, without creating new overlaps or violating the established left-to-right clone order. At a point where two overlapping clones cease to overlap, check if a new probe copy is required, and if so create one. It is easy to envision an interactive graphics system allowing the user to create feasible layouts in this way. Note that this does not create all feasible layouts, because for any feasible clone layout, there will be choices for how to place the probes.

Definition   The *interleaving* of a clone layout is a list of the clone pairs that overlap in the layout. An interleaving is called feasible if it is obtained from a feasible clone layout.

The set of all feasible interleavings can be represented in a succinct graphical manner, which is presented in [12]. The set of all feasible interleavings provides another representation of the set of all feasible clone layouts.

16.9. Physical mapping: last comments

Pooling

There is one further complication that we should mention. In STS-content mapping, one generally does not perform every STS/clone test. That is too expensive. Rather, clones are grouped into *pools*, and then each STS is individually tested against each pool of clones. A single test will show whether the STS occurs in at least one of the clones in a given pool but will not determine which specific clones it is in. In this way, STS-content mapping with pools of clones looks very much like radiation-hybrid mapping, except that the pools are constructed with a deliberate design whereas the "pools" in radiation-hybrid mapping are created by a random process.

Pooling strategies are either *static*, in which case the pools are fixed and tested once, or they are *adaptive* in that the results from a first round of pooled tests determines which pools should next be tried. However, the cost of creating pools is large, so that no more than two rounds are usually done, and even then the pools used are usually taken from a fixed set of choices. Clearly, pooling leads to problems in combinatorial design and statistical inference that go beyond the scope of this book. A brief discussion of pooling appears in [111].

Physical mapping provides an assay

Physical mapping techniques based on producing a layout of fragments have another important property that genetic maps lack: The physical map is accompanied by an ordered clone-library. The physical map is not just a symbolic representation of physical positions. The mapped fragments in the ordered clone-library can be used to quickly determine the rough location of a new piece of DNA obtained by some other experimental protocol. In a typical example, one might study a disease and find that a particular mRNA is produced in unafflicted people but not produced in people with the disease. That means that the disease is likely caused by a particular protein not being expressed. One wants to find the location of the gene for that protein to see precisely what differs between afflicted and unafflicted individuals. With a physical map, the mRNA can be turned into cDNA (see Section 11.8.3), which is then used to hybridize with the fragments in the ordered clone-library. The results of that experiment, combined with the physical map of the clone fragments, determine a rough location of the gene producing the protein of interest.

An excellent example of the use of a physical map, and an additional illustration of the interplay between (specialized) database searching and laboratory work, was the identification of a gene for early-onset Alzheimer's disease [47, 300]. A gene responsible for about 80% of the cases of early-onset Alzheimer's disease (the variant of the disease that affects relatively young people) had been found on chromosome 14 and cloned in early 1995. But the additional 20% remained unexplained. Shortly thereafter, genetic mapping (linkage) efforts identified the rough location on chromosome 1 of a gene linked to the additional cases. However, because genetic mapping generally only locates a gene to within one million bases, more physical methods needed to be employed to precisely locate the gene.

The researchers proceeded in the following way. They reasoned (or guessed) that the gene on chromosome 1 would be similar in sequence (locally at least) to the Alzheimer's gene on chromosome 14 already cloned and sequenced. They had available the sequences of a library of ESTs (see Section 3.5.1), and they matched the library sequences against the sequence of the Alzheimer's gene from chromosome 14. One EST matched well. That EST was 475 base pairs long when translated to the amino acid alphabet, and it aligned inside the chromosome 14 gene with an 80% match rate. At that point, they had an EST that might be part of the gene they sought, but they didn't know what part of the genome the EST was from (methods used to find ESTs do not usually determine their location). But using the selected EST as a probe in a map of YAC clones, they determined that the EST was located on a section of chromosome 1 previously deduced (by genetic linkage analysis) to contain the missing gene. Then, using PCR primers made from the EST (and a cDNA library), they were able to isolate a single fragment of length about two thousand base pairs that contained the missing gene. This fragment was then sequenced and (locally) aligned to the chromosome 14 gene (with 67% identity at the amino acid level), identifying a single putative gene on the fragment. Using PCR again, they then repeatedly extracted the interval of the putative gene from members of a family who were partially affected by early onset Alzheimer's disease. Comparing the sequences for this putative gene from the affected members with the sequences of the unaffected members, they found a consistent difference between the affected and the unaffected members, confirming that the gene had been found. Then, as usual, additional database searches were conducted to find proteins that have similar sequences to the deduced protein product of the newly found gene. The related proteins found from this search may help researchers understand the role of the deduced protein in causing Alzheimer's disease.

More combinatorics

Our discussion of physical mapping emphasized combinatorial problems in mapping and combinatorial, graph-theoretic approaches to solving those problems. This is in contrast to the large literature that views mapping problems in a statistical framework. The combinatorial view of mapping has been even more fully developed than discussed in this chapter, particularly by Ron Shamir and coauthors [181, 260] and by Richard Karp and coauthors [12, 13, 247, 264]. The graph-theoretic approach to current mapping problems was initiated with the paper by Waterman and Griggs [464] where certain mapping problems were cast as problems on interval graphs. Earlier work on the use of interval graphs in molecular biology (for early physical mapping methods) is contained in the book *Graphs and Genes* [327]. A special issue of *The Journal of Computational Biology* [367] focuses on mapping and sequencing, and many of the papers take a combinatorial approach.

16.10. An introduction to map alignment

In Section 7.14.6, we discussed the issue of searching in strings generated from a large alphabet and noted that *restriction-enzyme maps* can be treated as such strings. A restriction-enzyme map is really a string where every odd character is an integer (representing a distance) and every even character represents a particular type of restriction-enzyme. More generally, any map (physical or genetic) can be considered as a string where the odd characters are integers and the even characters are from some finite alphabet representing the mapped features.

We saw in Section 7.14 that, for exact matching, suffix arrays can be used to efficiently find all occurrences of a smaller map inside a larger map. This is a problem of real importance. But, maps typically contain errors, and we need some ways to solve *approximate matching* on maps and ways to define and compute the edit distance between two maps.

At one level, this problem is not new since maps can be considered as strings, and it is easy to choose sensible penalties for mismatches between integers. But there is greater subtlety in handling deletions. For example, the map x 125 y matches well in the interior of z 100 x 126 y and less well in the interior of z 100 x 60 y or z 100 x 66 y. That is, the mismatch of 125 with 126 carries a small penalty, but the mismatch of 125 with 60 or 66 carries a large penalty. It is easy to modify approximate matching methods to reflect those cases. But x 125 y might also be seen to match well in the interior of z 100 x 60 p 66 y, even though the penalty for mismatching 125 with 60 or with 66 is large. After all, if the character p were deleted, then the first string would align to x 126 y with only a small penalty. How should such deletions be modeled in a dynamic programming approach to map alignment?

One immediate approach is to rewrite the strings (maps) by converting each integer to *unary* notation. That is, replace a number d by d 1s. Then x 125 y becomes a string of 127 characters that occurs in z 10 x 60 p 66 y (a string of length 230) after just two edit operations: the deletion of character p from the long string and the deletion of a single 1 from 60 or 66. With such rewriting, the problem of finding approximate occurrences of a small map in a large one becomes exactly the problem of approximate matching in strings, and the edit distance between two maps becomes exactly the edit distance between the two rewritten strings. This idea of unary rewriting seems to solve the problem, but it has the disadvantage that it creates long strings. And because the dynamic programming computation of edit distance takes time proportional to the product of the string lengths, this approach is not efficient for large maps.

Alternative, efficient solutions to these map alignment or edit distance problems have been developed. The most efficient (worst-case) method runs in time $O(nm \log mn)$ [338], where n and m are the respective number of characters in the maps.

16.10.1. A nonunary dynamic programming approach to map alignment

In this section we will develop an earlier, less complex dynamic programming approach due to Waterman, Smith, and Katcher [465].

Definition   For the two maps in an alignment problem, let one be represented as a sequence of pairs (s_i, d_i), where s_i is a character representing the feature of the ith map site, and d_i denotes the distance of the site *from the left end* of the map. Let the second map be represented by a similar sequence of pairs (s'_j, d'_j).

Definition   The primitive operations that can be used to transform one map to the other are the insertion or deletion of feature sites or the change in the distance between sites in either map (i.e., changing the d_i or d'_j numbers).

Note that the substitution of one feature site to another is not allowed. At first that may seem to be a major restriction, but it is not unreasonable. In obtaining genetic or physical maps, a site may be missed or incorrectly inferred or displaced from its correct location, but it is very unlikely that a site of one type would be confused with a site of another type. In fact, many maps are constructed by finding the locations of each feature type separately and then combining the different maps. Each separate map might omit, misplace, or incorrectly deduce a site, but it can't deduce a site of a different type. Furthermore, the distances between sites are generally quite large, so it is unlikely that the resulting combined map would exchange a feature of one type for another.

The objective is to convert one map into the other with the lowest cost, where the cost of inserting or deleting a feature site is given by a parameter λ, and the cost of changing a distance by an amount t is βt.

A dynamic programming solution

Suppose the two maps have n and m sites, respectively, where $n > m$. First add two dummy sites numbered 0 to the left of each map, and add two dummy sites numbered $n + 1$ and $m + 1$, respectively, to the right of each map. Moreover, assume that $s_0 = s'_0$ and $s_{n+1} = s'_{m+1}$.

> **Definition**   Let $MD(i, j)$ be the minimum cost to convert the first map through site i into the second map through site j, *with the requirement that site i in the first map is aligned with site j in the second map.*

The requirement that sites i and j align in the definition of $MD(i, j)$ makes the meaning of $MD(i, j)$ different than the meaning of $D(i, j)$ used in computing edit distance. The objective is to compute $MD(n + 1, m + 1)$. We use the following recurrences:

$$MD(0, 0) = 0,$$

$$MD(0, j) = MD(n + 1, j) = MD(i, 0) = MD(i, m + 1) = \infty,$$

if $0 < j < m + 1$ or $0 < i < n + 1$. This imposes the constraint that the two 0 sites align, and that sites $n + 1$ and $m + 1$ align. When $s_i \neq s'_j$,

$$MD(i, j) = \infty,$$

due to the assumption that feature type s_i cannot be converted to feature type s'_j. When $s_i = s'_j$,

$$MD(i, j) = \min_{0 < k \leq i, 0 < l \leq j} [MD(i - k, j - l) + \lambda(l + k - 2) + \beta(|(d_i - d_{i-k}) - (d'_j - d'_{j-l})|)].$$

To better understand the last recurrence, recall that site i aligns with site j, so k indicates the number of sites to go back in the first map, and l indicates the number of sites to go back in the second map, in order to specify the last aligned pair of sites before i and j. Effectively, $k - 1$ sites before site i are removed in the first map, and $l - 1$ sites before site j are removed in the second map. By ranging over all values of k and l, the best alignment that aligns i and j will be found. And since pair 0, 0 has been included to handle the case where i and j are the first sites from the input maps to align, this recurrence will find the best alignment up through sites i and j.

The overall time to evaluate $D(n, m)$ is $O(n^2 m^2)$. Since the distance between sites is large (certainly larger than an average of n for a map with n sites), this time bound is better than that obtained by converting the strings into unary. As noted earlier, a faster, but more complex approach yields an $O(nm \log nm)$ time bound [338].

16.10.2. Extensions of the map alignment model

We argued above that one type of restriction-enzyme site is not easily confused for another type, and hence substitutions need not be handled by the alignment method. However, a more realistic model of restriction-enzyme maps would account for certain experimental errors not handled in the above model. One such experimental error is the reporting of only one enzyme site when there are actually two close sites for the same enzyme. Another experimental error is that of transposing the order of two close sites for two different enzymes. The paper by Miller, Ostell, and Rudd [323] tackles the transposition question under a somewhat different scoring scheme. Also, depending on how the map distances are experimentally obtained, it may be more realistic to let d_i be the distance between the sites i and $i - 1$ rather than the distance to site i from the left end.

16.11. Large-scale sequencing and sequence assembly

Some aspects of large-scale (genome-level) DNA sequencing have already been introduced in the book in different discussions of specific string or mapping problems. It is worthwhile trying to bring all these separate pieces together into a more comprehensive discussion of large-scale sequencing and its relationship to physical mapping. This is useful not only because of the importance of large-scale sequencing, but as a way to emphasize the role of various string algorithms.

The key problem today in sequencing a large string of DNA is that DNA is ultimately sequenced by a Maxam-Gilbert or Sanger-like method [111, 469] in which only a small amount of DNA can be sequenced in a single "read". That is, whether the sequencing is done by a fully automated machine or by a more manual method, the longest unbroken DNA substring that can be reliably determined in a single laboratory procedure is about 300 to 1000 bases long. For simplicity of exposition, we will use the figure of 400 bases as the standard read length, although people with "good hands" and newer machines claim to be able to reliably sequence much longer reads. A longer string can be used in the procedure (input to the machine or process), but only the initial 400 bases will be determined. Hence to sequence long strings or an entire genome, the DNA must be divided into many short strings that are individually sequenced and then used to assemble the sequence of the full string. The critical distinction between different large-scale sequencing methods is how the task of sequencing the full DNA is divided into manageable subtasks, so that the original sequence can be reassembled from sequences of length 400.

16.12. Directed sequencing

The most natural method, *directed sequencing*, was introduced in Exercises 61 and 63 of Chapter 7. We consider two variants of directed sequencing here: *primer walking* and *nested deletion*.

In primer walking (see also the discussion of chromosome walking on page 178), the left-most 400 bases of the large string are first sequenced. Knowing that substring allows one to produce PCR primers, which are used to make a copy of a section of the original string starting from near the right end of the 400 sequenced bases. At that point one has a substring of the original string starting from around base 400. The left-most 400 bases of that substring can then be sequenced, and the procedure can be iterated.

In nested deletion, one uses an enzyme (an exonuclease) to remove bases on the left end of the string. After the first 400 bases have been sequenced, the exonuclease removes the sequenced part (and hopefully only that part), so that the procedure can be iterated. Unfortunately, it seems that the use of the exonuclease to date is not as precise as desired, so that often too much or too little of the end is removed.

The main problem with directed sequencing is that it is inherently sequential and therefore slow. The sequential nature of the method can be overcome if several dispersed points are known in the DNA where directed sequencing can be initiated in parallel. A recent proposal (see Section 16.13.1) may make such direct parallel sequencing possible. The reader might also wonder about the obvious approach of cutting the long DNA string into smaller fragments that can each be directly sequenced. Unfortunately, the cutting process permutes the original order of the fragments, making such a pure directed strategy impossible. One approach to overcoming the permutation problem leads to shotgun sequencing, the main alternative to directed sequencing. We will discuss shotgun sequencing in Section 16.14.

A second critical problem for directed sequencing is that if for any reason the directed sequencing gets stuck at some point in the string, the string beyond that point cannot be sequenced unless a new starting point can be found. Directed sequencing often gets stuck or confused by repeated substrings, and DNA from higher organisms (mammals particularly) is highly repetitive (see Section 7.11.1). If the selected primer is in a repeated substring, it may hybridize with an unwanted part of the full string, in which case the PCR will copy the wrong part of the DNA. A complemented palindrome (see Section 7.11.1) causes problems for both nested deletion and primer walking because the two complemented segments base-pair with each other to form a *hairpin loop*, blocking the workings of the Maxam–Gilbert and Sanger sequencing methods. In the sequencing vernacular, directed sequencing cannot "read through" repetitive DNA. Hence directed sequencing is usually used for relatively small strings that contain no repeats. Directed sequencing is also used to close, or sequence across, small gaps left in a sequence found by shotgunning. We will return to that issue in Section 16.13.1.

16.13. Top-down, bottom-up sequencing: the picture using YACs

Because directed sequencing is not generally able to sequence large-scale, genome-sized DNA strings, more hierarchical or recursive approaches have been invented. Although the specific details can vary considerably, most large-scale sequencing efforts today follow a basic outline that starts with *physical mapping* of large clones and ends with *shotgun* sequencing of much smaller clones (typically *cosmids*).This is summarized by Maynard Olson.

> Following a period of competition between alternative sequencing strategies, a dominant technology has emerged for large-scale genomic sequencing: Clones the size of cosmids or larger are analyzed by random sampling (that is "shotgun" sequencing), implemented on commercial . . . sequencing instruments. The optimum size of the starting clones, the level of detail with which these clones should be mapped, and the extent to which the random sampling should be supplemented with more "directed" methods remain contentious. However, the important news is that the basic approach works in any of several well-tested variations. [352]

Accordingly, in our discussion of large-scale sequencing, we will mention some of the specific cloning vectors that are used today but not describe the full details of any real project. That would be much too complex and would obscure the description of the "dominant strategy" that has emerged. Figure 16.12 gives an overview of the process.

Our hypothetical large-scale project, sequencing say 100,000,000 base pairs, would start by creating a clone-library of fairly large clones, in the range of 100,000 to 1,000,000

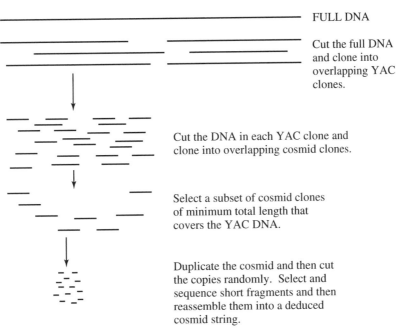

FULL DNA

Cut the full DNA
and clone into
overlapping YAC
clones.

Cut the DNA in each YAC clone and
clone into overlapping cosmid clones.

Select a subset of cosmid clones
of minimum total length that
covers the YAC DNA.

Duplicate the cosmid and then cut
the copies randomly. Select and
sequence short fragments and then
reassemble them into a deduced
cosmid string.

Figure 16.12: Overview of hypothetical large-scale sequencing. The top-down part constructs a YAC map of the DNA to be sequenced and then a cosmid map for each YAC in a covering set of YACs. A subset of cosmid clones is selected for sequencing, and each is sequenced by a shotgun (bottom-up) approach. Once the cosmids have been sequenced, they are used with the cosmid maps to assign characters to the YACs and hence to the full DNA. Possible extensions of this picture include performing radiation-hybrid mapping of DNA fragments larger than YAC inserts and restriction-enzyme mapping for maps more precise than cosmid maps, before sequencing is begun. Nothing in the figure is to scale.

base pairs each. These clones might be obtained from many copies of the target DNA (to be sequenced) by cutting the target DNA with an infrequently cutting restriction-enzyme. Each clone in the library is maintained by incorporating it into an appropriate *cloning vector* and inserting that vector into some biological system that replicates the DNA of the vector with its own DNA. For example, a *Yeast Artificial Chromosome* (*YAC*) is a cloning vector that can accommodate DNA inserts of length up to one million bases and that can be inserted and replicated in yeast cells [329, 335]. A YAC is made from a chromosome of *Saccharamyces cerevisiae* (common brewer's yeast) by extracting all the components of the chromosome that are essential for replication. When those components are spliced together (in the correct order and spatial separation) along with nonyeast DNA of an appropriate length, the resulting artificial chromosome can be inserted into a normal yeast cell and replicated along with the normal yeast DNA. In this way, each fragment of the library is maintained and reproduced in a YAC.

The next step is to choose at random a sufficient number of YACs in the library, with the expectation that the chosen YAC clones will cover the target DNA. Then fingerprints or STS data from the chosen YAC clones are obtained and used to deduce a physical map of these fragments.

Once the physical YAC map is finished, the procedure concentrates on each of the mapped YAC clones separately in order to obtain a more refined physical map inside that segment of DNA. The target DNA from each mapped YAC clone is replicated and then cut by a more frequently cutting restriction-enzyme, creating fragments of size say 40,000 base pairs long. A fragment of this size is typically incorporated into an artificial cloning vector called a *cosmid*, which itself is taken up by a virus (a *phage*) that infects *E. coli* bacteria. These phages inject the DNA they contain into *E. coli* cells, which then replicate that DNA along with their own DNA. A subset of these cosmid clones is chosen at random at sufficient density to expect that they will cover the YAC. The chosen cosmid clones are then mapped by physical mapping methods already discussed.

An alternative often exists to cutting the YACs. Often, an existing library may contain an unordered set of cosmids that cover the genome. One can pull out a subset of cosmids that cover the YAC of interest by hybridizing that full YAC to the fragments in the cosmid library. The cosmids that hybridize with the YAC are contained in it. These cosmids are then used to build a physical map of cosmids covering the YAC.

Other levels

Although not always part of the sequencing strategy, top-down physical mapping may start above the YAC level (for example, with a radiation-hybrid map) and continue below the cosmid level (for example, with a restriction-enzyme map of a frequent-cutting enzyme or with a genetic map). By picking a frequent-cutting restriction-enzyme, a highly refined physical map can be obtained in this way. Having a physical map of markers that are also used in genetic mapping (microsatellites for example) is very useful because it then allows the integration of genetic and physical maps. A feature that has been located relative to a genetic map can then be roughly located on the physical map.

YACs v. PACs v. BACs

Although the general logic of top-down physical mapping is not affected, we should note the existence of other cloning vectors that are becoming increasing popular [329]. These are the PACs and BACs, which are artificial chromosomes made from the *plasmid P*1, in the case of a PAC, and from *chromosomal bacterial DNA* in the case of a BAC. The insert size of these artificial chromosomes is less than a YAC, about 100,000 to 300,000 bases instead of one million. However, they appear to be easier to work with and seem not to be subject to the problems of chimerism and deletions that are common with YACs. Moreover, their smaller size is sometimes an advantage. For example, when trying to find a clone that spans a gap in a map, one looks for a clone that hybridizes with both ends of the known gap. A spanning YAC might be more easily found, but the size of the gap would be more precisely determined by finding a smaller BAC or PAC that spans the gap (say that fast ten times).

16.13.1. Is mapping necessary for sequencing?

Although the dominant strategy for large-scale sequencing typically begins with a mapping phase, the complete DNA sequence of two bacterial genomes and one archaeon were obtained [82, 162] using an approach that skipped mapping and went directly to shotgun sequencing. The larger of the two completed genomes was close to two million base pairs in length. The success of this approach surprised many people [350], but it is still unclear how extendable it will be to larger fragments or to nonbacterial DNA containing more noncoding and repetitive regions.

At the other extreme the success of mapless megabase sequencing of bacteria, and the development of BACs and PACs has lead to a recent proposal to bypass mapping in the human genome project. The proposed "BAC-PAC sequencing method" of Venter, Smith and Hood [442] works in the following way: First, a BAC (or PAC) library of the human genome is created, consisting of about 300,000 clones each with a human DNA insert of length around 150 kb. The library thus provides a 15-fold coverage of the genome and contains a BAC starting at roughly every 5 kb. The next step is to sequence about 500 bases from both ends of each cloned sequence, and to place each of these 600,000 sequences (along with a pointer to the BAC colony it comes from) into the computer. Each sequenced end is called a *sequence tagged connector* (*STC*).

To fully sequence a long region of human DNA, one extracts a BAC clone whose human insert lies in the region of interest (perhaps by hybridization of the region to the clones in the BAC library), and sequences the 150 kb insert by shotgun sequencing. Then, by computer, one examines the 600,000 STCs to find the rightmost (say) STC contained in the first sequenced BAC. This STC therefore identifies another BAC whose insert overlaps and extends the first sequence. By sequencing the second BAC insert and continuing in this way, one can sequence a long contiguous interval of DNA. This approach has been extended by J. Weber and E. Myers (see Genome Research, 7, 1997, p. 401–409 for the proposal, and p. 410–417 for a critique by Phil Green) who propose whole-genome human shotgun sequencing that avoids clone-by-clone sequencing. That proposal is being refined and implemented for the *Drosophila* genome at Celera Genomics corp., headed by Craig Venter, with $200 million backing from Perkin-Elmer (see Science, June 1998, p. 1540–1542).

16.13.2. Fragment selection for sequencing

We now return to discussing sequencing based on physical mapping. Once a physical map (for a single YAC) of overlapping cosmid clones is obtained, a subset of those clones are chosen for full DNA sequencing. If a clone is of length L, then roughly $L/400$ individual sequencing procedures will be required to sequence the clone. Hence the cost (in time and money) of sequencing a cosmid clone is roughly proportional to its length. This leads to a minor computational problem of selecting a subset of cosmid clones of minimum total length such that the selected cosmid clones cover the YAC clone they were obtained from. This subset is usually called the *minimum tiling path*.

One way to efficiently find a minimum tiling path is to cast the problem as a *shortest path* problem in a directed graph. Each clone from the cosmid map is represented by a node, with a directed edge from a node i to a node j if and only if cosmid clones i and j overlap, clone i starts to the left of the start of clone j, and clone i ends to the left of the end of clone j. Note that if one clone fully contains the other then there is no edge between them. The distance of the edge from i to j is set to the length of the overlap between clones i and j (i.e., the number of shared nucleotides). Any clone that starts at the extreme left end of the cosmid map is called a *source*, and any clone that ends at the extreme right end of the map is called a *sink*. Given this weighted directed (acyclic) graph, we have the following:

Theorem 16.13.1. *The nodes on any shortest (total distance) path in the graph from a source node to a sink node define a set of clones with smallest total length that covers the map. That set is a minimum tiling path.*

The proof of this is immediate and is left to the reader.

Since the graph is acyclic, a shortest path can be found in time linear in the number of edges, which is a speedup over the time needed to find a shortest path in a general graph.

16.13.3. Some real numbers

To get a more realistic picture of large-scale mapping, consider the status in January 1996 of the project to map human chromosome 16 done by a group at the Los Alamos National Laboratory [125]. Chromosome 16 is roughly 100 million base pairs long, about 3% of the human genome, and is estimated to contain around 3,000 genes. About 260 of these have been located so far. Roughly one thousand YAC and mega-YAC clones have been mapped to chromosome 16. These YACs cover about 99% of the chromosome, and only a few gaps remain in the YAC map. The cosmid map was derived from two clone-libraries containing about 40,000 cosmid clones in total. From these, about 4,000 clones were selected at random (with some filtration for GC-rich clones), providing about a tenfold coverage of chromosome 16. About 1,800 cosmids have been mapped to the chromosome, and these are organized into about 300 contigs covering over 80% of the chromosome. To date, about 300 STSs, 300 additional markers (ESTs, genes, cDNAs), as well as 1700 restriction-enzyme cutting sites have been mapped to chromosome 16.

16.14. Shotgun DNA sequencing

Having found a set of cosmid clones forming a minimum tiling path, the procedure finally turns to DNA sequencing. This is usually done by *shotgun* or *random* sequencing, summarized as follows: One first makes many copies of the DNA of interest (held in a cosmid in our hypothetical project) and then one cuts these copies by physical, enzymatic, or chemical means so that each copy is cut in (somewhat) random locations. The cutting produces fragments from the original DNA, but the process is such that the original order of the fragments is lost – one simply has a set of fragments without knowing where in the full string each came from. Still, the fragments from the different copies overlap each other so that if enough overlapping fragments are sequenced, the common patterns from overlapping sequenced fragments can be used to try to assemble the sequence of the full DNA string.

Actually, very few of the fragments from the cosmid will be sequenced, because the cosmid was replicated many times before those copies were cut and most procedures only sequence fragments in a fixed size range. (That range may be larger than 400, and often is, but only the first 400 bases can be read at one time.) So instead of sequencing all the fragments, the shotgun procedure randomly chooses fragments from the cosmids of an appropriate length and then it sequences the first 400 bases of each. The number of fragments picked must be sufficient to allow reconstruction of the full cosmid string from the overlapping sequenced parts of the fragments.

How many fragments should be picked to satisfy this expectation of sufficient overlap? Probabilistic analyses have been made [292], giving results that correspond well with empirically derived rules of thumb. The consensus based on current parameters is that one needs about a five- to tenfold coverage. That is, if the full cosmid string is of length L and the read length is 400, then the number of fragments picked should be between $5L/400$ and $10L/400$. Length-400 prefixes of that number of randomly chosen fragments should overlap sufficiently to allow reconstruction of the full cosmid string. In practice, the creation, cloning, and selection of fragments is not always uniform enough to completely reconstruct the cosmid, but this deviation from random coverage is due to molecular and biological reasons, not probabilistic reasons. Hence adding additional coverage does not always solve the problem, and other approaches, such as directed sequencing, must be employed where shotgun coverage is insufficient.

16.15. Sequence assembly

Having selected and sequenced a subset of partially overlapping strings (the first 400 bases of each chosen fragment), the problem now is to deduce the sequence of the full cosmid string. This is the algorithmic problem of shotgun *sequence assembly*. In the general approach we will describe, it is no longer relevant that each length-400 string is part of a longer chosen fragment,[6] and so for simplicity we assume here that each fragment is of length 400.

The specifics of different sequence assembly methods differ in various implementations, but they almost all follow a general outline first formalized in [362] and [363]. That outline contains three discrete steps: *overlap detection, fragment layout*, and *deciding the consensus* (often aided by multiple alignment). The first and last of these steps correspond to well-defined string problems.

16.15.1. Step one: overlap detection

The first step in the sequence assembly outline is to find, for each ordered pair of sequenced strings, how much a suffix of the first "matches" a prefix of the second. How is this suffix-prefix match formalized? If there were no sequencing errors, then for every ordered pair of strings S_1, S_2, we would compute the longest suffix of S_1 that exactly matches a prefix of S_2. (In Section 7.10 we presented a linear-time algorithm to find the longest suffix-prefix match for each ordered pair of strings in a set of strings.) But sequencing errors are a reality (even if they are only in the 1–5% range) and suffix-prefix matching must allow for approximate matches. Hence, most implementations define suffix-prefix matching by solving something like the following problem for each ordered pair of strings S_1, S_2:

Find a suffix of S_1 and a prefix of S_2 whose *similarity* is maximum over all suffix-prefix pairs of S_1 and S_2.

Here *similarity* is defined exactly as in Section 11.6, matches make a positive contribution to the objective function, and mismatches and spaces make negative contributions. As always, there is a question of picking the most useful weights and penalties, and we won't resolve that question here, but the algorithmic issue of how to find the most similar suffix-prefix pair from S_1 and S_2 should be clear.

To find the most similar suffix-prefix match, simply use the standard dynamic programming recurrences to compute the similarity between S_1 and S_2. However, in the dynamic programming table, initialize all entries in column zero to be all zeros (assuming S_1 is arrayed along the vertical axis of the table), and report the maximum value in row $|S_1|$. That value is the maximum similarity over all suffix-prefix pairs of S_1 and S_2. If the maximum value in row $|S_1|$ is in column j and the traceback in the dynamic programming table from that cell leads to row i in column one, then the best suffix-prefix pair consists of the suffix $S_1[i..n]$ and the prefix $S_2[1..j]$, assuming n is the length of S_1.

With lengths of S_1 and S_2 each around n, the best suffix-prefix match of the pair takes $\Theta(n^2)$ time. Thus, if there are k fragments, then the time to compute all the $k(k-1)$ best suffix-prefix matches is $\Theta(k^2 n^2)$. See Section 11.6.4 for more details.

Heuristic speedup

The computation of all best suffix-prefix matches (as formalized above) is actually a computational bottleneck in sequence assembly for the size problems of interest today. In some sequencing efforts it accounts for the majority of the computation time used in the project. But the purpose of computing the best suffix-prefix matches is to find ordered pairs with *large* similarity that specify candidates for overlapping, neighbor fragments. These candidate pairs are used in the second step of sequence assembly to lay out the fragments.

---

[6] In some recent approaches, that fact is actually relevant. We will consider one in Exercise 3.

Hence, we can speed up the first step of assembly if we can identify pairs of strings whose best suffix-prefix similarity is clearly not "sufficient" to be an attractive candidate.[7] The dynamic programming computation can then be skipped for these unattractive pairs.

How can unattractive pairs be efficiently identified? It is not difficult to think of a variety of approaches, and a certain amount of experimentation is needed to compare them. We will just mention one idea that has several variants. Since the error rate is fairly low (below 5%) and the coverage fairly high, two strings that have a "sufficient" overlap should have at least one "significantly" long *common substring* (see Section 7.4). What length is "significant" can again be determined by a probabilistic argument that is beyond the scope of this discussion. The point is that by computing the length of the longest common substring for each pair of chosen strings, we can recognize and exclude many pairs of strings that are unlikely to be overlapping neighbors in the full string. Recall from Exercise 8 in Chapter 7 that for k strings each of length n, the longest common substring for each of the $\binom{k}{2}$ pairs can be computed in $O(k^2 n)$ total time by the use of a generalized suffix tree. One can refine this approach to consider only common substrings that appear in the first quarter (say) of one of the two strings. This is a reasonable expectation of two neighboring overlapping strings and further reduces the number of pairs on which the full dynamic programming must be executed. Chen and Skiena [100] have done an in-depth examination of exact matching methods for overlap detection. They report vast speedups over dynamic programming with only a small decline in quality (see Section 7.15).

A generalization of the exact-matching approach would use something like BLAST to find regions of high *local* similarity in the pair of strings. The idea again is that one expects neighboring strings to have several "good-sized" regions of high local similarity. Whatever the specific method, by screening out unlikely neighbors the dynamic programming computation can be limited to a smaller number of pairs of attractive strings. Techniques of this type have dramatically reduced the overall computation time needed for sequence assembly.

16.15.2. Step two: substring layout

There are different approaches to layout, some of which are algorithmically quite sophisticated, but most practical implementations follow some variant of a *greedy* approach, similar to the way that fragments are laid out in physical mapping projects. In fact, one can view the DNA sequence of a substring as a very precise *ordered fingerprint* and follow the kind of reasoning employed in laying out physical maps.

In detail, the string pair with highest scoring suffix-prefix match is first chosen and merged, fixing the extent of its overlap. Then the next highest scoring pair is selected and merged, resulting in either one contig containing three strings or in two separate contigs containing four strings (see Figure 16.13). Since the suffix-prefix overlaps are determined by the similarity criteria, spaces are allowed in the suffix-prefix matches. Therefore, as successive cosmids are added into a contig, additional spaces may have to be inserted into the added string to be consistent with spaces in previously inserted strings (see Figure 16.14). This is exactly the same issue that arises in multiple alignment, and in effect, one is constructing here a highly structured multiple alignment.

Notice that, in this approach, one does not compute the best suffix-prefix overlap of a fixed *contig* with one of the fragments or with another contig. That might giver a better multiple alignment, but this would be computationally expensive and is not usually done

---

[7] Sufficiency depends on the coverage, on the length of the read, and on the type and distribution of errors. This kind of analysis is usually probabilistic and is outside the scope of this book.

Figure 16.13: a. The first contig. b and c show the two different possibilities after the merge of the next best suffix-prefix pair.

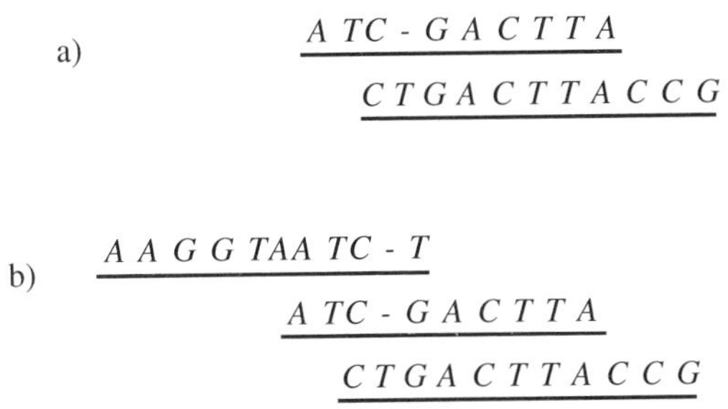

a)

$$\frac{A\ TC\ \text{-}\ G\ A\ C\ T\ T\ A}{C\ T\ G\ A\ C\ T\ T\ A\ C\ C\ G}$$

b)

$$\frac{A\ A\ G\ G\ TAA\ TC\ \text{-}\ T}{A\ TC\ \text{-}\ G\ A\ C\ T\ T\ A}$$
$$C\ T\ G\ A\ C\ T\ T\ A\ C\ C\ G$$

Figure 16.14: a. The first contig. The alignment in that pair of strings has one inserted space. The next best suffix-prefix match matches the suffix *ATCT* with the prefix *ATCG* as shown in b. That pairwise alignment does not contain a space, but a space is inserted into the third string when the string is added to the contig. That new space is inherited from the alignment of the first two strings.

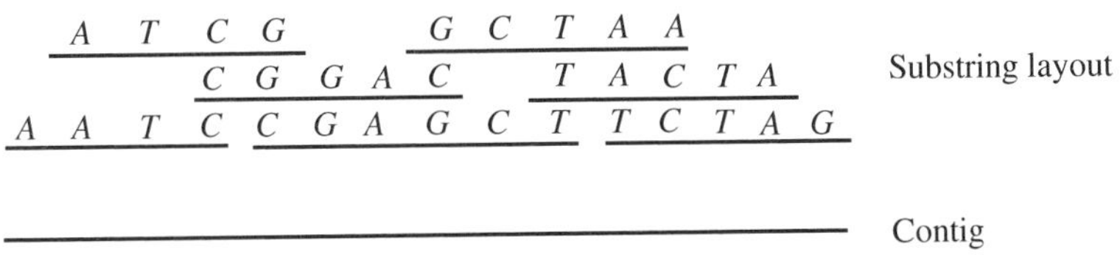

$$\frac{A\quad T\quad C\quad G}{\begin{array}{cccc}C&G&G&A&C\end{array}}$$
$$\frac{G\quad C\quad T\quad A\quad A}{T\quad A\quad C\quad T\quad A}\quad \text{Substring layout}$$
$$\overline{A\ \ A\ \ T\ \ C\ \ C\ \ G\ A\ \ G\ \ C\ \ T\ \ T\ C\ \ T\ A\ \ G}$$

Contig

Figure 16.15: The layout of sequenced strings created in step two assigns each character to a unique position in the cosmid. Hence the layout imposes a column structure on the characters of the sequenced strings. In this figure there are no spaces in the alignment, although spaces are possible in general.

in practice. Instead, all the layout decisions are based on the pairwise suffix-prefix scores computed in step one, and any inconsistencies caused by the layout are left to be resolved in step three. Hence step two involves no additional dynamic programming computation, and this step goes very quickly in practice.

16.15.3. Step three: deciding the consensus

The substring layout created in step two is now used to create a single *deduced cosmid* DNA string. That is, the bases in the cosmid are determined in this step. Exactly how this is done varies in different methods but each follows a similar theme. The substring layout has (implicitly) assigned each character in each sequenced substring to a particular cosmid position (see Figure 16.15). If all the characters assigned to a particular cosmid position are the same, then there is nothing to do. But if there is disagreement, then a choice has to be made to select one particular character or to indicate that the disagreement is too large to determine the base in that position. We will outline several approaches for illustration but remind the reader that existing systems each resolve this issue in somewhat different ways.

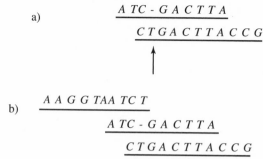

Figure 16.16: This figure shows the remerging following the merge history shown earlier. Figure a shows the first contig with one space in the alignment. The profile of that alignment contains one column with both a space and the character *T*. (The profile of each other column consists of a single character.) When the third string is aligned to the profile (instead of being aligned to only a single string), no space is inserted. The resulting alignment is shown in b. That alignment makes it more convincing that *T* should be the character deduced for the column containing the space.

The simplest approach is to report the frequency of each character in each column (a profile) and let the user decide how to use that information. Alternately, a *consensus* sequence can be compiled by taking the *plurality* character in each column, provided the plurality is above some preset threshold. At the other extreme, one can identify *windows* of the substring layout where the disagreements are large. For each such window, the substrings contained in that window can then be multiply aligned by some method, and the consensus string from that multiple alignment can be reported. (Recall the discussion in Section 14.7 of multiple alignment with the consensus objective function.) Doing multiple alignment in a window of high disagreement allows some characters to be changed and others to be moved a bit. A more rigorous approach to adjusting the layout and determining the bases is contained in [273], [269], and [267].

An intermediate approach is to *remerge* the substrings in the layout according to the same merge history determined in the layout step, but now each new substring is merged into its contig using an alignment of the new substring with a *profile* (see the first part of Section 14.3.1) of the substrings already in the contig. (See Figure 16.16) This kind of remerging is not as computationally intense as a full multiple alignment, and it only requires at most $n - 1$ profile alignments if there are n substrings. Nonetheless, it should give a better agreement of the merged string with the contig compared to the agreement obtained from the original layout, since that layout is based on the alignment of the new string with *only one* of the substrings in the contig. Once all the substrings are remerged, a single consensus string is created from that layout.

16.16. Final comments on top-down, bottom-up sequencing

Backing up: compiling the whole sequence

Having sequenced strings of length 400 and having deduced the sequence of each of the chosen cosmids, the hypothetical, large-scale sequencing project enters a *backup* phase that reports the sequence for the full DNA string. First, in each chosen YAC, the deduced strings from the chosen cosmids are assembled using the cosmid map of the YAC. All that is involved in this step is an application of the base-deducing step of assembly, already discussed above. Then the deduced YAC strings are used with the YAC map to deduce the full target DNA string.

In reality, there is some iteration between the top-down and the bottom-up phases of this entire procedure.

Problems and future directions in the shotgun approach

There are two persistent problems that arise in shotgun sequencing. First, the final deduced string obtained often needs additional correction. Usually, human technicians oversee the output from the sequencing protocol and the assembly program, looking for unreliable regions in the deduced string. These regions are then either resequenced or checked by some other laboratory method. One proposed verification method involves *sequencing by hybridization*, to be discussed in Section 16.18. It is still a matter of active research to more fully automate these last steps of spotting and correcting difficulties in the deduced string.

The second major problem in higher organisms is the frequent occurrence of repetitive substrings. The problem is particularly severe when the repeat is longer than the typical read length (400 base pairs). Repeats cause problems for the shotgun method that are similar to the problems discussed (in Section 16.12) for directed sequencing. However, since shotgun sequencing works on the entire string in parallel, it can deduce parts of the string on both sides of a repeat even if it can't correctly deduce all the repeated regions. This is in contrast to directed sequencing, which might not be able to continue past the first difficult repeat it encounters. Moreover, using shotgun methods, problems caused by repeats can often be more easily detected and handled by alternative laboratory methods. One indication of the existence of repeats is that the final deduced cosmid string is shorter than the true known length of the cosmid. A more informative indication of repeats is that the layout from step two contains a region where many more than the expected number of substrings overlap and/or contains a region where too few overlaps occur. It may also happen that a gap in the layout could be closed by redistributing some of the substrings that the greedy algorithm had placed in one of the high overlap regions.

Another problem that has been related in practice to shotgun sequencing is the common occurrence of chimeric clones in YACs. This is not strictly a problem of the shotgun methodology, but it is often included in such discussions because shotgun sequencing is usually used in large project where YACs are the cloning vector. A related problem is the occurrence of deletions in YACs that also seem to be common.

Recently, a number of people have attempted to incorporate available side information in the fragment assembly phase of the shotgun method. Side information often consists of knowing particular pairs of fragments that must overlap or that must not overlap, of knowing how the fragments relate to a genetic map or to some other lower resolution physical map, of knowing a range of physical distances between pairs of fragments, or of just knowing some partial order on the left-right layout of the fragments. It is hoped that by incorporating all of this side information, the number of feasible assemblies will be very small and the resulting computational problems will therefore be greatly simplified. A formal effort along these lines is reported in [343].

16.17. The shortest superstring problem

Having discussed mapping and sequencing, which comprise real front-line problems in molecular biology, we now step back a bit to discuss a more abstract pure string problem that is partly motivated by the sequence assembly problem in shotgun sequencing.

16.17.1. Introduction to superstrings

Definition   Given a set of k strings $\mathcal{P} = \{S_1, S_2, \ldots, S_k\}$, a *superstring* of the set $\mathcal{P}$ is a single string that contains every string in $\mathcal{P}$ as a substring.

For example, a concatenation of the strings of $\mathcal{P}$ in any order gives a trivial superstring of $\mathcal{P}$. For a more interesting example, let $\mathcal{P} = \{abcc, efab, bccla\}$. Then $bcclabccefab$ is a superstring of $\mathcal{P}$, and $efabccla$ is another, shorter superstring of $\mathcal{P}$. Generally, we are interested in finding superstrings whose length is small.

Definition   For set $\mathcal{P}$, let $S^*(\mathcal{P})$ denote the *shortest* superstring of $\mathcal{P}$. When the set $\mathcal{P}$ is clear by context, we will often use S^* in place of $S^*(\mathcal{P})$.

The problem of finding the shortest superstring may have applications in data compression (see Exercises 25 and 26) but is mostly motivated by one approach to the sequence assembly problem in shotgun sequencing. Each string in the set $\mathcal{P}$ models one of the sequenced DNA fragments created by the shotgun sequencing protocol (Section 16.14). The assembly problem is to deduce the originating DNA string S from the set of sequenced fragments $\mathcal{P}$. Without sequencing errors, the originating string S is a superstring of $\mathcal{P}$ and, under some assumptions, S is likely to be a shortest superstring of $\mathcal{P}$. In that case, a shortest superstring of $\mathcal{P}$ is a good candidate for the originating string S.

In reality, real sequence data always contain errors and omissions, and studies on shotgun data suggest that the shortest superstring is often shorter than the originating string, especially when the string contains regions of repeated substrings. Consequently, the shortest superstring problem is an abstract algorithmic problem that only loosely models the real sequence assembly task. Still, work on the superstring problem is justified in the same way that work on model organisms or *in vitro* research is justified in molecular biology. Techniques developed by studying an abstracted, simplified problem (or simple organism) are often important steps to obtaining practical results about more complex or realistic systems.

Finding the shortest superstring of a set of strings is known to be NP-hard [171], and so an efficient algorithm to find the exact optimal is thought to be unlikely. Moreover, the shortest superstring problem is MAX-SNP hard [69], and so it is also thought that there is no polynomial-time algorithm that can approximate the optimal to within an *arbitrary, but predetermined* constant. Nonetheless, there are polynomial-time methods that approximate the optimal to within the specific error bounds of four and of three (and better). We will examine in detail a method that guarantees an error bound of three, that is, a fast method that finds a superstring whose length is guaranteed to be no more than three times the length of the shortest superstring. This bound was established by Blum, Jiang, Li, Tromp, and Yannakakis [69]. The method we present here is derived from their solution but is somewhat different in detail, so that the two algorithms may produce different superstrings.

Since the publication of [69], the error bound has been reduced in a series of papers, although all the algorithms in these papers follow similar high-level ideas. The general approach in [69] was first refined by Teng and Yao [434] who found an algorithm that guarantees an error bound of at most 2.89. The guaranteed bound was reduced to $2\frac{50}{63}$ by Kosaraju, Park, and Stein [283] and to 2.75 by C. Armen and C. Stein [30]. A reduction to 2.67 was achieved by Jiang and Jiang [246] and then reduced to 2.596 by Breslauer, Jiang, and Jiang [77]. An alternative reduction to 2.67 was also achieved by Armen and Stein [29]. The best improvement to date, by Sweedyk, reduces the bound to 2.5 [427].

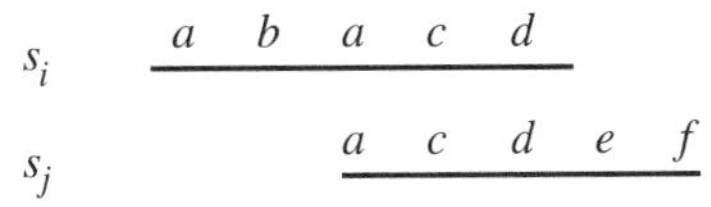

Figure 16.17: The overlap of strings $S_i = abacd$ and $S_j = acdef$. $pref(S_i, S_j) = ab$, $p(S_i, S_j) = 2$, and $ov(S_i, S_j) = 3$.

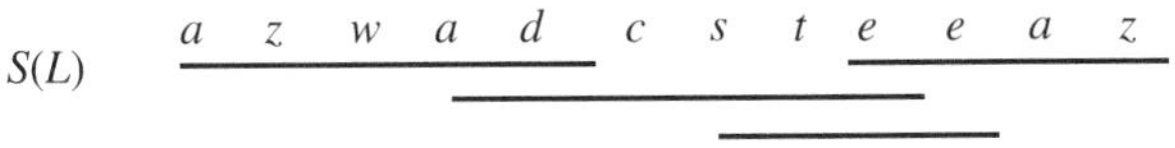

Figure 16.18: String $S(L)$ formed from four strings S_{o_1}, S_{o_2}, S_{o_3}, and S_{o_4} defined in the text. $pref(S_{o_1}, S_{o_2}) = azw$, $pref(S_{o_2}, S_{o_3}) = adc$, and $pref(S_{o_3}, S_{o_4}) = st$.

Improvements below a factor of three are outside the scope of this book. However, the results we discuss in depth capture much of the high-level approach and the spirit of the improved results (but not the details).

A related problem of interest is to find the shortest superstring of a set of strings when each string in the set can be used in either the forward or reverse orientation. This *orientation-free superstring problem* is motivated by the fact that in real sequence assembly the DNA fragments come from both strands of the DNA string, but one does not know the origin of any specific fragment (see Exercise 2). In the context of the superstring problem, this translates to the problem where one is given a set of substrings without knowing the orientation of any given string. The paper by Jiang, Li, and Du [248] considers this case. It details a factor-of-four approximation method and states that a factor-of-three is also obtainable.

Basic definitions

Throughout the discussion of superstrings, we assume that no string in $\mathcal{P}$ is a substring of any other string in $\mathcal{P}$. Any such substrings can be efficiently detected and removed (see Exercise 7.20 in Chapter 7). It is easy to see that after removing these substrings from $\mathcal{P}$, a shortest superstring of the remaining strings is also a shortest superstring of the original set. Hence, there is no loss in assuming that $\mathcal{P}$ is substring free.

Definition For any string S, let $|S|$ denote the length of S. For a *set* of strings $\mathcal{P}$, $\|\mathcal{P}\|$ denotes the total length of the strings in $\mathcal{P}$.

Definition For two strings S_i and S_j, let $ov(S_i, S_j)$ be the *length* of the longest match (overlap) of a suffix of string S_i and a prefix of string S_j. We define $p(S_i, S_j) = |S_i| - ov(S_i, S_j)$ and define *pref* (S_i, S_j) as the prefix of S_i of length $p(S_i, S_j)$.

Stated differently, *pref* (S_i, S_j) is the prefix of S_i ending at the start of the longest suffix-prefix match of S_i and S_j. See Figure 16.17 for an example.

Definition Let $L = o_1, o_2, o_3, \ldots, o_t$ be an ordered subset of the integers 1 through t. L defines the subset $\{S_{o_1}, S_{o_2}, \ldots, S_{o_t}\} \subseteq \mathcal{P}$ of t strings from $\mathcal{P}$, and it also defines the following string $S(L)$:

$$S(L) = pref(S_{o_1}, S_{o_2})pref(S_{o_2}, S_{o_3}) \ldots pref(S_{o_{t-1}}, S_{o_t})S_{o_t}.$$

$S(L)$ is the concatenation of the nonoverlapping prefixes of the pairs of adjacent string in L, followed by the full string S_{o_t}. For example, if $S_{o_1} = azwad$, $S_{o_2} = adcste$, $S_{o_3} = stee$, $S_{o_4} = eeaz$, then $S(L) = azwadcsteeaz$. (See Figure 16.18) We now develop two important facts about string $S(L)$.

Lemma 16.17.1. *The length of string $S(L)$ is* $\sum_{i=1}^{t-1} p(S_{o_i}, S_{o_{i+1}}) + |S_{o_t}|.$

The Lemma is illustrated in Figure 16.18. Its proof is immediate from the definitions.

Lemma 16.17.2. *Let L and $S(L)$ be defined as above. Then string $S(L)$ is a superstring* (*not necessarily shortest*) *of the strings* $\{S_{o_1}, S_{o_2}, \ldots, S_{o_t}\}$.

An examination of Figure 16.18 makes Lemma 16.17.2 almost self-evident, but a formal proof is still required.

PROOF OF LEMMA 16.17.2   The proof is by induction on i, starting from $i = t$ downward to $i = 1$. String S_{o_t} is a substring of $S(L)$ since it is explicitly included at the end of $S(L)$. Suppose the claim holds down to $i + 1$. We must show that S_{o_i} is also a substring of $S(L)$. Since $pref(S_{o_i}, S_{o_{i+1}})$ is explicitly included in $S(L)$, we only need to show that the remaining part of S_{o_i} is contained in $S(L)$ immediately following $pref(S_{o_i}, S_{o_{i+1}})$. This is immediate if $S_{o_i} = pref(S_{o_i}, S_{o_{i+1}})$ (i.e., if S_{o_i} and $S_{o_{i+1}}$ have no overlap). Otherwise, let r_{o_i} denote the suffix of S_{o_i} starting immediately after $pref(S_{o_i}, S_{o_{i+1}})$. By the fact that no string in $\mathcal{P}$ is a substring of another, r_{o_i} is a prefix of $S_{o_{i+1}}$. Now by the inductive hypothesis, $S_{o_{i+1}}$ is a substring of $S(L)$, so r_{o_i} is contained in $S(L)$ starting immediately following $pref(S_{o_i}, S_{o_{i+1}})$, and the induction is complete. $\square$

16.17.2. The objective function for superstrings

Setting the list L to be a permutation Π of all the integers 1 through k, Lemma 16.17.2 implies the following:

Corollary 16.17.1. For the set $\mathcal{P}$ of k strings, any permutation Π of the integers 1 through k specifies a superstring $S(\Pi)$ of $\mathcal{P}$.

Conversely, if S^* is a shortest superstring of the set $\mathcal{P}$, then S^* must be $S(\Pi)$ for some permutation Π. To see this, note that since $\mathcal{P}$ is substring free, a string S_i can begin to the left of a string S_j in the superstring S^* if and only if S_i also ends to the left of S_j in S^*. Further, no two strings begin or end at the same point in S^*. Hence the desired permutation Π (such that $S^* = S(\Pi)$) is obtained by the order that the left endpoints of strings in $\mathcal{P}$ appear in S^*. Each adjacent pair of strings in S^* must overlap by the most possible or else S^* could be shortened, and hence $S^* = S(\Pi)$. As an example, consider again the strings in Figure 16.18. If $\mathcal{P}$ is the ordered set $\{eeaz, adcste, azwad, stee\}$, then the permutation shown in Figure 16.18 is $\Pi = 3, 2, 4, 1$.

By Lemma 16.17.1, if permutation Π is $o_1, o_2, o_3, \ldots, o_k$, then

$$|S(\Pi)| = \sum_{i=1}^{k-1} p(S_{o_i}, S_{o_{i+1}}) + |S_{o_k}|.$$

It follows that the shortest superstring is defined by the permutation $\Pi = o_1, o_2, o_3, \ldots, o_k$ minimizing

$$\sum_{i=1}^{k-1} p(S_{o_i}, S_{o_{i+1}}) + |S_{o_k}|.$$

But,

$$p(S_{o_i}, S_{o_{i+1}}) = |S_{o_i}| - ov(S_{o_i}, S_{o_{i+1}}),$$

so

$$|S(\Pi)| = \sum_{i=1}^{k-1} p(S_{o_i}, S_{o_{i+1}}) + |S_{o_k}| = \sum_{i=1}^{k} |S_{o_i}| - \sum_{i=1}^{k-1} ov(S_{o_i}, S_{o_{i+1}})$$

$$= \|\mathcal{P}\| - \sum_{i=1}^{k-1} ov(S_{o_i}, S_{o_{i+1}}).$$

Hence the shortest superstring can also be defined by the permutation maximizing

$$\sum_{i=1}^{k-1} ov(S_{o_i}, S_{o_{i+1}}).$$

The latter definition may be a bit subtle because it is possible for more than two strings to overlap at a single point (as in Figure 16.18), yet the summation only includes terms for overlaps between pairs of successive strings in the permutation. We will primarily use the latter maximization criterion for the shortest superstring objective function.

16.17.3. Cyclic strings and cycle covers

Definition   Given a (linear) string S we define the *cyclic string* $\phi(S)$ to be a copy of S with the modification that the last character of S is considered to precede the first character of S. In other words, link the beginning and end of S together to make the cyclic string $\phi(S)$.

We will use ϕ to refer to the cyclic string $\phi(S)$ when the base string S is understood from context or is not critical.

Definition   The length of cyclic string $\phi(S)$, denoted $|\phi(S)|$, equals $|S|$, the length of string S.

Definition   Let $inf(S)$ be an infinite repetition of string S. A string α is said to *map* to cyclic string $\phi(S)$ if α is a substring of $inf(S)$. Clearly, if α is a substring of $inf(S)$ then one occurrence of α starts in some position from 1 through $|S|$ of $inf(S)$.

For example, if $S = abac$, then the string $acabacabacabaca$ maps to $\phi(S)$.

Note that a string that maps to some cyclic string ϕ can have arbitrarily more than $|\phi|$ characters. A physical way to think about cyclic strings is to imagine a wheel containing stamp pad characters that spell out ϕ. Then imagine that those characters are inked on a stamp pad. Any string (no matter how long) obtained by rolling the inked wheel on paper is a string that maps to ϕ.

Definition   Define a *cycle cover* $C(\mathcal{P})$ of $\mathcal{P}$ as a set of cyclic strings such that every string in $\mathcal{P}$ maps to at least one of the cyclic strings in $C(\mathcal{P})$. When $\mathcal{P}$ is clear by context, we will use C in place of $C(\mathcal{P})$.

Definition   The *length* of the cycle cover $C(\mathcal{P})$ is denoted by $\|C\|$ and equals the sum of the lengths of its cyclic strings. That is, $\|C\| = \sum_{\phi \in C} |\phi|$.

Definition   Let $C^*(\mathcal{P})$ denote the *minimum* length cycle cover of $\mathcal{P}$.

Any superstring S of $\mathcal{P}$ trivially defines a cycle cover of the same length (containing the single cyclic string $\phi(S)$). Therefore,

Lemma 16.17.3. *The length of the shortest superstring S^* of set $\mathcal{P}$ is at least $\|C^*(\mathcal{P})\|$.*

We will see that the problem of finding a cycle cover of minimum total length can be solved efficiently. This fact is the basis for all the published approximation methods for the shortest superstring problem. Before discussing any specific method, we examine how to obtain a superstring from a cyclic cover.

16.17.4. How cycle covers define superstrings

Given a cycle cover $C(\mathcal{P})$ of $\mathcal{P}$, we need to partition the strings of $\mathcal{P}$ into subsets that map to the same cyclic string in the cover, but because a string in $\mathcal{P}$ might map to more than one cyclic string of $C(\mathcal{P})$, there may be a choice in how this partition is created, as detailed next.

Definition   If string $S \in \mathcal{P}$ maps to only one cyclic string ϕ in $C(\mathcal{P})$ then we *associate* S with ϕ. Otherwise, we arbitrarily *associate* S with one of the cyclic strings of $C(\mathcal{P})$ that S maps to.

Definition   For a cyclic string $\phi \in C^*(\mathcal{P})$, let $\mathcal{P}_\phi$ denote the subset of strings of $\mathcal{P}$ that are associated with ϕ. Suppose $|S_\phi| = t$.

For any cyclic string $\phi(S)$, we number the characters of ϕ starting with the first character of S. With that numbering, each string in $\mathcal{P}_\phi$ maps to ϕ starting at one of the positions 1 through $|S|$. Furthermore, since no string in $\mathcal{P}$ is a substring of another, no two strings in $\mathcal{P}_\phi$ map to the same starting point.

Definition   Let $L_\phi = o_1, o_2, o_3, \ldots, o_t$ be the indices of the strings in $\mathcal{P}_\phi$ in order of their starting positions in ϕ. Be careful not to confuse these indices (which identify an ordering of a particular subset of $\mathcal{P}$) with their actual starting positions in ϕ.

For example, consider the string $S = azwadcstee$. The cyclic string $\phi(S)$ is shown at the top of Figure 16.19. Each string in the ordered set $\Pi = \{eeaz, adcste, azwad, stee\}$ maps to $\phi(S)$. The starting positions of each of these four strings are marked and numbered in the figure. L_ϕ is the permutation 3, 2, 4, 1, and the string $S(L_\phi)$ is shown at the bottom of the figure.

By Lemma 16.17.2, string $S(L_\phi) = pref(S_{o_1}, S_{o_2})pref(S_{o_2}, S_{o_3}) \ldots pref(S_{o_{t-1}}, S_{o_t})S_{o_t}$ is a superstring of the set $\mathcal{P}_\phi$. Further, if L'_ϕ is any circular shift of list L_ϕ then the string $S(L'_\phi)$ is also a superstring of $\mathcal{P}_\phi$. For example, if $L'_\phi = \{4, 1, 3, 2\}$ then $S(L'_\phi)$ is also a superstring of $\mathcal{P}_\phi$, and its length is one greater than the superstring $S(L_\phi)$ described in Figure 16.19. This happens because $ov(S_2, S_4) = 3 = ov(S_1, S_3) + 1$.

Definition   If o_i is the last index in list L'_ϕ (a circular shift of L_ϕ), then the string S_{o_i} is called the *final string* of $S(L'_\phi)$.

The next lemma follows from definitions and the fact that any string that maps to ϕ can be mapped so that its starting position is between 1 and $|\phi|$.

Lemma 16.17.4. *If S_{o_i} is the final string of $S(L'_\phi)$, then the length of $S(L'_\phi)$ is exactly* $|\phi| + ov(S_{o_i}, S_{o_{i+1}}) \leq |\phi| + |S_{o_i}|$, *where $t + 1$ is taken to be 1.*

Given Lemma 16.17.4, the circular shift minimizing the length of the resulting superstring $S(L'_\phi)$ is obtained by choosing the index o_i to minimize $ov(S_{o_i}, S_{o_{i+1}})$. We will need this fact later in the factor-of-three approximation.

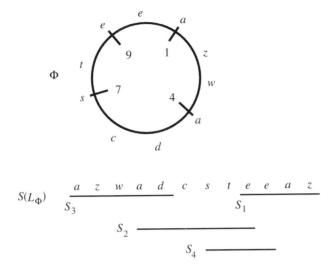

Figure 16.19: Cyclic string $\phi(S)$ for $S = azwadcstee$. String $S(L_\phi)$ formed from the four strings of $\mathcal{P}$ associated with ϕ.

16.17.5. Factor-of-four approximation

We now describe an algorithm producing a superstring that is never more than four times the optimal length.

Factor-of-four approximation algorithm

1. Find a minimum length cycle cover $C^*(\mathcal{P})$ of $\mathcal{P}$ and associate each string $S \in \mathcal{P}$ with exactly one cycle in $C^*(\mathcal{P})$ that S maps to. {We will discuss later how to efficiently find $C^*(\mathcal{P})$.}

2. For every cyclic string ϕ in $C^*(\mathcal{P})$, form the ordered list of indices L_ϕ and create the string $S(L_\phi)$, a superstring of $\mathcal{P}_\phi$. Let $\mathcal{P}'$ be the set of superstrings obtained in this step.

3. Concatenate the strings of $\mathcal{P}'$ (in any order) to obtain a superstring H of $\mathcal{P}$.

Clearly, H is a superstring of $\mathcal{P}$ since each $S(L_\phi)$ is a superstring of the strings associated with cyclic string ϕ, and $C^*(\mathcal{P})$ is a cycle cover of $\mathcal{P}$. Each string $S(L_\phi) \in \mathcal{P}'$ (found in step 2) has one *final string*. Let $\mathcal{P}_f$ be the set of these final strings. Then using Lemma 16.17.4,

Lemma 16.17.5. *Superstring H has length less than or equal to* $\|C^*\| + \sum_{S \in \mathcal{P}_f} |S|$.

Error analysis of the algorithm

Preparation for the analysis

We first develop a fact (the *Overlap Lemma*) that is a (or *the*) central tool in the analysis of all superstring approximation algorithms developed to date. The Overlap Lemma in turn requires a very important fact (the *GCD Theorem*) concerning periodic strings. The definition of a period of a string is given on page 40.

Theorem 16.17.1. (The GCD Theorem). *If string S has two periods of lengths p and $q \leq p$, and $|S| \geq p + q$, then S has a period of length $\gcd(p, q)$.*

PROOF We first show that S has a period of length $p - q$ (i.e. that character $S(i) = S(i + p - q)$ for all values of i in the range 1 to $|S| - p + q$). There are two cases: either

$i \leq q$ or $i > q$. In the first case, $i + p \leq |S|$ by the assumption that $|S| \geq p + q$; so character $i + p$ exists and equals character i since S has a period of length p. Now $q \leq p$, so character $i + p - q$ also exists and must equal character $i + p$ since S has a period of length q. Hence, in the first case, $S(i) = S(i + p) = S(i + p - q)$. In the second case, when $i > q$, symmetric reasoning establishes that characters $i - q$ and $i - q + p$ exist and must each equal character i. So $S(i) = S(i + p - q)$ in both cases, and S has a period of length $p - q$. It is also true, trivially, that $|S| \geq q + p - q$.

Now each step of Euclid's algorithm (for computing the greatest common divisor of two arbitrary integers x and y) recursively reduces the problem of computing $gcd(x, y)$ to the problem of computing $gcd(x, x - y)$, for $x > y$. And, Euclid's algorithm ends with $gcd(x, y)$. Hence when S has periods p and q, and $|S| \geq p + q$, the execution of Euclid's algorithm with input (p, q) will maintain at each step two integers x and y that are each period lengths of S. In particular, the final integer $x = y = gcd(x, y)$ is a period length of S, proving the claim that S has a period of length $gcd(p, q)$. □

Lemma 16.17.6. (The Overlap Lemma). *Let ϕ and ϕ' be any two cyclic strings in $C^*(\mathcal{P})$, and let α and α' be two strings (not necessarily in $\mathcal{P}$) that map to ϕ and ϕ', respectively. Then the longest suffix-prefix match between α and α' has length less than $|\phi| + |\phi'|$.*

Before proving the lemma, recall that α can be arbitrarily larger than ϕ, and the same is true for α' and ϕ'. So this lemma is not a triviality. When α is larger than the cyclic string ϕ it maps to, α wraps at least once around ϕ and so has a period equal to $|\phi|$.

PROOF OF LEMMA 16.17.6   Assume the lemma is not true, so the length of the α, α' overlap is at least $|\phi| + |\phi'|$. Let α'' denote the suffix-prefix match of α and α'. We first show that $|\phi| \neq |\phi'|$.

Clearly, α'' maps to both ϕ and ϕ' since α maps to ϕ and α' maps to ϕ', and α'' is a substring of both. By assumption $|\alpha''| \geq |\phi| + |\phi'|$, so α'' must wrap completely around ϕ and also around ϕ'. Hence if $|\phi| = |\phi'|$, ϕ and ϕ' both consist exactly of the first $|\phi|$ characters of α'', and hence ϕ is a cyclic rotation of ϕ' (see Exercise 1 on page 11). But then all strings that map to ϕ also map to ϕ', so ϕ could be removed from the cover. This is impossible in a cycle cover of minimum total length, so $|\phi| \neq |\phi'|$.

Since α'' maps both to ϕ and ϕ' and wraps around each of them, α'' must be a semiperiodic string with periods $|\phi|$ and $|\phi'|$. Furthermore, since the length of α'' is assumed to be at least $|\phi| + |\phi'|$, we can apply the GCD Theorem 16.17.1, which says that α'' has a period equal to $gcd(|\phi|, |\phi'|)$. Since $|\phi| \neq |\phi'|$, $gcd(|\phi|, |\phi'|) < \max(|\phi|, |\phi'|)$. Assume that maximum is $|\phi|$. However, $gcd(|\phi|, |\phi'|)$ *evenly* divides $|\phi|$, so ϕ consists of repeated *full* copies of a string of length $gcd(|\phi|, |\phi'|)$. Since $gcd(|\phi|, |\phi'|) < |\phi|$, ϕ must contain at least two of those full copies. Now consider the cyclic string γ obtained from ϕ by removing one of these full copies. Clearly, all strings that map to ϕ also map to the smaller cyclic string γ. But that contradicts the assumption that $C^*(\mathcal{P})$ is a minimum length cycle cover of $\mathcal{P}$; hence $|\alpha''| < |\phi| + |\phi'|$ and the lemma is proved. □

Theorem 16.17.2. *The superstring H has total length less than four times the length of the shortest superstring $S^*(\mathcal{P})$.*

PROOF   Recall that $\mathcal{P}_f \subseteq \mathcal{P}$ is the set of final strings obtained in step 2 of the algorithm. By definition, any two strings in $\mathcal{P}_f$ map to two different cyclic strings of $C^*(\mathcal{P})$. Clearly, the shortest superstring of $\mathcal{P}$ is at least as long as the shortest superstring of $\mathcal{P}_f$. By the Overlap Lemma (16.17.6), any two strings that map to two different cyclic strings ϕ and ϕ' in $C^*(\mathcal{P})$ can overlap by at most $|\phi| + |\phi'|$ characters. Therefore, if the ordered list $L = o_1, \dots, o_r$ defines the shortest superstring of the set of strings in $\mathcal{P}_f$, then $\sum_{i=1}^{r-1} ov(S_{o_i}, S_{o_{i+1}}) < 2\|C^*\|$. It follows that the shortest superstring of $\mathcal{P}_f$ (and hence of $\mathcal{P}$) has length greater than $\sum_{S \in \mathcal{P}_f} |S| - 2\|C^*\|$. That is, $|S^*(\mathcal{P})| > \sum_{S \in \mathcal{P}_f} |S| - 2\|C^*\|$, so $\sum_{S \in \mathcal{P}_f} |S| < 3|S^*(\mathcal{P})|$ (since $\|C^*\| \le |S^*(\mathcal{P})|$). However, by Lemma 16.17.5, $|H| \le \sum_{S \in \mathcal{P}_f} |S| + \|C^*\|$, so $|H| \le \sum_{S \in \mathcal{P}_f} |S| + |S^*(\mathcal{P})|$ and $|H| \le 4|S^*(\mathcal{P})|$. $\square$

The reader should appreciate the centrality of the Overlap Lemma in the analysis of the factor-of-four algorithm. The Lemma is also central in the factor-of-three algorithm presented next. Hence, it should be intuitive that improvements in the Overlap Lemma or its application could lead to reductions in the guaranteed error bounds. Such an approach is taken in [246] and [77], where it is shown (essentially, but not exactly) that the overlap of α and α' can be limited to $\frac{2}{3}(|\phi| + |\phi'|)$ rather than $|\phi| + |\phi'|$. A related approach is followed in [29]. The details are too involved for this discussion. The factor-of-three improvement that we discuss next follows a different idea, but improvements in the Overlap Lemma have been applied to reduce that bound as well.

16.17.6. Improvement to a factor of three

The motivating idea is to ask: Why does the factor-of-four algorithm just concatenate the strings in $\mathcal{P}'$? Instead, why not take the strings in $\mathcal{P}'$ as input to another execution of the algorithm (steps 1 and 2)? If that execution finds overlaps between strings in $\mathcal{P}'$ then it should produce a shorter superstring than is obtained by merely concatenating the strings in $\mathcal{P}'$.

Unfortunately, we don't gain anything by executing steps 1 and 2 on $\mathcal{P}'$. The problem is that the shortest cycle cover of $\mathcal{P}'$ will be the same as the shortest cycle cover of $\mathcal{P}$ because each string in $\mathcal{P}'$ would just wrap up again to form the cyclic string of C^* that it was derived from. We leave the formal proof of this as an exercise. So no progress would be made by running steps 1 and 2 again on $\mathcal{P}'$. However, we will now establish that if every cyclic string in the cover of $\mathcal{P}'$ has at least two strings that are associated with it, then (with the use of another small idea) there will be additional overlaps between strings in $\mathcal{P}'$ leading to a provable shortening of the resulting superstring.

Definition Define a *nontrivial* cycle cover of $\mathcal{P}'$ as a cycle cover of $\mathcal{P}'$ where at least two strings in $\mathcal{P}'$ are associated to every cyclic string in the cover.

We will discuss later how to find a minimum-length nontrivial cycle cover of a set of strings. We now present an improved approximation algorithm assuming such a cover can be found. Although stated somewhat differently, the first two steps are the same as in the previous (factor-of-four) algorithm.

Factor-of-three approximation algorithm

1. Find a minimum length cycle cover $C^*(\mathcal{P})$ of $\mathcal{P}$ and associate each string $S \in \mathcal{P}$ with exactly one cycle in $C^*(\mathcal{P})$ that S maps to.

2. For every cyclic string ϕ in $C^*(\mathcal{P})$, form the list of indices L_ϕ and create the string $S(L_\phi)$, a superstring of the strings associated with ϕ. Let $\mathcal{P}' = \{S_1', S_2', S_3', \dots, S_r'\}$ denote the set of superstrings obtained.

3. Find a minimum-length *nontrivial* cycle cover $C^*(\mathcal{P}')$ of the set of strings $\mathcal{P}'$.

4. For each cyclic string $\bar\phi$ in $C^*(\mathcal{P}')$, let $L_{\bar\phi} = o_1, o_2, \dots, o_t$ be the indices of the strings of $\mathcal{P}'$ (not the strings of $\mathcal{P}$) associated with $\bar\phi$, in order of their starting positions in $\bar\phi$. By

definition of a nontrivial cycle cover, t is at least two. Each cyclic shift of $L_{\bar{\phi}}$ defines a superstring of the strings that map to $\bar{\phi}$. Find the cyclic shift that gives the shortest of those superstrings. That is, choose $o_i \in L_{\phi'}$ to minimize $ov(s'_{o_i}, s'_{o_{i+1}})$, and shift $L_{\bar{\phi}}$ to make s'_{o_i} the final string. Call the resulting ordering of strings $L'_{\bar{\phi}}$, and form the string $S(L'_{\bar{\phi}})$, which is a superstring of the strings in $\mathcal{P}'$ associated with $\bar{\phi}$.

5. Concatenate the superstrings created in step 4 to form the string $H'(\mathcal{P})$.

Clearly, $H'(\mathcal{P})$ is a superstring of $\mathcal{P}'$ and hence of $\mathcal{P}$.

Error analysis

Definition   For any set of strings Q and any superstring q of Q, the *compression* of Q achieved by q is defined to be $\|Q\| - |q|$ and is written $comp(q)$.

Lemma 16.17.7. *If the set Q contains k strings $\{q_1, q_2, \ldots, q_k\}$ and Π is a permutation of the integers 1 to k, then the compression of Q achieved by superstring $S(\Pi)$ is $\sum_{i=1}^{k-1} ov(S_{o_i}, S_{o_{i+1}})$. Further, the shortest superstring of Q is the superstring achieving the maximum compression of Q.*

The proof of Lemma 16.17.7 is essentially contained in the discussion of objective functions in Section 16.17.2 (see page 429). A formal proof is left as an exercise.

Definition   Let $S^*(\mathcal{P}')$ denote the shortest superstring of $\mathcal{P}'$ and let $comp^* = \|\mathcal{P}'\| - |S^*(\mathcal{P}')|$ denote the amount of compression of $\mathcal{P}'$ achieved by $S^*(\mathcal{P}')$.

The following lemma bounds $comp^*$ from above. Although this may at first seem to be the wrong direction to be of interest, the bound actually turns out to be of use.

Lemma 16.17.8. $comp^* \leq 2\|C^*(\mathcal{P})\|$. *That is, the compression achieved by the shortest superstring of $\mathcal{P}'$ is at most $2\|C^*(\mathcal{P})\|$, where $C^*(\mathcal{P})$ is the minimum-length cycle cover of $\mathcal{P}$ computed in step 1 of the algorithm.*

PROOF   Suppose $\mathcal{P}'$ has r strings. The shortest superstring of $\mathcal{P}'$ is defined by some permutation $o_1, o_2, \ldots, o_r$ of the integers from 1 to r, and $comp^* = \sum_{i=1}^{r-1} ov(S'_{o_i}, S'_{o_{i+1}})$. Each string in $\mathcal{P}'$ is a string $S(L_\phi)$ obtained from a distinct cyclic string $\phi \in C^*(\mathcal{P})$, and each such $S(L_\phi)$ maps to the cyclic string ϕ from which it was obtained. Hence by the Overlap Lemma (16.17.6), the maximum possible overlap between two strings in $\mathcal{P}'$, obtained from cyclic strings ϕ and ϕ' say, is less than $|\phi| + |\phi'|$. It follows that $comp^* = \sum_{i=1}^{r-1} ov(S'_{o_i}, S'_{o_{i+1}}) < 2\|C^*(\mathcal{P})\|$.   □

Definition   For a cycle cover C of $\mathcal{P}'$, the *compression* of $\mathcal{P}'$ achieved by C is defined as $\|\mathcal{P}'\| - \|C\|$ and is written $comp(C)$. Let ϕ be a cyclic string in C and let Q be a set of strings that are associated with ϕ. Then the compression of Q achieved by ϕ is defined as $\|Q\| - |\phi|$.

Lemma 16.17.9. $|H'(\mathcal{P})| \leq |S^*(\mathcal{P}')| + comp^*/2$.

PROOF   Recall that $C^*(\mathcal{P}')$ is the minimum-length nontrivial cycle cover of $\mathcal{P}'$. By the definition of compression and the fact that $\|C^*(\mathcal{P}')\| \leq |S^*(\mathcal{P}')|$, it follows that $\|\mathcal{P}'\| - comp(C^*(\mathcal{P}')) = \|C^*(\mathcal{P}')\| \leq |S^*(\mathcal{P}')| = \|\mathcal{P}'\| - comp^*$. Hence $comp(C^*(\mathcal{P}')) \geq comp^*$.

Recall that in step 3 of the algorithm S'_{o_i} denotes the final string of $S(L'_{\bar{\phi}})$. $C^*(\mathcal{P}')$ is a nontrivial cover, so at least two strings of $\mathcal{P}'$ are associated with $\bar{\phi}$. Further, o_i was chosen to minimize $ov(S'_{o_i}, S'_{o_{i+1}})$. It follows that $ov(S'_{o_i}, S'_{o_{i+1}})$ is at most half the

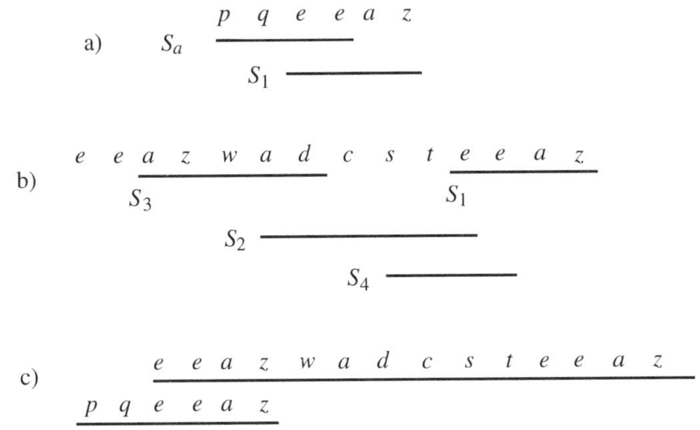

a) S_a $\underset{S_1}{\underline{\quad}}$ $p\quad q\quad e\quad e\quad a\quad z$

b) $\underset{S_3}{\underline{\quad}}$ $e\quad e\quad a\quad z\quad w\quad a\quad d\quad c\quad s\quad t\quad e\quad e\quad a\quad z$ $\underset{S_1}{\underline{\quad}}$

S_2 _____

S_4 _____

c) $e\quad e\quad a\quad z\quad w\quad a\quad d\quad c\quad s\quad t\quad e\quad e\quad a\quad z$

$p\quad q\quad e\quad e\quad a\quad z$

Figure 16.20: Expansion of $S^*(\mathcal{P}_f)$. String $S_1 = eeaz$ is the final string and $S_3 = azwad$ is the first string, S_l, in $S(L_\phi)$ shown in Figure 16.19. If $b = 1$, so that $eeaz$ is the last string in $S^*(\mathcal{P}_f)$, and $S_a = pqeea$ say, then S_a overlaps $pref(S_1, S_3)S(L_\phi)$ by exactly the same amount it overlaps S_3. The length of the expansion is exactly $|\phi|$.

compression achieved by $\bar\phi$. But $ov(S'_{o_i}, S'_{o_{i+1}})$ is precisely the additional length that is added to $|\bar\phi|$ to obtain string $S(L'_{\bar\phi})$ from $\bar\phi$. Now the compression of $\mathcal{P}'$ achieved by $H'(\mathcal{P})$ is the compression of $\mathcal{P}'$ achieved by $C^*(\mathcal{P}')$ minus those added lengths needed to create $H'(\mathcal{P})$ from $C^*(\mathcal{P}')$. Hence, $comp(H'(\mathcal{P})) \geq comp(C^*(\mathcal{P}')) - comp(C^*(\mathcal{P}'))/2 = comp(C^*(\mathcal{P}'))/2 \geq comp^*/2$. In summary, $comp(H'(\mathcal{P})) \geq comp^*/2$.

Using the facts established above, we finally have $|H'(\mathcal{P})| = \|\mathcal{P}'\| - comp(H'(\mathcal{P}) \leq \|\mathcal{P}'\| - comp^*/2 = \|\mathcal{P}'\| - comp^* + comp^*/2 = |S^*(\mathcal{P}')| + comp^*/2$.   □

Lemma 16.17.10. $|S^*(\mathcal{P}')| \leq |S^*(\mathcal{P})| + \|C^*(\mathcal{P})\| \leq 2|S^*(\mathcal{P})|$.

PROOF   To prove this lemma, we will create a superstring of $\mathcal{P}'$ whose length is at most twice the length of $S^*(\mathcal{P})$, the shortest superstring of $\mathcal{P}$.

Recall that $\mathcal{P}_f$ is the set of final strings obtained from the cycle cover $C^*(\mathcal{P})$, and let $S^*(\mathcal{P}_f)$ be a shortest superstring of $\mathcal{P}_f$. Since $\mathcal{P}_f \subseteq \mathcal{P}$, $S^*(\mathcal{P}_f)$ has length at most $|S^*(\mathcal{P})|$. We will create a superstring for $\mathcal{P}'$ in the following way: Let S be any string from $\mathcal{P}_f$ and let ϕ be the cycle in $C^*(\mathcal{P})$ that S is associated with. Therefore, S is the final string in $S(L_\phi)$, and $S(L_\phi)$ is a member of $\mathcal{P}'$. We will "expand" the copy of S in $S^*(\mathcal{P}_f)$ to contain $S(L_\phi)$, while still maintaining the overlaps found in $S^*(\mathcal{P}_f)$. When this has been done for each string S in $\mathcal{P}_f$, the resulting string will be a superstring for $\mathcal{P}'$. We now describe how to expand the strings.

Let $S_b \in \mathcal{P}_f$ be the last string in $S^*(\mathcal{P}_f)$ and let $S_a \in \mathcal{P}_f$ be the string to its immediate left in $S^*(\mathcal{P}_f)$. Let $\phi \in C^*(\mathcal{P})$ be the cyclic string that S_b is associated with, and let $S(L_\phi) \in \mathcal{P}'$ be the string created in step 2 from ϕ with final string S_b. Let S_l denote the first string in $S(L_\phi)$, and now form the string $pref(S_b, S_l)S(L_\phi)$. Since ϕ is a cyclic string, and S_b is the final string of $S(L_\phi)$, string S_b is both a prefix and a suffix of string $pref(S_b, S_l)S(L_\phi)$. (For an example, see Figure 16.20) Hence, S_a overlaps $pref(S_b, S_l)S(L_\phi)$ by at least as much as it overlaps S_b. Furthermore, the length of $pref(S_b, S_l)S(L_{c_b})$ is exactly $|\phi| + |S_b|$. So we can replace S_b with $pref(S_b, S_l)S(L_\phi)$ in $S^*(\mathcal{P}_f)$ and overlap S_a by as much as before. The net increase in the length of the resulting string is exactly $|\phi|$. By iterating this argument for each string in $\mathcal{P}_f$, moving left in $S^*(\mathcal{P}_f)$ with each iteration, we obtain a superstring for $\mathcal{P}'$ that has length at most $|S^*(\mathcal{P}_f)| + \|C^*(\mathcal{P})\| \leq |S^*(\mathcal{P})| + \|C^*(\mathcal{P})\| \leq 2|S^*(\mathcal{P})|$.

That superstring certainly has length no less than the shortest superstring of $\mathcal{P}'$, and the lemma is proved. $\square$

Theorem 16.17.3. $|H'(\mathcal{P})| < 3|S^*(\mathcal{P})|$.

PROOF   This follows immediately by combining the three previous lemmas. $\square$

16.17.7. Efficient implementation

Step 1 of both the factor-of-four and factor-of-three approximation algorithms finds a cycle cover of $\mathcal{P}$ of minimum total length. Such a cover can be found by reducing the problem to an instance of the "assignment problem". If $\mathcal{P}$ contains k strings, then input to an instance of the assignment problem is a k by k matrix A, where entry $A(i, j)$ is some number. A *complete assignment* in A is a set M of k cells from A such that every row and every column contains exactly one cell from A. The *weight of assignment M* is the sum of the numbers in the cells of M.

To see the relationship between a complete assignment and a cycle cover of $\mathcal{P}$, define a *cycle* of M as a set of distinct indices $\{i_1, i_2, \ldots, i_t\}$ such that $A(i_1, i_2)$ is in M, $A(i_2, i_3)$ is in $M, \ldots, A(i_t, i_1)$ is in M. Then a complete assignment defines a set of cycles of M that include the indices 1 to k exactly once. That is, suppose $A(i_1, i_2)$ is in M; if $A(i_2, i_1)$ is also in M then a cycle of the two indices $\{i_1, i_2\}$ is defined. Otherwise, $A(i_2, i_3)$ is in M for $i_3 \neq i_2$, etc. until finally $A(i_t, i_1)$ is in M for some index i_t. All the indices in this sequence then form a cycle of M. Since an index appears exactly once as a row index and exactly once as a column index in the cells of a complete assignment, no index can be in more than one cycle defined by a complete assignment. But because every index must be in some cycle, the cycles defined by M include each index exactly once.

By letting each index i in the assignment problem correspond to string S_i in $\mathcal{P}$, each cycle in M defines a cyclic string ϕ. By setting $A(i, j)$ to $ov(S_i, S_j)$, for each pair (i, j), a cycle cover of $\mathcal{P}$ with minimum total length is derived from a maximum weight complete assignment. The proof is left as an exercise. Note that, with the definition of an assignment, $A(i_1, i_1)$ can be in the assignment and hence will create a separate, *trivial* cycle of M.

There are several well-known algorithms that find a maximum weight complete assignment in $O(k^3)$ time when the weights are arbitrary [294]. However, when each $A(i, j)$ is set to $ov(S_i, S_j)$, a greedy assignment method runs in $O(k^2 \log k)$ time and also finds a maximum weight complete assignment. That greedy method hence finds a minimum-length cycle cover faster than the general assignment method does. These claims are explored in Exercises 13, 14, and 15.

Nontrivial cycle cover   Step three of the factor-of-three approximation calls for finding a nontrivial cycle cover of minimum-length. That can be accomplished in $O(k^3)$ time also by setting $A(i, i) = -\infty$, and then finding the resulting maximum weight complete assignment. Those values force a complete assignment that never assigns an index to itself.

How to form the matrix efficiently   To determine the elements $A(i, j)$ in matrix A, we need to compute the value $ov(S_i, S_j)$ for each ordered pair of strings S_i, S_j in $\mathcal{P}$. This can be done efficiently. In Section 7.10 we showed how to compute all k^2 of the needed suffix-prefix matches in $O(\|\mathcal{P}\| + k^2)$ time. Its use in the superstring problem was the main motivation for that result.

16.18. Sequencing by hybridization

There is an interesting and important special case of the superstring problem that is (in its idealized form) simpler than the general superstring problem. That special case of the superstring problem arises in a proposed method to sequence DNA, called *sequencing by hybridization* (*SBH*). Variants of the initial proposal are currently under extensive experimental study and partial commercial development (see [368], [134], [135], and [98]). The description given here of the experimental aspects of SBH will be a gross oversimplification but will serve to illustrate the essential flavor of the method and introduce several (idealized) computational problems involved with SBH.

Idealized sequencing by hybridization

In the sequencing by hybridization method one first creates a physical matrix (called a *chip*) of 4^k cells, for a fixed number k. Each cell holds a known distinct copy of one of the 4^k DNA strings of length k. (How the matrix is produced is an interesting story in itself [15, 163].) Then, many copies are made of the target DNA string to be sequenced and a radioactive or flourescent tag is attached to each of these copies. The set of target DNA copies is then put in contact with the matrix, with the result that a copy of the target will hybridize to a particular k-length string χ if and only if χ is a substring of the target DNA string. Each k-length string χ is affixed to an individual cell, and hence any target copy that hybridizes to χ gets affixed to that cell as well. When all the nonhybridized DNA copies are washed off the matrix, the cells of the matrix containing tagged strings specify the set of k-length substrings of the target DNA. The locations of the tagged cells are easily determined by machine, and since each cell corresponds to a known k-length string, the output of the experiment is a complete listing $\mathcal{L}$ of all the k-length substrings that appear in the target DNA. Note, however, that no substring is included in $\mathcal{L}$ more than once, no matter how many times it appears in the target string. That list $\mathcal{L}$ is the input to the following computational problem:

Definition The *SBH problem* is to determine as much as possible about the target DNA string S from the list $\mathcal{L}$ of all k-length substrings that appear in S. In particular, if possible, uniquely determine the original string S from list $\mathcal{L}$.

The number k is chosen long enough (in most proposals it is around ten) so that under *random* conditions no substring is likely to appear more than once in the target string S. Thus, in the *idealized* version of the SBH problem, we assume that no k-length substring appears more than once in S. Clearly, the target string S is a common superstring of each of the substrings in $\mathcal{L}$, and with the assumption that no substring in $\mathcal{L}$ appears more than once, S is the shortest common superstring of the set $\mathcal{L}$. However, the set $\mathcal{L}$ has more structure than an arbitrary instance of the superstring problem, because any two consecutive k-length substrings in S overlap by $k-1$ nucleotides. That high level of overlap can be exploited to obtain a very efficient and elegant way to model and attack the idealized problem. In fact, the *SBH* problem reduces to questions about finding Euler paths in a directed graph. This reduction was developed by P. Pevzner [366].

16.18.1. Reduction to Euler paths

Given the list $\mathcal{L}$ of all the k-length substrings in the target string S, construct a directed graph $G(\mathcal{L})$ as follows: Create 4^{k-1} nodes, each labeled with a distinct $(k-1)$-length DNA string. For each string χ in $\mathcal{L}$, direct an edge from the node labeled with the left-most

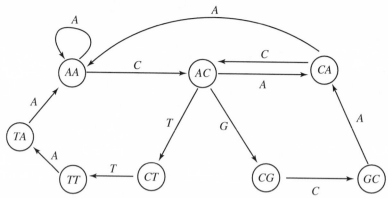

Figure 16.21: Directed graph $G(\mathcal{L})$ derived from the list $\mathcal{L} = AAA, AAC, ACA, CAC, CAA, ACG, CGC, GCA, ACT, CTT, TTA, TAA.$

$k-1$ characters of χ to the node labeled with the right-most $k-1$ characters of χ. That edge is labeled with the right-most character of χ. Note that some nodes of $G(\mathcal{L})$ may not touch any edges, and these nodes can be removed.

For an example, see Figure 16.21.

Definition   An Euler *path* in a directed graph G is a directed path that traverses every edge in G exactly once. An Euler *tour* is an Euler path that starts and ends at the same node.

A path in $G(\mathcal{L})$ *specifies* a string S in the following way: String S begins with the label of the first node on the path and follows thereafter with the concatenation, in order, of the labels on the edges it traverses. For example, the Euler path $\{AC, CA, AC, CG, GC, CA, AA, AC, CT, TT, TA, AA, AA\}$ specifies the string $S = ACACGCAACTTAAA$. Notice that all the three-tuples listed in Figure 16.21 occur in S and that no other three-tuples appear there. We can now formalize these observations.

Definition   A string S is called *compatible* with $\mathcal{L}$ if and only if S contains every substring in $\mathcal{L}$ and (assuming that $\mathcal{L}$ contains k-length substrings) S contains no other substrings of length k.

Theorem 16.18.1.   *A string S is compatible with $\mathcal{L}$ if and only if S is specified by an Euler path in $G(\mathcal{L})$.*

The proof of the theorem is immediate from the definitions of $\mathcal{L}$ and $G(\mathcal{L})$ (and is left to the reader).

Corollary 16.18.1.   Assume that each substring in $\mathcal{L}$ occurs in the target DNA string exactly once. Then the dataset $\mathcal{L}$ unambiguously determines the target DNA string if and only if $G(\mathcal{L})$ has a *unique* Euler path.

Corollary 16.18.2.   There is a one-one correspondence between Euler paths in $G(\mathcal{L})$ and strings that are compatible with $\mathcal{L}$. That is, each compatible string corresponds to a distinct Euler path, and each Euler path specifies a distinct compatible string.

If the data were collected correctly and no k-length substring appears in the target more than once (as assumed), then $G(\mathcal{L})$ must have at least one Euler path specifying the target DNA string. The difficulty arises when there is more than one Euler path in $G(\mathcal{L})$, but even in that case, it might be possible to extract some information about the target string from the set of possible Euler paths in $G(\mathcal{L})$. Therefore (ideally) the task of reconstructing the target string from the dataset $\mathcal{L}$ reduces to a problem concerning Euler paths in directed graphs.[8]

As can be expected, the idealized version of the SBH problem rarely arises in practice. As always, there is error in the data, and when repetitive DNA is likely, it is not reasonable to assume away repeated substrings in $\mathcal{L}$. For a full discussion of the sequencing by hybridization approach and the computational problems involved, see [368] and [135].

16.18.2. Continuity of compatible strings

In the case that there is more than one Euler path through the edges of the graph, there is more than one way to construct a compatible string from the hybridization data $\mathcal{L}$. That ambiguity complicates the problem of finding the target DNA string, but it might still be possible to extract useful information about the true target or about the set of all compatible strings. For example, it may be that there are certain substrings contained in all the compatible strings. Those substrings therefore are in the target string, and the ambiguities are only in the remaining parts of the string. Hence one would like to study the structure of the *set* of all compatible strings, or equivalently, the set of all Euler paths in $G(\mathcal{L})$.

There are a couple of ways of getting at the set of all Euler paths in a graph. One approach is based on a relationship between the set of all spanning trees and the set of all Euler paths of a graph (described in [144]), coupled with the fact that the set of all spanning trees of a graph can be conveniently enumerated. We will not take that approach here. Another approach is based on a conjecture due to Ukkonen [440] that was proved (in a broader context) by Pevzner [372]. In that approach, we transform any compatible string to any other compatible string by a series of simple string operations, such that each intermediate string produced along the way is itself a compatible string. This provides a kind of continuity among the set of strings that are compatible with the hybridization data.

Definition   Let S be the string $\alpha\beta\gamma\delta\epsilon$, where each Greek symbol denotes a substring of S, and where substrings β and δ are nonempty. A *string rotation* in S is an operation on S that produces the string $\alpha\delta\gamma\beta\epsilon$. That is, the nonempty substrings β and δ are rotated around the (possibly empty) substring γ.

For example, recall the compatible string $S = ACACGCAACTTAAA$ from the previous example. Consider the particular choice $S = AC\,A\,C\,GCA\,ACTTAAA$, where spaces separate the substring choices, and so $\alpha = AC, \beta = A, \gamma = C, \delta = GCA$, and $\epsilon = ACTTAAA$. With those choices, the string rotation in S creates the string $S' = AC\,GCA\,C\,A\,ACTTAAA$. As verified in Figure 16.21, S' is also specified by an Euler tour in $G(\mathcal{L})$, and hence S' is also a compatible string. We will prove below that this is not an accident.

Note that a string rotation is a generalization of a *cyclic rotation* (see Exercise 1 on page 11). In particular, a cyclic rotation of a string S is a string rotation in which α, γ, and ϵ are all empty, so $S = \beta\delta$. Then, the string $\delta\beta$ results from a cyclic rotation of S. We can now state the following continuity theorem for compatible strings.

Theorem 16.18.2. (SBH continuity theorem). *Let S and S' be any two distinct strings that are compatible with given hybridization data. S can be transformed to S' by a series*

---

[8]   In fact, each graph $G(\mathcal{L})$ is a subgraph of the well-known de Bruin graphs, so questions about Euler paths in $G(\mathcal{L})$ are related to questions about Euler paths in de Bruin graphs (see [419] for an interesting discussion of Euler paths in de Bruin graphs).

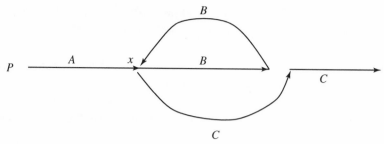

Figure 16.22: Cartoon of an Euler path P corresponding to string S. The subpaths P and P' are identical until point x.

of string rotations, such that each intermediate string in the series is itself compatible with the hybridization data.

PROOF   Let P and P' be the Euler paths in $G(\mathcal{L})$ that specify strings S and S', respectively. We will transform P to P' via a series of intermediate Euler paths, using operations that correspond to string rotations.

To begin, if P and P' have different first nodes, we will cyclicly rotate S so that the resulting string S'' is specified by an Euler path P'' that does have the same first node as P'. To prove this is possible, note that if $G(\mathcal{L})$ does not contain an Euler tour but does contain an Euler path, then $G(\mathcal{L})$ has a unique node with more in edges than out edges (and a unique node with more out edges than in edges). Hence, in this case, that unique node must start *every* Euler path in $G(\mathcal{L})$. So if P and P' do begin with different nodes, there must be an Euler *tour* in $G(\mathcal{L})$. But then the number of in edges must be the same as the number of out edges at *every* node, and every Euler path must in fact be an Euler tour. In particular, P must be an Euler tour.

Define P'' as the Euler tour (path) that starts with the same node v that P' starts with and then traverse the edges of $G(\mathcal{L})$ in the same (cyclic) order that P does. The string S'' described by P'' is a cyclic rotation of string S. In detail, let α be the prefix of S specified by path P up to the point where P first enters v, and let γ be the remainder of S (specified by the remainder of P). Path P'' specifies the string $S'' = \gamma\alpha$. Therefore, one initial cyclic rotation of S creates a string with a corresponding Euler path that starts at the same node where P' starts.

Given the above argument, we will assume here that P and P' start with the same first node. Let A denote the initial subpath (possibly containing no edges) on which P and P' are identical, and let x be the endpoint of A. That is, P and P' agree until point x but continue on two different edges e and e', respectively, out of x. (The point of the preceding paragraph was to establish that node x exists.) Since edge e' is not part of subpath A, but must be part of the whole path P, path P must return to node x some time after edge e is traversed. Let B denote the subpath of P starting with edge e and ending just before edge e' is traversed. Let C denote the remaining subpath of P (see Figure 16.22).

We claim that subpaths B and C meet at least once, say at a point y (see Figure 16.23). To see this, note that Euler path P' leaves subpath A along edge e', which is on subpath C, but P' must later traverse edge e, which is on subpath B. Hence there must be a first point, designated y, along C where C visits a point on subpath B.

The existence of y implies that we can create an Euler path P'' that agrees with P' on subpath A and also on edge e'. Path P'' starts with subpath A, then traverses the first part of C until reaching point y, then follows with the second part of B from point y to

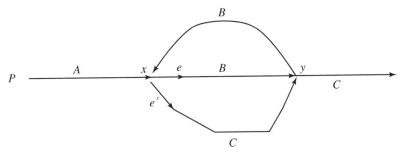

Figure 16.23: Subpaths B and C must meet at a point denoted y.

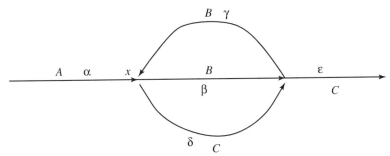

Figure 16.24: The relationship of Euler paths P and P'' to strings S and S''. A path rotation corresponds to a string rotation.

point x, then continues with the first part of B until y, and finally finishes with the second part of C. Path P'' is an Euler path containing an initial subpath that agrees with P' for at least one edge more than P does. Repeating this argument as many times as needed, we create a series of Euler paths whose initial subpaths agree with increasingly long initial subpaths of P.

The transformation from P to P'' is completely general, since we assumed only that P and P' are Euler paths in the graph. We call this transformation a *path rotation*. It follows that any Euler path P in G can be transformed to any other Euler path P' in G by a series of path rotations, where each intermediate path P'' generated in this way is itself an Euler path. To finish the proof of the theorem, we need to relate a path rotation to a string rotation.

How do the strings S and S'' generated from P and P'' relate to each other? Given the existence of point y, we decompose S into five substrings: $S = \alpha\beta\gamma\delta\epsilon$, where α is the substring of S generated along subpath A, β is the substring of S generated along subpath B until point y is reached, γ is the substring of S generated along the remaining part of subpath B, δ is the substring of S generated along C until point y, and ϵ is the remaining part of S generated along subpath C. (See Figure 16.24.) With this notation, Euler path P'' generates the string $S'' = \alpha\delta\gamma\beta\epsilon$. Hence S'' is created from S by a single *string rotation*. Each path rotation induces a string rotation and the theorem follows.   □

16.18.3. Last comments on SBH

Theorem 16.18.2 is very elegant and can be further exploited to give information about the target DNA string (see [204] and [210] for results of this type). We again, however, emphasize that the theorem concerns an idealized version of the SBH problem that only loosely models reality.

Although originally proposed as a method for fully sequencing strings of DNA, recent research and commercial development[9] of matrix hybridization methods and hybridization chips have been directed at more modest goals. Three applications seem the most promising: *sequence verification and correction*, sequence-based *genetic diagnostics*, and *monitoring of gene expression*.

The idea behind the first application is that one might do quick-and-dirty sequencing using either directed or shotgun methods, and then do an independent run of SBH to try to spot errors in the sequence. "Quick-and-dirty" can have three meanings: Using individual sequence reads well beyond the normal 300–600 base length, where the error rate is high; performing sequencing without redundancy; or performing sequencing without internal checks for consistency or error-spotting. The quick-and-dirty sequence would be checked to see if it is compatible with the SBH data. If not, then the SBH data and its set of solutions might be used to suggest places where the first sequence needs to be further investigated.

The second application, to genetic diagnostics, is based on cataloging known mutations causing genetic diseases. Knowing the "normal" sequence for a given gene, and knowing the common disease-causing mutations in that gene, one can design a hybridization chip that recognizes if the input DNA deviates from the normal sequence, and if so, by which common mutation (for example, see [98]). This is a cheap alternative to completely sequencing an individual's gene in order to see which, if any, of the known mutations of the gene is carried by that individual. Along the same lines, a catalog of the DNA of known pathogens (or any set of organisms) can be encoded onto a chip, so that the SBH solution could very quickly identify which particular organism is in hand. In the case of bacterial and viral diseases, this approach may be much faster and cheaper than existing pathogen culture/identification methods.

The third application provides a tool to study the central question of gene activation and control: When (under what circumstances) and where (in which cells) are particular genes activated? This is a critical question in biology that remains mostly unanswered. Hybridization matrices can be used to cheaply monitor which genes are active (being transcribed) by monitoring the messenger RNA that is being produced in a given cell or organ at any particular time. The idea is that each gene of interest will have a distinctive "signature" or hybridization pattern, making its mRNA or cDNA easily recognizable. One problem with this approach is that the amount of mRNA produced is not always proportional to the amount of protein being produced – some mRNA is highly stable and can be reused to produce more than one protein molecule, whereas other mRNA is easily degraded.

16.19. Exercises

1. Prove Theorem 16.13.1 on page 419.

2. The discussion of shotgun sequence assembly was a simplification of the real situation encountered in practice, and many details were omitted. There is one complication, however, that should be noted. The DNA to be sequenced starts as doubled-stranded DNA, but only a single-stranded DNA molecule can be sequenced by common methods. So the two strands are separated and then both strands are cut randomly. The result is that the set of fragments contains pieces from both DNA strands, and present methods can not

---

[9] There are two companies primarily focused on SBH technology: Affymetrix and Hyseq, both located in Silicon Valley.

determine which strand a fragment came from. Each fragment is sequenced from its $5'$ end to its $3'$ end, but on one strand that sequencing is "left to right" while on the other strand it is "right to left". Therefore, the sequence assembly problem is really to simultaneously assemble the two complementary sequences of the two strands, given a set of fragments covering both strands. Again, the complication is that one doesn't know which strand any particular fragment comes from. Clearly, you do not want to enumerate, in an exponential procedure, the strand choice for each fragment. Explain how to extend sequence assembly techniques discussed in the book to handle this problem. Explain the connection of this problem to the *orientation-free superstring problem* mentioned on page 427.

3. Recent efforts in sequence and fragment assembly have focused on obtaining and using additional side information that makes the assembly problem easier to solve. One type of side information is to learn, for some pairs of fragments, which of the pair is to the right of the other. Another type of side information specifies, for some pairs of fragments, a rough distance from one fragment to the other. This information can be obtained by using parts of the two fragments as PCR primers on the original full string and then measuring the length of the amplified substring. A third idea, called *double-barreled shotgunning*, is to sequence or fingerprint *both ends* of each fragment. The benefit is that one has two fingerprints for each fragment, and their relative positions are fixed and known. Discuss how this information can be incorporated into fragment assembly methods and how much it helps.

4. **Added sequencing coverage**

   For additional reliability, sequencing projects sometimes find and sequence more than one set of DNA fragments that cover the DNA string of interest. Each nucleotide in the DNA of interest must be in several covers, but no DNA fragment should be in more than one cover. That is, the covers should be fragment disjoint. We formalize this as follows: Given a parameter k (the desired *coverage*), select a subset of fragments with minimum total length so that k fragment-disjoint covers can be constructed from the set. This problem can be solved efficiently as a *minimum-cost network flow* problem on a directed graph. Flesh out the details and prove the correctness of this approach.

   There are many well-known ways and packaged programs to solve the min-cost flow problem. We don't know if any are presently used in practice for this fragment selection problem.

5. Prove Theorem 16.8.4 on page 410.

6. In the tightest clone layout solution, two clones were permitted to be placed arbitrarily close to one another, as long as there was some nonzero space between them. Discuss how the problem and solution change if the layout is required to have a fixed minimum distance between the left ends of any two clones.

7. Does Lemma 16.17.6 hold if $|\phi| + |\phi'|$ is replaced by $\max(|\phi|, |\phi'|)$? Give a proof or a counterexample. Why would this help, if it were true?

8. In the factor-of-three approximation algorithm for the superstring problem and its analysis, we assumed (without stating it) that the set $\mathcal{P}'$ is substring free. Prove that it is.

9. Suppose we could compute a minimum length cycle cover in which every cycle has at least three strings map to it. Would that reduce the error bound from a factor of three? If so, what would the bound be?

10. Explain why a minimum length cycle cover of the set $\mathcal{P}'$ (without the requirement that the cover be nontrivial) would just recreate the minimum length cycle cover of $\mathcal{P}$.

11. Prove Lemma 16.17.7 on page 434.

12. Fully justify the claim that a minimum length cycle cover of set $\mathcal{P}$ can be found by computing a maximum weight complete assignment where $A(i, j) = ov(S_i, S_j)$.

13. Greedy assignment

The dominant time step in the two superstring approximation algorithms described in the text is the time needed to compute the maximum weight complete assignment. In general that can be done in $O(k^3)$ time. However, when the weights come from suffix-prefix overlaps, the maximum weight assignment and the cycle cover can be found faster and more simply by a *greedy method*.

Greedy assignment

Input: A k by k matrix A.
Output: A complete assignment M.
1. Set M to the empty set and declare all cells of A available.
2. While there is an available cell in A do
Begin
2.1 Pick the cell (i, j) in A with the largest weight $A(i, j)$ among all available cells.
2.2 Put cell (i, j) in M and declare all cells in row i and all cells in column j to be unavailable.
End
3. Output the complete assignment M.

Clearly, the algorithm runs in $O(k^2 \log k)$ time (needing only to sort the edge weights), which is an improvement over the general assignment algorithm.

Prove that the algorithm produces a complete assignment.

14. The Monge inequality

Let u, u', v, and v' be four nodes in a complete, weighted bipartite graph $G = (N_1, N_2, E)$ such that u and u' are in N_1 and v and v' are in N_2. (A complete bipartite graph is one where there is an edge between every node in the node set N_1 and every node in the node set N_2.) Suppose, without loss of generality, that $A(u, v) \geq \max[A(u, v'), A(u', v), A(u', v')]$.

If $A(u, v) + A(u', v') \geq A(u, v') + A(u', v)$ then these four nodes are said to satisfy the *Monge inequality*. A complete, weighted bipartite graph is said to satisfy the Monge inequalities if the Monge inequality is satisfied for any two arbitrary nodes from N_1 together with any two arbitrary nodes from N_2.

Prove that when $A(u, v) = ov(S_u, S_v)$, for every pair (u, v), are used as the weights in a complete bipartite graph, then the graph satisfies the Monge inequalities.

15. In general, the greedy assignment will not produce a complete assignment of maximum weight. However, if the weights on a complete bipartite graph satisfy the Monge inequalities and all the weights are nonnegative, then the greedy assignment is a maximum weight matching. Prove this claim.

16. Show that the greedy assignment method will not always find a nontrivial complete assignment of maximum weight?

17. For arbitrary edge weights, the greedy algorithm will find a complete assignment whose weight is within one half the maximum weight of any assignment. Prove this.

18. Suppose we modify the factor-of-three approximation algorithm so that the greedy algorithm were used to find a nontrivial cycle cover. This would speed up the algorithm, but because the greedy algorithm is not guaranteed to find the maximum weight nontrivial cycle cover, the error bound of three is not guaranteed.

Using the error bound of one half established in the preceding problem, what (simple) approximation bound on superstring length is guaranteed if the greedy algorithm is used to find nontrivial cycle covers?

19. We explained both the factor-of-three algorithm and the factor-of-four algorithm in terms of the assignment problem, although a greedy assignment method can be used in the factor-of-four algorithm. In the original literature, the factor-of-four algorithm was explained without reference to the assignment problem. Instead, a greedy string merging algorithm was used. A *merge* of two strings α and β is defined as the string $pref(\alpha, \beta)\beta$. *The greedy merging algorithm* finds a superstring of $\mathcal{P}$ as follows:

Copy set $\mathcal{P}$ to set $\mathcal{S}$.

Repeat

Find a pair of strings α, β in $\mathcal{S}$ with the longest overlap (i.e., longest suffix-prefix match). Replace α and β in $\mathcal{S}$ with the merge of α and β.

Until $\mathcal{S}$ contains only a single string.

Problem: Show that the needed suffix-prefix matches of pairs of strings in $\mathcal{S}$ can be determined by the suffix-prefix matches of pairs of strings in the original set $\mathcal{P}$.

20. Show that the greedy merging algorithm can be implemented in $(k^2 \log k + |\mathcal{P}|)$ time.

21. Show that the greedy merging algorithm creates a superstring of $\mathcal{P}$ achieving a compression of $\mathcal{P}$ that is at least half the compression achieved by the shortest superstring of $\mathcal{P}$. This fact was first established in [431] and [436].

22. By relating the greedy merging algorithm to the greedy assignment algorithm, prove that the greedy merging algorithm produces a superstring that is never more than four times the length of the shortest superstring.

23. The GCD Theorem discussed in the book (page 431) can be strengthened as follows:

Theorem 16.19.1. *If string S has two periods of length p and q, and $|S| \geq p+q-gcd(p,q)$, then S has a period of length gcd(p,q).*

Prove this stronger version of the GCD theorem.

24. Recall the definition of a *semiperiodic* string given in Section 3.2.1. Prove the following claim:

If no string in $\mathcal{P}$ is semiperiodic, then the factor-of-four approximation algorithm on page 431 produces a superstring of $\mathcal{P}$ whose length is at most twice that of the shortest superstring.

25. Most papers on the shortest superstring problem list text compression as an application of the shortest superstring problem. The idea, apparently, is that one can replace and store a set of n strings $\mathcal{S}$ by a superstring of $\mathcal{S}$, along with $2n$ pointers into the superstring that indicate the start and end of each string in $\mathcal{S}$. Of course, since no practical algorithms are known for computing the shortest superstring, the compression achieved in practice will be less than optimal.

As an alternative to computing superstrings, describe a simple method to compress the n strings in $\mathcal{S}$ by using a minimum length cycle cover of $\mathcal{S}$. Explain how to recover the original strings from the compressed strings.

Recall that the minimum length cycle cover of $\mathcal{S}$ can be computed efficiently and that it is no larger (and usually smaller) than the shortest superstring of $\mathcal{S}$ (which cannot be computed in practice). In light of these facts, is text compression an interesting or convincing application of the superstring problem? (Problem contributed by Paul Stelling.)

26. The implication from the previous problem is that text compression is an inappropriate (phony) application of the shortest common superstring problem. But suppose one does use the minimum length cycle cover of $\mathcal{S}$ to get a compressed representation of a set of strings. Each cyclic string in the cover will itself be represented as a *linear string* created by cutting the cyclic string at some point. Since the choice of the cutting points is arbitrary,

maybe one can choose cutting points so that the resulting set of linear strings has a short superstring. That would allow further compression of the cyclic strings and hence of the original set of strings.

Essentially, the above idea leads to the following problem: Given a set of linear strings, you may rotate any of the strings by any chosen amount. How should the strings be rotated so that the resulting common superstring is shortest? Or how should they be rotated to obtain the best, efficiently computed guarantee on the length of the shortest superstring?

Explore and discuss the above problem. Consider first the case of two linear strings. That special case has a simple solution. In general, can you prove a limit on the amount of compression possible compared to the length of the minimum cycle cover when linear strings can be rotated?

27. Prove Theorem 16.18.1 on page 438.

28. It is easy to determine whether a directed connected graph G has an Euler path. In the context of the *SBH* problem, this allows us to efficiently determine if the data are consistent with some string. But to be of most utility, we want to know if the data specify one *unique* string. Give an efficient (linear-time) algorithm to determine whether a directed graph G has a *unique* Euler path.

29. Suppose the hybridization experiments can determine how many times each substring in $\mathcal{L}$ occurs in the target string. How can this information be incorporated into the solution method?

17

Strings and Evolutionary Trees

The dominant view of the evolution of life is that all existing organisms are derived from some common ancestor and that a new species arises by a splitting of one population into two (or more) populations that do not cross-breed, rather than by a mixing of two populations into one. Therefore, the high level history of life is ideally organized and displayed as a *rooted, directed tree*. The extant species (and some of the extinct species) are represented at the leaves of the tree, each internal node represents a point when the history of two sets of species diverged (or represents a common ancestor of those species), the length and direction of each edge represents the passage of time or the evolutionary events that occur in that time, and so the path from the root of the tree to each leaf represents the evolutionary history of the organisms represented there. To quote Darwin

...the great Tree of Life fills with its dead and broken branches the crust of the earth, and covers the surface with its ever-branching and beautiful ramifications. [119]

This view of the history of life as a tree must frequently be modified when considering the evolution of viruses, or even bacteria or individual genes, but remains the dominant way that high-level evolution is viewed in current biology. Hundreds (may be thousands) of papers are published yearly that depict deduced evolutionary trees. Several biological journals are primarily focused on such trees and the methods to obtain them.

Increasingly, the methods to create evolutionary trees are encoded into computer programs and the data used by these programs are based, directly or indirectly, on molecular sequence data. In this chapter we discuss some of the mathematical and algorithmic issues involved in evolutionary tree construction and the interrelationship of tree construction methods to algorithms that operate on strings and sequences.

Biological controversies

In the world of biological systematics (or classification) and evolutionary biology, there are three major competing theories of what classification trees should mean and how they should be constructed: *evolutionary taxonomy*, *phenetics* or *numerical taxonomy*, and *cladistics*. Debate between proponents of these three theories is intense and often bewildering to people outside of evolutionary biology. Although the technical side of tree building may appear to be a matter of pure graph theory and combinatorial optimization, the fundamental issues that determine the validity of these methods are sometimes discussed in terms more suited for religion. An insightful and (fairly) neutral account of some of these tree battles can be found in [380]; a mathematical treatment of some of these issues can be found in [148], [149], and [151]. And of course, the larger debate on how to reconstruct evolutionary history goes beyond tree building, with the argument that some important aspects of molecular evolution are not treelike at all (for example, see [430]).

447

Tree-building algorithms

Despite the debates on the biological foundations of trees and tree building, most tree-building *algorithms* can be classified into two broad categories: *distance-based* methods and *maximum-parsimony* methods.[1]

In *distance-based* methods, the input to the problem consists of evolutionary distance data (such as edit distance from sequences, melting temperature from DNA hybridizations, the strength of antibody cross reactions, etc.), and the goal is to reconstruct a weighted tree whose pairwise distances "agree" with the given evolutionary distances. When the distance data are *ultrametric*, the problem has an elegant solution that will be detailed in Section 17.1. When the data are not ultrametric but are *additive*, then the problem has an efficient solution, described in Section 17.4, based on the algorithm for ultrametric distances. However, the real problems arise when the data are not additive. Then, one has to find a tree whose distances "best approximate" the given data. Different definitions of "approximate" have been suggested in the computational biology literature, and there are a variety of (usually heuristic) algorithms that build trees based on these different definitions. But on whole, the problem remains open as there is no approach that both leads to a provably efficient algorithm and that follows a completely accepted definition of a "good approximation".

Maximum-parsimony methods take a different approach and do not reduce biological data to distances. Instead, parsimony methods are *character-based* methods that work directly on *character data*, very often using aligned sequence data.[2] The goal is to build a tree, with the input taxa at the leaves and inferred taxa at the internal nodes, that minimizes the total amount or cost of the "mutations" implied by that evolutionary history. (*Minimizing mutations* is often described in the literature as *maximizing parsimony*.) When the input taxa are each represented by a molecular sequence, the maximum parsimony problem generalizes the phylogenetic alignment problem discussed in Chapter 14. The parsimony problem seeks a leaf-labeled tree whose phylogenetic alignment has the minimum cost over all possible trees.

The seminal paper in algorithmic approaches to building phylogenetic trees is by Fitch and Margoliash [160]. The reader interested in the philosophy and mechanics of real tree building, as practiced in current molecular and evolutionary biology, should look at [380] for philosophy, at [223, 429] for mechanics, at [164] for a combination, and at [127] for insightful comments. Applications of these methods to specific phylogenetic and taxonomic studies are ubiquitous and are often reported in the popular press (for a recent example, see [78]). For detailed popular press articles (concerning the ancestry of the Giant Panda and of the Red Wolf), see [351], and [470]. For illustrations from representative research-level papers, see [207], [314], and [131].

String and tree algorithms

Even from this brief description of distance and parsimony methods it should be clear that string algorithms of the type considered in this book play an important role in the ultimate success or failure of methods for evolutionary inference. This conclusion is immediate in

---

[1] Felsenstein [150] writes "Two computational methods have dominated the reconstruction of molecular phylogenies: parsimony and distance. The parsimony method finds the evolutionary tree that requires the fewest changes to nucleotides to explain evolution of the observed sequences. Distance methods compute a table of pairwise numbers of differences between sequences and try to fit this to expected pairwise distances computed from the tree".

[2] The word "character" here means an observable trait or characteristic, which sometimes is a DNA or amino acid character, but need not always be one.

the case of maximum parsimony methods that operate directly on raw sequences or on characters derived from sequences (possibly multiply aligned). It is less immediate in the case of a distance method, since those methods operate on numbers. However, often the input to a distance-based method is the output from some particular string algorithm that finds pairwise distances or computes multiple alignments.[3] Thus the model of distance or multiple alignment that is embedded in the algorithm (global, local, gap type, weighted, unweighted, sum-of-pairs, consensus, variable parameter settings, etc.) and the quality of the algorithmic solution can both have a significant impact on the reliability of the final evolutionary tree.

In the next sections we discuss several idealized versions of distance-based and maximum-parsimony tree-building problems. The discussion focuses on the "elegant and the proven" rather than the "realistic and the practical", and the results may seem too abstracted from reality to be of direct practical use. Moreover, we concentrate on *combinatorial* aspects of tree building, even though statistical issues have had a very large practical and theoretical role in the field [148, 149, 151], and maximum-likelihood methods are becoming more widely used. Nonetheless, the problems and results we discuss serve to introduce the field, and cleanly expose the underlying ideal-world models that motivate and drive most practical tree-building methods.

17.1. Ultrametric trees and ultrametric distances

17.1.1. Introduction

We begin by discussing *ultrametric trees and distances*, constructs that can be used in evolutionary reconstruction when the data "perfectly fit" certain strong assumptions. Even when the data are not perfect, ultrametric trees arise implicitly in many numerical-based tree reconstruction methods. Ultrametric trees, or approximations of them, can be used to deduce both the branching patterns of evolutionary history and some measure of the time that has passed along each branch.

Definition   Let D be a symmetric n by n matrix of real numbers. An *ultrametric tree* for D is a rooted tree T with the following properties:

1. T contains n leaves, each labeled by a unique row of D.
2. Each internal node of T is labeled by one entry from D and has at least two children.
3. Along any path from the root to a leaf, the numbers labeling internal nodes *strictly decrease*.
4. For any two leaves i, j of T, $D(i, j)$ is the label of the least common ancestor of i and j in T.

Thus, an ultrametric tree for D (if there is one) is a compact representation of the matrix D. For an example, see Figure 17.1.

Definition   A *min-ultrametric tree* for D is a rooted tree T with all the properties of an ultrametric tree, except that property 3 is changed to the following: In a min-ultrametric tree, along any path from the root to a leaf, the labels of the internal nodes must *strictly increase*.

There is no accepted term for what we have defined as "min-ultrametric", and this deficiency is sometimes a source of confusion. Note that not every matrix D necessarily

---

[3] In some reconstruction studies, the pairwise distances used are the distances of induced pairwise alignments taken from a *multiple* alignment of all the sequences.

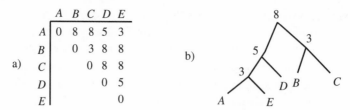

Figure 17.1: a. A symmetric matrix D. b. An ultrametric tree for matrix D.

has an ultrametric or a min-ultrametric tree representing it. An ultrametric tree, or a min-ultrametric tree, has at most $n - 1$ internal nodes, so if D has more than $n - 1$ distinct values, then there cannot be an ultrametric or a min-ultrametric tree for D.

The main mathematical and algorithmic problems are to characterize the conditions under which a matrix D can be represented by an ultrametric or a min-ultrametric tree and to develop an efficient algorithm for building these trees when possible. We will examine both of these problems in detail after relating ultrametric trees to evolutionary trees.

17.1.2. Evolutionary trees as ultrametric trees

If the true evolutionary history of n taxa forms a rooted directed tree T, with the extant taxa at the leaves of the tree, then each internal node v of T represents a *divergence event* or a historical branching. A divergence event is a point in time when the evolutionary histories of (at least) two taxa, say i and j, diverge. For simplicity, we will say that v is the point when "i and j diverge". However, this shorthand does not suggest that i is an ancestor of j, or j is an ancestor of i, or both are descendants of some other (possibly extinct) taxa. It simply says that before point v, i and j shared a common evolutionary history.

Suppose that in addition to knowing the topology of the tree (or *branching order* in the lexicon of *cladistics*), one knows the time (absolute or relative) that each divergence event occurred. Written at the nodes of the tree, those times must strictly increase along any path from the root. Moreover, if node v is the least common ancestor of leaves i and j in T, then the label on node v is the time when i and j diverged. Thus, T is a *min-ultrametric* tree for the n by n matrix D where, for each pair of leaves i and j, $D(i, j)$ is the time that i and j diverged. So a true evolutionary history (a directed tree plus times of divergence) forms a min-ultrametric tree for pairwise divergence-time data.

Equivalently, if we know the true evolutionary history, we can label each node v in T by the time that has passed since the divergence event represented by v. With that labeling, the numbers strictly decrease along any path from the root. Thus if $D(i, j)$ now represents the time since i and j diverged, then this second labeling of T makes T an ultrametric tree for D. As we will see below, real biological data more usually approximate *time since* divergence rather than *time of* divergence, so it is more natural to concentrate on ultrametric trees.

Now the real problem is that we don't know true evolutionary history, either trees or divergence times. Rather, we want to infer a plausible history from data reflecting time since divergence. Knowing that a true evolutionary history should form an ultrametric tree, the goal is to construct ultrametric trees from the time-since-divergence data. That raises an immediate question: Even if there is an ultrametric tree T for a matrix D of divergence data, how confident can we be that T actually captures the true evolutionary history we seek, or even captures some of it? We will address that question, as well as the original

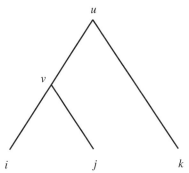

Figure 17.2: Suppose that D has an ultrametric tree. This figure shows the generic subtree containing leaves i, j, k. The subtree will generally contain other nodes that are not shown. Node v is the least common ancestor of i and j, and node u is the least common ancestor of the three leaves i, j, and k. Since this is an ultrametric tree for D, the number written at u must be strictly larger than the number at v. By definition, the number at v is $D(i, j)$ and the number at u is $D(i, k) = D(j, k)$. Hence these three numbers satisfy the condition that the maximum of $D(i, j)$, $D(i, k)$, and $D(j, k)$ is not unique. Indices i, j, k are arbitrary, so the picture is general and shows that if D has an ultrametric tree then D is an ultrametric matrix.

algorithmic question of how to build ultrametric trees, in the next section. After that, we will return to the issue of how to find biological data of greatest use in building ultrametric trees.

17.1.3. How to test for an ultrametric tree

In this section we state and prove the major theorem concerning ultrametric trees and distances and develop an efficient algorithm to build ultrametric trees when possible.

> **Definition**   A symmetric matrix D of real numbers defines an *ultrametric distance* if and only if for every three indices i, j, and k, there is a tie for the maximum of $D(i, j)$, $D(i, k)$, and $D(j, k)$. That is, the maximum of those three numbers is *not unique*. Similarly, D defines a *min-ultrametric distance* if and only if for each triple i, j, and k, there is a tie for the minimum of $D(i, j)$, $D(i, k)$, and $D(j, k)$.

When D defines an ultrametric distance, we will often say that D is an ultrametric *matrix*. Similarly, we may refer to D as a min-ultrametric matrix.

It is easy to see that if D has an ultrametric (or min-ultrametric) tree, then D is an ultrametric (or min-ultrametric) matrix. See Figure 17.2 for more explanation. The converse statement is less obvious. We next show that it is also true.

Theorem 17.1.1. *A symmetric matrix D has an ultrametric tree (or a min-ultrametric tree) if and only if D is an ultrametric (or min-ultrametric) matrix.*

PROOF   The *only-if* part of this theorem was observed above in Figure 17.2, so we will prove here that if D is an ultrametric matrix, then there is an ultrametric tree for D. This proof is constructive and leads to an efficient algorithm for constructing ultrametric trees when possible. The case of min-ultrametric matrices and trees is analogous and is left to the reader.

To build an ultrametric tree T from an ultrametric matrix D, first focus on a single leaf, say leaf i. For simplicity, assume that $D(i, i) \neq D(i, j)$ for all $j \neq i$.[4] If there are d distinct entries in row i of D, then any ultrametric tree T for D must contain a path from

---

[4] The definition of "ultrametric" usually requires $D(i, i) = 0$, but for generality we have relaxed that requirement in the definition.

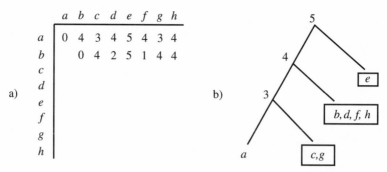

Figure 17.3: a. Two rows of a symmetric matrix D. The row for taxon a is used to obtain the path to leaf a shown in panel b. The node numbers on that path partition the remaining taxa as shown.

the root to leaf i with exactly d nodes. Moreover, each node on this path must be labeled by one of the d distinct entries in row i, and those labels must appear in decreasing order on the path. So the nodes and the labels on the path to leaf i are determined only by the entries in row i of D. Additionally, any internal node v on that path labeled $D(i, j)$ must be the least common ancestor of leaf i and leaf j. This fixes where leaf j must appear in T relative to the path to leaf i (see Figure 17.3). In this way, the path to leaf i partitions the remaining $n - 1$ leaves (the nodes other than i) into $d - 1$ classes. Call this partition $\mathcal{D}$. Leaves j and k are together in the same class of $\mathcal{D}$ if and only if $D(i, j) = D(i, k)$. It follows that each class in $\mathcal{D}$ is defined by a distinct node on the path to i. In particular, the node defining the class containing j is the node labeled with $D(i, j)$.

Given the partition $\mathcal{D}$ defined by the path to i, we would like to find a separate ultrametric tree for each of those $d - 1$ classes in $\mathcal{D}$ and then connect each resulting subtree to the node on the path to i that defines its class. That is, we would like to solve the ultrametric tree problem recursively on each class and then connect these trees to form the ultrametric tree for the full matrix D. We will now show that if D is an ultrametric matrix, then this approach works correctly.

Focus on the class defined by internal node v and let leaf j be contained in this class. Let l be some other leaf and consider three cases: l is in the same class as j, l is in a class located between leaf i and node v, or l is in a class between node v and the root of T. For example, in Figure 17.3, when i is a and j is b, then $l = d$ falls into the first case, $l = g$ falls into the second case, and $l = e$ falls into the third case. In the first case, $D(i, j) = D(i, l)$, so $D(j, l) \leq D(i, j)$ since D is an ultrametric matrix. That means that if a ultrametric subtree containing leaves j and l is attached at v, then $D(j, l)$ is correctly represented in the new tree, and the tree has the required property that node numbers strictly decrease along any path from the root. (If $D(j, l) = D(i, j)$ then node v will have degree greater than two, as happens when $l = h$ in Figure 17.3.) In the second case, $D(i, l) > D(i, j)$, so $D(j, l) = D(i, l)$. Therefore, if an ultrametric tree for the class containing j is connected at v, then $D(j, l)$ will be correctly written at the least common ancestor of leaves j and l. In the third case $D(i, l) < D(i, j)$, so $D(j, l) = D(i, j)$ and v must be the least common ancestor of j and l. Hence in all three cases, the ultrametric tree of the class defined by v can be correctly attached to v. Therefore, when D is an ultrametric matrix, this recursive approach correctly constructs an ultrametric tree for D. $\square$

Additional consequences

The proof of Theorem 17.1.1 immediately establishes several important facts.

Theorem 17.1.2. *If D is an ultrametric matrix, then the ultrametric tree for D is unique.*

To see this, note that in proving the correctness of the recursive approach, we saw that the classes in $\mathcal{D}$ defined by the path to node i were *forced*. That is, the nodes and their labels on the path to i must appear in every ultrametric tree for D. Moreover, the partition defined by that path, and the position of each specific class, was also forced and would appear in the same order in any ultrametric tree for D. Uniqueness is immediately implied by these facts and will be important in a later section. The uniqueness of a min-ultrametric tree can be established in the same way.

We should note one possible source of confusion about uniqueness. Ultrametric trees are sometimes defined without requiring that the node numbers *strictly* decrease along every path from the root. If this "strictness" is not required (contrary to the definitions we use here) and some node in the ultrametric tree has more than two children, then another ultrametric tree can be obtained by replacing that single node by a chain. In that case, Theorem 17.1.2 does not hold. However, even if "strictness" is not required, then the set of ultrametric trees obtained for D are very related in a way that we leave the reader to work out.

The fact that the ultrametric tree for an ultrametric matrix is unique answers the question posed earlier in Section 17.1.2: How informative is an ultrametric tree derived from D? That is, we are trying to deduce an unknown evolutionary tree T that is thought to have given rise to the observed data in D. How does the tree derived from D relate to the unknown T? The answer is that *if* the data in D are truly proportional to the node numbers labeling the unknown evolutionary tree T, then the ultrametric tree derived from D *must be T*. That is a very strong, and desirable, property.

Another consequence of the proof of Theorem 17.1.1 is

Theorem 17.1.3. *If D is an ultrametric matrix, then an ultrametric tree for D can be constructed in $O(n^2)$ time.*

Note in second printing: Although correct, I was mistaken in thinking that the given algorithm has an easy $O(n^2)$ time implementation. See http://www.cs.ucdavis.edu/~gusfield/ultraerrat/ultraerrat.html for an alternate combinatorial algorithm whose running time is very clearly $O(n^2)$.

It is also easy to establish that one can determine if D is an ultrametric matrix in $O(n^2)$ time. We leave that as an exercise.

17.1.4. How are ultrametric data obtained?

Theorem 17.1.2 implies that ultrametric time-since-divergence data are very powerful in reconstructing evolutionary history, both the tree topology and relative divergence times. But what hope is there that ultrametric data can be obtained in practice?

The molecular clock theory

The *molecular clock theory*, proposed in the early 1960s by Emile Zuckerkandl and Linus Pauling [489], [see also 279], states that for any given protein, *accepted mutations* in the amino acid sequence for the protein (and in the underlying DNA) occur at a constant rate. The term *accepted* here means those mutations that still allow the protein to function properly (i.e., that are not lethal to the protein; see Section 15.7.2). The implication is that the number of accepted mutations occurring in any time interval is *proportional* to the length of that interval. Hence once the clock has been calibrated, the length of an unknown interval can be measured by the number of accepted mutations that occur in that interval. Of course, this theory is affected somewhat by granularity and is only asserted to hold over "long-enough" time intervals. In addition, the rate of *accepted* mutations is different for different proteins (i.e., they have different clocks). For example, *hemoglobin* mutates faster than *cytochrome c*, but both are relatively stable (and are very similar in all mammals) compared to other proteins such as *fibrinopeptides* [11]. In fact, different parts of a protein may evolve at different rates [127], so care must be taken when invoking the clock. Underlying the molecular clock theory is the assumption that mutations (accepted or not) in all DNA occur at some constant rate. Differences in rates of accepted mutations result from differences in how constrained particular proteins are (by natural selection at the organism level or by physical chemistry at the molecular level).

A great deal of ink (and blood) has been spilt over the molecular clock, and the theory is intimately tied to another major controversy in evolutionary biology – the *neutral theory of molecular evolution*, developed by Motoo Kimura [275] and others. That theory concerns the relative importance of natural selection versus random genetic drift in evolution. For two very readable discussions on the neutral theory, one by Kimura and one by Stephen Jay Gould, see [274] and [188].

It should be clear how the molecular clock simplifies the task of collecting ultrametric data. Let A and B be two taxa that both make and utilize the "same" protein (for example hemoglobin). Suppose one can determine that k accepted mutations have occurred in the DNA or amino acid sequences for the hemoglobins of A and B since the time that A and B diverged. (By the molecular clock theory, it follows that roughly $k/2$ separate accepted mutations occurred in each of the two evolutionary histories since their point of divergence.) If the number of accepted mutations (or some number proportional to it) can be deduced in this manner for each pair of n taxa being studied, then those $\binom{n}{2}$ numbers are related proportionally to the respective times since divergence of those pairs of taxa. Those $\binom{n}{2}$ numbers then perfectly satisfy the requirements for an ultrametric tree, and by Theorem 17.1.2, that tree recaptures the *true* evolutionary history of the taxa. So the molecular clock theory reduces the problem of finding time-since-divergence data to the problem of finding the number of mutations between two taxa. That reduction is very appealing, but how are mutation data obtained?

Laboratory-based methods

The first methods used to estimate the number of accepted mutations between two taxa generally used physical/chemical means. One approach takes the DNA of the two taxa, denatures it (by heating) so that the double strands separate, mixes the single strands from the two DNA sources together to allow them to hybridize, and then checks the temperature at which the hybrid strands separate. The idea is that the more similar the two DNA sources are, the stronger the hybridization and the higher the temperature must be in order to separate the two strands. The separating temperature of two taxa A and B is then assumed (after some adjustments) to be proportional to the total number of accepted mutations that have occurred in the time since A and B diverged.

The highly simplified description of an experimental procedure is essentially what was done in the classic, massive study of bird evolution [406, 407] by Sibley and Ahlquist. Moreover, the Sibley and Ahlquist data had a strong "self-verifying" property – the data were nearly ultrametric.[5] In turn, the ultrametric nature of their data gave a strong boost to

---

[5] In the language of evolutionary biology, the test to see if the data are ultrametric is called the *relative rate test*, which was suggested by V. Sarich and A. Wilson. Ahlquist and Sibley write, "We have found that the DNA clock seems to tick at the same average rate in all lineages of birds. The evidence comes from a procedure known as the relative rate test ... The relative rate test compares any three species ..." [406].

the molecular clock theory since ultrametric data are hard to explain without accepting a molecular clock and also believing that the obtained data correctly reflect true evolutionary history.[6]

Sequence-based methods

More current methods for estimating the numbers of accepted mutations are based directly on DNA or amino acid sequences. For two taxa A and B, one estimates the number of accepted mutations that have occurred since their divergence by examining differences in the DNA or amino acid sequences coding for proteins common to both A and B. Usually (see [127]) this estimate is related to *edit distance*. (This should come as no surprise; it had to be edit distance or some related concept to make this topic appropriate for a book on string algorithms.)

In actuality, edit distance has to be adjusted because a given nucleotide position might mutate several times in the course of evolutionary history. Also, certain DNA mutations (particularly in the third nucleotide of a codon) do not change the amino acid specified by the DNA. But after adjusting for these factors (see [127]), the edit distance between existing DNA or protein sequences for A and B is used to estimate the total number of accepted mutations that have occurred since the divergence of A and B. Those pairwise edit distance numbers form a matrix D, and if D is ultrametric then its tree is used as a hypothesis for the true evolutionary history of the taxa.

To recap, under the molecular clock theory, edit distance can be used to estimate the number of accepted mutations that have occurred since the time any two taxa diverged. That number should then be proportional to the actual time since their divergence. Hence, given a set of n taxa and the $\binom{n}{2}$ edit distances between the pairs of taxa, an ultrametric tree (if one exists) for the data should give a possible evolutionary history explaining the data. Moreover, when an ultrametric tree exists for D, it is the only ultrametric tree that exists for D. The conclusion is that if pairwise edit distances can be used to obtain numbers proportional to the true pairwise time-since-divergence data (as the molecular clock theory suggests), then the unique ultrametric tree for those data should be the true evolutionary tree with the correct topology and proportionally correct edge lengths. Furthermore, if the absolute time of divergence is known for any pair (say from fossil data) then the molecular clock can be "set", allowing proportionally correct times to be converted to correct absolute times. This is the ideal case.

Final comments

Most often, real data are not ultrametric and even ultrametric data do not necessarily reflect true times since divergence. Still, being ultrametric is a property of data that is not likely to occur by chance, so when data are ultrametric, or nearly so, this is taken as strong evidence for the molecular clock theory and as strong evidence that the data do capture true evolutionary history.

In response to data that are not ultrametric, one might consider the problem of perturbing the data by "the smallest" amount possible so that the resulting data become ultrametric. If the perturbations are not too large, then the data may still support the molecular clock theory. Several variants of this perturbation problem have been studied. Consider the

---

[6] We should note, however, that the chemistry behind the work of Sibley and Ahlquist has been severely questioned [395], along with some of their record keeping, and the reliability of their results is no longer clear. An alternate view of bird evolution appears in [147].

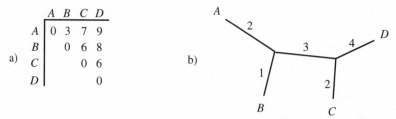

Figure 17.4: a. A symmetric matrix D. b. An additive tree for matrix D. Note that T contains some unlabeled nodes.

problem when the perturbed data are required to be ultrametric and the perturbations can only decrease the initial values. There is a solution to the problem where each value is (simultaneously) decreased by as little as possible so that the end result is ultrametric. This will be explored in Exercise 21. When the changes can be both positive as well as negative, then there is an efficient algorithm that creates ultrametric data and minimizes the maximum change to any single initial value. If only increases are allowed, then the problem is NP-hard. The last two results are due to Farach, Kannan, and Warnow [145].

17.2. Additive-distance trees

17.2.1. Introduction

Ultrametric data are the Holy Grail of phylogenetic reconstruction – when time-since-divergence data are ultrametric, the belief is that the true evolutionary history can be reconstructed. But this is mostly an idealized abstraction, and real data are rarely ultra-metric. What can be done in this case? A weaker requirement on evolutionary-distance data is that they be *additive*.

> **Definition**   Let D be a symmetric n by n matrix where the numbers on the diagonal are all zero and the off-diagonal numbers are all strictly positive. Let T be an edge weighted tree with at least n nodes, where n distinct nodes of T are labeled with the rows of D. Tree T is called an *additive tree* for matrix D if, for every pair of *labeled* nodes (i, j), the path from node i to node j has total weight (or distance) exactly $D(i, j)$.

That is, T encodes the matrix D in that every entry $D(i, j)$ equals the distance in T from node i to node j. For example, see Figure 17.4.

> **Additive tree problem**   Given a symmetric matrix D with diagonal entries equal to zero and other entries strictly positive, find an additive tree for D, or determine that none exists.

When the data in D reflect some evolutionary distance between pairs of taxa, such as weighted edit distance or other estimates of time since divergence, an additive tree for D gives one possible reconstruction of evolutionary history. The tree shows a branching pattern and gives edge distances (reflecting time) consistent with the data in D. However, since the tree is undirected, it does not indicate ancestral relations or the direction of evolution. Those have to be deduced by separate means.

It is easy to establish that if D is ultrametric and $D(i, i) = 0$ for each i, then D is also additive. In fact, such an ultrametric matrix D can be characterized as follows: D is ultrametric if there is an additive tree T for D and a node v in T such that all leaves in T have the same distance to v. However, it is not true that when D is additive it must be ultrametric. Thus the condition that distance data be additive is a weaker requirement for evolutionary validity than the requirement that it be ultrametric. Still, the additive tree problem, like the ultrametric problem, is a very idealized problem since data are rarely additive, due to either small errors in the data or large problems in the evolutionary model. One large problem is that evolution (of plants perhaps, and bacteria certainly) is not always divergent (i.e., tree-like). Genetic material can merge through *horizontal transfer* (see [430] for example), causing evolutionary histories to merge rather than diverge.

When data are not perfectly additive but the tree model is still valid, then one looks for a tree whose pairwise distances deviate from the distances in D as "little as possible". This is a large area of research with different suggested definitions of deviation and many suggested (heuristic) algorithms. For a practical treatment of this problem see [429]. For a theoretical method with a *provable* analysis, see [145] and [6]. For NP-completeness results see [120].

17.2.2. Algorithms for the additive tree problem

There are several $O(n^2)$-time algorithms in the literature that solve the problem of reconstructing, if possible, an additive tree from an n by n matrix (e.g., [46, 118, 218, 458]). Before these, the first polynomial-time solution appears to be in [83, 85], where the basic "four-point condition" was proved (see Exercise 31 in this chapter). That condition implies a direct $O(n^4)$-time algorithm for the problem. Interest in additive tree problems has arisen in many diverse fields, and many results have been rediscovered several times. A history and bibliography of much of this work appears in [49], where it is stated: "The works of various authors were remarkably little known one to another. Thus, for example, the four-point condition was published at least five times"! In fact, we now know of ten papers that give polynomial-time solutions to the additive tree problem.

Many of the published $O(n^2)$-time algorithms for the additive tree problem are similar and involve some amount of linear equation solving that obscures the true *combinatorial* nature of the additive tree problem. So we will not present any of those algorithms here. Instead, we will defer the algorithmic question until Section 17.4, where several evolutionary tree problems will be reduced to ultrametric tree problems. We will show there that the additive tree problem can be reduced in $O(n^2)$ time to an ultrametric tree problem, so that the given (combinatorial) algorithm for building an ultrametric tree can be adapted to building an additive tree. In this way, we solve the additive tree problem using only the *sorted order* of certain numbers derived from matrix D, without having to do any equation solving. But first, we discuss here another combinatorial result for a special case of the additive tree problem.

Compact additive trees

Compact additive tree problem   Given an n by n symmetric matrix D with diagonal entries equal to zero and other entries strictly positive, determine if there is an additive tree for D containing *exactly n* nodes.

For example, the tree shown in Figure 17.4 is an additive tree for D, but it is not a compact additive tree for D. In fact, no compact additive tree exists for that matrix, as we will establish shortly. However, the matrix shown in Figure 17.5 can be represented by a compact additive tree. The reader should be able to find it in the time it takes to turn a few pages.

	A	B	C	D	E	F
A	0	3	7	9	2	5
B		0	6	8	1	4
C			0	6	5	2
D				0	7	4
E					0	3
F						0

Figure 17.5: This matrix can be represented by a compact additive tree. Which one?

Definition   Given D, let $G(D)$ (or G for short) be the n-node complete graph where the nodes are labeled 1 through n and each edge (i, j) is given weight $D(i, j)$.

Theorem 17.2.1. *If there is a compact additive tree T for D, then T must be the unique minimum spanning tree of $G(D)$.*

PROOF   Let T be a compact additive tree for D, and let $e = (x, y)$ be any edge not in T. Now since the path in T from x to y must have total weight equal to $D(x, y)$, and all edge weights in T are strictly positive, $D(x, y)$ must be strictly larger than the weight of *every* edge on the x-to-y path in T. We will use this fact to show that edge (x, y) cannot possibly be in any minimum spanning tree of G. It will then follow that T must itself be a minimum spanning tree of G and that T is the only minimum spanning tree of G.

Suppose for contradiction that $e = (x, y)$ is in some minimum spanning tree $\overline{T}$ of G. Removing e from $\overline{T}$ divides the nodes of G into two connected components denoted S and $\overline{S}$. Since x is in S and y is in $\overline{S}$, there must be some edge e' on the x-to-y path in T that crosses from S to $\overline{S}$. Clearly, e' is not in tree $\overline{T}$. By construction, e is the only edge in $\overline{T}$ to cross from S to $\overline{S}$. Now add edge e' to $\overline{T}$ and remove edge e. The result is again a spanning tree of G; call it T'. But the total edge weight of T' is less than $\overline{T}$ since edge e', whose weight is less than $D(x, y)$, has weight less than edge e, whose weight is $D(x, y)$. This contradicts the assumption that $\overline{T}$ is a minimum spanning tree of G. Hence e cannot be part of any minimum spanning tree of G, and the theorem is proved.   □

By Theorem 17.2.1, one can solve the compact additive tree problem in $O(n^2)$ time: Given D, (implicitly) construct $G(D)$ and then run any $O(n^2)$-time minimum spanning tree algorithm (see for example [294]) that (for simplicity of exposition) maintains a single growing tree. Assume inductively that the algorithm knows the path weights $d(i, j)$ for every pair of nodes (i, j) in the current tree. When a new edge (x, y) is added to the tree, and x is already in the tree, compute $d(i, y) = d(i, x) + D(x, y)$ for each node i in the tree. Then check that $d(i, y) = D(i, y)$. This takes $O(n)$ time per iteration and hence $O(n^2)$ time overall. If the algorithm finds that $d(i, j) \neq D(i, j)$ at any point, then there is no compact additive tree for D.

17.3. Parsimony: character-based evolutionary reconstruction

17.3.1. Introduction

In this section we introduce a different, *character-based* approach to reconstructing evolutionary history. In this approach, the input is a set of *attributes* called *characters* that the objects may possess.[7] If the input characters are chosen well, then the distribution of the attributes among the objects may be used to deduce partial evolutionary history in the form of an evolutionary tree. This tree provides the branching order of the history, but does not by itself establish the times of divergence events. Alternatively, character attributes can be used to form a *taxonomy* (a systematic classification) of the objects without any suggested historical importance. After discussing character-based reconstruction problems, we will relate this approach to ultrametric trees.

The algorithmic study of character-based reconstruction has been very active in recent years, and there are many results that could be discussed here. However, the intent is only to introduce character-based reconstruction. We therefore will only discuss a few very idealized and simplified versions of character-based (or maximum-parsimony) problems. We will discuss in detail only *binary-character* problems, where objects either do or do not have any particular character, although generalizations will be mentioned. The major focus will be on the (binary) *perfect phylogeny* problem, which is a special case of the general maximum-parsimony problem.

Definition   Let M be an n by m, 0-1 (binary) matrix representing n objects in terms of m characters or traits that describe the objects. Each character takes on one of two possible *states*, 0 or 1, and cell (p, i) of M has a value of one if and only if object p has character i.

Definition   Given an n by m binary-character matrix M for n objects, a *phylogenetic tree for M* is a rooted tree T with exactly n leaves that obeys the following properties:

1.   Each of the n objects labels exactly one leaf of T.

2.   Each of the m characters labels exactly one edge of T.

3.   For any object p, the characters that label the edges along the unique path from the root to leaf p specify all of the characters of p whose state is one.

In the example in Figure 17.6, the first matrix M has a phylogenetic tree T, but the second matrix M' does not.

The interpretation of a phylogenetic tree for M is that it gives an estimate of the evolutionary history of the objects (in terms of branching pattern, but not time), based on the following biological assumptions:

1.   The root of the tree represents an ancestral object that has none of the present m characters. That is, the state of each character is zero in the ancestral object.

2.   Each of the characters changes from the zero state to the one state exactly once and never changes from the one state to the zero state.

The key feature of a phylogenetic tree (without which there would be no interesting problem) is that each character labels *exactly one edge* of the tree. This corresponds to the second biological assumption above and represents the point in the evolutionary history of the objects when that character changes from its zero state to its one state. Hence any objects below that edge definitely have that character.

Finding a set of characters that obey the assumptions is often a difficult (and controversial) task in evolutionary research. It is therefore worth discussing the issue of where character data actually come from.

---

[7] Note that the word "character" in this context does not refer to a member of an alphabet but to an attribute or a trait of an object.

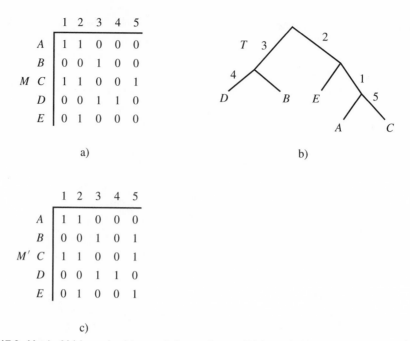

		1	2	3	4	5
	A	1	1	0	0	0
	B	0	0	1	0	0
M	C	1	1	0	0	1
	D	0	0	1	1	0
	E	0	1	0	0	0

a)

b)

		1	2	3	4	5
	A	1	1	0	0	0
	B	0	0	1	0	1
M'	C	1	1	0	0	1
	D	0	0	1	1	0
	E	0	1	0	0	1

c)

Figure 17.6: Matrix *M* (shown in a) has a phylogenetic tree *T* (shown in b). However, matrix *M'* (shown in c) has no phylogenetic tree.

17.3.2. Where do character data come from?

In the case of biological objects (generally species), the characters used have traditionally been *morphological* features or traits of the objects. These morphological characters may be gross features such as "possessing a backbone" or may be very fine features only understood by specialists studying those organisms. An excellent recent example of an evolutionary tree based purely on binary characters appears in [231]. There, an early history of bird radiation is deduced from binary characters such as "has feathers" or "loss of postorbital bone". But characters can also be based on DNA or protein sequences possessed by the different species. For example, whether or not the amino acid sequence for a given protein contains a specific substring makes a perfectly good binary character.

Characters based on morphology can be problematic because, under selection pressures, similar morphological features could arise independently, or a feature could be gained and lost and then regained. For example, it is thought that wings have evolved independently several times. Behavioral traits are equally problematic. For example, the character "walks on knuckles" is a trait possessed by both chimpanzees and gorillas but not (generally) by humans.[8] Therefore, if this features is used to build a phylogenetic tree, chimpanzees and gorillas should be found in a subtree that does not contain humans. However, contrary to that reasoning, current thinking has humans and chimpanzees branching together away from gorillas.

Nonfunctional substrings in DNA generally make better characters for character-based evolutionary reconstruction than do morphological or behavioral traits. Substrings of DNA outside of protein coding regions (genes) and outside of regulatory regions mutate semi-randomly, so that a long randomly mutated substring in the DNA of one species is very unlikely to be found in the DNA of (evolutionarily) distant species. But if the mutation rate is slow enough for the span of history being studied, the random substring (or vestiges of

---

[8] Thanks to Gary Churchill for this example.

it) may be recognized in more closely related species. Thus character-based evolutionary reconstruction from certain stretches of DNA should be increasingly important as more genomic DNA sequences are obtained.

One increasingly important type of character used in evolutionary studies is the specific nucleotide in a fixed position of a DNA sequence. Generally, analogous DNA sequences from a number of species are first multiply aligned. Each column of that alignment then specifies a single character, which can take on one of four states, A, T, C, or G. The state of character i, for any given species p, is the nucleotide in position p, i of the multiple alignment. In this case the characters are not binary, but DNA sequences are sometimes used as binary characters by grouping A with G (the purines) and grouping C with T (the pyrimidines). Similarly, multiply aligned amino acid sequences are sometimes used as characters having twenty states (or fewer, if characters are grouped).

Another type of *binary* character of potential importance is based on whether or not the expression of a particular protein is *regulated* by another particular protein. (For introductory discussions on regulation and three common structures in regulatory proteins – zinc fingers, helix-turn-helix motifs, and leucine zippers – see [378] and [54].) It is increasingly understood that regulation and control of protein expression (when and in what cells a gene for a particular protein is expressed) is as important (or maybe more important) in species differentiation than differences in the protein sequences. For example, mice and humans make essentially the same complement of proteins, and for many particular proteins, there is little difference in the amino acid sequences between the human and the mouse analogs of the protein.[9] What makes a mouse different from a human may be determined more by differences in protein regulation (which is reflected in DNA sequences outside of protein coding regions) than by differences in the corresponding protein sequences. Recent work [208] supports this view, as reflected in a review [311] of that work: "genes can gain new developmental functions not just through alterations in their protein sequences but also by acquiring new regulatory sequences that alter where and when they are expressed". Evolutionary history therefore may be better understood by examining changes in protein regulation (which proteins act to enhance or repress the expression of which other proteins) than by examining changes in amino acid sequence. Moreover, the character "protein A acts to enhance (or repress) the expression of protein B" is a *binary* character, and hence such binary characters may be of additional utility as more regulatory data are obtained.

Finally, we mention an implicit binary character used [45] to tackle the controversial relationship among fungi, plants, and animals. The question is: Are fungi more closely related to plants or animals? In the approach of [45], a multiple alignment of certain protein sequences from fungi, plants, and animals was first computed. This multiple alignment contained gaps, and many of the aligned strings shared the same gap(s). The width, regularity, and frequency of the shared gaps was such that each gap could be treated as a binary character, although that language is not used in [45]. Relative to the multiple alignment, each string either contained or didn't contain a specific gap. These "binary characters" were then used (implicitly) in [45] to build an evolutionary tree that placed fungi closer to animals than to plants. Similarly, one can identify discrete gaps in the alignment of the RNA sequences represented in Figure 11.6 on page 237 (and in the more complete figure in [123]). These gaps can be used as binary characters to construct an evolutionary tree of the viruses containing those sequences.

Given these examples of binary characters, we now turn to the algorithmic question of how to build evolutionary trees from binary characters.

---

[9] I've heard one biologist refer to mice as "furry, little people".

$$
\begin{array}{c}
\begin{array}{ccccc} 2 & 1 & 3 & 5 & 4 \\ 1 & 2 & 3 & 4 & 5 \end{array}
\end{array}
$$

$$
\bar{M} \quad
\begin{array}{c} A \\ B \\ C \\ D \\ E \end{array}
\left[
\begin{array}{ccccc}
1 & 1 & 0 & 0 & 0 \\
0 & 0 & 1 & 0 & 0 \\
1 & 1 & 0 & 1 & 0 \\
0 & 0 & 1 & 0 & 1 \\
1 & 0 & 0 & 0 & 0
\end{array}
\right.
$$

Figure 17.7: Matrix $\bar{M}$ resulting from sorting the columns of the matrix M shown earlier in Figure 17.6. (Each column is treated as a binary number.) The first row of numbers above $\bar{M}$ indicates the original character name (column) of each character in M. The second row of numbers gives the new name for each character.

17.3.3. Perfect Phylogeny

Perfect phylogeny problem   Given the n by m, 0-1 matrix M, determine whether there is a phylogenetic tree for M, and if so, build one.

We will solve the perfect phylogeny problem with a very simple $O(nm)$-time algorithm, where each comparison operation and each reference to M takes one time unit.

For the algorithm and its proof of correctness, it will be convenient to first reorder the columns of M. Considering each column of M as a binary number (with the most significant bit in row 1), sort these m numbers into decreasing order, placing the largest number in column 1. Let $\bar{M}$ denote the reordered matrix M. For an example, see Figure 17.7. Certainly, M has a phylogenetic tree if and only if $\bar{M}$ does. From this point on, each character will be named by the column it occupies in $\bar{M}$. Hence a character named j will be to the right (in $\bar{M}$) of any character named i if and only if $i < j$.

Definition   For any column k of $\bar{M}$, let O_k be the set of objects with a one in column k (i.e., objects that have character k).

Clearly, if O_k strictly contains O_j then column (character) k must be to the left of column j in matrix $\bar{M}$ (see Figure 17.7). Also, any duplicate copies of a column are placed together in a consecutive block of columns of $\bar{M}$. Now we can state the major theorem and the basis for the efficient solution to the perfect phylogeny problem.

Theorem 17.3.1. *Matrix $\bar{M}$ (or M) has a phylogenetic tree if and only if for every pair of columns i, j, either O_i and O_j are disjoint or one contains the other.*

PROOF   First, suppose that T is a phylogenetic tree for $\bar{M}$ and consider two characters i and j. Let e_j be the edge of T on which character i changes from zero state to one state, and let e_j be the similar edge for character j. All of the objects that possess character i (or j) are found at the leaves of T below edge e_i (or edge e_j). One of four cases must hold: 1. $e_i = e_j$, 2. e_i is on the path from the root of T to e_j, 3. e_j is on the path from the root of T to e_i, or 4. the paths to the two respective edges diverge before reaching either e_i or e_j. In case 1, the objects possessing character i and j are identical (so $O_i = O_j$); in case 2, the objects possessing character j must all possess character i (so $O_j \subset O_i$); case 3 is symmetric to case 2 and so $O_i \subset O_j$; and in case 4, $O_i \cap O_j = \emptyset$. In all cases, either O_i and O_j are disjoint or one contains the other. This completes one half of the proof.

We will prove the other half by construction. Consider objects p and q, and let k be the largest (right-most in $\bar{M}$) character that p and q both possess. We claim that if p possesses a character $i < k$, then q must also contain character i. To see this, note that O_i and

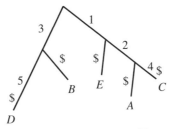

Figure 17.8: The keyword tree for the strings defined by matrix $\bar{M}$ shown in Figure 17.7. This tree defines a perfect phylogeny for $\bar{M}$. The reader can check that when the column numbers of $\bar{M}$ are translated back to their original numbers in M, then the tree above is the phylogenetic tree shown in Figure 17.6.

O_k intersect (since p possesses both characters); so, by assumption, O_i contains O_k and hence q also possesses character i. Since p and q are arbitrary, if q possesses a character $i < k$ then p must also possess that character. In summary, $\bar{M}(p, i) = \bar{M}(q, i)$ for every character $i \leq k$, and $\bar{M}(p, j) = \bar{M}(q, j)$ for $j > k$ if and only if $\bar{M}(p, j) = M(q, j) = 0$.

To finish the proof, we take advantage of our knowledge of keyword trees (from Section 3.4). We label an object p with a string consisting of the characters, in the order they appear in $\bar{M}$, that object p possesses. As usual, we also append an end-of-string character, say $, that is outside the original alphabet. Hence no resulting string is the prefix of any other string. For example, object A from Figure 17.7 is given the string 1, 2, $, and object D is given the string 3, 5, $. From the previous paragraph, the string for two objects p and q must be identical to some character k and thereafter have no characters (in either the string or phylogenetic sense) in common. It follows that the keyword tree (without failure links) for the n strings derived from the n objects in $\bar{M}$ specifies a perfect phylogeny for the matrix $\bar{M}$ (see Figure 17.8). To obtain a perfect phylogeny from the keyword tree, simply remove the $ symbols from edges in the tree.   □

Nonconstructive proofs of Theorem 17.3.1 appear in a number of places (for example, [139], [141], and [140]). Note that an algorithm based on a straightforward implementation of Theorem 17.3.1 would take $\Omega(nm^2)$ time just to determine if M has a phylogeny. The first $O(nm)$-time perfect phylogeny algorithm was given in [200], but the proof given above suggests a different $O(nm)$-time algorithm.

17.3.4. An $O(nm)$-time algorithm for the perfect phylogeny problem

1. Consider each column of M as a binary number. Using radix sort [10], sort these numbers in decreasing order, placing the largest number in column 1. Call the new matrix $\bar{M}$ and name each character by its column position in $\bar{M}$.

2. For each row p of $\bar{M}$, construct the string consisting of the characters, in sorted (increasing) order, that p possesses.

3. Build the keyword tree T for the n strings constructed in step 2.

4. Test whether T is a perfect phylogeny for M.

The algorithm can be implemented in $O(nm)$ time by using radix sort with pointers to avoid all but the last column permutations (see [10] for details on $O(nm)$-time radix sort). All other operations in the algorithm are trivially done in that time bound. Note that radix sort does not necessarily sort the columns by the *number* of 1's in the column. Such a sort could be used to solve the perfect phylogeny problem, but would not run in $O(nm)$ time.

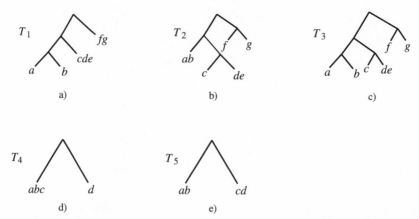

Figure 17.9: Trees T_1 and T_2 are compatible; they are refined by T_3. Trees T_4 and T_5 are not compatible.

17.3.5. Tree compatibility: an application of perfect phylogeny

An extension of the perfect phylogeny problem that has received considerable attention is how to determine whether two (or more) differing phylogenetic trees describe a "consistent" evolutionary history, and if so, how to combine the trees into a single phylogenetic tree incorporating all the known history. Problems of this type are of central importance in building trees from real data, because different tree-building methods and different computer packages will most often give trees that differ in some detail. In addition, an evolutionary tree is usually built for a set of taxa based on the comparison of a single protein or a single position in aligned protein sequences, but very often the resulting tree will be different depending on which particular protein or position is used. It is now generally accepted that several trees, each from a different protein or position, must be built and be shown to be "generally consistent" before the implied evolutionary history is considered reliable. There are several different approaches to defining and extracting "a consistent" history. Since our purpose here is only to introduce the topic and relate it to perfect phylogeny, only a single formalism will be discussed.

Definition   A phylogenetic tree T' is a *refinement* of T if T can be obtained by a series of contractions of edges of T.

If T' refines T, then T' agrees with all the evolutionary history displayed in T, while displaying additional history not contained in T.

Let T_1 and T_2 be two phylogenetic trees for a set of n objects. We will assume that T_1 and T_2 are both in "reduced form", that is, both are binary trees, and no node except the root can have exactly one child.

Definition   Trees T_1 and T_2 are *compatible* if there exists a phylogenetic tree T_3 refining both T_1 and T_2 (see Figure 17.9).

Tree compatibility problem   Given trees T_1 and T_2, determine whether the two trees are compatible, and if so, produce a refinement tree T_3.

Let M_1 be a 0-1 matrix with one row for each object and one column for each internal node j in T_1. Entry (i, j) of M_1 has value one if and only if the leaf for object i is found below node j. That is, column j of M_1 records the objects found in the subtree of T_1 rooted at node j. Matrix M_2 is similarly defined for T_2, and matrix M_3 is the matrix formed by the union of the columns of M_1 and M_2. Then the following theorem holds:

Theorem 17.3.2. *T_1 and T_2 are compatible if and only if there is a phylogenetic tree for M_3. Further, a phylogenetic tree T_3 for M_3 is a refinement of both T_1 and T_2.*

We leave the proof of this theorem as an exercise. For details see [142] and [143].

Theorem 17.3.2 reduces the compatibility problem to the perfect phylogeny problem. We will see below that the perfect phylogeny problem can be reduced to a min-ultrametric problem, and so the compatibility problem can also be seen as a disguised version of an ultrametric problem. Theorem 17.3.2 also implies an $O(n^2)$-time algorithm for the compatibility problem, using the algorithm described for the perfect phylogeny problem. However, the problem can be solved more directly in $O(n)$ time. Details of this are in [200] and again are left as an exercise.

17.3.6. Generalized perfect phylogeny

We have discussed the perfect phylogeny problem in detail when each character can take on two states. The *generalized phylogeny problem* allows a character to take on more than two states. In that case, a perfect phylogeny for M is a directed tree T where each object labels exactly one leaf of T as before, but now edges are labeled with *character-state transitions*. That is, the label applied to an edge is an *ordered triple* (c, x, y) indicating that character c changes from state x to state y along that edge.

As in the binary case, we specify the starting state for each character at the root node and require that the path from the root to a leaf labeled p must describe the character states of object p: The ending states specified by the changes on the path to p must correctly specify the character states of object p. The critical constraint now, for any state y of any character c, is that there can be at most one edge where the state of character c changes to y. That is, there can be at most one edge labeled with a triple that begins with c and that ends with y.

The definition of phylogenetic tree in the case of binary characters is easily seen as a special case of this more general definition. In the binary case, a character label c on an edge is just an abbreviation for the triple $(c, 0, 1)$. We should note that the description of a generalized perfect phylogeny given here is somewhat nonstandard but equivalent to the more commonly used definitions found, for example, in [258]. The definition here is used to make the continuity between the binary and the general case more explicit.

The generalized perfect phylogeny problem   Given a character matrix M where each character may take on up to r states, determine if there is a perfect phylogeny for M, and if so, construct one.

It is beyond the scope of this book to discuss the generalized problem in any detail. It was first proposed in a paper by Buneman [84] in 1974 and the issue remained open for almost twenty years. Kannan and Warnow [258] first established that the problem has a polynomial time solution in terms of n and m, if r is fixed at three or four. Independently, Dress and Steele [133] gave a polynomial-time solution for r fixed at three. However, if r is variable (specified in the input to the problem) then the generalized perfect phylogeny problem is NP-complete [72, 418]. Agarwalla and Fernandez-Baca [7] subsequently established that if r is any fixed number, then the generalized perfect phylogeny problem can be solved in polynomial time in terms of n, m. Of course, r appears in the exponent of the worst-case time bound. More recently, Kannan and Warnow [259] simplified that solution and considered a way to represent all the solutions to any given problem instance. That approach may prove to be very valuable for computing statistical estimates of the "significance" of any given tree or of any edge in the tree.

17.4. The centrality of the ultrametric problem

Although the four tree problems we have considered in detail – ultrametric, additive, (binary) perfect phylogeny, and tree compatibility – are quite different in appearance, they are actually strongly related. In fact, all of these problems are solvable as ultrametric tree problems. We have already seen that the tree compatibility problem reduces to the (binary) perfect phylogeny problem. The main work in this section is to show how to reduce the additive tree problem and the (binary) perfect phylogeny problem to ultrametric tree problems. In the case of the additive tree problem, this reduction will also give an efficient algorithm for solving the additive tree problem.

17.4.1. The additive tree problem viewed as an ultrametric problem

We will show how to reduce the additive tree problem to an ultrametric problem in $O(n^2)$ time, by creating a matrix D' that is ultrametric if and only if matrix D is additive. To introduce the idea of the reduction, assume that D is additive and that an additive tree T for D is known. Also, assume without loss of generality that each of the n taxa in D labels a *leaf* of T.[10] We will label the nodes of T with particular numbers to create an ultrametric tree, and this tree will expose the idea of the desired reduction.

Let v be the row of D containing the largest entry in D, and let m_v be that maximum entry. Hence, over all nodes in T, node v has the maximum distance to any leaf in T. Now root T at node v, creating a directed tree. We want to "stretch" leaf edges (edges that end at a leaf) so that v is equidistant to each leaf in the resulting tree. To do that for each leaf i, simply add $m_v - D(v, i)$ to the distance on the edge in T into leaf i. The result is a rooted, edge-weighted tree T', where the distance from v to any leaf is exactly m_v, and where each internal node is equidistant to any leaf in its subtree. (See Figure 17.10 a, b.) Note that only distances on leaf edges were changed. Now label each node of T' with the (unique) distance from it to any of the leaves in its subtree. Those labels are nonincreasing and can therefore be used to define an ultrametric matrix D', where $D'(i, j)$ is the label at the least common ancestor of leaves i and j in T'.

How the reduction works on the matrix

In the above exposition, matrix D' depends on T', which depends on T, which comes from D, and this defines the reduction from additive matrix D to ultrametric matrix D'. But we want a reduction from D to D' directly without having to build T or T'. To see the idea of a direct reduction, consider two leaves i and j of T and let w be their least common ancestor. We can deduce their (equal) distance to w in T', without explicitly knowing T'.

Let x be the distance in T from the root v to node w, and let y be the distance from node w to leaf i. The distance from w to i (or j) in T' is exactly $y + m_v - D(v, i)$. But what is y? Distance y is exactly $D(v, i) - x$, and $2x$ is easily seen, from Figure 17.11, to be $D(v, i) + D(v, j) - D(i, j)$. Similar reasoning can be used to obtain the distance from w to j. Hence, the following lemma holds:

---

[10] This requires that we relax the *strictness* condition in the definition of an ultrametric, allowing two adjacent nodes on a path from the root to have equal node labels.

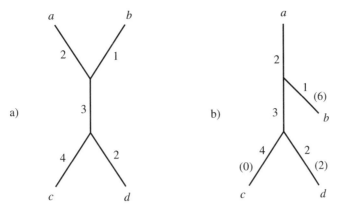

Figure 17.10: a. An edge-weighted tree with four labeled leaves. Node v is node a, and m_v is nine. b. The tree rooted at node a; the numbers in parentheses are the values $m_a - D(a, i)$ for each leaf i that are added to the weight of each leaf edge. Then each leaf is at distance nine from the root. Also, each internal node is equidistant from each leaf in its subtree.

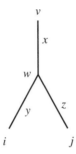

Figure 17.11: A schematic of distances in tree T. The distance from the root v to w is x, and the distance from w to leaf i (or j) is y (or z). Therefore, $D(v,i) = x+y$, $D(v, j) = x+z$, and $D(i, j) = y+z$. It follows that $2x = (x + y) + (x + z) - (y + z) = D(v, i) + D(v, j) - D(i, j)$.

Lemma 17.4.1. *Without knowing T or T' explicitly, we can deduce that $D'(i, j) = m_v + (D(i, j) - D(v, i) - D(v, j))/2$.*

Given Lemma 17.4.1, we have

Theorem 17.4.1. *If D is an additive matrix, then D' is an ultrametric matrix, where $D'(i, j) = m_v + (D(i, j) - D(v, i) - D(v, j))/2$.*

PROOF The proof requires assembling the pieces established above. $D'(i, j) = y + m_v - D(v, i)$, $y = D(v, i) - x$, and $x = D(v, i) + D(v, j) - D(i, j)$. Substituting and cancelling equal terms establishes the theorem. $\square$

Hence, if we are given a matrix D and want to establish whether D is additive, we can create matrix D' and test if D' is ultrametric. If not, then D is not additive. But what of the converse?

Theorem 17.4.2. *If matrix D' is ultrametric, then matrix D is additive.*

PROOF Let T'' be the ultrametric tree for matrix D'. (We don't use T' here since it was defined from T, which is unknown.) First, assign weights to edges of T'' so that the path from any leaf i to an ancestor node w has a distance equal to the number labeling node w. (The label on each leaf is zero.) To do this, just assign to each edge (p, q) the absolute

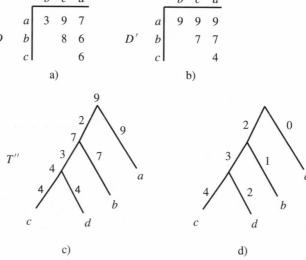

Figure 17.12: a. The distance matrix D obtained from the tree in Figure 17.10a. The largest entry has value 9 and is in row a. b. The derived ultrametric matrix D'. c. The ultrametric tree T'' along with the derived edge weights. d. The resulting tree after $m_a - D(a, i)$ is subtracted from leaf edges. The original tree is recovered after contracting the zero-weight edge to leaf a.

difference between the number written at node p and the number written at q. The resulting path-distance between a pair of leaves (i, j) is exactly twice the number written at the least common ancestor of i and j. However, because T'' is an ultrametric tree for matrix D', that distance is exactly $2 \times D'(i, j) = 2 \times m_v + D(i, j) - D(v, i) - D(v, j)$. Therefore, if for each leaf i, we now "shrink" the leaf edge into i by $m_v - D(v, i)$, then the resulting path between leaves i and j will have distance exactly $D(i, j)$. That creates an additive tree for D from an ultrametric tree for D'. □

In summary, matrix D is additive if and only if D' is ultrametric. Moreover, if D is additive, then the additive tree T for D can be created as follows:

Additive tree algorithm   Create matrix D' from D and construct the ultrametric tree T'' from D'. Next, assign a distance to each edge equal to the absolute difference between the node labels of its endpoints. Then, for each leaf i, subtract $m_v - D(v, i)$ from the distance on the edge into leaf i. The resulting tree is an additive tree for matrix D.

For an example, see Figure 17.12.
All the steps of the algorithm are easily implemented in $O(n^2)$ time; hence

Theorem 17.4.3. *An additive tree for an additive matrix can be constructed in $O(n^2)$ time.*

At heart, this algorithm relies on the algorithm for building an ultrametric tree (Section 17.1.3), and the heart of that algorithm involves only sorting and partitioning of numbers. Hence, when fully implemented along these lines, an additive tree can be constructed for D by a combinatorial algorithm (rather than a numerical one) that only sorts and partitions numbers held in matrix D'.

17.4.2. The perfect phylogeny problem viewed as an ultrametric problem

Having shown how to solve the additive tree problem as an ultrametric tree problem, we now show how to solve the (binary) perfect phylogeny problem as an ultrametric tree problem.

Definition Given the n by m character-matrix M, define the following n by n matrix D_M: For each pair of objects p, q, set $D_M(p, q)$ to be the *number* of characters that objects p and q both possess. That is, $D_M(p, q)$ equals the number of columns i such that $M(p, i) = M(q, i) = 1$.

The first relationship between M and D_M is given in the following lemma:

Lemma 17.4.2. *If M has a perfect phylogeny, then D_M is a min-ultrametric matrix.*

PROOF Let T be the perfect phylogeny for M. We will convert T to a min-ultrametric tree for matrix D_M, establishing that D_M is min-ultrametric. First, write a zero at the root node of T. Then process the nodes of T in a top-down manner, successively writing, at each node v, the number written at v's parent plus the number of characters labeling the edge into v from its parent node. The result is that if p and q are objects labeling leaves below v, then the number written at v is the number of characters in common between p and q (i.e., $D_M(p, q)$). Moreover, these numbers are strictly increasing along any path from the root. Hence the tree T together with these node numbers create a min-ultrametric tree for matrix D_M. □

Lemma 17.4.2 establishes a necessary condition for M to have a perfect phylogeny. It does not, however, establish a sufficient condition. In fact, D_M may be ultrametric even though M does not have a perfect phylogeny. We leave it to the reader to find an example of this. However, one can still solve the perfect phylogeny problem using ultrametric machinery. Recall that if a matrix is min-ultrametric, then the min-ultrametric tree for that matrix is unique. So if M has a perfect phylogeny, then the min-ultrametric tree derived from matrix D_M must have the same topology as the perfect phylogeny T obtained directly from M by the perfect phylogeny algorithm. Moreover, if one knows the topology of T without knowing the edge labels (characters of M), it is a simple matter to correctly assign characters to edges of T to obtain the full perfect phylogeny for M. This suggests the following method to determine if M has a perfect phylogeny:

Perfect phylogeny via ultrametrics

1. Create matrix D_M from M as above.
2. Try to create a min-ultrametric tree T' from D_M. If D_M has no min-ultrametric tree, then M has no perfect phylogeny.
3. If D_M has a min-ultrametric tree T', then try to label the edges of T' with the m characters of M, converting T' into a perfect phylogeny for M. If this fails, then M has no perfect phylogeny. Otherwise the converted T' is the perfect phylogeny T.

In this way, the perfect phylogeny problem can be viewed as a special kind of ultrametric problem and solved using ultrametric machinery. It follows also that the tree compatibility problem can be viewed as a kind of ultrametric problem as well. However, this view is only for conceptual clarity, and it is more efficient to solve the perfect phylogeny and compatibility problems by algorithms specialized to those problems.

17.5. Maximum parsimony, Steiner trees, and perfect phylogeny

We have mentioned "the maximum-parsimony problem" several times and stated that the perfect phylogeny problem is a special case of it, but we have not explained this or given a clean definition of the maximum-parsimony problem. To do that, we introduce a class of problems known as *Steiner tree* problems and then view the maximum-parsimony problem on character data as a Steiner tree problem on a graph that is a *hypercube*. We first restrict attention to binary characters.

17.5.1. Basic definitions

Definition   Let $G = (N, E)$ be an undirected graph on node set N and edge set E, with a nonnegative weight $w(i, j)$ on each edge (i, j) in G. Let $X \subseteq N$ be a given subset of nodes. A *Steiner tree ST* for X is any connected subtree of G that contains all the nodes of X, although it may contain nodes in $N - X$ as well. The *weight* of a Steiner tree ST is the sum of the weights of the edges in ST, and is denoted $W(ST)$. Given G and X, the *weighted Steiner tree problem* is to find the Steiner tree of minimum total weight. When all edges have weight one, the problem is called the *unweighted* Steiner tree problem.

Definition   A *hypercube* of dimension d is an undirected graph with 2^d nodes, where the nodes are labeled with the integers between 0 and $2^d - 1$. Two nodes in the hypercube are adjacent if and only if the binary representation of their labels differs in exactly one bit.

Definition   The *weighted Steiner tree problem on hypercubes* is the weighted Steiner tree problem where the graphs are hypercubes. The *unweighted* Steiner tree problem is the special case when all edges have weight one.

Until now we have loosely and vaguely defined the maximum-parsimony problem as the problem of reconstructing evolutionary history with the fewest number of mutations. Equipped with the language of hypercubes and Steiner trees, we can now give a precise definition of "the maximum-parsimony problem". We will do this for the case of binary characters.

Definition   For a set of input taxa X described in terms of d binary characters, *the maximum-parsimony problem* is the *the unweighted Steiner tree problem* on a d-dimensional hypercube. Each element in set X is described by a d-length binary vector and hence describes one node in the hypercube.

The d-dimensional hypercube has one node for each of the 2^d possible objects that can be described by a d-length binary vector. Two possible objects that differ by a single mutation event (a change in the state of a single character) correspond to two adjacent nodes in the hypercube. Therefore, there is a Steiner tree connecting a given set of nodes X using l edges if and only if there is a corresponding phylogenetic tree that involves l character-state mutations.

The generalization of this viewpoint to the case when characters can take on more than two states is left as an exercise.

The Steiner interpretation of perfect phylogeny

We can now precisely relate the perfect phylogeny problem to the maximum-parsimony problem. If the input to the maximum-parsimony problem contains d binary characters and each character is nontrivial (in that it is contained by some but not all of the taxa), then the maximum-parsimony tree must have at least d mutations (i.e., places where the state of one character changes). Thus the perfect phylogeny problem can be viewed as the question: Does the maximum-parsimony problem have a solution whose cost equals exactly d, the obvious lower bound? The generalized perfect phylogeny problem can be viewed in the same way, where the lower bound is based on the fact that if a character has r states, then there must be at least $r - 1$ places in the maximum-parsimony tree where that character changes state.

So from the viewpoint of the Steiner tree problem and maximum parsimony, the perfect phylogeny problem asks whether the optimal Steiner tree must have a cost greater than the obvious lower bound. There is no known efficient solution to the Steiner tree problem on unweighted graphs generally, or even on hypercubes in particular. However, the polynomial time solutions to the generalized perfect phylogeny problem show that when r is fixed, this special question about Steiner trees can be answered in polynomial time.

17.5.2. Approximations to maximum parsimony

The unweighted Steiner tree problem on hypercubes has been shown to be NP-hard [165]. Usually, the NP-hardness of an unweighted problem immediately implies the NP-hardness on the weighted version of the problem. After all, the unweighted case is simply the weighted case with all the weights set to one. However, the case of the Steiner tree problem on hypercubes is different, and the result in [165] does *not* imply the NP-hardness of the weighted Steiner tree problem on hypercubes. We leave it to the reader to puzzle out why this might be. The answer, along with a proof that the weighted version is in fact NP-hard, is contained in [198].

Although the weighted Steiner tree problem on hypercubes is NP-hard, an efficient algorithm to approximate it to within a factor of less than two does exist. This follows because the weighted Steiner tree problem on any graph can be efficiently approximated within an error bound of $11/6$ [62, 484]. Prior to that result, it was known that the minimum spanning tree can be used to obtain a Steiner tree whose total weight is less than twice the weight of the optimal Steiner tree [284]. Specialized to the case of the hypercube, the method in brief is the following: a. compute the distance $d(i, j)$ in the hypercube between each pair of input objects i and j in set X (in the case of binary characters, this is just the Hamming distance); b. form a complete, undirected graph K_X with one node for each object in X and one edge between each pair of nodes, and assign weight $d(i, j)$ to each edge (i, j) in K_X; c. compute a minimum spanning tree T of K_X (where each edge in T corresponds to a path in the hypercube); d. re-expand each edge in T to its original corresponding path in the hypercube; e. superimpose the paths found in step d to form graph G', and then find any spanning tree of G'. The result is a Steiner tree of X with total weight less than twice that of the optimal.

The above description is only for conceptual purposes, and in the case of binary character data, this computation can be done without explicitly embedding the problem in an actual hypercube. That is quite important for efficiency. The better approximations given in [62] and [484] can also be adapted for the maximum-parsimony problem without needing the hypercube explicitly. We leave that as an exercise for the reader.

17.6. Phylogenetic alignment, again

The previous sections focused on methods to deduce evolutionary history from sequence data, and this is the direction of evolutionary deduction that has received most of the attention in the popular press (for example, the "Eve" episode [481, 86, 175, 31][11]). But deduction in the opposite direction is also important. That is, if the evolutionary history of a set of taxa is well established from fossil or morphological data, one can use that history to shed light on the *molecular* evolution of homologous sequences, one from each of the taxa in the set.

There are several ways to formalize this approach to studying molecular evolution. One particular formalization is again the *phylogenetic alignment problem* discussed earlier in Section 14.8: One is given a tree with a known sequence at every leaf, and the problem is to determine sequences at the internal nodes to minimize the total evolutionary (edit) distance on the edges. In more detail, once sequences at internal nodes have been assigned, the distance of an edge is the edit distance between the two sequences that label the endpoints of that edge. The distance of a tree is the sum of the distances of its edges. The problem is then to determine sequences at the internal nodes to minimize the distance of the tree.

In the earlier discussion of phylogenetic alignment (Section 14.8), the purpose of using the established tree was to guide a multiple alignment of the extant sequences. The multiple alignment was consistent with the labeled tree, and so the tree implicitly determined which pairs of sequences should be emphasized in the alignment. However, once that alignment of the extant sequences was obtained, the sequences from the internal nodes could be removed. Here, the view is reversed. The valued information from the phylogenetic alignment is precisely those deduced internal sequences, as they represent the hypothesized ancestral sequences that gave rise to the current, known sequences. In this way, the phylogenetic alignment problem is a particular case of the parsimony problem. If edit distance is a realistic way to model evolutionary distance between sequences (i.e., the cost of mutations, etc.), and if the globally optimal phylogenetic alignment properly captures a process that actually operates by a series of local changes, then the optimal phylogenetic alignment may serve as a sensible hypothesis of evolutionary history.

The major technical aspects of the phylogenetic alignment problem were detailed in Section 14.8 and won't be repeated here. However, there is a particularly simple, widely used variant of phylogenetic alignment that has a direct meaning in the context of evolutionary inference, but not in the context of multiple alignment. Hence it is discussed in this chapter rather than in Chapter 14. The variant is called the *minimum mutation problem*. It has a simple, efficient solution due to W. Fitch and J. Hartigan [159, 216, 217]. That solution can also be incorporated into methods that tackle the full phylogenetic alignment problem (see Section 17.7).

17.6.1. The Fitch–Hartigan minimum mutation problem

Definition The *minimum mutation problem* is a variant of the phylogenetic alignment problem where, in addition to the tree and the strings labeling the leaves, the input contains a *multiple alignment* of those input strings.

---

[11] The Eve theory is that all presently living humans have a common female ancestor who lived about 200,000 years ago. All other human lineages (not descended from that Eve) have died out. The theory is primarily based on comparison of mitochondrial DNA sequences. Mitochondrial DNA is used partly because it is inherited only from the mother (making it easier to trace ancestry), partly because it mutates rapidly in humans (making it useful for examining short time periods), and partly because the mutations are generally neutral (no mutations are lethal or are beneficial to the organism). Although the Eve theory has received a great deal of popular press and has very strong supporters, parts of the early papers have been withdrawn (reflecting computational problems in that work), and its main thesis remains controversial.

This variant may be appropriate when the input to the phylogenetic alignment problem consists of protein strings so closely related that their multiple alignment is not in question. For example, in the classic cases of *cytochrome c* or *hemoglobin*, the amino acid sequences from different mammals have very similar lengths, and a biologically meaningful multiple alignment is easily obtained using a small number of spaces. From this alignment, each amino acid in each string is assigned to a specific position relative to a chosen reference string. The evolutionary history of the strings is then viewed as the combined *but independent* evolutionary history of each amino acid *position*. This is the underlying model used to construct PAM units and PAM matrices (see Sections 15.7 and 17.6.2).

By assigning each amino acid to a specific position, and by assuming that the evolutionary history of each position is independent of the other positions, the phylogenetic alignment problem becomes greatly simplified. Each column in the multiple alignment can be solved separately. Essentially, the strings assigned to the interior nodes must conform to the given multiple alignment. Hence the phylogenetic alignment problem reduces to separate instances of the problem where each input string consists of just a *single character*. The permitted input characters are the characters in the original alphabet augmented by a character used to represent a space. We now examine this simple variant of the problem, but only the version in which each mismatch has a cost of one and each match has a cost of zero. The generalization of the problem to general costs is immediate and is left as an exercise.

The minimum mutation problem for a single position   Given a rooted tree with n nodes (not necessarily binary) and a single character labeling each leaf, the *minimum mutation* problem is to label each interior node of the tree with a single character so as to minimize the number of edges whose endpoints have different labels.

This problem is easily solved with dynamic programming. For any node v, let T_v be the subtree rooted at node v and let $C(v)$ be the cost of the optimal solution to the mutation problem restricted to subtree T_v alone. For every character x, let $C(v, x)$ be the cost of the best labeling of subtree T_v when node v is *required* to be labeled with character x. Let v_i denote the ith child of node v. The base cases specify the values of $C(v)$ and $C(v, x)$, for every leaf v of T, and every character x in the alphabet. For each leaf v, $C(v) = 0$ and $C(v, x) = 0$ if x is the input character that labels leaf v, and $C(v, x) = \infty$ if x is not the input character that labels v. When v is an internal node, then

$$C(v) = \min_x C(v, x)$$

and

$$C(v, x) = \sum_i [\min(C(v_i) + 1, C(v_i, x))].$$

The recurrences are evaluated starting from the base cases. This corresponds to working bottom-up from the leaves of the tree, since the recurrences for any node v can be evaluated after, and only after, the recurrences for all of v's children have been evaluated. As usual in dynamic programming, the full specification of the optimal solution is determined in a backtrack phase after all the $C(v, x)$ values have been computed. In the backtrack phase, the algorithm first sets the character at the root r to be the character x such that $C(r) = C(r, x)$. It then traverses the tree top-down. If the character at node v has been set to y, then the character at its child v_i is set to y if $C(v_i) + 1 > C(v_i, y)$; otherwise the character at v_i is set to the character x such that $C(v_i) = C(v_i, x)$.

It is immediately apparent that the recurrences are correct and that they can be evaluated bottom-up in $O(n\sigma)$ time, where σ is the size of the alphabet. When general mismatch and match costs are used, the solution can be generalized and solved in $O(n\sigma^2)$ time. (See Exercise 36.) Exercise 37 explores the original Fitch–Hartigan solution to the minimum mutation problem. Their solution is different than the one presented here.

17.6.2. Phylogenetic alignment used to compute PAM matrices

The construction of Dayhoff PAM matrices was discussed in Section 15.7.4, but one detail of the method was omitted. In Section 15.7.4 we said that amino acid substitutions were counted [122] from aligned pairs of protein sequences with known, small PAM distances. In fact, the counts came from pairs of known sequences augmented with additional inferred sequences obtained by solving the *phylogenetic alignment problem* on 71 evolutionary trees [122]. The leaves of each tree were labeled with extant protein sequences, and ancestral sequences were inferred and assigned to the internal nodes of each tree. Then amino acid substitution data were collected from the pairs of sequences that label neighboring nodes in the tree. In this way, additional pairs of sequences were obtained, extending the available data; these pairs were even more similar to each other than were the extant pairs at the leaves. Dayhoff et al. [122] state: "By comparing observed sequences with inferred ancestral sequences, rather than with each other, a sharper picture of accepted point mutation is obtained".

17.7. Connections between multiple alignment and tree construction

Although we have mostly considered the tasks of tree building and multiple alignment separately, many approaches in the biological literature integrate or iterate these two tasks. This was mentioned in Section 14.10.2, where we noted that the history of merges determined by cluster-based multiple alignment methods also are used to define evolutionary trees. Progressive alignment methods formalize this dual task of building evolutionary trees and constructing multiple alignments.

Other methods use multiple alignment as a step in computing evolutionary trees or do the opposite, using evolutionary trees as a step in constructing multiple alignments, or they iterate between those two tasks. For example, the Fitch–Hartigan algorithm can be iterated with algorithms for the general phylogenetic alignment problem (where no alignment of the input is given). After a phylogenetic alignment has been computed for a tree T (by whatever method) it may be improved in the following way: First, compute a multiple alignment $\mathcal{M}(T)$ that is consistent with T. By Theorem 14.6.1, the optimal pairwise alignments corresponding to edges of T are preserved in $\mathcal{M}(T)$. Next, remove from $\mathcal{M}(T)$ the inferred sequences corresponding to the internal nodes of T. Then use the resulting multiple alignment of the original sequences as input to the Fitch–Hartigan method. The result is a phylogenetic alignment on T that is no worse, and will generally be better, than the previous one. That phylogenetic alignment can be used to create a final multiple alignment, or it can be left in the tree form as a hypothesized evolutionary history.

The above approach can also be adapted to try to construct a phylogenetic alignment where the string at any internal node v is desired to be roughly equidistant from each of the strings at the leaves of v's subtree. Such a phylogenetic alignment would be much more consistent with the molecular clock thesis than is a purely lifted alignment.

Another common way that evolutionary tree building is iterated with multiple align-ment is to first multiply align the input set of strings; then compute the *induced pairwise distances* given by the multiple alignment; and then use those distances to build evo-lutionary trees with distance-based methods. The opposite approach is to first build an evolutionary tree T from pairwise edit distances computed directly from the strings; next, solve the phylogenetic alignment problem for T; then compute the multiple alignment consistent with the labeled tree T; and finally remove the strings corresponding to the internal, inferred strings. Of course, one can iterate those two approaches. Moreover, at the points in the method where both a multiple alignment and a tree are in hand, one can solve the phylogenetic alignment problem using the Fitch–Hartigan method, as described above.

17.8. Exercises

1. Give a simple algorithm that can determine in $O(n^2)$ time if an n by n matrix D is an ultrametric matrix. You do not need to use the constant-time least common ancestor method developed in Chapter 8.

2. The following is an alternative definition of an ultrametric matrix: Recall the definition of graph $G(D)$ from Section 17.2.2, and let MD be the minimum spanning tree of $G(D)$. D is ultrametric if and only if, for every pair of nodes i, j, the maximum weight edge in MD on the path between i and j has weight $D(i, j)$. Prove this.

3. Prove that if D is an ultrametric matrix, then there exists a minimum spanning tree of $G(D)$ consisting of a single path of n nodes.

4. When the "strictness" condition is removed from the definition of a ultrametric tree, there can be more than one ultrametric tree for a given matrix D. Nevertheless, the variation in permitted trees is not great. Precisely describe the class of ultrametric trees that are possible for a single matrix D.

5. Complete the details to establish that a min-ultrametric tree is unique, using the definition of min-ultrametric in the text (i.e., with the "strictness" condition included).

6. Is it true that if there is an additive tree for matrix D then that tree is unique?

7. Without using the reduction from the additive tree problem to the ultrametric tree problem, devise an $O(n^2)$-time algorithm that takes in an n by n matrix D and determines if there is an additive tree for D. Of course, the algorithm should produce a tree when there is one.

8. Show that if a matrix D is ultrametric and $D(i, i) = 0$ for each i, then D is also additive.

   Show that the converse is not true. What about the case of min-ultrametric matrices?

9. Suppose a matrix D is additive but perhaps not representable by a compact additive tree. Let T be the minimum spanning tree of the data from D and consider an arbitrary pair of rows i and j from D. Is it possible that the distance in T from i to j is less than $D(i, j)$? Can this type of observation be used to construct an $O(n^2)$-time algorithm to build additive trees?

10. Let M be a matrix containing the $\binom{n}{2}$ pairwise edit distances from n sequences. We know that if M is an additive matrix, then the pairwise edit distances can be represented by the interleaf distances on an edge-weighted additive tree with n leaves. But what about the converse situation? Suppose one is given an edge-weighted tree with n leaves. Is it always possible to create n sequences and a one-to-one mapping of the sequences and leaves so that, for any pair of leaves (i, j) and their associated sequences S_i, S_j, the distance between i and j equals the edit distance between S_i and S_j? Now address the same question for ultrametric trees.

11. Suppose that M is represented by a sparse-matrix data structure. That is, for each object p, the characters that p possesses are represented in a linked list. Let l be the total length of these n lists. Give an $O(l)$-time algorithm that determines if M has a perfect phylogeny and constructs one if possible.

12. Recall the definition of the consecutive-ones property for rows from Section 16.5.1. The analogous consecutive-ones property for columns requires that all the ones in every column be in a contiguous block. It is known [217] that if M has a phylogenetic tree then the rows of M can be permuted so that every column has the consecutive-ones property. Prove this.

    The consecutive-ones property allows a visually nicer presentation of the matrix. An $O(n^2 m)$ algorithm to reorder the rows was given in [217]. Linear-time, but fairly complex, algorithms are known for creating, when possible, the consecutive-ones property [74], but for matrices with a phylogenetic tree, there is a much simpler linear-time method. Find it.

13. Is it true that if a matrix M has the consecutive-ones property then there must be a perfect phylogeny for M?

14. Using Theorem 17.3.1, establish that, in worst case, every entry in M must be examined just to determine whether or not M has a perfect phylogeny.

15. In the binary character perfect phylogeny problem discussed in the text, it was assumed that each character is in state zero at the root of the tree. This assumption is often too strong. However, once choices for the root states are made, the states can be relabeled so that the root states are all zero, and the resulting matrix can be tested to see if it has a phylogenetic tree. Of course, if M has a perfect phylogeny, then it has one where the root (ancestor node) is labeled with any chosen row of M. However, it may not seem natural to place one of the input objects at the root of the tree. A more natural choice for the root state of a character j is the *majority* state: Assign the root state of character j to be 1 if and only if $|O_j| \geq n/2$. It has been shown [139] that if there is any choice of root states that leads to a phylogenetic tree, then the majority choice for each character will also lead to one.

16. Give a simple linear-time algorithm to determine whether two phylogenies are isomorphic. The fact that the leaves have distinct labels is helpful.

17. The following problem was recently discussed in an active exchange on `bionet.mol-bio.evolution`, an Internet newsgroup concerned with evolution: How can one tell if two binary trees in New Hampshire format are the same tree? The discussion on the net did not fully resolve the issue (in computer science worst-case terms) although one cubic-time (or more) method was suggested, and a very practical, but exponential worst-case, method was also suggested. A simple linear-time method is possible.

    Let T be a binary tree where each leaf has a unique label. A New Hampshire encoding of T is a string that describes T. It can be obtained by a depth-first traversal of T as follows: The string is accumulated left to right with all symbols appended at the right end; the first time a nonleaf node is visited, append a left parenthesis to the growing string; when a leaf is visited, append its label; the first time the traversal backs up to a node, append a comma; and when the traversal backs up to a node for the second time, append a right parenthesis. For example, the tree shown in Figure 17.8 has New Hampshire code of $((D, B), (E, (C, A)))$. Note that although the code uniquely identifies the tree, the code is not unique because of choices allowed during the depth-first traversal. Stated another way, even if T is kept in a data structure that specifies a particular left and a particular right child for each nonleaf node, the choice of which child is left and which is right may differ in any two separate copies of T. Therefore, even if the traversal was required to be an in-order traversal (left child visited before the right child), two different strings could be created for the two different copies of T. This gives rise to the problem addressed on the net of determining if two New Hampshire codes, which are not identical strings, actually specify the same tree.

Show how to use the solution to the previous problem on tree isomorphism to solve this problem in linear time.

An alternative approach is to pick a "canonical" New Hampshire encoding with the property that any tree can be encoded in only one way. If an arbitrary New Hampshire encoding for the tree can be efficiently converted to the canonical one, then the problem of comparing two codes is solved by converting each to its respective canonical code and then checking for equality of the codes. An obvious canonical encoding is the "lexicographic-least" one. For example, the code above would be $(((A, C), E), (B, D))$.

Give a precise definition of "lexicographic-least" so that each tree has a unique lexicographic-least New Hampshire encoding. Then give a simple linear-time algorithm that converts any New Hampshire code to its lexicographic-least one. The fact that the labels are distinct is helpful.

18. Prove Theorem 17.3.2 on page 465.

19. Develop an $O(n)$-time algorithm to determine if two trees (each of size $O(n)$) are compatible.

20. The Hamming distance of two binary vectors is the number of positions at which the two vectors differ. For a binary character matrix M with n rows, let D be the n by n matrix of the pairwise Hamming distances. Show that if M has a perfect phylogeny then D is additive. What can be established about the converse? Can this relationship be used to efficiently reduce the perfect phylogeny problem to the additive tree problem?

21. Let D be an n by n matrix of numbers that is not ultrametric. We want to *decrease* the values in D so that the resulting numbers are ultrametric. Consider the following procedure: Build a complete, labeled graph G on n nodes where edge (i, j) has weight $D(i, j)$; form a minimum spanning tree T from G; let $D'(i, j)$ be the largest edge weight on the path from i to j in T. Prove that $D'(i, j) \leq D(i, j)$ and that matrix D' is ultrametric.

Let D'' be any ultrametric matrix such that $D''(i, j) \leq D(i, j)$ for each pair (i, j). Prove that $D''(i, j) \leq D'(i, j)$.

Explain how this solves the problem of decreasing the values by "the smallest amount" to create an ultrametric matrix, for all interpretations of "smallest amount".

22. Consider the problem of increasing matrix values by the "smallest amount" to create a min-ultrametric matrix. What can be done?

23. Is it possible to efficiently reduce the ultrametric problem to the perfect phylogeny problem?

24. In the reduction from the additive tree problem to the ultrametric tree problem, we selected a node v whose maximum distance to any other node was maximum. Show that it is possible to reduce the additive tree problem to an ultrametric tree problem, using *any* arbitrary node v. The specifics of the reduction are different, but much of the logic is the same and the running time is again $O(n^2)$.

25. Following the reduction given on page 468, describe a self-contained algorithm that tests if a matrix is additive and if so constructs an additive tree.

26. Show how to efficiently implement step 3 of the *perfect phylogeny via ultrametrics* algorithm described in Section 17.4.2 (page 469).

27. Generalize the Steiner tree viewpoint of maximum parsimony and perfect phylogeny (Section 17.5) to characters that can take on more than two states.

28. Prove that the minimum spanning tree method of [284], sketched in Section 17.5.2, gives a Steiner tree whose weight is at most twice that of the optimal, for any graph. Explain how this can be used in the parsimony problem without explicitly creating a hypercube.

STRINGS AND EVOLUTIONARY TREES

29. In the hypercube, there is a very simple method to create the optimal Steiner tree of any three given nodes. Find that method. Try to extend the idea to four nodes and explain why it fails (assuming it does).

30. Consider an undirected tree T with n leaves labeled by distinct numbers. For a given set of four leaves, consider the smallest subtree of T that connects those four leaves, and delete (contract) any internal node that has degree equal to two. The resulting tree is called the *quartet tree* for those four leaves. The quartet tree is either a tree with four leaves, four edges, and one internal node of degree four or a tree with four leaves, five edges, and two internal nodes, each of degree three.

Considering leaf labels, show that there are three nonisomorphic quartet trees of the second kind.

Show how to reconstruct T uniquely from its set of quartet trees. Try to devise an efficient algorithm for this reconstruction.

31. A symmetric nonnegative matrix D with zero values on the diagonal satisfies the *four point condition* if and only if for each ordered choice of four rows, x, y, z, and t, $D(x, y) + D(z, t) \leq \max(D(x, z) + D(y, t), D(x, t) + D(y, z))$.

Prove that matrix D is additive if and only if it satisfies the four point condition.

32. Consider a rooted binary tree T with n leaves labeled by distinct numbers. For a given set of three leaves, consider the smallest connected subtree of T that connects those three leaves, and delete (contract) any internal node that has only one child. We call the resulting tree the *triple tree* of those three leaves. A triple tree has three leaves and two internal nodes, each with two children. Essentially, the triple tree specifies which two of the three leaves have a common ancestor below the least common ancestor of all three leaves.

Considering leaf labels, show that there are three nonisomorphic triple trees.

Show that one can uniquely reconstruct the rooted tree T from the set of triple trees.

33. If, instead of being given all the triple trees at once, one can adaptively decide which triple trees to ask for, show that T can be reconstructed from only $O(n^2)$ triple trees. Next try for an adaptive algorithm that runs in $O(n \log n)$ time. The solution appears in [256].

34. Suppose you are now given a set of triple trees that may or may not be consistent with a binary rooted tree. Consider the complexity of the problem of determining whether the set is consistent with a rooted binary tree.

35. Read the Steiner tree paper(s) ([62], [484]) that give an approximation method with an $11/6$ error bound. Then show how that method can be applied to the parsimony problem without explicitly creating a hypercube.

36. Prove the correctness and time analysis of the dynamic programming solution to the minimum mutation problem. Show how to solve the minimum mutation problem in $O(n\sigma^2)$ time when arbitrary match and mismatch scores are used.

37. In the text we solved the minimum mutation problem for a single position via dynamic programming. However, the original, well-known Fitch–Hartigan method does not use dynamic programming. The original algorithm is the following:

The original Fitch–Hartigan algorithm for a single position

Make two passes through the tree, a bottom-up pass and then a top-down pass. During the bottom-up pass, recursively construct the *equality set* S_v, for each interior node v. A character x is in S_v if and only if x appears in the equality sets of v's children as many times as any other character appears in those sets. For example, if v is a node whose children are all leaves, then S_v contains the characters that appear at those leaves as many times as any other character appears. Note that an equality set might only contain a single character. Note also that, in constructing S_v, the algorithm only considers how many times a character appears in the equality sets of v's children, not how many times it appears in the equality sets of nodes below those children.

The top-down pass assigns a character to each node. To start, if there is only one character x in the equality set at the root, then assign the root the character x. Otherwise, arbitrarily assign the root a character from its equality set. Now working top-down, assign each interior node v a character according to the following rule:

> If the equality set at node v contains a character x that has been assigned as the character of v's parent, then assign v the same character x. Otherwise, arbitrarily assign v a character from v's equality set.

Note that if there is only one character in v's equality set, then the rule will assign that character to node v.

Problems: Prove the correctness of the original Fitch–Hartigan algorithm, and analyze its running time. Can it be generalized to handle general mismatch costs the way that dynamic programming can? What advantages does the original Fitch–Hartigan method have over the dynamic programming solution? How can the method be extended to generate all the optimal solutions to the minimum mutation problem?

38. When there are multiple optimal solutions to the minimum mutation problem, one might want a solution that assigns to any internal node v a string that is roughly equidistant to the strings at the leaves of its subtree. Why is such a solution desirable? Examine the problem of constructing such an optimal solution.

39. In relation to the previous problem, find an efficient solution to the following problem: Given two strings S_1 and S_2, construct a string S' minimizing $|D(S', S_1) - D(S', S_2)|$, such that $D(S', S_1) + D(S', S_2) = D(S_1, S_2)$.

18

Three Short Topics

18.1. Matching DNA to protein with frameshift errors

In Section 15.11.3, we discussed the canonical advice of translating any newly sequenced gene into a derived amino acid sequence to search the protein databases for similarities to the new sequence. This is in contrast to searching DNA databases with the original DNA string. There is, however, a technical problem with using derived amino acid sequences. If a single nucleotide is missing from the DNA transcript, then the reading frame of the succeeding DNA will be changed (see Figure 18.1). A similar problem occurs if a nucleotide is incorrectly inserted into the transcript. Until the correct reading frame is reestablished (through additional errors), most of the translated amino acids will be incorrect, invalidating most comparisons made to the derived amino acid sequence.

Insertion and deletion errors during DNA sequencing are fairly common, so frameshift errors can be serious in the subsequent analysis. Those errors are in addition to any substitution errors that leave the reading frame unchanged. Moreover, informative alignments often contain a relatively small number of exactly matching characters and larger regions of more poorly aligned substrings (see Section 11.7 on local alignment). Therefore, two substrings that would align well without a frameshift error but would align poorly with one can easily be mistaken for regions that align poorly due only to substitution errors. Therefore, without some additional technique, it is easy to miss frameshift errors and hard to correct them.

How can DNA sequences be productively compared to amino acid sequences in the presence of both frameshift and substitution errors? We give both a theoretical and a practical answer to this question. The primary approach is based on the technique of *inexact matching of a string to a network*. This technique will also be used in the discussion of gene finding in Section 18.2.

18.1.1. Matching a string to a network

In Sections 11.4 and 11.6.3, the weighted edit distance problem (for strings S_1 and S_2) was represented as a shortest path problem in a directed, acyclic grid derived from the two input strings S_1 and S_2. In the grid, each diagonal edge represented either a match or a substitution, each horizontal edge represented the insertion of a character into S_1, and each vertical edge represented the deletion of a character from S_1. Therefore, the weighted edit distance problem was viewed as the problem of finding a minimum cost path from node $(0, 0)$ to node (n, m) that specifies ("spells out") string S_1, when errors are allowed at specified costs.

Edit distance is only one of many string problems that can be productively represented as a problem of inexactly matching a string to a path in a directed, acyclic graph. To define the general setting, let G be a directed, acyclic graph, called a *network*, with a designated

Figure 18.1: a. DNA sequence with derived amino acid sequence below it. b. The sequence after deletion of the fourth nucleotide (G) and the new derived amino acid sequence with a frameshift error.

start node s and stop node t and with a single character labeling each edge. Thus, each s to t path *specifies* (or spells out) a particular string.

Definition   For a network G as above, we say that a string S' is in G if S' is specified by some s-to-t path in G. Then the *network-matching* problem is: Given an input string S, find a string S' in G whose (weighted) edit distance to S is minimum over all strings in G.

The network matching problem can be efficiently solved via a set of straightforward dynamic programming recurrences. Let $D(i, v)$ denote the minimum (weighted) edit distance between the prefix $S[1..i]$ and any string specified by a path from node s to node v in graph G. For a node v, let $E(v)$ be the set of edges in G directed into node v, and for an edge $e = (u, v) \in E(v)$, let $T(i, e)$ be the cost of transforming character $S(i)$ to the character labeling edge e. Generally, $T(i, e)$ will be zero if the two characters match and will be positive if they mismatch. Let $I(e)$ be the cost of inserting the character on edge e into S, and let $DL(i)$ be the cost of deleting character $S(i)$ from S.

For $i > 0$ and $v \neq s$, $D(i, v)$ is the minimum taken over three possibilities:

$$\min_{e=(u,v)\in E(v)} [D(i - 1, u) + T(i, e)],$$

$$\min_{e=(u,v)\in E(v)} [D(i, u) + I(e)],$$

$$D(i - 1, v) + DL(i).$$

The needed base cases are $D(i, s)$ for $i \geq 0$ and $D(0, v)$ for $v \neq s$. Clearly, $D(0, s) = 0$, and $D(i, s)$ for $i > 0$ is just the cost of deleting the first i characters from S. The edit distance $D(0, v)$ is obtained by repeated application of the second recurrence above to the nodes in G. We leave the details as an exercise. We also leave the proof of correctness of these recurrences and the following theorem as exercises. At this point in the book, these should be straightforward.

Theorem 18.1.1. *If (acyclic) network G has edge set E, then the network-matching problem can be solved via dynamic programming in $O(|S||E|)$ time.*

We now return to the motivating problem of matching DNA to protein with frameshift errors.

18.1.2. DNA/protein matching cast as network matching

Let S be a newly sequenced DNA string and let P be an amino acid sequence that is to be aligned with S. Let $\mathcal{P}$ be the set of all the back-translations of P into DNA strings that encode P. The following is one way to formalize the problem of matching a DNA string

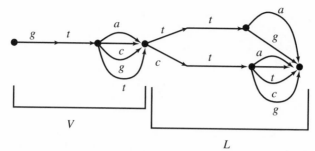

Figure 18.2: The subnetwork for the amino acid pair *VL*.

S to a protein string P in the face of frameshift errors:

DNA to protein matching   Find the DNA string $S' \in \mathcal{P}$ whose (weighted) edit distance to S is minimum over all strings in $\mathcal{P}$. The edit distance of S and S' then is taken as the measure of similarity of S and P.

This formalization seems to correctly capture the problem, but since the size of $\mathcal{P}$ is exponential in the length of $|P|$, can this conceptual approach be efficiently implemented? It can certainly be implemented inefficiently. In fact, there are computer packages that do DNA to protein comparisons using this back-translation approach, explicitly generating set $\mathcal{P}$ and then aligning S to each string in $\mathcal{P}$. That is not the solution we seek.

To efficiently implement the back-translation approach, we build a network G based on P, and then solve the network matching problem for S and G. For each amino acid in P, construct a subnetwork that specifies all of the DNA codons for that amino acid. Then connect these subnetworks in series, creating the network G. The set of strings in G are exactly the set of strings in $\mathcal{P}$. As an example, the subnetwork for the amino acid pair VL is shown in Figure 18.2. Finally, find the string in G with minimum edit distance to the DNA string S.

Since there are at most six codons for any amino acid and each consists of three nucleotides, network G has $O(|P|)$ edges. Hence, the problem of finding the string $S' \in \mathcal{P}$ with the minimum edit distance to S is solved in $O(|S||P|)$ time. That is the same time it would take to align S to P if they were both strings over the same alphabet.

The above approach to DNA/protein matching, based on network matching, was essentially worked out in [364]. A sublinear expected time elaboration of it was outlined in [94]. A different, practical approach to DNA/protein matching is outlined in Exercise 5, and two formalizations of that approach are outlined in Exercises 6 and 7 at the end of the chapter. Finally, Claverie [103] uses a very different, heuristic approach to the problem. It is based on using an amino acid substitution matrix that reflects possible frameshift errors in the underlying DNA. For another effort to align DNA to protein, for somewhat different purposes, see [219]. For an in-depth examination of several methods, see [485].

18.2. Gene prediction

With the recent ability and determination to sequence large amounts of DNA, the problem of identifying genes contained in newly sequenced DNA has become of critical importance. This problem arises in two ways. First, the various genome projects are producing a flood of new, relatively unstudied DNA sequences. Identifying the genes in these sequences is one of the first steps in annotating and characterizing the new sequences. Second, in positional cloning, one may sequence a large stretch of DNA that is expected to contain a disease-related gene (see page 397). Then one searches that sequence to find regions that have the "look and feel" of a gene. In the case of eukaryotes, the problem is compounded by introns that interrupt the coding regions of genes, so one has to distinguish between the introns and exons in a predicted gene.

The gene-finding problem in eukaryotes is more complex than in prokaryotes and is generally divided into two tasks: finding *candidate exons* in a long DNA sequence (believed to contain a gene) and selecting a subset of these candidates to form the predicted gene.

The task of finding candidate exons is outside the scope of this book and we will only mention some of the ideas used. The most basic is to identify all the maximal *open reading frames* (*ORFs*) in the sequence. An open reading frame is a substring that contains no stop codons when read in a single reading frame (see the glossary for an alternate definition of ORF). If the proper reading frame is not known, then all three (or six) possible reading frames are examined. Additional information can then be used to predict if an ORF contains an exon. For example, the length of the ORF is a key indicator. Coding regions are significantly longer than the average maximal ORF. Also, the frequencies with which different synonymous codons are used in exons is often different from their frequencies in noncoding regions. This is called *codon bias*. The use of codon bias to distinguish coding from noncoding regions is made more effective by the fact that different organisms exhibit different codon biases. Other indicators are based on particular motifs that one expects to find in coding regions. Methods to identify candidate exons usually produce a large number of overlapping candidates but also provide a score that reflects the "likelihood" that each identified candidate is a true exon. In [412], a score is reported for every substring of the long string. Exon-finding methods often have the most difficulty identifying short exons and in precisely locating exon-intron boundaries (splice sites). A survey of some of the indicators and methods is contained in [412]. See also [172], [174], [156], and [192].

The second task of selecting a "good" subset of nonoverlapping candidate exons to form the predicted gene is called *exon assembly*. In a more complex version, candidate introns can also be identified. In that case, the exon assembly problem is one of finding nonoverlapping, alternating candidate exons and candidate introns that cover the sequence.

18.2.1. Exon assembly

We consider the simple version of exon assembly, namely selecting a nonoverlapping subset of exon candidates, while ignoring intron candidates.

Since each identified exon candidate is given a score representing its goodness, the exon assembly problem may seem to be solved by finding a set of nonoverlapping candidates of highest total score. That is precisely the one-dimensional chaining problem discussed in Section 13.3. However, it is reported in [174] that this approach does not work well, and that claim makes sense. Each exon candidate in the optimal chain may look like an exon (based on statistics from many exons in different organisms), but the chaining method does not reflect information on the correlated inclusion of subsets of exons in genes. A related, more sophisticated approach to problems like exon assembly appears in [277].

An alternative approach, due to Gelfand, Mironov, and Pevzner [173, 174], is to use the power of sequence comparison and database search (i.e., the *first fact* of sequence comparison) to solve a problem they call *the spliced-alignment problem*. In that paper, a candidate exon is called a *block*, and we will use that term here. Their approach is to find a subset of nonoverlapping blocks whose concatenation is highly similar to a sequence from a known gene in the database. Because we are considering eukaryotes, the database sequences are cDNA sequences, and hence their intron sequences have been removed (see Section 11.8.3).

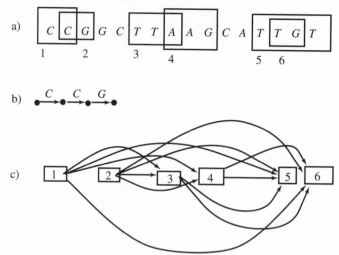

Figure 18.3: a. Sequence S and six blocks. b. The subpath for block B_1. c. A representation of the network G for S. Each box represents the subpath for the corresponding block, but the actual path is omitted. Only the edges connecting those subpaths are shown.

Definition   Let S denote a genomic DNA sequence, let $\mathcal{B} = \{B_1, B_2, \ldots, B_k\}$ be the ordered set of blocks identified in S, and let T be a cDNA sequence. Let S' denote a string obtained by concatenating ordered, nonoverlapping blocks from $\mathcal{B}$. String S' is called a *chain*. The *spliced alignment problem* is to find the chain S^* with greatest similarity to T. "Similarity" is used here exactly as discussed in earlier chapters, allowing the use of specific scoring schemes to best model gene recognition.

Given a database of gene sequences, the spliced alignment problem is solved for S and each sequence T in the database. The chain corresponding to the best spliced alignment is then used for the exon assembly of S.

Solving spliced alignment via network alignment

The spliced alignment problem of S and T is easily solved by aligning string T to a network G derived from the set of blocks $\mathcal{B}$. However, the objective function used is to maximize similarity rather than minimize edit distance (see Exercise 4). For each block B_i in $\mathcal{B}$, create a subpath in G with $|B_i|$ edges. Each of these edges is labeled with a single character, so the path spells out the substring B_i. For any pair of blocks B_i, B_j, connect the last node of the path for B_i by a directed, unlabeled edge to the first node of the path for B_j if and only if B_i ends strictly to the left of the start of B_j. (See Figure 18.3) Then the optimal alignment of string T to this network solves the spliced alignment problem.

If $\mathcal{B}$ has k blocks containing a total of b characters, then network G has $\Theta(b)$ edges in the paths derived from the individual blocks plus $O(k^2)$ edges connecting the paths. Hence, by Theorem 18.1.1, the spliced alignment problem can be solved in $O(|T|b + |T|k^2)$ time. However, the network G created for spliced alignment has additional structure that allows a reduction in the running time to $O(|T|b + |T|k)$ time. That reduction is considered in Exercise 11 of this chapter.

Another approach to finding exon/intron splice sites, due to Laub and Smith [293] also uses the power of sequence databases, but avoids dynamic programming. Rather, it looks for a high density of good matches between short substrings in the target DNA, and sequences in the exon database.

18.3. Molecular computation: computing with (not about) DNA strings

The idea of using biomolecules in computers and computing devices has been extensively explored in the past decade [67, 68]. Prototype hybrid computers based on properties of particular proteins have been built and special-purpose computing devices and components made from organo-electronic materials are being commercially developed. Some of these efforts represent a real turning of the tables. Instead of computing *about* strings, one computes *with* strings (polymers such as protein and DNA). One effort has received a great deal of attention and, coming out of the theoretical computer science community, has been particularly attractive to many theoreticians.

In fall of 1994, Leonard Adleman [4] showed how to solve a small (eight node) instance of the Hamilton path problem by setting up a laboratory procedure based on recombinant DNA techniques. This experiment suggested that such DNA-based computing, exploiting the inherent parallelism in DNA-based biochemistry, might be able to solve larger instances of hard problems than is presently possible through the use of conventional computers. His work received a great deal of attention in both the scientific and popular press, including a major article in the *New York Times* [281]. Other people followed up on Adleman's initial work, suggesting additional ways of using DNA to tackle hard computation problems. The most widely cited of those papers is due to Richard Lipton [305], who described how to use DNA to solve instances of the classic *Satisfiability* problem. We will describe here Lipton's paper rather than Adleman's, because it is simpler to describe and has a somewhat more general feel to it.

The Satisfiability problem (SAT) Given a formula in *conjunctive normal form* (*CNF*), determine if there is a setting of the variables (true or false) that makes the entire formula evaluate to true.

A formula in CNF consists of clauses made up from n variables denoted $x_1, x_2, \ldots, x_n$. An occurrence of a variable x_i, or an occurrence of its negation $\overline{x_i}$, is called a *literal*. Each clause consists of literals connected by **OR** operators (written $\vee$) and enclosed in a matching pair of parentheses. The clauses are connected by **AND** operators (written $\wedge$), and the entire string is called a *formula*. For example, $(x_1 \vee \overline{x_2}) \wedge (x_2 \vee x_3) \wedge (\overline{x_1} \vee \overline{x_3})$ is a formula in CNF with three clauses constructed from three variables and two literals per clause.

The Satisfiability problem is to set the true/false values of the variables to make the entire formula true under the following rules: For any variable x_i, every occurrence of x_i must have the same value, and that value must be the opposite value of every occurrence of $\overline{x_i}$. The **OR** of two values is true if and only if at least one of the two values is true; the **AND** of two values is true if and only if both of the values are true. A formula therefore is made true if and only if at least one literal in each clause is set true under the constraint that a variable must have the opposite setting as its negation.

The Satisfiability problem for formulas with only two literals per clause can be solved efficiently, but the problem is NP-complete for formulas with as few as three literals per clause [171]. Therefore, there is no known efficient (worst-case) algorithm for the Satisfiability problem, and it is believed that no efficient algorithm exists.[1] Being NP-complete not only makes SAT (or Hamilton path) an interesting test case for DNA computing, but

---

[1] However, it should be remembered that NP-completeness is a theory about asymptotic, worst-case phenomena and important instances of an NP-complete problem may still be solvable in practice.

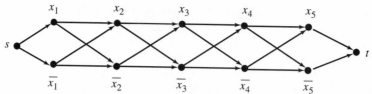

Figure 18.4: An acyclic directed graph that represents all the ways to choose true/false settings for the variables x_1, x_2, x_3, x_4, and x_5. Each s to t path represents one way to set the variables.

it makes a practical solution for SAT very valuable. The reason is that any instance of any problem in the class NP, a vast class of problems, can be efficiently reduced (on a conventional computer) to an instance of SAT, with the property that the solution of that SAT instance solves the original problem instance. So a method that can solve the Satisfiability problem in practice for large CNF formulas would have an importance well beyond the significance of the SAT problem itself.

18.3.1. Lipton's approach to the Satisfiability problem

Conceptually, Lipton's approach to the Satisfiability problem is complete brute force: Enumerate all 2^n ways to set the n variables and then test each enumerated setting to see if it makes the formula true. The novelty is that with DNA technology, the enumeration and the evaluations can be done in a massively parallel way. Now we turn to the details.

The method consists of an enumeration stage and an evaluation stage. The enumeration stage can be executed without knowing any specific CNF formula. For a chosen setting of a parameter m, the enumeration stage synthesizes 2^m short DNA strings, each representing a distinct way that m variables can be set true or false. Once synthesized, those strings can be used to solve the Satisfiability problem for any specific CNF formula with $n \leq m$ variables.

The enumeration stage

It is useful to represent all the 2^m possible settings for m variables $(x_1, \ldots, x_m)$ as a directed acyclic graph with $2m + 2$ nodes. An example is shown in Figure 18.4. The graph consists of m columns of two nodes each, plus a left start node s and right stop node t. The top node of every column i represents the choice of setting variable x_i true, and the bottom node of column i represents the choice of setting variables x_i false. There is an edge from node s to each of the two nodes in column one, an edge from each of the nodes in column m to node t, and an edge from each of the two nodes in column i to each of the two nodes in column $i + 1$, for i from 1 to $m - 1$. Clearly, there is a one-to-one correspondence between the 2^m ways that the m variables can be set and the 2^m directed paths in the graph from node s to node t. If a path goes through the top (bottom) node of column i, then x_i is set true (false).

Lipton's method synthesizes one DNA string for each directed path in s to t, as follows: First, create a random DNA string of length twenty say, for every node of the graph. Such a small DNA string is called an *oligo*. The strings for the top m nodes are called T-oligos, the strings for the bottom m nodes are called F-oligos, and the strings for nodes s and t are called *start* and *stop* oligos, respectively. These oligos are collectively called *node-oligos*.

Next, create a DNA string for each edge in the graph. These strings, called *linker-oligos*, are also of length twenty, but they are not random. For an arbitrary edge from node u to

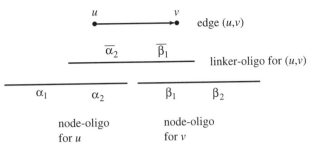

Figure 18.5: The (u, v) linker-oligo hybridizes to the node-oligos for nodes u and v. $\alpha_1, \alpha_2, \beta_1$ and β_2 represent oligos of length ten each. $\overline{\alpha_2}$ and $\overline{\beta_1}$ represent oligos that are complementary to α_2 and β_1 respectively.

node v, the linker-oligo for (u, v) is created as follows:[2] The first ten nucleotides are the *complements* ($A \leftrightarrow T$, $C \leftrightarrow G$) of the last ten nucleotides of the node-oligo for node u; the last ten nucleotides are the complements of the first ten nucleotides of the node-oligo for v. The point is that the linker-oligo for (u, v) can hybridize with the node-oligos for u and v, holding those two oligos together to create a length-forty oligo (see Figure 18.5).

Next, create "a large number" of copies of all the node and linker oligos. The exact amount is technology dependent, but at a minimum there must be 2^m copies of the start and stop oligos, 2^{m-1} copies of each node-oligo, and 2^{m-2} copies of each linker-oligo.

There is one essential constraint on the selection of oligos. The oligo for a node u must only be able to hybridize on its left end to the two linker-oligos for the two edges into u, and it must only be able to hybridize on its right end with the two linker-oligos for the two edges out of u. Similarly, the linker-oligo for (u, v) should only be able to hybridize with nodes u and v, and these hybridizations must occur on their intended ends. Thus, no ten-length suffix of one node-oligo can be a prefix of another node-oligo. One expects that random oligos will obey these constraints, but if they don't, with experimentation one can create or select $2m + 2$ oligos so that no pair can hybridize. By using experimentally derived oligos for the node-oligos, and then creating linker-oligos as described above, the entire set of oligos should obey the hybridization constraints.

In the next step of the enumeration stage, all the copies of all the oligos are mixed together and allowed to hybridize under the constraints just described. This yields DNA strings of length $20m + 20$ bases, where in each string, the first and last ten bases are single stranded, and the rest of the string is double stranded. In each double stranded string, one strand contains only node-oligos and the other strand contains only (offset) linker-oligos. We call the former strand the *value string*. Each value string therefore corresponds to a directed path from s to t in the graph and hence corresponds to a choice for setting the m variables $x_1, x_2, \ldots, x_m$. At this point, the two strands cannot be separated because neighboring node-oligos on the value string are held together by the offset linker-oligos on the other strand. So as a final step, for each double-stranded molecule, neighboring oligos on the same strand are ligated (chemically joined) so that the two strands can be separated. The net result is that each value string is now a stable, single-stranded DNA molecule that will be used in the evaluation stage. In general, the method will require many copies of each value string, but the actual number is again technology dependent.

Note that the value strings can be created without knowing any specific CNF formula. They can be used for any CNF formula with $n \leq m$ variables.

---

[2] There is also a technical point involving DNA orientation that is omitted in this description.

The evaluation stage

Once a specific CNF formula with $n \leq m$ variables is given, the Satisfiability problem for that formula is solved in k steps, where k is the number of clauses of the formula. We assume that the variables in the formula are named $x_1, x_2, \ldots, x_n$.

The evaluation stage starts with the full collection (soup) of value strings created in the enumeration stage. (The strands made up of linker-oligos are also in the soup but can be ignored or removed.) At each successive step, some number of value strings (possibly none) will be deleted from the soup, and the "surviving" value strings will be passed on to the next step. In detail, at each step another clause of the formula is processed, and all the value strings in the soup whose variable settings make that clause *false* are removed from the soup. (With standard technology, some strings that make the clause true, i.e., that "satisfy" the clause, are also removed, which is one reason that many copies of each value string are needed.) The surviving strings that remain in the soup are then used in the next step. For example, if the first clause processed is $(x_2 \vee \overline{x_3})$, then any value strings in the soup that contain the T-oligo for x_2 or contain the F-oligo for x_3 survive the step. All other value strings are deleted. Hence at each step, the surviving strings represent variable settings that satisfy all the clauses processed so far. When all the k clauses have been processed, one simply determines if there are any surviving value strings in the soup. If yes, then the variables can be set to make the given CNF formula true; otherwise the variables cannot be set to make it true.

Biotechnical details

Lipton's original paper did not specify the biotechnical details to implement the evaluation stage. Here we engage in speculation on how the method might be implemented.[3]

Prior to knowing any specific CNF formula, one creates $2m$ additional length-twenty oligos, one for each possible truth setting of m variables. For variable x_i, one new oligo is made complementary to the T-oligo for x_i, and one new oligo is made complementary to the F-oligo for x_i. Each of these $2m$ new oligos is replicated and embedded in its own nylon filter. This creates $2m$ nylon filters, each embedded with copies of the oligo that can hybridize to one specific node-oligo.

The $2m$ nylon filters can be used to process any clause containing up to m literals. We describe the method by showing how to process $(x_2 \vee \overline{x_3})$. First, all the value strings that have survived so far (before $(x_2 \vee \overline{x_3})$ is processed) are passed through the filter embedded with the oligo that can hybridize to the T-oligo for x_2. Consider a value string that contains the T-oligo for x_2. That T-oligo can hybridize and stick to the oligo in the first filter. However, with the right prodding, all strings that do not contain the full T-oligo for x_2 can be made to pass through the filter. (In the process, some strings containing the T-oligo for x_2 are also lost.) The strings that do not stick to the first filter are then passed through a filter embedded with the oligo that can hybridize to the F-oligo for x_3. As before, many of the strings containing the F-oligo for x_3 will stick there, but no strings not containing the F-oligo for x_3 will stick. The strings that do not stick to either filter are not needed further, while the strings stuck to the two filters are washed free and mixed together. These are the surviving value strings that represent settings of the variables

---

[3] The details are surely fanciful, but the point should be clear. If one can easily think up plausible-sounding laboratory procedures to implement the method (at least for small instances), then biotech professionals should be able to find implementations that really work.

making $(x_2 \vee \overline{x_3})$ true. Hence, after this evaluation step, each surviving string represents a setting that satisfies all the clauses processed so far.

When all the clauses have been processed, one determines if the soup is empty or whether it actually contains some surviving strings. One way to do this is to use the start and stop oligos as PCR primers to generate a large number of copies of any surviving strings. PCR will quickly make enough copies so that the product (if there is any) will be visible on an electrophoresis gel or be detected by some other method. If there are surviving value strings, one can also efficiently determine (through a computational technique called *self-reduction*) a choice of variable settings to make the formula true. How to do that using filters and PCR is left as an exercise.

A CNF formula with k clauses containing a total of l distinct literals can be processed in only k steps, using a total of l filter passes. Hence even though the underlying logic of Lipton's method involves exhaustive enumeration and evaluation of all possible truth settings, the use of DNA introduces massive parallelism into the process, raising the hope that it can be made practical.

18.3.2. Critique

The idea of using DNA to solve computational problems is certainly intriguing and elegant, and DNA does provide a massive parallelism far beyond what is available on existing silicon-based computers. However, there are many technological hurdles to overcome, and one huge fundamental problem with the particular methods discussed here.

The fundamental problem is that the function 2^n is exponential whether it counts time or molecules. Worse, 2^n is the *minimum* number of DNA strings needed to make Lipton's SAT method work. Similarly, it has been estimated [302] that Adleman's Hamilton path method applied to a 70-city problem with moderate edge density $(\Theta(n \log n))$ would require 10^{25} kilograms of DNA just to make one copy of each needed string! My back-of-the-envelope calculations show a far greater amount of DNA is needed. In an examination of several new developments in nonstandard computing, a news article in Science [180] states that a year after his initial work,

> Adleman now calculates that using the same algorithm to solve "even a problem of modest size" – involving, say, 50 or 100 cites – would require "tons of DNA". Later, Penn State's Beaver found that an apparently promising algorithm for factoring a 300-digit number actually called for "an ocean [of DNA] the size of the universe".

The minimum amount of required DNA for Lipton's SAT method is more modest (for a 70-variable formula, the estimated minimum amount of DNA ranges from a few grams to a few thousand kilograms, depending on various technical assumptions). But even this zone of relative practicality is quickly exhausted. Even if a 70-variable formula needs only a few grams, a formula with just 100 variables will have a minimum DNA requirement of millions of kilograms. And 100 is not a large number, especially for problem instances that are created by reducing other NP-complete problems to SAT.

So the *base, raw* idea of brute-force enumeration is not going to work beyond modest problem sizes, even with the massive parallelism that DNA affords. Even ignoring technical issues in laboratory economics and DNA chemistry (and these are huge problems in themselves), the fundamental problem of exponential growth means that the enumerative methods suggested to date will be of importance for very small problem niches at best. The question becomes, even if DNA chemistry can be made to work at the gram or kilogram level, are there important instances of SAT that are too large for a silicon computer to solve and yet small enough that DNA computing will be practical?

Rather than trying to provide practical solutions for NP-complete problems, DNA computing based on pure enumeration may prove more valuable in speeding up solutions to problems already considered tractable. Or, one might find applications where the underlying algorithm only needs to enumerate a small fraction of the solution space. Algorithms of this type are found in the realm of *randomized algorithms*, where often the key observation is that a particular event occurs with high enough probability so that only a "small" amount of random sampling is required. Only time (and money) will tell where the initial ideas for DNA computing will lead. Many additional ideas have been proposed (for example, see [53]) since the first papers of Adelman and Lipton. Despite the impracticality of most of the proposals to date, some practical molecular computing method in the future might trace its origins to these early ideas. Hopefully though, future contributions will be more fully evaluated before the popular press declares that computer scientists are "astonished" at the "revolutionary advances".

18.4. Exercises

1. Prove the correctness and time analysis of the dynamic programming recurrences for the network matching problem of Section 18.1.1 (page 480).

2. Explain the validity of modeling the DNA/protein matching problem as a network matching problem (as detailed in Section 18.1.2).

3. The solution to the DNA/protein matching problem developed in Section 18.1 essentially encodes a *global* alignment of the derived amino acid sequence to a protein sequence. Since local alignment is generally of greater biological importance, rework the method to solve the local alignment version of the problem.

4. Reformulate the network matching problem to maximize similarity rather than minimize edit distance. State the dynamic programming recurrences.

5. One practical approach to DNA/protein matching is based on a direct use of local alignment. Let S be the DNA sequence to be compared to protein sequence P. Assume the correct orientation of S is known, but its reading frame is not. Translate S into three amino acid sequences, S_1, S_2, and S_3, one for each choice of reading frame. Then locally align each of these three sequences to P and find all the local alignments that are above some threshold of similarity (see Sections 11.7.3 and 13.2). The result will be three sets of good local alignments. If a frameshift error occurs in a region of the DNA that has high similarity to a region of P (when translated in the correct frame), then this local alignment will be part of one of the three sets. A frameshift error in that region shows up as two adjacent, good local alignments in two of the sets. By looking for such changes, one can piece together a good overall alignment of S and P by hand. Discuss the validity and practicality of this approach compared to the network-based approach given in Section 18.1.

6. The approach given in the previous exercise can be automated. Once the three sets of good local alignments have been obtained, the problem is to find a high scoring subset of nonoverlapping local alignments from those sets. This is exactly the two-dimensional chaining problem discussed in Section 13.3. Explain this in detail, and discuss any advantages or disadvantages in solving DNA/protein matching in this way.

7. Guan and Uberbacher [195] follow the general approach to DNA/protein matching given in Exercise 5, but they automate the approach by using dynamic programming. We describe their method to find a single best local alignment between a DNA sequence and a protein sequence.

Conceptually, solve (in parallel) the three problems of locally aligning the prefixes of S_1, S_2, and S_3 to prefixes of P, using the normal local alignment recurrences. In each table, the value of any cell is obtained using values from three of its neighboring cells in its own table. In addition, however, appropriate cell values from the other two tables are also considered, and these can be used along with a penalty to reflect a frameshift error. Make this idea precise. Write out the recurrences in detail and determine the running time of the algorithm. Then consider the problem of finding many good local alignments of S to P.

8. Generalizing the problem of DNA/protein alignment, consider the problem of aligning two derived amino acid sequences P_1 and P_2. The problem now is to find a DNA back-translation of P_1, call it S_1, and a DNA back-translation of P_2, call it S_2, whose edit distance is minimum over all such pairs of back-translated sequences. Give the most efficient solution you can for this problem.

9. Given two DNA strings, find the minimum number of changes to the two strings (or minimum cost of those changes) so that the two new strings translate to the same amino acid string. Does this problem have an efficient solution? How might this problem arise in practice?

10. Modify the network model used for exon assembly (Section 18.2.1) to include candidate introns as well as blocks. That is, the assembled gene must consist of nonoverlapping, alternating blocks and candidate introns that cover the candidate coding region. The goodness of the solution is judged both by how well the concatenated blocks match a sequence in the database and how well the selected candidate introns look like introns. Explain any needed changes to the dynamic programming solution to standard network matching.

11. The running time for the spliced alignment problem can be reduced from $O(|T|b + |T|k^2)$ time to $O(|T|b + |T|k)$ time. The improved method is in the same spirit as the solution to the one-dimensional chaining problem detailed in Section 13.3. It follows a left-to-right scan of S, but it requires additional dynamic programming effort whenever the left end of a block is encountered. Instead of simply looking up a single value for the block, the method must evaluate dynamic programming recurrences to consider the value of using that block in an alignment with string $T[1..l]$, for each position l. For any block B_i, let $C(B_i, l)$ be the value of the best alignment between $T[1..l]$ and any chain containing a subset of blocks from the set B_1 to B_i. If B_h is the last block to end before block B_i begins, then for any position l in T, $C(B_i, l)$ is the maximum of $C(B_h, l)$ and is the alignment value obtained by using all of block B_i in an alignment ending at position l of T. It is the latter value that has to be computed using dynamic programming when the left of B_i is reached.

Flesh out the details of the method, prove its correctness, and analyze its running time.

12. Explain in detail why in Lipton's method every strand containing only linker-oligos can be ignored. Explain how they are automatically discarded when the first clause is evaluated.

13. As described in Section 18.3.1, Lipton's method for the Satisfiability problem only determines whether there is a satisfying assignment of the formula. Even when the formula can be satisfied, the method does not explicitly describe how the variables should be set. One can of course sequence any DNA strings obtained at the end of the method, but that is highly undesirable.

Give a simple extension of Lipton's method (using the biotech operations described in the text) to efficiently determine how to set the variables to satisfy the formula, when the formula is satisfiable. One approach uses the common technique of *self-reduction*.

19

Models of Genome-Level Mutations

19.1. Introduction

String search, edit, and alignment tools have been extensively used in studies of molecular evolution. However, their use has primarily been aimed at comparing strings representing single genes or single proteins. For example, evolutionary studies have usually selected a single protein and have examined how the amino acid sequence for that protein differs in different species. Accordingly, string edit and alignment algorithms have been guided by objective functions that model the most common types of mutations occurring at the level of a single gene or protein: point mutations or amino acid substitutions, single character insertions and deletions, and block insertions and deletions (gaps).

Recently, attention has been given to mutations that occur on a scale much larger than the single gene. These mutations occur at the *chromosome* or at the *genome* level and are central in the evolution of the whole genome. These larger-scale mutations have features that can be quite different from gene- or protein-level mutations. With more genome-level molecular data becoming available, larger-scale string comparisons may give insights into evolution that are not seen at the single gene or protein level.

The guiding force behind genome evolution is "duplication with modification" [126, 128, 301, 468]. That is, parts of the genome are duplicated, possibly very far away from the original site, and then modified. Other genome-level mutations of importance include *inversions*, where a segment of DNA is reversed; *translocations*, where the ends of two chromosomes (telomeres) are exchanged; and *transpositions*, where two adjacent segments of DNA exchange places. These kinds of large-scale mutations are collectively called *genome rearrangements*.

19.1.1. Genome rearrangements give new evolutionary insights

Studying genome rearrangements may allow certain evolutionary deductions that aren't possible from gene-level comparisons. In many organisms, genome-level mutations occur more slowly than gene-level mutations. This allows one to compare molecular data from species that diverged a very long time ago. For example, to deduce the divergence order of three species, one traditionally does three pairwise similarity computations using a protein common to the three species. Then, provided that the similarity values are "consistent", one reasons that the pair with highest similarity diverged from a common ancestor, which had earlier diverged from the third species. But if divergence times are far enough in the past, the accumulated gene mutations may be so great that all the three pairwise similarity values may be essentially indistinguishable and unusable to reliably determine a fine-grain history of divergence. In contrast, more slowly occurring genome rearrangements may still allow detailed historical deductions. Even though a gene has mutated at the DNA level, one can still recognize it as the same gene in two different species from its protein product

492

and/or function. This allows one to compare the *order* that genes appear on a chromosome in the different species and to use differing gene order to deduce evolutionary history far back in time.

At the other end of the continuum, there is DNA, which is extremely stable and which doesn't accumulate many mutations at the gene level, but which does exhibit genome rearrangements. This is true, for example, of plant mitochondrial DNA:[1]

> ... the *Brassica* mitochondrial genome may be viewed as a collection of unchanging sequences whose relative arrangement is extremely fluid. [355].

A related example where genome rearrangements can be informative is the mammalian X-chromosome:

> The X-chromosome genes have been less mobile during evolution than autosomal genes. A gene found on the X-chromosome in one mammalian species is highly likely to be found on the X-chromosome in virtually all other mammals. [317]

This allows one to study the *gene order* of the X-chromosome genes across the mammals.

Work on algorithms for genome rearrangements only began a few years ago, but already there is an impressive literature (see [272], [270], [271], [40], [39], [209], [211], [212], [213], [154], [61], [261], [38], and [390]). We detail here one of the first results, illustrating the flavor of much of the work being done.

19.2. Genome rearrangements with inversions

In this section, we describe some of the original work of Kececioglu and Sankoff [271, 272] on genome rearrangements caused by *inversions* but no other types of rearrangements. This single focus is partly for conceptual simplicity, as a way to begin to tackle the computational issues, but there is DNA (plant mitochondrial DNA for example [355]) where inversions alone are the dominant form of mutation.

To begin, we assume that the genes on the chromosome are distinguishable so that they can be given distinct integer labels. This is a reasonable starting assumption, since most genes do occur as a single copy, and even genes that have mutated at the DNA level can be identified (from the remaining sequence similarity) and given the same integer. Therefore, the genes are numbered 1 through n on the original chromosome (this ordering is called the *identity permutation*) before inversions have taken place. The chromosome after inversions is represented by a *permutation* of the integers from 1 to n (see Figure 19.1).

Now, given two permutations representing the gene orders of a chromosome in two species, we would like to define an *inversion distance* between the two species. This distance should reflect molecular-level inversions that would transform one gene order into the other. That is, we would like to know how the second permutation could arise (or did arise) from the first permutation by a series of natural molecular-level inversions. Unfortunately, a complete model of inversions in DNA is not known, and no one knows how such historical reconstruction should be done, or what the most meaningful definition of *inversion distance* is. However, because inversions are relatively rare in chromosomes

---

[1] This point can be very confusing because mitochondrial DNA sequence in animal species mutates much faster than nuclear DNA and therefore is commonly used to establish a fine-grain molecular clock for very recent time intervals [468].

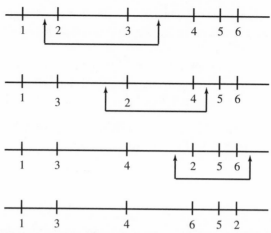

Figure 19.1: Each line represents a "chromosome" with six genes. Three inversions transform the top chromosome to the bottom one. Two arrows enclose the segment to be inverted. The first chromosome is represented by the identity permutation 1, 2, 3, 4, 5, 6; the second by 1, 3, 2, 4, 5, 6; the third by 1, 3, 4, 2, 5, 6; and the last one is represented by 1, 3, 4, 6, 5, 2.

and are seemingly random in the segments selected for inversion, the *parsimony* criteria have usually been invoked.

> **Definition** Under the parsimony criteria, the *inversion distance* between two permutations is defined as the *minimum* number of inversions capable of transforming one permutation into the other.

A reduction

The general problem of computing the inversion distance between two arbitrary permutations is equivalent to the problem of transforming a *single* permutation to the *identity* permutation. This comes from picking one of the permutations as the reference, giving it the identity permutation, and relabeling the other permutation accordingly. With this reduction, the general problem of computing inversion distance becomes the problem of computing the minimum number of inversions capable of transforming a permutation into the identity permutation. That problem has been shown to be NP-hard [87].

We will present a simple method [271, 272] that transforms any input permutation Π to the identity permutation, using a number of inversions that never exceeds twice the inversion distance.

19.2.1. Definitions and initial facts

Because the order of genes on a chromosome is represented by a permutation, we first need some terminology to discuss permutations.

> **Definition** Let $\Pi = \pi_1, \pi_2, \pi_3, \ldots, \pi_n$ represent a permutation of the integers 1 through n, where π_i is the number in the ith position.

For example, if $\Pi = 3, 2, 4, 1$ then $\pi_1 = 3, \pi_2 = 2, \pi_3 = 4$, and $\pi_4 = 1$. Note that π_{i+1} is the number to the right of π_i in Π, whereas $\pi_i + 1$ is the number that is one greater than π_i.

Definition   A *breakpoint* in Π occurs between two numbers π_i and π_{i+1}, for $1 \leq i \leq n - 1$, if and only if $|\pi_i - \pi_{i+1}| \neq 1$. Further, there is a breakpoint at the front of Π if $\pi_1 \neq 1$ and there is a breakpoint at the end of Π if $\pi_n \neq n$.

For example, consider $\Pi = 3, 2, 4, 5, 1$. There are breakpoints at both the front and end of Π, between 2 and 4, and between 5 and 1. But there is no breakpoint between 3 and 2, even though they are out of order.

Definition   Let $\phi(\Pi)$ denote the number of breakpoints in permutation Π.

Lemma 19.2.1. *The inversion distance of any permutation Π is at least $\phi(\Pi)/2$.*

PROOF   A single inversion operation can reduce the number of breakpoints by at most two, removing a breakpoint at either end of the inverted segment. Since the identity permutation contains no breakpoints, it takes at least $\phi(\Pi)/2$ inversion operations to transform a permutation with $\phi(\Pi)$ breakpoints to the identity permutation.   □

Definition   A *strip* in Π is a *maximal* subinterval of Π that contains no breakpoints.

For example, $\Pi = 1, 5, 6, 7, 4, 3, 2$ contains three strips. One is 1 by itself, the next is 5, 6, 7, and the third is 4, 3, 2.

Definition   A strip is called *increasing* if the numbers in it are successively increasing, and it is called *decreasing* if the numbers are successively decreasing. A strip consisting of a single number is defined to be *decreasing*.

For example, the first and last of the three strips in the previous example are decreasing, whereas the second strip is increasing. It is easy to establish that every strip is either increasing or decreasing. Decreasing strips will play a central role in the heuristics developed next.

19.2.2. The heuristics

We first develop an algorithm that never uses more than $2\phi(\Pi)$ inversion operations on any permutation Π. Hence, the number of inversions it uses is never more than four times the number used by the optimal algorithm.

Lemma 19.2.2. *Suppose Π is a nonidentity permutation containing no decreasing strips. Then there exists an inversion in Π that does not increase the number of breakpoints, such that the resulting permutation contains a decreasing strip of length at least two.*

PROOF   Since there are no decreasing strips, every strip is increasing and hence every strip must have at least two numbers. Further, unless Π is already the identity permutation, there must be a strip having a breakpoint at each end. The inversion of that strip creates a decreasing strip without increasing the number of breakpoints.   □

Lemma 19.2.3. *If Π contains a decreasing strip, then there is an inversion that decreases the number of breakpoints by at least one.*

PROOF   Consider the decreasing strip with the smallest number π_i contained in any decreasing strip. By definition π_i is at the right end of its strip. Now π_{i+1} can't be $\pi_i - 1$ (or else $\pi_i - 1$ would be in a decreasing strip), and it can't be $\pi_i + 1$ (for if it were, π_i would be in an increasing, not a decreasing strip). Hence there must be a breakpoint between π_i and π_{i+1}. By similar reasoning, there must be a breakpoint between the number $\pi_i - 1$ and the number to its immediate right. Now either $\pi_i - 1$ is located to the right or to the

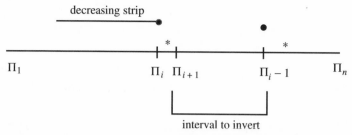

Figure 19.2: Element π_i is the smallest number contained in any decreasing strip. There is a breakpoint (shown by a ∗) after π_i and after $\pi_i - 1$. Inverting the interval between π_{i+1} and $\pi_i - 1$ reduces the number of breakpoints by at least one.

left of π_i. In the first case, we can invert the interval starting with π_{i+1} and ending with $\pi_i - 1$ (inclusive) to bring π_i together with $\pi_i - 1$, removing a breakpoint and extending the decreasing strip containing π_i (see Figure 19.2). That inversion might create (or might remove) a breakpoint at the other end, but there was already one there before the inversion. Hence the number of breakpoints falls by at least one and perhaps by two.

The second case, when $\pi_i - 1$ is to the left of π_i, is symmetric and its proof is left to the reader. □

Lemmas 19.2.2 and 19.2.3 immediately suggest an algorithm using a number of inversions at most four times the inversion distance. The algorithm is:

First heuristic

While there is a breakpoint in the permutation do
begin
If there is a decreasing strip, then find and invert one that reduces the number of breakpoints. {Justified by Lemma 19.2.3}.
Else (if there is no decreasing strip), find and invert an increasing strip (creating a decreasing strip) without increasing the number of breakpoints. {Justified by Lemma 19.2.2}.
end;

Theorem 19.2.1. *The first heuristic uses at most* $2\phi(\Pi)$ *inversion operations to transform* Π *to the identity permutation. Hence the number of inversions it uses is never more than four times the inversion distance.*

PROOF Using the first heuristic, the number of breakpoints falls by at least one after every other inversion. Hence the number of inversions is twice the number of initial breakpoints, or $2\phi(\Pi)$. The theorem follows since the inversion distance is at least $\phi(\Pi)/2$. □

Improving the guarantee

If every inversion that reduces the number of breakpoints also leaves a decreasing strip, then the number of inversions used by the first heuristic would be at most $\phi(\Pi)$ and so would never be more than twice the inversion distance. Unfortunately, at times there are no such nice inversions. However, the situation is only a little more complex.

Lemma 19.2.4. *Let* Π *be a permutation with a decreasing strip. Suppose that there is no inversion that both reduces the number of breakpoints and that leaves a decreasing strip. Then there is an inversion in* Π *that reduces the number of breakpoints by two.*

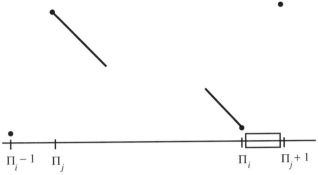

Figure 19.3: Element $\pi_i - 1$ must be to the left of π_j, which must be to the left of π_i, which must be to the left of $\pi_j + 1$.

PROOF Consider the decreasing strip with the smallest number π_i contained in any decreasing strip. In the proof of Lemma 19.2.3, we showed that if $\pi_i - 1$ is to the right of π_i then inverting the interval from starting with π_{i+1} and ending with $\pi_i - 1$ (inclusive) brings π_i together with $\pi_i - 1$, removing a breakpoint. Moreover, this inversion leaves the decreasing strip that number π_i resides in, violating the premise of the Lemma. Hence the number $\pi_i - 1$ must be to the *left* of π_i.

Now consider the decreasing strip with the *largest* number π_j contained in any decreasing strip. By an argument symmetric to the one for π_i, number $\pi_j + 1$ must be to right of π_j. Furthermore, π_j must be to the left of π_i, for if it were not, then the inversion of the interval starting with $\pi_i - 1$ and ending immediately before π_i would reduce the number of breakpoints by one and still leave the existing decreasing strip containing π_j, contradicting the premise of the lemma. Also, π_j must be to the right of $\pi_i - 1$, for if it were located to the left of both π_i and $\pi_i - 1$, then inverting the same interval as before would reduce the number of breakpoints and leave a decreasing strip, again contradicting the premise. Similarly, $\pi_j + 1$ must be to the right of π_i. Figure 19.3 shows the situation to this point.

We now claim that number $\pi_j + 1$ must immediately follow π_i. That is, $\pi_j + 1$ must be π_{i+1} and the box shown in Figure 19.3 must be empty. For contradiction, suppose that there are numbers in the box. There cannot be any decreasing strips in the box, for if there were, the inversion of the familiar interval starting with $\pi_i - 1$ and ending with π_i would reduce the number of breakpoints by one and still leave the decreasing strip in the box, contradicting the premise of the lemma. So if the box has anything in it, it must contain an increasing strip. However, then the inversion of the interval starting with π_j and ending just before $\pi_j + 1$ would reduce the number of breakpoints by one and convert the increasing strip in the box to a decreasing strip, which again would be a contradiction. Hence the box must be empty.

A symmetric argument shows that number $\pi_i - 1$ must immediately precede π_j. It follows that the inversion of the interval starting with π_j and ending with π_i removes a breakpoint on both ends and hence reduces the number of breakpoints by two. □

Lemma 19.2.4 suggests the following improvement to the heuristic algorithm:

Second heuristic

While there is a breakpoint in the permutation do
begin
If there is an inversion that reduces the number of breakpoints and leaves a decreasing strip, execute it.
Else (if there is no inversion that decreases the number of breakpoints and leaves a

decreasing strip), then find an inversion that decreases the number of breakpoints by two, and execute it. Then find and invert an increasing strip without increasing the number of breakpoints.

end;

Theorem 19.2.2. *The second heuristic uses at most $\phi(\Pi)$ inversion operations to transform Π to the identity. Hence it uses at most twice the number of inversions used by the optimal algorithm.*

PROOF In the second heuristic, the number of breakpoints falls by at least one whenever the algorithm executes an inversion that leaves a decreasing strip. When the inversion does not leave a decreasing strip, then it decreases the number of breakpoints by two and is followed by an inversion that doesn't increase the number of breakpoints but does create a decreasing strip. Consequently, the number of inversions is at most $\phi(\Pi)$, the number of original breakpoints. The theorem then follows because any algorithm must do at least $\phi(\Pi)/2$ inversion operations. □

By a more careful analysis and by making the algorithm "look ahead" more than one inversion, Bafna and Pevzner [40] have extended the second heuristic and reduced the error bound from 2 to 1.75.

19.3. Signed inversions

Kececioglu and Sankoff [272] also introduced a *signed* version of the inversion problem. In the *signed inversion* problem, each number in a permutation has a sign (+ or −) that changes every time the number is contained in an inverted interval. For example, when the middle three numbers in $+5, -2, -4, +3, -1$ are inverted, the result is $+5, -3, +4, +2, -1$. The problem is to use the minimum number of inversion operations to transform the given initial signed permutation into the identity permutation where all numbers have positive sign.

The signed inversion problem better models DNA evolution than does the unsigned version. Signed permutations model the fact that DNA is double stranded and that a gene is expressed (or works) on only one of the two strands. When DNA is inverted, it is also *reflected*, so that the inverted segment changes the strand it is on. Hence, a transformation of one gene order to another must not only create the correct order, it must put each gene on the correct strand. The use of signs encodes this requirement. An initial signed permutation is created from two gene orders in the following manner: The unsigned permutation Π is obtained exactly as described by the reduction used for unsigned permutations. To assign the signs, give a positive sign to some arbitrary number in Π, say 1. Then each other number i in Π is given a negative sign if and only if gene i is on the same strand as gene 1 in exactly one of the two gene orders. Thus, a negative sign indicates that gene i must be in an odd number of inversions to end up on the correct strand, whereas a positive sign indicates that gene i must be in an even number of inversions.

The signed inversion problem may seem harder than the unsigned problem, since both the order and the signs have to be transformed. But the problem is actually easier. With appropriate modification of the definitions of increasing and decreasing strips (to allow for the sign) it was shown in [272] that the second heuristic is unchanged and Theorem 19.2.2 still holds for the signed inversion problem (see Exercise 1). That bound was then reduced in [40] to 1.5, and Hannenhalli and Pevzner [211] showed, in contrast to the unsigned version, that the signed inversion problem can be solved in polynomial time. The time bound was then reduced to almost $O(n^2)$ [61], and later to $O(n^2)$ [261].

The signed version of the problem will become more relevant in the future, as many more genome-scale molecular sequences become available. However, today, most of the known gene orders are known without orientation, and hence the unsigned version best models most of the existing data. In fact, algorithmic problems concerning genome rearrangements where even the gene order is not known have been studied [154, 390], and these problems model much of the existing data on large-scale genome rearrangements.

19.4. Exercises

1. In a signed permutation Π, an *adjacency* is defined as a pair of consecutive numbers of the form $+i, +(i+1)$ or $-(i+1), -i$. A breakpoint is defined as occurring between any two consecutive numbers that do not form an adjacency. Also, there is a breakpoint at the front of Π unless the first number is $+1$, and there is a breakpoint at the end of Π unless the last number is $+n$.

   Specify definitions for increasing and decreasing strips so that when the second heuristic algorithm is applied to the signed inversion problem, all the lemmas continue to hold, as does Theorem 19.2.2.

2. Suppose string S' is created from string S by a single inversion. Is there a standard string alignment model that could be used to align S and S' in a way that highlights the inversion? If so, explain the need for new algorithms to study inversions.

3. **Evolution by recombination** In some circumstances, a population of organisms can rapidly evolve by a process of repeated meiotic cross-overs and recombinations. Algorithmic issues in reconstructing the history of such recombinations have been studied in [268]. The following exercises outline the most basic result.

   Let S_1 and S_2 be two strings of length n each. An *equal cross-over* of S_1 and S_2 at position i can generate either of two new strings, $S_1[1..i]S_2[i+1..n]$ or $S_2[1..i]S_1[i+1..n]$.

   Give an $O(n)$-time algorithm that takes in three strings, S_1, S_2, and S_3, and determines if there is a position i such that S_3 can be generated from S_1 and S_2 by an equal cross-over.

   Now consider the problem with the modification that S_3 can be smaller than S_1 or S_2.

4. Let S be a set of k equal length strings of total length m. Give an $O(m)$-time algorithm to preprocess set S so that when any three strings from S are specified, one can determine in *constant time* whether the third string can be generated from the first two strings by an equal cross-over at some position.

   Hint: Use a suffix tree.

   Give a simpler algorithm that uses $O(k^2 + m)$ preprocessing time.

5. Now we model the evolution of some population via crossing over events. Initially the population consists of two strings that generate a third string by an equal cross-over. At that point, some pair of those three (maybe the original pair) generate a fourth string by an equal cross-over. At each step, some pair of the existing set generates a new string by an equal cross-over. The initial pair of strings is called the *protopair* for the set of k strings S generated in this way.

   Consider the situation where the ending population S is known, but the history of S is unknown. The computational problem is to reconstruct a plausible history of S. Assume that the $O(m)$ preprocessing from the previous problem has been done. Give an algorithm that can take in a subset S' of S and a string S_i in S and determine in $O(k)$ time if there is a pair in S' that can generate S_i. This algorithm is used in the next task.

Assume again that the $O(m)$ preprocessing has been done. Given a set S of k strings and a designated pair S_1, S_2 in S, give an $O(k^2)$-time algorithm to determine if S_1, S_2 is a protopair for S.

6. With the results from the previous exercises one can test in $O(m + k^4)$ time if the set S contains a protopair. Just do the preprocessing and then test each pair in S to see if it is a protopair. This time can be reduced to $O(m + k^3)$ because for each string $S_i \in S$, there is at most one other string in S that could join S_i to form a protopair for S. Explain why, and explain how to efficiently find the $k/2$ candidate protopairs.

Epilogue – Where Next?

In this book I have tried to present fundamental ideas, algorithms, and techniques that have a wide range of application and that will likely remain important even as the present-day interests change. I have also tried to explain the fundamental reasons why computations on strings and sequences are productive in biology and will remain important even as the specific applications change. But with only 500 pages (a mere 285,639 words formed from 1,784,996 characters), there are certain algorithmic methods and certain present and anticipated applications that I could not cover.

Additional techniques

For additional pure computer science results on exact matching, the reader is referred to *Text Algorithms* by M. Crochemore and W. Rytter [117]. That book goes more deeply into several pure computer science issues, such as periodicities in strings and parallel algorithms. For a survey of many string searching algorithms and inexact matching methods, see *String Searching Algorithms* by G. Stephen [421]. For additional topics in computational molecular biology, particularly probabilistic and statistical questions about strings and sequences, see *An Introduction to Computational Biology* by M. Waterman [461]. For another introduction to combinatorial and string problems in computational molecular biology, see *Introduction to Computational Molecular Biology*, by J. Setubal and J. Meidanis [402]. For topics in computational molecular biology more focused on issues of protein structure, see the chapter *Computational Molecular Biology* by A. Lesk in [297]. For specific open technical problems in sequence-oriented computational biology, see [371] by P. Pevzner and M. Waterman.

Where is computational molecular biology going?

It's going where large-scale molecular biology is going and where the data are found. For a broad view of the future, see [291] by Eric Lander. For an example of the future, see [474].

Where are the good problems in computational molecular biology and where will they be in the future? I don't know for sure, but I distinguish between *technology-driven* problems and *biology-driven* problems.

The fragment assembly problem in shotgun sequencing is a good example of a technology-driven problem. Technology-driven problems tend to be the easiest ones to formalize, and the immediate importance of the formalizations are often easily assessed. Hence technology-driven problems are attractive to computer scientists entering computational biology. However, biotechnology changes quickly and many computational problems are very sensitive to small changes, so good technology-driven problems can appear and disappear quickly. It's easy to imagine the day when alternative large-scale sequencing technology will make the fragment assembly problem irrelevant. (In fact, the end of shotgun sequencing has been predicted, and hoped for, many times in the past).

Technology-driven problems will always be a part of computational biology, but to work on such problems, you must maintain a current understanding of the technology. It is not safe to look for technology-driven problems in old textbooks.

Examples of a biology-driven problems are: find functionally significant motifs in a family of protein sequences; develop techniques to detect alternate genetic codes; develop techniques to identify the extent of horizontal gene and intron transfer; or develop techniques to help understand the role of DNA repeats in genome evolution. These problem statements are vague, and specific formal, technical problems are needed. For example, in the motif problem, what is the definition of a motif?; can a motif be described by a regular expression?; are motifs likely to be highlighted by a multiple alignment?; are significant motifs likely to be more compressible than random substrings?; what is a family?; is it defined by similar function, structure, evolutionary history, general sequence similarity, etc.?

Biology-driven problems are harder to formalize, and specific formalizations (and the abstract models they are built on) may be harder to defend or evaluate, but well-formalized biology-driven problems may have a much longer lifespan than technology-driven problems. This is because the underlying biological issues are more stable than is biotechnology. Moreover, it may be easier to make general predictions about what kinds of biology-driven problems will arise in the future. There are predictable trends that will generate specific biology-driven questions as more sequences and protein structures are learned.

One major trend is that more *correlated* sequence data will be available, driven by the growing emphasis on studying large-scale, complex, genome-level, interactions between genes, between proteins, and between genes and proteins. Correlation takes numerous specific forms: correlation between protein sequences and protein structures (which leads to homology modeling of protein structure); correlation between sequence motifs and features over long stretches of genomic DNA (which addresses questions of long-range interactions); correlation of similarities over many related sequences (for which multiple string comparison and alignment will remain important); correlations between related genes in different organisms (which leads to questions about gene order and genome evolution); correlations between sequences from different organisms (to further exploit knowledge gained from model organisms, or to study horizontal gene transfer); correlation between sequences and functions collected from specific cell lines and tissues over time and under varying conditions (which addresses the question of which genes are expressed where and when and how that expression is correlated with development, disease, or environmental stress); and correlations between polymorphic sequence features (collected over an entire population) and the appearance of diseases or differences in function. Already there are projects to monitor the gene expression in specific cells over time by capturing all the mRNAs produced and counting how often a particular gene is expressed. This kind of study will someday also be performed at the protein level. Studies like those will generate vast amounts of data that will have to be analyzed by computer: "We are now going to have expression maps of 100,000 different genes. Good luck figuring that out!" [349].

Consequently, the biological focus may change from single gene or single protein analysis to whole pathway or network analysis, or to whole genome analysis, or to multicell analysis, or to multiloci linkage or disease analysis, or to whole population analysis, or to multiple organism analysis. But despite the change in focus, and the increase in the complexity of the systems being studied, sequence data and sequence analysis will remain central in computational biology for a long time. The big-picture issue in computational molecular biology, the issue behind most specific, formal, biology-driven string problems, is how to "do" or augment "real biology" by exploiting sequence data. Sequence analysis will be with us long after sequencing becomes routine. The challenge for computational molecular biology is to fully develop and elaborate the power of the sequence-centric approach. The challenge to string algorithmists is to formalize and solve the computational string problems created by that approach.

How should computer scientists proceed in computational biology?

The advice to computer scientists is twofold. First, learn as much real biology as possible – read biology texts and journals (*Science*, *Nature*, and *Trends in Genetics* are particularly good for a broad view), attend biology talks and conferences, talk extensively to biologists, and seriously consider computational problems and abstract models already defined by biologists. But, second, do not be limited by the already formalized computational problems and established models, and do not be discouraged if you cannot immediately incorporate the full biological complexity into your formalized problems.

After immersion into real biology, try to frame and explore your own questions (particularly biology-driven questions) guided by your understanding of the biology, by the goal of ultimately solving the "full problem", and by your understanding of what is computationally feasible and infeasible. Computer science will make the most serious contribution to biology after the emergence of a large community of people who understand both fields and who know the strengths and needs of both. I am not suggesting that you ignore problems posed by biologists, but rather that you augment or modify them with your own questions. To quote from Leroy Hood "As computer scientists come in knowing more and more biology, they will chart their own course. Once they get into it, we don't tell them what to do or where to go. They just take off" [280].

In the long run, a community of biology-educated computer scientists will make the largest impact by proceeding the same way that molecular biology proceeds – by picking problems and research approaches that best suit available and potential techniques; by developing new techniques that one believes will be valuable, even if they do not perfectly fit existing problems; by focusing on model organisms (model computational problems); by framing manageable, somewhat simplified problems where progress can be made; and by working as a community to build on each others' results, incrementally adding in more realistic features of the biology-driven problem. That indirect and incremental approach is well accepted in both computer science and biology, but it will generate many silly, overly idealized computational problems along the way, and these may be disappointing to some people. However, this approach is realistic. It will be more likely to succeed than will premature frontal attacks on the most difficult computational problems for the same reasons that fundamental research on model organisms in biology is often a more productive, but more indirect way to obtain practical insights to problems arising in more complex organisms. It is instructive to realize that less than one percent of all known micro-organisms have been successfully cultured in the lab, creating a huge bias in what laboratory biologists have focussed on. And yet, studies of that one percent have lead to profound discoveries. Looking intensely under a lamppost, while also trying to expand its beam, is often both a necessary and a successful strategy.

I offer one more piece of advice to theoretical computer scientists. Focus more on the *biological quality* of a computation, and not exclusively on speed and space improvements. Because the formalization of a biology-driven problem may be difficult, the biological quality of the computed results must be examined, possibly requiring successive changes in the formal problem. The skills of computer scientists can be very important in iterating this process (finding practical solutions to each of the succeeding formalized problems). Making each step practical may be more important than optimally speeding up any given step.

In summary, learn real biology, talk extensively to biologists, and work on problems of known importance to biology. But, in addition, try to become your own biological consultant. Guided by real biology, frame your own manageable, technical questions. Be willing to take the criticism that many of these questions will be too idealized to be of immediate practical use, knowing that in the long run, this incremental approach (with its many failures) is how a research community best addresses difficult problems. Most of all, be curious, have fun, and appreciate the wonderful opportunities you have to work in such an exciting and important field.

Bibliography

[1] *Trends in Genetics*, vol. 11, no. 3, 1995. See the entire issue.

[2] Detecting dinosaur DNA. Four technical comments by various authors in *Science*, May 26, 1995, pages 1191–93.

[3] K. Abrahamson. Generalized string matching. *SIAM J. Comput.*, 16:1039–51, 1987.

[4] L. Adleman. Molecular computation of solutions to combinatorial problems. *Science*, 266:1021–24, 1994.

[5] P. Agarwal and D. States. The repeat pattern toolkit (RPT): Analyzing the structure and evolution of the C. elegans genome. *Proceedings of the Second International Conference on Intelligent Systems for Molecular Biology*, pages 1–9, 1994.

[6] R. Agarwala, V. Bafna, M. Farach, B. Narangyan, M. Paterson, and M. Thorup. On the approximability of numerical taxonomy: fitting distances with trees. *Proc. 7th ACM-SIAM Symp. on Discrete Algs.*, pages 365–72, 1996.

[7] R. Agarwala and D. Fernandez-Baca. A polynomial-time algorithm for the perfect phylogeny problem when the number of character states is fixed. *SIAM J. Comput.*, 23:1216–24, 1994.

[8] A. Aho. Algorithms for finding patterns in strings. In J. van Leeuwen, editor, *Handbook of Theoretical Computer Science* vol. A, pages 257–300, MIT Press/Elsevier, 1990.

[9] A. Aho and M. Corasick. Efficient string matching: an aid to bibliographic search. *Comm. ACM*, 18:333–40, 1975.

[10] A. Aho, J. Hopcroft, and J. Ullman. *Design and Analysis of Computer Algorithms*. Addison-Wesley, Reading, MA, 1974.

[11] B. Alberts, D. Bray, J. Lewis, M. Raff, K. Roberts, and J. Watson. *Molecular Biology of the Cell*, 3rd ed. Garland Press, New York, 1994.

[12] F. Alizadeh, R. Karp, L. Newberg, and D. Weisser. Physical mapping of chromosomes: a combinatorial problem in molecular biology. *Algorithmica*, 13:52–76, 1995.

[13] F. Alizadeh, R. Karp, D. Weisser, and G. Zweig. Physical mapping of chromosomes using unique probes. *J. Comp. Biol.*, 2:159–84, 1995.

[14] L. Allison and C. N. Yee. Minimum message length encoding and the comparison of macro-molecules. *Bull. Math. Biology*, 52:431–53, 1990.

[15] J. Alper. Putting genes on chips. *Science*, 264:1400, 1994.

[16] S. Altschul. Amino acid substitution matrices from an information theoretic perspective. *J. Mol. Biol.*, 219:555–65, 1991.

[17] S. Altschul, M. Boguski, W. Gish, and J. Wooton. Issues in searching molecular sequence databases. *Nature Genetics*, 6:119–29, 1994.

[18] S. Altschul and B. W. Erickson. Optimal sequence alignment using affine gap costs. *Bull. Math. Biology*, 48:603–16, 1986.

[19] S. Altschul, W. Gish, W. Miller, E. W. Myers, and D. Lipman. A basic local alignment search tool. *J. Mol. Biol.*, 215:403–10, 1990.

[20] S. Altschul and D. Lipman. Trees, stars, and multiple sequence alignment. *SIAM J. Appl. Math*, 49:197–209, 1989.

[21] S. F. Altschul and W. Gish. Local alignment statistics. In R. F. Doolittle, editor, *Methods in Enzymology Vol. 266. Computer Methods for Macromolecular Sequence Analysis*, pages 460–80. Academic Press, 1996.

[22] A. Amir, G. Benson, and M. Farach.

Alphabet-independent two-dimensional matching. *SIAM J. Comput.*, 23:313–23, 1994.

[23] A. Anderson and S. Nilsson. Efficient implementation of suffix trees. *Software Practice and Experience*, 25:129–41, 1995.

[24] A. Apostolico. The myriad virtues of subword trees. In A. Apostolico and Z. Galil, editors, *Combinatorics on Words*, pages 85–96. Springer-Verlag, Nato ASI series vol. 112, 1985.

[25] A. Apostolico. Improving the worst-case performance of the Hunt-Szymanski strategy for the longest common subsequence of two strings. *Information Processing Lett.*, 23:63–69, 1986.

[26] A. Apostolico and R. Giancarlo. The Boyer-Moore-Galil string searching strategies revisited. *SIAM J. Comput.*, 15:98–105, 1986.

[27] A. Apostolico and C. Guerra. The longest common subsequence problem revisited. *Algorithmica*, 2:315–36, 1987.

[28] V. L. Arlazarov, E. A. Dinic, M. A. Kronrod, and I. A. Faradzev. On economic construction of the transitive closure of a directed graph. *Dokl. Acad. Nauk SSSR*, 194:487–88, 1970.

[29] C. Armen and C. Stein. A 2 2/3-approximation algorithm for the shortest superstring problem. *Proc. 7th Symp. on Combinatorial Pattern Matching. Springer LNCS 1075*, pages 87–101, 1996.

[30] C. Armen and C. Stein. A 2 3/4-approximation algorithm for the shortest superstring problem, technical report from Computer Science Dept., Dartmonth College, 1994.

[31] F. Ayala. The myth of Eve: molecular biology and human origins. *Science*, 270:1930–36, 1995.

[32] S. Baase. *Computer Algorithms,* 2nd ed. Addison-Wesley, Reading, MA, 1988.

[33] D. Bacon and W. Anderson. Multiple sequence alignment. *J. Mol. Biol.*, 191:153–61, 1986.

[34] R. Baeza-Yates and G. Gonnet. All-against-all sequence matching. Unpublised note.

[35] R. Baeza-Yates and G. Gonnet. A new approach to text searching. *Comm. ACM*, 35:74–82, 1992.

[36] R. A. Baeza-Yates and C. Perleberg. Fast and practical approximate string matching. *Proc. 3rd Symp. on Combinatorial Pattern Matching. Springer LNCS 644*, pages 185–92, 1992.

[37] V. Bafna, E. Lawler, and P. Pevzner. Approximation algorithms for multiple sequence alignment. *Proc. 5th Symp. on Combinatorial Pattern Matching. Springer LNCS 807*, pages 43–53, 1994.

[38] V. Bafna and P. Pevzner. Sorting by reversals: genome rearrangements in plant organelles and evolutionary history of X chromosome, report CSE-94-032. Technical report, Penn. State, 1994.

[39] V. Bafna and P. Pevzner. Sorting by transpositions. *Proc. 6th ACM-SIAM Symp. on Discrete Algs.*, pages 614–21, 1995.

[40] V. Bafna and P. Pevzner. Genome rearrangements and sorting by reversals. *SIAM J. Comput.*, 25:272–89, 1996.

[41] A. Bairoch. PROSITE: a dictionary of sites and patterns in proteins. *Nucl. Acids Res.*, 20:2013–18, 1992.

[42] A. Bairoch and P. Bucher. PROSITE: recent developments. *Nucl. Acids Res.*, 22:3583–89, 1992.

[43] B. Baker. A theory of parameterized pattern matching: algorithms and applications. *Proc. of the 25th ACM Symp. on the Theory of Computing*, pages 71–80, 1993.

[44] T. Baker. A technique for extending rapid exact-match string matching to arrays of more than one dimension. *SIAM J. Comput.*, 7:533–41, 1978.

[45] S. Baldauf and J. Palmer. Animals and fungi are each other's closest relatives: congruent evidence from multiple proteins. *Proc. of the Natl. Acad. Science*, 90:11, 558–62, 1993.

[46] H. J. Bandelt. Recognition of tree metrics. *SIAM J. Discrete Math*, 3:3–6, 1990.

[47] M. Barinaga. Missing Alzheimer's gene found. *Research News in Science*, 269:917–18, August 1995.

[48] D. Barlow. Gametic imprinting in mammals. *Science*, 270:1610–13, 1995.

[49] J. P. Barthelemy and A. Guenoche. *Trees and Proximity Representations*. Wiley, New York, 1991.

[50] G. J. Barton. Computer speed and sequence comparison (letter to the editor). *Science*, 257:1609, 1992.

[51] G. J. Barton and M. J. Sternberg. Evalu-

ation and improvements in the automatic alignment of protein sequences. *Protein Engineering*, 1:89–94, 1987.

[52] G. J. Barton and M. J. Sternberg. A strategy for rapid multiple alignment of protein sequences: confidence levels from tertiary structure comparisons. *J. Mol. Biol.*, 198:327–37, 1987.

[53] E. Baum, D. Boneh, P. Kaplan, R. Lipton, J. Reif, and N. Seeman (eds). Second annual meeting on DNA based computers. Dimacs Workshop at Princeton University June 1996.

[54] T. Beardsley. Smart genes. *Scientific American*, pages 86–95, August 1991.

[55] S. Begley and A. Rogers. It's all in the genes: in computational biology, scientists track elusive DNA stands through databases. *Newsweek*, September 5, 1994, p. 64. Quote attributed to Greg Lennon.

[56] S. A. Benner, M. A. Cohen, and G. H. Gonnet. Response to Barton's letter: Computer speed and sequence comparison. *Science*, 257:1609–10, 1992.

[57] S. A. Benner, M. A. Cohen, and G. H. Gonnet. Empirical and structural models for insertions and deletions in the divergent evolution of proteins. *J. Mol. Biol.*, 229:1065–82, 1993.

[58] D. Benson. Digital signal processing methods for biosequence comparison. *Nucl. Acids Res.*, 18:3001–6, 1990.

[59] D. Benson. Fourier method for biosequence analysis. *Nucl. Acids Res.*, 18:6305–10, 1991.

[60] G. Benson. A space efficient algorithm for finding the best non-overlapping alignment score. *Proc. 5th Symp. on Combinatorial Pattern Matching. Springer LNCS 807*, pages 1–14, 1994.

[61] P. Berman and S. Hannenhali. Fast sorting by reversal. *Proc. 7th Symp. on Combinatorial Pattern Matching. Springer LNCS 1075*, pages 168–85, 1996.

[62] P. Berman and V. Ramaiyer. Improved approximations for the Steiner tree problem. *J. Algorithms*, 17:381–408, 1994.

[63] P. Bieganski. *Genetic sequence data retrieval and manipulation based on generalized suffix trees*. PhD thesis, Univ. Minnesota, Dept. Computer Science, 1995.

[64] P. Bieganski, J Riedl, J. V. Carlis, and E. R. Retzel. Generalized suffix trees for biological sequence data: applications and implementation. In *Proc. of the 27th Hawaii Int. Conf. on Systems Sci.*, pages 35–44. IEEE Computer Society Press, 1994.

[65] P. Bieganski, J Riedl, J. V. Carlis, and E. R. Retzel. Motif explorer – a tool for interactive exploration of amino acid sequence motifs. In *Proc. of the First Pacific Symp. on Biocomputing*, pages 705–6. IEEE Computer Society Press, 1996.

[66] R. Bird. Two dimensional pattern matching. *Information Processing Lett.*, 6:168–70, 1977.

[67] R. Birge. Protein-based optical computers and memories. *IEEE Computer*, 25, November 1992.

[68] R. Birge. Protein-based computing. *Scientific American*, 272:90–95, 1995.

[69] A. Blum, T. Jiang, M. Li, J. Tromp, and M. Yannakakis. Linear approximation of shortest superstrings. *J. ACM*, 41:634–47, 1994.

[70] A. Blumer, J. Blumer, D. Haussler, D. Ehrenfeucht, A. Chen, and M. T. Seiferas. The smallest automaton recognizing the subwords of a text. *Theor. Comp. Sci.*, 40:31–55, 1985.

[71] A. Blumer, J. Blumer, D. Haussler, R. McConnell, and D. Ehrenfeucht. Complete inverted files for efficient text retrieval and analysis. *J. ACM*, pages 578–95, 1987.

[72] H. Bodlaender, M. Fellows, and T. Warnow. Two strikes against perfect phylogeny. *Proc. of the 19th Int. Colloq. on Automata, Languages and Programming*, pages 273–83, 1992.

[73] M. S. Boguski and G. D. Schuler. Establishing a human transcript map. *Nature Genetics*, 10:369–71, 1995.

[74] K. Booth and G. Lueker. Testing for the consecutive ones property, interval graphs and graph planarity testing using pq-tree algorithms. *J. Comp. Sys. Sci.*, 13:333–79, 1976.

[75] R. S. Boyer and J. S. Moore. A fast string searching algorithm. *Comm. ACM*, 20:762–72, 1977.

[76] S. E. Brenner. Network sequence retrieval. *Trends in Genetics*, 11, 1995, pages 247–48.

[77] D. Breslauer, T. Jiang, and Z. Jiang. Rota-

tions of periodic strings and short super-strings. Preprint, May 30, 1996.

[78] W. J. Broad. Clues to fiery origins of life sought in hothouse microbes. *New York Times*, May 9, 1995.

[79] J. Brody. Multiple sclerosis. Will anyone recovered from it please communicate with patient. *New York Times*, May 3, 1995.

[80] M. Browne. Critics see humbler origin in "dinosaur" DNA. *New York Times*, June 20, 1995.

[81] D. L. Brutlag, J. P. Dautricourt, S. Maulik, and J. Relph. Improved sensitivity of biological sequence database searches. *Comp. Appl. Biosciences*, 6:237–45, 1990.

[82] C. Bult, C. R. Woese, and J. C. Venter et al. Complete genome sequence of the methanogenic archaeon, Methanococcus jannaschii. *Science*, 273:1058–72, 1996.

[83] P. Buneman. The recovery of trees from measures of dissimilarity. In D. G. Kendall and P. Tautu, editors, *Mathematics in the Archaeological and Historical Sciences*, pages 387–85. Edinburgh Univ. Press, 1971.

[84] P. Buneman. A characterization of rigid circuit graphs. *Discrete Math*, 9:205–12, 1974.

[85] P. Buneman. A note on metric properties of trees. *J. Combinatorial Theory (B)*, 17:48–50, 1974.

[86] R. Cann, M. Stoneking, and A. Wilson. Mitochondrial DNA and human evolution. *Nature*, 325:31–36, 1987.

[87] A. Caprara. Sorting by reversals is difficult. In *Proc. of RECOMB 97: The first international conference on computational molecular biology*, pages 75–83. ACM Press, 1997.

[88] H. Carrillo and D. Lipman. The multiple sequence alignment problem in biology. *SIAM J. Appl. Math*, 48:1073–82, 1988.

[89] R. Cary and G. Stormo. Graph-theoretic approach to RNA modeling using comparitive data. In *Proc. of Intelligent Systems in Mol. Biol.*, pages 75–80, 1995.

[90] D. Casey, C. Cantor, and S. Spengler. *Primer on Molecular Genetics*. US Dept. Energy, Human Genome Program, Washington, DC, 1992.

[91] C. Caskey, R. Eisenberg, E. Lander, and J. Straus. Hugo statement on patenting of DNA. *Genome Digest*, 2:6–9, 1995.

[92] S. Chan, A. Wong, and D. Chiu. A survey of multiple sequence comparison methods. *Bull. Math. Biology*, 54:563–98, 1992.

[93] W. I. Chang and J. Lampe. Theoretical and empirical comparions of approximate string matching algorithms. *Proc. 3rd Symp. on Combinatorial Pattern Matching. Springer LNCS 644*, pages 175–84, 1992.

[94] W. I. Chang and E. L. Lawler. Sublinear expected time approximate string matching and biological applications. *Algorithmica*, 12:327–44, 1994.

[95] K. Chao, W. Pearson, and W. Miller. Aligning two sequences within a specified diagonal band. *Comp. Appl. Biosciences*, 2:6–9, 1995.

[96] K. M. Chao. Computing all suboptimal alignments in linear space. *Proc. 5th Symp. on Combinatorial Pattern Matching. Springer LNCS 807*, pages 1–14, 1994.

[97] K. M. Chao, R. Hardison, and W. Miller. Recent developments in linear-space alignment methods: a mini survey. *J. Comp. Biol.*, 1:271–91, 1994.

[98] M. Chee and P. A. Fodor et al. Accessing genetic information with high-density DNA arrays. *Science*, 274:610–614, 1996.

[99] E. Cheever, G. Christian Overton, and D. Searls. Fast Fourier Transform–based correlation of DNA sequences using complex plane encoding. *Comp. Appl. Biosciences*, 7:143–54, 1991.

[100] T. Chen and S. Skiena. Trie-based data structures for sequence assembly. Department of Computer Science, Stony Brook N.Y., Preprint, July 11, 1996.

[101] C. Chothia. One thousand families for the molecular biologist. *Nature*, 357:543–44, 1992.

[102] N. Christofides. Worst-case analysis of a new heuristic for the travelling salesman problem. Report 338, Grad. School Industrial Administration, Carnegie-Mellon, 1976.

[103] J. M. Claverie. Detecting frame shifts by amino acid sequence comparsion. *J. Mol. Biol.*, 234:1140–57, 1993.

[104] A. Clift, D. Haussler, R. McConnell, T. D. Schneider, and G. Stormo. Sequence landscapes. *Nucl. Acids Res.*, 14:141–58, 1986.

[105] A. Cobbs. Fast approximate matching using suffix trees. *Proc. 6th Symp. on Combinatorial Pattern Matching. Springer LNCS 937*, pages 41–54, 1995.

[106] F. E. Cohen. Folding the sheets: using computational methods to predict the structure of proteins. In E. Lander and M. S. Waterman, editors, *Calculating the Secrets of Life*, pages 236–71. National Academy Press, 1995.

[107] R. Cole. Unpublished manuscript.

[108] R. Cole. Tight bounds on the complexity of the Boyer-Moore pattern matching algorithm. *SIAM J. Comput.*, 23:1075–91, 1994.

[109] B. Commentz-Walter. A string matching algorithm fast on average. *Proc. of the 6th Int. Colloq. on Automata, Languages and Programming*, pages 118–32, 1979.

[110] G. Cooper. *Oncogenes*. Jones and Bartlett, Boston, MA, 1990.

[111] N. Cooper. *The Human Genome Project*. Univ. Science Books, Mill Valley, CA, 1994.

[112] T. Cormen, C. Leiserson, and R. Rivest. *Introduction to Algorithms*. MIT Press and McGraw Hill, Cambridge, MA, 1992.

[113] D. Cox, M. Burmeister, E. R. Price, S. Kim, and R. Myers. Radiation hybrid mapping: a somatic cell genetic method for constructing high-resolution maps of mammalian chromosomes. *Science*, 250:245–50, 1990.

[114] M. Crochemore. An optimal algorithm for computing repetitions in a word. *Information Processing Lett.*, 12:244–50, 1981.

[115] M. Crochemore. Transducers and repetitions. *Theor. Comp. Sci.*, 45:63–86, 1985.

[116] M. Crochemore and W. Rytter. Usefulness of the Karp-Miller-Rosenberg algorithm in parallel computations on strings and arrays. *Theor. Comp. Sci.*, 88:59–82, 1991.

[117] M. Crochemore and W. Rytter. *Text Algorithms*. Oxford Univ. Press, New York, 1994.

[118] J. Culberson and P. Rudnicki. A fast algorithm for constructing trees from distance matrices. *Information Processing Lett.*, 30:215–20, 1989.

[119] C. R. Darwin. *The Origin of Species*. John Murray, London, 1859.

[120] W. H. Day. Computational complexity of inferring phylogenies from dissimilarity matrices. *Bull. Math. Biol.*, 49:461–67, 1987.

[121] M. O. Dayhoff. Computer analysis of protein evolution. *Scientific American*, pages 87–94, July 1969.

[122] M. O. Dayhoff, R. M. Schwartz, and B. C. Orcutt. A model of evolutionary change in proteins. *Atlas of Protein Sequence and Structure*, 5:345–52, 1978.

[123] N. J. Deacon, J. Mills, et al. Genomic structure of an attenuated quasi species of HIV-1 from a blood transfusion donor and recipients. *Science*, 270:988–91, 1995.

[124] A. Dembo and S. Karlin. Strong limit theorems of empirical functions for large exceedances of partial sums of i.i.d. variables. *Ann. Prob.*, 19:1737–55, 1991.

[125] N. A. Doggett, R.K. Moyzis, et al. An integrated physical map of human chromosome 16. *Nature*, 377:335–65, 1995.

[126] R. F. Doolittle. Similar amino acid sequences: Chance or common ancestry? *Science*, 214:149–59, 1981.

[127] R. F. Doolittle. *Of Urfs and Orfs: A primer on How to Analyze Derived Amino Acid Sequences*. University Science Books, Mill Valley, CA, 1986.

[128] R. F. Doolittle. Redundancies in protein sequences. In G. Fasman, editor, *Prediction of Protein Structure and the Principles of Protein Conformation*, pages 599–624. Plenum, New York, 1989.

[129] R. F. Doolittle. Searching through sequence databases. In R. F. Doolittle, editor, *Methods in Enzymology Vol. 183. Molecular Evolution: Computer Analysis of Protein and Nucleic Acid Sequences*, pages 99–110. Academic Press, New York, 1990.

[130] R. F. Doolittle. What we have learned and will learn from sequence databases. In G. Bell and T. Marr, editors, *Computers and DNA*, pages 21–31. Addison-Wesley, Reading, MA, 1990.

[131] R. F. Doolittle, R. F. Feng, S. Tsang, G. Cho, and E. Little. Determining divergence times of the major kingdoms of

living organisms with a protein clock. *Science*, 271:470–77, 1996.

[132] R. F. Doolittle, M. Hunkapiller, L. E. Hood, S. Devare, K. Robbins, S. Aaronson, and H. Antoniades. Simian sarcoma virus onc gene v-sis, is derived from the gene (or genes) encoding a platelet-derived growth factor. *Science*, 221:275–77, 1983.

[133] A. Dress and M. Steel. Convex tree realizations of partitions. *Appl. Math. Lett.*, 5:3–6, 1993.

[134] R. Drmanac, L. Hood, R. Crkvenjakov, et al. DNA sequence determination by hybridization: a strategy for efficient large-scale sequencing. *Science*, 260:1649–52, 1993.

[135] R. Drmanac, I. Labat, I. Brukner, and R. Crkvenjakov. Sequencing of megabase plus DNA by hybridization: theory of the method. *Genomics*, 4:114–28, 1989.

[136] D. Eppstein. Sequence comparison with mixed convex and concave costs. *J. Algorithms*, 11:85–101, 1990.

[137] D. Eppstein, Z. Galil, R. Giancarlo, and G. F. Italiano. Sparse dynamic programming I: linear cost functions. *J. ACM*, 39:519–45, 1992.

[138] D. Eppstein, Z. Galil, R. Giancarlo, and G. F. Italiano. Sparse dynamic programming II: convex and concave cost functions. *J. ACM*, 39:516–67, 1992.

[139] G. Estabrook, C. Johnson, and F. McMorris. An idealized concept of the true cladistic character. *Math. Bioscience*, 23:263–72, 1975.

[140] G. Estabrook, C. Johnson, and F. McMorris. An algebraic analysis of cladistic characters. *Discrete Math.*, 16:141–47, 1976.

[141] G. Estabrook, C. Johnson, and F. McMorris. A mathematical foundation for the analysis of cladistic character compatibility. *Math. Bioscience*, 29:181–87, 1976.

[142] G. Estabrook and F. McMorris. When are two qualitative taxonomic characters compatible? *J. Math. Bioscience.* 4:195–200, 1977.

[143] G. Estabrook and F. McMorris. When is one estimate of evolutionary relationships a refinement of another? *J. Math. Bioscience*, 10:367–73, 1980.

[144] S. Even. *Graph Algorithms*. Computer Science Press, Mill Valley, CA, 1979.

[145] M. Farach, S. Kannan, and T. Warnow. A robust model for finding evolutionary trees. *Algorithmica*, 13:511–20, 1995.

[146] M. Farach, M. Noordewier, S. Savari, L. Shepp, A. Wyner, and J. Ziv. On the entropy of DNA: algorithms and measurements based on memory and rapid convergence. *Proc. 6th ACM-SIAM Symp. on Discrete Algs.*, pages 48–57, 1995.

[147] A. Feducia. Explosive evolution in tertiary birds and mammals. *Science*, 267:637–38, 1995.

[148] J. Felsenstein. Numerical methods for inferring evolutionary trees. *Quart. Rev. Biol.*, pages 379–404, 1982.

[149] J. Felsenstein. Parsimony in systematics: biological and statistical issues. *Annu. Rev. Ecol. Syst.*, 14:313–33, 1983.

[150] J. Felsenstein. Perils of molecular introspection. *Nature*, 335:118, 1988.

[151] J. Felsenstein. Phylogenies from molecular sequences: inference and reliability. *Annu. Rev. Genetics*, 22:521–65, 1988.

[152] J. Felsenstein, S. Sawyer, and R. Kochin. An efficient method for matching nucleic acid sequences. *Nucl. Acids Res.*, 10:133–39, 1982.

[153] D. Feng and R. F. Doolittle. Progressive sequence alignment as a prerequisite to correct phylogenetic trees. *J. Mol. Evol.*, 25:351–60, 1987.

[154] V. Ferretti, J. Nadeau, and D. Sankoff. Original synteny. *Proc. 7th Symp. on Combinatorial Pattern Matching. Springer LNCS 1075,* pages 149–67, 1996.

[155] J. W. Fickett. Fast optimal alignment. *Nucl. Acids Res.*, 12:175–80, 1984.

[156] J. W. Fickett. Orfs and genes: How strong a connection? *J. Comp. Biol.*, 2:117–23, 1995.

[157] M. Fischer and M. Paterson. String-matching and other products. In R. M. Karp, editor, *Complexity of Computation*, pages 113–25. SIAM-AMS Proc., 1974.

[158] V. Fischetti, G. Landau, J. Schmidt, and P. Sellers. Identifying periodic occurrences of a template with applications to protein structure. *Information Processing Lett.*, 45:11–18, 1993.

[159] W. M. Fitch. Toward defining the course of evolution: minimum changes for a specific tree typology. *Syst. Zoology*, 20:406–16, 1971.

[160] W. M. Fitch and E. Margoliash. The construction of phylogenetic trees. *Science*, 155:279–84, 1967.

[161] W. M. Fitch and T. F. Smith. Optimal sequence alignments. *Proc. Natl. Academy Science*, 80:1382–86, 1983.

[162] R. Fleischmann, H. Smith, J. C. Venter, et al. Whole-genome random sequencing and assembly of *Haemophilus influenzae rd*. *Science*, 269:496–512, 1995.

[163] S. Fodor, J. Read, M. Pirrung, L. Stryer, A. Lu, and D. Solas. Light-directed spatially addressable parallel chemical synthesis. *Science*, 251:767–73, 1991.

[164] P. L. Foley, C. J. Humphries, I. L. Kitching, R. W. Scotland, D. J. Siebert, and D. M. Williams. *Cladistics: A Practical Course in Systematics*. Oxford Univ. Press, Oxford, 1992.

[165] L. R. Foulds and R. L. Graham. The Steiner problem in phylogeny is NP-complete. *Advances Appl. Math.*, 3:43–49, 1982.

[166] M. L. Fredman. Algorithms for computing evolutionary similarity measures with length independent gap penalties. *Bull. Math. Biol.*, 46:553–66, 1984.

[167] M. L. Fredman, J. Komlos, and E. Szemeredi. Storing a sparse table with O(1) worst case access time. *J. ACM*, 31:61–68, 1984.

[168] Z. Galil. On improving the worst case running time of the Boyer-Moore string searching algorithm. *Comm. ACM*, 22:505–8, 1979.

[169] Z. Galil and K. Park. Alphabet-independent two-dimensional witness computations. *SIAM J. Comput.*, 25:907–35, 1996.

[170] Z. Galil and R. Giancarlo. Speeding up dynamic programming with applications to molecular biology. *Theor. Comp. Sci.*, 64:107–18, 1989.

[171] M. Garey and D. Johnson. *Computers and Intractability*. Freeman, San Francisco, 1979.

[172] M. S. Gelfand. Prediction of function in DNA sequence analysis. *J. Comp. Biol.*, 2:87–115, 1995.

[173] M. S. Gelfand, A. A. Mironov, and P. A. Pevzner. Gene recognition via spliced alignment. *Proc. of the Nat. Academy of Science*, pages 9061–66, 1996.

[174] M. S. Gelfand, A. A. Mironov, and P. A. Pevzner. Spliced alignment: a new approach to gene recognition. *Proc. 7th Symp. on Combinatorial Pattern Matching. Springer LNCS 1075*, pages 141–58, 1996.

[175] A. Gibbons. Mitochondrial Eve: wounded, but not dead yet. *Science*, 257:873–75, 1992.

[176] R. A. Gibbs. Pressing ahead with human genome sequencing. *Nature Genetics*, 11:121–25, 1995.

[177] R. A. Gibbs, P. N. Nguyen, and C. T. Caskey. Detection of single DNA base differences by competitive oligonucliotide priming. *Nucl. Acids Res.*, 17:2437–48, 1989.

[178] R. Giegerich and S. Kurtz. From Ukkonen to McCreight and Weiner: A unifying view of linear-time suffix tree construction. *Algorithmica*, to appear.

[179] W. Gilbert. Towards a paradigm shift in biology. *Nature*, 349:99, 1991.

[180] J. Glanz. Computer scientists rethink their discipline's foundations. *Science*, 269:1363–64, 1995.

[181] M. C. Golumbic, H. Kaplan, and R. Shamir. On the complexity of DNA physical mapping. *Advances Appl. Math.*, 15:251–61, 1994.

[182] G. Gonnet and R. Baeza-Yates. An analysis of the Karp-Rabin string matching algorithm. *Information Processing Lett.*, 34:271–74, 1990.

[183] G. H. Gonnet, M. A. Cohen, and S. A. Benner. Exhaustive matching of the entire protein sequence database. *Science*, 256:1443–45, 1992.

[184] G. H. Gonnet and R. Baeza-Yates. *Handbook of Algorithms and Data Structures*, 2nd ed. Addison-Wesley, Reading, MA, 1991.

[185] A. Gorbalenya and E. Koonin. Helicases: amino acid sequence comparisons and structure-function relationships. *Current Opinion Structural Biol.*, 3:419–29, 1993.

[186] O. Gotoh. An improved algorithm for matching biological sequences. *J. Mol. Biol.*, 162:705–8, 1982.

[187] O. Gotoh. Alignment of three biological sequences with an efficient traceback. *J. Theor. Biol.*, 121:327–37, 1986.

[188] S. J. Gould. Through a lens, darkly. *Nat-*

ural History, pages 16–24, September 1989.

[189] E. Green and P. Green. Sequence-tagged site (STS) content mapping of human chromosomes: theoretical considerations and early experiences. In *PCR Methods and Applications*, pages 77–90. Cold Spring Harbor Lab. Press, 1991.

[190] P. Green, D. Lipman, D. Hillier, R. Waterston, D. States, and J. M. Claverie. Ancient conserved regions in new gene sequences and the protein databases. *Science*, 259:1711–16, 1993.

[191] D. S. Greenberg and S. Istrail. Physical mapping by STS hybridization: algorithmic strategy and the challenge of software evaluation. *J. Comp. Biol.*, 2:219–73, 1995.

[192] M. Gribskov and J. Devereux. *Sequence Analysis Primer*. Stockton Press, New York, 1991.

[193] M. Gribskov, R. Luthy, and D. Eisenberg. Profile analysis. In R. F. Doolittle, editor, *Methods in Enzymology Vol. 183. Molecular Evolution: Computer Analysis of Protein and Nucleic Acid Sequences*, pages 146–59. Academic Press, New York, 1990.

[194] M. Gribskov, A. McLachlan, and D. Eisenberg. Profile analysis detection of distantly related proteins. *Proc. Natl. Academy Science*, 88:4355–58, 1987.

[195] X. Guan and E. C. Uberbacher. Alignment of DNA and protein sequences containing frameshift errors. *Comp. Appl. Biosciences*, 12:31–40, 1996.

[196] L. J. Guibas and A. M. Odlyzko. A new proof of the linearity of the Boyer-Moore string searching algorithm. *SIAM J. Comput.*, 9:672–82, 1980.

[197] S. Gupta, J. Kececioglu, and A. Schaffer. Making the shortest-paths approach to sum-of-pairs multiple sequence alignment more space efficient in practice. *Proc. 6th Symp. on Combinatorial Pattern Matching. Springer LNCS 937*, pages 128–43, 1995.

[198] D. Gusfield. The Steiner tree problem in phylogeny. Technical Report No. 334, Yale Univ. Computer Science Dept., 1984.

[199] D. Gusfield. An increment-by-one approach to suffix arrays and trees. Techni-

cal Report CSE-90-39, UC Davis, Dept. Computer Science, 1990.

[200] D. Gusfield. Efficient algorithms for inferring evolutionary history. *Networks*, 21:19–28, 1991.

[201] D. Gusfield. Efficient methods for multiple sequence alignment with guaranteed error bounds. *Bull. Math. Biol.*, 55:141–54, 1993.

[202] D. Gusfield. Simple uniform preprocessing for linear-time string matching. Technical Report CSE-96-5, UC Davis, Dept. Computer Science, 1996.

[203] D. Gusfield, K. Balasubramanian, and D. Naor. Parametric optimization of sequence alignment. *Algorithmica*, 12:312–26, 1994.

[204] D. Gusfield, R. Karp, L. Wang, and P. Stelling. Graph traversals, genes and matroids: an efficient special case of the travelling salesman problem. *Proc. 7th Symp. on Combinatorial Pattern Matching. Springer LNCS 1075,* pages 304–19, 1996.

[205] D. Gusfield and P. Stelling. Parametric and inverse-parametric sequence alignment with XPARAL. In R. F. Doolittle, editor, *Methods in Enzymology Vol. 266. Computer Methods for Macromolecular Sequence Analysis*, pages 481–94. Academic Press, New York, 1996.

[206] D. Gusfield and L. Wang. New uses for uniform lifted alignments. Technical Report, CSE-96-4, UC Davis, Dept. Computer Science, 1996.

[207] K. Halanych, J. D. Bacheller, A. M. Aguinaldo, S. M. Liva, D. M. Hillis, and J. A. Lake. Evidence from 18s ribosomal DNA that lophophorates are protostome animals. *Science*, 267:1641–43, 1995.

[208] M. Hanks, W. Wurst, L. Anson-Cartwright, A. Auerback, and A. Joyner. Rescue of the en-1 mutant phenotype by replacement of en-1 with en-2. *Science*, 269:679–82, 1995.

[209] S. Hannenhalli. Polynomial-time algorithm for computing translocation distance between genomes. *Proc. 6th Symp. on Combinatorial Pattern Matching. Springer LNCS 937*, pages 162–76, 1995.

[210] S. Hannenhalli, W. Fellows, H. Lewis, S. Skiena, and P. Pevzner. Positional sequencing by hybridization. *Comp. Appl. Biosciences*, 12:19–24, 1996.

[211] S. Hannenhalli and P. Pevzner. Transforming cabbage into turnip: polynomial algorithm for sorting signed permutations by reversals. *Proc. of the 27th ACM Symp. on the Theory of Computing*, pages 178–89, 1995.

[212] S. Hannenhalli and P. Pevzner. Transforming mice into men: polynomial algorithm for genomic distance problem. *Proc. of the 36th IEEE Symp. on Foundations of Comp. Sci.*, pages 581–92, 1995.

[213] S. Hannenhalli and P. Pevzner. To cut or not to cut: applications of comparitive physical maps in molecular evolution. *Proc. 7th ACM-SIAM Symp. on Discrete Algs.*, pages 304–13, 1996.

[214] D. Harel and R. E. Tarjan. Fast algorithms for finding nearest common ancestors. *SIAM J. Comput.*, 13:338–55, 1984.

[215] R. Harper. World Wide Web resources for the biologist. *Trends in Genetics*, 11, 1995.

[216] J. Hartigan. Minimum mutation fits to a give tree. *Biometrics*, 29:53–65, 1972.

[217] J. Hartigan. *Clustering Algorithms*. Wiley, New York, 1975.

[218] J. Hein. An optimal algorithm to reconstruct trees from additive distance data. *Bull. Math. Biol.*, 51:597–603, 1989.

[219] J. Hein and J. Stovlbaek. Combined DNA and protein alignment. In R. F. Doolittle, editor, *Methods in Enzymology Vol. 266. Computer Methods for Macromolecular Sequence Analysis*, pages 402–27. Academic Press, New York, 1996.

[220] J. G. Henikoff and S. Henikoff. BLOCKS database and its applications. In R. F. Doolittle, editor, *Methods in Enzymology Vol. 266. Computer Methods for Macromolecular Sequence Analysis*, pages 88–105. Academic Press, New York, 1996.

[221] S. Henikoff and J. G. Henikoff. Automated assembly of protein blocks for database searching. *Nucl. Acids Res.*, 19:6565–72, 1991.

[222] S. Henikoff and J. G. Henikoff. Amino acid substitution matrices from protein blocks. *Proc. Natl. Academy Science*, 89:10,915–19, 1992.

[223] D. Hillis and C. Moritz (eds.). *Molecular Systematics*. Sinauer Associates, Sunderland, MA, 1990.

[224] D. S. Hirschberg. Algorithms for the longest common subsequence problem. *J. ACM*, 24:664–75, 1977.

[225] J. Hodgkin, R. H. Plasterk, and R. H. Waterston. The nemotode *Caenorhabditis elegans* and its genome. *Science*, 270:410–14, 1995.

[226] F. Hodson, O. Kendall, and P. Tautu (eds.). *Mathematics in the Archaeological and Historical Sciences*. Edinburgh University Press, Edinburgh, Scotland, 1971.

[227] R. Holliday. A different kind of inheritance. *Scientific American*, pages 60–73, June 1989.

[228] J. Hopcroft and J. Ullman. *Introduction to Automata Theory, Languages and Computation*. Addison-Wesley, Reading, MA., 1979.

[229] N. Horspool. Practical fast searching in strings. *Software Practice and Experience*, 10:501–6, 1980.

[230] P. Horton and E. Lawler. An analysis of the efficiency of the A^* algorithm for multiple sequence alignment. Unpublished note, 1994.

[231] L. Hou, L. Martin, Z. Zhou, and A. Feduccia. Early adaptive radiation of birds: Evidence from fossils from notheastern china. *Science*, 274:1164–67, 1996.

[232] X. Huang and W. Miller. A time-efficient, linear-space local similarity algorithm. *Adv. Appl. Math.*, 12:337–57, 1991.

[233] T. J. P. Hubbard, A. M. Lesk, and A. Tramontano. Gathering them into the fold. *Nature Structural Biology*, 4:313, April 1996.

[234] T. J. Hudson, E. S. Lander, et al. An STS-based map of the human genome. *Science*, 270:1945–54, 1995.

[235] R. Hughey. Parallel sequence comparison and alignment. *Proc. Int. Conf. Application-Specific Array Processors*, IEEE Computer Society Press, July 1995.

[236] L. Hui. Color set size problem with applications to string matching. *Proc. 3rd Symp. on Combinatorial Pattern Matching. Springer LNCS 644*, pages 227–40, 1992.

[237] A. Hume and D. M. Sunday. Fast string searching. *Software–Pract. Exper.*, 21:1221–48, 1991.

[238] J. W. Hunt and T. G. Szymanski. A fast al-

gorithm for computing longest common subsequences. *Comm. ACM*, 20:350–53, 1977.

[239] R. Idury and A. Schaffer. Multiple matching of parameterized patterns. *Proc. 5th Symp. on Combinatorial Pattern Matching. Springer LNCS 807*, pages 226–39, 1994.

[240] R. W. Irving and C. B. Fraser. Two algorithms for the longest common subsequence of three (or more) strings. *Proc. 3rd Symp. on Combinatorial Pattern Matching. Springer LNCS 644*, pages 214–29, 1992.

[241] R. W. Irving and C. B. Fraser. Maximal common subsequences and minimal common supersequences. *Proc. 5th Symp. on Combinatorial Pattern Matching. Springer LNCS 807*, pages 173–83, 1994.

[242] S. Istrail. The chimeric clones problem. Technical Report, Sandia Natl. Lab., New Mexico, 1993.

[243] M. Ito, K. Shimiza, M. Nakanishi, and A. Hashimoto. Polynomial-time algorithms for computing characteristic strings. *Proc. 5th Symp. on Combinatorial Pattern Matching. Springer LNCS 807*, pages 274–88, 1994.

[244] G. Jacobson and K. P. Vo. Heaviest increasing/common subsequence problems. *Proc. 3rd Symp. on Combinatorial Pattern Matching. Springer LNCS 644*, pages 52–65, 1992.

[245] R. Jenson, M. C. King, and J. Holt, et al. BRCA1 is secreted and exhibits properties of a granin. *Nature Genetics*, 12:303–8, 1996.

[246] T. Jiang and Z. Jiang. Rotation of periodic strings and short superstrings. Preprint, 1995.

[247] T. Jiang and R. M. Karp. Mapping clones with a given ordering or interleaving. *Proc. 8'th ACM-SIAM Symp. on Discrete Algs.*, 1997.

[248] T. Jiang, M. Li, and D. Du. A note on shortest superstrings with flipping. *Information Processing Lett.*, 44:195–99, 1992.

[249] T. Jiang, L. Wang, and E. L. Lawler. Approximation algorithms for tree alignment with a given phylogeny. *Algorithmica*, 16:302–15, 1996.

[250] D. S. Johnson and C. H. Paadimitriou.

Performance guarantees for heuristics. In E. L. Lawler, J. K. Lenstra, A. H. G. Rinnooy Kan, and D. B. Shmoys, editors, *The Travelling Salesman Problem*, pages 145–80. Wiley-Interscience, New York, 1985.

[251] D. Joseph, J. Meidanis, and P. Tiwari. Determining DNA sequence similarity using maximum independent set algorithms for interval graphs. In *Proc. of the Third Scand. Workshop on Algorithm Theory. Springer LNCS 621*, pages 326–37, 1992.

[252] T. H. Jukes and C. R. Cantor. Evolution of protein molecules. In H. N. Munro, editor, *Mammalian Protein Metabolism*, pages 21–132. Academic Press, New York, 1969.

[253] J. Jurka. Human repetitive elements. In R. A. Meyers, editor, *Molecular Biology and Biotechnology*, pages 438–41. VCH Publishers, New York, 1995.

[254] J. Jurka. Origin and evolution of alu repetitive elements. In R. J. Maraia, editor, *The Impact of Short Interspersed Elements (SINEs) on the Host Genome*, pages 25–41, R. G. Landes, New York, 1995.

[255] J. Jurka, J. Walichiewicz, and A. Milosavljevic. Prototypic sequences for human repetitive DNA. *J. Mol. Evol.*, 35:286–91, 1992.

[256] S. Kannan, E. Lawler, and T. Warnow. Determining the evolutionary tree using experiments. *J. Algorithms*, 21:26–50, 1996.

[257] S. Kannan and E. Myers. An algorithm for locating non-overlapping regions of maximum alignment score. *SIAM J. Comput.*, pages 648–62, 1996.

[258] S. Kannan and T. Warnow. Inferring evolutionary history from DNA sequences. *SIAM J. Comput.*, 23:713–37, 1994.

[259] S. Kannan and T. Warnow. A fast algorithm for the computation and enumeration of perfect phylogenies when the number of character states is fixed. *Proc. 6th ACM-SIAM Symp. on Discrete Algs.*, pages 595–603, 1995.

[260] H. Kaplan, R. Shamir, and R. E. Tarjan. Tractability of parameterized completion problems on chordal and interval graphs: minimum fill-in and physical mapping. *Proc. 35th IEEE Symp. Found. Computer Science*, pages 780–91, 1994.

[261] H. Kaplan, R. Shamir, and R.E. Tarjan.

Faster and simpler algorithm for sorting signed permutations by reversals. *Proc. 8'th ACM-SIAM Symp. on Discrete Algs.*, 1997.

[262] S. Karlin and S. F. Altschul. Methods for assessing the statistical significance of molecular sequence features by using general scoring schemes. *Proc. Natl. Academy Science*, 87:2264–68, 1990.

[263] S. Karlin and S. F. Altschul. Applications and statistics for multiple high-scoring segments in molecular sequences. *Proc. Natl. Academy Science*, 90:5873–77, 1993.

[264] R. Karp. Mapping of the genome: Some combinatorial problems arising in molecular biology. *Proc. of the ACM Symp. on the Theory of Computing*, pages 278–85, 1993.

[265] R. Karp, R. Miller, and A. Rosenberg. Rapid identification of repeated patterns in strings, trees and arrays. *Proc. of the ACM Symp. on the Theory of Computing*, pages 125–36, 1972.

[266] R. Karp and M. Rabin. Efficient randomized pattern matching algorithms. *IBM J. Res. Development*, 31:249–60, 1987.

[267] J. Kececioglu. The maximum weight trace problem in multiple sequence alignment. *Proc. 4th Symp. on Combinatorial Pattern Matching. Springer LNCS 684*, pages 106–19, 1993.

[268] J. D. Kececioglu and D. Gusfield. Reconstructing a history of recombinations from a set of sequences. *Proc. 5th ACM-SIAM Symp. on Discrete Algs.*, pages 471–80, 1994.

[269] J. D. Kececioglu and E. Myers. Exact and approximation algorithms for the sequence reconstruction problem. *Algorithmica*, 13:7–51, 1995.

[270] J. D. Kececioglu and R. Ravi. Of mice and men: algorithms for evolutionary distances between genomes with translocation. *Proc. 6th ACM-SIAM Symp. on Discrete Algs.*, pages 604–13, 1995.

[271] J. D. Kececioglu and D. Sankoff. Efficient bounds for oriented chromosome inversion distance. *Proc. 5th Symp. on Combinatorial Pattern Matching. Springer LNCS 807*, pages 307–25, 1994.

[272] J. D. Kececioglu and D. Sankoff. Exact and approximation algorithms for sorting by reversal. *Algorithmica*, 13:180–210, 1995.

[273] J. D. Kececioglu. *Exact and approximation algorithms for the sequence reconstruction problem in computational biology*. PhD thesis, Univ. Ariz., Dept. Computer Science, 1990.

[274] M. Kimura. The neutral theory of molecular evolution. *Scientific American*, pages 98–126, November 1979.

[275] M. Kimura. *The Neutral Theory of Molecular Evolution*. Cambridge University Press, Cambridge, 1983.

[276] J. R. Knight and E. W. Myers. Approximate regular expression pattern matching with concave gap penalties. *Proc. 3rd Symp. on Combinatorial Pattern Matching. Springer LNCS 644*, pages 67–78, 1992.

[277] J. R. Knight and E. W. Myers. Superpattern matching. *Algorithmica*, 13:211–43, 1995.

[278] D. E. Knuth, J. H. Morris, and V. B. Pratt. Fast pattern matching in strings. *SIAM J. Comput.*, 6:323–50, 1977.

[279] L. F. Kolakowski, J. Leunissen, and J. E. Smith. Prosearch: Fast searching of protein sequences with regular expression patterns related to protein structure and function. *Biotechniques*, 13:919–21, 1992.

[280] G. Kolata. Biology's big project turns into challenge for computer experts. *New York Times*, June 11, 1996.

[281] G. Kolata. A vat of DNA may become fast computer of the future. *New York Times*, April 11, 1995.

[282] E. V. Koonin, S. F. Altschul, and P. Bork. BRCA1 protein products: functional motifs. *Nature Genetics*, 13:266–68, 1996.

[283] R. Kosaraju, J. Park, and C. Stein. Long tours and short superstrings. *Proc. of the 35th IEEE Symp. on Foundations of Comp. Sci.*, pages 166–77, 1994.

[284] L. Kou, G. Markowsky, and L. Berman. A fast algorithm for Steiner trees. *ACTA Informatica*, 15, 1981.

[285] A. Krogh, M. Brown, I. Mian, K. Sjolander, and D. Haussler. Hidden markov models in computational biology: Applications to protein modelling. *J. Mol. Biol.*, 235:1501–31, 1994.

[286] T. G. Krontiris. Minisatellites and human disease. *Science*, 269:1682–83, 1995.

[287] G. Landau and U. Vishkin. Introducing efficient parallelism into approximate string matching and a new serial algorithm. *Proc. of the ACM Symp. on the Theory of Computing*, pages 220–30, 1986.

[288] G. M. Landau and J. P. Schmidt. An algorithm for approximate tandem repeats. *Proc. 4th Symp. on Combinatorial Pattern Matching. Springer LNCS 684*, pages 120–33, 1993.

[289] G. M. Landau and U. Vishkin. Efficient string matching with k mismatches. *Theor. Comp. Sci.*, 43:239–49, 1986.

[290] G. M. Landau, U. Vishkin, and R. Nussinov. Locating alignments with k differences for nucleotide and amino acid sequences. *Comp. Appl. Biosciences*, 4:12–24, 1988.

[291] E. Lander. The new genomics: Global views of biology. *Science*, 274:536–39, 1996.

[292] E. Lander and M. Waterman. Genomic mapping by fingerprinting random clones: a mathematical analysis. *Genomics*, 2:231–39, 1988.

[293] M. T. Laub and D. W. Smith. Finding intron/exon splice junctions using INFO, Interuption Finder and Organizer. Preprint February 1997, U.C. San Diego, Dept. of Biology.

[294] E. L. Lawler. *Combinatorial Optimization: Networks and Matroids*. Holt, Rinehart, and Winston, New York, 1976.

[295] C. Lawrence, S. Altschul, M. Boguski, J. Liu, A. Neuwald, and J. Wooton. Detecting subtle sequence signals: a Gibbs sampling strategy for multiple alignment. *Science*, 262:208–14, 1993.

[296] A. M. Lesk, M. Levitt, and C. Chothia. Alignment of the amino acid sequences of distantly related proteins using variable gap penalties. *Protein Eng.*, 1:77–78, 1986.

[297] A. M. Lesk. Computational molecular biology. In A. Kent and J. G. Williams, editors, *Encyclopedia of Computer Science and Technology*, 31:101–65. Marcel Dekker, New York, 1994.

[298] M. Y. Leung, B. Blaisdell, C. Burge, and S. Karlin. An efficient algorithm for identifying matches with errors in multiple long molecular sequences. *J. Mol. Biol.*, 221:1367–78, 1991.

[299] V. I. Levenstein. Binary codes capable of correcting insertions and reversals. *Sov. Phys. Dokl.*, 10:707–10, 1966.

[300] E. Levy-Lahad, et al. Candidate gene for the chromosome 1 familial Alzheimer's disease locus. *Science*, 269:973–77, 1995.

[301] W. H. Li and D. Graur. *Fundamentals of Molecular Evolution*. Sinauer, Sunderland, MA, 1991.

[302] M. Linial and N. Linial. On the potential of molecular computing. Letter to *Science*, 268:481, April 28, 1995.

[303] D. Lipman, S. Altshul, and J. Kececioglu. A tool for multiple sequence alignment. *Proc. Natl. Academy Science*, 86:4412–15, 1989.

[304] D. J. Lipman and W. R. Pearson. Rapid and sensitive protein similarity searches. *Science*, 227:1435–41, 1985.

[305] R. J. Lipton. DNA solution of hard computational problems. *Science*, 268:542–44, 1995.

[306] C. D. Livingstone and G. J. Barton. Identification of functional residues and secondary structure from protein multiple alignment. In R. F. Doolittle, editor, *Methods in Enzymology Vol. 266. Computer Methods for Macromolecular Sequence Analysis*, pages 479–512. Academic Press, New York, 1996.

[307] M. Main and R. Lorentz. An $O(n \log n)$ algorithm for finding all repeats in a string. *J. Algorithms*, 5:422–32, 1984.

[308] U. Manber and G. Myers. Suffix arrays: a new method for on-line search. *SIAM J. Comput.*, 22:935–48, 1993.

[309] E. Marshall. Emphasis turns from mapping to large-scale sequencing. News article in *Science*, 268:1270–71, June 1995.

[310] H. Martinez. An efficient method for finding repeats in molecular sequences. *Nucl. Acids Res.*, 11:4629–34, 1983.

[311] J. Marx. Developmental biology: knocking genes in instead of out. *Science*, 269:636, 1995.

[312] W. J. Masek and M. S. Paterson. How to compute string-edit distances quickly. In D. Sankoff and J. Kruskal, editors, *Time Warps, String Edits, and Macromolecules: The Theory and Practice of Sequence Comparion*, pages 337–49. Addison Wesley, Reading, MA, 1983.

[313] W. J. Masek and M. S. Paterson. A faster algorithm for computing string edit

distances. *J. Comp. Sys. Sci.*, 20:18–31, 1980.

[314] M. Max, P. McKinnon, K. Seidenman, R. Barret, M. Apllebury, J. Takahashi, and R. Margolskee. Pineal opsin: a nonvisual opsin expressed in chick pineal. *Science*, 267:1502–6, 1995.

[315] J. Maynard Smith. *Evolutionary Genetics*. Oxford Univ. Press, Oxford, 1989.

[316] M. McClure, T. Vasi, and W. Fitch. Comparitive analysis of multiple protein-sequence alignment methods. *Mol. Biol. Evolution*, 11:571–92, 1994.

[317] E. McConkey. *Human Genetics: The Molecular Revolution*. Jones and Bartlett, Boston, MA, 1993.

[318] E. M. McCreight. A space-economical suffix tree construction algorithm. *J. ACM*, 23:262–72, 1976.

[319] W. McGinnis and M. Kuziora. The molecular architects of body design. *Scientific American*, pages 58–66, February 1994.

[320] H. W. Mewes and K. Heumann. Genome analysis: pattern search in biological macromolecules. *Proc. 6th Symp. on Combinatorial Pattern Matching. Springer LNCS 937*, pages 261–85, 1995.

[321] W. Miller. An algorithm for locating a repeated region. Unpublished manuscript.

[322] W. Miller and E. W. Myers. Sequence comparison with concave weighting functions. *Bull. Math. Biol.*, 50:97–120, 1988.

[323] W. Miller, J. Ostell, and K. Rudd. An algorithm for searching restriction maps. *Comp. Appl. Biosciences*, 6:247–52, 1990.

[324] A. Milosavljevic. Discovering dependencies via algorithmic mutual information: a case study in DNA sequence comparisons. *Machine Learning*, 21:35–50, 1995.

[325] A. Milosavljevic and J. Jurka. Discovering simple DNA sequences by the algorithmic significance method. *Comp. Appl. Biosciences*, 9:407–11, 1993.

[326] A. Milosavljevic and J. Jurka. Discovery by minimal length encoding: a case study in molecular evolution. *Machine Learning*, 12:69–87, 1993.

[327] B. G. Mirkin and S. N. Rodin. *Graphs and Genes*. Springer-Verlag, New York, 1984.

[328] A. Mirzanian. A halving technique for the longest stuttering subsequence problem. *Information Processing Lett.*, 26:71–75, 1987.

[329] A. Monaco and Z. Larin. YACs, BACs, PACs and MACs: artificial chromosomes as research tools. *Trends in Genetics*, 12:280–86, 1994.

[330] J. Monod. *Chance and Necessity; An Essay on the Natural Philosophy of Modern Biology*. Knopf, New York, 1971.

[331] R. Motwani and P. Raghavan. *Randomized Algorithms*. Cambridge University Press, Cambridge, 1995.

[332] R. K. Moyzis. The human telomere. *Scientific American*, pages 48–55, April 1991.

[333] K. Mullis. The unusual origin of the polymerase chain reaction. *Scientific American*, pages 56–65, April 1990.

[334] M. Murata, J. Richardson, and J. Sussman. Simultaneous comparison of three protein sequences. *Proc. Natl. Academy Science*, 82:3073–77, 1985.

[335] A. Murray and J. Szostak. Artificial chromosomes. *Scientific American*, pages 62–68, November 1987.

[336] S. Muthukrishnan. Detecting false matches in string matching algorithms. *Proc. 4th Symp. on Combinatorial Pattern Matching. Springer LNCS 684*, pages 164–78, 1993.

[337] E. Myers. Algorithmic advances for searching biosequence databases. In S. Suhai, editor, *Computational Methods in Genome Research*, pages 121–35. Plenum Press, New York, 1994.

[338] E. Myers and X. Huang. An $O(n^2 \log n)$ restriction map comparison and search algorithm. *Bull. Math. Biol.*, 54:599–618, 1992.

[339] E. W. Myers. A Four-Russians algorithm for regular expression pattern. Technical Report, Univ. Ariz., 1988.

[340] E. W. Myers. A Four-Russians algorithm for regular expression pattern matching. *J. ACM*, 39:430–48, 1992.

[341] E. W. Myers. An O(nd) difference algorithm and its variations. *Algorithmica*, 1:251–66, 1986.

[342] E. W. Myers. A sublinear algorithm for approximate keyword searching. *Algorithmica*, 12:345–74, 1994.

[343] E. W. Myers. Towards simplifying and ac-

curately formulating fragment assembly. *J. Comp. Biol.*, 2:275–90, 1995.

[344] E. W. Myers and W. Miller. Optimal alignments in linear space. *Comp. Appl. Biosciences*, 4:11–17, 1988.

[345] E. W. Myers and W. Miller. Row replacement algorithms for screen editors. *ACM TOPLAS*, 11:33–56, 1989.

[346] D. Naor and D. Brutlag. On near-optimal alignments in biological sequences. *J. Comp. Biol.*, 1:349–66, 1994.

[347] S. B. Needleman and C. D. Wunsch. A general method applicable to the search for similarities in the amino acid sequence of two proteins. *J. Mol. Biol.*, 48:443–53, 1970.

[348] S. Nicol, et al. Genetic identification of a hantavirus associated with an outbreak of acute respiratory illness. *Science*, 262:914–17, 1993.

[349] R. Nowak. Entering the postgenome era. *Science*, 270:368-71, 1995.

[350] R. Nowak. Venter wins sequencing race – twice. *Science*, 268:1273, 1995.

[351] S. O'Brien. The ancestry of the Giant Panda. *Scientific American*, pages 102–16, November 1987.

[352] M. V. Olson. A time to sequence. *Science*, 270:394–96, 1995.

[353] J. Ott. *Analysis of Human Genetic Linkage*. Johns Hopkins University Press, Baltimore, MD, 1991.

[354] M Overduin, T. Harvey, S. Bagby, K. Tong, P Yau, M. Takeichi, and M. Ikura. Solution structure of the epithelial cadherin domain responsible for selective cell adhesion. *Science*, 267:386–89, 1995.

[355] J. Palmer and L. Herbon. Plant mitochondrial DNA evolves rapidly in structure, but slowly in sequence. *J. Mol. Evol.*, 28:87–89, 1988.

[356] W. R. Pearson. Rapid and sensitive sequence comparisons with FASTP and FASTA. In R. F. Doolittle, editor, *Methods in Enzymology Vol. 183. Molecular Evolution: Computer Analysis of Protein and Nucleic Acid Sequences*, pages 63–69. Academic Press, New York, 1990.

[357] W. R. Pearson. Searching protein sequence libraries: comparison of the sensitivity and selectivity of the Smith-Waterman and FASTA algorithms. *Genomics*, 11:635–50, 1991.

[358] W. R. Pearson. Comparison of methods for searching protein sequence databases. *Protein Science*, 4:1145–60, 1995.

[359] W. R. Pearson. Effective protein sequence comparison. In R. F. Doolittle, editor, *Methods in Enzymology Vol. 266. Computer Methods for Macromolecular Sequence Analysis*, pages 227–58. Academic Press, New York, 1996.

[360] W. R. Pearson. Protein sequence comparison and protein evolution. Tutorial T6 of Intelligent Systems in Mol. Biol., Cambridge, England, July 1995.

[361] W. R. Pearson and David J. Lipman. Improved tools for biological sequence comparison. *Proc. Natl. Academy Science*, 85:2444–48, 1988.

[362] M. Peltola, H. Soderlund, J. Tarhio, and E. Ukkonen. Algorithms for some string matching problems arising in molecular genetics. *Proc. of the 9th IFIP World Computer Congress*, pages 59–64, 1983.

[363] M. Peltola, H. Soderlund, and E. Ukkonen. Sequaid: a DNA sequence assembly program based on a mathematical model. *Nucl. Acids Res.*, 12:307–21, 1984.

[364] M. Peltola, H. Soderlund, and E. Ukkonen. Algorithms for the search of amino acid patterns in nucleic acid sequences. *Nucl. Acids Res.*, 14:99–107, 1986.

[365] E. Pennisi. Research news: premature aging gene discovered. *Science*, 272:193–94, April 1996.

[366] P. Pevzner. l-Tuple DNA sequencing: computer analysis. *J. Biomol. Structure Dynamics*, 7:63–73, 1989.

[367] P. Pevzner (ed.). Combinatorial methods for DNA mapping and sequencing, 1995. Vol. 2, no. 2. Special issue of *J. Comp. Biol.*

[368] P. Pevzner and R. Lipshutz. Towards DNA sequencing chips. *Proc. 19th Symp. on Math. Found. of Comp. Sci. Springer LNCS 841*, pages 143–58, 1994.

[369] P. Pevzner and M. Vingron. Multiple sequence comparison and *n*-dimensional image reconstruction. *Proc. 4th Symp. on Combinatorial Pattern Matching. Springer LNCS 684*, pages 243–53, 1993.

[370] P. Pevzner and M. Waterman. Matrix longest common subsequence problem, duality and Hilbert bases. *Proc. 3rd Symp. on Combinatorial Pattern Match-*

ing. Springer LNCS 644, pages 79–89, 1992.

[371] P. Pevzner and M. S. Waterman. Open combinatorial problems in computational molecular biology. *Proc. of the 3'rd Israel Symposium on Computers and Systems. IEEE Computer Society Press*, 1995.

[372] P. A. Pevzner. DNA physical mapping and alternating Eulerian cycles in colored graphs. *Algorithmica*, 12:77–105, 1994.

[373] P. A. Pevzner and M. Waterman. Multiple filtration and approximate pattern matching. *Algorithmica*, 13:135–54, 1995.

[374] J. Posfai, A. Bhagwat, G. Posfai, and R. Roberts. Predictive motifs derived from cytosine methyltransferases. *Nucl. Acids Res.*, 17:2421–35, 1989.

[375] V. Pratt. Personal communication.

[376] V. Pratt. Applications of the Weiner repetition finder. Unpublished manuscript, 1973.

[377] S. Prusiner. The prion diseases. *Scientific American*, pages 48–57, January 1995.

[378] R. Rhodes and A. Klug. Zinc fingers. *Scientific American*, pages 56–65, February 1993.

[379] E. Rich and K. Knight. *Artificial Intelligence*. McGraw Hill, New York, 1991.

[380] M. Ridley. *Evolution and Classification: The Reformation of Cladism*. Longman, London, 1986.

[381] E. De Roberts, G. Oliver, and C. Wright. Homeobox genes and the vertibrate body plan. *Scientific American*, pages 46–52, July 1990.

[382] M. Rodeh, V. R. Pratt, and S. Even. A linear algorithm for data compression via string matching. *J. ACM*, 28:16–24, 1981.

[383] J. B. Rosser and L. Schoenfield. Approximate formulas for some functions of prime numbers. *Illinois J. Math.*, 6:64–94, 1962.

[384] W. Rytter. A correct preprocessing algorithm for Boyer-Moore string searching. *SIAM J. Comput.*, 9:509–12, 1980.

[385] M. F. Sagot, A. Viari, and H. Soldano. Multiple sequence comparison: a peptide matching approach. *Proc. 6'th Symp. on Combinatorial Pattern Matching. Springer LNCS 937*, pages 366–85, 1995.

[386] P. Salamon and A. Konopka. A maximum entropy principle for distribution of local complexity in naturally occur-

ring nucleotide sequences. *Computers and Chemistry*, 16:117–24, 1992.

[387] D. Sankoff. Minimal mutation trees of sequences. *SIAM J. Appl. Math.*, 28:35–42, 1975.

[388] D. Sankoff and R. Cedergren. Simultaneous comparisons of three or more sequences related by a tree. In [360], pages 253–64.

[389] D. Sankoff and J. Kruskal (eds.). *Time Warps, String Edits, and Macromolecules: The Theory and Practice of Sequence Comparion*. Addison-Wesley, Reading, MA, 1983.

[390] D. Sankoff, V. Ferretti, and J. Nadeau. Conserved segment identification. In *Proc. of RECOMB 97: The first international conference on computational molecular biology*, pages 252–56. ACM Press, 1997.

[391] C. Sapienza. Parental imprinting of genes. *Scientific American*, pages 52–60, October 1990.

[392] B. Schatz. Information retrieval in digital libraries: Bringing search to the net. *Science*, 275:327–34, 1997.

[393] B. Schieber and U. Vishkin. On finding lowest common ancestors: simplifications and parallelization. *SIAM J. Comput.*, 17:1253–62, 1988.

[394] R. T. Schimke. Gene amplification and drug resistance. *Scientific American*, 243:60–69, 1980.

[395] C. Schmid and J. Marks. DNA hybridization as a guide to phylogeny: chemical and physical limits. *J. Mol. Evol.*, 30:237–46, 1990.

[396] J. Schmidt. All highest scoring paths in weighted grid graphs and its application to finding all approximate repeats in strings. *SIAM J. Comput.*, to appear.

[397] G. Schuler, J. Epstein, H. Ohkawa, and J. Kans. *Entrez:* molecular biology database and retrieval system. In R. F. Doolittle, editor, *Methods in Enzymology Vol. 266. Computer Methods for Macromolecular Sequence Analysis*, pages 141–62. Academic Press, New York, 1996.

[398] G. D. Schuler, S. F. Altschul, and D. J. Lipman. A workbench for multiple alignment construction and analysis. *Proteins: Structure, Function and Genetics*, 9:180–90, 1991.

[399] G.D. Schuler, E.S. Lander, and T.J. Hudson et al. A gene map for the human genome. *Science*, 274:540–46, 1996.

[400] R. Schwarz and M. Dayhoff. Matrices for detecting distant relationships. In M. Dayhoff, editor, *Atlas of Protein Sequences*, pages 353–58. Natl. Biomed. Res. Found., 1979.

[401] R. Sedgewick. *Algorithms,* 2nd ed.. Addison-Wesley, Reading, MA, 1988.

[402] J. Setubal and J. Meidanis. *Introduction to Computational Molecular Biology*. PWS, Boston, 1997.

[403] C. Shih and R. Weinberg. Isolation of a transforming sequence from a human bladder carcinoma cell line. *Cell*, pages 161–69, 1982.

[404] Y. Shiloach. Fast canonization of circular strings. *J. Algorithms*, 2:107–21, 1981.

[405] S. K. Shyue, D. Hewett-Emmett, H. Sperling, D. Hunt, J. Bowmaker, J. Mollon, and W. H. Li. Adaptive evolution of color vision of genes in higher primates. *Science*, 269:1265–67, 1995.

[406] C. G. Sibley and J. E. Ahlquist. Reconstructing bird phylogeny by comparing DNA's. *Scientific American*, pages 82–92, February 1986.

[407] C. G. Sibley and J. E. Ahlquist. *Phylogeny and Classification of Birds*. Yale Univ. Press, New Haven, CT, 1990.

[408] D. Slonim, L. Kruglyak, L. Stein, and E. Lander. Building human genome maps with radiation hybrids. In *Proc. of RECOMB 97: The first international conference on computational molecular biology*, pages 277–86. ACM Press, 1997.

[409] P. D. Smith. Experiments with a very fast substring search algorithm. *Software – Pract. Exper.*, 21:1065–74, 1991.

[410] P. D. Smith. On tuning the Boyer-Moore-Horspool string searching algorithm. *Software – Pract. Exper.*, 24:435–36, 1994.

[411] T. F. Smith and M. S. Waterman. Identification of common molecular subsequences. *J. Mol. Biol.*, 147:195–97, 1981.

[412] E. E. Snyder and G. D. Stormo. Identification of protein coding regions in genomic DNA. *J. Mol. Biol.*, 284:1–18, 1995.

[413] J. L. Spouge. Improving sequence-matching algorithms by working from both ends. *J. Mol. Biol.*, 181:137–38, 1985.

[414] J. L. Spouge. Speeding up dynamic programming algorithms for finding optimal lattice paths. *SIAM J. Appl. Math.*, 49:1552–66, 1989.

[415] J. L. Spouge. Fast optimal alignment. *Comp. Appl. Biosciences*, 7:1–7, 1991.

[416] R. Staden. Screening protein and nucleic acid sequences against libraries of patterns. *J. DNA Sequencing Mapping*, 1:369–74, 1991.

[417] P. Steeg. Granin expectations in breast cancer? *Nature Genetics*, 12:223–25, 1996.

[418] M. Steel. The complexity of reconstructing trees from qualitative characters and subtrees. *J. Classification*, 9:91–116, 1992.

[419] S. Stein. The mathematician as explorer. *Scientific American*, pages 149–63, May 1961.

[420] J. Steitz. Snurps. *Scientific American*, pages 56–63, June 1988.

[421] G. A. Stephen. *String Searching Algorithms*. World Scientific, Singapore, 1994.

[422] M. J. Sternberg. PROMOT: A fortran program to scan protein sequences against a library of known motifs. *Comp. Appl. Biosciences*, 7:257–60, 1991.

[423] J. A. Storer. *Data Compression: Methods and Theory*. Computer Science Press, Rockville, MD, 1988.

[424] B. S. Strauss. Book review: DNA repair and mutagenesis. *Science*, 270:1511–13, 1995.

[425] D. M. Sunday. A very fast substring search algorithm. *Comm. ACM*, 33:132–42, 1990.

[426] D. Suzuki, A. Griffiths, J. Miller, and R. Lewontin. *An Introduction to Genetic Analysis,* 3rd ed. Freeman, New York, 1986.

[427] E. S. Sweedyk. *A 2 1/2 appoximation algorithm for shortest common superstring*. PhD thesis, Univ. Calif., Berkeley, Dept. Computer Science, 1995.

[428] M. B. Swindells. Commentary: Finding your fold. *Protein Eng.*, 7:1–3, 1994.

[429] D. L. Swofford and G. L. Olsen. Phylogeny reconstruction. In D. M. Hillis and C. Moritz, editors, *Molecular Systematics*, pages 411–501. Sinauer, Sunderland, MA, 1990.

[430] M. Syvanen. Horizontal gene transfer: ev-

idence and possible consequences. *Annu. Rev. Genetics*, 28:237–61, 1994.

[431] J. Tarhio and E. Ukkonen. A greedy approximation algorithm for constructing shortest common superstrings. *Theor. Comp. Sci.*, 57:131–45, 1988.

[432] W. R. Taylor. Hierarchical method to align large numbers of biological sequences. In R. F. Doolittle, editor, *Methods in Enzymology Vol. 183. Molecular Evolution: Computer Analysis of Protein and Nucleic Acid Sequences*, pages 456–74. Academic Press, New York, 1990.

[433] W. R. Taylor. Multiple protein sequence alignment: algorithms and gap insertion. In R. F. Doolittle, editor, *Methods in Enzymology Vol. 266. Computer Methods for Macromolecular Sequence Analysis*, pages 343–67. Academic Press, New York, 1996.

[434] S. H. Teng and F. Yao. Approximating shortest superstrings. *Proc. of the 34th IEEE Symp. on Foundations of Comp. Sci.*, pages 158–65, 1993.

[435] E. Trifonov and V. Brendel. *Gnomic: A Dictionary of Genetic Codes*. VCH Press, Deerfield, FL, 1986.

[436] J. Turner. Approximation algorithms for the shortest common superstring problem. *Information Comput.*, 83:1–20, 1989.

[437] E. Ukkonen. Approximate string-matching over suffix trees. *Proc. 4th Symp. on Combinatorial Pattern Matching. Springer LNCS 684*, pages 228–42, 1993.

[438] E. Ukkonen. On-line construction of suffix-trees. *Algorithmica*, 14:249–60, 1995.

[439] E. Ukkonen. Algorithms for approximate string matching. *Information Control*, 64:100–18, 1985.

[440] E. Ukkonen. Approximate string matching with q-grams and maximal matches. *Theor. Comp. Sci.*, 92:191–211, 1992.

[441] L. Urdang. *Random House Dictionary (College Edition)*. Random House, New York, 1968.

[442] J. C. Venter, H. O Smith, and L. Hood. A new strategy for genome sequencing. *Nature*, 381:364–66, 1996.

[443] M. Vingron. Multiple sequence alignment and applications in molecular biology. Preprint 91-12, Univ. Heidelberg, 1991.

[444] M. Vingron and P. Argos. A fast and sensitive multiple sequence alignment algorithm. *Comp. Appl. Biosciences*, 5:115–21, 1989.

[445] M. Vingron and P. Argos. Determination of reliable regions in protein sequence alignments. *Protein Eng.*, 7:565–69, 1990.

[446] M. Vingron and P. Argos. Motif recognition and alignment for many sequences by comparison of dot-matrices. *J. Mol. Biol.*, 218:33–43, 1991.

[447] M. Vingron and M. Waterman. Sequence alignment and penalty choice. *J. Mol. Biol.*, 235:1–12, 1994.

[448] G. Vogt, T. Etzold, and P. Argos. An assessment of amino acid exchange matrices in aligning protein sequences. *J. Mol. Biol.*, 294:816–31, 1995.

[449] G. von Heijne. *Sequence Analysis in Molecular Biology: Treasure Trove or Trivial Pursuit*. Academic Press, New York, 1987.

[450] B. A. Voyles. *The Biology of Viruses*. Mosby, St. Louis, MO, 1993.

[451] G. Wagner. E-cadherin: a distant member of the immunoglobulin superfamily. *Science*, 267:342, 1995.

[452] M. M. Waldrop. On-line archives let biologists interrogate the genome. *Science*, 269:1356–58, 1995.

[453] L. Wang and D. Gusfield. Improved approximation algorithms for tree alignment. *Proc. 7th Symp. on Combinatorial Pattern Matching. Springer LNCS 1075*, pages 220–33, 1996.

[454] L. Wang and T. Jiang. On the complexity of multiple sequence alignment. *J. Comp. Biol.*, 1:337–48, 1994.

[455] M. Waterman. Multiple sequence alignment by consensus. *Nucl. Acids Res.*, 14:9095–102, 1986.

[456] M. Waterman, R. Arratia, and D. Galas. Pattern recognition in several sequences: consensus and alignment. *Bull. Math. Biol.*, 46:515–27, 1984.

[457] M. Waterman and M. Perlwitz. Line geometries for sequence comparisons. *Bull. Math. Biol.*, 46:567–77, 1984.

[458] M. Waterman, T. Smith, M. Singh, and W. Beyer. Additive evolutionary trees. *J. Theor. Biol.*, 64:199–213, 1977.

[459] M. Waterman and M. Vingron. Rapid and accurate estimates of statistical significance for sequence data base searches. *Proc. Natl. Academy Science*, 91:4625–28, 1994.

[460] M. S. Waterman. Sequence alignments in the neighborhood of the optimum with general application to dynamic programming. *Proc. Natl. Academy Science*, 80:3123–24, 1983.

[461] M. S. Waterman. *Introduction to Computational Biology: Maps, Sequences, Genomes*. Chapman and Hall, London, England, 1995.

[462] M. S. Waterman, M. Eggert, and E. Lander. A new algorithm for best subsequence alignments with application to tRNA-rRNA comparisons. *J. Mol. Biol.*, 197:723–28, 1987.

[463] M. S. Waterman, M. Eggert, and E. Lander. Parametric sequence comparisons. *Proc. Natl. Academy Science*, 89:6090–93, 1992.

[464] M. S. Waterman and J. R. Griggs. Interval graphs and maps of DNA. *Bull. Math. Biol.*, 48:189–95, 1986.

[465] M. S. Waterman, T. F. Smith, and H. L. Katcher. Algorithms for restriction maps. *Nucl. Acids Res.*, 12:237–42, 1984.

[466] M. S. Waterman. Efficient sequence alignment algorithms. *J. Theor. Biol.*, 108:333–37, 1984.

[467] M. S. Waterman and R. Jones. Consensus methods for DNA and protein sequence alignment. In R. F. Doolittle, editor, *Methods in Enzymology Vol. 183. Molecular Evolution: Computer Analysis of Protein and Nucleic Acid Sequences*, pages 221–37. Academic Press, New York, 1990.

[468] J. Watson, M. Gilman, J. Witkowski, and M. Zoller. *Recombinant DNA*, 2nd ed. Scientific American Books, San Fransisco, CA, 1992.

[469] J. Watson, N. Hopkins, J. Roberts, J. Steitz, and A. Weiner. *Molecular Biology of the Gene*, 4th ed. Benjamin Cummings, Menlo Park, CA, 1987.

[470] R. Wayne and J. Gittleman. The problematic Red Wolf. *Scientific American*, pages 36–39, July 1995.

[471] R. Weinberg. A molecular basis of cancer. *Scientific American*, pages 126–42, May 1983.

[472] R. Weinberg. Finding the anti-oncogene. *Scientific American*, pages 44–51, September 1988.

[473] P. Weiner. Linear pattern matching algorithms. *Proc. of the 14th IEEE Symp. on Switching and Automata Theory*, pp. 1–11, 1973.

[474] J. Weinstein and K. Paul et al. An information-intensive approach to the molecular pharmacology of cancer. *Science*, 275:343–49, 1997.

[475] S. C. West. The processing of recombination intermediates: mechanistic insights from studies of bacterial proteins. *Cell*, 76:9–15, 1994.

[476] R. White and J. M. Lalouel. Chromosome mapping with DNA markers. *Scientific American*, 238:40–48, 1988.

[477] W. J. Wilbur. On the PAM matrix model of protein evolution. *Mol. Biol. Evol.*, 2:434–37, 1985.

[478] N. Williams. Closing in on the complete yeast genome sequence. *Science*, 268:1560–61, 1995.

[479] N. Williams. Europe opens institute to deal with gene data deluge. *Science*, 269:630, 1995.

[480] N. Williams. How to get databases talking the same language. *Science*, 275:301–2, 1997.

[481] A. Wilson and R. Cann. The recent African genesis of humans. *Scientific American*, pages 68–73, April 1992.

[482] S. Wu and U. Manber. Fast text searching allowing errors. *Comm. ACM*, 35:83–91, 1992.

[483] S. Wu, U. Manber, G. Myers, and W. Miller. An O(np) sequence comparison algorithm. *Information Processing Lett.*, 35:317–23, 1990.

[484] A. Zelikovsky. An 11/6 approximation algorithm for the Steiner tree problem in graphs. *Information Processing Lett.*, 46:79–83, 1993.

[485] Z. Zhang, W. R. Pearson, and W. Miller. Aligning a DNA sequence with a protein sequence. In *Proc. of RECOMB 97: The first international conference on computational molecular biology*, pages 337–43. ACM Press, 1997.

[486] R. Zimmer and T. Lengauer. Fast and numerically stable parametric alignment of biosequences. In *Proc. of RECOMB 97: The first international conference on*

computational molecular biology, pages 344–53. ACM Press, 1997.

[487] J. Ziv and A. Lempel. A universal algorithm for sequential data compression. *IEEE Trans. on Info. Theory*, 23:337–43, 1977.

[488] J. Ziv and A. Lempel. Compression of individual sequences via variable length coding. *IEEE Trans. on Info. Theory*, 24:530–36, 1978.

[489] E. Zuckerkandl and L. Pauling. Molecular disease, evolution and genic heterogeneity. In M. Kash and B. Pullman, editors, *Horizons in Biochemistry*, pages 189–225. Academic Press, New York, 1962.

[490] M. Zuker. Suboptimal sequence alignments in molecular biology, alignment with error analysis. *J. Mol. Biol.*, 221:403–20, 1991.

Glossary

This glossary defines terms that are used in several parts of the book and that may be new to the reader. The definitions are stated to the level of precision needed for this book, and some may be incomplete compared to a glossary for a biology text. Very familiar terms are omitted, as are specialized terms that are only used in the same section of the book where they are defined. When a particular term is discussed or defined more completely in the body of the book, the page where that discussion begins is given in parentheses after the definition.

allele One of two or more forms that a substring of DNA (often a gene) can take on.

algorithm A high level description of a mechanistic way to solve a problem or compute a function. The description must lay out the logic of the method, but should avoid many low-level programming details needed to implement it on a computer. Often, in the biological literature, "algorithm" and "program" are used interchangeably, but this is not correct. A program contains all the implementation detail needed to make the method work in a specific computer programming language on specific computers. That level of detail usually obscures the logic and the ideas behind the algorithm.

alpha helix Helical structure in protein. The alpha helix is one of two common secondary structures in protein. The other is the beta sheet. (pages 248, 361)

amino acid Molecules that form the building blocks of proteins. There are twenty common amino acids found in proteins. Each amino acid is coded in DNA by a "codon" (see genetic code and codon). (page 14)

amino acid alphabet A twenty-character alphabet consisting of the characters A, C, D, E, F, G, H, I, K, L, M, N, P, Q, R, S, T, V, W, Y, each representing one of the twenty amino acids coded for by DNA.

amino acid substitution matrix A matrix specifying the scores to use for character-specific matches and mismatches. The two most widely used classes of amino acid substitution matrices are the PAM matrices and the BLOSUM matrices. (pages 381 and 386)

analogous protein See homologous protein.

antibody A protein that binds to a foreign molecule. Also known as an **immunoglobulin**. They are used in nature to identify and destroy foreign bodies. They are used in biochemistry as assays to identify specific proteins.

antigen A molecule that is recognized as foreign and causes a response by antibodies.

autosome Any chromosome other than a sex chromosome. In humans, this is any chromosome other than the X and Y chromosomes.

base In the context of molecular biology, a base refers to a single nucleotide (DNA or RNA). The DNA bases are abbreviated A, T, C, and G. The RNA bases are abbreviated A, U, C, and G. See nucleotide.

base pair Two bases (nucleotides) that bond to form **complementary** pairs. In DNA these are A and T; C and G. In RNA they are A and U; C and G. These pairs are also called WATSON–CRICK pairs.

beta sheet One of two common secondary structures in protein. The other is the alpha helix.

Big-Oh O and related notation For two non-negative functions $f(n)$ and $g(n)$, if $\lim_{n\to\infty} \frac{f(n)}{g(n)} < \infty$, then we say $f(n) = O(g(n))$. Intuitively (but not exactly) this means that $f(n)$ grows no faster than $g(n)$, ignoring small values of n and any multiplicative constants.

If $\lim_{n\to\infty} \frac{g(n)}{f(n)} < \infty$, then we say $f(n) = \Omega(g(n))$. Intuitively (but not exactly) this means that $f(n)$ grows at least as fast as $g(n)$, ignoring small values of n and any multiplicative constants.

If $f(n) = O(g(n))$ and $f(n) = \Omega(g(n))$, then we say that $f(n) = \Theta(g(n))$. Intuitively (but not exactly) this means that $f(n)$ grows at the same rate as $g(n)$, ignoring small values of n and any multiplicative constants. See [32] or [112] for more detail.

bioinformatics A made-up word that refers to a wide range of computational efforts in the human, and other, genome projects. It encompasses data base development, communication software, sequence analysis, development of automated laboratory notebooks, development of standards, and more. The great virtue of the term is that it has no previous meaning so that it can be almost anything. The downside is that people who do informatics are called "bioinformaticists". Although the term is very inclusive, many people identified as bioinformaticists self-label themselves as being in "computational molecular biology".

catalyst A molecule that enables or accelerates a specific chemical reaction. Proteins are the primary catalysts in the cell, but RNA can act as a catalyst as well.

cDNA Complementary or copy DNA. This is DNA made from messenger RNA (through reverse transcription). When the mRNA is from eukaryotes, the cDNA lacks the introns that are contained in the DNA substring from which the mRNA is transcribed. (page 238)

chromosome Large linear or circular structure made of DNA in complex with certain proteins. The hereditary information of an organism is organized into a number of chromosomes.

cladistics A particular systematic approach to building evolutionary trees and interpreting their meaning. Generally, the cladistic approach attempts to reconstruct the branching pattern (order) of the evolutionary tree but not estimate the passage of time on any particular branch.

clone A population of organisms, genes, or molecules consisting of many exact copies of the originating entity. As a verb, "to clone" means to make many copies. To "clone a sequence" also refers to the process of extracting a specific DNA sequence or gene from a larger body of DNA.

clone library See DNA library.

cloning vector An organism, often a virus or a plasmid, or more recently an artificial chromosome (for example, a YAC), that carries a fragment of DNA into a host cell that can clone (replicate) the inserted DNA fragment along with its own DNA. (page 416)

coding region An interval of DNA containing the codons that specify a particular protein.

In Eukaryotes, the coding region is broken into disjoint exons. However, coding regions for different proteins can sometimes overlap; hence an interval should be defined as a coding region for a specific protein. (page 238)

codon A string of three nucleotides in DNA or mRNA that specifies a particular amino acid. See also **genetic code** and **translation**. Two codons are called "synonymous" if they specify the same amino acid. See genetic code. (page 14)

complementary base In DNA, the complementary base pairs are A, T and C, G. In RNA they are A, U and C, G. See also **base pairs**.

complementary nucleotide sequence A nucleotide sequence S is the complementary sequence of S' if each base in S is the complementary base of the corresponding base of S'.

cosmid An artificially constructed cloning vector capable of holding inserts of lengths around 40,000 bases. (pages 416 and 419)

crossing over The process, during meosis, where two homologous chromosomes align together, break, and exchange fragments of DNA to form two hybrid chromosomes. Each hybrid chromosome contains alternating intervals of DNA from the two original chromosomes. Normally in a pair of exchanged fragments, each fragment in the pair is the same length as the other. But in **unequal crossing over** one fragment is larger than the other, so the two new chromosomes are of unequal length. See **recombination**.

denaturation A large disruption of the natural structure of protein or DNA due to heating or exposure to chemicals. In double-stranded DNA, denaturation often refers to splitting apart of the two strands into two single strands. In protein, it often refers to destroying the three-dimensional structure of the protein.

Depth-First Search (dfs) A very common recursive algorithm used to explore a graph. In this book it is only used to explore trees. Depth-first search first visits the root of the tree (the first "current node") and then successively executes a *dfs* computation on the tree rooted at each child of the current node. When all those recursive executions are completed, the search backs up from the current node to its parent node. See [10], [32], [112] or [401] for more details.

diploid An organism is diploid if it contains two copies of each chromosome. The copies need not be identical, as each can contain different alleles. In humans, most cells other than germ-line cells (sperm and egg) are diploid. See also **haploid**.

DNA   A chain of nucleotides (bases) in a single molecule. The bases of DNA are **adenine (A), thymine (T), cytosine (C)**, and **guanine (G)**. DNA is the basic carrier of genetic information. DNA usually consists of two strands of complementary nucleotide sequences that are base paired (Watson-Crick) to each other. Hence, in describing the two strands, one typically only specifies one of the strands and its orientation. DNA in humans forms a linear chain, but DNA can also form a circular molecule (see **plasmid**). (page 14)

DNA library   A physical collection of unordered cloned fragments of DNA, possibly cDNA obtained from mRNA. The fragments can come from the entire genome, but often the fragments in a particular DNA library come from a particular tissue or chromosome region and represent only a subset of the entire DNA of an organism.

DNA sequencing   The process of determining the complete base pair sequence of a target DNA string. (page 415)

Drosophila melanogaster   Fruit fly commonly used in genetic studies.

enhancer sequence   A DNA sequence that binds to regulatory proteins. This binding affects the rate at which certain intervals of DNA are transcribed, and hence it affects the rate at which certain proteins are made (expressed). The concept of an enhancer sequence is often confused with a **promoter** sequence, but unlike the latter, the enhancer sequence can be very far away from the DNA that it regulates. The opposite of an enhancer is a **repressor**.

electrophoresis   A process in which molecules (DNA, RNA, or protein) with different properties (such as charge, length, or size) are separated. The molecules are put into a gel and an electric field (sometimes alternating) is applied. Molecules with different properties move through the gel at different rates in response to the field.

Escherichia coli (E. coli)   Bacteria found in the human gut. E. coli is one of the widely studied model organisms in genetics and molecular biology.

enzyme   A protein catalyst, working to accelerate specific chemical reactions in the cell.

EST (expressed sequence tag)   An STS derived from a cDNA molecule. Therefore, an EST is an STS that is found in a gene, rather than in a noncoding region of DNA. (page 61)

eukaryote   An organism whose DNA is enclosed in a nucleus. Includes all "higher-order" life. Often in this book, the important feature of a eukaryote is that its genes are broken into alternating exons and introns. Contrast with **prokaryote**.

exon   In Eukaryotes, genes are typically broken up into alternating regions of exons and introns. The exons contain information that is represented in the messenger RNA (mRNA) and ultimately used to produce a protein (or sometimes RNA). (page 237)

exon-intron splice site   A point in a eukaryotic gene where an intron adjoins an exon. Nature knows how to identify and splice out the introns from transcribed RNA, but no complete explanation of how splice sites are recognized is known. (pages 238, 248)

expression   A gene is expressed if the protein it codes for is produced. Sometimes we also say a protein is expressed.

extant organism   The word "extant" means currently existing. I've found that this term is not well known to computer scientists, and I know of no computer science setting where it is used. It is a standard word in evolutionary biology.

fushi tarazu   One of the coolest names for a mutant allele in *Drosophila*, an organism where many mutants have cool names. Japanese for "missing a stripe".

gene   A contiguous interval of DNA that contains the information needed to code for a protein, or less often, for some RNA. Genes form the basic units of heredity.

gene regulatory protein   A protein that binds to DNA to affect the expression of a gene.

genetic code   The correspondence between specific nucleotide triplets (**codons**) and specific amino acids. Since there are $4^3 = 64$ codons but only 20 standard amino acids, some amino acids are coded by more than one codon. Therefore, the genetic code is said to be **degenerate**. For example, the amino acid alanine is specified by codons GCA, GCC, GCG, GCU. There are also three codons that do not code for any amino acid, but usually specify the end of the polypeptide. These are called **stop codons**. (page 14)

genome   The entire genetic information contained in an organism.

genotype   The description of a specific organism in terms of its genome. This is opposed to **phynotype**, which is a description of an organism in terms of its expressed features.

germ cells   In humans, sperm or egg cells.

These cells are haploid, containing only a single copy of each chromosome. See also **somatic cells**.

globular protein A protein that forms a round shape.

haploid A cell containing only one copy of each chromosome.

hemoglobin The primary functional protein of red blood cells. It is involved with binding and transporting oxygen.

homeobox A DNA sequence to which regulatory proteins bind, affecting major features of the organism's development. These sequences are highly conserved across many species. (page 230)

homologous chromosomes The two copies of the same chromosome in a diploid organism.

homologous protein Two proteins that are related by common evolutionary history. For example, human cytochrome c is homologous to duckbill platypouse cytochrome c. However, sometimes the term is used even when the shared history is unclear. Two proteins in different organisms that have common biological or chemical features (function, structure, motifs) are often referred to as "homologous", but the word "analogous" seems more appropriate, expressing similarity of the proteins without any implied cause. Unfortunately, in some subareas of biology the phrase "analogous protein" has been defined to mean that the two proteins are similar due to *convergent* evolution, i.e., they are known to have no significant shared history. This deviation from normal English robs biology of an important common word. Sometimes biologists will use the phrase "same protein" (in quotes) to capture the normal English meaning of "analogous protein". In this book, I have tried to use the term "homologous protein" only when a shared history is implied.

homology A similarity due to common evolutionary history. "Sequence homology" is sometimes used interchangeably with "sequence similarity", although the later term does not imply a common evolutionary history. A phrase such as "degree of homology" is commonly used, but it is not meaningful under the strict definition given above.

hybridization The base-pairing (bonding) of two complementary DNA or RNA molecules. Hybridization of a single stranded **probe** to a longer sequence of DNA is often used to determine where the complement of the probe resides in the longer sequence.

intron See **exon**.

linkage The degree to which two markers in DNA are inherited together. Linkage provides a crude reflection of physical distance. The closer two markers are on a single chromosome, the more likely they are to be inherited together and not separated by a crossing-over (or recombination) event.

linkage map Same as genetic map.

leucine zipper A common motif for a transcription factor. (page 62).

ligate To join together two molecules.

ligand A molecule that binds to a specific location on another molecule.

marker A feature of a chromosome that can be detected.

meiosis The process during which germ cells (sperm or egg) are created from parent cells. Four haploid cells are created from one diploid cell during two rounds of cell division. Typically, during meiosis each pair of homologous chromosomes come together, cross over/recombine, and replicate, creating four hybrid chromosomes.

messenger RNA (mRNA) The template RNA that codes for a single protein. Each codon of the mRNA specifies a single amino acid in the protein sequence. The mRNA is processed by a **ribosome**, which, with the aid of **transfer RNA**, strings together the prescribed amino acids of the protein.

mitochondria Bacteria-sized organelle inside eukaryotic cells. It is the energy center of the cell and contains it own DNA. It is an unsolved puzzle how it came to co-exist with eukaryotic cells. Mitochondrial DNA is inherited only from one's mother, and mitochondrial DNA sequences are often used to infer evolutionary history.

nucleic acid Essentially a chain of nucleotides; an RNA or DNA molecule.

nucleotide Adenine, thymine, cytosine, or guanine in DNA; Adenine, uricil, cytosine, or guanine in RNA. See base.

oligonucleotide (or **oligo** for short) A short nucleic acid chain (DNA or RNA).

oncogene A gene that acts to promote the development of cancer, when active. In contrast, an **antioncogene** is a gene that normally acts to suppress the development of cancer, allowing cancer when inactive.

open reading frame (ORF) A substring in DNA that contains no stop codons (*UAA, UAG,* or *UGA*) when read in a single reading frame. Sometimes an ORF is alternately defined as a substring in DNA between the start codon *AUG* and the first stop codon found in the same reading frame as the start codon. That definition is more appropriate for prokaryotes and is confusing for eukaryotes, because eukaryotic genes are sometimes described as containing several ORFS [192], meaning several exons. A single exon does satisfy the first definition of an ORF, but need not satisfy the alternate definition. However, there seems to be no absolute standard, and two different chapters of [192] use the two different definitions, even though both chapters concern eukaryotic genes. Identifying the open reading frames is often the first step in trying to locate genes in a stretch of anonymous DNA.

PAM Short for "point accepted mutation" or "percent accepted mutations". PAM is used as a unit of evolutionary distance and also as an identifier of specific amino acid substitution matrices. (page 381)

palindromic sequence in DNA A DNA string that becomes a palindrome in the normal English use of the word after half of its string is replaced by its complementary sequence. (page 138)

polymerase chain reaction (PCR) A process used to make copies (or amplify) the DNA in an interval of DNA defined between two **primer** sequences. The DNA can be *in vitro* (i.e., outside of a living cell), and the two primer sequenes are chosen by the experimenter. However, the primers must be located within an "acceptable" distance from each other. PCR proceeds in cycles, and in each cycle the number of copies of the defined interval of DNA roughly doubles. The ability to quickly and cheaply make essentially unlimited quantities of DNA between any two (relatively close) and user-defined points on a DNA molecule, has revolutionized the practice of molecular biology. PCR has been described as "being to genes what Gutenberg's printing press was to the written word" [90]. The inventor of PCR, Kary Mullis, was awarded the Nobel Prize in chemistry in 1994. A very amusing account by Mullis of that invention appears in [333].

phenotype The total, observable set of features of an organism. See also **genotype** for contrast.

phylogeny The tree-like evolutionary history of a set of taxa. (Chapter 17)

plasmid A circular molecule of DNA found outside the normal genome of the organism. Typically found in bacteria. Often used as a cloning vector. (page 12).

point mutation A single change of one nucleotide (in DNA or RNA) or one amino acid residue.

polypeptide A long chain of linked amino acids. Sometimes used for "protein".

polymere A long linear molecule made up of identical or similar subunits.

positional cloning A fairly recent approach to identifying disease-related genes. First, using genetic mapping, find an interval in the genome likely to contain the gene; then sequence parts of the interval in disease afflicted individuals and unafflicted relatives to find systematic differences that indicate the desired gene has been found. (page 397)

prokaryote An organism whose DNA is not enclosed in a nucleus. Contrast with **eukaryote**.

primary structure The sequence description of a molecular (DNA, RNA, or amino acid) string.

promoter A substring in DNA upstream of a gene, where the RNA polymerase (which helps transcribe the DNA to RNA during transcription) binds to the DNA.

polymerase An enzyme that facilitates the creation of a nucleic acid molecule complementary to an existing single-stranded nucleic acid molecule (called a template).

protease An enzyme that cuts protein.

protein A polypeptide, a chain of linked amino acids. The structural material and workhorse molecule of the cell (DNA proposes, but protein disposes). Enzymes are protein catalysts, working to enable or accelerate chemical reactions in the cell. Until the late 1940s and early 1950s it was generally believed that proteins would also be found to be the molecule encoding hereditary information. It was very surprising when DNA was shown to play that role.

proto-oncogene A normal gene that can be converted to an oncogene by one of several mechanisms.

purine In the context of this book, a purine is either adenine (A) or guanine (G). How do you remember this? Biochemistry students at U. C. Davis (the AGgies) are taught the mnemonic: *AGgies are PURe.*

pyrimadine In the context of this book, a pyrimadine is either cytosine (C) or thymine (T).

rate-limiting step One of several favorite phrases used in biochemistry but not in computer science. In computer science we would call it a "bottleneck".

reading frame One of three places to start

reading when translating a string from the DNA alphabet into the amino acid alphabet. If the direction of the string is also not established, then one often refers to six reading frames. (page 14)

recombination A general term for several mechanisms resulting in the breaking or cutting and resplicing of intervals of DNA. Occurs both naturally (for example, during meiosis via crossing-over) and in the laboratory.

repressor A protein that reduces (or eliminates) the expression of another protein by interfering with the transcription of its DNA. A typical mechanism is for the repressor to bind close to the DNA encoding the repressed protein, hence interfering with its transcription.

residue A single unit in a polymere. Used both for a single nucleotide in a DNA molecule or a single amino acid in a protein.

restriction enzyme An enzyme that cuts DNA at locations containing specific short (usually palindromic) DNA sequences.

retrovirus A virus with an RNA genome that must be transcribed to a double-stranded DNA molecule in order to reproduce.

reverse transcription The process of creating a double-stranded DNA molecule encoding a sequence from a single-stranded RNA molecule.

reverse transcriptase Enzyme that creates a double-stranded DNA "copy" of a single-stranded RNA molecule.

ribosome A complex made of RNA and protein where the protein defined by a messenger RNA is synthesized.

RNA ribonucleic acid molecule. See **nucleic acid**.

satellite DNA Short DNA substrings that are highly repetitive in a genome. These are further subdivided into mini- and micro-satellite by the length of the repeated substring. (pages 140 and 138)

sequence tagged site (STS) Roughly, a short DNA sequence that occurs only once in the genome. More exactly, a pair of PCR primers within a bounded distance, with the property that PCR succeeds using those primers at only one location in the genome. STSs provide markers throughout the genome, but they need not be located in genes. See also **EST**. (pages 61 and 398)

silent mutation A mutation in a DNA codon that does not change the specified amino acid. Most often, a silent mutation is in the third

nucleotide in the codon. For example, TCN codes for the amino acid serine, where N is any of the four DNA nucleotides. So a mutation in the third nucleotide is a silent mutation.

start codon Codon that signals the start of a sequence to be translated to protein. Frequently *AUG*, but it can vary in different organisms.

stop codon Codon that signals the end of a sequence to be translated to protein. Frequently *UAA*, *UAG* or *UGA*.

TATA box The name given to the commonly occurring substring "TATA" that appears in the promotor region of many genes.

telomere The DNA forming each end of a chromosome. It contains highly repetitive short substrings. (page 138)

transcription The process by which an RNA molecule is synthesized complementary to the DNA in a gene. In eukaryotes, both the introns and the exons are transcribed into the RNA. The introns are later spliced out, creating an mRNA.

transcription factor General term for a protein that aids in initiating or regulating transcription. See also **leucine zipper** or **zinc finger**. (page 62).

transfer RNA (tRNA) The RNA molecule that transports a specific amino acid to a growing amino acid chain, as directed by a specific codon in the mRNA.

translation The process by which a protein is synthesized, according to the "blueprint" given by a messenger RNA molecule.

Yeast Artificial Chromosome (YAC) An artificially created cloning vector used to hold DNA sequences up to one or two million bases long. (page 416)

You aren't expected to absorb this Phrase used by biologists in talks when displaying a 35-mm photographic slide containing unreadable DNA or amino acid sequences.

You aren't expected to absorb this Phrase used by computer scientists in talks when displaying an overhead transparency containing unreadable C or C++ code.

You aren't expected to absorb this Phrase used by mathematicians or statisticians in talks when filling the blackboard with inscrutable equations.

zinc finger A common motif for a transcription factor. See **transcription factor** (page 62)

Index